Advances in Bio-Based and Biodegradable Polymeric Composites

Advances in Bio-Based and Biodegradable Polymeric Composites

Editor

Evgenia G. Korzhikova-Vlakh

MDPI • Basel • Beijing • Wuhan • Barcelona • Belgrade • Manchester • Tokyo • Cluj • Tianjin

Editor
Evgenia G. Korzhikova-Vlakh
Institute of Macromolecular
Compounds
Russian Academy of Sciences
St. Petersburg
Russia

Editorial Office
MDPI
St. Alban-Anlage 66
4052 Basel, Switzerland

This is a reprint of articles from the Special Issue published online in the open access journal *Polymers* (ISSN 2073-4360) (available at: www.mdpi.com/journal/polymers/special_issues/bio_based_polym_compos).

For citation purposes, cite each article independently as indicated on the article page online and as indicated below:

LastName, A.A.; LastName, B.B.; LastName, C.C. Article Title. *Journal Name* **Year**, *Volume Number*, Page Range.

ISBN 978-3-0365-6492-0 (Hbk)
ISBN 978-3-0365-6491-3 (PDF)

Contents

About the Editor

Evgenia G. Korzhikova-Vlakh

Evgenia G. Korzhikova-Vlakh studied chemistry at Saint-Petersburg State University (SPBU, Russia) and then obtained her PhD in the Institute of Macromolecular Compounds, Russian Academy of Sciences (IMC RAS). Since 2012, she has been an Associated Professor at SPBU, and since 2016, she has been a leading scientist, head of the Laboratory of Polymer sorbents and carriers for biotechnology, in IMC RAS. Her research has focused on the development and study of polymeric materials for bioanalysis, biocatalysis, solid-phase extraction, drug delivery, and regenerative medicine. Dr. Korzhikova-Vlakh has published more than 100 scientific papers in peer-reviewed international journals.

polymers MDPI

Editorial

Preface to "Advances in Bio-Based and Biodegradable Polymeric Composites"

Evgenia G. Korzhikova-Vlakh

Institute of Macromolecular Compounds, Russian Academy of Sciences, Bolshoy pr. 31, 199004 St. Petersburg, Russia; vlakh@mail.ru

The use of polymers for various purposes is increasing every year. In this regard, the production of polymeric materials from natural renewable sources, as well as the obtention of biodegradable polymeric materials and their recycling, comes to the fore in the packaging and disposable products industry. In addition, bio-based and biodegradable materials are becoming more and more prominent in medical practice as implants, regenerative and wound-healing materials, etc. Thus, modern environmental requirements, as well as the introduction of new technologies in biomedicine, pose the challenge to develop innovative "green" polymeric materials with the desired properties. The doping of polymeric materials with various fillers to prepare the composites makes it possible to adjust and improve their properties, such as mechanical and physicochemical characteristics, biological performance, degradation rate, conductivity, etc., in a targeted manner. The combination of bio-based and biodegradable polymers with fillers/reinforcements of various chemistries, dimensions, and geometries can lead to a higher level of diversity provided by composite materials as compared to just polymeric ones.

Given the growing interest in polymeric composite materials, this Special Issue summarizes the latest advances in science and technology of bio-based and biodegradable composites, including the production, characterization, and application of such composites, as well as theoretical studies in the field. This Special Issue contains a number of original articles and reviews. The first group of papers reports on the use of poly (lactic acid), poly (ε-caprolactone), poly (butylene succinate), poly (vinyl alcohol), chitosan, pectin, lignin-epoxide, and corn starch as matrix polymers, and nanocellulose, bacterial cellulose, wood fibers, multiwalled carbon nanotubes, and silver, $CaCO_3$ and ZnO nanoparticles as fillers. In turn, reviews are focused on overviewing the modification of cellulose and waste natural polymers as fillers, the production of conductive polymer composites for biomedical and biosensor technologies, and advances in the production of polymer composite packaging materials with antibacterial properties.

Conflicts of Interest: The author declares no conflict of interest.

Citation: Korzhikova-Vlakh, E.G. Preface to "Advances in Bio-Based and Biodegradable Polymeric Composites". *Polymers* **2022**, *14*, 2926. https://doi.org/10.3390/polym14142926

Received: 15 July 2022
Accepted: 18 July 2022
Published: 20 July 2022

polymers

MDPI

Review

A Review on Antimicrobial Packaging from Biodegradable Polymer Composites

Siti Hasnah Kamarudin [1,*], Marwah Rayung [2], Falah Abu [1], So'bah Ahmad [1], Fatirah Fadil [1], Azrena Abdul Karim [1], Mohd Nurazzi Norizan [3], Norshahida Sarifuddin [4], Mohd Shaiful Zaidi Mat Desa [5], Mohd Salahuddin Mohd Basri [6], Hayati Samsudin [7] and Luqman Chuah Abdullah [8]

1 School of Industrial Technology, Faculty of Applied Sciences, Uitm Shah Alam, Shah Alam 40450, Malaysia; falah@uitm.edu.my (F.A.); sobah@uitm.edu.my (S.A.); fatirahfadil@uitm.edu.my (F.F.); rena184@uitm.edu.my (A.A.K.)
2 Department of Chemistry, Faculty of Science and Technology, Universiti Putra Malaysia, Serdang 43400, Malaysia; marwahrayung@yahoo.com
3 Centre for Defence Foundation Studies, Universiti Pertahanan Nasional Malaysia (UPNM), Kem Perdana Sungai Besi, Kuala Lumpur 57000, Malaysia; mohd.nurazzi@gmail.com
4 Department of Manufacturing and Materials Engineering, International Islamic University Malaysia, Jalan Gombak, Kuala Lumpur 53100, Malaysia; norshahida@iium.edu.my
5 Faculty of Chemical Engineering Technology and Process, Universiti Malaysia Pahang, Lebuhraya Tun Razak, Gambang 26300, Malaysia; shaiful@ump.edu.my
6 Department of Process and Food Engineering, Faculty of Engineering, Universiti Putra Malaysia, Serdang 43400, Malaysia; salahuddin@upm.edu.my
7 Food Technology Division, School of Industrial Technology, Universiti Sains Malaysia, Pulau Pinang 11800, Malaysia; hayatis@usm.my
8 Department of Chemical and Environmental Engineering, Faculty of Engineering, Universiti Putra Malaysia, Serdang 43400, Malaysia; chuah@upm.edu.my
* Correspondence: sitihasnahkam@uitm.edu.my

Citation: Kamarudin, S.H.; Rayung, M.; Abu, F.; Ahmad, S.; Fadil, F.; Karim, A.A.; Norizan, M.N.; Sarifuddin, N.; Mat Desa, M.S.Z.; Mohd Basri, M.S.; et al. A Review on Antimicrobial Packaging from Biodegradable Polymer Composites. *Polymers* **2022**, *14*, 174. https://doi.org/10.3390/polym14010174

Academic Editor: Evgenia G. Korzhikova-Vlakh

Received: 8 November 2021
Accepted: 24 November 2021
Published: 2 January 2022

Abstract: The development of antimicrobial packaging has been growing rapidly due to an increase in awareness and demands for sustainable active packaging that could preserve the quality and prolong the shelf life of foods and products. The addition of highly efficient antibacterial nanoparticles, antifungals, and antioxidants to biodegradable and environmentally friendly green polymers has become a significant advancement trend for the packaging evolution. Impregnation of antimicrobial agents into the packaging film is essential for impeding or destroying the pathogenic microorganisms causing food illness and deterioration. Higher safety and quality as well as an extended shelf life of sustainable active packaging desired by the industry are further enhanced by applying the different types of antimicrobial packaging systems. Antimicrobial packaging not only can offer a wide range of advantages, but also preserves the environment through usage of renewable and biodegradable polymers instead of common synthetic polymers, thus reducing plastic pollution generated by humankind. This review intended to provide a summary of current trends and applications of antimicrobial, biodegradable films in the packaging industry as well as the innovation of nanotechnology to increase efficiency of novel, bio-based packaging systems.

Keywords: antimicrobial packaging; biodegradable; natural fibre; polymer composite; sustainable

1. Introduction

Packaging is a billion global industry and plays a significant role for essential items for consumer goods ranging from basic chemicals to household and personal care products, drinks, foods, medical devices, and much more. The value of the packaging industry is highly expanding due to competitiveness in making commodities and luxury packaging. To date, the applications of plastics in the packaging sectors have been increasing at a fast speed due to their benefits of being commercially low cost and possessing intrinsic characteristics of plastic films in packaging industries. The most frequent plastic films used in the development of the packaging industry include polypropylene (PP), low-density polyethylene (LDPE), linear low-density polyethylene (LLDPE), poly(vinyl chloride) (PVC), and poly(ethylene terephthalate) (PET). The unique properties of plastics such as low cost and superior processability and having good barriers, mechanical properties, good sealing characteristics, and high transparency make them a favorable material [1]. In addition, they can be totally recycled and are lightweight alternatives to traditional, non-recyclable materials due to their super functionality [2–4]. Despite all the listed usefulness and benefits, the use of plastics as base materials in the packaging system suffered from the limitations of the materials themselves, such as the harmful chemicals and waste that packaging leaves behind. The wide usage of plastic packaging has caused serious plastic waste disposal problems, which, in turn, create massive environmental pollution [5]. In 2018, the World Wildlife Fund also reported that China, Indonesia, Malaysia, the Philippines, Thailand, and Vietnam contributed around 60% of the estimated 8 million tonnes of plastic that enter the world's oceans every year [6]. This threat to the environment is basically due to the significant level of highly toxic emissions, composting management issues, and alteration in carbon dioxide cycle [7]. Furthermore, disposed packaging plastics in many countries are rarely recycled due to technical issues and socio-economic constraints. For example, in China, there is only about 20% of recycled plastic waste as compared with 1 million tons of plastic generated [8]. Moreover, a huge proportion of the used plastic materials is either deposited in landfills or contributes to litter everywhere, surrounding the environment, which ends up putting stress and strain on the environmental balance. The alternative way to minimize the waste contributed by plastic is to use compounds from nature. Therefore, this phenomenon has stimulated the attention of many researchers to develop sustainable, active packaging material [9]. Therefore, the design of the packaging should consider not only shelf-life, cost, and protection, but also user-friendliness and environmental sustainability [10].

The consumption of food packaging was said to have increased during this pandemic (Figure 1) [11]. A comparison of different regions shows that the consumption of food packaging before and after the Covid-19 pandemic vary strongly. Apparently, Indonesia has contributed to a large amount of food packaging consumption before the pandemic caused by Covid-19. During the pandemic, Hong Kong passed as the highest region consuming food packaging. Because of the pandemic, there is a high concern regarding the hygiene and safety aspects by customers. Most people have resorted to their last option of buying bulk stocks of groceries or having their meals taken away. According to the Agriculture and Horticulture Development Board (AHDB) in its 2020 article on *'Takeaway food performance during Covid-19'*, the pandemic effect has urged people to switch from dine-in to takeaway-delivery due to social distancing recommendations.

One of the important safety aspects related to food packaging is its influence towards the microbial shelf life of food. In the environment where we live, there are millions of microbes, most of which are not visible to the human eye. The microbes, such as viruses and bacteria, have a very simple composition and replicate very quickly. For instance, a single bacteria can generate up to 500 new bacteria within 3 h through binary fission. Some of these bacteria and viruses may cause infections, ranging from mild to deadly diseases. Throughout human existence, dangerous microbes have been a source of horrifying epidemics such as plague, cholera, tuberculosis, etc. Although they are invisible, microbes continue to cause health problems, especially to the respiratory, digestive, and nervous systems. The types of diseases found today have been extremely difficult to prevent and cure due to high levels of antimicrobial resistance. Microbes can be transmitted in the following ways: (1) Coughing and sneezing, (2) Breathing contaminated air, (3) Contact with infected people by shaking hands, and (4) Contact with the infected objects or contaminated surfaces, water, or food. The threat posed by bacteria has inspired numerous researchers to research and develop unique antimicrobial plastic packaging for farm, food, and cosmetics.

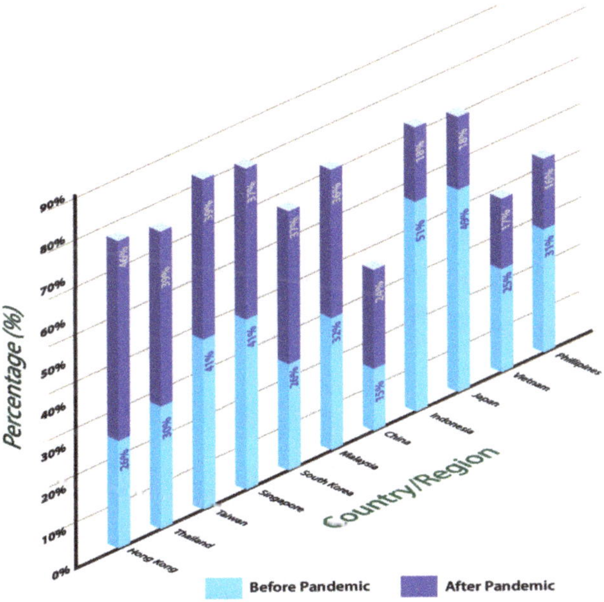

Figure 1. COVID-19's impact on consumers ordering food, APAC 2020, by country or region. Modified from [11].

The reactions of bacteria, enzymes, molds, and some microorganisms towards the surrounding humidity and temperature on different types of foods also contribute to food spoilage in the food packaging [12–15], as displayed in Figure 2 [16]. Figure 2 shows that *Shewanella putrefaciens*' growth rate on fresh fish was the highest compared to Pseudomonas spp. and *Brochothrix thermosphacta*, which was around 0.5 per hour at 20 °C. On the other hand, bacteria and yeast on cooked and cured pork products showed the lowest growth rate, which was less than 0.1 per hour at around 12 °C. However, *Monascus ruber* (fungus) showed a unique growth rate, which started when the temperature was at 20 °C with less than 0.2 per hour and rose gradually to more than 0.6 per hour. Some researchers did their research on how to lower this carbon footprint [17,18].

There are a number of limitations in current packaging, which are non-sustainable production, legislation, cost pressures, and consumer education. Helping consumers to understand the importance of packaging, whether it is in food, drink, or medicine or giving economical access to products they need every day to make their life easier, safer, and more confident not only about the products they buy but the role of the packaging it is served in helps ensure those products maintain freshness, quality, and efficacy. Thus, poor barrier properties to water, vapor, and gases are the important critical issues in packaging. Fresh products like vegetables and fruits need to be packaged in an oxygen-permeable membrane environment, whereas processed products do not require much transfer. Another challenge faced by many producers is speed to market. A shorter research and development (R&D) process is needed for the development of packaging, which is around 9 months instead of the 12 to 18 months for the current packaging development cycle. Food waste reduction as well as new consumer experiences in new consumption occasions for the benefits of consumers and protecting the food quality are among the biggest challenges for current packaging right now. Additionally, bringing new, innovative products and at the same time maintaining sustainability goals and profitability goals both for consumers and the packaging company are the reforms that need to be made. The excessive growth of microorganisms because of contamination and temperature abuse, the high degrees of nutritional qualities to the oxidation, and laws of nutritional qualities to the interaction with extreme factors are among examples of food quality and safety issues.

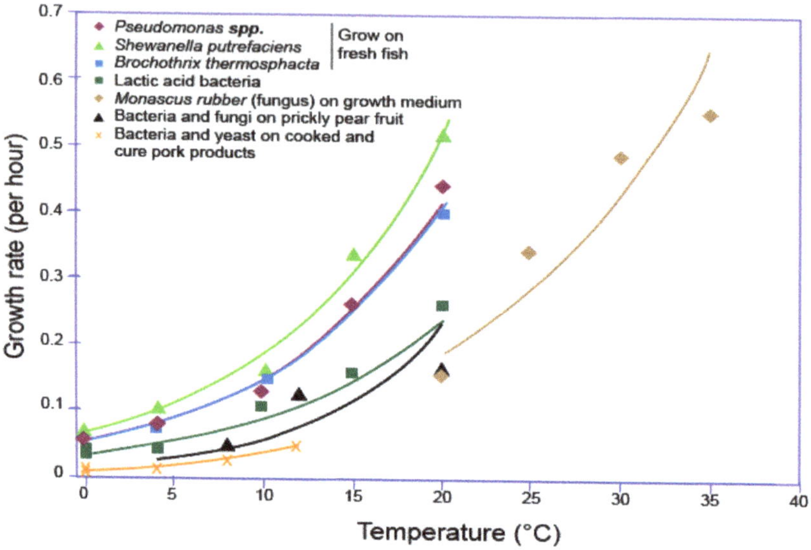

Figure 2. Growth rate of various types of microbes depending on the time. Modified from [16].

Antimicrobial packaging was introduced to combat this problem so that the shelf-life storing of the food can be extended, reducing food waste [19,20]. Apparently, antimicrobial agents had been applied to be incorporated with the food packaging [21–25] The antimicrobial properties in antimicrobial agents have made them become suitable to be incorporated with food packaging [26–28]. According to Rhim et al., antimicrobial-function nanocomposites were found effective for minimizing the growth of contaminant pathogens that exist after the post-processing, extending food shelf-life and enhancing food protection [29].

The usage of green polymers together with nanoparticles such as silver nanoparticles (AgNPs) as an antimicrobial agent is common in food packaging industries [30,31] since the characteristics of silver nanoparticles are good enough to make them a widely used nanofiller for making packaging because of their antimicrobial properties. Fillers are substances that are applied to regular packaging products, typically in low percentages, to improve the performance of the original content. In a composite, it is basically the mixture of the regular packaging material and a filler [32]. From the collective reviews that were done, it can be summed up that those researchers mostly used silver nanoparticles (AgNPs) as their antimicrobial agent because of their effective antimicrobial performance, low toxicity, and high thermal stability, and they have continued to gain attention since. In addition, it is possible to significantly increase the stability and mechanical strength of poly-saccharide films by adding AgNPs [33]. Evidently, a few articles that have been published in the field of bio-composites and bio-nano composites agreed on the potent properties of silver nanoparticles in making antimicrobial food packaging with a longer shelf-life [31,34,35].

Even though biopolymers are environmentally friendly and considered as most fascinating packaging materials, the industrial applications are restricted due to several factors such as their oxygen/water vapor barriers, thermal resistance, and other mechanical properties associated with costs [36]. In order to encounter these challenges and urge the industrial applications of biopolymers for packaging materials, there is the requirement for advanced research to effectively improve their stability, quality, nutritional values, and microbial resistance. Moreover, the barrier properties need to be intensified. Biodegradable polymers containing starch/cellulose fibres are most likely to make a solid growth in applications. Numerous approaches for elevating the properties and performance of antimicrobial packaging materials, such as chemical and physical modifications, polymeric blending, and nanocomposites, have indicated a bright potential for many types of applications.

Hence, more advanced research tools and huge investment are required to obtain fully sustainable materials with antimicrobial activity and effective alternatives for the existing ones. The enhancement of a moisture barrier and mechanical and other properties of biodegradable polymer will benefit the significant innovation in these packaging materials. Moreover, an increment in the use of biodegradable packaging must be intensified by more composting infrastructure. The development of specialised recycling procedures for these types of packaging should be considered. Despite all the advantages related to the use of silver nanoparticles with biodegradable polymers for sustainable packaging and a safer environment, several important constraints are minimizing the toxicity and environmental risks impacted from the packaging waste containing these nanoparticles.

Both functional and technical gaps have been the limitation barriers toward the development and applications of antimicrobial packaging materials in industries. Several limitations include vapor and air barriers, the stability of antimicrobial agents under processing conditions, and the low processability of bioplastics' toxicity as well as the changes in mechanical properties of the packaging materials. Accordingly, further research work should be focused on filling the void linking the antimicrobial actions to microbial growth kinetics in the packaged foods in both lab and industrial approaches. Close collaboration between both academic and industrial players could be an effective alternative to filling the gap between commercial aspects and research. Synergism and blending of nanocomposites would be the core tools as the useful strategies for improving antimicrobial performances for improving antimicrobial packaging and preventing some of the limits encountered during activity. This would contribute to the initial essays on the research and development of antimicrobials' packaging.

In addition, a forecast of market demand shows that the estimated global market growth for antimicrobial packaging was exponentially increased, as indicated from a growing Compound Annual Growth Rate (CAGR) value from 2020 until 2024. The contribution in the increasing economic value of these antimicrobial packaging products from biodegradable polymer composites is driven by the growing awareness of the consumers towards the consumption of sustainable and green packaging. Consumers are now consciously aware of the possible threat coming from synthetic food preservatives to human health, as some are potentially transformed into carcinogenic agents, thus indirectly helping in reducing the dependency on the consumption of synthetic preservatives. The use of these antimicrobial packagings from biodegradable polymer composites will be greatly beneficial in accelerating the transition towards preservative-free food products. The vast potential of antimicrobial packaging in sustaining the freshness of some selected highly perishable food including meat and poultry, seafood, fruits and vegetables, baked goods, and cheese and dairy-based products has contributed to the rise in the market demand, thus creating growth profitability for the players operating in the global market.

This review focused on the summary of current trends and applications of antimicrobial biodegradable films in the packaging industry as well as the innovation of nanotechnology to provide high efficiency of novel, bio-based packaging systems. For that reason, the influence of attractive product packaging plays an important role in the consumer purchasing decision. Most consumers are looking into new, added value possessed in the advanced packaging technology over the traditional packaging. The ideal antimicrobial packaging materials should be equipped with intelligent indicators' technology to measure certain crucial conditions, such as temperature, pH value, and humidity, to show the degree of bacterial contamination developed in packaged food throughout its shelf life. Universal protocol standards are needed not only to evaluate their antimicrobial activity against common food-borne bacteria and maintaining food quality, but also to meet consumer sensory preferences. The alternative packaging asserts to perform similarly as conventional packaging in terms of achieving expected shelf life of food, durability, sealing strength, printability, and flexibility. The integration of these responsive technologies into food packaging will provide a massive impact in the food processing industries, to fulfil the growing demand for packaged, ready-to-eat foods that are distinguished as a primary driver of future packaging trends.

2. Antimicrobial Packaging Agents

Antimicrobial packaging can be produced by the addition of antimicrobial agents such as chitosan and essential oil into the systems [37,38]. The incorporation of antimicrobial agents can be done by several techniques such as direct addition, encapsulation, coating, or grafting into or onto the matrix [39]. Meanwhile, polymeric materials are often used in conventional packaging systems. These polymers can be obtained from various sources and they can be classified based on their biodegradability. A current trend has been focusing on replacing non-biodegradable polymers to biodegradable polymers due to environmental concern, legislative rules, and consumer demands for green products. In view of this, the antimicrobial packaging has also been shifted to produce products from natural resources for both the host polymer and the antimicrobial agents. The types, advantages, and their limitations will be discussed in the following subtopic.

Various types of antimicrobial agents have been investigated for their potential applications in the antimicrobial packaging, and some of the examples are presented in Figure 3. Each of the antimicrobial agents has a unique mechanism and reacts differently to different types of microorganisms. In this case, the types of antimicrobials sometimes set restrictions for their applications. In general, these compounds can be classified into synthetic and natural classes based on their sources and physiologies. The synthetic antimicrobial agents can be further categorized as organic and inorganic. The literature reports various types of synthetic antimicrobial agents such as metallic nanoparticles (Ag, Cu, S) [40], oxide nanoparticles (ZnO, TiO_2, CuO) [41], clay nanoparticles (bentonite, cloisite, montmorilonitrile) [42], chelating agents [43], volatile compounds (SO_2, ClO_2, ethanol) [44], organics acids and their salts [45], etc. [25,46]. The type of antimicrobial agents selected may differ depending on for which application the packaging material is being used. Basically, synthetic organic compounds containing ethylene diamine tetraacetic acid (EDTA), parabens, fungicides, and other chemicals are the major antimicrobial compounds used in the food packaging industry [47]. In particular, metal and metal oxides are potential antibacterial agents, but issues concerning their long-term impact on the environment and human health remain unresolved. On the other hand, the use of silver-based antimicrobial packaging for food purposes is expected to grow and has been used in countries such as Japan and the United States [48].

Figure 3. Classification of antimicrobial agents. Modified from [49].

There were two ways of synthesizing the AgNPs in biological methods, which are intracellular or extracellular [50–52]. For the extracellular process, it involves trapping the metal ions on the outer surface of the cells. Then, it aims to reduce the silver in the presence of biomolecules. On the other hand, for the intracellular, the process takes place inside the microbial cells [53]. Extracellular synthesis was shown to be preferable in most studies compared to intracellular due to its simplicity, lower cost, and preferred large-scale production [51,53,54]. Additionally, some of the advantages of using this biological method are that it is environmentally friendly, does not produce any toxic residue, and is cost-effective [53–59]. Figure 4 shows the illustration of biological synthesis of silver nanoparticles using plant extraction.

Next, chemical synthesis of silver nanoparticles was also reviewed [53,55,59]. In this method, chemical reduction was discussed [53,55]. Reducing, stabilizing, or capping agents were used in this method. The size distribution of the produced silver nanoparticles was stated to be influenced by these agents [57]. The simple equipment used and the convenience of the chemical reduction are the benefits of using this method [60]. Nonetheless, this method is not widely preferred due to the requirement of using harmful chemicals such as sodium citrate, borohydride, potassium bitartrate, and sodium dodecyl, which are very toxic [53,61]. Additionally, this method was mentioned to yield toxic by-products during the process [59]. Figure 5 shows chemical synthetization of silver nanoparticles.

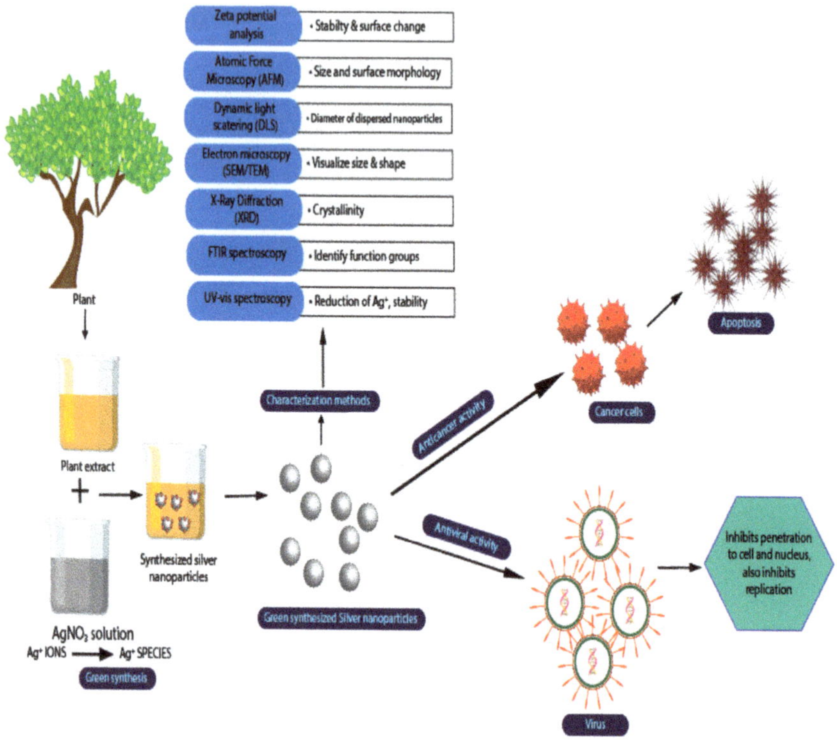

Figure 4. Biological synthesis of silver nanoparticles using plant extraction.

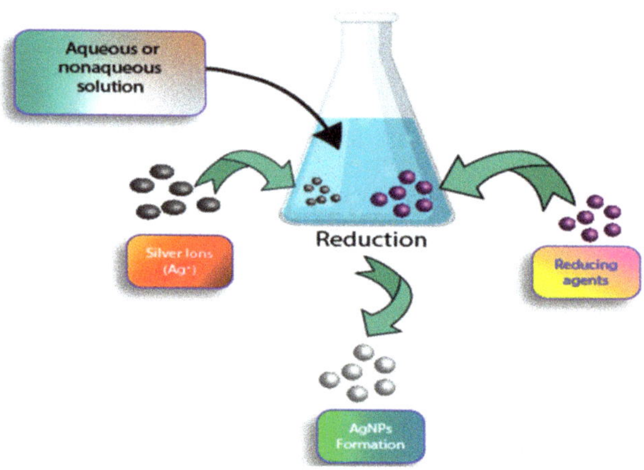

Figure 5. Chemical synthetization of silver nanoparticles through chemical reduction.

The positive attributes, such as being biocompatible, safe, and non-toxic to the environment, can be especially effective in food packaging, while posing no hazard to humans [62]. Natural antimicrobial agents come from various sources and they can be obtained from animals, plants, or microbial sources [63]. Some commonly used antimicrobial agents are from animal sources such as proteins (lactoferrin, ovotransferrin), enzymes (lysozyme, lactoperoxidase), and polysaccharides (chitosan). Additionally, microbial products such as nisin, pediocins, and bacteriocins are often used as antimicrobial agents in packaging. In a similar way, plant-derived antimicrobial agents are usually obtained in a form of essential oil or extract. Some of the most effective natural extracts are ginger, garlic, oregano, thyme, cinnamon, clove, coriander, and more. The presence of active compounds in these materials such as flavanols, terpenes, anthocyanins, phenolic acids, tannins, and stilbenes is responsible for the antimicrobial effects for specific microorganisms. Moreover, they could provide additional health benefits such as being nutritional supplements [48]. From this point of view, the use of antimicrobial agents from plant-derived sources could be an excellent choice especially for food packaging purposes. At present, the challenge related to the plant-derived antimicrobial compounds is due to their loss during high-temperature processing and reduced antimicrobial efficiency [64]. Other factors that restrict their application are the production cost and sometimes the strong aroma produced [63].

2.1. Polymeric Matrix Used in Antimicrobial Packaging

The selection of a polymeric material is dependent on the intended use on application and highly depends upon the properties of the polymer matrix [65]. Polymers such as poly(ethylene terephthalate) (PET), polyethylene (PE), poly(vinyl chloride) (PVC), polypropylene (PP), polystyrene (PS), and others have been investigated in this field. PVC is one of the examples used for polymers for packaging in the world. It has several advantages including flexibility, toughness, light weight, and ease of processing. Studies on PVC loaded with silver nanoparticles [66], zinc [67], and orange essential oil [68] have been reported. PET is another type of polymer that is useful and has the potential to be as good as PVC. This polymer has a good mechanical strength and toughness [69]. Further, PE is another widely used polymer. Low-density polyethylene (LDPE) is the cheapest among other polymers. Meanwhile, PP with high grade has a high melting point; thus, it is suitable for high-temperature packaging [47]. These materials are often used because of their good properties and relatively low cost [70]. Table 1 shows some examples of petroleum-based polymers that have been studied as a host with a variety of antimicrobial agents for packaging materials.

Table 1. Petroleum-based polymers for antimicrobial packaging.

Host Polymer	Antimicrobial Agent	Preparation Method	Targeted Organism	Ref.
PVC	Ag-NP	Solvent casting	*B. subtilis*, *A. niger*, and *F. solani*	[66]
PVC	PHE-Zn	Solvent casting	*E. coli* and *S. aureus*	[67]
PVC	Orange essential oil	Solvent casting	*E. coli* and *S. aureus*	[68]
PET	Ag-NP	Melt blending	*E. coli* and *Z. Bailii*	[71]
PET	LDH-p-hydroxybenzoate	Coating	*Salmonella* spp. and *C. jejuni*	[69]
PET	ZnO, TiO_2	Melt blending	-	[72]
LDPE	ZnO	Melt blending	*E. coli*	[73]
LDPE	Ag-NP	Melt blending	*E. coli*, *S. aureus*, *E. faecalis*, and *Salmonella enterica*	[74]
LDPE	Thymol	Solvent casting	*E. coli* and *Salmonella enterica*	[75]
PE	Carvacrol and menthol	Coating	*E. coli*, *S. aureus*, *L. innocua*, and *S. cervicea*	[76]
PP	Sorbic acid	Extrusion molding	*E. coli* and *S. aureus*	[77]
PP	Oregano EO	Melt blending	*B. thermosphacta*	[78]
PP	Carvacrol	Melt compounding	*E. coli and A. alternata*	[79]
PS	GO-p(VBC)	Solvent casting	*B. cereus*, *P. aeruginosa*, and *fungus candida*	[80]
PS	ZnO-NP $CaCO_3$-NP TiO_2-NP	Encapsulation	*S. aureus*, *P. aeruginosa*, *C. albicans*, and *A. niger*	[81]

Nanoparticle (NP); Pentaerythritol p-hydroxybenzoate ester-based zinc metal alkoxides (PHE-Zn); Layered double hydroxide para-hydroxybenzoate (LDH-p-hydroxybenzoate); Essential Oil (EO); graphene oxide/poly(4-vinylbenzyl chloride), GP(VBC).

2.2. Antimicrobial Packaging from Bio-Based Polymers

Despite the excellent performance of the petroleum-based polymers, they possess several limitations. One of the main constraints is their non-biodegradable properties that can cause short- and long-term pollution [69]. To address this issue, much attention has been given by the researchers and industries in developing antimicrobial packaging from bio-based polymers [82]. The use of bio-based polymers in place of traditional petroleum-based polymers could avoid the disposal problem and produce products that are environmentally friendly, safer, and non-toxic. Additionally, these materials give advantages in the sense that they are renewable and available abundantly in nature. The general classification of bio-based polymers is depicted in Figure 6. They can be broadly classified into three categories, which are (1) polymers directly extracted from biomass sources, (2) polymers chemically synthesized from bio-derived monomers, and (3) polymers produced directly by microorganisms [83]. Polymers such as poly(lactic acid) (PLA), starch, cellulose, and chitosan are gaining more favor in the antimicrobial packaging.

Figure 6. Classification of bio-based polymers based on their origin.

PLA is a type of biodegradable and renewable polymer that has been studied extensively for antimicrobial packaging. PLA is obtained from two major pathways: ring opening of lactide or direct polycondensation of lactic acid, a monomeric precursor derived from renewable resources. The monomer was produced from the fermentation process of sugar feedstock such as dextrose or chemical synthesis. Sugar feedstock can be obtained in two ways: firstly, directly from sources (sugar cane, sugar beets) or secondly, through conversion of starch from corn, potato, wheat, rice, or agricultural waste [70]. PLA has mechanical properties that are almost similar to commercial thermoplastics like PET, making it possible to be applied in a wide range of products [84]. Starch is a promising biodegradable and biocompatible polymer in the packaging industry. It is non-toxic and readily available. However, starch has a strong hydrophilic behavior, thus making it sensitive to moisture [85]. Meanwhile, cellulose is the most abundant renewable material, and it can exist in various forms upon modification. In packaging, cellulose can act as a filler or host polymer [86]. Chitosan is another attractive polymer that has been frequently investigated for antimicrobial packaging. Chitosan is generated from the deacetylation reaction of chitin. Chitosan possesses antimicrobial properties and, thus, can be used as a host and antimicrobial agent at the same time [47]. Table 2 shows some examples of bio-based polymers reported for antimicrobial packaging in recent years. They are being used with a variety of antimicrobial agents, both natural and synthetic, with various types of preparation methods involving solvent casting and encapsulation and being targeted for different types of microorganisms. In a way, bio-based polymers show a great potential in antimicrobial packaging; however, their relatively high cost compared to the traditional polymers somehow limits their applications on a larger scale.

Table 2. Antimicrobial packaging systems utilizing bio-based polymers.

Host Polymer	Antimicrobial Agent	Preparation Method	Targeted Organism	Ref.
PLA	ZnO, MgO, TiO_2	Solvent casting	*E. coli*	[70]
PLA	ZnO	Solvent casting	*E. coli* and *L. monocytogenes*	[87]
PLA	TV-EO, EEP	Solvent casting	*S. aureus* and *Penicillium* sp.	[88]
Starch	Nisin and Natamycin	Solvent casting	*B. cereus* and *A. niger*	[89]
Starch	Ferulic acid, Cinnamic acid	Melt blending	*E. coli* and *L. innocua*	[90]
Starch	Carvacrol, montmorillonite	Solvent casting	*E. Coli*	[62]
Carrageenan	Orange essential oil, Trehalose	Solvent casting	*S. aureus*, *E. coli* and *C. albicans*	[91]
κ-Carrageenan	Olive leaves extract	Solvent casting	*E. coli*	[92]
κ-Carrageenan	CuS-NP	Solvent casting	*S. aureus* and *E. coli*	[93]
Nanocellulose	Nisin	Solvent casting	*L. monocytogenes*	[94]
Carboxymethyl Cellulose	Curcumin, Zinc Oxide	Solvent casting	*L. monocytogenes* and *E. coli*	[86]
Nanocellulose	Anthocyanin, Oregano essential oil	Solvent casting	*L. monocytogenes* and *E. coli*	[95]
Gelatin	Bacteriophages	Solvent casting	*S. aureus*	[96]
Gelatin	Curcumin	Solvent casting	*E. coli* and *L. monocytogenes*	[97]
Gelatin	Pomegranate peel powder	Solvent casting	*S. aureus*, *L. monocytogenes* and *E. coli*	[98]
Pectin	Copaiba oil	Solvent casting	*S. aureus* and *E. coli*	[99]
Pectin	Ag-NP	Solvent casting	*E.coli* and *Salmonella* Typhimurium	[100]
Pectin-Alginate	Carvacrol	Encapsulation	*E. coli*	[101]
Alginate	Sulphur-NP	Solvent casting	*E. coli* and *L. monocytogenes*	[102]
Alginate-Chitosan	ZnO-NP	Coating	-	[103]
Alginate-Chitosan	Nisin	Encapsulation	*L. monocytogenes*	[104]
Chitosan-Starch	Grapefruit seed extract	Solvent casting	*A. niger*	[85]
Chitosan	Proanthocyanidins	Solvent casting	*E. coli*, *Salmonella*, *S. aureus*, and *L. monocytogenes*	[105]
Chitosan-Agar	Ag-NP	Solvent casting	*P. aeruginosa*, *E. coli*, and *S. aureus*	[106]
Agar	Ag-NP	Solvent casting	*A. hydrophilla*	[107]
Agar- Carboxymethyl Cellulose	Ag-MMT	Solvent casting	*B. subtilis* and *E. coli*	[108]

Thymus vulgaris essential oil (TV-EOs); ethanolic extract of Mediterranean propolis (EEP); Silver modified montmorillonite (Ag-MMT).

3. Types of Antimicrobial Packaging

The antibacterial, antifungal, and antioxidant activities can be prompted by the main polymer used for packaging or by addition of numerous components from natural agents (bacteriocins, essential oils, natural extracts, etc.) to synthetic agents, both organic and inorganic (Ag, TiO_2 nanoparticles, ZnO, synthetic antibiotics, etc.) [46].

This review on antimicrobial packaging for various applications was supported with bibliometric analysis as a systematic approach. Data used in the present study were retrieved on 8 June 2021 from Scopus. Data from June 2021 onwards were not considered in this study for data consistency. Presently, to this writing, the keyword search analysis in Scopus on the query string (TITLE-ABS ("antimicrobial packaging")) AND TITLE-ABS (food*) AND PUBYEAR < 2021 OR PUBDATETXT (("January 2021" OR "February 2021" OR "March 2021" OR "April 2021" OR "May 2021")) AND (EXCLUDE (PUBYEAR, 2022)) AND (LIMIT-TO (LANGUAGE, "English")) resulted in 306 documents (Figure 7) wherein 195 were research articles, 56 were book chapters, 33 were review works, 18 were conference papers, and 4 were books (8 June 2020).

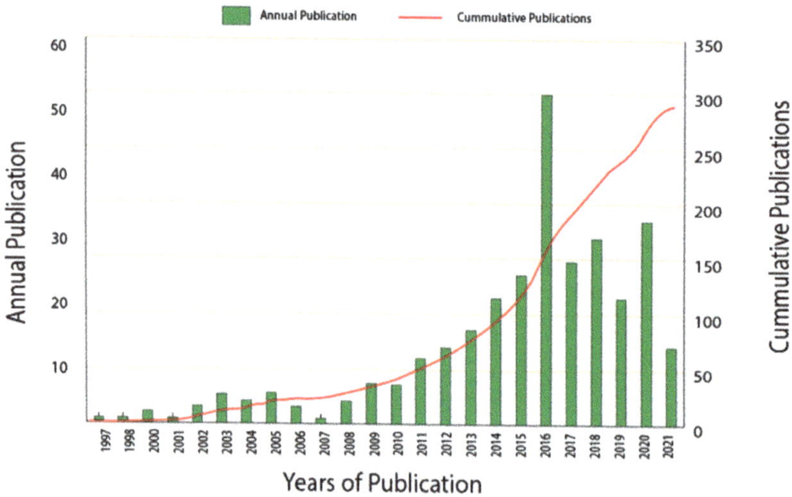

Figure 7. Annual and cumulative publications on antimicrobial packaging for various applications.

There are several forms of antimicrobial packaging, which are (1) addition of sachets/pads containing volatile antimicrobial agents into packages, (2) incorporation of volatile and non-volatile antimicrobial agents directly into polymers, (3) coating or adsorbing antimicrobials onto polymer surfaces, (4) immobilization of antimicrobials to polymers by ion or covalent linkages, and (5) use of polymers that are inherently antimicrobial [25,109].

Overall, the antimicrobial packaging strategy is classified into two groups, either direct or indirect contact between antimicrobial surface and the preserved food [46]. Table 3 briefly explains the definition, types, and function of the antimicrobial packaging strategies.

Table 3. Description of antimicrobial packaging strategies.

Strategies	Definition	Types	Function
Antimicrobial sachet or pad	The most common type of antimicrobial packaging. The sachets or pads that contain antimicrobial packaging are attached, enclosed, or loose in the interior of a package.	Three types of antimicrobial agents added in the sachets or pads are oxygen absorbers, moisture absorbers, and ethanol vapor generators.	To prevent oxidation, inhibit growth of molds, and lower water activity.
Direct integration in polymer	Any polymer used for packaging is incorporated with antimicrobial agents.	Edible films incorporated with nisin, lysozymes, antimicrobial enzymes (lactoferrin and lactoperoxidase), antimicrobial peptides (magainins, cecropins, natural phenols, antioxidants), metals (copper), and zeolites substituted by 1–3% silver incorporated into polyethylene, polypropylene, nylon, and butadiene styrene.	To disrupt the enzymatic activity of microbial cells and to prevent surface growth in packages.
Antimicrobial coating	Applying antimicrobial coatings on the polymer surfaces such as films, wax paper, and cellulose casing.	Waxes, fungicides, sorbic acid, polyethylene films coated with nisin/methylcellulose, poultry coated with nisin/zinc	For wrapping or packaging materials.
Immobilization of antimicrobials to polymers by ionic or covalent linkages	Ionic and covalent immobilization of antimicrobials onto polymers with the presence of functional groups and spacer molecules that link antimicrobial agents to polymers surfaces.	Antimicrobial agents with functional groups are peptides, enzymes, polyamines, and organic acids, whereas antimicrobial compounds with functional groups are enzymes, peptides, polyamines ethylene vinyl acetate, ethylene methyl acrylate, ethylene acrylic acid, ethylene methacrylic acid, ionomer, nylon, polystyrene, etc.	To reduce antimicrobial activity per unit area such as in proteins and peptides due to change in conformation and denaturation by solvents.
Inherently antimicrobial polymer	Cationic polymers that are inherently antimicrobial, and physical modification of polymers were used in films and coatings.	Chitosan and poly-1-lysine polymers films and coatings, polyamide films treated with UV irradiation.	It acts as a barrier between the nutrients contained and microorganisms to protect them from fungal degradation.

Firstly, the most common strategy is by having the antimicrobial sachet or pad with antimicrobial substance inside a sachet and added to the food packaging [46,110]. The antimicrobial compounds are released from the sachets into the headspace of packaging or to the surface of food products and subsequently inhibit the growth of food-borne pathogens [111]. The most popular antimicrobial agents for active packaging include nisin, chitosan, potassium sorbate, silver substituted zeolite, and essential oils [112].

Secondly is the inclusion or embedding of antimicrobials directly into the interior of the polymer films. In this method, the antimicrobial compounds are inside polymer films and introduced during the manufacturing process of these films [111]. The materials used in edible films should be Generally Recognized as Safe (GRAS) and may be eaten with food [113]. Thirdly is by covering the polymer surfaces with a layer of antimicrobial. The antimicrobial agents are coated onto the surfaces of the polymer films [98]. Then, the antimicrobial substance would either evaporate into the headspace or migrate into the food through diffusion [110].

The following antimicrobial packaging strategy is immobilization of antimicrobials in the polymers using ion or covalent linkages. This method needs (1) antimicrobial agents with functional groups that can be linked to the polymers and (2) antimicrobial compounds containing functional groups such as enzymes, peptides, and polyamines [98]. Lastly is the permanent existence of antimicrobial polymers. Some polymers used to construct films inherently have antimicrobial properties themselves [111]. For example, chitosan is categorized as an active food packaging material because of its inherent antimicrobial properties and capacity to carry various active components [114].

4. Performance of Antimicrobial Packaging

As mentioned before, antimicrobial packaging is frequently prescribed as a combination of antimicrobial material and agents based on a specific matrix, which, in turn, leads to different types of packaging functions and uses. The review of correlation analysis between antimicrobial properties in terms of its use and advantages toward its application in food safety is presented in Table 4. What stands out in this table is the general pattern of each antimicrobial agent such as volatile gas form [115], silver compound [96], sanitizer and fungicide [116], plant extract [115], plant essential oil [115,117–120], enzyme [121], chitosan [122,123], bacteriocin [124,125], and inorganic nanoparticle [126,127] in controlling the growth of microorganisms, which, in turn, leads to indicators of prolonged shelf life of food, which is the essence of food safety. An important finding that emerged from the data was that antimicrobial agents can inhibit the growth of pathogenic microorganisms in food such as *Bacillus cereus*, *Escherichia coli*, *Listeria monocytogenes*, *Salmonella* spp., *and Staphylococcus aureus* [128] (Table 4). In addition, a further striking factor to emerge from the table is how antimicrobial material also shows a similar ability to protect from microbial growth by adding value by controlling the moisture migration and nutrient oxidation [115]. Taken together, these trends suggest that there is an association between antimicrobial properties toward superior food safety and longest shelf life.

The next part of the review was concerned with the performance of antimicrobial packaging towards environmental impact, as shown in Figure 8. Looking at Figure 8, it is apparent that the antimicrobial packaging showed a positive impact on the environment as well as in the ecosystem cycle. One concept that emerged during the extensive review was antimicrobial agents and material that normally come under renewable raw material provide added value in ensuring ecosystem sustainability. This is because, if antimicrobial packaging is made from a combination of antimicrobial agents and materials that are based on renewable raw materials, it may accelerate the biodegradation process and further stabilize the ecosystem balance. These results suggest that antimicrobial packaging is not only able to show the ability of controlling the growth of microorganisms and prolonging the shelf life of food, which is very important in food safety, but also has a positive effect on the environment [129].

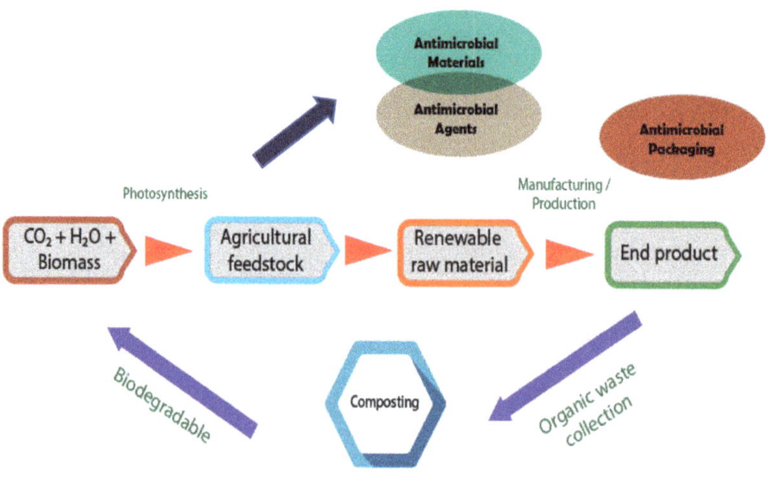

Figure 8. Interrelated antimicrobial packaging towards environmental impact.

Table 4. Use, Advantages, and Applications of Different Types of Antimicrobial Properties in Antimicrobial Packaging for Food Safety.

Type of Antimicrobial Packaging Properties	Example	Use	Commercial Product	Advantages	Application	Ref.
			Antimicrobial Agents			
Volatile gas form	Chlorine dioxide, ethanol and sulfur dioxide	In sachets/pads that are attached to the internal part of the package	• Maicrogarde™, (BarrierSafe International Inc., Lake Forest, IL, USA)	• Initiates the solid-state dry reaction, subsequently producing chlorine dioxide that diffuses throughout the package to inhibit microbial contamination and control odor • Inhibit aerobic total viable count	Iceberg lettuce	[115]
			• Ethicap™ (Freund Corp., Tokyo, Japan)	• Able to retard mold growth	Bakery and dried fish products	
Silver compound		Inhibit a wide range of microorganisms, bacteria, and mold by disrupting the microbial enzymes activities	• Zeomic™	• Able to control the growth of gram-positive bacteria, gram-negative bacteria, and fungi	Chopping board, food packaging film, and glove and lunch box	[109]
			• Aglon® • Novaron® • Cleanaid™		Food packaging	
Sanitizer and fungicide			• Mircoban®	• Inhibited the growth of *S. Typhimurium* (ATCC 14028), *S. aureus* (ATCC 12598), *B. thermosphacta* (ATCC 11509), *B. subtilis* (ATCC 605), *E.coli* (ATCC 25922), *S. flexneri* (ATCC 12022), and several strains of *E.coli* O157:H7.	Packaging of meat	[116]
Plant extract	Wasabi extract		• Wasapower™	• Volatile allyl isothiocyanate (AIT) play inhibits bacteria such as *E.coli*, *S.aureus*, fungi *A. niger*, and *P.italucum*.	Sushi products	[115]
Plant essential oils	Linalool, thymol, carvacrol, clove oil, cinnamaldehyde, basil essential oil			• Inhibit microorganism growth through disturbance of the cytoplasmic membrane, disrupting the proton motive force, electron flow, active transport, and inhibition of protein synthesis.	Food packaging	[115]
	Grapefruit seed extract			• Inhibit the growth of pathogenic bacteria such as *L. monocytogenes*, *E.coli*, and *S. Typhimurium*	Packaging ground beef	[118]
	Oregano essential oil and citral			• Reduce the number of *E. coli*, *S. enteric*, and *L. monocytogenes* in salad	Packaging salad	[119]
	Allyl isothiocyanate (AIT)			• Effective against *E. coli*	Ground meat patties	[120]

Table 4. *Cont.*

Type of Antimicrobial Packaging Properties	Example	Use	Commercial Product	Advantages	Application	Ref.
Plant extract	Garlic oil			• Inhibit the growth of microbial on sprout • Effectively reduce the number of gram-positive and gram-negative bacteria • Inhibit the growth *S. aureus* and *B. cereus*	Sprout	[117,130]
Enzyme	Lysozyme			• Effective on *E. coli* O157:H7 and *S. typhimurium*	Beef patties	[121]
Chitosan				• Reduce the number of *E. coli* and *S. aureus*	Food packaging,	[131]
				• Reduce population of total aerobic count in pork	Vacuum packaging of refrigerated grilled pork	[132]
	Coated on plastic film, incorporated with 1% oregano oil and clove essential oil			• Control cheese exhibiting *L. monocytogenes*, *S. aureus*, and *E. coli O157:H7*	Vacuum- sealed cheese	[123]
	Incorporated with nisin and Thymus kotschyanus essential oil			• Inhibition level of *L. monocytogenes*, *E. coli*, *S. aureus*, and *S. Typhimurium*	Food packaging	[122]
Bacteriocin	Enterocins A and B			• Control of proliferation of *L. monocytogenes*	Oyster and beef	[125]
	Pedicin			• Reduction of *L. monocytogenes*	Raw chicken	[124]
Inorganic Nanoparticles	Titanium dioxide (TiO_2)			• Inactivate microorganism by oxidizing the polyunsaturated phospholipids' component of the cell membrane • Reduction of *E. coli* and *Pseudomonas* spp.	Food packaging	[126]
	Zinc oxide (ZnO)			• Exhibited *E. coli* and *S. aureus*	Food packaging	[127]
Biodegradable materials	Edible biopolymer Food-grade additives		Antimicrobial Material • Protein, lipids, and polysaccharides • Plasticizers • Colorant • Flavors • Emulsifiers • antioxidants	• Protected from microbial growth, moisture migration, and nutrient oxidation	Packaging of nuts, candies, and fruits	[115]

Many studies have been conducted on the mechanism of silver nanoparticles as an active packaging ingredient in packaging [53,55,58,133–139]. Figure 9 shows the possible mechanism of silver nanoparticles' action in the antimicrobial food packaging. The mechanism of the AgNPs were said to impede the cell wall synthesis in the cell [139]. Daneshniya et al. (2020) [140] reported that the AgNPs in the range of 1–10 nm in size attached to the cell membrane and disrupted the membrane functions such as permeability and respiration. The silver nanoparticles were believed to be penetrating the bacteria cell and causing further wreckage by associating with the thiol groups from the respiratory chain proteins and transport proteins such as DNA, glutathione (GSH), and thioredoxin, which led to hindering their functions [137]. The damage towards the thiol groups was also stated to be due to the release of silver ions (Ag+) from the silver nanoparticles that are very reactive when they are reacted with the cell membrane that was negatively charged [41]. The reaction resulted in further involvement in the bactericidal effect of silver nanoparticles and led to cell death [135,137]. However, the specific action of silver nanoparticles' mechanism was still unclear throughout these extensive studies.

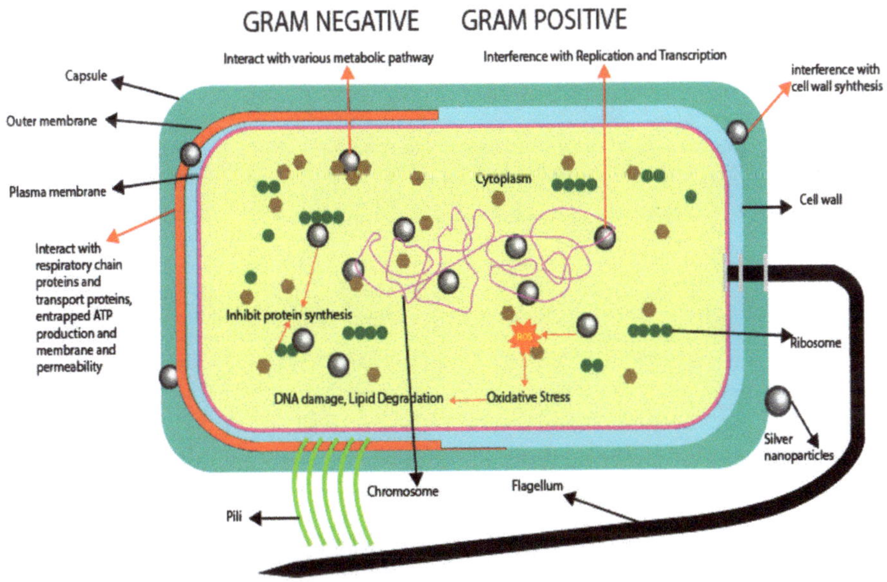

Figure 9. Possible mechanism of silver nanoparticles towards microbes.

All organic and inorganic compounds that were widely studied in the past research such as chitosan, chitin, titanium oxide (TiO_2), and copper (Cu) have shown great antimicrobial effects toward the bacteria, microorganisms, and enzymes, as mentioned in Table 5. Table 5 depicts the widely used material of antimicrobial agents in producing the antimicrobial food packaging.

Table 5. Advantages and disadvantages of widely used materials of antimicrobial agent for food packaging industries.

Antimicrobial Agent	Advantages	Disadvantages
Silver	• Act as a catalyst instead of chemically reacted with microorganisms in their destruction, and microorganism cannot resist them [141]. • Eliminate the risk of genetic mutations of microorganisms due to direct use of toxins [141]. • Can be combined with both degradable and nondegradable biomaterials, resulting in improving the permeability of the film, quality of the product, and mechanical properties of the coating [54,141]. • Stable in very high temperature compared to other compounds [142].	• Can cause particles to migrate from packaging to the food if it is used at a high level [143,144].
Titanium oxide	• High stability, extensive range of antibiosis [145]. • Biologically inactive, demonstrates quite low toxicity, thus low risk to human [146]. • Shows no absorption or tissue storage of TiO_2 and no hazardous effects for occupational workers and public health [147].	• All molecular sizes of TiO_2 and crystal forms (anatase and rutile) might cause phototoxicity due to reactive species (ROS) under UV radiations [147]. • Reactive oxygen species (ROS), such as hydrogen peroxide (H_2O_2), hydroxyl radicals, and superoxide lead, which can lead to the oxidative stress pathway. This is one of the ways in which TiO_2 and Ag NPs exert their toxic effects and interrupt the life cycle of Drosophila through the ROS generation enhancement and DNA damage that led to related adverse consequences [148].
Copper	• Inhibits or declines bacteria, viruses, and fungi growth [141] • Increases the film thermal stability and mechanical properties [141] • Can inhibit survival of microorganisms [149]	• Toxic, especially at the nanometre dimension [141] • Nanoscale size Cu increases their reaction since the surface atoms are increased, which make them highly reactive sites, causing severe antimicrobial behavior and toxicity [141,150]
Chitosan	• Water-soluble cationic polymer due to positive charge on its amino groups [151]. • Polycationic, non-toxic, biocompatible, and biodegradable [152–155]. • Soluble in dilute acids in pH less than 6.0–6.5, for example, acetic acid, formic acid, lactic acid, and HCI [156] • Good mechanical properties and can be consumed along with the product in the package [157,158]	• Insoluble at neutral and higher pH due to the D-glucosamine [156,159]
Chitin	• Renewable, biocompatible, biodegradable, and non-toxic compounds [160–162]. • Abundant [151] • Antioxidant [163]	• Highly hydrophobic, thus insoluble in water and even organic solvent [151]
Lysozyme	• Naturally present in avian eggs and mammalian milk [160] • Cost effective [164] • Showed high activity towards Gram-negative bacteria and moderately effective against Gram-positive bacteria [165]	• Showed no action towards yeasts or fungi [165]

5. Issues Related to Antimicrobial Packaging

Application of antimicrobial packaging systems based on biopolymers incorporated with different bioactive agents possesses immense potential for improving the food quality and safety along with a possible increment in shelf life. As mentioned earlier, a variety of bioactive substances, both synthetic and natural, such as essential oils, antimicrobial peptides, enzymes, etc., have been investigated and applied in antimicrobial packaging systems. Several investigations on the subject have indicated the potential of antimicrobial packaging systems in effectively inhibiting the targeted spoilage microorganisms, employing a suitable combination of biopolymer and a bioactive compound to produce an antimicrobial film [166].

Despite all the above advantages of antimicrobial packaging, there are some challenges and limitations, which should be discussed and overcome. One of the main challenges is health issues and risks regarding the safety and migration of nanoparticles of antimicrobial agents. The possibility of inhalation by the respiratory system, skin penetration through skin nodes, and unintentional migration and ingestion of nanoparticles by the digestive system might badly affect human health.

5.1. Safety Issues

Numerous studies have found that nanoparticles of antimicrobial agents are effectively proven in enhancing the barrier, mechanical, and antimicrobial properties of antimicrobial packaging when appropriate amounts of antimicrobial agents are incorporated into packaging materials. Figure 10 illustrates how nanoparticles of antimicrobial agents can improve the barrier properties as compared with pure polymer materials. Nanoparticle and pure polymer matrix properties are among the most important factors that determine the properties of the resulting composite. For food packaging applications, nanocomposites that have been studied the most are clay and polymer nanocomposites, while bio-based polymers that have been studied the most are PLA. These nanomaterials will intensify the water and serve as moisture-repellent properties of food packaging materials.

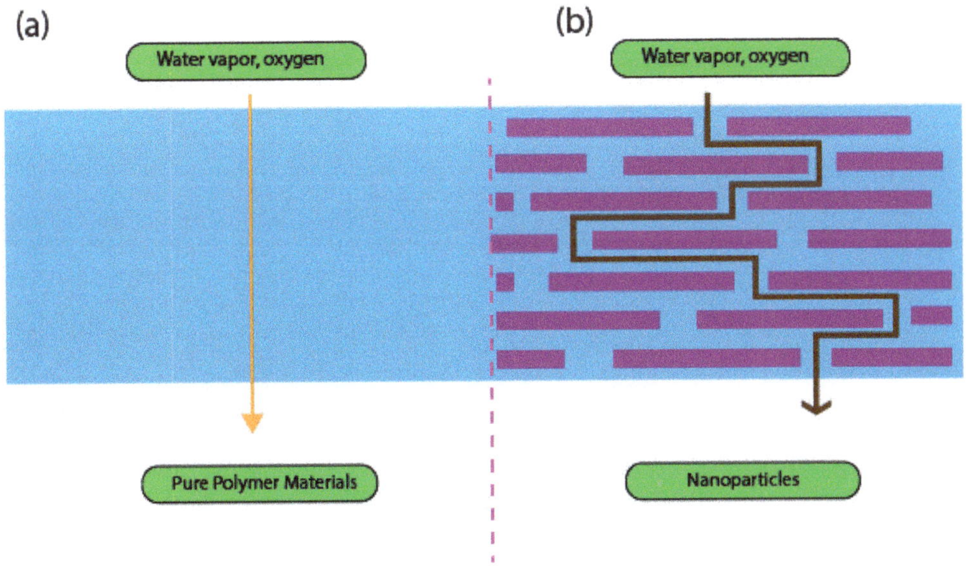

Figure 10. Water vapor and oxygen passing through (**a**) pure polymer materials and (**b**) nanoparticles.

However, there are a few limitations and issues that need to be reconsidered. In terms of migration, nanoparticles are susceptible to migrate from packaging into the food, which depend on nanomaterial characteristics such as size, concentration, shape, and dispersion. Other than those, there are environmental factors (temperature, mechanical stress), food condition (composition and pH), polymer properties (viscosity and structure), and contact duration. These will bring limitations and potentially result in adverse health effects. It has been reported that some nanoparticles can cause intracellular damage, pulmonary inflammation, and vascular diseases [167]. Thus, a detailed toxicological analysis is needed to explain the risks.

There are three types of migrations of substances into food: (1) overall migration limit (OML), which evaluates the total weight of extracted substances, which is non-specific and above the limit that is allowed to be penetrate into food; (2) specific migration limit, which measures the concentration of the material-specific restricted substances based on their toxicological risks using advanced detecting assays; and (3) maximum permitted quantity (QM), which measures the maximum level of the residual given substance that can migrate from a material into foodstuffs or simulants. To ensure the overall quality of the plastics, the overall migration to a food of all substances together may not exceed the OML, which, for polymers of about 60 mg/kg of food (or food simulant) or 10 mg/dm2 of the contact material, will usually be used for inertness of the substances. Regulation No. (EU) 10/2011 from Plastics Regulation and No. (EU) 2016/1416 from the European Commission-published Commission Regulation ensure the safety of plastic materials with the use of migration limits, which specify the maximum amount of substances allowed to migrate to food. They impose the permitted value of 5 to 25 mg zinc per kg food (25 to 5 mg/kg food) for food contact items based on the SML consideration. Furthermore, 40 mg/day of zinc daily consumption for the human body is the restricted amount level for food contact materials by the National Institutes of Health [168]. A nanocomposite containing 0.5 g/L 0.5 g/L ZnO NPs is the permitted level of value of migration [169]. There was a study conducted by Bumbudsanpharoke et al. who experimented and discovered the migration of Zn^{2+} from LDPE-ZnO nanocomposite films, in which the level of migrated Zn^{2+} (3.5 mg L^{-1}) was considered safe for human health due to a lower value than the specific migration limit provided by European Plastics Regulation (EU No. 10/2011) [170]. Examples for Specific Migration Limits are Polycyclic Aromatic Hydrocarbons (PAH) from carbon black and Bisphenol A (BPA) from polycarbonate plastic, most of the time known to be carcinogens. However, except in some cases, the level of migrated Zn^{2+} increased despite the migrated level being lower than the maximum migration limit based on the National Institutes of Health for food contact materials with the existence of essential oil in the nanocomposite [171].

5.2. Production Cost

Antimicrobials of nanoparticles are diversely used in packaging materials due to their advanced properties for industrial purposes. Most common uses of antimicrobial properties are Ag, TiO_2, and ZnO of antimicrobial packaging systems. These types of nanoparticles are used in the lab and the production cost could be considered as way more affordable than the production cost for the real industry, in which the production costs required are 10 times more to be useful as the original one. The prices of antimicrobial agents are way more expensive in industrial scale, and scaling up the packaging for nanocomposites demands cutting-edge technologies, which may amplify the final cost, thereby reducing the market acceptability [172,173]. Apart from that, antimicrobial agents are frequently developed for a specific food and do not provide the same results with other types of food; thus, the price will be more expensive to buy several antimicrobial agents for several types of food.

5.3. Strong Aroma, Flavor, and Color

Essential oils of natural antimicrobial agents such as carvacrol, ginger/garlic oil, linalool, clove oil, thymol, basil, and cinnamaldehyde possess a high intensity of off-flavors. These types of essential oils have high antibacterial properties but have a strong smell and flavor, which inhibits the original flavor of the food, which represents the critical challenges for the food industry. Moreover, they carry a striking color. It was mentioned by [174] Bhullar et al. in 2015 that around 85–99% of essential oils contain phenolic and hydrophilic volatile terpenoids, which cause a generation of intense reddish color to the films. Furthermore, they have a sharp flavor, which restricts their applications in the food packaging industry constituents [174].

6. Conclusions and Future Perspective

The review presented here summarized comprehensive available information on the recent development of antimicrobial packaging, especially in food packaging industries. This review introduced a brief background on the concept of antimicrobial packaging and their principles, followed by the main components of the antimicrobial packaging composition. The discussions were narrowly focused into the types of antimicrobial packaging, the applications, the implementations of recent discoveries, and strategies aiming to curb microbial growth through innovative antimicrobial packaging. Among the demonstrated potential applications, their massive use in food packaging has received considerable interest compared to others. The reviewed research work from the literature offers evidence in favor of antimicrobial packaging use to control food quality over targeted perishable products, generally, and the current plan to execute the mass production of antimicrobial packaging in real food systems, specifically. The antimicrobial packaging synergistically made of selected green polymers incorporated with certain chemical agents, natural agents, or probiotics have been shown to be effective to address issues on antimicrobial activity and plastic pollution towards sustainable development. The strong ground supported by the regulatory authorities, the commitments from industry players, and the growing public awareness are pacing the anticipation toward the use of antimicrobial packaging. The strategies of hybridizing those inexpensive, abundant natural polymers with functional additives will enhance the polymeric properties in order to satisfy the stringent requirement set by the packaging industry. At the time of writing, countless efforts were made to accelerate the mass production of antimicrobial packaging throughout technological advancement. However, there are a few challenges that are faced during the replacement transition from conventional petroleum-based plastic packaging towards antimicrobial packaging materials. The consideration towards the suitably formulated components between various antimicrobial agents and polymeric matrices needs to be really understood. For instance, some of the potential antimicrobial agents such as essential oils might also experience a high loss rate due to rapid volatilization due to several causes. Oxidative and polymerization processes may result in a loss of quality and pharmacological properties. A slow and sustained release of the essential oils will be useful to maintain food quality due to the presence of a high concentration of essential oils trapped in the packaging. Further, in-depth research is required to limit volatile loss and to sustain the durability and efficiency of the fabricated antimicrobial packaging materials at their optimum.

The advanced technology offered in the innovative antimicrobial packaging also has countered the resistance phenomenon in microbes to conventional processing technologies. Despite the excellent antimicrobial activity in controlling the microbial contamination by reducing the growth rate and extending the lag period of targeted foodborne pathogens, the depth of evaluation of the migration of active antimicrobial agents throughout the packaging needs to be accentuated. The importance of preventing the migration of active substances throughout the packaging materials has drawn attention from consumers and regulatory authorities, in regard to human health due to the fact that some can cause irritation due to cytotoxic effects while others can be allergens. Migration of undesirable substances must be strictly under the limit established by regulations to protect the safety of the consumers. For nanoparticles-embedded packaging, the specific toxicological tests are of the utmost necessity for future studies to clarify that prolonged consumption of packaged food from these innovative packaging materials are safe to humans, without long-term side effects. The application of nanoparticles into the food packaging needs to have a concise guide and should be carefully assessed prior to being available on the market.

Despite having many outstanding properties and a realm of possibilities for antimicrobial agents for the packaging industry for retarding microbial growth and improving the shelf life of foods, more comprehensive research is still a requirement, considering the abovementioned limitations. Otherwise, the advantages of a prolonged shelf life may come at the expense of major unforeseen health repercussions.

Apart from that, the possible incoming threats to both terrestrial and aquatic ecosystems and the adverse effects of these antimicrobial substances-embedded packaging to long-term environmental impact should be considered. The disposal issue regarding the probability of the packaging containing nanoparticles and their subsequent breakdown, which could result in the release of unstable forms of chemical compounds into our natural ecosystems, should be highlighted. More future research should be focused on fully biodegradable polymers such as blends of starch and others for their high-efficiency usage in food packaging. Biopolymers are prominent candidates to be modified or combined with an antimicrobial substance to develop the antimicrobial systems with applications in several fields and in good directions to reach these goals.

Author Contributions: Writing—Reviewing and Editing, S.H.K., M.R., F.A., S.A., F.F., A.A.K., M.N.N., N.S., M.S.Z.M.D., M.S.M.B., H.S. and L.C.A. All authors have read and agreed to the published version of the manuscript.

Funding: This research received no external funding.

Institutional Review Board Statement: Not applicable.

Informed Consent Statement: Not applicable.

Data Availability Statement: The data that support the findings of this study are available on request from the corresponding author.

Acknowledgments: The authors gratefully acknowledge the technical and financial support from the Universiti Teknologi MARA (UiTM). This research was funded by IIUM-UMP-UiTM Sustainable Research Collaboration Grant 2020 (SRGC), 600-RMC/SRC/5/3 (035/2020).

Conflicts of Interest: The authors declare no conflict of interest.

References

1. Sangroniz, A.; Zhu, J.B.; Tang, X.; Etxeberria, A.; Chen, E.Y.X.; Sardon, H. Packaging materials with desired mechanical and barrier properties and full chemical recyclability. *Nat. Commun.* **2019**, *10*, 3559. [CrossRef]
2. Fu, Y.; Dudley, E.G. Antimicrobial-coated films as food packaging: A review. *Compr. Rev. Food Sci. Food Saf.* **2021**, *20*, 3404–3437. [CrossRef] [PubMed]
3. Kamarudin, S.H.; Abdullah, L.C.; Aung, M.M.; Ratnam, C.T. Thermal and structural analysis of epoxidized jatropha oil and alkaline treated kenaf fiber reinforced poly(Lactic acid) biocomposites. *Polymers* **2020**, *12*, 2604. [CrossRef]
4. Shireesha, Y.; Nandipati, G.; Chandaka, K. Properties of hybrid composites and its applications: A brief review. *Int. J. Sci. Technol. Res.* **2019**, *8*, 335–341.
5. Karthi, N.; Kumaresan, K.; Sathish, S.; Gokulkumar, S.; Prabhu, L.; Vigneshkumar, N. An overview: Natural fiber reinforced hybrid composites, chemical treatments and application areas. *Mater. Today Proc.* **2019**, *27*, 2828–2834. [CrossRef]
6. Dashtizadeh, Z.; Abdan, K.; Jawaid, M.; Khan, M.A.; Behmanesh, M.; Dashtizadeh, M.; Cardona, F.; Ishak, M. Mechanical and thermal properties of natural fibre based hybrid composites: A review. *Pertanika J. Sci. Technol.* **2017**, *25*, 1103–1122.
7. Mishra, S.; Mohanty, A.K.; Drzal, L.T.; Misra, M.; Parija, S.; Nayak, S.K.; Tripathy, S.S. Studies on mechanical performance of biofibre/glass reinforced polyester hybrid composites. *Compos. Sci. Technol.* **2003**, *63*, 1377–1385. [CrossRef]
8. Venkata Reddy, G.; Venkata Naidu, S.; Shobha Rani, T. Impact properties of kapok based unsaturated polyester hybrid composites. *J. Reinf. Plast. Compos.* **2008**, *27*, 1789–1804. [CrossRef]
9. Thwe, M.M.; Liao, K. Effects of environmental aging on the mechanical properties of bamboo-glass fiber reinforced polymer matrix hybrid composites. *Compos. Part A Appl. Sci. Manuf.* **2002**, *33*, 43–52. [CrossRef]
10. Tsampas, S.A.; Greenhalgh, E.S.; Ankersen, J.; Curtis, P.T. Compressive failure of hybrid multidirectional fibre-reinforced composites. *Compos. Part A Appl. Sci. Manuf.* **2015**, *71*, 40–58. [CrossRef]
11. Statistica. Impact of Coronavirus (COVID-19) on Consumers Ordering Take Away Food Post-Pandemic in the Asia Pacific Region in 2020, by Country or Region. 2021. Available online: https://www.statista.com/statistics/1111688/apac-covid-19-impact-on-consumers-ordering-food-post-pandemic-by-country-or-region/ (accessed on 25 November 2021).

12. Tran, L.Q.N.; Fuentes, C.; Verpoest, I.; Van Vuure, A.W. Tensile Behavior of Unidirectional Bamboo/Coir Fiber Hybrid Composites. *Fibers* **2019**, *7*, 62. [CrossRef]
13. Nurazzil, N.M.; Khalina, A.; Sapuan, S.M.; Laila, A.M.D.; Rahmah, M. Curing behaviour of unsaturated polyester resin and interfacial shear stress of sugar palm fibre. *J. Mech. Eng. Sci.* **2017**, *11*, 2650–2664. [CrossRef]
14. Norrrahim, M.N.F.; Kasim, N.A.M.; Knight, V.F.; Halim, N.A.; Shah, N.A.A.; Noor, S.A.M.; Jamal, S.H.; Ong, K.K.; Wan Yunus, W.M.Z.; Farid, M.A.A.; et al. Performance evaluation of cellulose nanofiber reinforced polymer composites. *Funct. Compos. Struct.* **2021**, *3*, 024001. [CrossRef]
15. Norrrahim, M.N.F.; Ariffin, H.; Yasim-Anuar, T.A.T.; Hassan, M.A.; Ibrahim, N.A.; Yunus, W.M.Z.W.; Nishida, H. Performance evaluation of cellulose nanofiber with residual hemicellulose as a nanofiller in polypropylene-based nanocomposite. *Polymers* **2021**, *13*, 1064. [CrossRef]
16. Hammond, S.T.; Brown, J.H.; Burger, J.R.; Flanagan, T.P.; Fristoe, T.S.; Mercado-Silva, N.; Nekola, J.C.; Okie, J.G. Food Spoilage, Storage, and Transport: Implications for a Sustainable Future. *Bioscience* **2015**, *65*, 758–768. [CrossRef]
17. Ilyas, R.A.; Sapuan, S.M.; Mohd Nurazzi, N.; Faiz Norrrahim, M.N.; Ibrahim, R.; Atikah, M.S.N.; Huzaifah, M.R.M.; Radzi, A.M.; Izwan, S.; Noor Azammiah, A.M.; et al. Chapter 3—Macro to nanoscale natural fiber composites for automotive components: Research, development, and application. In *Biocomposite and Synthetic Composites for Automotive Applications*; Sapuan, S.M., Ilyas, R.A., Eds.; Woodhead Publishing Series in Composites Science and Engineering; Woodhead Publishing: Sawston, UK, 2021; pp. 51–105, ISBN 978-0-12-820559-4.
18. Ilyas, R.A.; Sapuan, S.M.; Norrrahim, M.N.F.; Yasim-Anuar, T.A.T.; Kadier, A.; Kalil, M.S.; Atikah, M.S.N.; Ibrahim, R.; Asrofi, M.; Abral, H.; et al. Nanocellulose/starch biopolymer nanocomposites: Processing, manufacturing, and applications. In *Advanced Processing, Properties, and Applications of Starch and Other Bio-Based Polymers*; Al-Oqla, F.M., Sapuan, S.M., Eds.; Elsevier Inc.: Amsterdam, The Netherland, 2020; pp. 65–88.
19. Mtui, G.Y.S. Recent advances in pretreatment of lignocellulosic wastes and production of value added products. *Afr. J. Biotechnol.* **2009**, *8*, 1398–1415.
20. Zakaria, M.R.; Fujimoto, S.; Hirata, S.; Hassan, M.A. Ball milling pretreatment of oil palm biomass for enhancing enzymatic hydrolysis. *Appl. Biochem. Biotechnol.* **2014**, *173*, 1778–1789. [CrossRef]
21. Ramamoorthy, S.K.; Skrifvars, M.; Persson, A. A review of natural fibers used in biocomposites: Plant, animal and regenerated cellulose fibers. *Polym. Rev.* **2015**, *55*, 107–162. [CrossRef]
22. Abdul Khalil, H.P.S.; Davoudpour, Y.; Islam, M.N.; Mustapha, A.; Sudesh, K.; Dungani, R.; Jawaid, M. Production and modification of nanofibrillated cellulose using various mechanical processes: A review. *Carbohydr. Polym.* **2014**, *99*, 649–665. [CrossRef] [PubMed]
23. Jonoobi, M.; Harun, J.; Tahir, P.M.; Shakeri, A.; Saifulazry, S.; Makinejad, M.D. Physicochemical characterization of pulp and nanofibers from kenaf stem. *Mater. Lett.* **2011**, *65*, 1098–1100. [CrossRef]
24. Chen, W.; Yu, H.; Liu, Y.; Hai, Y.; Zhang, M.; Chen, P. Isolation and characterization of cellulose nanofibers from four plant cellulose fibers using a chemical-ultrasonic process. *Cellulose* **2011**, *18*, 433–442. [CrossRef]
25. Alemdar, A.; Sain, M. Isolation and characterization of nanofibers from agricultural residues—Wheat straw and soy hulls. *Bioresour. Technol.* **2008**, *99*, 1664–1671. [CrossRef]
26. Jonoobi, M.; Harun, J.; Shakeri, A.; Misra, M.; Oksmand, K. Chemical composition, crystallinity, and thermal degradation of bleached and unbleached kenaf bast (*Hibiscus cannabinus*) pulp and nanofibers. *BioResources* **2009**, *4*, 626–639.
27. Norrrahim, M.N.F.; Ariffin, H.; Yasim-Anuar, T.A.T.; Ghaemi, F.; Hassan, M.A.; Ibrahim, N.A.; Ngee, J.L.H.; Yunus, W.M.Z.W. Superheated steam pretreatment of cellulose affects its electrospinnability for microfibrillated cellulose production. *Cellulose* **2018**, *25*, 3853–3859. [CrossRef]
28. Fahma, F.; Iwamoto, S.; Hori, N.; Iwata, T.; Takemura, A. Isolation, preparation, and characterization of nanofibers from oil palm empty-fruit-bunch (OPEFB). *Cellulose* **2010**, *17*, 977–985. [CrossRef]
29. Roy, S.; Rhim, J.W. Carboxymethyl cellulose-based antioxidant and antimicrobial active packaging film incorporated with curcumin and zinc oxide. *Int. J. Biol. Macromol.* **2020**, *148*, 666–676. [CrossRef]
30. Yasim-Anuar, T.A.T.; Ariffin, H.; Norrrahim, M.N.F.; Hassan, M.A.; Tsukegi, T.; Nishida, H. Sustainable one-pot process for the production of cellulose nanofiber and polyethylene/cellulose nanofiber composites. *J. Clean. Prod.* **2019**, *207*, 590–599. [CrossRef]
31. Sharip, N.S.; Yasim-Anuar, T.A.T.; Norrrahim, M.N.F.; Shazleen, S.S.; Nurazzi, N.M.; Sapuan, S.M.; Ilyas, R.A. *A Review on Nanocellulose Composites in Biomedical Application*; CRC Press: Boca Raton, FL, USA, 2020; ISBN 9780429327766.
32. Norrrahim, M.N.F.; Ariffin, H.; Yasim-Anuar, T.A.T.; Hassan, M.A.; Nishida, H.; Tsukegi, T. One-pot nanofibrillation of cellulose and nanocomposite production in a twin-screw extruder. *IOP Conf. Ser. Mater. Sci. Eng.* **2018**, *368*, 012034. [CrossRef]
33. Fareez, I.M.; Jasni, A.H.; Norrrahim, M.N.F. Nanofibrillated Cellulose Based Bio-phenolic Composites. In *Phenolic Polymers Based Composite Materials*; Jawaid, M., Asim, M., Eds.; Springer: Singapore, 2021; pp. 139–151, ISBN 9789811589324.
34. Faiz Norrrahim, M.N.; Mohd Kasim, N.A.; Knight, V.F.; Mohamad Misenan, M.S.; Janudin, N.; Ahmad Shah, N.A.; Kasim, N.; Wan Yusoff, W.Y.; Mohd Noor, S.A.; Jamal, S.H.; et al. Nanocellulose: A bioadsorbent for chemical contaminant remediation. *RSC Adv.* **2021**, *11*, 7347–7368. [CrossRef]
35. Jones, D.; Ormondroyd, G.O.; Curling, S.F.; Popescu, C.-M. Chemical compositions of natural fibres. In *Advanced High Strength Natural Fibre Composites in Construction*; Fan, M., Fu, F., Eds.; Woodhead Publishing: Duxford, UK, 2017; pp. 23–58, ISBN 9781782424468.

36. Sabbagh, F.; Muhamad, I.I. Production of poly-hydroxyalkanoate as secondary metabolite with main focus on sustainable energy. *Renew. Sustain. Energy Rev.* **2017**, *72*, 95–104. [CrossRef]
37. Ke, C.L.; Deng, F.S.; Chuang, C.Y.; Lin, C.H. Antimicrobial actions and applications of Chitosan. *Polymers* **2021**, *13*, 904. [CrossRef]
38. Wińska, K.; Mączka, W.; Łyczko, J.; Grabarczyk, M.; Czubaszek, A.; Szumny, A. Essential oils as antimicrobial agents—Myth or real alternative? *Molecules* **2019**, *24*, 2130. [CrossRef] [PubMed]
39. Sfiligoj, M.; Hribernik, S.; Stana, K.; Kree, T. Plant Fibres for Textile and Technical Applications. In *Advances in Agrophysical Research*; IntechOpen: London, UK, 2013.
40. Sánchez-López, E.; Gomes, D.; Esteruelas, G.; Bonilla, L.; Lopez-Machado, A.L.; Galindo, R.; Cano, A.; Espina, M.; Ettcheto, M.; Camins, A.; et al. Metal-based nanoparticles as antimicrobial agents: An overview. *Nanomaterials* **2020**, *10*, 292. [CrossRef]
41. Nguyen, V.T.; Vu, V.T.; Nguyen, T.H.; Nguyen, T.A.; Tran, V.K.; Nguyen-Tri, P. Antibacterial activity of TiO_2- and ZNO-decorated with silver nanoparticles. *J. Compos. Sci.* **2019**, *3*, 61. [CrossRef]
42. Ebrahimi, H.; Abedi, B.; Bodaghi, H.; Davarynejad, G.; Haratizadeh, H.; Conte, A. Investigation of developed clay-nanocomposite packaging film on quality of peach fruit (*Prunus persica* Cv. Alberta) during cold storage. *J. Food Process. Preserv.* **2018**, *42*, e13466. [CrossRef]
43. Mohamed, N.A.; Abd El-Ghany, N.A.; Fahmy, M.M.; Khalaf-Alla, P.A. Novel polymaleimide containing dibenzoyl hydrazine pendant group as chelating agent for antimicrobial activity. *Int. J. Polym. Mater. Polym. Biomater.* **2018**, *67*, 68–77. [CrossRef]
44. Chen, M.; Chen, X.; Ray, S.; Yam, K. Stabilization and controlled release of gaseous/volatile active compounds to improve safety and quality of fresh produce. *Trends Food Sci. Technol.* **2020**, *95*, 33–44. [CrossRef]
45. Gómez-García, M.; Sol, C.; De Nova, P.J.G.; Puyalto, M.; Mesas, L.; Puente, H.; Mencía-Ares, Ó.; Miranda, R.; Argüello, H.; Rubio, P.; et al. Antimicrobial activity of a selection of organic acids, their salts and essential oils against swine enteropathogenic bacteria. *Porc. Health Manag.* **2019**, *5*, 32. [CrossRef]
46. Bouasker, M.; Belayachi, N.; Hoxha, D.; Al-Mukhtar, M. Physical characterization of natural straw fibers as aggregates for construction materials applications. *Materials* **2014**, *7*, 3034–3048. [CrossRef]
47. Smole, M.S.; Hribernik, S. Grass Fibers, Physical Properties. In *Encyclopedia of Earth Sciences Series*; Gliński, J., Horabik, J., Lipiec, J., Eds.; Springer: Dordrecht, The Netherlands, 2011; pp. 333–334, ISBN 9789048135851.
48. González, O.M.; Velín, A.; García, A.; Arroyo, C.R.; Barrigas, H.L.; Vizuete, K.; Debut, A. Representative hardwood and softwood green tissue-microstructure transitions per age group and their inherent relationships with physical-mechanical properties and potential applications. *Forests* **2020**, *11*, 569. [CrossRef]
49. Aloui, H.; Khwaldia, K. Natural Antimicrobial Edible Coatings for Microbial Safety and Food Quality Enhancement. *Compr. Rev. Food Sci. Food Saf.* **2016**, *15*, 1080–1103. [CrossRef]
50. Allemang, R.; De Clerck, J.; Niezrecki, C.; Blough, J.R. (Eds.) *Topics in Modal Analysis I, Volume 5*; Proceedings of the 30th IMAC, A Conference on Structural Dynamics, 2012; Springer: New York, NY, USA, 2012; ISBN 9783319007793.
51. Srinivasa, C.V.; Bharath, K.N. Impact and hardness properties of areca fiber-epoxy reinforced composites. *J. Mater. Environ. Sci.* **2011**, *2*, 351–356.
52. Saba, N.; Jawaid, M.; Sultan, M.T.H. *An Overview of Mechanical and Physical Testing of Composite Materials*; Elsevier Ltd.: Amsterdam, The Netherlands, 2018; ISBN 9780081022924.
53. Rahmat, M. Dynamic Versus Static: Evolving Mechanical Characterisation. Available online: https://researchoutreach.org/articles/dynamic-versus-static-evolving-mechanical-characterisation/ (accessed on 8 September 2021).
54. FAO Data. Available online: https://www.fao.org/faostat/en/#data (accessed on 21 April 2021).
55. Ramesh, M. Flax (*Linum usitatissimum* L.) fibre reinforced polymer composite materials: A review on preparation, properties and prospects. *Prog. Mater. Sci.* **2019**, *102*, 109–166. [CrossRef]
56. Van De Velde, K.; Kiekens, P. Thermoplastic polymers: Overview of several properties and their consequences in flax fibre reinforced composites. *Polym. Test.* **2001**, *20*, 885–893. [CrossRef]
57. Neves, A.C.C.; Rohen, L.A.; Mantovani, D.P.; Carvalho, J.P.R.G.; Vieira, C.M.F.; Lopes, F.P.D.; Simonassi, N.T.; Luz, F.S.D.; Monteiro, S.N. Comparative mechanical properties between biocomposites of Epoxy and polyester matrices reinforced by hemp fiber. *J. Mater. Res. Technol.* **2020**, *9*, 1296–1304. [CrossRef]
58. Ranalli, P.; Venturi, G. Hemp as a raw material for industrial applications. *Euphytica* **2004**, *140*, 1–6. [CrossRef]
59. Pejić, B.M.; Kramar, A.D.; Obradović, B.M.; Kuraica, M.M.; Žekić, A.A.; Kostić, M.M. Effect of plasma treatment on chemical composition, structure and sorption properties of lignocellulosic hemp fibers (*Cannabis sativa* L.). *Carbohydr. Polym.* **2020**, *236*, 116000. [CrossRef]
60. *Michael Carus European Hemp Industry: Cultivation, Processing and Applications for Fibres, Shivs, Seeds and Flowers*; European Industrial Hemp Association: Brussels, Belgium, 2017; pp. 1–9. Available online: https://eiha.org/media/2016/05/16-05-17-European-Hemp-Industry-2013.pdf (accessed on 15 September 2021).
61. Gupta, M.K.; Srivastava, R.K.; Bisaria, H. Potential of Jute Fibre Reinforced Polymer Composites: A review. *Int. J. Fiber Text. Res.* **2015**, *5*, 30–38.
62. Rohit, K.; Dixit, S. A review—Future aspect of natural fiber reinforced composite. *Polym. Renew. Resour.* **2016**, *7*, 43–60. [CrossRef]
63. Thyavihalli Girijappa, Y.G.; Mavinkere Rangappa, S.; Parameswaranpillai, J.; Siengchin, S. Natural Fibers as Sustainable and Renewable Resource for Development of Eco-Friendly Composites: A Comprehensive Review. *Front. Mater.* **2019**, *6*, 226. [CrossRef]

64. Thomas, J. Kenaf: Nature's Little-Known Wonder. The Asean Post, 11 January 2019. Available online: https://theaseanpost.com/article/kenaf-natures-little-known-wonder (accessed on 21 April 2021).
65. Leao, A.L.; Souza, S.F.; Cherian, B.M.; Frollini, E.; Thomas, S.; Pothan, L.A.; Kottaisamy, M. Agro-based biocomposites for industrial applications. *Mol. Cryst. Liq. Cryst.* **2010**, *522*, 18–27. [CrossRef]
66. Odusote, J.; Kumar, V. Mechanical Properties of Pineapple Leaf Fibre Reinforced Polymer Composites for Application as Prosthetic Socket. *J. Eng. Technol.* **2016**, *6*, 24–32. [CrossRef]
67. Todkar, S.S.; Patil, S.A. Review on mechanical properties evaluation of pineapple leaf fibre (PALF) reinforced polymer composites. *Compos. Part B Eng.* **2019**, *174*, 106927. [CrossRef]
68. Kumar Sinha, A.; Narang, H.K.; Bhattacharya, S. Evaluation of Bending Strength of Abaca Reinforced Polymer Composites. *Mater. Today Proc.* **2018**, *5*, 7284–7288. [CrossRef]
69. Abaca Natural Fiber. ABC Oriental Rug & Carpet Cleaning Co. Available online: https://www.abc-oriental-rug.com/abaca-natural-fiber.html (accessed on 21 April 2021).
70. Ranum, P.; Peña-Rosas, J.P.; Garcia-Casal, M.N. Global maize production, utilization, and consumption. *Ann. N. Y. Acad. Sci.* **2014**, *1312*, 105–112. [CrossRef]
71. Husseinsyah, S.; Marliza, M.Z.; Selvi, E. Biocomposites from polypropylene and corn cob: Effect maleic anhydride grafted polypropylene. *Adv. Mater. Res.* **2014**, *3*, 129–137. [CrossRef]
72. Costa, L.A.S.; Assis, D.d.J.; Gomes, G.V.P.; Da Silva, J.B.A.; Fonsêca, A.F.; Druzian, J.I. Extraction and Characterization of Nanocellulose from Corn Stover. *Mater. Today Proc.* **2015**, *2*, 287–294. [CrossRef]
73. De Andrade, M.R.; Nery, T.B.R.; de Santana e Santana, T.I.; Leal, I.L.; Rodrigues, L.A.P.; de Oliveira Reis, J.H.; Druzian, J.I.; Machado, B.A.S. Effect of cellulose nanocrystals from different lignocellulosic residues to chitosan/glycerol films. *Polymers* **2019**, *11*, 658. [CrossRef]
74. Yu, M.; Huang, R.; He, C.; Wu, Q.; Zhao, X. Hybrid Composites from Wheat Straw, Inorganic Filler, and Recycled Polypropylene: Morphology and Mechanical and Thermal Expansion Performance. *Int. J. Polym. Sci.* **2016**, *2016*, 2520670. [CrossRef]
75. United States Department of Agriculture. Cereal, Grasses, and Grains. Available online: https://www.fs.fed.us/wildflowers/ethnobotany/food/grains.shtml (accessed on 21 April 2021).
76. Krepker, M.; Shemesh, R.; Danin Poleg, Y.; Kashi, Y.; Vaxman, A.; Segal, E. Active food packaging films with synergistic antimicrobial activity. *Food Control* **2017**, *76*, 117–126. [CrossRef]
77. Panthapulakkal, S.; Zereshkian, A.; Sain, M. Preparation and characterization of wheat straw fibers for reinforcing application in injection molded thermoplastic composites. *Bioresour. Technol.* **2006**, *97*, 265–272. [CrossRef]
78. Alemdar, A.; Sain, M. Biocomposites from wheat straw nanofibers: Morphology, thermal and mechanical properties. *Compos. Sci. Technol.* **2008**, *68*, 557–565. [CrossRef]
79. Krepker, M.; Prinz-Setter, O.; Shemesh, R.; Vaxman, A.; Alperstein, D.; Segal, E. Antimicrobial carvacrol-containing polypropylene films: Composition, structure and function. *Polymers* **2018**, *10*, 79. [CrossRef]
80. Morais, J.P.S.; Rosa, M.D.F.; De Souza Filho, M.D.S.M.; Nascimento, L.D.; Do Nascimento, D.M.; Cassales, A.R. Extraction and characterization of nanocellulose structures from raw cotton linter. *Carbohydr. Polym.* **2013**, *91*, 229–235. [CrossRef]
81. Perelshtein, I.; Applerot, G.; Perkas, N.; Wehrschuetz-Sigl, E.; Hasmann, A.; Guebitz, G.; Gedanken, A. CuO-cotton nanocomposite: Formation, morphology, and antibacterial activity. *Surf. Coat. Technol.* **2009**, *204*, 54–57. [CrossRef]
82. Chand, N.; Fahim, M. Cotton reinforced polymer composites. In *Tribology of Natural Fiber Polymer Composites*; Chand, N., Fahim, M., Eds.; Woodhead Publishing: Cambridge, UK, 2008; pp. 129–161.
83. Jayaweera, C.D.; Karunaratne, D.W.T.S.; Bandara, S.T.S.; Walpalage, S. Investigation of the effectiveness of nanocellulose extracted from Sri Lankan Kapok, as a filler in Polypropylene polymer matrix. In Proceedings of the 2017 Moratuwa Engineering Research Conference (MERCon), Moratuwa, Sri Lanka, 29–31 May 2017; pp. 1–6.
84. Macedo, M.J.P.; Silva, G.S.; Feitor, M.C.; Costa, T.H.C.; Ito, E.N.; Melo, J.D.D. Composites from recycled polyethylene and plasma treated kapok fibers. *Cellulose* **2020**, *27*, 2115–2134. [CrossRef]
85. Mani, G.K.; Rayappan, J.B.B.; Bisoyi, D.K. Synthesis and Characterization of Kapok Fibers and its Composites. *J. Appl. Sci.* **2012**, *12*, 1661–1665. [CrossRef]
86. Chang, F.; Lee, S.H.; Toba, K.; Nagatani, A.; Endo, T. Bamboo nanofiber preparation by HCW and grinding treatment and its application for nanocomposite. *Wood Sci. Technol.* **2012**, *46*, 393–403. [CrossRef]
87. Han, S.; Yao, Q.; Jin, C.; Fan, B.; Zheng, H.; Sun, Q. Cellulose Nanofibers From Bamboo and Their Nanocomposites with Polyvinyl Alcohol: Preparation and Characterization. *Polym. Polym. Compos.* **2016**, *39*, 2611–2619. [CrossRef]
88. Shah, A.U.M.; Sultan, M.T.H.; Jawaid, M.; Cardona, F.; Talib, A.R.A. A Review on the Tensile Properties of Bamboo Fiber Reinforced Polymer Composites. *BioResources* **2016**, *11*, 10654–10676.
89. Ghaderi, M.; Mousavi, M.; Yousefi, H.; Labbafi, M. All-cellulose nanocomposite film made from bagasse cellulose nanofibers for food packaging application. *Carbohydr. Polym.* **2014**, *104*, 59–65. [CrossRef]
90. Verma, D.; Gope, P.C.; Maheshwari, M.K.; Sharma, R.K. Bagasse fiber composites-A review. *J. Mater. Environ. Sci.* **2012**, *3*, 1079–1092.
91. Mokhena, T.C.; Mochane, M.J.; Motaung, T.E.; Linganiso, L.Z.; Thekisoe, O.M.; Songca, S.P. Sugarcane Bagasse and Cellulose Polymer Composites. In *Sugarcane—Technology and Research*; IntechOpen: London, UK, 2018.

92. Rowell, R.M. Mixed morphology nanocrystalline cellulose from sugarcane bagasse fibers/poly(lactic acid) nanocomposite films: Synthesis, fabrication and characterization. *Iran. Polym. J.* **2017**, *26*, 125–136.
93. Rowell, R.M. Natural fibres: Types and properties. In *Properties and Performance of Natural-Fibre Composites*; Woodhead Publishing Limited: Cambridge, UK, 2008; pp. 3–66, ISBN 9781845692674.
94. Rana, M.; Hao, B.; Mu, L.; Chen, L.; Ma, P.C. Development of multi-functional cotton fabrics with Ag/AgBr-TiO_2 nanocomposite coating. *Compos. Sci. Technol.* **2016**, *122*, 104–112. [CrossRef]
95. Shah, D.U.; Schubel, P.J.; Licence, P.; Clifford, M.J. Determining the minimum, critical and maximum fibre content for twisted yarn reinforced plant fibre composites. *Compos. Sci. Technol.* **2012**, *72*, 1909–1917. [CrossRef]
96. Ghalia, M.A.; Abdelrasoul, A. *Compressive and Fracture Toughness of Natural and Synthetic Fiber-Reinforced Polymer*; Elsevier Ltd.: Amsterdam, The Netherlands, 2018; ISBN 9780081022924.
97. Faruk, O.; Bledzki, A.K.; Fink, H.P.; Sain, M. Progress report on natural fiber reinforced composites. *Macromol. Mater. Eng.* **2014**, *299*, 9–26. [CrossRef]
98. Swolfs, Y.; Verpoest, I.; Gorbatikh, L. Recent advances in fibre-hybrid composites: Materials selection, opportunities and applications. *Int. Mater. Rev.* **2019**, *64*, 181–215. [CrossRef]
99. Schneider, K.; Lauke, B. Determination of compressive properties of fibre-reinforced polymers in the in-plane direction according to ISO 14126. Part 2: A critical investigation of failure behaviour. *Appl. Compos. Mater.* **2007**, *14*, 177–191. [CrossRef]
100. Garat, W.; Le Moigne, N.; Corn, S.; Beaugrand, J.; Bergeret, A. Swelling of natural fibre bundles under hygro- and hydrothermal conditions: Determination of hydric expansion coefficients by automated laser scanning. *Compos. Part A Appl. Sci. Manuf.* **2020**, *131*, 105803. [CrossRef]
101. Mokhothu, T.H.; John, M.J. Review on hygroscopic aging of cellulose fibres and their biocomposites. *Carbohydr. Polym.* **2015**, *131*, 337–354. [CrossRef] [PubMed]
102. Zafeiropoulos, N.E. *Interface Engineering of Natural Fibre Composites for Maximum Performance*; Zafeiropoulos, N.E., Ed.; Woodhead Publishing Limited: Cambridge, UK, 2011; ISBN 9781845697617.
103. Engelund, E.T.; Thygesen, L.G.; Svensson, S.; Hill, C.A.S. A critical discussion of the physics of wood-water interactions. *Wood Sci. Technol.* **2013**, *47*, 141–161. [CrossRef]
104. Fredriksson, M. On wood-water interactions in the over-hygroscopic moisture range-mechanisms, methods, and influence of wood modification. *Forests* **2019**, *10*, 779. [CrossRef]
105. Placet, V.; Cisse, O.; Boubakar, M.L. Influence of environmental relative humidity on the tensile and rotational behaviour of hemp fibres. *J. Mater. Sci.* **2012**, *47*, 3435–3446. [CrossRef]
106. Zhang, D.; Milanovic, N.R.; Zhang, Y.; Su, F.; Miao, M. Effects of humidity conditions at fabrication on the interfacial shear strength of flax/unsaturated polyester composites. *Compos. Part B Eng.* **2014**, *60*, 186–192. [CrossRef]
107. Ashik, K.P.; Sharma, R.S.; Guptha, V.L.J. Investigation of moisture absorption and mechanical properties of natural/glass fiber reinforced polymer hybrid composites. *Mater. Today Proc.* **2018**, *5*, 3000–3007. [CrossRef]
108. Jumahat, A.; Soutis, C.; Jones, F.R.; Hodzic, A. Fracture mechanisms and failure analysis of carbon fibre/toughened epoxy composites subjected to compressive loading. *Compos. Struct.* **2010**, *92*, 295–305. [CrossRef]
109. Jegan, M.M.; Sathishkumar, T.P.; Anand, V.; Arun Ganesh, R.; Logesh Raja, C.; Mohanraj, K. Investigation of Fracture Behavior on Glass & Natural Fibre Reinforced Epoxy Composites. *Int. J. Eng. Res. Technol.* **2019**, *7*, 13–17.
110. Khan, R. Fiber bridging in composite laminates: A literature review. *Compos. Struct.* **2019**, *229*, 111418. [CrossRef]
111. Greco, F.; Leonetti, L.; Lonetti, P.; Nevone Blasi, P. Crack propagation analysis in composite materials by using moving mesh and multiscale techniques. *Comput. Struct.* **2015**, *153*, 201–216. [CrossRef]
112. Soutis, C. Compression failure of composite laminates. In *Modeling Damage, Fatigue and Failure of Composite Materials*; Elsevier: Amsterdam, The Netherlands, 2016; Volume 1, pp. 197–211, ISBN 9781782422860.
113. Yao, L.; Cui, H.; Sun, Y.; Guo, L.; Chen, X.; Zhao, M.; Alderliesten, R.C. Fibre-bridged fatigue delamination in multidirectional composite laminates. *Compos. Part A Appl. Sci. Manuf.* **2018**, *115*, 175–186. [CrossRef]
114. Ma, Q.; Rejab, M.R.M.; Siregar, J.P.; Guan, Z. A review of the recent trends on core structures and impact response of sandwich panels. *J. Compos. Mater.* **2021**, *55*, 2513–2555. [CrossRef]
115. Caseri, W.R. Nanocomposites of polymers and inorganic particles: Preparation, structure and properties. *Mater. Sci. Technol.* **2006**, *22*, 807–817. [CrossRef]
116. Maleque, M.A.; Belal, F.Y.; Sapuan, S.M. Mechanical properties study of pseudo-stem banana fiber reinforced epoxy composite. *Arab. J. Sci. Eng.* **2007**, *32*, 359–364.
117. Mohd Nurazzi, N.; Khalina, A.; Mohd Sapuan, S.; Rahmah, M. Development of sugar palm yarn/glass fibre reinforced unsaturated polyester hybrid composites. *Mater. Res. Express* **2018**, *5*, 045308. [CrossRef]
118. Asim, M.; Jawaid, M.; Abdan, K.; Ishak, M.R.; Alothman, O.Y. Effect of Hybridization on the Mechanical Properties of Pineapple Leaf Fiber/Kenaf Phenolic Hybrid Composites. *J. Renew. Mater.* **2018**, *6*, 38–46. [CrossRef]
119. Khan, T.; Sultan, M.T.H.; Shah, A.U.M.; Ariffin, A.H.; Jawaid, M. The Effects of Stacking Sequence on the Tensile and Flexural Properties of Kenaf/Jute Fibre Hybrid Composites. *J. Nat. Fibers* **2021**, *18*, 452–463. [CrossRef]
120. Yorseng, K.; Rangappa, S.M.; Pulikkalparambil, H.; Siengchin, S.; Parameswaranpillai, J. Accelerated weathering studies of kenaf/sisal fiber fabric reinforced fully biobased hybrid bioepoxy composites for semi-structural applications: Morphology, thermo-mechanical, water absorption behavior and surface hydrophobicity. *Constr. Build. Mater.* **2020**, *235*, 117464. [CrossRef]

121. Ismail, A.S.; Jawaid, M.; Sultan, M.T.H.; Hassan, A. Physical and mechanical properties of woven kenaf/bamboo fiber mat reinforced epoxy hybrid composites. *BioResources* **2019**, *14*, 1390–1404.
122. Radzi, A.M.; Sapuan, S.M.; Jawaid, M.; Mansor, M.R. Mechanical performance of Roselle/Sugar palm fiber hybrid reinforced polyurethane composites. *BioResources* **2019**, *13*, 6238–6249. [CrossRef]
123. Rahman, M.R.; Hamdan, S.; Jayamani, E.; Kakar, A.; Bin Bakri, M.K.; Yusof, F.A.B.M. Tert-butyl catechol/alkaline-treated kenaf/jute polyethylene hybrid composites: Impact on physico-mechanical, thermal and morphological properties. *Polym. Bull.* **2019**, *76*, 763–784. [CrossRef]
124. Pappu, A.; Pickering, K.L.; Thakur, V.K. Manufacturing and characterization of sustainable hybrid composites using sisal and hemp fibres as reinforcement of poly (lactic acid) via injection moulding. *Ind. Crop. Prod.* **2019**, *137*, 260–269. [CrossRef]
125. Jawaid, M.; Abdul Khalil, H.P.S.; Abu Bakar, A. Woven hybrid composites: Tensile and flexural properties of oil palm-woven jute fibres based epoxy composites. *Mater. Sci. Eng. A* **2011**, *528*, 5190–5195. [CrossRef]
126. Senthil Kumar, K.; Siva, I.; Rajini, N.; Winowlin Jappes, J.T.; Amico, S.C. Layering pattern effects on vibrational behavior of coconut sheath/banana fiber hybrid composites. *Mater. Des.* **2016**, *90*, 795–803. [CrossRef]
127. Asim, M.; Jawaid, M.; Paridah, M.T.; Saba, N.; Nasir, M.; Shahroze, R.M. Dynamic and thermo-mechanical properties of hybridized kenaf/PALF reinforced phenolic composites. *Polym. Compos.* **2019**, *40*, 3814–3822. [CrossRef]
128. Chandrasekar, M.; Shahroze, R.M.; Ishak, M.R.; Saba, N.; Jawaid, M.; Senthilkumar, K.; Kumar, T.S.M.; Siengchin, S. Flax and Sugar Palm Reinforced Epoxy Composites: Effect of Hybridization on Physical, Mechanical, Morphological and Dynamic Mechanical Properties. *Mater. Resear* **2019**, *6*, 105331. [CrossRef]
129. Cavalcanti, D.K.K.; Banea, M.D.; Neto, J.S.S.; Lima, R.A.A.; da Silva, L.F.M.; Carbas, R.J.C. Mechanical characterization of intralaminar natural fibre-reinforced hybrid composites. *Compos. Part B Eng.* **2019**, *175*, 107149. [CrossRef]
130. Balasubramanian, K.; Rajeswari, N.; Vishvak, R. Influence of sequential positioning of fibre mats in enhancing the properties of natural fibre mat hybrid composite. *Mater. Today Proc.* **2019**, *28*, 1144–1148. [CrossRef]
131. Siakeng, R.; Jawaid, M.; Ariffin, H.; Sapuan, S.M. Mechanical, dynamic, and thermomechanical properties of coir/pineapple leaf fiber reinforced polylactic acid hybrid biocomposites. *Polym. Compos.* **2019**, *40*, 2000–2011. [CrossRef]
132. Kumar, R.; Singh, R.P.; Kataria, R. Study on mechanical properties of fabricated hybrid natural fibre polymeric composites. *Int. J. Mater. Prod. Technol.* **2020**, *60*, 73–91. [CrossRef]
133. Manral, A.; Ahmad, F.; Chaudhary, V. Static and dynamic mechanical properties of PLA bio-composite with hybrid reinforcement of flax and jute. *Mater. Today Proc.* **2019**, *25*, 577–580. [CrossRef]
134. Ramesh, P.; Prasad, B.D.; Narayana, K.L. Effect of fiber hybridization and montmorillonite clay on properties of treated kenaf/aloe vera fiber reinforced PLA hybrid nanobiocomposite. *Cellulose* **2020**, *27*, 6977–6993. [CrossRef]
135. Kanitkar, Y.M.; Kulkarni, A.P.; Wangikar, K.S. Characterization of Glass Hybrid composite: A Review. *Mater. Today Proc.* **2017**, *4*, 9627–9630. [CrossRef]
136. Sanjay, M.R.; Arpitha, G.R.; Yogesha, B. Study on Mechanical Properties of Natural—Glass Fibre Reinforced Polymer Hybrid Composites: A Review. *Mater. Today Proc.* **2015**, *2*, 2959–2967. [CrossRef]
137. Adesina, O.T.; Jamiru, T.; Sadiku, E.R.; Ogunbiyi, O.F.; Beneke, L.W. Mechanical evaluation of hybrid natural fibre–reinforced polymeric composites for automotive bumper beam: A review. *Int. J. Adv. Manuf. Technol.* **2019**, *103*, 1781–1797. [CrossRef]
138. Saleem, A.; Medina, L.; Skrifvars, M.; Berglin, L. Hybrid Polymer Composites of Bio-Based Bast Fibers with Glass, Carbon and Basalt Fibers for Automotive Applications—A Review. *Molecules* **2020**, *25*, 4933. [CrossRef] [PubMed]
139. Saleem, A. Development and Characterisation of Bast and Basalt Fibre Hybrid Polymer Composites for Automotive Applications. Ph.D. Thesis, University of Borås, Faculty of Textiles, Borås, Sweden, 2021.
140. Daneshniya, M.; Maleki, M.H.; Nezhad, H.J.; Keshavarz Bahadori, N. Investigating the Application of Silver Nanoparticles in Active Food Packaging: Antimicrobial Properties and Synthesis Methods Bioactive peptides View project Probiotics View project. *Chem. Res. J.* **2020**, *5*, 28044.
141. Jothibasu, S.; Mohanamurugan, S.; Vijay, R.; Lenin Singaravelu, D.; Vinod, A.; Sanjay, M.R. Investigation on the mechanical behavior of areca sheath fibers/jute fibers/glass fabrics reinforced hybrid composite for light weight applications. *J. Ind. Text.* **2020**, *49*, 1036–1060. [CrossRef]
142. Sapuan, S.M.; Aulia, H.S.; Ilyas, R.A.; Atiqah, A.; Dele-Afolabi, T.T.; Nurazzi, M.N.; Supian, A.B.M.; Atikah, M.S.N. Mechanical properties of longitudinal basalt/woven-glass-fiber-reinforced unsaturated polyester-resin hybrid composites. *Polymers* **2020**, *12*, 2211. [CrossRef]
143. Patel, N.; Patel, K.; Gohil, P.; Chaudhry, V. Investigations on Mechanical Strength of Hybrid Basalt/Glass Polyester Composites. *Int. J. Appl. Eng. Res.* **2018**, *13*, 4083–4088.
144. Raghavendra Rao, H.; Varada Rajulu, A.; Ramachandra Reddy, G.; Hemachandra Reddy, K. Flexural and compressive properties of bamboo and glass fiber-reinforced epoxy hybrid composites. *J. Reinf. Plast. Compos.* **2010**, *29*, 1446–1450. [CrossRef]
145. Retnam, B.S.J.; Sivapragash, M.; Pradeep, P. Effects of fibre orientation on mechanical properties of hybrid bamboo/glass fibre polymer composites. *Bull. Mater. Sci.* **2014**, *37*, 1059–1064. [CrossRef]
146. Sivakumar, D.; Ng, L.F.; Selamat, M.Z. Sivaraos Investigation on fatigue life behaviour of sustainable bio-based fibre metal laminate. *J. Mech. Eng.* **2017**, 123–140.
147. Vasumathi, M.; Murali, V. Effect of alternate metals for use in natural fibre reinforced fibre metal laminates under bending, impact and axial loadings. *Procedia Eng.* **2013**, *64*, 562–570. [CrossRef]

148. Vieira, L.M.G.; dos Santos, J.C.; Panzera, T.H.; Rubio, J.C.C.; Scarpa, F. Novel fibre metal laminate sandwich composite structure with sisal woven core. *Ind. Crop. Prod.* **2017**, *99*, 189–195. [CrossRef]
149. Chandrasekar, M.; Ishak, R.; Sapuan, S.; Leman, Z.; Jawaid, M.; Shahroze, R.M. Fabrication of Fibre Metal Laminate with Flax and Sugar Palm Fibre based Epoxy Composite and Evaluation of their Fatigue Properties. *J. Polym. Mater.* **2019**, *35*, 463–473. [CrossRef]
150. Schultheisz, C.R.; Waas, A.M. Compressive Failure of Composites, Part 1: Testing and Micromechanical Theories. *Prog. Aerosp. Sci.* **1996**, *32*, 1–42. [CrossRef]
151. Mohammed, I.; Talib, A.R.A.; Hameed Sultan, M.T.; Jawaid, M.; Ariffin, A.H.; Saadon, S. Mechanical properties of Fibre-metal laminates made of natural/synthetic fibre composites. *BioResources* **2018**, *13*, 2022–2034. [CrossRef]
152. Krishnasamy, P.; Rajamurugan, G.; Thirumurugan, M. Performance of fiber metal laminate composites embedded with AL and CU wire mesh. *J. Ind. Text.* **2020**. [CrossRef]
153. Ng, L.F.; Sivakumar, D.; Zakaria, K.A.; Bapokutty, O.; Sivaraos. Influence of kenaf fibre orientation effect on the mechanical properties of hybrid structure of fibre metal laminate. *Pertanika J. Sci. Technol.* **2017**, *25*, 1–8.
154. Feng, N.L.; Malingam, S.D.; Ping, C.W. Mechanical characterisation of kenaf/PALF reinforced composite-metal laminates: Effects of hybridisation and weaving architectures. *J. Reinf. Plast. Compos.* **2021**, *40*, 193–205. [CrossRef]
155. Chai, G.B.; Manikandan, P. Low velocity impact response of fibre-metal laminates—A review. *Compos. Struct.* **2014**, *107*, 363–381. [CrossRef]
156. Cai, R.; Jin, T. The effect of microstructure of unidirectional fibre-reinforced composites on mechanical properties under transverse loading: A review. *J. Reinf. Plast. Compos.* **2018**, *37*, 1360–1377. [CrossRef]
157. Ishak, N.M.; Malingam, S.D.; Mansor, M.R.; Razali, N.; Mustafa, Z.; Ghani, A.F.A. Investigation of natural fibre metal laminate as car front hood. *Mater. Res. Express* **2021**, *8*, 025303. [CrossRef]
158. Sivakumar, D.; Ng, L.F.; Zalani, N.F.M.; Selamat, M.Z.; Ab Ghani, A.F.; Fadzullah, S.H.S.M. Influence of kenaf fabric on the tensile performance of environmentally sustainable fibre metal laminates. *Alex. Eng. J.* **2018**, *57*, 4003–4008. [CrossRef]
159. Nurazzi, N.M.; Demon, S.Z.N.; Halim, N.A.; Mohamad, I.S.; Abdullah, N. Carbon nanotubes-based sensor for ammonia gas detection—An overview. *Polimery* **2021**, *66*, 175–186. [CrossRef]
160. Norizan, M.N.; Moklis, M.H.; Ngah Demon, S.Z.; Halim, N.A.; Samsuri, A.; Mohamad, I.S.; Knight, V.F.; Abdullah, N. Carbon nanotubes: Functionalisation and their application in chemical sensors. *RSC Adv.* **2020**, *10*, 43704–43732. [CrossRef]
161. Nurazzi, N.M.; Asyraf, M.R.M.; Khalina, A.; Abdullah, N.; Sabaruddin, F.A.; Kamarudin, S.H.; Ahmad, S.; Mahat, A.M.; Lee, C.L.; Aisyah, H.A.; et al. Fabrication, functionalization, and application of carbon nanotube-reinforced polymer composite: An overview. *Polymers* **2021**, *13*, 1047. [CrossRef]
162. Ahmadi, M.; Zabihi, O.; Masoomi, M.; Naebe, M. Synergistic effect of MWCNTs functionalization on interfacial and mechanical properties of multi-scale UHMWPE fibre reinforced epoxy composites. *Compos. Sci. Technol.* **2016**, *134*, 1–11. [CrossRef]
163. Sarker, F.; Karim, N.; Afroj, S.; Koncherry, V.; Novoselov, K.S.; Potluri, P. High-Performance Graphene-Based Natural Fiber Composites. *ACS Appl. Mater. Interfaces* **2018**, *10*, 34502–34512. [CrossRef]
164. Sarker, F.; Potluri, P.; Afroj, S.; Koncherry, V.; Novoselov, K.S.; Karim, N. Ultrahigh Performance of Nanoengineered Graphene-Based Natural Jute Fiber Composites. *ACS Appl. Mater. Interfaces* **2019**, *11*, 21166–21176. [CrossRef] [PubMed]
165. Monteiro De Lima, A.; García Del Pino, G.; Valin Rivera, J.L.; Chong, K.B.; Bezazi, A.; Costa De Macêdo Neto, J.; De Las Graças Da Silva Valenzuela, M.; Dehaini, J.; Valenzuela Díaz, F. Characterization of Polyester Resin Nanocomposite with Curauá Fibers and Graphene Oxide [Caracterización de nanocompuestos de resina de poliéster con fibras de curauá y óxido de grafeno]. *Rev. Cienc. Técnicas Agropecu.* **2019**, *28*, 1–10.
166. Costa, U.O.; Nascimento, L.F.C.; Garcia, J.M.; Bezerra, W.B.A.; Fabio da Costa, G.F.; da Luz, F.S.; Pinheiro, W.A.; Monteiro, S.N. Mechanical properties of composites with graphene oxide functionalization of either epoxy matrix or curaua fiber reinforcement. *J. Mater. Res. Technol.* **2020**, *9*, 13390–13401. [CrossRef]
167. Han, S.O.; Karevan, M.; Bhuiyan, M.A.; Park, J.H.; Kalaitzidou, K. Effect of exfoliated graphite nanoplatelets on the mechanical and viscoelastic properties of poly(lactic acid) biocomposites reinforced with kenaf fibers. *J. Mater. Sci.* **2012**, *47*, 3535–3543. [CrossRef]
168. Chaharmahali, M.; Hamzeh, Y.; Ebrahimi, G.; Ashori, A.; Ghasemi, I. Effects of nano-graphene on the physico-mechanical properties of bagasse/polypropylene composites. *Polym. Bull.* **2014**, *71*, 337–349. [CrossRef]
169. Chen, J.; Huang, Z.; Lv, W.; Wang, C. Graphene Oxide Decorated Sisal Fiber/MAPP Modified PP Composites: Toward High-Performance Biocomposites. *Polym. Compos.* **2017**, *39*, E113–E121. [CrossRef]
170. Bumbudsanpharoke, N.; Choi, J.; Park, H.J.; Ko, S. Zinc migration and its effect on the functionality of a low density polyethylene-ZnO nanocomposite film. *Food Packag. Shelf Life* **2019**, *20*, 100301. [CrossRef]
171. Sapiai, N.; Jumahat, A.; Mahmud, J. Mechanical properties of functionalised CNT filled kenaf reinforced epoxy composites. *Mater. Res. Express* **2018**, *5*, 045034. [CrossRef]
172. Wu, G.; Gu, Y.; Hou, X.; Li, R.; Ke, H.; Xiao, X. Hybrid nanocomposites of cellulose/carbon-nanotubes/polyurethane with rapidly water sensitive shape memory effect and strain sensing performance. *Polymers* **2019**, *11*, 1586. [CrossRef]

173. Nabinejad, O.; Sujan, D.; Rahman, M.E.; Liew, W.Y.H.; Davies, I.J. Hybrid Composite Using Natural Filler and Multi-Walled Carbon Nanotubes (MWCNTs). *Appl. Compos. Mater.* **2018**, *25*, 1323–1337. [CrossRef]
174. Prasob, P.A.; Sasikumar, M. Viscoelastic and mechanical behaviour of reduced graphene oxide and zirconium dioxide filled jute/epoxy composites at different temperature conditions. *Mater. Today Commun.* **2019**, *19*, 252–261.

polymers

MDPI

Review

Polyaniline/Biopolymer Composite Systems for Humidity Sensor Applications: A Review

Yuriy A. Anisimov [1], Richard W. Evitts [2], Duncan E. Cree [3,*] and Lee D. Wilson [1,*]

1 Department of Chemistry, University of Saskatchewan, 110 Science Place (Room 156 Thorvaldson Building), Saskatoon, SK S7N 5C9, Canada; yuriy.anisimov@usask.ca
2 Department of Chemical and Biological Engineering, University of Saskatchewan, 57 Campus Drive, Saskatoon, SK S7N 5A9, Canada; richard.evitts@usask.ca
3 Department of Mechanical Engineering, University of Saskatchewan, 57 Campus Drive, Saskatoon, SK S7N 5A9, Canada
* Correspondence: duncan.cree@usask.ca (D.E.C.); lee.wilson@usask.ca (L.D.W.); Tel.: +1-306-966-3244 (D.E.C.); +1-306-966-2961 (L.D.W.)

Abstract: The development of polyaniline (PANI)/biomaterial composites as humidity sensor materials represents an emerging area of advanced materials with promising applications. The increasing attention to biopolymer materials as desiccants for humidity sensor components can be explained by their sustainability and propensity to absorb water. This review represents a literature survey, covering the last decade, which is focused on the interrelationship between the core properties and moisture responsiveness of multicomponent polymer/biomaterial composites. This contribution provides an overview of humidity-sensing materials and the corresponding sensors that emphasize the resistive (impedance) type of PANI devices. The key physicochemical properties that affect moisture sensitivity include the following: swelling, water vapor adsorption capacity, porosity, electrical conductivity, and enthalpies of adsorption and vaporization. Some key features of humidity-sensing materials involve the response time, recovery time, and hysteresis error. This work presents a discussion on various types of humidity-responsive composite materials that contain PANI and biopolymers, such as cellulose, chitosan and structurally related systems, along with a brief overview of carbonaceous and ceramic materials. The effect of additive components, such as polyvinyl alcohol (PVA), for film fabrication and their adsorption properties are also discussed. The mechanisms of hydration and proton transfer, as well as the relationship with conductivity is discussed. The literature survey on hydration reveals that the textural properties (surface area and pore structure) of a material, along with the hydrophile–lipophile balance (HLB) play a crucial role. The role of HLB is important in PANI/biopolymer materials for understanding hydration phenomena and hydrophobic effects. Fundamental aspects of hydration studies that are relevant to humidity sensor materials are reviewed. The experimental design of humidity sensor materials is described, and their relevant physicochemical characterization methods are covered, along with some perspectives on future directions in research on PANI-based humidity sensors.

Keywords: humidity sensor; polyaniline; biopolymer; hydration; adsorption; electrical conductivity; response/recovery time; hysteresis; sensing mechanism; hydrophilicity/hydrophobicity

Citation: Anisimov, Y.A.; Evitts, R.W.; Cree, D.E.; Wilson, L.D. Polyaniline/Biopolymer Composite Systems for Humidity Sensor Applications: A Review. *Polymers* **2021**, *13*, 2722. https://doi.org/10.3390/polym13162722

Academic Editor: Evgenia G. Korzhikova-Vlakh

Received: 11 July 2021
Accepted: 8 August 2021
Published: 14 August 2021

1. Introduction

To advance the field of multicomponent polyaniline (PANI)/biomaterial composites for humidity-sensing devices, a review of past studies on related developments enables the identification of knowledge gaps in the field. This contribution provides an overview of recent studies that highlight humidity-sensing PANI materials and their key structural and physicochemical properties. The mechanism of polyaniline-water vapor interaction will be presented along with physicochemical properties of composites related to pore structure, flexibility, hydrophilicity, surface area, conductivity and others. Suitable synthetic methodologies will enable the design of new (bio)polymer systems from known, accessible and naturally abundant components. This information can assist in advancing practical moisture-sensitive systems via a bottom-up approach, by incorporating tunable properties to yield improved materials for use in moisture-measurement systems. This review will address several objectives, as follows:

1. To provide a literature overview of PANI/biopolymer sensor materials;
2. To provide insight on the moisture-sensing mechanism of PANI and its hybrid composites that contain biopolymer systems;
3. To develop a rational approach for the design of PANI-based materials that are suitable for humidity-sensing applications.

Figure 1a shows a statistical survey of the literature related to PANI-based humidity sensors over the last decade. The total number of research works (about 100) is limited, which suggests that this field is positioned well for further study and future development. Figure 1b shows a statistical survey of various materials that could serve as potential humidity sensors. In contrast to conventional ceramics [1–9], biopolymer materials are highly promising since they are (a) sustainable and environmentally friendly, (b) low cost, and (c) relatively easy to prepare. The search results, shown in Figure 1a, used the following keyword search terms: "polyaniline + humidity sensors", whereas Figure 1b employed "polyaniline + biopolymer + composites".

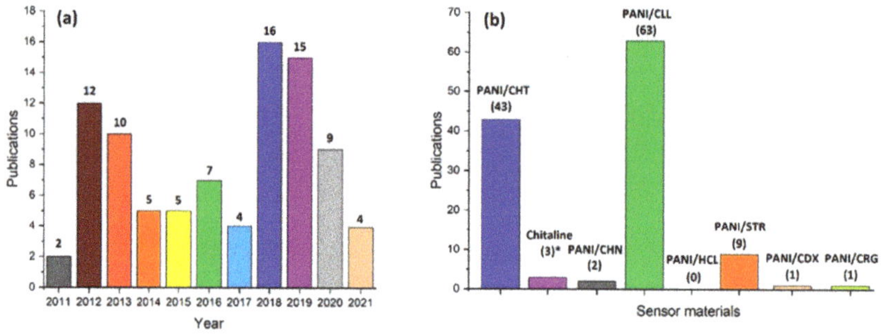

Figure 1. Statistical survey on recent publications: (**a**) polyaniline-based humidity sensors in the last decade; (**b**) total publications to date for polyaniline-biopolymer composites (according to the search queries stated in the graph). Source of data: Scopus; search time: 2 July 2021. Acronyms: CHT—chitosan, CHN—chitin, CLL—cellulose, HCL—hemicellulose, STR—starch, CDX—cyclodextrin, CRG—carrageenan. Chitaline was searched by a plain request "chitaline".

Some authors refer to PANI/CHT composites to as "chitaline" or "PANI-*g*-CHT" and will be used interchangeably herein. The relevance of "chitaline" materials relates to the uncertainty in the structure of such PANI/CHT composites, which will be discussed later. Figure 1 illustrates that chitosan and cellulose are among the most commonly investigated biopolymers used in conjunction with PANI. Currently, there is no humidity sensor device based on PANI/CHT technology. However, there is one known humidity sensor based on chitin, the native biopolymer precursor of chitosan [10], and some sensors are based on PANI/CLL composites [11–13]. For other PANI/biopolymer composites, there is a dearth of information, and limited information on such humidity sensors is available in the open scientific literature. Hence, there is a need to provide coverage of such materials in line with the objectives stated above.

The regulation of humidity plays a significant role in the maintenance of human health and providing comfortable conditions in many aspects of daily life. Relative humidity (RH) is related to the ratio of the partial pressure of water vapor to the total pressure of the saturated vapor at a given temperature and is often expressed in %. Humidity control within indoor environments, such as commercial and residential buildings, plays a role in maintaining the quality of infrastructure and consumable goods. For example, moisture can affect the stability of paintings, computers, plants, food, fruits, vegetables, egg incubation, coatings, chemicals, etc. Moisture-responsive materials have been tested and designed over the decades to monitor humidity in order to prevent the degradation of various items. They range from a simple human hair to sophisticated ceramics [1–9] or conductive polymer materials [10,14–16]. Among these systems, advances in the development of conductive organic polymers have resulted in lightweight, compact, and cost-effective materials that have widened the conventional scope of application beyond traditional ceramic and carbonaceous materials [17]. In this category, PANI has promoted ongoing research interest due to its intrinsic conductivity, where it can be used exclusively as a humidity-responsive material. However, to enhance its properties, PANI can be prepared in the form of composites by the addition of biopolymer additives, such as starch [18–20], cellulose [21–23] and chitosan [24–26].

A humidity sensor is synonymous with a hygrometer, where early hygrometers were based on human hair due to its ability to stretch and shrink according to the changes in the humidity [27]. About half a century ago there were popular "psychrometric charts", which looked like two-dimensional tables with a "dry" and "wet" thermometer reading [28]. Modern humidity sensors are electronic devices that measure both moisture content (RH, %) and temperature, as these two physical quantities are linked. There is no need to compare thermometer readings as the value of RH (%) readily appears on an electronic display, due to the alteration of electrical currents. The aim of such technology is to precisely detect changes in humidity and convert them into RH values. Humidity sensors can be divided into three types, as outlined in Table 1.

Table 1. Common humidity sensor types: resistive, capacitive, and thermal [29].

Properties \ Type	Resistive	Capacitive	Thermal
Basis	Conducting moisture-absorbing film	Two conducting plates with a hygroscopic dielectric located in between	Two thermal sensors: one encased in dry air (or N_2), where another is exposed to humid air
Measurand	Resistance	Capacitance	Temperature
RH range (%)	5–90	Full range (0–100)	Measures the absolute humidity in g/m^3
Temperature range (°C)	−40–100	−50–150	90–300
Accuracy	Moderate	High	Depends on the temperature, where a value above 90 °C is optimal
Price	Low to moderate	High	Varies
Potential applications	Smart food packaging, automotive devices, residential environments	Medicine, scientific research	Industry, machinery; clothes, textile and wood drying

As shown in Table 1, the most appropriate sensors for general daily life are resistive and capacitive types since they operate at room temperature and allow the measurement of RH with good accuracy. Resistive sensors are practical, cost-effective, and suitable for everyday use, and can be subdivided into two categories: polymer-based sensors and ceramic sensors [30]. Capacitive sensors possess the ability to measure RH with higher precision, but they are relatively expensive. Thermal sensors are applicable for industrial use in high-temperature conditions. While ceramic sensors are popular and widely used, polymer-based humidity sensors are used less frequently, in spite of the lower manufacturing cost of polymer-based sensors. In turn, this has driven the interest in further research and development for these materials.

The moisture-sensing mechanism is generally based on the diffusion and mobility of hydrogen (H^+) or hydroxide (OH^-) ions (also known as the Grotthuss mechanism) [31].

Water is an amphoteric molecule that can dissociate into H^+ and OH^- ions, especially onto the surface of moisture-responsive materials. Hydrogen ions are attracted by the electron-rich O-atoms, which can further interact with water molecules and transform them into hydronium (H_3O^+) ions:

$$H_2O + H^+ \rightarrow H_3O^+ \quad (1)$$

H^+ will then be released into the liquid layer located on the adsorption surface layer, and the cycle will repeat continuously:

$$H_3O^+ \rightarrow H_2O + H^+ \quad (2)$$

In this case, hydrogen ions are the dominant carriers of the electric charge on the superficial aqueous layers [32]. Another possible mechanism of the electric charge transport process is also based on the transmission of hydroxide ions [33]. When a water molecule eliminates H^+, it contributes to the protonation of an adjacent H_2O molecule, resulting in the formation of OH^-, as follows:

$$2H_2O \rightarrow H_3O^+ + OH^- \quad (3)$$

Then, the OH^- ion interacts with the next water molecule, inducing the chain of the OH^- transmission process:

$$OH^- + H_2O \rightarrow H_2O + OH^- \quad (4)$$

The mechanism described above demonstrates that electrical conductivity is a key requirement for moisture-responsive materials. In order to attract water molecules, the material should be hydrophilic and can be prepared in the form of a thin film as part of the design of a humidity sensor device [34].

The moisture-responsive nature of a material used in a resistive (impedance) humidity sensor can be divided into three categories, based on the most abundant material component: polymers, ceramics and polymer/ceramic materials, as outlined in Table 2.

Table 2 represents three classes of material types (polymers, ceramics and polymer/ceramic) [1–9,38–49]. Additional classification becomes more complicated as some materials can be attributed to multiple classes of materials. For instance, classification by a conduction mechanism (electronic or ionic) is not that obvious since most materials exhibit both mechanisms, which depend on their formulation or ambient RH level [50]. A typical humidity sensor may include a polymer (conducting or electrolyte) and inorganic compound (salt, oxide, metal, perovskite etc.), or their combinations may function via an electronic–ionic sensing mechanism. Detailed consideration of Table 2 reveals that polymer components may be classified as biomaterials or synthetics, which can be used in conjunction with a conductive polymer (e.g., PANI/CHN [10]) or a ceramic component (e.g., cellulose nanofibers/ZnO [42]). The use of biopolymers has attracted greater interest due to their relative availability, sustainability, and low cost. Recently, humidity sensors based on chitin [10], cellulose [51,52], chitosan [44,53], starch [54–56], cyclodextrin [57] and carrageenan [58] have been reported.

Table 2 highlights some examples of PANI-based humidity sensors reported in the literature, where such sensors are found to consist of variable components: (1) PANI/biopolymer; (2) PANI/metal; (3) PANI/metal oxide; other various permutations and combinations of components are possible. Recent review articles [59,60] enumerate a few dozen types of humidity and other PANI-based sensors. Such types of PANI-based devices are characterized by various parameters: (1) response time; (2) recovery time; and (3) RH interval, in which they have functional properties. The optimal response and recovery times are those having characteristics below 1-min intervals across variable RH (0 to 100%). Sensor materials are not always functional across the full RH range, where many fall within a narrow band or the 5–95% range. For example, a PANI/Co(0) humidity sensor performed with an 8 s response time and 60 s recovery time, while the electric conductivity was dependent on the magnitude of water vapor adsorption [61].

Table 2. Various types of resistive humidity sensors.

Material Type	Classification	Sub-Classification	Examples
Polymer	Polymer-electrolyte	Quaternary ammonium, sulfonate and phosphonium salts	AEPAB/styrene [35] PEDOT:PSS/GO [36] VBTPC/*n*-butylacrylate [37]
	Conducting polymers	Polyaniline, polypyrrole, poly(ortho-phenylenediamine), polythiophene	PANI/CHN [10] PPy/graphene oxide [14] PoPD/PANI [15] P3HT/$Cu_3(BTC)_2$ [16]
Ceramic	Electronic/ionic conduction mechanism	Perovskites	$MnPS_3$ [1], $BaTiO_3$ [2], $Ba_{0.5}Sr_{0.5}TiO_3$ [3]
		Thick film ceramics	Gd-doped CeO_2 [4], Fe_2O_3/SiO_2 [5]
		Cation-doped ceramics	Na^+/K^+-doped Ga_2O_3 [6], La^{3+}-doped $BaSnO_3$ [7]
		Thin-film ceramics	$Mn_{1.2}Co_{1.5}Ni_{0.3}O_4$ thin film [8], WO_3/TiO_2 thin films [9]
Polymer/Ceramic	Electronic/ionic conduction mechanism	Polymer/metal	PVP/Ag [38], PVA/Ag [38], PVP/Au [39], PANI/PTFMA/Ag [40]
		Polymer/metal oxide	PVA/SnO_2 [41], CLNF/ZnO [42], PANI/CuO [43], CHT/$CuMn_2O_4$ [44], PDMS/Armalcolite [45]
		Polymer/inorganic salt	PANI/$MgCrO_4$ [46] PPy/$Sr_3(AsO_4)_2$ [47] SPEEK/$CaCl_2$ [48] PVA-PAA/NbC [49]

AEPAB—[2-(acryloyloxy)ethyl] dimethyl propyl ammonium bromide; PEDOT—poly(3,4-ethylenedioxythiophene); PSS—polystyrene sulfonate; VBTPC—(vinylbenzyl) tributyl phosphonium chloride; PPy—polypyrrole; PoPD—poly(o-phenylenediamine); P3HT—poly(3-hexylthiophene-2,5-diyl); BTC—1,3,5-benzenetricarboxylate; PVP—polyvinylpyrrolidone; PVA—polyvinyl alcohol; PTFMA—polytrifluoromethyl aniline; CLNF—cellulose nanofibers; PDMS—polydimethylsiloxane; SPEEK—sulfonated poly (ether ketone); PAA—polyacrylic acid.

Table 2 provides limited coverage of ceramic humidity sensors since they are among the most abundant and widespread sensor types reported, as described in some recent reviews [31,62,63]. While ceramics are not the focus of the current study, the sensing mechanism that is drawn from the field of ceramic materials has relevance to polymer-based humidity sensors.

The development of a suitable humidity sensor requires a specific material composition to achieve humidity sensitivity behavior and high electrical conductivity. A possible synthetic polymer candidate is PANI, whose electric conductivity has been demonstrated widely in other studies [59]. In conjunction with a suitable hydrophilic polymer, PANI can be tailored as a candidate material for humidity sensors. Biopolymer components, such as cellulose, starch, or chitosan, may serve a structural role or act as the hydrophilic counterpart in such sensors. An important and necessary condition is that the material should be prepared as a thin film that has suitable film-forming characteristics, which is a key feature of humidity-sensing device fabrication.

Polyaniline is an interesting synthetic polymer with unique properties that have been known for more than 150 years; its facile synthesis occurs via an oxidation process [64]. Multiple factors are known to influence the electrical conductivity of PANI, which affords tailoring of its properties for sensing applications. PANI has numerous oxidation (colored forms; cf. Figure 2 and Table 3) and protonated forms (pK_a values) that endow variable electrical conductivity for diverse sensor applications.

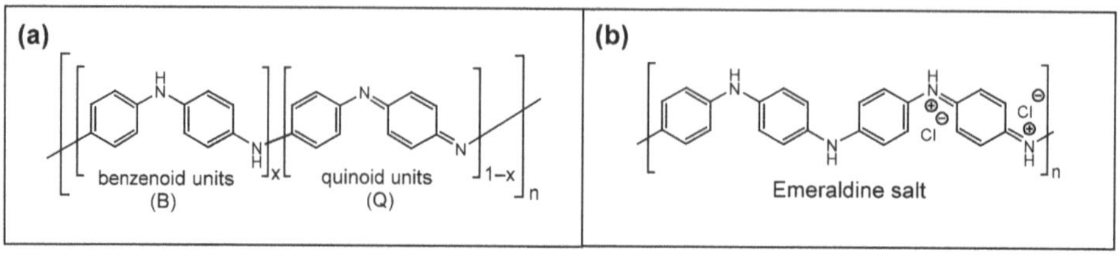

Figure 2. Structures of PANI: (**a**) general structure showing benzenoid (B) and quinoid (Q) units in its base form; and (**b**) emeraldine salt (PANI-ES). Adapted from [64,65]. The x-value is defined as the fraction of benzenoid (B) and quinoid (Q) units, as follows: -$(B)_x$-$(Q)_{1-x}$- for a given degree of polymerization (n), according to panel (**a**).

Table 3. PANI in various forms and oxidation states *.

x Value	Oxidation State	Name	Base Form	Color	Salt Form	Color
1	fully reduced	leucoemeraldine	PANI-LB	colorless	PANI-LS	light yellow
0.5	half-oxidized	emeraldine **	PANI-EB	blue	PANI-ES	green
0	fully oxidized	pernigraniline	PANI-PB	violet	PANI-PS	dark blue

* Data obtained from [64–67]. ** The most stable emeraldine form of PANI is highlighted in grey for emphasis.

The half-oxidized form of PANI (emeraldine) is the most stable and is chiefly used for PANI-based sensor materials, where grey shading is used to emphasize its importance. Both leucoemeraldine and pernigraniline forms are unstable and exist only under certain conditions. Among the forms of PANI, only the PANI-ES is electrically conductive [65], and this constitutes the main structural form for PANI-based humidity sensors herein, unless noted otherwise.

The electrical conductivity depends on the surface area and porosity [62], and various dopant anions, such as chloride, sulfate, polystyrene sulfonate, etc. [68]. In addition, water vapor affects conductivity [69], and that enhances the efficiency of such PANI-based sensor materials.

The synthesis of PANI in variable morphological forms, such as nanosheets [70], nanofibers [71], nanorods [72], nanocoils [73], nanotubes and nanoflowers [74], have been reported, where the morphology depends on the nature of the synthetic conditions. Some structural forms possess high crystallinity with high conductivity, which lends to their utility for electronic devices, including humidity sensors. For example, a higher melting temperature leads to greater chain disorder, whereas ordered polymers are typically obtained at lower temperatures or at slower rates of reagent addition. The reaction time is a key parameter for the preparation of PANI materials. For example, the role of stirring for 48 h and longer has led to the formation of highly agglomerated units, such as nanosheets and nanoflowers [75].

As outlined above, PANI can form hybrid composites with various biopolymers that contain oxygen and nitrogen heteroatoms, such as cellulose, chitosan, starch, and alginate, which can enhance the hydrophilicity and conductivity of the material. Conductivity depends on RH, where the molar fraction of the biopolymer plays a key role in the electrochemical performance of the composite material. This creates a synergistic effect that is due to the π-π * conjugation of PANI, which can affect sensitivity behavior [76]. The synergism between PANI and a second component (mainly the metal ions) has been reported by Singh and Shukla [59]. Surface interactions are critical for producing a sufficient electrical signal for humidity sensing. Judicious selection of biopolymer components and their ratios may yield high-quality humidity sensors with minimal response and recovery times. In some cases, a third matrix component can be incorporated, as outlined for a ternary humidity sensor material that contains PANI/cellulose/PVA [12].

To devise a good humidity sensor, several conditions should be considered. Firstly, the composition of the material may include a conductive polymer (PANI) and a hydrophilic (bio)polymer (such as CHT). Secondly, the materials should display sensitivity to humidity, and the components should be characterized prior to the fabrication of the sensor. Thirdly, the components should have moderately high conductivity, which suggests the utility of materials with lower resistivity values. Such criteria as moisture response and conductivity can be investigated for the pristine components and compared with their blends or composites. While there are no commercial PANI/biopolymer sensors available on the market at the time of writing this review, there are many types of commercial ceramic-based sensors available (cf. Table 2 for selected examples). To further the discussion of PANI/biopolymer composites, an outline of the important design features for good humidity sensors is required. Once the key design elements are outlined, they can be translated to PANI/biopolymer systems to achieve *state-of-the-art* humidity sensor materials, in line with the objectives of this review.

2. Key Features of Ceramic Humidity Sensors

In Table 2, ceramic humidity sensors are mainly constituted by metal oxides, perovskites, sulfides, or combinations of such systems. The effective detection of vapor phases is based on their fundamental physicochemical properties. Firstly, the ability of vapor to adsorb onto a ceramic surface is feasible, based on the surface chemistry of the adsorption sites, where the magnitude of adsorption can vary over a range of humidity values. Secondly, the structure of a humidity-sensing material is unique, due to its surface area and pore structure, according to grain size and grain boundaries. Thirdly, the electrical (e.g., conductivity/resistivity, capacitance) and mechanical (i.e., Young's modulus) properties are subject to change under variable degrees of water vapor adsorption, especially the electrical properties [63].

2.1. *Water Vapor Adsorption vs. Electrical Properties*

The value of water vapor adsorption is proportional to the environmental RH [77]. Hereafter, an increase in RH implicitly assumes an increase in the extent of adsorption, where there are two types of humidity-sensing (resistive and capacitive) devices, as outlined in Table 1. The conductive mechanism for resistive sensors involves electronic or ionic mechanisms or a combination of these effects (cf. Table 2). This implies that the electric charge is born by electrons and/or ions that result in greater conductivity (decreased resistivity) of a sample with greater RH, according to the Grotthuss mechanism (cf. Equations (1)–(4), Section 1). For capacitive sensors, the magnitude of capacitance grows with increasing RH. Their mechanism is different from the resistive one and is based on changes in the dielectric constant of a material upon altering RH [78]. Typical resistance/impedance, vs. RH or capacitance, vs. RH curves, are shown in Figure 3.

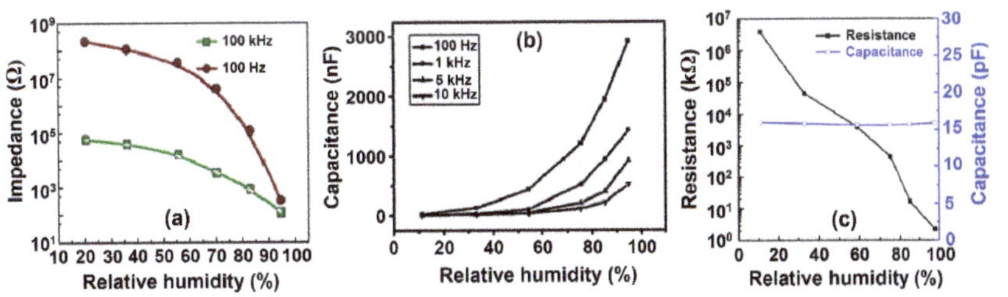

Figure 3. Panel (**a**): impedance (real) vs. RH dependence for a ZnO-based resistive humidity sensor. Reprinted with permission from [79]. Copyright 2013, Elsevier. Panel (**b**): capacitance vs. RH curve for ZnO-based capacitive humidity sensor. Reprinted with permission from [80]. Copyright 2013, Elsevier. Panel (**c**): comparative resistance and capacitance vs. RH curves for a mesoporous resistive $LaFeO_3$ sensor. Reprinted with permission from [81]. Copyright 2013, Elsevier.

Figure 3c shows an exponential decrease of resistance and growth of capacitance that is RH-dependent. Notably, the capacitance of a resistive-type sensor stays constant over the wide range of humidity. In addition, the use of higher alternating current frequencies led to lower resistance and capacitance.

2.2. Pore Structure, Grains and Grain Boundaries

A pore is a cavity located inside the grain (intragranular pore) or a space between the different grains (intergranular pore). A grain is a single structural unit of material, whereas the grain boundary is a surface that separates grains from the intergranular space. Pores, grains and grain boundaries form the unique structure of a material and determine its electrical and mechanical properties [82]. A network of intra- and intergranular pores represents a system with interpenetrating capillary channels that can potentially accommodate absorbed water molecules, together with the electrical charge carriers (electrons and ions). Therefore, the presence of pores is vital for effective humidity sensor materials, whereas the pore sizes should be in a range that does not compromise the material mechanical stability. Figure 4 below represents two sensing materials of similar composition, with variable grain and pore structure [83].

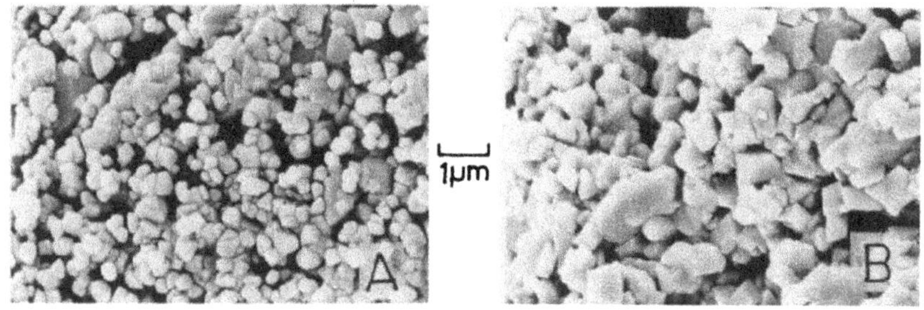

Figure 4. SEM images of porous ceramics: (**A**) $MgCr_2O_4$; (**B**) $7MgCr_2O_4{\cdot}3TiO_2$. Reprinted with permission from [83]. Copyright 1980, The American Ceramic Society.

Figure 4 highlights an illustrative example of how the porosity of a sample can be tuned. Pristine magnesium chromite has smaller grains (hence, longer grain boundaries) and smaller intergranular pores (Figure 4A); whereas the same sample, modified by TiO_2, acquires larger grains, shorter grain boundaries and larger intergranular pores. Intragranular pores are not visible in this image; however, they were measured by mercury intrusion porosimetry. This system exhibits a trend that is similar to intergranular pores [83], where the resistivity of composites was reported to depend on the TiO_2 content. In turn, the addition of TiO_2 boosts the resistivity ca. 200-fold, whereas the further addition of TiO_2 results in a resistivity drop by ca. three orders of magnitude (cf. Table 4).

Table 4. Pore structure, resistivity, and conductivity of $MgCr_2O_4$–TiO_2 sintered composites [83].

Specimen	Grain Size (μm)	Pore Size (nm)	Surface Area (m^2/g)	Resistivity (Ω·cm)	Conductivity (S/cm)
$MgCr_2O_4$	0.2	100	1.6	1.3×10^{10}	7.7×10^{-11}
7 $MgCr_2O_4{\cdot}3TiO_2$	1	270	0.3	2.6×10^{12}	3.8×10^{-12}
3 $MgCr_2O_4{\cdot}7TiO_2$	4	350	0.1	1.0×10^{9}	1.0×10^{-9}

Conductivity is calculated as an inverse resistivity, which shows a reverse trend in values. The grain boundary is proportional to the surface area, which becomes shorter when the grains and pores increase in size. These experimental data lend support that larger pores constitute a capillary network that allows for the transfer of larger amounts of water, and a larger number of charge carriers, which results in greater conductivity overall.

In summary, the presence of larger pores yields higher conductivity. The concept of tunable "pore size–conductivity" can be translated from ceramics to PANI/biopolymer humidity sensor materials, along with the creation of particle grain interfaces through the presence of defects in the structure of the composite material.

3. Structure and Physicochemical Properties of PANI/Biopolymer Composites

3.1. PANI/Chitosan Binary Composites

Chitin is a naturally abundant biopolymer derived from the exoskeletons of crustaceans, insects, and fungal biomass. The controlled deacetylation of CHN using chemical or enzymatic processes yields CHT. CHT is a biocompatible, non-toxic, eco-friendly, biodegradable and relatively low-cost biopolymer. Due to its limited mechanical strength in the dry state, CHT materials can be prepared as composites, along with other polymers such as PANI. CHT contains numerous functional groups (hydroxyl and amine) that can readily undergo further covalent modification or alteration via composite formation with PANI.

PANI/CHT binary composites are normally prepared by the chemical polymerization of aniline, with subsequent grafting onto chitosan in acidic media [84–88]. Since the conductivity of PANI depends strongly on its oxidation state, synthesis is often directed at the preparation of conductive emeraldine salt (PANI-ES) forms. Salt formation is aided by the use of dopants such as HCl or acetic acid, where some studies report the use of an intermediary neutralization step with NaOH (leading to a de-doped emeraldine base (PANI-EB) form), with subsequent re-doping upon the addition of an acid to obtain conductive PANI-ES [77,85,86]. Polymerization is carried out in the presence of an ammonium persulfate (APS) initiator. In general, aniline and powdered CHT are combined and treated with APS; however, one study reports the use of a prefabricated solid-phase CHT film that was submerged in an aniline-APS solution [89]. A generalized synthetic procedure is illustrated in Figure 5.

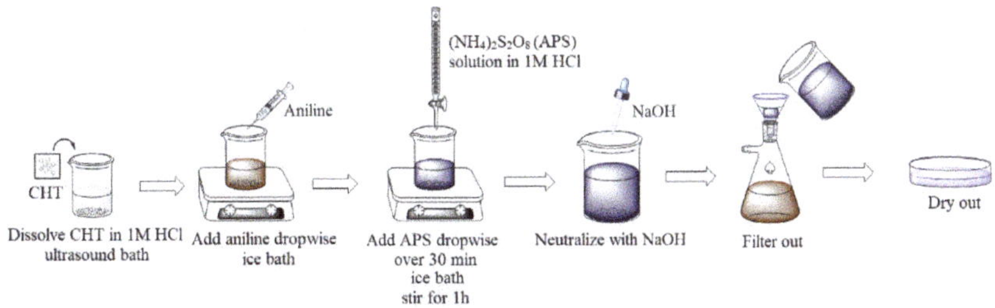

Figure 5. Generalized synthetic procedure for PANI/CHT binary composites.

Although numerous papers reported comparable synthetic procedures, the structural features of PANI/CHT composites are variable. The bonding interactions between PANI and CHT have been described in the literature [77,90–97], which include noncovalent (hydrogen bonding) interactions shown in Figure 6a, along with covalent grafting of chitosan units onto PANI via covalent bonding (cf. Figure 6b). Such grafted composites (PANI-g-CHT or chitaline) have been reported for some years, where the first record dates back to 1989 [94], and were later reintroduced in 2009 [96], 2013 [97] and 2016 [91]. Notably, the authors based their interpretation on FTIR spectral results. In 2009 [96], ^{1}H NMR results were reported, but there was no correlation of the signatures with covalent bonding via the NH group (Figure 6b).

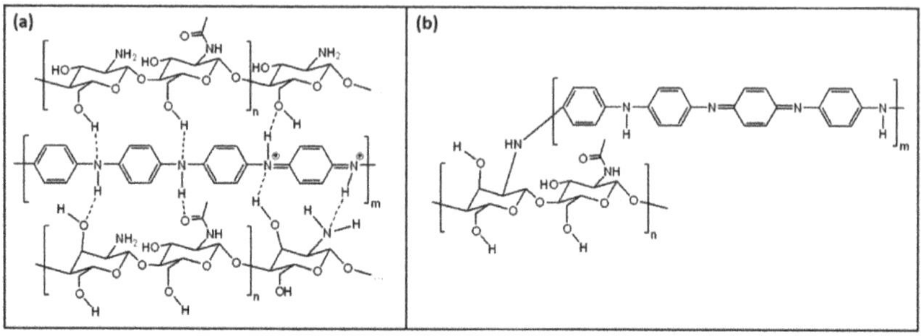

Figure 6. Schematic structures of chemical bonding between PANI and CHT units: (**a**) hydrogen bonding (PANI/CHT); (**b**) covalent bonding via grafting (PANI-g-CHT; "chitaline"). Adapted with permission from the results presented in [77,90–97].

Therefore, the predominant structural form of PANI/CHT (cf. Figure 6a,b) remains an open question, where future research will establish the structure of such composites according to the nature of the synthetic conditions.

It is critically important to know about the structural features of such PANI/CHT composites since hydration processes are inferred to impact on the nature of hydrogen-bonded versus covalently bound systems. During the humidity-sensing process, a water molecule that emerges amidst the H-bonded network will tend to form its own hydrogen bond interactions with PANI or CHT. In turn, this will result in an alteration of the PANI/CHT hydrogen-bonded structure. The dissociation of the hydrogen bonds of PANI/CHT will create variable accessibility for water on the (bio)polymer surface sites and pores upon the influx of water, which may result in interpenetrating capillary channels. These interpolymer channel structures will facilitate electron and proton transfer within the composite, leading to greater conductivity. In contrast, networks containing covalent bonds are static relative to H-bonded systems, which are less amenable to the influx of water. This implies that covalently grafted composites may possess fewer capillary channels and reduced conductivity. Therefore, the hydrogen-bonded network structure in Figure 6a is likely a more favorable ensemble to achieve the desired properties for a humidity sensor material.

The morphology of composites may vary depending on the type of dopant. For PANI/CHT copolymers, several doping acids have been used (acetic acid, CH_3COOH; hydrochloric acid, HCl; and sulfuric acid, H_2SO_4) [84,85]. The morphology of such composites was investigated using scanning electron microscopy (SEM), as shown in Figure 7. Figure 7 reveals that the morphology is dependent on the type of Brønsted acid used during synthesis. In the case of the H_2SO_4 dopant, the PANI/CHT composite surface appears to be coarse and random [84], while the surfaces using HCl and CH_3COOH dopants allowed for better-organized structures. Samples prepared in HCl acquired forms with a granular morphology and a high surface area [85], whereas the use of CH_3COOH as the dopant resulted in a rose-leaf form of PANI/CHT composite [84]. A complex or rough surface morphology by the judicious choice of acid during synthesis leads to enhanced textural and adsorption properties of samples, as a result of the variable textural properties due to acid-driven templation effects.

Figure 7. SEM images of PANI/CHT composites (in their powdered form) that were synthesized in various acid media: (**a**) CH_3COOH, (**b**) HCl, (**c**) H_2SO_4 medium. Panels (**a**,**c**) reprinted with permission from [84]. Copyright 2011, Elsevier. Panel (**b**) reprinted with permission from [85]. Copyright 2018, Wiley Online Library.

The electrical conductivity of PANI/CHT composites is influenced by their morphology, crystallinity, and water vapor adsorption properties. The morphology affects the electrical conductivity of the composites. For example, PANI/CHT composites were prepared in powder and film forms, where various authors used SEM to describe how the morphology was related to the conductivity [84,85,87,89]. These studies revealed that the size and distribution of particles for a powdered sample influenced conductivity, where the smoothness and presence of pores and cracks in films played an important role. A smaller particle size leads to greater conductivity [84], whereas surface irregularities cause a decrease in conductivity [85]. In general, the literature indicates that PANI/CHT films consist of highly smooth surfaces with few defects, which supports their utility for the fabrication of smart devices and sensors. The structure of PANI/CHT composites possesses a semi-crystalline structure, where the degree of crystallinity of PANI/CHT composites is known to affect their electrical conductivity. Crystallinity determines the spatial arrangement of ions in the lattice, which also affects the ion-pair formation and the mobility of charge carriers [98]. Increased crystallinity leads to samples with higher conductivity [84]. The crystallinity of various PANI/CHT composites ranges from 3.4 to 11%, which differs according to the mass ratio of components, dopant counter-ions and substituted/unsubstituted PANI.

XRD results revealed that all PANI/CHT composites generally show a characteristic, sharp XRD line at 2θ-values between 20° and 26°, which relate to the amorphous regions of CHT and PANI, respectively [84,85,89]. The degree of crystallinity is normally calculated as the peak area ratio (area of the crystalline domains/(crystalline area + amorphous area) [84]. Previous reports [85,89] for PANI/CHT composites indicate that such materials possess higher crystallinity with respect to pristine PANI but are lower than that of pure CHT. The electrical conductivity values of PANI/CHT samples have been reported in the range of 10^{-6} to 10^{-3} S/cm [77,84,87]. The water vapor adsorption by PANI/CHT composites shows an increase as the RH levels rise, resulting in greater electrical conductivity. Since PANI/CHT is a binary composite, the addition of the component that adsorbs water results in greater adsorption of water. By comparison, if the additive component does not absorb water, the extent of water uptake may not rise incrementally [77]. Although there are limited reports on PANI/CHT humidity sensors, there was a study outlining a sensor based on PANI and chitin [10]. CHN differs from CHT, due to the presence of N-acetyl groups at C-2, as compared with variable levels of deacetylation at C-2 (glucosamine), according to Figure 8. In other words, the degree of acetylation for chitin is 100%, while that of chitosan is generally 50% or lower. Therefore, research on PANI/CHN composites is also relevant to PANI/CHT materials as a result of their structural similarity, along with PANI/cellulose composites, due to the variable functionality at C-2 for these biopolymers (cf. Figure 8).

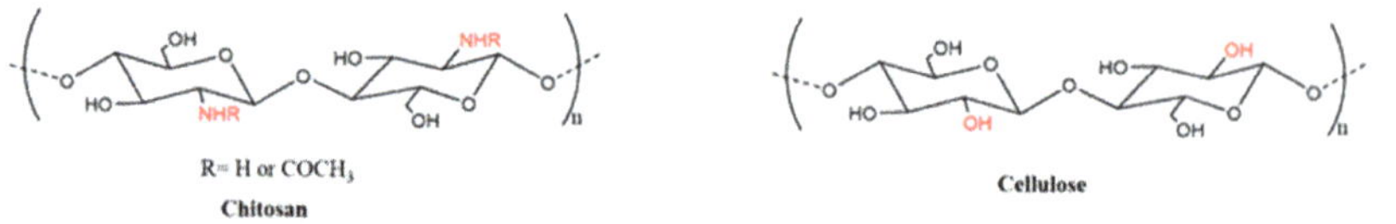

Figure 8. Chemical structures of chitosan and cellulose. Reprinted with permission from Ref. [99]. Copyright 2021, MDPI.

The thermal stability (decomposition temperature) of PANI/CHT composites falls in a range-between PANI and CHT. For example, decomposition of PANI chains and rupture of benzene rings occur above 400 °C [87,100], while the degradation of chitosan chains takes place at ca. 290 °C [101]. The decomposition of PANI/CHT copolymers occurs at 330–350 °C [77,87]. The increased stability of copolymers can be attributed to the formation of intra- and intermolecular hydrogen bonds between PANI and the functional (amino, hydroxyl, acetyl) groups of chitosan [102]. On the other hand, stability also depends on the relative content of each (bio)polymer in the composite. For example, greater PANI and less CHT yield a PANI/CHT composite that will decompose at a temperature closer to that for PANI (400 °C). Similarly, greater CHT content in the composite and less PANI results in a decomposition temperature nearer to that of CHT (290 °C). This trend concurs with theoretical expectations, based on a linear combination of the thermal properties of each polymer, according to their respective heat capacity.

3.2. Humidity Sensing Mechanism of PANI/CHT Composites

The electrical resistance of PANI/CHN composites decreases linearly with a rise in humidity [10]. This suggests that the electrical conductivity exhibits a linear growth since the conductance is inversely proportional to resistance. A rise in electrical conductivity results from the greater mobility of the dopant ion, which is related to the unfolding polymer chains of PANI. In turn, the curling and uncurling (association/dissociation; cf. Figure 6a) of polymer chains depend on the level of swelling. For instance, when the polymer is hydrated (at a high RH), it physically swells, where the coiled polymer chains uncoil to allow for the transport of an electric charge. CHT and chitin are known to swell tremendously [103–105], which enhances the unfolding of polymer chains in the composite. This may suggest that, to a certain degree, the addition of CHT or CHN to PANI will increase its conductivity for wet conditions, despite the non-conductive nature of CHT. Excess amounts of CHT will attenuate the overall conductivity of the composite. The geometry of polymer chains becomes more favorable for the charge transfer, as the (bio)polymer chains are aligned in parallel, which leads to greater conductivity [106]. The linear response shape (conductivity versus % RH) is advantageous, as it can be used in amplifiers (circuits that increase the amplitude of the electric signal) for converting measurands into the measurable values of RH.

On the other hand, elevated humidity delivers water molecules to PANI chains and results in the greater intrinsic conductivity of PANI. This is achieved by the interchain electron transfer, as well as by the greater mobility of dopant ions [107]. As mentioned above, the polymerization of aniline occurs with the aid of a strong oxidant, ammonium persulfate (APS), and therefore PANI contains oxidized, half-oxidized and non-oxidized (reduced) units. Oxidized units are characterized by the quinoid forms (=N–; Q), whereas the reduced fraction consists of benzenoid units (–NH–; B), and the half-oxidized PANI is a mixture of Q/B forms (=N– and –NH–) [108]. Nitrogen atoms serve as the source of available electron pairs, which accounts for the p-type doping of PANI. Lone electron pairs on N-groups of PANI become protonated, according to Equations (5) and (6):

$$\text{–NH– + H}^+ \rightarrow \text{–NH}_2{}^+\text{–} \quad (5)$$

$$\text{=N– + H}^+ \rightarrow \text{=NH}^+\text{–} \quad (6)$$

The electron transfer model [109–113] shows that the electron is recurrently hopping between the oxidized and reduced forms of PANI (although it seems to be proton hopping; the proton cannot transfer without interaction with donor electrons). Supporting evidence is provided by NMR results [113], which indicate that such electron transfer occurs in the presence of water:

$$\text{-NH}_2{}^+\text{-} + H_2O \rightarrow \text{-NH-} + H_3O^+ \quad (7)$$

Mohamed et al. reported that the addition of CHT to PANI changed the ratio of B:Q units (in favor of Q units) [100]. The Q-units contribute to a more linear and rigid structure of PANI/CHT composites that may enhance the proton-hopping mechanism laterally between the chains of PANI and CHT (cf. Figure 9).

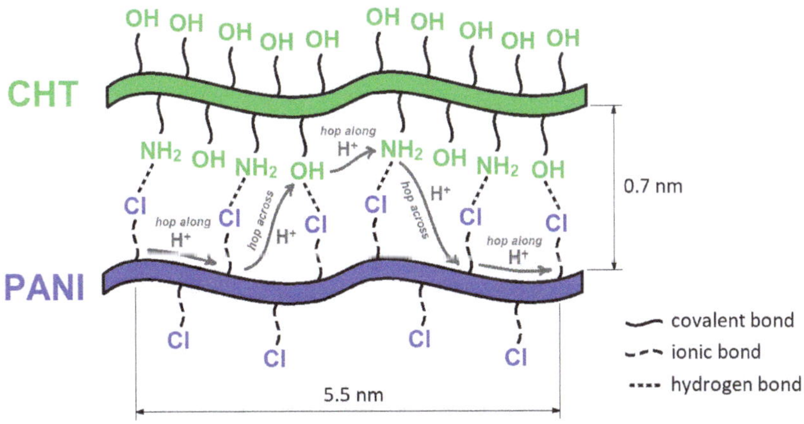

Figure 9. Illustrative view of the proton-hopping mechanism in PANI/CHT composites. The estimated dimensions of PANI/CHT chains are based on the covalent radii of the atoms.

Figure 9 illustrates two possible ways of proton transfer within the composite—along the polymer chain ("hop along") and between the polymer chains ("hop across"). It can be realized more fully if the polymer chains of both PANI and CHT are oriented in parallel, which implies a higher fraction of Q forms in the PANI backbone.

The hydrogen bond network between PANI and CHT accounts for the important role of absorbed water in the humidity-sensing mechanism for PANI. PANI/CHT composite materials are prone to absorbing larger amounts of water versus pristine PANI [77], despite their ability to serve as suitable humidity sensors. Overall, the sensing mechanism can be described as the electron transfer that occurs by the proton exchange, via H_2O.

Currently, there are no known PANI/CHT humidity sensors available commercially. The impetus for the fabrication of such sensor materials relates to their low cost, structural versatility, and sustainability. In 2011, Li et al. fabricated an H_2 gas humidity sensor [93]. The PANI/CHT nanocomposite sensor was exposed to various levels of hydrogen in an air stream, where they observed a change in resistance. The authors concluded that a hydrogen-sensing mechanism was supported by H_2O-polymer interactions, which were caused by the reaction between H_2 and O_2 from the air. The response and recovery times of such sensors were slow. However, this feature is commonplace for PANI/biopolymer sensors, as discussed in Section 4.1 below. In general, data from [93] relate to the abovementioned sensing mechanism (cf. Figure 9), which suggests that PANI/CHT composites warrant further investigation.

3.3. PANI/Cellulose Binary Composites

Polyaniline-cellulose composites are attractive materials for electrical device applications, due to their high electrical conductivity and amenability for the fabrication of conductive nanocomposites. They can be produced from different types of cellulose, such as microcrystalline cellulose [114,115], nanocellulose [116,117], bacterial cellulose [118,119], cotton cellulose [120,121] and various cellulose derivatives (e.g., cellulose acetate [122,123], carboxymethyl cellulose [13,124] and others [125]). Similar to chitosan, cellulose can form composites that are stabilized by hydrogen bonding and other electrostatic interactions [126]. This leads to the functionalization of PANI to afford new forms of polymer composites. As in chitosan, PANI/cellulose composites are prepared via the oxidative polymerization of aniline using APS in an acidic media [127].

3.3.1. PANI/Bacterial Cellulose Composites

Bacterial or microbial cellulose has gained attention for humidity sensor use over the last decade [118,119,128–136]. This type of cellulose is derived from the bacteria cultivated under appropriate conditions. Bacterial cellulose (BC) is attractive due to its outstanding mechanical properties, porosity, and electrical conductivity [137]. For instance, H. Kim et al. [128] produced PANI/BC composites with a conductivity that reached 0.02 S/cm, which is almost 10 to 10^4 times greater than the same value of PANI/CHT composites [77,87,92]. In another study, Alonso et al. [129] also compared the electrical conductivity of PANI/BC membranes synthesized via in situ and ex situ polymerization. The authors concluded that in situ polymerization favored composites with higher PANI contents, leading to materials with higher conductivity. SEM results for pure BC and PANI/BC composites are shown in Figure 10 [129].

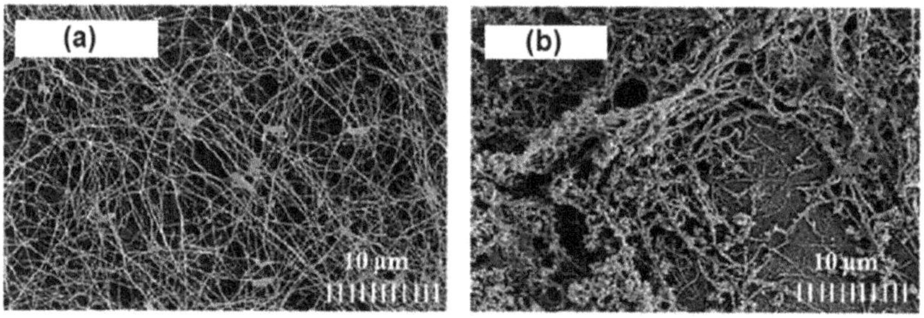

Figure 10. SEM micrograph of (**a**) bacterial cellulose; (**b**) PANI/BC composite. Reprinted with permission from [129]. Copyright 2018, Elsevier.

Figure 10a displays bacterial cellulose in its pristine state after drying. It can clearly be seen that pure BC contains an intricate network of nanofibers connected by physical joints. The gaps between individual fibers are much larger than their fiber diameter, which illustrates their outstanding porosity. After modification by PANI, the fiber network surface is altered, as shown in Figure 10b. The newly formed composite has a unique morphology resembling that of native BC, but mostly consisting of granules and flakes. The surface of a composite becomes more compact, and pores are being obstructed by incorporated PANI, which implies a decreased distance between the PANI and BC units. The SEM results support the hypothesis that favorable interactions occur between the functional groups of BC and PANI, in line with the role of hydrogen-bonding effects (cf. Figures 10 and 11a).

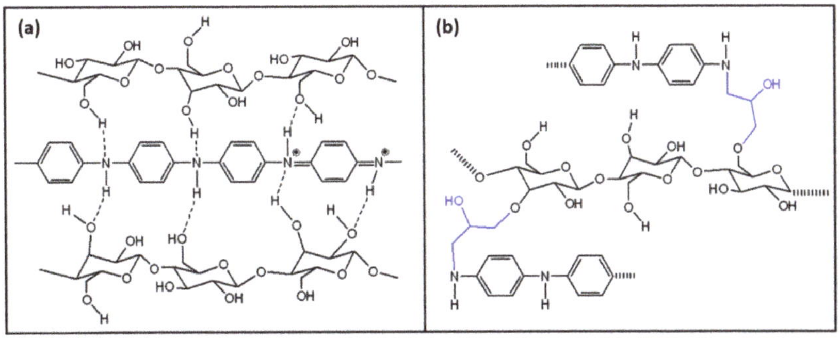

Figure 11. Schematic representation of polymer structures: (**a**) PANI/BC composite; (**b**) PANI/*g*EBC composite (cross-linker is highlighted in blue). Based on the results reported in [114,118,119,126]. EBC—epoxy-modified bacterial cellulose.

Similar to chitosan, cellulose-containing samples exhibit hydrogen bonding between BC and PANI (cf. Figure 11a). However, this was concluded based on Fourier transform infrared (FTIR) results, which are likely to yield unequivocal interpretations on the structure of the composite fiber structure [114,119,126]. Some complementary methods (UV-Vis, PXRD and TGA) were also applied, but serve to provide indirect evidence of chemical interactions, in contrast to the use of NMR spectroscopy, as outlined in one published report indicating that the "... composite has the characteristic features of both polyaniline and cellulose" [126]. Another group performed the grafting of cellulose onto PANI using epoxychloropropane as a cross-linker. They propose the structure shown in Figure 11b, according to spectral characterization by FTIR and X-ray photoelectron spectroscopy (XPS) [118].

The thermal stability of PANI/BC composites correlates with the IR spectral data, where the spectra of PANI/BC composites show weak interactions between –OH groups of cellulose and –NH groups of PANI. In addition, weak interactions between the β-glycosidic oxygen of BC and –NH groups of PANI also provide support for a hydrogen bond formation that is similar to the results for PANI/chitosan composites [92]. Interestingly, hydrogen bond formation between PANI and BC causes weakening of the intermolecular hydrogen bonding between cellulose chains, as reflected in the thermogravimetry profiles for PANI/BC [138]. The onset degradation temperature of the binary composites resulted in lower values than that of the pristine cellulose (224 °C vs. 276 °C). Nevertheless, the thermal stability of PANI/BC composites is still sufficiently high to favor humidity measurements in ambient conditions. The relatively high electric conductivity of PANI/BC indicates the potential suitability of such composites for humidity sensing.

3.3.2. PANI/Carboxymethylcellulose (CMC) Composite

A PANI/CMC composite material was developed as a working humidity sensor [13]. The key parameters for humidity sensor performance were given: the response time (10 s) and recovery time (90 s) were reported between RH 25–75%. At RH levels above 75%, the response became saturated. The properties for PANI/CMC compare favorably with PANI/chitin humidity sensor materials [10]. The response time was ca. 30 s, with a measurable humidity that ranged between 10% and 100% RH, where the humidity absorption mechanism for the PANI/CMC composite occurred via the curling and uncurling of the (bio)polymer chains, [13] similar to that described for chitosan above (cf. Section 3.2).

The structure of a generalized PANI/CMC composite is sketched in Figure 12a, where it can be seen that it is similar to PANI/BC and PANI/CHT composites in terms of the hydrogen bonding arrangements between carboxyl or carboxylate groups of CMC and the NHR groups of PANI [124]. These interactions were confirmed by IR spectroscopy, similar to structurally related biopolymers, such as cellulose and chitosan. The authors have pointed out that CMC (sodium salt) requires greater (alkaline) pH to prompt its greater dissociation into $-COO^-$ and Na^+. After that, the negatively charged carboxylate anion is electrostatically attracted to the positively charged anilinium cation during the polymerization process, resulting in a greater conductivity of the PANI/CMC composites. Indeed, the conductivity of pristine PANI ($6.3 \cdot 10^{-4}$ S/cm) proved to be twice as low as a PANI/CMC composite ($12.5 \cdot 10^{-4}$ S/cm) [124].

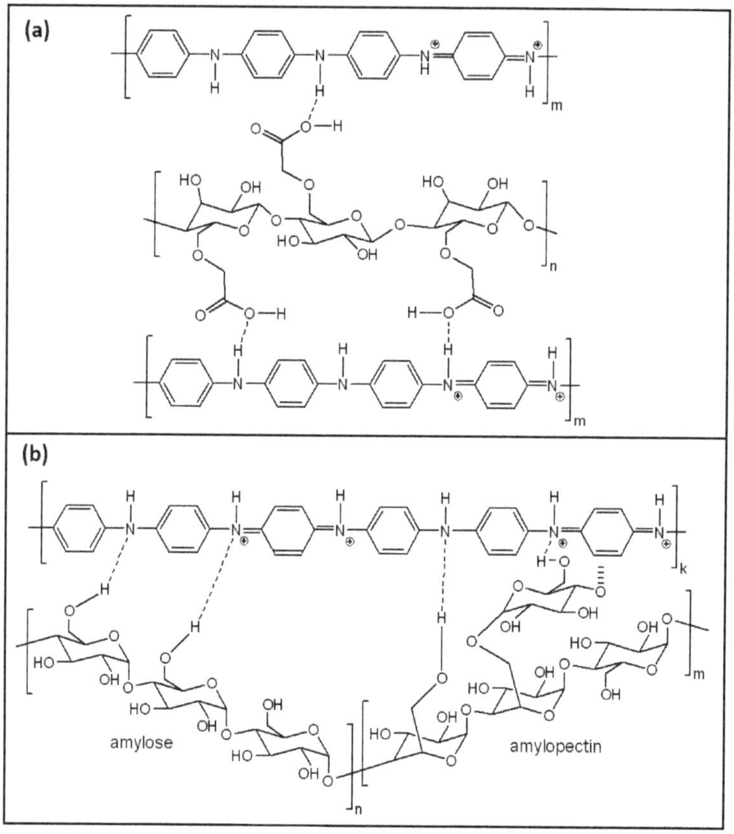

Figure 12. Schematic representation of polymer structures: (**a**) PANI/CMC composite; and (**b**) PANI/STR composite. The terms "m" and "n" refer to the degree of polymerization for the amylopectin and amylose segments, respectively. Based on the results from [124,139].

3.4. PANI/CLL Ternary Composites

A remarkable humidity sensor, consisting of PANI, nanofibrillated cellulose (NFC), and PVA, was reported in 2019 [12]. This composite has limited properties in the case of PANI, such as hygroscopicity and flexibility, which improved upon the addition of NFC and PVA, respectively. Similar to other PANI/CLL composites, it forms a hydrogen-bonded structure that likely becomes more complex upon the addition of PVA. Based on IR spectral data, Anju et al. [12] proposed that the PANI chains are sandwiched between NFC and PVA chains, according to the following illustration (cf. Figure 13).

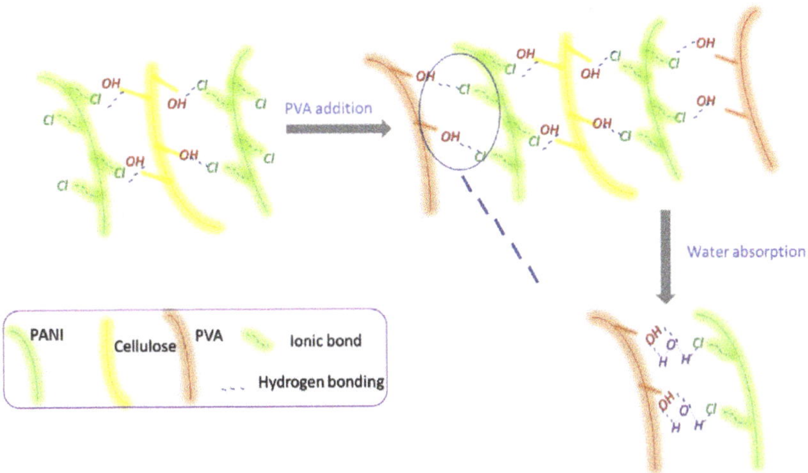

Figure 13. An illustration of the conceptualized structure of a PANI/NFC/PVA composite, to account for its humidity sensitivity. Reprinted with permission from [12]. Copyright 2019, Elsevier.

The humidity-sensitive mechanism occurs due to the intercalation of water molecules between the chains of PANI and PVA. The hydrogen bonds of PANI-PVA break, with the subsequent formation of PANI-H_2O and H_2O-PVA hydrogen bonds. The important features of this sensor are discussed further (cf. Section 4.1).

3.5. PANI/Starch Binary Composites

PANI/starch (PANI/STR) composites are normally synthesized via the in situ polymerization of aniline with starch under acidic conditions [139–141]. The synthesis is similar to that for PANI/CHT and PANI/CLL composites, meaning that various PANI/biopolymer composites can be obtained by the same method reported for PANI/CHT systems. However, other studies report a different method using post-polymerization of PANI, blended with starch solutions [142–144]. The latter is an ex situ method that is in contrast to the conventional approach via the in situ polymerization of PANI with biopolymers.

The structure of PANI/STR composites is portrayed in Figure 12b. It can be seen that the amylose and amylopectin units of starch are connected to PANI via hydrogen bonding, similar to chitosan and cellulose. Such types of interactions were confirmed by an FTIR spectral study, where a detailed spectral assignment for the respective PANI/STR signatures was reported [139]. Interestingly, this study reported an FTIR characterization of PANI/CHT and PANI/CMC composites, since starch, cellulose and chitosan possess certain structural similarities, viz. OH-groups, pyranose rings and glycoside linkages. The reason for comparing PANI/biopolymer signatures with those of pristine PANI in its leucoemeraldine (LE) form (according to a reported ratio of aniline:oxidizer = 1:1.25 [139]) is unclear, whereas the preparation of PANI was reported in its emeraldine (EB) form [145].

SEM images revealed that PANI/STR composites exhibit greater porosity versus the PANI/CMC and PANI/CHT composites [139]. This may lead to higher water vapor adsorption, with an improved sensing performance of the starch-based composites. However, the humidity-sensing properties of PANI/STR materials [139–144] have not been widely reported in the open literature.

3.6. Concluding Remarks

PANI/biopolymer composites share certain similarities, both in the modality of their preparation and in the corresponding physicochemical properties. The literature mainly covers the in situ polymerization for preparation of PANI/biopolymer composites, whereas ex situ methods were reported for composites that contain cellulose and starch. It appears that both methods lead to hydrogen-bonded materials, while there are some limited reports that indicate the role of covalent grafting between PANI and the biopolymer system. It remains unclear whether there is a particular way to direct a certain type of composite structure, but the occurrence of an H-bonded network is important for humidity-sensing properties. This relates to the ability of water molecules to undergo dynamic formation and the dissociation of hydrogen bonds between PANI with the biopolymer system as a function of variable humidity.

There are some reports of working PANI/polymer humidity sensors described in the literature. This includes a range of PANI/biopolymer systems, as follows: PANI/CHN humidity sensor (2010) [10]; PANI/CMC humidity sensor (2016) [13]; and a recent 2019 report [12] for a comparative study of humidity sensing by ternary composites (PANI/nanofibrillated cellulose/PVA and PANI/carbon nanofibers/PVA). The latter will be discussed in further detail in Section 4. Overall, the research and applications of humidity sensors based on PANI/biopolymer composites are still at the early stages of research. However, it is noteworthy to compare such PANI/biopolymer systems with other types of well-known conventional humidity sensors, such as carbon-based materials. For example, PANI/carbon humidity sensors represent a variation on the conceptual design of PANI/biopolymer systems to form biopolymer-based sensors, as outlined in Table 5.

Table 5. Comparative features of PANI/biopolymer and PANI/carbon humidity sensors.

Feature	PANI/Biopolymer	PANI/Carbon
Conductivity of components	One is non-conductive	Both are conductive
Hydrophilicity	Biopolymer is hydrophilic	Carbon is hydrophobic
Type of interactions	H- or covalent bonding	van der Waals interactions
Size of a fiber	Biopolymer is thin	Carbon is thick

As seen from Table 5, these are two extreme cases of moisture-responsive materials, which serve the same goal—to enable humidity sensing. Section 4 will provide an outline of carbon-based humidity sensors.

4. Carbon-Based Humidity Sensors

Carbon nanomaterials are represented by the various types of carbon allotropes, such as nanotubes and nanofibers, as well as graphene and carbon black. These materials stand out due to the superior properties that make them perfect candidates for various electronic devices, including humidity sensors. Carbon nanomaterials can acquire various structural hierarchies: one-dimensional (fibers), two-dimensional (films) and three-dimensional (monoliths) shapes that can be used for humidity sensing. Among their various properties that support their utility in the design of sensor materials are their large specific surface area, which allows for the design of sensor materials with high sensitivity and quick response times [146].

In a similar manner, ceramic-, carbon-, and biopolymer-based sensors can work according to resistive or capacitive transduction principles [147,148]. Similar to ceramic sensors, carbon-based capacitive sensors dominate over the resistive-type sensors in terms of flexibility and response/recovery times [146]. Carbon materials can be modified in a similar way to PANI by the addition of hygroscopic and conductive polymers, to yield hybrid materials for humidity sensing. The top features of the most relevant examples for PANI/carbon and PANI/biopolymer sensors are discussed in the section below (cf. Section 4.1).

4.1. PANI/Carbon vs. PANI/Biopolymer Sensors

Multiwall carbon nanotubes (MWCNT) have been used in combination with PANI to fabricate and characterize a humidity sensor [149]. Another example of a sensor reported for the detection of humidity and ammonia was fabricated using carbon nanofibers (CNFs),

PANI and PVA [12]. Other types of gas sensors were recently reported that include an H_2S sensor based on carbon aerogel (CA) and PANI [150], and a CO_2 sensor comprising graphene (G)/PANI [151].

Table 6 illustrates several key features of carbon- and biopolymer-based sensors:

1. Response time (how rapidly the sensor responds to humidity);
2. Recovery time (how rapidly the sensor returns to its initial state after response);
3. Hysteresis (how the sensor recovers fully after each measurement).

Table 6. Comparative features of carbon- and biopolymer-based sensors.

Material	Type of Sensor	Response Time (s)	Recovery Time (s)	Hysteresis Error (%)	Ref.
		PANI/carbon sensors			
PANI/MWCNT	Resistive	60	140	0.5	[149]
PANI/CNF/PVA	Capacitive	41	50	1	[12]
PANI/CA (H_2S)	Resistive	1	960	—	[150]
PANI/G (CO_2)	Resistive	81	20	—	[151]
		PANI/biopolymer sensors			
PANI/NFC/PVA	Capacitive	47	58	5	[12]
PANI/CHN	Resistive	30	180	18–20 *	[10]
PANI/CMC	Resistive	10	90	2	[13]

* Depends on the number of cycles.

In Table 5, biopolymer-based sensors exhibit competitive properties that compare favorably to carbon-based sensors, despite the differing properties of these materials. Nanofibrillated cellulose sensors show a slightly slower response when compared with CNFs, whereas some biopolymer sensors show an even faster response (such as chitin- and CMC-based materials). One conspicuous drawback of biopolymer-based sensors is that they have greater hysteresis error, as evidenced by their incomplete recovery after a humidity measurement.

Figure 14a–c illustrates the hysteresis loops of carbon- and biopolymer-based composites. A hysteresis loop is shaped by the adsorption profile (lower line) and the desorption profile (upper line) evident in Figure 14a,b. In Figure 14c, a wider hysteresis loop is noted that reveals the sample recovers to a lesser extent, and there is a greater hysteresis error. Figure 14 shows a gradual increase in the level of hysteresis from (a) to (b) to (c) at 1%, 5%, and 18%, respectively. The level of hysteresis (H_e, %) was calculated by either Equation (8) or (9):

$$H_e = \pm\frac{\Delta R}{2\Delta R_0} * 100\% \quad (8)$$

$$H_e = \pm\frac{\Delta C}{2\Delta C_0} * 100\% \quad (9)$$

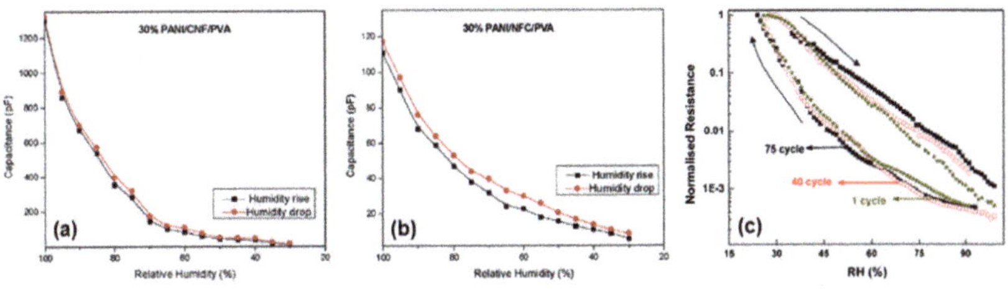

Figure 14. Hysteresis response curves for various composites: (**a**) PANI/CNF/PVA; (**b**) PANI/NFC/PVA; and (**c**) PANI/CHN. Panels (**a**,**b**) reprinted with permission from [12]. Copyright 2019, Elsevier. Panel (**c**) reprinted with permission from [10]. Copyright 2010, Elsevier.

The difference in resistance (ΔR) or difference in capacitance (ΔC) depends on the resistive/capacitive sensor type between adsorption and desorption values. The full-scale resistance (ΔR_0) or capacitance (ΔC_0) range of a measurement is given in [149].

It is noteworthy that after multiple work cycles, the hysteresis error increases in Figure 14c, which indicates sensor fatigue [10]. Hence, durability is another important feature of a desirable sensor.

Nanocarbon materials generally possess higher porosity than conventional biopolymers. For example, carbon aerogel materials possess enormous porosity, whereby such types of sensors exhibit an extremely fast response and very slow recovery (cf. Table 6) [150].

As discussed in Table 6 (cf. Section 4.1), carbon sensors share similarities in their performance when compared with biopolymer sensors, irrespective of differences in their hydrophilicity (cf. Table 5, Section 3.6). Despite the hydrophobic nature of activated carbon, the utility of carbon-only humidity sensors has been reported [152,153]. The application of asymmetric porous carbon films [153] was described for humidity-sensing applications. Upon exposure to an electrical current, its surface becomes oxidized by oxygen from the air, generating hydrophilic –COOH groups, which enable proton hopping via water bridges, as shown in Figure 15.

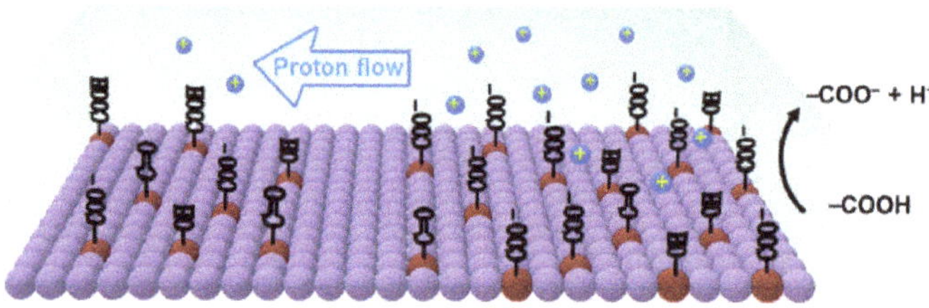

Figure 15. The proton-transfer mechanism in functionalized porous carbon film. Reprinted with permission from [153]. Copyright 2016, Wiley Online Library.

The sensing mechanism in Figure 15 correlates well with proton hopping in PANI/CHT chains (cf. Figure 9) and the flow of water in hydrophobic channels (cf. Figure 16), which will be discussed in the following section.

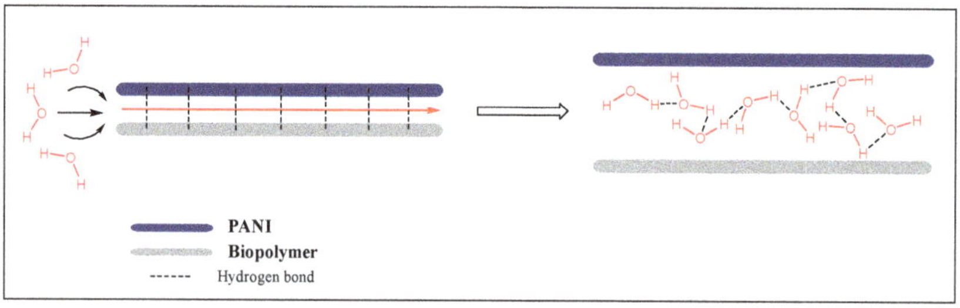

Figure 16. A conceptual structural model of a hydrogen-bonded composite in an anhydrous form (initial state; **left**) and accompanying dissociation of the polymer complex (final state; **right**) at elevated relative humidity levels.

5. Experimental Strategies for the Study of Hydration

It can be inferred from the previous discussion that hydration phenomena play a key role in the structure–function properties of moisture-responsive materials, because the adsorption of water activates labile OH^- and H^+ ions via the Grotthuss mechanism (cf. Equations (1)–(4), Section 1; and Figure 9, Section 3.2), which governs the sensitivity of such humidity sensors. This section provides a selective overview of hydration studies since a full treatment extends beyond the scope of this review. Instead, aspects of hydration phenomena will be covered that are relevant to the functional properties of the humidity sensor materials, as described herein for heterogeneous systems. The molecular-level details of hydration processes (solid–liquid; solid-vapor) are not particularly well understood in complex macromolecular systems, especially in the case of starch and cellulose biopolymers, as reported by Dehabadi et al. [154]. This section will discuss the challenges related to experimental studies of hydration processes using selected techniques as described above (cf. Section 1 and Table 5, Section 3.6). It can be understood that a hydrophilic biopolymer is necessary for improving the moisture-responsiveness of PANI-based sensor materials, due to the key role of hydrogen bonding in binary composites such as PANI with a biopolymer (cf. Figures 6a, 11a, 12 and 13, *vide infra*). Hence, it is necessary to gain a molecular-level understanding of how water uptake affects the structure and properties of the PANI/biopolymer composites. In Section 3.1, the influx of water was inferred to disrupt the intermolecular PANI/CHT hydrogen bonds within the binary composite, versus the formation of H_2O/H_2O H-bonds. Due to its small size and versatile binding strength, water is an efficient competitor in hydrogen-bonded systems. In turn, the uptake of water by the composite may serve to disrupt the noncovalent interactions of the composite material, which results in an "*unzipping*" of the hydrogen bond network of the PANI/biopolymer complex, as illustrated in Figure 16.

Figure 15 builds upon the concept illustrated in Figure 9, where the proton-hopping mechanism occurs across the chain length of the PANI/CHT composite. As a result, the conductivity of hydrated composite will rise (cf. see also Section 3.2; and Figure 24 in Section 6). When the sample reverts to its dry state, water molecules will be desorbed, along with the reformation of intermolecular PANI/biopolymer H-bonds. The process is akin to a *ziplock*-type process, where the "*unzipped*" form corresponds to the dissociated complex and the "*zipped*" form relates to the noncovalent association complex between the polymer units (cf. Figure 16). The processing of *zipping–unzipping* will occur in a cyclic manner, along with the hydration and dehydration processes of the (bio)polymer composite. The entire cycle is referred to as hysteresis (cf. Table 6, Section 4.1), where the sensor material undergoes the maximum possible number of hysteresis cycles that determine its durability. However, the initial state will not fully recover, due to an incomplete loss of solvent, leading to the partial loss of its initial shape, which can reach ca. 20% for biopolymer-based materials (cf. Table 6). This corresponds to the scenario described in Section 3.1. (cf. Figure 6), where two possible structures of PANI/CHT composites are illustrated: hydrogen-bonded and covalent-bonded networks. Hydrogen bonding is likely responsible for the "*zip-unzip*" mechanism, whereas covalent bonding is more likely to prevent or attenuate such a process, based on a consideration of bonding in such covalent networks. The presence of partial covalent bonding may account for the role of hysteresis phenomena, since the unrecoverable loss of structure upon desorption of water that may relate to the role of covalent bonds within such types of composites. Therefore, there may be possible contributions arising from both types of structures, as illustrated in Figure 6a,b (cf. Section 3). For systems that adopt hydrogen bonding solely, this provides an account of the greater conductivity at a higher RH ("*unzipped*" form). In the case of proton-exchange membranes, the role of hydrogen bonding is well established, and is further demonstrated by the electrochemical impedance spectroscopy results at variable humidity, shown in Figure 17 [155].

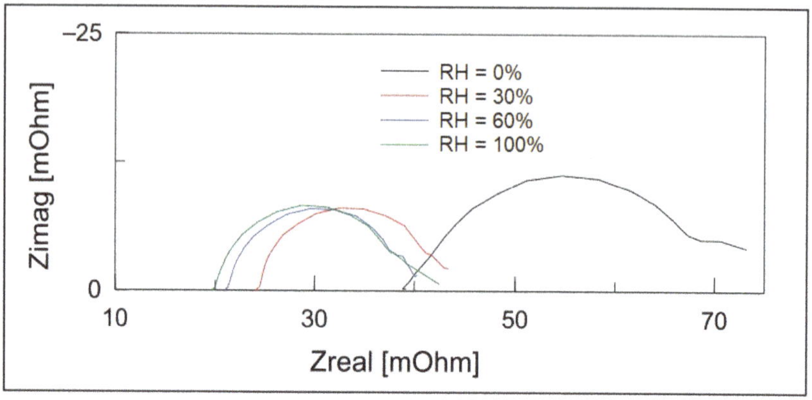

Figure 17. Nyquist plot (imaginary vs. real impedance) derived from electrochemical impedance spectroscopy (EIS) for a proton-exchange membrane in a fuel cell. Reprinted with permission from [155]. Copyright 2012, Elsevier.

Z_{real} is directly proportional to the resistivity of the sample, where the curved line at a higher RH is shifted left (lower resistivity), which corresponds to the higher conductivity of the sample. This bolsters the concept that water provides a conduit for hydrogen ions to start hopping laterally between the polymer chains or along the polymer chain length (cf. Figure 9).

Depending on the nature of the interactions between two hydrophilic polymer chains, the resulting structure may combine to yield a composite with an even more hydrophilic character. However, in the case of a composite that results in effective hydrogen bonding between subunits, the complex may result in the formation of hydrophobic domains, such as cavities or channels, allowing the migration of water to flow that supports transport of the charge carriers. This general phenomenon can be understood based on "hydrophobic effects". The latter is distinguished from hydrophobic interactions, and is discussed further below. A detailed understanding of the kinetics and thermodynamics of hydration is required to address the aforementioned knowledge gap. Blokzijl et al. [156] introduced the term "hydrophobic hydration", which implies that hydrophobicity and hydrophilicity may work in opposition but are closely interconnected, accounting for distinguishable processes related to hydration.

Hydrophobicity provides information about the limited solubility of nonpolar solutes in water and their subsequent aggregation. Thermodynamically, hydrophobicity is defined on the basis of the energetics of transfer from nonpolar compounds, from an apolar solvent to a water environment [156]. However, the fundamental basis of the term "hydrophobicity" has a variable interpretation, as outlined in a recent study on hydrophobic effects [157] that categorizes such phenomena into "classical" (entropy-driven) and "non-classical" (enthalpy-driven) processes. A study on cyclodextrin (CDX) hydration showed that enthalpic contributions dominate the entropy gain by the water molecules [158]. Cyclodextrins are amphiphilic oligomers of glucose, with an apolar cavity and hydrophilic functional groups external to the cavity that resemble features of cellulose, starch and chitosan. Hence, PANI/biopolymer materials will likely have an enthalpic driving force upon changes in the hydration of the apolar and/or hydrophilic domains of such macromolecular systems.

The term "hydrophobic hydration" in the literature appears to be misunderstood [156–159]. In short, it refers to the assembling of water molecules around a nonpolar solute. The process of hydration is described as (1) the formation of a cavity in the bulk of water, (2) insertion of the solute in this cavity with subsequent solute-solvent interactions and (3) rearrangement of water molecules in close proximity to the nonpolar compound. The formation of apolar domains upon the interaction of PANI with biopolymer components is anticipated for hydrogen-bonded complexes between complementary (bio)polymer units (cf. Figures 6a and 11a, Section 3). The apolar surface area is likely to depend on the ratio of PANI and the biopolymer components, along with the morphology adopted by the system.

Hydrophobic effects have been thoroughly studied for CDX (cyclodextrin) complexes [160–163]. Three types of CDX structures (α, β, and γ) differ by the number of glucose subunits (6, 7, and 8, respectively). Each structure contains a cavity, which increases in size in the sequence $\alpha \rightarrow \beta \rightarrow \gamma$. The size of the apolar cavity is important, as it may occlude more or fewer H_2O molecules. For example, when the void is too small (less than 0.9 nm) [164], it is entropically unfavorable to fill with water where the full complement of hydrogen bonding is achieved, as compared with bulk water. In contrast, when the void becomes larger, water molecules penetrate inside the cavity and form hydrogen-bonded clusters. These clusters are more stable when they form a hydrogen-bonded network. If the water molecules escape from the cavity, the optimal interactions can be restored, where the driving force is called "high-energy water" [157]. This scenario is likely realized in medium-sized cavities (1.0 to 1.1 nm), where the driving force (referred to as enthalpic force) is directed toward the external environment. Although hydrophobic interactions seem to be well studied in the literature, their mechanisms are not completely resolved. Given the uncertainty on hydrophobic mechanisms, in PANI/biopolymer systems they appear even more complex, in view of the complex polymer structure of such composite materials.

It follows that the concept of hydrophobicity for PANI/biopolymer composites for humidity sensor materials is important, due to the role of cavities and channels, similar to that described for cyclodextrins by Blokzijl et al. [156]. Based on Figure 16, the presence of hydrophobic channels provides a conduit for water that prompts the water molecules to assemble and move as a whole. In other words, water has higher kinetic mobility ("dynamic water"), whereas hydrogen-bonded water is more static and serves as a poor conductor. Dynamic water (synonymous with the abovementioned "high-energy water") exists in a different thermodynamic state that tends toward escape from the cavity. This is absolutely necessary to make the sample electrically conductive. To sum up, two hydrophilic polymers become more hydrophobic overall upon association but contribute to electrical conductivity. This controversial statement highlights the complexity of hydration phenomena, warranting further study to address this knowledge gap.

Hydration based on solute-water interactions can be studied via practical and theoretical approaches. A practical approach includes thermodynamic methods, such as calorimetry, along with spectroscopic characterization (e.g., NMR, Raman, XRD, SAXS etc.). NMR diffusion experiments in the solid state represent a *state-of-the-art* method for studying hydration processes. In the work reported by Thiessen et al. (2018) [165], hydration was studied from the perspective of kinetics, using double- and zero-quantum-filtered (DQF/ZQF) deuterium NMR. The study was directed at evaluating proton migration and detecting hydrophilic pockets in Nafion membranes. T_1 and T_2 relaxation times were calculated for trapped water as it migrated through these membranes (Figure 18). The equations that describe the line-shapes in Figure 18 using relaxation times can be found in [165] (cf. Equations (1) and (2) in [165]).

Differential scanning calorimetry (DSC) allows for estimating the enthalpy of the hydration/dehydration processes of composites. As mentioned above, enthalpy may supersede entropy as the thermodynamic driving force. Thus, the determination of enthalpy in such processes can be quite meaningful for studying hydration effects. Calorimetric studies allow for the determination of the heat flow at certain temperatures (thermal events), where each thermal event bears enthalpic information for the state of hydration of the composite. In turn, the estimation of the energetics in molar (kJ/mol) or specific (J/g) quantities can be estimated, along with the coordination number of water molecules. Water molecules adsorb onto PANI or CHT, according to the number of adsorption sites borne by each monomer unit. For example, each monomer unit of PANI possesses a single adsorption site (NH^+, or acidic nitrogen) [166], whereas CHT may have several types of adsorption sites (-OH and -NH groups and polymer-chain end groups) [167]. This will allow for an estimate of the amount of water associated with each PANI/composite moiety, and how it depends on the relative composition of the precursors in the composite.

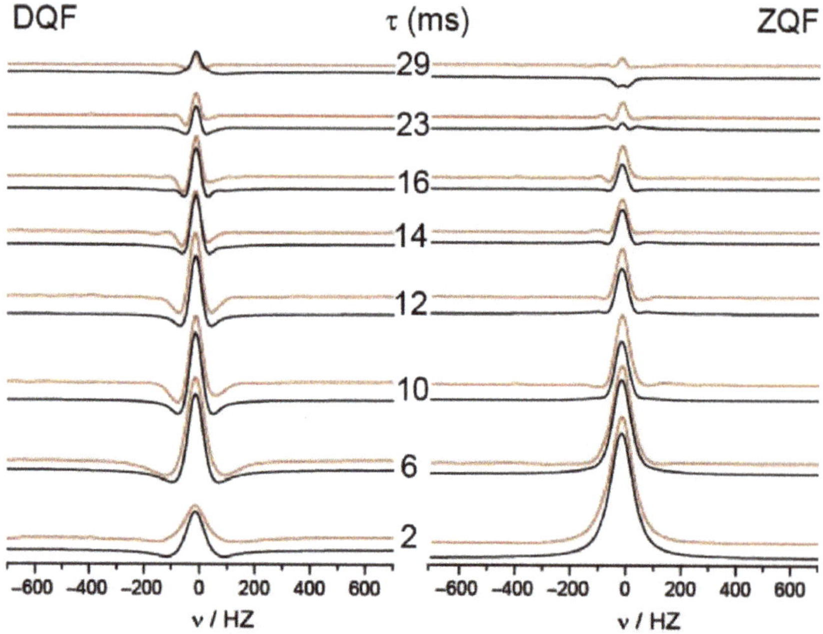

Figure 18. Deuterium DQF/ZQF NMR spectra of a Nafion membrane, hydrated to 15 D_2O molecules per sulfonate group. Reprinted with permission from [165]. Copyright 2018, Elsevier.

Interestingly, water in PANI can be detected even after being submitted to a drying cycle that relates to the so-called "fixed" and "mobile" states of water [166]. Upon drying a PANI composite at a temperature range from 30 to 300 °C for a known duration (15 min), DSC confirmed that the first thermal event was slightly higher (ca. 100 °C), indicating evaporative water loss [168]. In addition to hydrogen bonding and other non-covalent interactions, such as electrostatic forces or hydrophobic effects upon hydration, can be proven by characterization of the enthalpies of transition in calorimetric studies.

The DSC curves of PANI/CHT/PVA (PVA-polyvinyl alcohol) composites were described by Anisimov et al. [77]. Evidence of unique hydration in micropore domains and occluded water in composites was shown by thermal events that extended up to 200 °C, where an increase in the chitosan content revealed a growth in the level of incorporated water, which supports the assumption of an increase in the number of hydrophilic adsorption sites.

As shown in Figure 19, the SEM results revealed the morphological features of hydrated polyanionic cellulose composites. The extent of hydration inevitably leads to a change in the morphology [169].

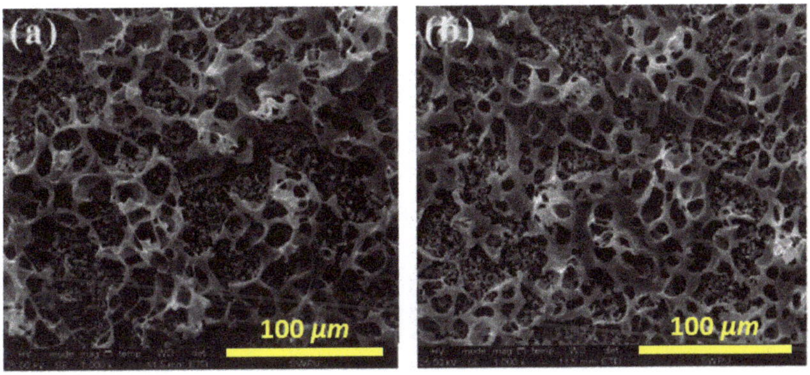

Figure 19. SEM images of the polyanionic cellulose-reinforced cement, with different degrees of hydration: (**a**) smaller; and (**b**) larger. Reprinted with permission from [169]. Copyright 2020, Elsevier.

Figure 19 shows that when the composite sample is exposed to moisture for a longer period of time, more regular network structures emerge. The initial porous structure of calcium-silicate-hydrate cement fills with polymer particles of polyanionic cellulose, and finally, these particles form a fusion membrane that causes a regular network structure together with the hydrated products [169]. Another work [77] revealed that PANI/CHT/PVA composites exhibited exponential growth in their adsorption capacities with respect to increased exposure to moisture.

Small-angle X-ray scattering (SAXS) is a suitable technique to study the variation in crystallinity as a function of the hydration state. For example, potato starch particles before and after hydration were studied using SAXS, where changes in the crystallinity were supported by plotting the scattered intensity I versus the magnitude of the scattering vector, q (cf. Figure 20) [170].

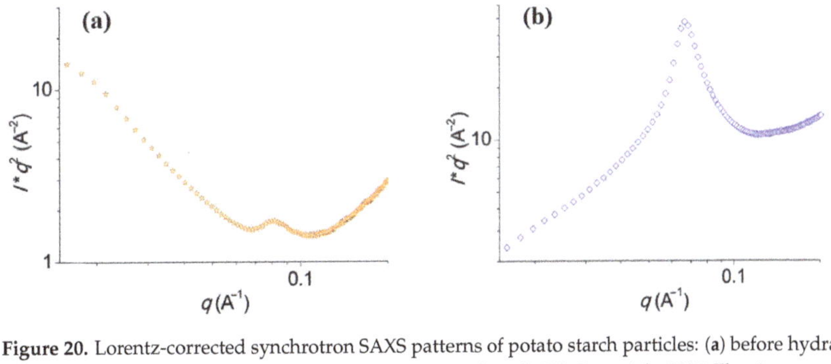

Figure 20. Lorentz-corrected synchrotron SAXS patterns of potato starch particles: (**a**) before hydration; and (**b**) after hydration. Adapted with permission from [170]. Copyright 2017, Elsevier.

The SAXS signature of dry potato starch (Figure 20a) is ascribed to the lamellar semi-crystalline structure of starch [170]. After hydration, the lamellar peak became more intense and underwent a slight shift to the left, from 0.09 to 0.07 $Å^{-1}$. This occurred due to the change from a glassy nematic state to a smectic state of starch [171]. Starch helices typically form starch crystallites; however, the alignment of starch units during hydration may change the crystalline starch types. For further understanding of how these crystalline types change, wide-angle X-ray scattering (WAXS) is required [170]. Along with the determination of crystalline structure, SAXS can also aid in determining the attractive or repulsive interactions between particles. For instance, electrostatic/non-electrostatic interactions can be calculated through the structure factor (S), which is derived from scattered intensity (I). Such types of interactions are significant for understanding the structure-function properties in PANI-based composites and may also give a clue about their humidity-sensing mechanism. When a hydrogen-bonded composite is formed, it becomes more ordered and crystalline. After the infiltration of water, it becomes less crystalline. This yields a dynamic system, with higher conductivity in its amorphous (wet) state and lower conductivity in its crystalline (dry) state.

The molecular-level details concerning the nature and mechanism of the hydration of PANI-based materials are not fully understood and remain an area that warrants further research via complementary experimental approaches. Due to challenges related to interpreting molecular-level details from experiments, computational studies may serve to provide insight into the nature of the hydration processes, energetics and kinetics of the system. Theoretical approaches that are based on molecular dynamics and computational methods can be used to gain insight into the hydration properties of biopolymers during the adsorption process [172]. Computational studies have been carried out for biopolymer/dye systems that provide valuable insight into the role of hydration during adsorption processes (cf. Figure 5 from [172]).

Water absorption by PANI is a dual process that occurs within the internal microstructure and outer surface sites of the composite. Water molecules may be bound to the acidic sites of PANI via hydrogen bonding, where a proposed model [173] illustrates that PANI is surrounded by several hydration shells that include 5, 10, and 20 molecules of water around a single unit of PANI (cf. Figure 21). The respective hydration shells were simulated with the aid of density functional theory (DFT) calculations and the B3LYP/3-21G** model. The results showed the build-up of water at the NH-groups of PANI that yield enhanced conductivity.

Figure 21 schematically shows the water molecules that accumulate at the hydrophilic sites of PANI. The model indicates that four H_2O molecules are attached to the NH-groups of PANI, and one molecule for the H-atom at the meta-position of a benzenoid ring of PANI. These five molecules form the first hydration shell. The second shell is formed by 10 molecules of water, which occur via H-bonding to water in the first shell. The third shell accumulates 20 H_2O molecules, each of which is bound to the H_2O of the previous shell. An illustration of the original ball-and-stick model obtained by DFT calculations was reported by Alghunaim in 2019 (cf. Figures 3–5 in [173]).

To gain further insight on the role of hydration processes in humidity sensor materials, a wide range of complementary experimental and computational methods can be used, as outlined in the selected examples above. A selected overview of recent studies that focus on hydration-related studies in diverse types of chemical systems is summarized in Table 7.

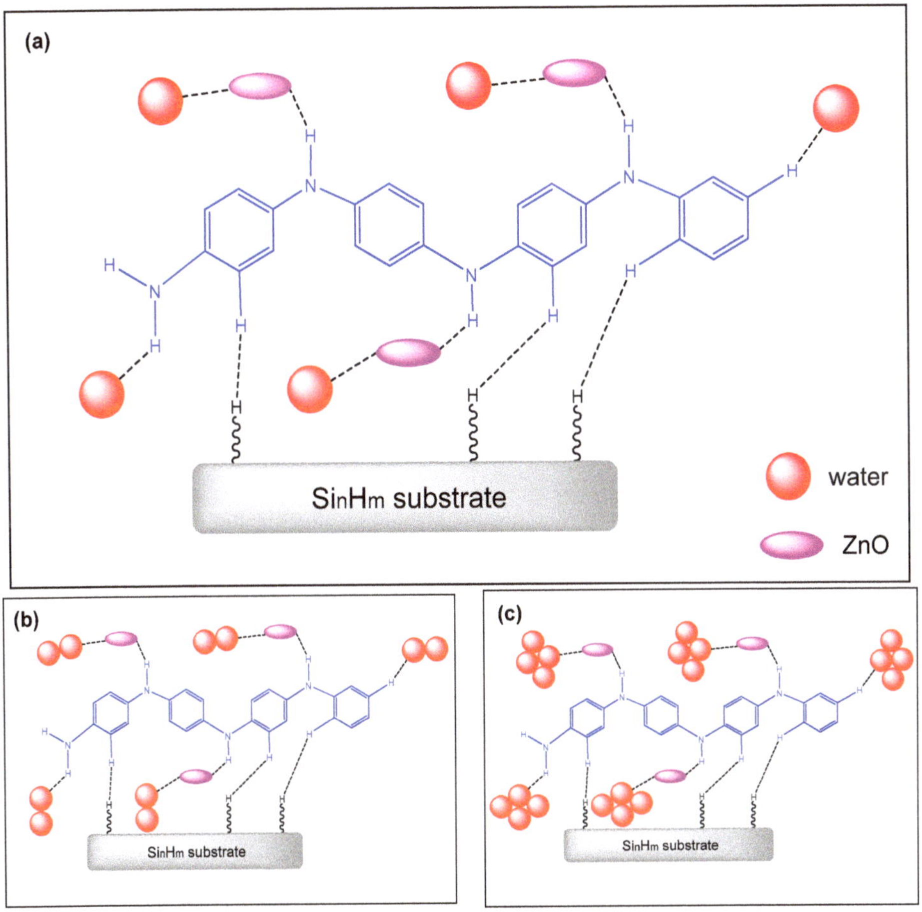

Figure 21. B3LYP/3-21G** calculated structural model for a hydrated PANI-Si-ZnO composite: (**a**) first, (**b**) second, and (**c**) third hydration shells of the PANI-Si-ZnO composites with 5, 10, and 20 molecules, respectively. Based on DFT results from [173].

Table 7. Selected experimental studies for the study of hydration processes.

Category	Experimental Method	Application	Ref.
Spectroscopy	^{2}H NMR diffusion	Detection of hydrophilic pockets	[165]
	^{1}H NMR	Intramolecular H-bonding by calculation of J-coupling constants	[174]
	SEM	Morphology as a function of variable hydration	[169]
	SAXS/WAXS	Crystallinity as a function of variable hydration	[170]
	NMR crystallography	Noncovalent interactions, detection of labile H atoms	[175]
	Raman spectroscopy: D_2O spectral probe	Bound-water fraction by detection of HOD uncoupled oscillators via isotopic dilution	[154]

Table 7. *Cont.*

Category	Experimental Method	Application	Ref.
Calorimetry	DSC	Enthalpies of hydration and dehydration (solid-vapor or solid-liquid systems)	[77,168]
	ITC	Solvent binding enthalpy (solid-liquid system)	[176]
	Immersion calorimetry	Immersion enthalpies in water (liquid or vapor phase)	[177]
Thermodynamic	Water vapor adsorption isotherms	Surface area and pore volume at variable temperature for solid-vapor systems	[177,178]
	Equilibrium dye adsorption	Use of dye probes to estimate the hydrophile-lipophile character of adsorbent and water adsorption capacity	[77]
	GIST	Enthalpy and entropy contributions to the free energy of solvation	[179]
	Hydration distribution model	Thermodynamics of pocket hydration	[180]
Computational	Quantum-Mechanical DFT	Accumulation of water molecules at hydrophilic sites	[173]
		Correlation of dye-based adsorption with hydration	[172]
	Physical models	Derivation of the equations for water binding (free energies versus chemical potentials and activities)	[176]

ITC—isothermal titration calorimetry; GIST—grid inhomogeneous solvation theory.

6. Material Design Approach for Unique Hydration Properties

PANI is an ideal polymer matrix for materials with unique hydration properties. Several factors should be considered in the design of a suitable humidity sensor device at each level of fabrication, such as the sensitive layer, electrodes, substrate layer, etc. The first element is the sensitive layer, which is responsive toward moisture. PANI itself has a limited hygroscopic character and moderate hydrophilicity; however, its electrical conductivity depends on the extent of hydration. For a PANI-based composite to be a good humidity-sensing material it may be modified with a secondary material in order to enhance its hydration properties. Additional blending with a third component will afford the desired property of being sufficiently hygroscopic, as reported by Anju et al. [12].

In addition to adding other polymer-based materials to PANI to improve its hydration properties, PANI can be functionalized with different chemical groups. The humidity-sensing properties can be precisely tuned, depending on the functional groups introduced. These functional groups can be either hydrophilic or hydrophobic. Lyuleeva et al. [181] proposed that hydrophilic methacrylic acid (MAA) and hydrophobic *tert*-butyl acid (*t*-BMA) functional groups can be grafted onto a moisture-responsive material, as a way of tuning the humidity-sensing performance of the materials.

Among the multitude of functionalized polyanilines are halogenated PANI [182,183], carboxylic, sulfonic and sulfamic derivatives of PANI [184], and butylthio-functionalized PANI [185]. Both the protonated and non-protonated forms of PANI will undergo changes in band gaps, highest occupied molecular orbital (HOMO), and lowest unoccupied molecular orbital (LUMO). Therefore, its chemical and electronic properties will be altered, where such alteration can be precisely tailored to the required humidity performance of a sensor. The polarity and size of the functional group are the primary criteria that impart sensitive properties through appropriate functionalization. However, these changes will not significantly influence the intrinsic properties of PANI, such as conductivity [186]. This allows for the tuning of PANI to an adequate level of hydrophilicity, while maintaining its conductive properties. A hydrophobic material with good porosity can also adsorb water, examples of which include cyclodextrins and porous activated carbon (cf. carbon aerogel from Section 4.1).

The second element is the lower layer located beneath the sensitive layer. This should be a substrate, typically made of glass, alumina, or silicone since they are insulators [187], along with embedded electrodes (cf. Figure 22).

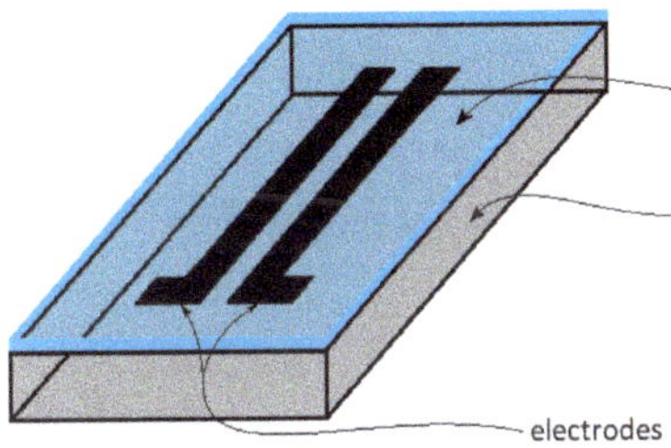

Figure 22. Simplified sensor configuration, with the substrate, electrodes, and sensing layer above them.

The third important component is the electrodes, which can be made from ultrathin metal plates. In general, copper, nickel, gold, or aluminum serve as good electrical conductors for electrodes. The ultimate humidity sensor can be fabricated according to a resistive or capacitive mode [188] since the humidity response can be measured via either resistance [10] or capacitance [12]. The modality of measurement depends on several factors: (1) the content of PANI; (2) the range of measurable relative humidity; and (3) the temperature of the environment.

PANI/biopolymer films should serve as a promising component for resistive-type humidity sensors since their change in resistance as a function of RH can be precisely measured. The preparation of samples in the form of rectangular or circular shapes depends on the type of resistivity measurements, where two modalities are often employed: two-terminal (2T) and four-terminal (4T) sensing (cf. Figure 23).

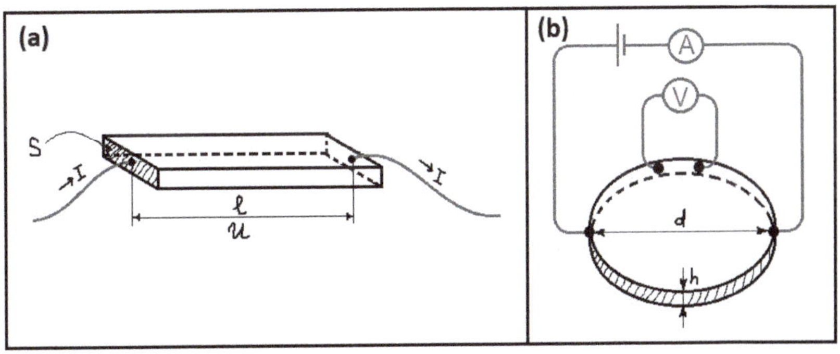

Figure 23. Schematic representation of two types of resistivity measurements: (**a**) two-point (2T) probe; and (**b**) four-point (4T) probe methods.

The two-point sensing technique is the simplest and more established method to measure electrical resistivity, one that is based on the voltage drop and current measurement across a sample (Figure 23a). It works well when the sample has high resistivity (e.g., low conductivity, as reported for PANI/biopolymers [189] (cf. Table 9 also for additional examples, *vide supra*). The net equation for resistivity (ρ) is [190]:

$$\rho = \frac{US}{Il} \tag{10}$$

S and l are the cross-sectional area and length of a conductor, respectively. The ratio U/I is measured as a slope on the current-voltage graph. When the conductor behaves as a semiconductor, the trend for U/I may differ from a straight-line relationship, where some corrections may be required [77].

The four-point sensing technique (also known as a 4T Wenner probe [191]) is used to conduct more accurate measurements, especially for the evaluation of the sheet resistivity of thin films [192]. This method is based on four electrodes, each pair of which bears current (A) and voltage (V) separately (Figure 23b). This method is also convenient since the resistivity of regular circular samples does not depend on the diameter (d), but only on thickness (h); the resistivity can be calculated, if resistance R is known, by Equation (11):

$$\rho = Rh \tag{11}$$

This method is applicable for samples with moderate and low resistivity and is especially useful for samples that experience a sudden drop in resistivity or a boost in conductivity under high RH [190,193]. It is clearly seen that resistivity depends on the thickness of films, which in turn affects the response time of a sensor. Packirisamy et al. [188] fabricated a polyimide-based resistive humidity sensor and showed the difference between two samples of variable thickness (cf. Figure 8 in [188]).

The response curves were fitted according to Equation (12):

$$R = \frac{C}{RH - RH_c} + R_0 \tag{12}$$

where R is the resistance in MOhm, RH is relative humidity (%), and RH_c is the so-called "cut-off" humidity, which indicates a lower RH limit, after which the sensor behaves as an insulator; C is a fitting constant, and R_0 is residual resistance due to impurities.

To test the sample as a humidity sensor, it is placed into a chamber according to Figure 24. The chamber is connected to a hygrometer in order to measure RH inside the compartment, and to a potentiostat, denoted as "Keithley" in Figure 24. The vacuum pump ensures an air and water vapor outlet exhaust if needed. The vaporizer nebulizes moisture into a chamber, which can be replaced by a saturated salt solution for simplicity, as described in a recent review by Singh and Shukla [59] (cf. Figure 12 in [59]). A fan ensures the uniform distribution of water vapor around the chamber. The change in conductivity/resistivity at different RH values is measured with a potentiostat.

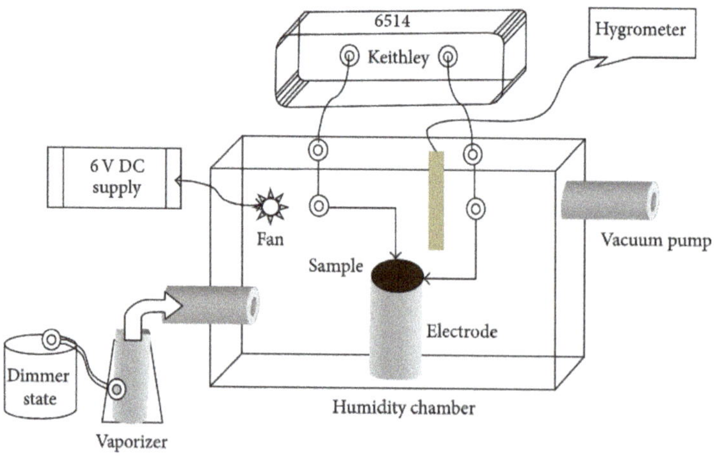

Figure 24. Schematic sketch of a humidity-sensing setup. Reprinted with permission from [194]. Copyright 2014, Hindawi.

It is understood that the cut-off humidity constitutes 41.8% for polyimide sensors (cf. Figure 7 in [188]), which implies that the lowest operational RH for such samples is near 45%. The practical working range of the RH is an important feature of humidity sensor materials, as described above. To optimize the desired humidity response and to fabricate the appropriate moisture-sensitive material, a central composite design [195] and Box–Behnken methodology [196] offer a reliable statistical approach. These methods are designated for multifactorial optimization; however, the Box-Behnken design (BBD) is preferred when the number of variables is three or greater [197]. BBD allows for calculating the minimum number of experiments *(N)* that are needed for such an optimization, where N can be calculated by Equation (13) [198]:

$$N = 2k(k - 1) + C_o \qquad (13)$$

where k is a number of factors (3 in the given case) and C_o is the number of central points (equals to 1). Consequently, to optimize a humidity response, 13 experiments are required. Each experiment is denoted as a point in the middle of a cube edge, and the last point is at the center of the cube (cf. Figure 25) [197].

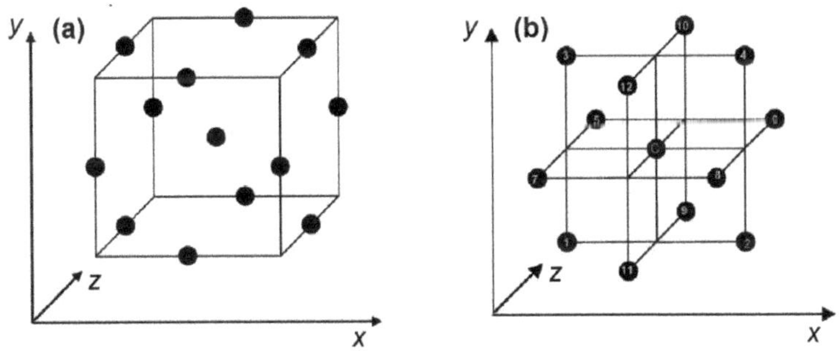

Figure 25. Two different representations for a 3-factorial Box-Behnken design: (**a**) a cube; and (**b**) a three interlocking 2^2 factorial design. Reprinted with permission from [197]. Copyright 2007, Elsevier.

The advantage of the Box–Behnken design is that it enables the selection of the optimal conditions of an experiment, excluding extreme cases, which results in the prevention of failed experiments. Each level of condition is coded as –1, 0 and 1, which can be interpreted as "low", "moderate" and "high" values, respectively. Table 8 displays such conditions, if it is assumed that the humidity-responsive material would operate approximately at room temperature.

Table 8. Numerically encoded conditions of the experiment.

	–1	0	1
PANI (% *w/w*)	25	50	75
RH, %	35	65	95
T, °C	15	25	35

All possible points for the BBD (12 points + C—central point), avoiding extrema conditions, are shown in detail elsewhere (cf. Table 1 in [197]).

The BBD was employed successfully for various types of polymer-based sensors' design [199], optimization of mechanical properties [200], and the study of electrochemical processes [201] among various other examples. The great advantage of BBD is that it enables the alteration of multiple factors simultaneously and the prediction of a resulting response for such changes. In contrast to that, the one-factor-at-a-time (OFAT) approach [202] is slow and time-consuming. Another advantage is that BBD will aid in the determination of the optimal conditions of the experiment with high precision, which may not be achieved by the regular OFAT method.

7. Discussion and Knowledge Gaps

Although the coverage of the scientific literature reported for various humidity-sensing materials appears to be broad in scope, there are limited reports on PANI/biopolymer composites as potential humidity sensors. Various review articles [30,64] indicate that the humidity-sensing performance of PANI-based composites have received good coverage for PANI in combination with metal additives (e.g., Co, Cd, Ag), metal oxides (e.g., TiO_2, WO_3, V_2O_5) and metal chalcogenides (e.g., CdS, CdSe). Some of these metals that have been developed as sensors may suffer from limitations due to material cost and the relative availability of materials, in some cases. Therefore, there is a need to find alternatives that are more cost-effective and sustainable, such as biopolymers and oligomers: chitin, chitosan, cellulose, starch, cyclodextrin, or carrageenan. Table 9 provides a summary of relevant data for PANI/biopolymer humidity-sensing materials. The results as shown reveal that such materials possess features such as porosity, moisture absorption, electrical conductivity, and response-recovery times, which are among the most important design considerations for humidity sensor devices. However, many entries in this table remain unreported, which reveals the need for further study in this field.

Table 9. PANI/biopolymer humidity sensing materials and their relevant properties.

Material	Porosity		Electrical Conductivity (S/cm)	Response Time (s)	Recovery Time (s)	Ref.
	BET Surface Area (m^2/g)	Pore Diameter (nm)				
PANI/NFC/PVA	17	17.1	—	47	58	[12]
PANI/CNF/PVA	34	21.2	—	41	50	[12]
PANI/CHT/PVA	—	—	$3.9 \cdot 10^{-6}$	—	—	[77]
PANI/NFC/PLA	—	—	—	—	—	[203]
PANI/CHN	—	—	$2.2 \cdot 10^{-5}$	120	—	[10]
PANI/CMC	—	—	$7.6 \cdot 10^{-4}$	10	90	[13,124]
PANI/MCC	—	—	$1.25 \cdot 10^{-2}$	40	60	[204]
PANI/NC	—	—	0.65	—	—	[116]
PANI/BC	—	—	0.12	—	—	[118]
PANI/EBC	—	—	1.1	—	—	[118]
PANI/EBC/PAM	—	—	1.4	—	—	[118]
PANI/CA	—	—	$4.0 \cdot 10^{-3}$	—	—	[123]

NFC—nanofibrillated cellulose, CNF—carbon nanofibers, PVA—polyvinyl alcohol, CHT—chitosan, PLA—polylactic acid, CHN—chitin, CMC—carboxymethylcellulose, MCC—microcrystalline cellulose, NC—nanocellulose, BC—bacterial cellulose, EBC—epoxy-modified bacterial cellulose, PAM—polyacrylamide, CA—cellulose acetate.

For instance, the electrical conductivity of many PANI/biopolymer samples is known, whereas the PANI/cellulose samples have the highest conductivity among the various biopolymer composites, as compared with those containing chitosan and chitin. It may appear that chitin/chitosan is not suitable for humidity sensing. However, the conductivity of the entire composite depends on the component ratio, and on the presence of other (additive) components, including the degree of PANI-doping. A knowledge gap in the role of multi-factorial effects remains to be studied, which can be addressed through statistical optimization using methods such as the Box–Behnken design.

There are knowledge gaps for other relevant humidity-sensing parameters. For example, the porosity for the PANI/biopolymer composites has not been widely studied, especially under relevant environmental conditions. In the case of studies other than SEM imaging, an estimate of the porosity (microporous, mesoporous, and macroporous) of such materials by this approach is prone to uncertainty, particularly for samples in an anhydrous state that are likely to undergo solvent swelling. The surface accessibility of pores and the average pore diameter can be estimated from the Brunauer-Emmett-Teller (BET) analysis of vapor adsorption isotherms. In contrast to rigid carbonaceous or inorganic solids, biopolymers can undergo incremental swelling, with greater levels of water uptake (vide supra). This is in contrast to ceramic materials that are rigid in nature, and their structure is relatively stable regardless of the moisture content or RH. Hysteresis in a sensor device is also important, as discussed in Section 4.1. Although biopolymers usually have greater hysteresis effects relative to ceramic materials, the performance still resides within an acceptable range.

To design a good humidity sensor, the response and recovery times are key parameters that have not been widely studied. Moisture absorption, calculated as a percentage increase by mass, represents a potential topic area with limited reported research. In the case of the design of humidity sensor materials, it is a key feature that governs the sensitivity and efficiency parameters. Finally, as PANI composites are relatively new materials for use as humidity sensors, the literature reviewed herein highlights the need for further studies on humidity sensor design and their functional properties. In particular, this statement refers to the use of soft (biopolymers) versus hard (ceramic materials), in conjunction with PANI.

8. Summary and Future Outlook

In conclusion, PANI is a preferred candidate for humidity sensor materials with unique hydration properties. It has an electrical conducting mechanism that relies on the role of molecular interactions between water and the hydrophilic sites of PANI. The conductivity of PANI is highly dependent on the ambient RH, which is a key factor to consider in the fabrication of such devices. To enhance the hydrophilic properties of PANI, it can be modified by the formation of composites with biopolymers, such as chitosan, cellulose, starch, cyclodextrin and other types of polysaccharides that facilitate the hydrogen-bonded composites. The formation of these composites ultimately affect the structure and surface chemical properties of the system. The moisture-responsive mechanism of PANI/biopolymer systems depend on two factors: (i) electron/proton hopping between water molecules and acidic nitrogen groups of PANI; and (ii) on the morphology of the composites, which can be altered under the influence of relative humidity (curling–uncurling of PANI/biopolymer chains). The kinetics of this process have not been fully reported to date, despite the importance of kinetic factors that affect the performance of fast-response humidity sensor devices. In practice, the hydration properties of PANI/biopolymer materials can be investigated with the aid of complementary methods that provide insight into the thermodynamics, kinetics, and structural aspects of the humidity-sensing process. Computational methods are anticipated to provide further insight on the structure of hydrated PANI-based composites, through the deployment of methods such as molecular dynamics and DFT calculations (cf. Table 7).

PANI-based composites are normally obtained via oxidative polymerization in acidic media. The nature of the dopant (acid) plays an important role as a template in the resulting morphology and humidity-sensing performance of the composites. A secondary component (biopolymer) results in a synergistic effect for improving the sensitivity and performance of PANI but may affect the electronic properties (conductivity) of PANI, according to structural considerations. The conductivity of PANI is an intrinsic property that is influenced by the functionalization of PANI via physical or chemical means.

The configuration of a humidity sensor consists of at least two components: (i) a sensitive layer that may comprise a PANI-based composite, and (ii) a substrate layer with embedded electrodes, where the humidity response is measured as impedance (resistance) or capacitance at variable RH. To optimize the humidity-sensing performance, multifactorial design approaches have been employed, such as the Box–Behnken design. The latter is desirable and serves to guide experimental design for the optimization of multi-parameter systems.

Nevertheless, the literature on PANI-based moisture-responsive materials is at an early stage, in comparison with conventional ceramic and carbonaceous sensor materials. A perspective drawn from this contribution recommends that further studies be directed at gaining an improved understanding of humidity-sensing mechanisms for PANI/biopolymer-based nanocomposites. Dedicated experimental and computational studies are anticipated to achieve this goal, through the use of complementary experimental methods. The holistic approaches and methodologies outlined herein are anticipated to provide a molecular-level understanding of the kinetics and thermodynamics of hydration for such systems. In turn, emerging materials such as PANI-based biopolymer nanocomposites are envisioned to occupy an increasingly important role in the future development of sensor technology that employs PANI-based materials for diverse applications.

Author Contributions: Writing—original draft preparation, Y.A.A.; writing—review and editing, L.D.W., D.E.C. and R.W.E.; supervision, L.D.W. and D.E.C.; project administration, L.D.W.; funding acquisition, L.D.W. All authors have read and agreed to the published version of the manuscript.

Funding: L.D.W. acknowledges the support by the Government of Canada through the Natural Sciences and Engineering Research Council of Canada (Discovery Grant Number: RGPIN 04315-2021).

Institutional Review Board Statement: Not applicable.

Informed Consent Statement: Not applicable.

Data Availability Statement: Not applicable.

Acknowledgments: Y.A.A. acknowledges the University of Saskatchewan for partial support of this work through the scholarship and teaching assistant program in the Department of Chemistry.

Conflicts of Interest: The authors declare no conflict of interest.

References

1. Kumar, R.; Jenjeti, R.N.; Sampath, S. Bulk and few-layer 2d, p-MnPS_3 for sensitive and selective moisture sensing. *Adv. Mater. Interfaces* **2019**, *6*, 1900666. [CrossRef]
2. He, Y.; Liu, X.; Wang, R.; Zhang, T. An excellent humidity sensor with rapid response based on $BaTiO_3$ nanofiber via electrospinning. *Sens. Lett.* **2011**, *9*, 262–265. [CrossRef]
3. Farahani, H.; Wagiran, R.; Urban, G.A. Perovskites as surface-assisted room temperature protonic conductor humidity sensor. In Proceedings of the 2019 IEEE International Conference on Sensors and Nanotechnology, Montreal, QC, Canada, 27–30 October 2019. [CrossRef]
4. Almara, L.; Tarancóna, A.; Andreua, T.; Torrella, M.; Hub, Y.; Dezanneaub, G.; Morata, A. Mesoporous ceramic oxides as humidity sensors: A case study for gadolinium-doped ceria. *Sens. Actuators B* **2015**, *216*, 41–48. [CrossRef]
5. Yuan, Q.; Li, N.; Geng, W.; Chi, Y.; Tu, J.; Li, X.; Shao, C. Humidity sensing properties of mesoporous iron oxide/silica composite prepared via hydrothermal process. *Sens. Actuators B* **2011**, *160*, 334–340. [CrossRef]
6. Wang, D.; Lou, Y.; Wang, R.; Wang, P.; Zheng, X.; Zhang, Y.; Jiang, N. Humidity sensor based on Ga_2O_3 nanorods doped with Na^+ and K^+ from GaN powder. *Ceram. Int.* **2015**, *41*, 14790–14797. [CrossRef]
7. Prajapati, M.J.; Vardhan, R.V.; Mandal, S. Effect of lanthanum on the phase evolution of perovskite barium stannate synthesized through polymerized complex method. *Ceram. Int.* **2019**, *45*, 17420–17428. [CrossRef]
8. Ga, A.; Zhao, Q.; He, D.; Zhao, Y.; Chang, A. Rapid response humidity sensor based on DC magnetron sputtered $Mn_{1.2}Co_{1.5}Ni_{0.3}O_4$ thin film. *Mater. Lett.* **2020**, *271*, 127685. [CrossRef]
9. Huang, Y.; Wang, M.; Wen, Z.; Ma, J.; Zhang, J.; Hou, Z. Progress on preparation and application of nanostructured WO_3/TiO_2 composite thin films. *J. Artif. Cryst.* **2020**, *49*, 144–151.
10. Ramaprasad, A.T.; Rao, V. Chitin-polyaniline blend as humidity sensor. *Sens. Actuators B* **2010**, *148*, 117–125. [CrossRef]
11. Putri, N.P.; Kusumawati, D.H.; Widiyanti, N.; Munasir. Synthesis of polyaniline/cellulose composite as humidity sensor. *IOP Conf. Ser. J. Phys. Conf. Ser.* **2018**, *997*, 012009. [CrossRef]
12. Anju, V.P.; Jithesh, P.R.; Narayanankutty, S.K. A novel humidity and ammonia sensor based on nanofibers/polyaniline/polyvinyl alcohol. *Sens. Actuators A* **2019**, *285*, 35–44. [CrossRef]
13. Kotresh, S.; Ravikiran, Y.T.; Raj Prakash, H.G.; Ramana, C.H.V.V.; Vijayakumari, S.C.; Thomas, S. Humidity sensing performance of spin coated polyaniline–carboxymethyl cellulose composite at room temperature. *Cellulose* **2016**, *23*, 3177–3186. [CrossRef]
14. Wang, M.; Zhang, D.; Yang, A.; Wang, D.; Zong, X. Fabrication of polypyrrole/graphene oxide hybrid nanocomposite for ultrasensitive humidity sensing with unprecedented sensitivity. *J. Mater. Sci. Mater. Electron.* **2019**, *30*, 4967–4976. [CrossRef]
15. Zhou, T.; Wu, S.; Cai, J.; Ruan, W. Rapid humidity sensors based on poly (o-phenylenediamine-co-aniline) spherical nanoparticles. *Polym. Bull.* **2020**, *77*, 1095–1105. [CrossRef]

16. Jang, Y.J.; Jung, Y.E.; Kim, G.W.; Lee, C.Y.; Park, Y.D. Metal–organic frameworks in a blended polythiophene hybrid film with surface-mediated vertical phase separation for the fabrication of a humidity sensor. *RSC Adv.* **2019**, *9*, 529–535. [CrossRef]
17. Cherrington, R.; Liang, J. Chapter 2–Materials and deposition processes for multifunctionality. In *Design and Manufacture of Plastic Components for Multifunctionality*; Elsevier: Walthem, MA, USA, 2016; pp. 19–52.
18. Ragab, E.; Shaban, M.; Abdel Khalek, A.; Mohamed, F. Design and characterization of PANI/starch/Fe_2O_3 biocomposite for wastewater remediation. *Int. J. Biol. Macromol.* **2021**, *181*, 301–312. [CrossRef] [PubMed]
19. Yeamin, M.B.; Islam, M.M.; Chowdhury, A.-N.; Awual, M.R. Efficient encapsulation of toxic dyes from wastewater using several biodegradable natural polymers and their composites. *J. Clean. Prod.* **2021**, *291*, 125920. [CrossRef]
20. Arrieta-Almario, A.; Mendoza-Fandiño, J.; Palencia-Luna, M. Composite material elaborated from conducting biopolymer cassava starch and polyaniline. *Rev. Mex. Ing. Química* **2019**, *19*, 707–715. [CrossRef]
21. Hadi Salehi, M.; Golbaten-Mofrad, H.; Jafari, S.H.; Goodarzi, V.; Entezari, M.; Hashemi, M.; Zamanlui, S. Electrically conductive biocompatible composite aerogel based on nanofibrillated template of bacterial cellulose/polyaniline/nano-clay. *Int. J. Biol. Macromol.* **2021**, *173*, 467–480. [CrossRef]
22. Ragazzini, I.; Gualandi, I.; Selli, S.; Polizzi, C.; Cassani, M.C.; Nanni, D.; Gambassi, F.; Tarterini, F.; Tonelli, D.; Scavetta, E.; et al. A simple and industrially scalable method for making a PANI-modified cellulose touch sensor. *Carbohydr. Polym.* **2021**, *254*, 117304. [CrossRef] [PubMed]
23. Dubey, N.; Arora, S. Surfactant assisted synthesis of pH responsive polyaniline-cellulose biocomposite for sensor applications. *Polym. Plast. Tech. Mat.* **2021**, *60*, 1135–1147.
24. Rathore, B.R.S.; Chauhan, N.P.S.; Rawal, M.K.; Ameta, S.C.; Ameta, R. Synthesis and characterization of chitosan-polyaniline-manganese dioxide nanocomposite for removal of methyl orange dye. *Asian J. Chem.* **2021**, *33*, 671–676. [CrossRef]
25. Jasenská, D.; Kašpárková, V.; Radaszkiewicz, K.A.; Capáková, Z.; Pacherník, J.; Trchová, M.; Minařík, A.; Vajd'ák, J.; Bárta, T.; Stejskal, J.; et al. Conducting composite films based on chitosan or sodium hyaluronate. Properties and cytocompatibility with human induced pluripotent stem cells. *Carbohydr. Polym.* **2021**, *253*, 117244. [CrossRef] [PubMed]
26. Rozova, E.Y.; Zoolshoev, Z.F.; Kuryndin, I.S.; Saprykina, N.N.; Elyashevich, G.K. Physicochemical properties and morphological features of modified chitosan/polyaniline composite films. *Russ. J. Phys. Chem.* **2021**, *95*, 193–198. [CrossRef]
27. Draper, J.W. *A Textbook on Chemistry*; Harper & Bros: New York, NY, USA, 1861; p. 55.
28. Wyszyński, P.; Przybylak, R. Variability of humidity conditions in the Arctic during the first International Polar Year, 1882–1883. *Polar Res.* **2014**, *33*, 23896. [CrossRef]
29. Lee, C.-Y.; Lee, G.B. Humidity Sensors: A Review. *Sens. Lett.* **2005**, *3*, 1–15. [CrossRef]
30. Harito, C.; Utari, L.; Putra, B.R.; Yuliarto, B.; Purwanto, S.; Zaidi, S.Z.J.; Bavykin, D.V.; Marken, F.; Walsh, F.C. Review—The development of wearable polymer-based sensors: Perspectives. *J. Electrochem. Soc.* **2020**, *167*, 037566. [CrossRef]
31. Farahani, H.; Wagiran, R.; Hamidon, M.N. Humidity sensors principle, mechanism, and fabrication technologies: A comprehensive review. *Sensors* **2014**, *14*, 7881–7939. [CrossRef]
32. Miyake, T.; Rolandi, M. Grotthuss mechanisms: From proton transport in proton wires to bioprotonic devices. *J. Phys. Condens. Matter* **2016**, *28*, 023001. [CrossRef]
33. Conway, B.E.; Bockris, J.O.; Linton, H. Proton conductance and the existence of the H_3O ion. *J. Chem. Phys.* **1956**, *24*, 834. [CrossRef]
34. Rafique, M.S.; Tahir, M.B.; Rafique, M.; Shakil, M. Chapter 12—Photocatalytic nanomaterials for air purification and self-cleaning. In *Nanotechnology and Photocatalysis for Environmental Applications*; Elsevier: Cambridge, MA, USA, 2020; pp. 203–219.
35. Gong, M.-S.; Kimb, J.-U.; Kim, J.-G. Preparation of water-durable humidity sensor by attachment of polyelectrolyte membrane to electrode substrate by photochemical crosslinking reaction. *Sens. Actuators B* **2010**, *147*, 539–547. [CrossRef]
36. Romero, F.J.; Rivadeneyra, A.; Becherer, M.; Morales, D.P.; Rodríguez, N. Fabrication and characterization of humidity sensors based on graphene oxide–PEDOT:PSS composites on a flexible substrate. *Micromachines* **2020**, *11*, 148. [CrossRef]
37. Gong, M.-S.; Lee, C.-W.; Joo, S.-W.; Cho, B.-K. Humidity-sensitive properties of phosphonium salt-containing polyelectrolytes. *J. Mater. Sci.* **2002**, *37*, 4615–4620. [CrossRef]
38. Mahapure, P.D.; Gosavi, S.W.; Aiyer, R.C. Studies on PVP, PVA and their nAg composites based humidity sensors. In Proceedings of the 2015 2nd International Symposium on Physics and Technology of Sensors, Pune, India, 8–10 March 2015.
39. Kanitkar, P.; Adhyapak, P.; Aiyer, R.; Mulik, U.; Amalnerkar, D. Synthesis and electrical characterization of Ag, Au-PVP nanocomposites for humidity sensing. *Sens. Lett.* **2012**, *10*, 932–940. [CrossRef]
40. Jeeva, A.; Vijayanand, P.S.; Ashokan, S.; Kojima, T.; Kato, S.; Deepalekshmi, P. A facile synthesis of poly(aniline-co-3-trifluoromethyl aniline) doped silver nanoparticles in micellar solution: Its humidity sensor application. *Polym. Sci. Ser. B* **2018**, *60*, 505–515. [CrossRef]
41. Hatamie, S.; Dhas, V.; Kale, B.B.; Mulla, I.S.; Kale, S.N. Polymer-embedded stannic oxide nanoparticles as humidity sensors. *Mater. Sci. Eng. C* **2009**, *29*, 847–850. [CrossRef]
42. Park, T.; Kim, N.; Kim, D.; Kim, S.-W.; Oh, Y.; Yoo, J.-K.; You, J.; Um, M.-K. An organic/inorganic nanocomposite of cellulose nanofibers and ZnO nanorods for highly sensitive, reliable, wireless, and wearable multifunctional sensor applications. *ACS Appl. Mater. Interfaces* **2019**, *11*, 48239–48248. [CrossRef] [PubMed]
43. Ashokan, S.; Jayamurugan, P.; Ponnuswamy, V. Effects of CuO and oxidant on the morphology and conducting properties of PANI:CuO hybrid nanocomposites for humidity sensor application. *Polym. Sci. Ser. B* **2019**, *61*, 86–97. [CrossRef]

44. Chani, M.T.S.; Karimov, K.S.; Khan, S.B.; Fatima, N.; Asir, A.M. Impedimetric humidity and temperature sensing properties of chitosan-$CuMn_2O_4$ spinel nanocomposite. *Ceram. Int.* **2019**, *45*, 10565–10571. [CrossRef]
45. Tripathy, A.; Pramanik, S.; Manna, A.; Shah, N.F.A.; Shasmin, H.N.; Radzi, Z.; Osman, N.A.A. Synthesis and characterizations of novel Ca-Mg-Ti-Fe-oxides based ceramic nanocrystals and flexible film of polydimethylsiloxane composite with improved mechanical and dielectric properties for sensors. *Sensors* **2016**, *16*, 292. [CrossRef] [PubMed]
46. Machappa, T.; Ambika Prasad, M.V.N. Humidity sensing behaviour of polyaniline/magnesium chromate ($MgCrO_4$) composite. *Bull. Mater. Sci.* **2012**, *35*, 75–81. [CrossRef]
47. Ganiger, S.K.; Murugendrappa, M.V. Lab scale study on humidity sensing and DC conductivity of polypyrrole/strontium arsenate ($Sr_3(AsO_4)_2$) ceramic composites. *Polym. Sci. Ser. B* **2018**, *60*, 395–404. [CrossRef]
48. Zhuang, Z.; Qi, D.; Ru, C.; Pan, J.; Zhao, C.; Na, H. Fast response and highly sensitive humidity sensors based on $CaCl_2$-doped sulfonated poly (ether ether ketone)s. *Sens. Actuators B* **2017**, *253*, 666–676. [CrossRef]
49. Hashim, A.; Hamad, Z.H. Fabrication and characterization of polymer blend doped with metal carbide nanoparticles for humidity sensors. *J. Nanostruct.* **2019**, *9*, 340–348.
50. Edwin Suresh Raj, A.M.; Mallika, C.; Swaminathan, K.; Sreedharan, O.M.; Nagaraja, K.S. Zinc(II) Oxide-zinc(II) molybdate composite humidity sensor. *Sens. Actuators B* **2002**, *81*, 229–236. [CrossRef]
51. Meng, X.; Yang, J.; Liu, Z.; Lu, W.; Sun, Y.; Dai, Y. Non-contact, fibrous cellulose acetate/aluminum flexible electronic-sensor for humidity detecting. *Compos. Commun.* **2020**, *20*, 100347. [CrossRef]
52. Li, J.; Zhang, J.; Sun, H.; Yang, Y.; Ye, Y.; Cui, J.; He, W.; Yong, X.; Xie, Y. An optical fiber sensor based on carboxymethyl cellulose/carbon nanotubes composite film for simultaneous measurement of relative humidity and temperature. *Opt. Commun.* **2020**, *467*, 125740. [CrossRef]
53. Zou, J.; Zhang, K.; Zhang, Q. Giant humidity response using a chitosan-based protonic conductive sensor. *IEEE Sens. J.* **2016**, *16*, 8884–8889. [CrossRef]
54. Yadav, C.G.; Sharma, G.; Singh, V.; Kumar, M.; Srivastava, N.; Kumar, S.; Gupta, V. Coupled mode surface plasmon resonance sensor: In situ detection of humidity with starch biofilm. *Opt. Quantum Electron.* **2018**, *50*, 11. [CrossRef]
55. Hashim, A.; Jassim, A. Novel of biodegradable polymers-inorganic nanoparticles: Structural, optical and electrical properties as humidity sensors and gamma radiation shielding for biological applications. *J. Bionanosci.* **2018**, *12*, 170–176. [CrossRef]
56. Hashim, A.; Habeeb, M.A.; Hadi, A. Synthesis of novel polyvinyl alcohol–starch-copper oxide nanocomposites for humidity sensors applications with different temperatures. *Sens. Lett.* **2017**, *15*, 758–761. [CrossRef]
57. Baatout, Z.; Teka, S.; Jaballah, N.; Sakly, N.; Sun, X.; Maurel, F.; Majdoub, M. Water-insoluble cyclodextrin membranes for humidity detection: Green synthesis, characterization and sensing performances. *J. Mater. Sci.* **2018**, *53*, 1455–1469. [CrossRef]
58. Wu, J.; Wu, Z.; Xu, H.; Wu, Q.; Liu, C.; Yang, B.-R.; Gui, X.; Xie, X.; Tao, K.; Shen, Y.; et al. An intrinsically stretchable humidity sensor based on anti-drying, self-healing and transparent organohydrogels. *Mater. Horiz.* **2019**, *6*, 595–603. [CrossRef]
59. Singh, P.; Shukla, S.K. Advances in polyaniline-based nanocomposites. *J. Mater. Sci.* **2020**, *55*, 1331–1365. [CrossRef]
60. Shoaie, N.; Daneshpour, M.; Azimzadeh, M.; Mahshid, S.; Khoshfetrat, S.M.; Jahanpeyma, F.; Gholaminejad, A.; Omidfar, K.; Foruzandeh, M. Electrochemical sensors and biosensors based on the use of polyaniline and its nanocomposites: A review on recent advances. *Microchim. Acta* **2019**, *186*, 465. [CrossRef]
61. Vijayan, A.; Fuke, M.; Hawaldar, R.; Kulkarni, M.; Amalnerkar, D.; Aiyer, R.C. Optical fibre based humidity sensor using Co-polyaniline clad. *Sens. Actuators B* **2008**, *129*, 106–112. [CrossRef]
62. Chen, Z.; Lu, C. Humidity sensors: A review of materials and mechanisms. *Sens. Lett.* **2005**, *3*, 274–295. [CrossRef]
63. Blank, T.A.; Eksperiandova, L.P.; Belikov, K.N. Recent trends of ceramic humidity sensors development: A review. *Sens. Actuators B* **2016**, *228*, 416–442. [CrossRef]
64. Sen, T.; Mishra, S.; Shimp, N.G. Synthesis and sensing applications of polyaniline nanocomposites: A review. *RSC Adv.* **2016**, *6*, 42196–42222. [CrossRef]
65. Do Nascimento, G.M. Spectroscopy of polyaniline nanofibers. In *Nanofibers*; Kumar, A., Ed.; InTech: Rijeka, Croatia, 2010; ISBN 978-953-7619-86-2.
66. Stejskal, J.; Kratochvíl, P.; Jenkins, A.D. Polyaniline: Forms and formation. *Collect. Czech. Chem. Commun.* **1995**, *60*, 1747–1755. [CrossRef]
67. Yang, Q.-H.; Hao, Q.; Lei, J.-P.; Ju, H.-X. Photoelectron-regulated redox reaction of polyaniline for visual detection of trace copper. *Chin. J. Anal. Chem.* **2017**, *45*, 1895–1902. [CrossRef]
68. Liu, J.; Agarwal, M.; Varahramyan, K.; Berney, E.S.; Hodo, W.D. Polymer-based microsensor for soil moisture measurement. *Sens. Actuators B* **2008**, *129*, 599–604. [CrossRef]
69. Song, X.; Qi, Q.; Zhang, T.; Wang, C. A humidity sensor based on KCl-doped SnO_2 nanofibers. *Sens. Actuators B* **2009**, *138*, 368–373. [CrossRef]
70. Pang, S.; Li, G.; Zhang, Z. Synthesis of polyaniline–vanadium oxide nanocomposite nanosheets. *Macromol. Rapid Commun.* **2005**, *26*, 1262–1265. [CrossRef]
71. Spain, E.; Kojima, R.; Kaner, R.B.; Wallace, G.G.; O'Grady, J.; Lacey, K.; Barry, T.; Keyes, T.E.; Forster, R.J. High sensitivity DNA detection using gold nanoparticle functionalised polyaniline nanofibres. *Biosens. Bioelectron.* **2011**, *26*, 2613–2618. [CrossRef] [PubMed]

72. Tamer, U.; Seçkin, A.İ.; Temur, E.; Torul, H. Fabrication of biosensor based on polyaniline/gold nanorod composite. *Int. J. Electrochem.* **2011**, *2011*, 869742. [CrossRef]
73. Rakhi, R.B.; Chena, W.; Alshareef, H.N. Conducting polymer/carbon nanocoil composite electrodes for efficient supercapacitors. *J. Mater. Chem.* **2012**, *22*, 5177–5183. [CrossRef]
74. Nobya, H.; El-Shazly, A.H.; Elkady, M.F.; Ohshima, M. Strong acid doping for the preparation of conductive polyaniline nanoflowers, nanotubes, and nanofibers. *Polymer* **2019**, *182*, 121848. [CrossRef]
75. Freitas, T.V.; Sousa, E.A.; Fuzari, G.C., Jr.; Arlindo, E.P.S. Different morphologies of polyaniline nanostructures synthesized by interfacial polymerization. *Mater. Lett.* **2018**, *224*, 42–45. [CrossRef]
76. Bae, J.; Shin, K.; Kwon, O.S.; Hwang, Y.H.; An, J.; Jang, A.; Kim, H.J.; Lee, C.-S. A succinct review of refined chemical sensor systems based on conducting polymer-cyclodextrin hybrids. *J. Ind. Eng. Chem.* **2019**, *79*, 19–28. [CrossRef]
77. Anisimov, Y.A.; Cree, D.E.; Wilson, L.D. Preparation of multi-component biocomposites and characterization of their physico-chemical and mechanical properties. *J. Compos. Sci.* **2020**, *4*, 18. [CrossRef]
78. Wang, J.; Wang, X.-H.; Wang, X.-D. Study on dielectric properties of humidity sensing nanometer materials. *Sens. Actuators B* **2005**, *108*, 445–449. [CrossRef]
79. Biswas, P.; Kundu, S.; Banerji, P.; Bhunia, S. Super rapid response of humidity sensor based on MOCVD grown ZnO nanotips array. *Sens. Actuators B* **2013**, *178*, 331–338. [CrossRef]
80. Wang, L.L.; Wang, H.Y.; Wang, W.C.; Li, K.; Wang, X.C.; Li, X.J. Capacitive humidity sensing properties of ZnO cauliflowers grown on silicon nanoporous pillar array. *Sens. Actuators B* **2013**, *177*, 740–744. [CrossRef]
81. Zhao, J.; Liu, Y.; Li, X.; Lu, G.; You, L.; Liang, X.; Liu, F.; Zhang, T.; Du, Y. Highly sensitive humidity sensor based on high surface area mesoporous $LaFeO_3$ prepared by a nanocasting route. *Sens. Actuators B* **2013**, *181*, 802–809. [CrossRef]
82. Bayhan, B.; Kavasoğlu, N. A study on the humidity sensing properties of $ZnCr_2O_4$–K_2CrO_4 ionic conductive ceramic sensor. *Sens. Actuators B* **2006**, *117*, 261–265. [CrossRef]
83. Nitta, T.; Terada, Z.; Hayakawa, S. Humidity-sensitive electrical conduction of $MgCr_2O_4$-TiO_2 porous ceramics. *J. Am. Ceram. Soc.* **1980**, *63*, 295–300. [CrossRef]
84. Yavuz, A.G.; Uygun, A.; Can, H.K. The effect of synthesis media on the properties of substituted polyaniline/chitosan composites. *Carbohydr. Res.* **2011**, *346*, 2063–2069. [CrossRef]
85. Lee, S.; Choi, D.; Son, Y. Hazardous acid detection based on chitosan-grafted-polyaniline copolymer. *Polym. Eng. Sci.* **2019**, *59*, E105–E110. [CrossRef]
86. Rajeev, K.K.; Kim, E.; Nam, J.; Lee, S.; Mun, J.; Kim, T.-H. Chitosan-grafted-polyaniline copolymer as an electrically conductive and mechanically stable binder for high-performance Si anodes in Li-ion batteries. *Electrochim. Acta* **2020**, *333*, 135532. [CrossRef]
87. Ratuchne, F.; Danczuk, M.; de Castro, E.G. Enhanced stability and conductivity of (polyaniline-chitosan) composites. *Orbital Electron. J. Chem.* **2018**, *10*, 239–246. [CrossRef]
88. Minisy, I.M.; Salahuddin, N.A.; Ayad, M.A. Adsorption of methylene blue onto chitosan–montmorillonite/polyaniline nanocomposite. *Appl. Clay Sci.* **2021**, *203*, 105993. [CrossRef]
89. Mohammadi, B.; Pirsa, S.; Alizadeh, M. Preparing chitosan-polyaniline nanocomposite film and examining its mechanical, electrical, and antimicrobial properties. *Polym. Polym. Compos.* **2019**, *27*, 507–517. [CrossRef]
90. Karunanithy, P.; Prasad, R.G.S.V.; Jakka, V.S.; Aparna, R.S.L.; Phani, A.R.; Prabhakara, G.S.; Ahmed, S.A. Enhanced antimicrobial activity of polyaniline grafted chitosan. *Adv. Sci. Eng. Med.* **2013**, *5*, 420–426. [CrossRef]
91. Yavuz, A.G.; Uygun, A.; Bhethanabotla, V.R. Substituted polyaniline/chitosan composites: Synthesis and characterization. *Carbohydr. Polym.* **2009**, *75*, 448–453. [CrossRef]
92. Abdi, Z.; Sedaghat, S. Synthesis and characterization of functionalized single-walled carbon nanotube/chitosan/polyaniline nanocomposite. *Int. J. Nano Dimens.* **2016**, *7*, 25–32.
93. Li, W.; Jang, D.M.; An, S.Y.; Kim, D.; Hong, S.-K.; Kim, H. Polyaniline-chitosan nanocomposite: High performance hydrogen sensor from new principle. *Sens. Actuators B* **2011**, *160*, 1020–1025. [CrossRef]
94. Yang, S.; Tirmizi, S.A.; Burns, A.; Barney, A.A.; Risen, W.M. Chitaline materials: Soluble chitosan-polyaniline copolymers and their conductive doped forms. *Synth. Met.* **1989**, *32*, 191–200. [CrossRef]
95. Ramaprasad, A.T.; Rao, V.; Sanjeev, G.; Ramanani, S.P.; Sabharwal, S. Grafting of polyaniline onto the radiation crosslinked chitosan. *Synth. Met.* **2009**, *159*, 1983–1990. [CrossRef]
96. Hosseini, S.H.; Simiari, J.; Farhadpour, B. Chemical and electrochemical grafting of polyaniline onto chitosan. *Iran. Polym. J.* **2009**, *18*, 3–13.
97. Zhao, S.; Xu, H.; Wang, L.; Zhu, P.; Risen, W.M.; Suggs, J.W. Synthesis of novel chitaline–silica aerogels with spontaneous Au and Ag nanoparticles formation in aerogels matrix. *Micropor. Mesopor. Mat.* **2013**, *171*, 147–155. [CrossRef]
98. Lee, G.-W.; Park, M.; Kim, J.; Lee, J.I.; Yoon, H.G. Enhanced thermal conductivity of polymer composites filled with hybrid filler. Compos. *Appl. Sci. Manuf.* **2006**, *37*, 727–734. [CrossRef]
99. Xue, C.; Wilson, L.D. An overview of the design of chitosan-based fiber composite materials. *J. Compos. Sci.* **2021**, *5*, 160. [CrossRef]
100. Mohamed, M.H.; Dolatkhah, A.; Aboumourad, T.; Dehabadi, L.; Wilson, L.D. Investigation of templated and supported polyaniline adsorbent materials. *RSC Adv.* **2015**, *5*, 6976–6984. [CrossRef]

101. Udoetok, I.A.; Wilson, L.D.; Headley, J.V. Self-assembled and cross-linked animal and plant-based polysaccharides: Chitosan–cellulose composites and their anion uptake properties. *ACS Appl. Mater. Interfaces* **2016**, *8*, 33197–33209. [CrossRef]
102. Cabuk, M.; Yavuz, M.; Unal, H.I. Electrokinetic properties of biodegradable conducting polyaniline-graft-chitosan copolymer in aqueous and non-aqueous media. *Colloids Surf. A Physicochem. Eng. Asp.* **2014**, *460*, 494–501. [CrossRef]
103. Casey, L.S.; Wilson, L.D. Investigation of chitosan-PVA composite films and their adsorption properties. *J. Geosci. Environ. Prot.* **2015**, *3*, 78–84. [CrossRef]
104. He, Y.; Zhu, B.; Inoue, Y. Hydrogen bonds in polymer blends. *Prog. Polym. Sci.* **2004**, *29*, 1021–1051. [CrossRef]
105. Wang, T.; Gunasekaran, S. State of water in chitosan–PVA hydrogel. *J. Appl. Polym. Sci.* **2006**, *101*, 3227–3232. [CrossRef]
106. Kulkarni, M.V.; Viswanath, A.K. Spectroscopic, thermal and electrical properties of sulphonic acids doped poly (o-anisidine) and their application as humidity sensor. *Sens. Actuators B* **2005**, *107*, 791–797. [CrossRef]
107. McGovern, S.T.; Spinks, G.M.; Wallace, G.G. Micro-humidity sensors based on a processable polyaniline blend. *Sens. Actuators B* **2005**, *107*, 657–665. [CrossRef]
108. MacDiarmid, A.G. *The Polyanilines: A Novel Class of Conducting Polymers*; Technical Report No. 1992-35; University of Pennsylvania: Philadelphia, PA, USA, 1992; pp. 1–28.
109. Feng, C.D.; Sun, S.L.; Wang, H.; Segre, C.U.; Stetter, J.R. Humidity sensing properties of nation and sol–gel derived SiO_2/Nafion composite thin films. *Sens. Actuators B* **1997**, *40*, 217–222. [CrossRef]
110. Nechtschein, M.; Santier, C.; Travers, J.P.; Chroboczek, J.; Alix, A.; Ripert, M. Water effects in polyaniline: NMR and transport properties. *Synth. Met.* **1987**, *18*, 311–316. [CrossRef]
111. Chiang, J.C.; MacDiarmid, A.G. Polyaniline: Protonic acid doping of the emeraldine form to the metallic regime. *Synth. Met.* **1986**, *13*, 193–205. [CrossRef]
112. Angelopoulos, M.; Ray, A.; McDiarmid, A.G. Polyaniline: Processability from aqueous solutions and effect of water vapor on conductivity. *Synth. Met.* **1987**, *21*, 21–30. [CrossRef]
113. Travers, J.P.; Nechtschein, M. Water effects in polyaniline: A new conduction process. *Synth. Met.* **1987**, *21*, 135–141. [CrossRef]
114. Abdi, M.M.; Liyana, R.; Md Tahir, P.; Heng, L.Y.; Sulaiman, Y.; Waheeda, N.F.; Abu Hassan, N. Physical and structural properties of polyaniline/microcrystalline cellulose nanocomposite. *AIP Conf. Proceed.* **2017**, *1901*, 100018.
115. Abdi, M.M.; Md Tahir, P.; Liyana, R.; Javahershenas, R. A surfactant directed microcrystalline cellulose/polyaniline composite with enhanced electrochemical properties. *Molecules* **2018**, *23*, 2470. [CrossRef]
116. Zhang, Z.; Chu, F. Study on conductivity of the composites of polyaniline/nano-cellulose. In *Applied Sciences in Graphic Communication and Packaging, Lecture Notes in Electrical Engineering 477*; Springer Nature: Singapore, 2018; pp. 877–883.
117. Razalli, R.L.; Abdi, M.M.; Tahir, P.M.; Moradbak, A.; Sulaiman, Y.; Heng, L.Y. Polyaniline-modified nanocellulose prepared from Semantan bamboo by chemical polymerization: Preparation and characterization. *RSC Adv.* **2017**, *7*, 25191–25198. [CrossRef]
118. Fei, G.; Wang, Y.; Wang, H.; Ma, Y.; Guo, Q.; Huang, W.; Yang, D.; Shao, Y.; Ni, Y. Fabrication of bacterial cellulose/polyaniline nanocomposite paper with excellent conductivity, strength, and flexibility. *ACS Sustain. Chem. Eng.* **2019**, *7*, 8215–8225. [CrossRef]
119. Pleumphon, C.; Thiangtham, S.; Pechyen, C.; Manuspiya, H.; Ummartyotin, S. Development of conductive bacterial cellulose composites: An approach to bio-based substrates for solar cells. *J. Biobased Mater. Bioenergy* **2017**, *11*, 321–329. [CrossRef]
120. Tissera, N.D.; Wijesena, R.N.; Rathnayake, S.; de Silva, R.M.; de Silva, K.M.N. Heterogeneous in situ polymerization of polyaniline (PANI) nanofibers on cotton textiles: Improved electrical conductivity, electrical switching, and tuning properties. *Carbohydr. Polym.* **2018**, *186*, 35–44. [CrossRef] [PubMed]
121. Silva, M.J.; Sanches, A.O.; Medeiros, E.S.; Mattoso, L.H.C.; McMahan, C.M.; Malmonge, J.A. Nanocomposites of natural rubber and polyaniline-modified cellulose nanofibrils. *J. Therm. Anal. Calorim.* **2014**, *117*, 387–392. [CrossRef]
122. Takano, T.; Tagaya, M.; Kobayashi, T. Dual-layer hollow fiber of polyaniline–cellulose acetate prepared with simple wet technique of chemical polymerization of aniline. *Polym. Bull.* **2013**, *70*, 3019–3030. [CrossRef]
123. De Paoli, M.A.; Duek, E.R.; Rodrigues, M.A. Poly(aniline)/cellulose acetate composites: Conductivity and electrochromic properties. *Synth. Met.* **1991**, *41–43*, 973–978. [CrossRef]
124. Megha, R.; Ravikiran, Y.T.; Kotresh, S.; Vijaya Kumari, S.C.; Raj Prakash, H.G.; Thomas, S. Carboxymethyl cellulose: An efficient material in enhancing alternating current conductivity of HCl doped polyaniline. *Cellulose* **2018**, *25*, 1147–1158. [CrossRef]
125. El-Sayed, N.S.; Abd El-Aziz, M.E.; Kamel, S.; Turky, G. Synthesis and characterization of polyaniline/tosylcellulose stearate composites as promising semiconducting materials. *Synth. Met.* **2018**, *236*, 44–53. [CrossRef]
126. Janaki, V.; Vijayaraghavan, K.; Oh, B.-T.; Ramasamy, A.K.; Kamala-Kannan, S. Synthesis, characterization and application of cellulose/polyaniline nanocomposite for the treatment of simulated textile effluent. *Cellulose* **2013**, *20*, 1153–1166. [CrossRef]
127. Saikia, A.; Karak, N. Cellulose nanofiber-polyaniline nanofiber-carbon dot nanohybrid and its nanocomposite with sorbitol based hyperbranched epoxy: Physical, thermal, biological and sensing properties. *Colloids Surf. A Physicochem. Eng. Asp.* **2020**, *584*, 124049. [CrossRef]
128. Kim, H.; Song, J.E.; Silva, C.; Kim, H.R. Production of conductive bacterial cellulose-polyaniline membranes in the presence of metal salts. *Text. Res. J.* **2020**, *90*, 1517–1526. [CrossRef]
129. Alonso, E.; Faria, M.; Mohammadkazemi, F.; Resnik, M.; Ferreira, A.; Cordeiro, N. Conductive bacterial cellulose-polyaniline blends: Influence of the matrix and synthesis conditions. *Carbohydr. Polym.* **2018**, *183*, 254–262. [CrossRef] [PubMed]
130. Jasim, A.; Ullah, M.W.; Shi, Z.; Lin, X.; Yang, G. Fabrication of bacterial cellulose/polyaniline/single-walled carbon nanotubes membrane for potential application as biosensor. *Carbohydr. Polym.* **2017**, *163*, 62–69. [CrossRef]

131. Alonso, E.; Faria, M.; Ferreira, A.; Cordeiro, N. Influence of the matrix and polymerization methods on the synthesis of BC/PANi nanocomposites: An IGC study. *Cellulose* **2018**, *25*, 2343–2354. [CrossRef]
132. Wan, Y.; Li, J.; Yang, Z.; Ao, H.; Xiong, L.; Luo, H. Simultaneously depositing polyaniline onto bacterial cellulose nanofibers and graphene nanosheets toward electrically conductive nanocomposites. *Curr. Appl. Phys.* **2018**, *18*, 933–940. [CrossRef]
133. Yue, L.-N.; Zheng, Y.-D.; Luan, J.; Yu, Y. Effects of water content on preparation and properties of bacterial cellulose/polyaniline composite gel-membranes. *Acta Polym. Sin.* **2014**, *9*, 1228–1237.
134. Zhou, Z.; Yang, Y.; Han, Y.; Guo, Q.; Zhang, X.; Lu, C. In situ doping enables the multifunctionalization of templately synthesized polyaniline@cellulose nanocomposites. *Carbohydr. Polym.* **2017**, *177*, 241–248. [CrossRef] [PubMed]
135. Marins, J.A.; Soares, B.G.; Dahmouche, K.; Ribeiro, S.J.L.; Barud, H.; Bonemer, D. Structure and properties of conducting bacterial cellulose-polyaniline nanocomposites. *Cellulose* **2011**, *18*, 1285–1294. [CrossRef]
136. Lee, H.-J.; Chung, T.-J.; Kwon, H.-J.; Kim, H.-J.; Yin Tze, W.T. Fabrication and evaluation of bacterial cellulose-polyaniline composites by interfacial polymerization. *Cellulose* **2012**, *19*, 1251–1258. [CrossRef]
137. Torres, F.G.; Commeaux, S.; Troncoso, O.P. Biocompatibility of bacterial cellulose based biomaterials. *J. Funct. Biomater.* **2012**, *3*, 864–878. [CrossRef]
138. Müller, D.; Mandelli, J.S.; Marins, J.A.; Soares, B.G.; Porto, L.M.; Rambo, C.R.; Barra, G.M.O. Electrically conducting nanocomposites: Preparation and properties of polyaniline (PANI)-coated bacterial cellulose nanofibers (BC). *Cellulose* **2012**, *19*, 1645–1654. [CrossRef]
139. Gautam, V.; Srivastava, A.; Singh, K.P.; Yadav, V.L. Vibrational and gravimetric analysis of polyaniline/polysaccharide composite materials. *Polym. Sci. A* **2016**, *58*, 206–219. [CrossRef]
140. Pandi, N.; Sonawane, S.H.; Gumfekar, S.P.; Kola, A.K.; Borse, P.H.; Ambade, S.B.; Guptha, S.; Ashokkumar, M. Electrochemical performance of starch-polyaniline nanocomposites synthesized by sonochemical process intensification. *J. Renew. Mater.* **2019**, *7*, 1279–1293. [CrossRef]
141. Alkhursani, S.A.; Ghobashy, M.M.; Madani, M. Radiation synthesis of organostarch as fluorescence label. *Asian J. Chem.* **2020**, *32*, 1799–1805. [CrossRef]
142. Sarma, T.K.; Chattopadhyay, A. Reversible encapsulation of nanometer-size polyaniline and polyaniline−Au-nanoparticle composite in starch. *Langmuir* **2004**, *20*, 4733–4737. [CrossRef] [PubMed]
143. Sammanan, B.; Sekar, J.R.; Thavasikan, J. Synthesis and characterization of polyacid doped conducting and non-conducting polymers: A comparative study. *Asian J. Chem.* **2018**, *30*, 799–803. [CrossRef]
144. Noreen, S.; Bhatti, H.N.; Iqbal, M.; Hussain, F.; Sarim, F.M. Chitosan, starch, polyaniline and polypyrrole biocomposite with sugarcane bagasse for the efficient removal of Acid Black dye. *Int. J. Biol. Macromol.* **2020**, *147*, 439–452. [CrossRef]
145. Sapurina, I.; Stejskal, J. The mechanism of the oxidative polymerization of aniline and the formation of supramolecular polyaniline structures. *Polym. Int.* **2008**, *57*, 1295–1325. [CrossRef]
146. Jian, M.; Wang, C.; Wang, Q.; Wang, H.; Xia, K.; Yin, Z.; Zhang, M.; Liang, X.; Zhang, Y. Advanced carbon materials for flexible and wearable sensors. *Sci. China Mater.* **2017**, *60*, 1026–1062. [CrossRef]
147. He, S.; Chen, P.; Qiu, L.; Wang, B.; Sun, X.; Xu, Y.; Peng, H. A mechanically actuating carbon-nanotube fiber in response to water and moisture. *Angew. Chem. Int. Ed.* **2015**, *54*, 14880–14884. [CrossRef]
148. Cheng, H.; Liu, J.; Zhao, Y.; Hu, C.; Zhang, Z.; Chen, N.; Jiang, L.; Qu, L. Graphene fibers with predetermined deformation as moisture-triggered actuators and robots. *Angew. Chem. Int. Ed.* **2013**, *52*, 10482–10486. [CrossRef]
149. Kundu, S.; Majumder, R.; Ghosh, R.; Pradhan, M.; Roy, S.; Singha, P.; Ghosh, D.; Banerjee, A.; Banerjee, D.; Chowdhury, M.P. Relative humidity sensing properties of doped polyaniline-encased multiwall carbon nanotubes: Wearable and flexible human respiration monitoring application. *J. Mater. Sci.* **2020**, *55*, 3884–3901. [CrossRef]
150. Bibi, A.; Rubio, Y.R.M.; Santiago, K.S.; Jia, H.-W.; Ahmed, M.M.M.; Lin, Y.-F.; Yeh, J.-M. H_2S-sensing studies using interdigitated electrode with spin-coated carbon aerogel-polyaniline composites. *Polymers* **2021**, *13*, 1457. [CrossRef] [PubMed]
151. Hashemi Karouei, S.; Milani Moghaddam, H.; Saadat Niavol, S. Characterization and gas sensing properties of graphene/polyaniline nanocomposite with long-term stability under high humidity. *J. Mater. Sci.* **2021**, *56*, 4239–4253. [CrossRef]
152. Li, C.; Yang, S.; Guo, Y.; Huang, H.; Chen, H.; Zuo, X.; Fan, Z.; Liang, H.; Pan, L. Flexible, multi-functional sensor based on all-carbon sensing medium with low coupling for ultrahigh-performance strain, temperature and humidity sensing. *Chem. Eng. J.* **2021**, *426*, 130364. [CrossRef]
153. Liu, K.; Yang, P.; Li, S.; Li, J.; Ding, T.; Xue, G.; Chen, Q.; Feng, G.; Zhou, J. Induced potential in porous carbon films through water vapor absorption. *Angew. Chem. Int. Ed.* **2016**, *55*, 8003–8007. [CrossRef]
154. Dehabadi, L.; Karoyo, A.H.; Wilson, L.D. Spectroscopic and thermodynamic study of biopolymer adsorption phenomena in heterogeneous solid−liquid systems. *ACS Omega* **2018**, *3*, 15370–15379. [CrossRef] [PubMed]
155. Miege, A.; Steffen, F.; Luschtinetz, T.; Jakubith, S.; Freitag, M. Real time water detection for adaptive control strategy in pemfc-systems. *Energy Procedia* **2012**, *29*, 431–437. [CrossRef]
156. Blokzijl, W.; Engberts, J.B.F.N. Hydrophobic effects. Opinions and facts. *Angew. Chem. Int. Ed. Engl.* **1993**, *32*, 1545–1579. [CrossRef]
157. Biedermann, F.; Nau, W.M.; Schneider, H.-J. The hydrophobic effect revisited—studies with supramolecular complexes imply high-energy water as a noncovalent driving force. *Angew. Chem. Int. Ed.* **2014**, *53*, 2–16. [CrossRef]

158. Smithrud, D.B.; Wyman, T.B.; Diederich, F. Enthalpically driven cyclophane–arene inclusion complexation: Solvent-dependent calorimetric studies. *J. Am. Chem. Soc.* **1991**, *113*, 5420–5426. [CrossRef]
159. Merzel, F.; Avbelj, F. Why do water molecules around small hydrophobic solutes form stronger hydrogen bonds than in the bulk? *BBA-Gen. Subj.* **2020**, *1864*, 129537. [CrossRef] [PubMed]
160. Rekharsky, M.V.; Inoue, Y. Complexation thermodynamics of cyclodextrins. *Chem. Rev.* **1998**, *98*, 1875–1918. [CrossRef] [PubMed]
161. Kimura, T.; Yukiyama, T.; Fujisawa, M. Thermodynamic properties of inclusion complexes of α-cyclodextrin + aliphatic nitriles ($H(CH_2)nCN$: n = 1–8) in aqueous solution. *J. Therm. Anal. Calorim.* **2012**, *108*, 695–704. [CrossRef]
162. Betzel, C.; Saenger, W.; Hingerty, B.E.; Brown, G.M. Topography of cyclodextrin inclusion complexes, part 20. Circular and flip-flop hydrogen bonding in beta-cyclodextrin undecahydrate: A neutron diffraction study. *J. Am. Chem. Soc.* **1984**, *106*, 7545–7557. [CrossRef]
163. Koehler, J.E.H.; Saenger, W.; van Gunsteren, W.F. The flip-flop hydrogen bonding phenomenon. Molecular dynamics simulation of crystalline β-cyclodextrin. *Eur. Biophys. J.* **1988**, *16*, 153–168. [CrossRef]
164. Vaitheeswaran, S.; Yin, H.; Rasaiah, J.C.; Hummer, G. Water clusters in nonpolar cavities. *Proc. Natl. Acad. Sci. USA* **2004**, *101*, 17002–17005. [CrossRef]
165. Thiessen, A.N.; Verbeek, W.; Gritter, K.; Ooms, K.J. Assessment of the sensitivity of DQF/ZQF ^2H NMR of D_2O for studying modified Nafion membranes at 20 °C and 80 °C. *Solid State Nucl. Magn. Reson.* **2018**, *93*, 1–6. [CrossRef]
166. Alix, A.; Lemoine, V.; Nechtschein, M.; Travers, J.P.; Menardo, C. Water absorption study in polyaniline. *Synth. Met.* **1989**, *29*, 457–462. [CrossRef]
167. Gocho, H.; Shimizu, H.; Tanioka, A.; Chou, T.-J.; Nakajima, T. Effect of polymer chain end on sorption isotherm of water by chitosan. *Carbohydr. Polym.* **2000**, *41*, 87–90. [CrossRef]
168. Aizamddin, M.F.; Roslan, N.C.; Kamarudin, M.A.; Omar, S.N.I.; Safian, M.F.; Halim, M.I.A.; Mahat, M.M. Study of conductivity and thermal properties of polyaniline doped with p-toluene sulfonic acid. *Malays. J. Anal. Sci.* **2020**, *24*, 413–421.
169. Ming Li, M.; Gu, S.; Jin, J.; Zheng, Y.; Xie, D. Research on the influence of polyanionic cellulose on the microstructure and properties of oil well cement. *Constr. Build. Mater.* **2020**, *259*, 119841.
170. Qiao, D.; Zhang, B.; Huang, J.; Xie, F.; Wang, D.K.; Jiang, F.; Zhao, S.; Zhu, J. Hydration-induced crystalline transformation of starch polymer under ambient conditions. *Int. J. Biol. Macromol.* **2017**, *103*, 152–157. [CrossRef] [PubMed]
171. Lopez-Rubio, A.; Htoon, A.; Gilbert, E.P. Influence of extrusion and digestion on the nanostructure of high-amylose maize starch. *Biomacromolecules* **2007**, *8*, 1564–1572. [CrossRef]
172. Chanajaree, R.; Sriuttha, M.; Lee, V.S.; Wittayanarakul, K. Thermodynamics and kinetics of cationic/anionic dyes adsorption on cross-linked chitosan. *J. Mol. Liq.* **2021**, *322*, 114507. [CrossRef]
173. Alghunaim, N.S. Effect of hydration on the electronic properties of Si/PANi/3ZnO nanocomposite. *J. Inorg. Organomet. Polym. Mater.* **2020**, *30*, 451–456. [CrossRef]
174. Sandilya, A.A.; Natarajan, U.; Priya, M.H. Molecular view into the cyclodextrin cavity: Structure and hydration. *ACS Omega* **2020**, *5*, 25655–25667. [CrossRef]
175. Grosu, I.A.; Filip, X.; Miclăuș, M.O.; Filip, C. Hydrogen-mediated noncovalent interactions in solids: What can nmr crystallography tell about? *Molecules* **2020**, *25*, 3757. [CrossRef]
176. Eggers, D.K.; Fu, S.; Ngo, D.V.; Vuong, E.H.; Brotin, T. Thermodynamic contribution of water in cryptophane host−guest binding reaction. *J. Phys. Chem. B* **2020**, *124*, 6585–6591. [CrossRef]
177. Hähnel, T.; Möllmer, J.; Klauck, M.; Kalies, G. Vapor adsorption and liquid immersion experiments on carbon molecular sieves. *Adsorption* **2020**, *26*, 361–373. [CrossRef]
178. Sabzevari, M.; Cree, D.E.; Wilson, L.D.; Cree, D.E. Gas and solution uptake properties of graphene oxide-based composite materials: Organic vs. inorganic cross-linkers. *J. Compos. Sci.* **2019**, *3*, 80. [CrossRef]
179. Kraml, J.; Kamenik, A.S.; Waibl, F.; Schauperl, M.; Liedl, K.R. Solvation free energy as a measure of hydrophobicity: Application to serine protease binding interfaces. *J. Chem. Theory Comput.* **2019**, *15*, 5872–5882. [CrossRef]
180. Tang, D.; Dwyer, T.; Bukannan, H.; Blackmon, O.; Delpo, C.; Barnett, J.W.; Gibb, B.C.; Ashbaugh, H.S. Pressure induced wetting and dewetting of the nonpolar pocket of deep-cavity cavitands in water. *J. Phys. Chem. B* **2020**, *124*, 4781–4792. [CrossRef] [PubMed]
181. Lyuleeva, A.; Helbich, T.; Bobinger, M.; Rieger, B.; Becherer, M.; Lugli, P.; Rivadeneyra, A. Functionalized and oxidized silicon nanosheets: Customized design for enhanced sensitivity towards relative humidity. *Sens. Actuators B* **2019**, *283*, 451–457. [CrossRef]
182. Wang, X.-H.; Wang, L.-X.; Jing, X.-B.; Wang, F.-S. Synthesis and characterization of chlorinated polyaniline. *Synth. Met.* **1995**, *69*, 149–150. [CrossRef]
183. Saidani, N.; Morallo, E.; Huerta, F.; Besbes-Hentati, S.; Montilla, F. Electrochemical synthesis of fluorinated polyanilines. *Electrochim. Acta* **2020**, *348*, 136329. [CrossRef]
184. Skopalová, K.; Capáková, Z.; Bober, P.; Pelková, J.; Stejskal, J.; Kašpárková, V.; Lehocký, M.; Junkar, I.; Mozetič, M.; Humpolíček, P. In-vitro hemocompatibility of polyaniline functionalized by bioactive molecules. *Polymers* **2019**, *11*, 1861. [CrossRef]
185. Chen, P.-Y.; Hung, H.-L.; Han, C.-C.; Chiu, H.-C. Correlation between nanoscale elasticity, semiconductivity, and structural order in functionalized polyaniline thin films. *Langmuir* **2020**, *36*, 4153–4164. [CrossRef]

186. Chen, X.P.; Jiang, J.K.; Liang, Q.H.; Yang, N.; Ye, H.Y.; Cai, M.; Shen, L.; Yang, D.G.; Ren, T.L. First-principles study of the effect of functional groups on polyaniline backbone. *Sci. Rep.* **2015**, *5*, 16907. [CrossRef]
187. Noboru Yamazoe, N.; Yasuhiro Shimizu, Y. Humidity sensors: Principles and applications. *Sens. Actuators* **1986**, *10*, 379–398. [CrossRef]
188. Packirisamy, M.; Stiharu, I.; Li, X.; Rinaldi, G. A polyimide based resistive humidity sensor. *Sens. Rev.* **2005**, *25*, 271–276. [CrossRef]
189. Harris, F.K. (Ed.) *Electrical Measurements*; Wiley: New York, NY, USA, 1952.
190. Singh, Y. Electrical resistivity measurements: A review. *Int. J. Mod. Phys. Conf. Ser.* **2013**, *22*, 745–756. [CrossRef]
191. Wenner, F. The four-terminal conductor and the Thomson bridge. *Bull. Bur. Stand.* **1912**, *8*, 559–610. [CrossRef]
192. Chandra, H.; Allen, S.W.; Oberloier, S.W.; Bihari, N.; Gwamuri, J.; Pearce, J.M. Open-source automated mapping four-point probe. *Materials* **2017**, *10*, 110. [CrossRef]
193. Chethan, B.; Raj Prakash, H.G.; Ravikiran, Y.T.; Vijayakumari, S.C.; Ramanad, C.H.V.V.; Thomas, S.; Kim, D. Enhancing humidity sensing performance of polyaniline/water soluble graphene oxide composite. *Talanta* **2019**, *196*, 337–344. [CrossRef]
194. Nagaraju, S.C.; Roy, A.S.; Kumar, J.B.P.; Anilkumar, K.R.; Ramagopal, G. Humidity sensing properties of surface modified polyaniline metal oxide composites. *J. Eng.* **2014**, *2014*, 925020. [CrossRef]
195. Lavrič, P.K.; Warmoeskerken, M.M.C.G.; Jocic, D. Functionalization of cotton with poly-NiPAAm/chitosan microgel. Part I. Stimuli-responsive moisture management properties. *Cellulose* **2012**, *19*, 257–271. [CrossRef]
196. Box, G.E.P.; Behnken, D.W. Some new three level designs for the study of quantitative variables. *Technometrics* **1960**, *2*, 455–475. [CrossRef]
197. Ferreira, S.L.C.; Bruns, R.E.; Ferreira, H.S.; Matos, G.D.; David, J.M.; Brandão, G.C.; da Silva, E.G.P.; Portugal, L.A.; dos Reis, P.S.; Souza, A.S.; et al. Box-Behnken design: An alternative for the optimization of analytical methods. *Anal. Chim. Acta* **2007**, *597*, 179–186. [CrossRef]
198. Salehi, R.; Irani, M.; Rashidi, M.-R.; Aroujalian, A.; Raisi, A.; Eskandani, M.; Haririand, I.; Davaran, S. Stimuli-responsive nanofibers prepared from poly(N-isopropylacrylamide-acrylamide-vinylpyrrolidone) by electrospinning as an anticancer drug delivery. *Des. Monomers Polym.* **2013**, *16*, 515–527. [CrossRef]
199. Hierlemann, A.; Weimar, U.; Kraus, G.; Schweizer-Berberich, M.; Göpel, W. Polymer-based sensor arrays and multicomponent analysis for the detection of hazardous organic vapours in the environment. *Sens. Actuators B* **1995**, *26–27*, 126–134. [CrossRef]
200. Srinivasa, P.C.; Ravi, R.; Tharanathan, R.N. Effect of storage conditions on the tensile properties of eco-friendly chitosan films by response surface methodology. *J. Food Eng.* **2007**, *80*, 184–189. [CrossRef]
201. Ruotolo, L.A.M.; Gubulin, J.C. A factorial-design study of the variables affecting the electrochemical reduction of Cr(VI) at polyaniline-modified electrodes. *Chem. Eng. J.* **2005**, *110*, 113–121. [CrossRef]
202. Czitrom, V. One-Factor-at-a-Time versus Designed Experiments. *Am. Stat.* **1999**, *53*, 126–131.
203. Li, S.; He, Y.; Guan, Y.; Liu, X.; Liu, H.; Xie, M.; Zhou, L.; Wei, C.; Yu, C.; Chen, Y. Cellulose nanofibril-stabilized Pickering emulsion and in situ polymerization lead to hybrid aerogel for high-efficiency solar steam generation. *ACS Appl. Polym. Mater.* **2020**, *2*, 4581–4591. [CrossRef]
204. Shukla, S.K. Synthesis of polyaniline grafted cellulose suitable for humidity sensing. *Indian J. Eng. Mater. Sci.* **2012**, *19*, 417–420.

polymers MDPI

Review

Modification of Cellulose Micro- and Nanomaterials to Improve Properties of Aliphatic Polyesters/Cellulose Composites: A Review

Mariia Stepanova and Evgenia Korzhikova-Vlakh *

Institute of Macromolecular Compounds, Russian Academy of Sciences, Bolshoy pr. 31, 199004 St. Petersburg, Russia; maristepanova@hq.macro.ru
* Correspondence: vlakh@hq.macro.ru

Abstract: Aliphatic polyesters/cellulose composites have attracted a lot attention due to the perspectives of their application in biomedicine and the production of disposable materials, food packaging, etc. Both aliphatic polyesters and cellulose are biocompatible and biodegradable polymers, which makes them highly promising for the production of "green" composite materials. However, the main challenge in obtaining composites with favorable properties is the poor compatibility of these polymers. Unlike cellulose, which is very hydrophilic, aliphatic polyesters exhibit strong hydrophobic properties. In recent times, the modification of cellulose micro- and nanomaterials is widely considered as a tool to enhance interfacial biocompatibility with aliphatic polyesters and, consequently, improve the properties of composites. This review summarizes the main types and properties of cellulose micro- and nanomaterials as well as aliphatic polyesters used to produce composites with cellulose. In addition, the methods for noncovalent and covalent modification of cellulose materials with small molecules, polymers and nanoparticles have been comprehensively overviewed and discussed. Composite fabrication techniques, as well as the effect of cellulose modification on the mechanical and thermal properties, rate of degradation, and biological compatibility have been also analyzed.

Keywords: microcrystalline cellulose; nanocrystalline cellulose; cellulose fibers; cellulose modification; aliphatic polyesters; polyhydroxyalkanoates; poly(lactic acid); poly(ε-caprolactone); poly(glycolic acid); poly(lactic acid-co-glycolic acid); poly(hydroxybutyrate); poly(butylene succinate); (bio)composites; "green" materials; mechanical properties; thermal properties; degradation; biocompatibility

Citation: Stepanova, M.; Korzhikova-Vlakh, E. Modification of Cellulose Micro- and Nanomaterials to Improve Properties of Aliphatic Polyesters/Cellulose Composites: A Review. *Polymers* **2022**, *14*, 1477. https://doi.org/10.3390/polym14071477

Academic Editor: Debora Puglia

Received: 28 February 2022
Accepted: 31 March 2022
Published: 5 April 2022

1. Introduction

In recent decades, aliphatic polyesters have attracted enormous interest as an alternative to plastics derived from petroleum [1]. Aliphatic polyesters are biocompatible, biodegradable, and have an excellent ability to a number of processing techniques allowing the production of electrospun nanofibers, films, filaments, nonwoven materials, 3D-printed materials of different shapes, molded and pressed materials, nanocomposite bulk materials, etc. [2–4]. Degradation to nontoxic products, the possibility of recycling, thermoplasticity, nontoxicity, comparability of some parameters with poly(ethylene terephthalate) (PET) and polypropylene (PP) [5–7], low flammability, smoke and refractive index, and dyeability [8] are among other positive features of aliphatic polyesters. In sum, these advantages make aliphatic polyesters very attractive polymers for obtaining biomedical (drug-delivery systems, suture threads, scaffolds for tissue engineering, etc.) [5,9–11] and environmentally friendly materials (packaging and disposable items such as clothing, tableware, etc.) [8,12,13]. However, their high hydrophobicity, insufficient thermal stability, and mechanical and barrier properties limit their wide application for technical purposes. The most powerful way to modify the properties of aliphatic polyesters is to obtain various composites [14]. In this case, the properties of the matrix polymer can be adjusted by the selection of a certain filler. For example, metals [15], carbon nanotubes [16], graphene [17] and its derivatives [18], ceramics [19], and different organic nanoparticles [20–22] are considered to improve the properties of interest.

Despite the variety of potential fillers, the most attention is paid to micro- and nanomaterials that are nontoxic and inexpensive, which makes it possible to produce "green" biocomposites on an industrial scale. Cellulose micro- and nanomaterials are among the most potential fillers for producing such environmentally friendly and biocompatible composites [23–26]. Excellent mechanical properties, a large specific surface area of cellulosic materials, the possibility to obtain them from the wastes of various industries, as well as biodegradability and biocompatibility make their application as reinforcing materials for a variety of areas, including biomedicine and obtaining "green" materials, relevant. However, the hydrophilicity of cellulose impairs significantly its dispersion in hydrophobic polyesters, which leads to cellulose aggregation, poor dispersion in the matrix polyester, and as a consequence, unsatisfactory material properties [27,28]. This obstacle can be overcome by modifying cellulose materials to improve their compatibility with hydrophobic polymer matrices, and as a result, to provide a more homogeneous dispersion of the filler.

In recent years, much attention has been paid to improving the compatibility of cellulose with aliphatic polyesters by covalent and noncovalent modification [29–38]. Modification of the cellulose surface, in turn, affects the properties of the cellulose filler and allows the properties of aliphatic polyester/cellulose composites to be adjusted in a wide range. Recently, several review articles devoted to the composites based on poly(lactic acid) and cellulose [39,40] with special focusing on the processing techniques [41,42] and biofiber's properties [43] have been published. Some reviews have partially discussed cellulose modification [39,44]; however, progress in this area has not been extensively overviewed.

In this comprehensive review, we have summarized the progress on the various approaches reported for the modifications of cellulose micro- and nanomaterials and the further preparation of composites with aliphatic polyesters. The different techniques such as adsorption, covalent modification with small molecules, grafting with polymers, and modification with inorganic and organic nanoparticles have been discussed. Unlike most reviews that consider only poly(lactic acid) (PLA), we have also included other aliphatic polyesters used to produce composites with modified cellulose, e.g., poly(glycolic acid), poly(ε-caprolactone), poly(hydroxybutyrate), poly(butylene succinate) and their copolymers. Furthermore, the effect of modification on various properties of composites, such as mechanical, thermal, degradation and biological ones, have been analyzed.

2. Cellulose Micro- and Nanomaterials

It is known that cellulose is the most abundant polysaccharide on our planet. Its main sources are primarily plants, including wood, as well as algae, tunicate, and metabolic products of some bacteria [45,46]. The highest cellulose content (more than 90%) is characteristic of "bacterial cellulose" (BC), while for other sources this value does not exceed 80% (plant—30–80%, tunicate—about 60%, algae—8–47%) [46–49]. The exception is mature cotton fibers, which consist almost entirely of cellulose (88.0–96.5%) [50]. Accordingly, BC and mature cotton are characterized by fewer impurities, such as lignin and hemicellulose, which are present in large amounts in plant and algae cellulose [46,50]. Another feature of BC is the presence of a finer mesh structure [46]. Furthermore, the degree of crystallinity for cellulose from different sources also varies quite a lot. Regardless of the source, cellulose is a linear homopolysaccharide and consists of β-1,4-glycosidic bonded anhydro-D-glucose units [49,51]. A large number of Van der Waals and hydrogen interactions are formed between and within the polymeric cellulose chains, which lead to the formation of three-dimensional hierarchical structures, the structural unit of which is an elementary fibril [52]. Elementary fibrils, also called elementary nanofibrils, are threadlike bundles of cellulose molecules consisting of alternating crystalline and a number of amorphous domains providing fiber flexibility [26]. Elementary fibrils due to aggregation are packed into microfibrils, which, in turn also aggregate, and this leads to the formation of cellulose fiber [53]. The widths of elementary fibrils and microfibrils range from 1.5 to 5 nm [54,55] and from 10 to 30 nm [49,54], respectively, and the width and length of microfibril aggregates can reach the order of 100 nm and tens of micrometers [49,51,53,55], respectively. The large number of hydroxyl groups (three reactive groups in each monomeric unit) and the supramolecular structure of cellulose determine its physical and chemical properties (insolubility in water and basic solvents, semicrystallinity, good mechanical properties, relative reactivity) [26,55–57].

Depending on the origin and the method of isolation, the degree of polymerization (DP) of cellulose and the molecular orientation of its chains can be different. For native cellulose, the most common crystalline structure is cellulose I, which under the influence of sodium-hydroxide solution or recrystallization changes to the most stable crystalline state, cellulose II. More details about the different forms of cellulose can be found elsewhere [49,51,55,58,59]. The degree of polymerization of cellulose ranges from a few hundred to several tens of thousands [46,49,58]. Given the structure of cellulose, cellulose objects can be produced as fibers, micro/nanofibrils, and micro/nanocrystals [26,44], which vary in degree of polymerization, crystallinity, and shape [60]. Figure 1 schematically demonstrates the general hierarchical structure and structure of a single polymer chain of cellulose with a list of the main cellulose micro- and nanomaterials obtained.

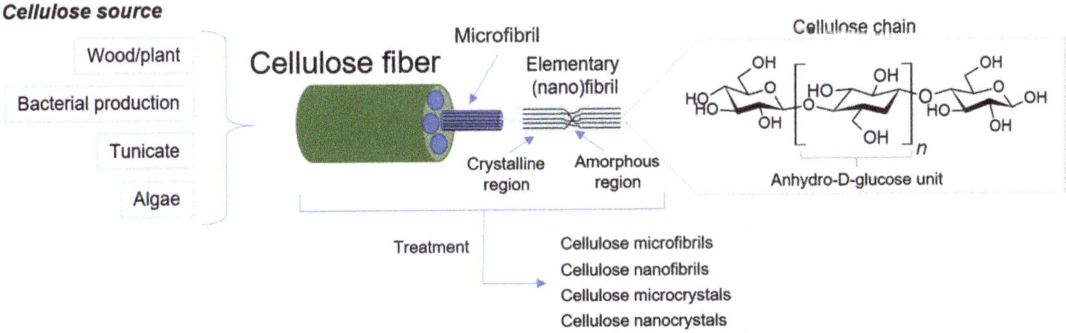

Figure 1. Cellulose from source to molecule and micro- and nanomaterials.

The nomenclature used to designate the various types of micro- and nanocellulose materials is currently ambiguous. Thus, cellulose nanocrystals (CNC) are called nanocrystalline cellulose (NCC), cellulose nanowhiskers (CNW), cellulose whiskers, nanocrystals, nanofibers, nanoparticles, nanorods, rod-like cellulose crystals, cellulose microcrystals (CMC), cellulose microcrystallites, cellulose microfibrils (CMF) [26,39,51,55]; cellulose microfibrils (CMF) are called microfibrillated cellulose (MFC), microfibrillar cellulose, nanofibrillated cellulose (NFC), cellulose nanofibrils (CNF) [26,55]; a synonym of microcrystalline cellulose (MCC) is whiskers [53]. Some time ago, the Technical Association of the Pulp and Paper Industry proposed to standardize the terminology used (nomenclature and abbreviation). According to the recommendations (WI 3021), depending on the dimensions (width (w) and length/width ratio (L/w)), cellulose materials are divided into: cellulose nanocrystals (CNC, w = 3–10 nm, L/w > 5), cellulose nanofibrils (CNF, w = 5–30 nm, L/w > 50), cellulose microcrystals (CMC, w = 10–15 μm, L/w < 2), cellulose microfibrils (CMF, w = 10–100 nm, L/w > 50) [61]. The main cellulose types found in the literature and used in the production of composite materials are shown in Figure 2.

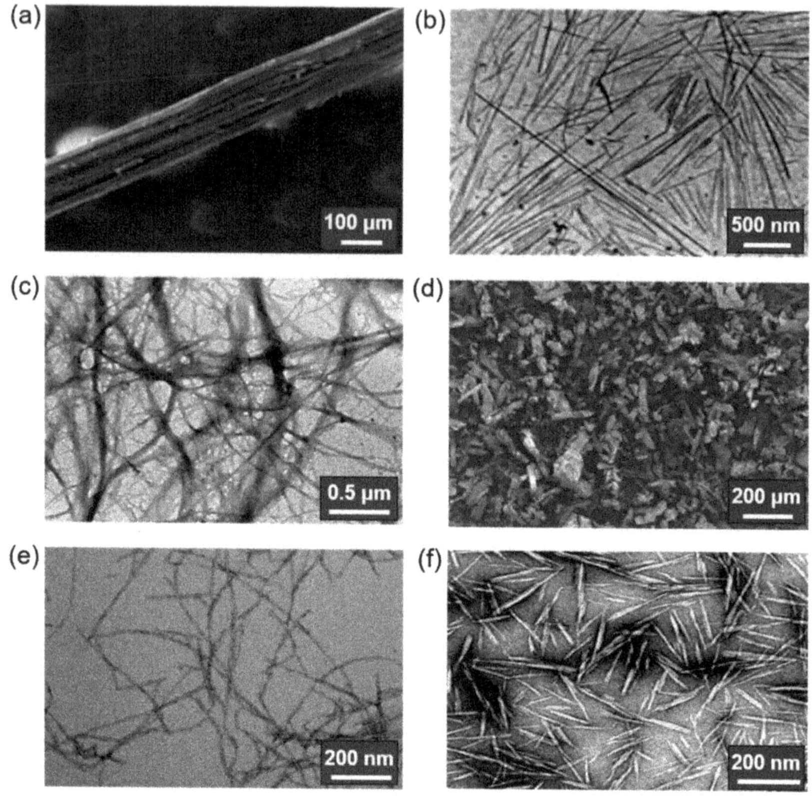

Figure 2. Overview of cellulose micro- and nanomaterials commonly used as fillers to prepare composite materials. Electron micrographs of (**a**) sisal fiber (scanning electron microscopy (SEM), reproduced from [62] under the terms of the Creative Commons CC BY license), (**b**) tunicate whiskers (transmission electron microscopy (TEM), reproduced from [63] with permission of American Chemical Society), (**c**) sugar beet CMF (TEM, reproduced from [64] with permission of Elsevier), (**d**) CMC, commercial (SEM, reproduced from [65] with permission of John Wiley & Sons, Inc), (**e**) wood CNF (TEM, reproduced from [66] with permission of American Chemical Society), (**f**) CNC from ramie fibers (TEM, reproduced from [67] under the terms of the Creative Commons CC BY license).

The size, type, and consequently the physical and chemical properties of the resulting cellulose materials are influenced by the source of origin, processing, and extraction method [44]. For instance, the use of mechanical action alone or its combination with chemical treatment of previously purified cellulose pulp/fibers (e.g., carboxymethylation or TEMPO-mediated oxidation) and/or enzymatic hydrolysis results in thin long flexible micro- (CMF) or nanofibrillar (CNF) structures with alternating crystalline and noncrystalline domains. In turn, acidic hydrolysis produces stiffer particles with a high degree of crystallinity (CMC and CNC), which are the result of the action of acid on both amorphous and crystalline domains. Thus, in the first case, the obtained cellulose micro- and nanofibrils retain the inherent semicrystallinity and high aspect ratio (L/w, over 50 for CMF and CNF) [39,46,49,55], while acid exposure reduces the number of defects in the structure and results in more highly crystalline materials with much lower L/w values (8 to 67 for CMC and CNC) [26,53,68].

Despite the existing terminology recommendations for cellulosic micro- and nanomaterials (see above), the use of terminology in the practice of current publications varies. Nevertheless, we have attempted to generalize the available data on the size of the various cellulose-based materials used. The preparation methods and summarized descriptions and characteristics of the obtained micro- and nanocellulose materials found in literature are presented in Table 1 [26–28,39,44,46,69,70].

Table 1. Cellulose micro- and nanomaterials.

Type	Fabrication [26,28,39,44]	Structure [26,27,39,46]	Size [26–28,46,69,70]
CMF	Mechanical treatment	Long thin flexible aggregates of elementary fibrils/microfibrils with amorphous and crystalline domains	Width 20–100 nm Length 0.5—several μm
CNF	Mechanical with/without chemical and/or enzymatic treatment	Long thin flexible structures with amorphous and crystalline domains	Width 2–100 nm Length 0.5—several μm
CMC	Hydrolysis with diluted inorganic acids with/without mechanical treatment	Rigid crystalline spherical or rod-shaped particles (large aggregates of nanocrystals)	10–200 μm
CNC	Hydrolysis with concentrated inorganic acids with mechanical and/or ultrasound treatment	Rigid whiskers, needle-like crystalline particles	Width 3–50 nm Length 100–500 nm (up to several μm for cellulose from algae, tunicate and BC)

Abbreviations: CNC: cellulose nanocrystals (nanocrystalline cellulose); CMC: cellulose microcrystals (microcrystalline cellulose); CNF: cellulose nanofibers; CMF: cellulose microfibers.

The degree of crystallinity and DP for all obtained materials depends largely on the source of the cellulose as well as the processing technique. The found values are very scattered. For example, DPs for micro- and nanoobjects in the literature range from 100 to 15,000 [46,70], and the degrees of crystallinity vary from a few dozen to more than 90% [26,27,55,70]. For instance, the degree of crystallinity for BC and tunicin (cellulose from tunicate) is 80–100%, for cellulose from algae it is more than 70%, and cellulose from plants it is 40–60% [39,46].

The source of cellulose also has a significant impact on its mechanical properties. Young's tensile modulus for cellulose fibers varies from 5 to 200 GPa [39,46,71]. The highest values from this range are typical for tunicin fibers (from 110 GPa), while for fibers from other origins the elastic modulus does not exceed 115–130 GPa [39,46]. Elongation at break and tensile strength of cellulose fibers are in the ranges of 1–30% and 0.2–1.2 GPa, respectively [26,71]. The application of cellulose-fiber treatments that help to reduce the amorphous components in the chain packing thereby leads to a decrease in DP, an increase in crystallinity and, as a result, an increase in the mechanical properties of the resulting cellulose material compared to the original fibers [39,55]. The theoretically calculated Young's modulus (E) of an ideal cellulose crystal (along the axis of the cellulose chain) is 167.5 GPa [72]. According to the published data, the practically identified Young's modulus values for micro/nanocrystals range from 60 to 220 GPa [26,39,44,46,49,73]; for micro/nanofibers from 14 to 84 GPa [39,46,73]; and for a single tunicin microfibril, a value of about 150 GPa has been found [74]. The established tensile-strength data are in the range of 1–10 GPa for CMC and CNC [39,44,46,58]; about 2–6 GPa for cellulose nanofibers [66]; and about 4–8% elongation at break for wood-cellulose CNF has been reported [46]. The above data indicate the excellent mechanical properties of these micro- and nanoscale cellulose materials. The characteristics of crystalline cellulose (density 1.5–1.6 g/cm^3) are close to—and in some cases significantly higher than—those of glass fibers used for composites (E about 70 GPa, density 2.6 g/cm^3), Kevlar (E 60–125 GPa, density 1.45 g/cm^3) and steel (E 200–220 GPa, density about 8 g/cm^3) [61].

In addition to the above subgroups, cellulose materials such as amorphous nanocellulose (ANC) and cellulose nanoyarn (CNY) are also mentioned in the literature. ANC are obtained from regenerated cellulose by acid hydrolysis and ultrasonic treatment and are generally elliptical particles 50–200 nm wide with DP of 60–70 with an amorphous structure and hence poor mechanical properties. CNYs are electrospun nanofibers; their width and DP range from 500–800 nm and 300–600 nm, respectively [26]. CNFs can be made into cellulose filaments by flow-focusing, wet-extrusion, or spinning processes, but their mechanical properties are also inferior to the more highly crystalline forms of micro- and nanocellulose materials [46,75]. Figure 3 shows these three forms of cellulosic materials.

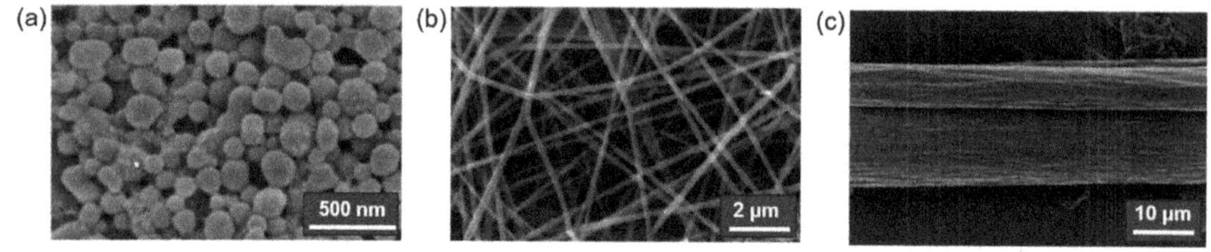

Figure 3. SEM images of (**a**) ANC from CMC (reproduced from [76] under the terms of the Creative Commons CC BY license), (**b**) CNY from carboxymethyl cellulose sodium salt with polyethylene oxide at ratio 1:1 (reproduced from [77] with permission of John Wiley & Sons), (**c**) cellulose filament from carboxymethylated CNF (reproduced from [78] under the terms of the Creative Commons CC BY license).

3. Aliphatic Polyesters

Aliphatic polyesters (polyhydroxyalkanoates) are biocompatible and biodegradable materials. Currently, they are of interest for the fabrication of biomedical and environmentally friendly materials to replace petroleum-based plastics. Polylactide, polyglycolide, polyhydroxybutyrate, polycaprolactone, poly(butylene succinate) and some copolymers based on them are among the widely considered aliphatic polyesters [79–83]. The chemical structures of key aliphatic polyesters are illustrated in Figure 4, while their key characteristics are summarized in Table 2.

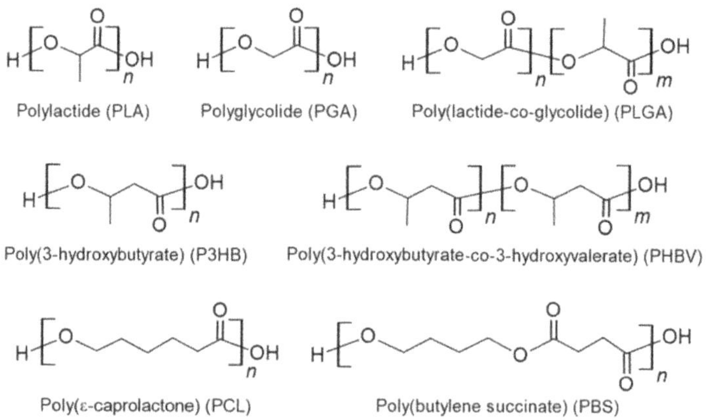

Figure 4. Structures of PLA, PGA, PLGA, P3HB, PHBV, PCL, and PBS.

Table 2. Some characteristics of commonly used aliphatic polyesters.

Characteristics	PLLA	PDLA	PDLLA	PGA	PHB	PCL	PBS
Crystallinity, (%)	Up to 40 (50, thermal treatment) [3,11,84,85]		Amorphous [11,84]	45–77 [86]	50–80 [79,87,88]	Up to 69 (up to 80 during degradation) [89,90]	35–45 [80,91]
Density, (g/cm^3)	1.29 [84]	1.25 [84]	1.25 [84]	1.50–1.71 [86,92]	1.26 [88]	1.07–1.20 [89]	1.23–1.26 [80,91]
T_g, (°C)	55–80 [84]	40–50 [84]	43–60 [11,84]	35–40 [86,93]	4–9 [79,87]	(−65)–(−54) [89,93]	(−45)–(−10) [91]
T_m, (°C)	170–200 [11,79,84]	120–150 [84]	120–170 [84]	220–230 [86,93]	165–185 [79,87]	55–70 [90,93,94]	90–120 [91]
E, (GPa)	2–4 [93]		1–3.5 [93]	6–7 [86,93]	2.5–3.5 [79,87,88]	0.21–0.44 [89,90]	0.03–0.71 [80]
Tensile strength, (MPa)	60–70 [93]		40 [93]	60–110 [86]	20–43 [79,87,88]	4–785 [89,90]	20–35 [80,91]
Elongation at break, (%)	2–6 [93]		1–2 [93]	1–20 [86,93]	5–10 [79,88]	20–4000 [89,93]	560 [91]
Solubility	$CHCl_3$, 1,4-dioxan, furan [84]	Ethyl lactate, ethyl acetate, THF, DMF, DMSO, xylene [84]	Aceton, $CDCl_3$, 1,4-dioxan, furan [84]	Hexafluoro-isopropanol (only for polymers with M < 45,000) [86,93]	Hot $CHCl_3$ and CH_2Cl_2 [87]	1,4-dioxane, 2-nitropropane, cyclohexanone, THF, toluene, benzene, $CHCl_3$, CCl_4, CH_2Cl_2 [90]	$CHCl_3$ [95]
Degradation	More than 2 years [93]		3–6 months [93]	From 6 weeks to 6 months [86 93,96]	4–12 weeks [97]	From several months to several years [89,98]	1–30% for 6 months [99,100]

Abbreviations: THF: tetrahydrofuran; DMF: dimethylformamide; DMSO: dimethyl sulfoxide. Physical quantities: T_g—glass-transition temperature; T_m—melting temperature; E—elastic modulus.

3.1. Poly(lactic acid)

Poly(lactic acid) (PLA) or polylactide is a synthetic, thermoplastic, biocompatible, and biodegradable aliphatic polyester derived from renewable lactic acid [7,11,84]. PLA can be produced by lactic-acid condensation or by the ring-opening polymerization (ROP) of lactide (lactic-acid cyclic dimer) [5,7,11,14,44]. Polycondensation of lactic acid allows the synthesis of only a low-molecular-weight polymer due to the side reaction of hydrolysis preventing the production of high-molecular-weight polymer chains. In contrast to polycondensation, ROP provides PLA with high molecular weight but requires the use of catalysts [7]. A combination of these methods is commonly used to produce PLA on an industrial scale [3]. In this case, a low-molecular-weight polymer is synthesized from lactic acid (2-hydroxypropionic acid) by polycondensation, then the formed polymer is depolymerized to form lactide, which is further used to produce PLA of high molecular weights by ROP [3,7].

The presence of optical isomers of lactic acid and lactide (L-lactic acid, D-lactic acid, L,L-lactide or L-lactide, D,D-lactide or D-lactide, D,L-lactide or meso-lactide) leads to obtaining PLA of four types: isotactic and optically active poly-L-lactide (PLLA) and poly-D-lactide (PDLA), syndiotactic and atactic optically inactive poly-D,L-lactide (PDLLA) [44,84,85]. Equimolar racemic mixture of L- and D-enantiomers of lactide (rac-lactide) is also designated as D,L-lactide [7,85]. Chemical structures of stereoisomers of monomers and PLA can be found elsewhere [44].

The molecular structure (enantiopure) of PLA and heat treatment affect its crystallinity [3,84]. Optically inactive PDLLA is amorphous, whereas stereoregular PLLA and PDLA are capable of homocrystallization [84], forming α- (highest thermodynamic stability), β- or γ-crystalline forms depending on composition [3,6]. Blending PLLA and PDLA leads to their cocrystallization and the formation of a stereocomplex with a different crystal structure, characterized by an increase in melting temperature (T_m) by about 50 °C relative to the homopolymer PLLA or PDLA [6,101]. The parameters of the crystalline forms of PLA can be found in detail elsewhere [6]. Stereochemistry has a tremendous influence on the supramolecular structure of PLA. The presence of more than 10 mol% of the links in the polymer chain different from the basic optical form leads to a significant decrease in crystallinity [84,102]. Branching of the polymer chain also leads to a rather significant decrease in the crystallinity and glass-transition temperature (T_g) of PLA [102]. The crystallinity of PLA is also affected by the molecular weight of the polymer, various treatments, the introduction of nucleation agents, plasticizers into the matrix, and for final products, PLA crystallization can be initiated by temperature annealing [3,6,103].

Crystallinity, as well as the parameters listed above, determine the physical (thermal, rheological, barrier, etc.) and mechanical properties, as well as the degradability of PLA [3,7,44,104,105]. The high degree of crystallinity of polylactide leads to excellent thermal and mechanical properties [44]. However, a high degree of crystallinity is not always necessary and is determined by the application of the final polymer product. For example, the rapid crystallization can complicate stretching the product by blow molding, can diminish the optical transparency of the product, such as a bottle, and can increase the degradation time of the polymer, which may limit its use for some biomedical applications. At the same time, the presence of thermal stability due to high crystallinity is very important for products formed by injection molding [44]. It has been reported that the increase in molecular weight and crystallinity is accompanied by an increase in viscosity and softening point, so that the behavior of PLA in the melt becomes similar to polystyrene [85]. The thermal stability of PLA is similar to poly(vinyl chloride) (PVC), but significantly inferior to PP, polyethylene (PE), polystyrene (PS), and PET [85].

The glass-transition temperature and melting point of PLA are important parameters for predicting the material properties [84]. Both are influenced by the molecular weight of the polymer. T_g and T_m increase sharply when the number-average molecular weight (M_n) increases to 80,000 and 100,000, respectively, and then remain almost unchanged [102]. As optical purity decreases with a constant molecular weight, a decrease in glass-transition temperature is observed. Moreover, PDLA is characterized by lower T_g values than PLLA with the same molecular weight [3,102]. In turn, T_m is more influenced by the amorphous state of PLA than T_g, due to the lack or complete absence of a crystalline phase [102].

The commercially available type of PLA produced on a large scale is mainly PLLA [6,84], since about 90% of all lactic acid is produced from renewable sources by microorganisms as L-isomer [84]. Thus, a commercial polylactide is a semicrystalline polymer with a T_g of 55 to 65 °C and a T_m of 140 to 180 °C depending on the amount of the D-enantiomer impurity [94]. In comparison with petrochemical-based plastics, PLA has a slow crystallization rate, low impact strength, low thermal resistance, and low glass-transition temperature and fragility [11,94]. For instance, to consider the substitution of PET by PLA for packaging fabrication, the barrier properties of PLA need to be improved. Typically, aliphatic polyesters with molecular weights greater than 60,000 are used for packaging, agricultural, and biomedical applications [11].

3.2. Poly(glycolic acid)

Poly(glycolic acid) or polyglycolide (PGA) and its copolymer with PLA (PLGA) is among the most widely studied and used polymers [84,86]. PGA is a semicrystalline, biodegradable, biocompatible aliphatic polyester that differs from PLA in the absence of a methyl group in the monomer unit (glycolic acid residue) [86,93]. PGA can be obtained by polycondensation of glycolic acid (difficult to obtain high molecular weights), ROP of glycolide (more economical, but pure monomer is required), and solid-phase polycondensation of halogen acetates (low degree of polymerization) [86]. The synthesis conditions determine PGA molecular weight, crystallinity, T_m and T_g, and terminal groups. The growth of the molecular weight of PGA contributes to an increase in crystallinity, mechanical properties and a decrease in the biodegradation rate. Acceptable mechanical properties of PGA are achieved at molecular weights greater than 30,000 [86]. Due to the high degradation rate of PGA, its synthesis is more difficult and expensive than for PLA.

Mechanical, thermal, degradation properties and density of PGA are determined by molecular weights, dispersity, and crystallinity. PGA is characterized by high crystallinity. The most common crystallization degree is 45–55%, but 77% has also been reported. Due to the stabilized crystal cage, PGA has a high melting point (220–230 °C) and poor solubility (soluble only in highly fluorinated solvents such as hexafluoroisopropanol) [86]. The glass-transition temperature (T_g) of PGA is higher than the ambient temperature, but close to human body temperature (35–40 °C), which makes the material elastic when introduced into the human body (e.g., implantation) [86,93]. PGA is characterized by poor thermal stability because T_m is very high and close to the degradation temperature [90,92]. The lack of solubility in conventional organic solvents and the narrow processing window of PGA melt create a problem in obtaining products based on it [93]. At the same time, the supramolecular structure of PGA provides excellent mechanical properties [86,93]. For example, the elastic modulus (*E*) of PGA is higher than that of other synthetic biodegradable polymers (PLLA, PDLLA, poly-ε-caprolactone) and is 6–7 GPa [86,93]. The high density of PGA (1.50–1.71 g/cm^3), due to the molecular-packing features, provides high gas-tight properties of the polymer, exceeding this parameter of polyethylene terephthalate (PET) by 100 times [86,92].

Currently, PGA is of great interest for renewable industry and biomedical applications due to its thermal properties, biocompatibility, biodegradation, nontoxicity, excellent mechanical characteristics, and low gas permeability. The obstacles of PGA application are overcome by making PGA-based copolymers and composites [86,93,106]. For example, by copolymerizing glycolide and various enantiomers of lactide and varying their ratios, the properties of the resulting PLGA (stiffness, crystallinity, melting point, and biodegradation rate) can be set. For example, PLGA demonstrates mechanical properties similar to those of human calcareous bone. In addition, PLGA is widely used as implants, micro- and nanoparticles for drug delivery [107,108].

3.3. *Poly(hydroxybutyrate)*

Poly(3-hydroxybutyrate) (P3HB, PHB) is a thermoplastic, biodegradable, semicrystalline, linear microbial aliphatic polyester [5,8,79,87]. Biosynthesis in cells of natural/transgenic plants and bacterial fermentation, including the use of genetically modified microorganisms, are the main ways for the production of P3HB [8,79,87]. In particular, P3HB is produced by prokaryotic microorganisms from sugar-based media (agricultural industrial wastes, hydrolysates of some wood) and other carbon sources in the form of inclusion bodies, which serve as intracellular bacterial depots storing carbon and energy [5,79,87]. Microorganisms may accumulate up to 40–50% of P3HB from the dry-cell mass, and in the case of *Alcaligenes eutrophus* the accumulation may reach up to 96% of the dry cell mass [87]. Depending on the conditions and isolation forms, the resulting P3HB can have different characteristics (molecular weight, crystallinity, mechanical properties and ability to biodegrade) [8,79].

P3HB exhibits optical activity due to a chiral central carbon, and the main natural configuration is poly(D-3-hydroxybutyrate) [87]. The stereostructure and tacticity of P3HB can be specified by chemical synthesis, obtaining isotactic, syndiotactic, or atactic PHB [8,88,109]. The number of monomeric units in P3HB can vary in different range: (1) over than 10,000 for P3HB produced as cytosolic inclusions of bacteria, (2) 100–300 for P3HB from cell membranes and (3) up to 30 monomeric units for P3HB for other natural sources, including human tissues [8]. P3HB with a number of monomeric units greater than 1000 (molecular weight greater than 100,000) can be obtained chemically from β-butyrolactone [8,109].

The linear structure of the P3HB chain ensures its high crystallinity with the presence of an amorphous phase in addition to the crystalline phase [79]. The crystallinity of P3HB can vary in a wide range from 50 to 80%, and as with PLA, has a significant effect on the mechanical properties [79,87]. P3HB is generally characterized as a strong and stiff polymer with low thermal stability and low crystallization rate. Secondary crystallization of P3HB occurs at room temperature with the formation of amorphous lamellae, leading to polymer brittleness [87,110]. P3HB has piezoelectric properties and is also characterized by good resistance to acids, bases, and ultraviolet radiation [87]. In addition, P3HB has better barrier properties than PP, PE, PVC, and PET, and is characterized by some other properties similar to or superior to those of PP and PE [79]. In addition to these advantages, P3HB is a biocompatible and nontoxic polyester, which makes it a promising environmentally friendly alternative to petrochemical polymers, and also demonstrates suitability for various tissue-engineering and other biomedical applications (scaffolds, surgical threads, drug-delivery systems, surgical mesh, etc.).

The low thermal stability is due to the close values of the melting and degradation temperatures of P3HB, which leads to a narrow heat-treatment window. Thus, thermal degradation of the polymer melt can occur during processing [110,111]. This problem can be partially solved by using a random copolymer of 3-hydroxybutyric and 3-hydroxyvaleric acids (poly(3-hydroxybutyrate-co-3-hydroxyvalerate) or poly(3-hydroxybutyric acid-co-3-hydroxyvaleric acid) abbreviated as PHBV or PHBHV. PHBV as well as P3HB homopolymer is characterized by high crystallinity, biocompatibility, biodegradability in different environments, good barrier properties, nontoxicity, UV stability, similarity to P3HB in solubility and chemical stability, hydrophobicity, and low impact resistance and fragility. Unlike P3HB, PHBV has a lower T_m, higher surface tension and flexibility [111]. Thus, PHBV appears to be technologically more attractive and of interest as materials for biomedical applications, agriculture, and packaging materials, and has been developed on an industrial scale [88]. PHBV copolymer can be produced by various microorganisms [111], including recombinant strains, in amounts up to 80% of dry-cell weight, and PHBV composition can vary over a wide range depending on substrate composition [111].

In addition to PHBV, lower melting-point values are observed for another copolymer, namely poly(3-hydroxybutyrate-co-4-hydroxybutyrate) (P(3HB-co-4HB)). Thus, an increase in the number of 4-hydroxybutyrate units (from 0 to 38 mol%) provides a significant decrease in T_m (from 176 to 54 °C), and with a further increase in the proportion of 4HB the melting temperature of copolymers practically does not change. P(3HB-co-4HB) is a thermoplastic biodegradable aliphatic polyester produced by bacterial fermentation. The ratio of monomeric units is largely determined by the substrate used. Moreover, during biosynthesis the production of a mixture of copolymer compositions with a wide range of monomer compositions is observed, including the presence of P4HB, which significantly affects the characteristics of isolated polymers [112]. Regarding P3HB, P4HB is a relatively new material, which can also be obtained by polycondensation of 4-hydroxybutyric acid or ROP γ-butyrolactone. P4HB is also nontoxic, biocompatible, biodegradable, UV-resistant, demonstrates relatively good barrier properties, and exhibits optical activity, and therefore can be used for biomedical and packaging applications [79].

3.4. *Poly(ε-caprolactone)*

Poly-ε-caprolactone (also called, polycaprolactone, PCL) is a biodegradable, hydrophobic, semicrystalline synthetic aliphatic polyester whose monomer unit is built from 6-hydroxyhexanoic acid (6-hydroxycaproic acid) [89,98]. PCL can be synthesized by polycondensation of 6-hydroxyhexanoic acid, but because of the equilibrium nature of this process and the need to remove the water formed during the reaction, obtaining a polymer with a high degree of polymerization and molecular weight values above 10,000 is challenging [90]. Thus, the most promising method is the ROP of ε-caprolactone, which allows the production of PCL with low dispersity and high molecular weights. However, this method of synthesis requires the use of catalysts, often based on metals, which can possess toxic effects [89]. The ε-caprolactone monomer may be obtained by oxidation of cyclohexanol by peracetic acid (industrial method of production) [11], and also by microorganisms as one of the intermediate products [89]. PCL can be synthesized in a wide range of molecular weights. The crystallinity degree of PCL can be up to 69%, and it decreases with the growth in molecular weight [89,98]. It was reported that the nonenzymatic hydrolytic degradation process increased the crystallinity of the PCL sample from the initial 45% to nearly 80% [113]. As with other crystalline aliphatic polyesters, the molecular weight and corresponding degree of crystallinity of PCL have a significant impact on its physical, mechanical, thermal, and degradation properties [89,98]. For example, PCL with M_n below 15,000 has a low viscosity and forms very brittle materials, while polymer with M_n from 25,000 to 90,000 is characterized by desirable mechanical and rheological properties [90]. A rapid thermal degradation is observed for PCL at temperatures above 170 °C, but its low melting (55 to 70 °C) and glass-transition temperatures (−65 to −54 °C) allow its easy processing, which distinguishes PCL from other biodegradable aliphatic polyesters with higher T_m, such as PLA, PGA, their copolymers, PHB and its copolymers [79,90]. In addition to higher thermal stability, PCL in comparison with its biodegradable analogues is characterized by higher viscoelastic properties [98]. Thus, the listed characteristics along with variable viscosity make PCL very technological, suitable, and promising for melt processing such as melt extrusion, electrospinning, injection molding, and 3D printing [90].

PCL is strongly inferior in its mechanical properties to other aliphatic polyesters, and the materials made from it—depending on the architecture, and especially porous ones—are characterized by an even lower load-bearing capacity. This disadvantage imposes significant limitations on the use of PCL materials. This problem can be solved by obtaining copolymers with PCL or composites based on it [90,98]. It should be noted that PCL can be blended with many polymers (PVC, polycarbonates, PLA, PLGA, and several others to form mechanically compatible composites. This property of PCL opens up many opportunities to regulate mechanical, biodegradation, and biological properties for a variety of tasks [89,90,98]. Finally, in comparison to PLA, the low degradation rate of PCL also contributes to the minimal formation of physiological problems caused by pH shifts in the environment during the biodegradation of PCL [90,98].

3.5. *Poly(butylene succinate)*

Poly(butylene succinate) (PBS) is a synthetic, biocompatible, semicrystalline, thermoplastic, biodegradable aliphatic polyester [80,91]. PBS is prepared by the polycondensation of succinic acid (or dimethylsuccinate) with 1,4-butanediol. The monomers (succinic acid and 1,4-butanediol) can be obtained from renewable or fossil-based resources [91]. Safe and accessible microwave radiation can be used to synthesize PBS, resulting in reduced reaction times and increased yields. PBS can be synthesized with high molecular weights and is a commercial product, but its cost is higher compared to common petrochemical plastics. PBS is characterized by two types of crystalline structures (α- and β-form). A β-crystalline structure is observed for the material in the state of deformation [80]. The degree of crystallinity for PBS is 35–45% [91]. Mechanical and thermal properties, as well as biodegradation, depend on the molecular weight of the polymer and its crystallinity [80,91]. PBS is characterized by flexibility and tensile strength close to that of PE and PP [91]. Some physical properties of PBS are comparable to PET. PBS is characterized by good thermal properties. The glass-transition temperature of this polymer is considerably lower than room temperature and ranges from −45 to −10 °C. The melting point of PBS is higher than for PCL but lower in comparison with PLA, PHB, PHBV, and PGA, and varies in the range of 90–120 °C. In view of the above, for PBS various methods of processing are applicable: extrusion, thermoforming, injection molding, etc. These properties, along with biodegradation, distinguish PBS from polyolefins [91]. However, PBS is characterized by disadvantages such as low melt viscosity, slow crystallization rate, gas tightness, and relative brittleness [80,91]. To improve these characteristics, as well as to vary the rate of decomposition and reduce the cost of materials, it is possible to obtain copolymers and composites based on PBS [91]. Furthermore, the introduction of plasticizers in the PBS matrix can improve the rheological properties of this polymer and reduce brittleness [80]. Biodegradability, environmental friendliness, chemical resistance, transparency, physical and mechanical properties, recyclability, and processability allow its application in various fields, first of all in packaging and disposable tableware, but also in textiles, automotive industry, agriculture, forestry, fisheries, medicine, etc.

4. Modification of Cellulose Micro- and Nanomaterials

4.1. Adsorption

Physical modification, or adsorption, is one of the simplest and oldest techniques to modify cellulose nanomaterials [114]. To date, there are a lot of publications describing the modification of cellulose nanomaterials with small molecules and polymers. Selected examples are summarized in Table 3. Basically, the modifying agents are surface-active molecules, or surfactants. Being amphiphilic, they serve as intermediates between hydrophilic cellulose and hydrophobic polyesters. The hydrophilic fragments of surfactants interact with cellulose hydroxyls, while other parts of the (macro)molecule surround the surface like "brushes", preventing the direct interaction between cellulose fibers/particles in nonpolar surroundings.

Among the small molecules, ethoxylated nonylphenol phosphate ester and cetyltrimethylammonium bromide (CTAB) are the most widely used compounds. Li et al. showed that the modification of CNC with low-molecular surfactants did not affect the size of nanocrystals and their distribution [115]. As for modification with polymeric surfactants, such polymers as lignin, poly(vinyl alcohol) (PVA), poly(vinyl acetate) (PVAc), poly(N-vinylpyrrolidone) (PVP), and poly(ethylene glycol) (PEG) are widely applied to enhance the compatibility of cellulose nanomaterials with aliphatic polyesters.

The most common technique for the physical modification of cellulose is the adding of surfactants to aqueous cellulose dispersion, followed by freeze-drying [116–118]. The success of modification of cellulose via adsorption was testified by several groups by such methods for structure characterization as Fourier transform infrared (FTIR) spectroscopy [115,118–120] and solid-state ^{13}C NMR spectroscopy [116]. Besides these, a decrease in turbidity also can indirectly testify the surface modification. In particular, Gois et al. also evaluated the turbidity of the dispersions of the neat CNC and CNC modified with different PEGs or Pluronic VR L44 in chloroform [117]. The authors found that pure CNC began to precipitate after 2 min, while the modified CNC started to reduce its turbidity only after 3–5 min, depending on surfactant.

Table 3. Selected studies on modification of cellulose micro- and nanomaterials via adsorption of small molecules and polymers.

Type of Cellulose	Modifier	Filler Content (wt%)	Matrix Aliphatic Polyester	Processing/ Design ofComposites	Characterization Methods	Refs.
CNC	Ethoxylated nonylphenol phosphate ester	5	PLA/PHB	Melt blending/ Films	TEM, FTIR, XRD, TGA and DSC	[119]
CNW	Ethoxylated nonylphenol phosphate ester	5	PLA	Extrusion + Hot pressing/ Strips	GPC, SEM, TEM, DMA and tensile tests	[114]
CNC	Ethoxylated nonylphenol phosphate ester	1 or 5	PLA/Ag NPs	Casting/Films; Electrospinning/Mats	TEM, FE-SEM, AFM, DSC, DMTA and tensile tests	[121,122]
CNC	CTAB	0.5, 1, 3 or 5	PLA	Hot pressing/Strips	FTIR, UVis, TEM, SEM, TGA, DTG and tensile tests	[115]
CNC	CTAB	1–3	PLA/rGO	Hot pressing/Sheets	FTIR, XRD, AFM, FE-SEM, TGA, DTG, WVP, WAXD, mechanical and MTT-tests	[118]
CNC	Decamethylene dicarboxylic dibenzoyl hydrazide	1 or 3	PLLA	Torque rheometry or Casting/Films	FTIR, XPS, AFM, SEM, TGA, WAXD, DSC, DMA and tensile tests	[123]
Cellulose fibers	Dopamine	40	PLA	Extrusion + hotmolding	FTIR, SEM, XRD, DSC, TGA and mechanical tests	[124]
CNF	Lignin	1, 3 or 5	PLA	Extrusion/Filaments	FTIR, SEM, DSC, DMA, tensile tests	[125]
CNC	Lignin	0.3–2.5	PLA	Compression molding/Disks	Optical microscopy, SEM, DSC, DMA and rheological tests	[104]
Cellulose fibers	Lignin and tannin	35	PLA	Compression molding/Sheets	NMR, FTIR, SEM, TGA, DMA, water sorption, SBS and flexural tests	[116]
CNC	Poly(vinyl alcohol)	1	PLA, PLA/PEG	Casting/Films	ATR-FTIR, XRD, TGA, DSC, mechanical tests	[126]
CNC	Poly(N-vinylpyrrolidone)	5, 9 or 15	PCL	Casting/Films	DLS, BET, SEM, POM, mechanical tests, molecular dynamics simulation	[127]
CNC	PEG300, PEG-1000, PEG monooleate, Pluronic VR L44	3	PLA	Casting/Films	Turbidity measurements, AFM, TGA, mechanical tests	[117]
CNC	Poly(vinyl acetate), poly(ethylene glycol)	2.4 or 4.8	PHB, PHBV	Melt blending/Films	FTIR, POM, TEM, SEM, AFM, TGA, DSC, mechanical tests	[120]
CNC	PEG (after oxidation with TEMPO)	1–5	PLA/rGO	Casting/Films	FTIR, XRD, SEM, TEM, DMA, TGA, DSC, WVP, tensile and MTT-tests, antioxidant activity	[128]

Methods: TEM: transmission electron microscopy; SEM: scanning electron microscopy; FE-SEM: field-emission scanning electron microscopy; AFM: atomic force microscopy; FTIR spectroscopy: Fourier transform infrared spectroscopy; XRD: X-ray diffraction; GPC: gel-permeation chromatography; TGA: thermogravimetric analysis; DTG: differential thermogravimetric analysis; DSC: differential scanning calorimetry; DMA: dynamic-mechanical analysis; DMTA: dynamic-mechanical thermal analysis; UVis: ultraviolet–visible spectroscopy (transparency); XPS: X-ray photoelectron spectroscopy; WVP: water-vapor permeability; WAXS/WAXD: wide-angle X-ray scattering/diffraction; MTT-test: (3-[4,5-dimethylthiazol-2-yl]-2,5 diphenyl tetrazolium bromide) test; NMR: nuclear magnetic resonance; SBS: short beam shear; ATR-FTIR: attenuated total reflection FTIR; DLS: dynamic light scattering; BET: Brunauer–Emmett–Teller; POM: polarized optical microscopy. Abbreviations: Pluronic: triblock copolymer of poly(ethylene oxide) and poly(propylene oxide); rGO: reduced graphene oxide; CTAB: cetyltrimethylammonium bromide; PEG: poly(ethylene glycol).

Modification of the cellulose nanomaterials with surfactants leads to diminishing the agglomeration of cellulose nanomaterials, and as result to their better distribution in the matrix of hydrophobic polyesters. The homogenous distribution of the filler in the matrix polymer has been observed by many authors [114,116,127].

The properties of modified cellulose and its composite materials are strongly influenced by the chemical nature of the modifying agent. For example, cellulose fibers modified with lignin had no effect on the degradation temperature, while coating with tannin led to a decrease in the fiber degradation temperature [116]. In the latter case, such behavior can be explained by the lower intrinsic thermal resistance of tannin molecules compared to cellulose and lignin. No effect on degradation temperature was observed either for PLA composites filled with CNC bearing adsorbed CTAB [115]. A discussion of the dependence of the mechanical properties of composites on cellulose modification is presented below (see Section 5.2).

In general, the advantages of adsorption as a modification tool are its simplicity and its ability to vary the modifier over a wide range. In turn, the possibility of a leakage of adsorbed molecules from the surface when dispersing the modified cellulose in nonpolar solvents to prepare composites by solution casting or precipitation is a main disadvantage of this approach. As a result, leakage of the modifier from the cellulose surface may affect the properties of the prepared composites, for example, not improving the mechanical properties.

4.2. Covalent Modification with Small Molecules

Cellulose covalent modification is limited by the reactions of its functional groups, namely hydroxyls. In particular, depending on modifier agents, ester, urethane, or silyl ether bonds can be formed due to chemical reactions of cellulose hydroxyls (Figure 5).

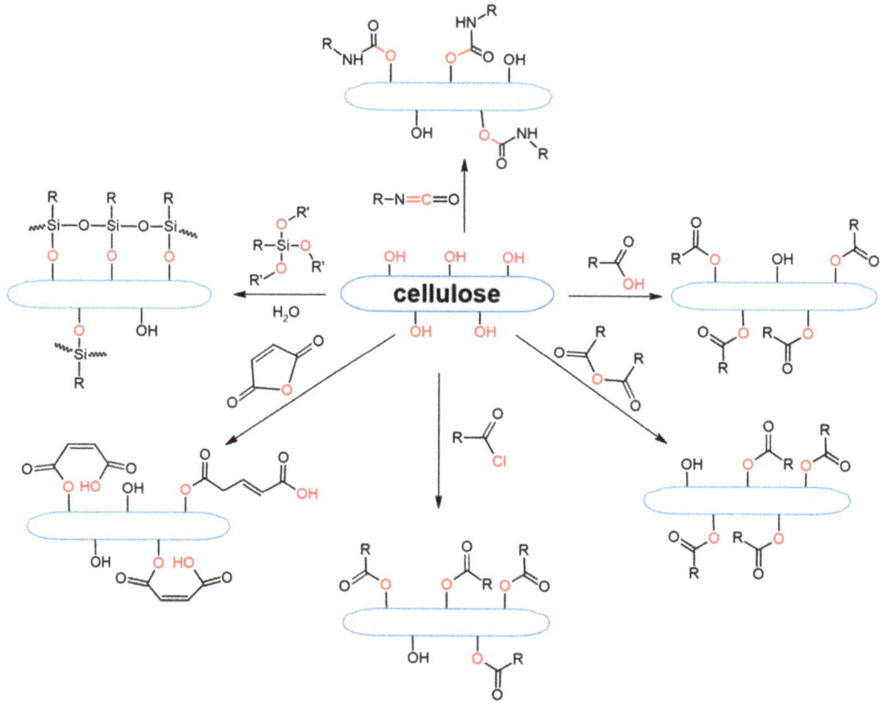

Figure 5. Scheme of cellulose-modification pathways with small molecules (esterification, acylation, silanization, modification with isocyanates).

4.2.1. Ester Bond Formation

Ester bonds are formed as result of the esterification, transesterification, or acylation of hydroxyls by anhydrides of carbonic acids or acyl chlorides.

a. Esterification and Transesterification

Esterification and transesterification are very simple ways to modify cellulose. Esterification is based on the reaction between hydroxyl and carboxyl groups under strong-acid catalysis (Fischer esterification) to produce ester plus water. The main drawback of this reaction is its reversible nature. In general, to shift the balance toward the direct reaction, the hydroxyls must be in excess relative to the carboxylic acid. The sulfuric acid used in the reaction not only acts as a catalyst, but also serves as a dehydrating agent that binds the released water. Besides sulfuric acid, the esterification can be catalyzed by 4-dimethylaminopyridine (DMAP) as an acyl-transfer catalyst (Steglich esterification).

In the case of cellulose micro- and nanomaterials, the esterification is the second most popular method after adsorption [129–133]. Some of the selected studies are summarized in Table 4. Such carboxylic acids as acetic, butanoic, valeric, dodecanoic, oleic, methacrylic, benzoic, etc., as well as fatty acids have been utilized to hydrophobize cellulose for its further use as a filler for aliphatic polyester matrices. To increase the efficiency of heterogeneous esterification, the modified method involves the activation of cellulose-surface hydroxyls with thionyl chlorides (in organic media) followed by the displacement of thionyl with an appropriate carboxylic acid in the presence of pyridine as a catalyst [134–136].

The main methods used for qualitative confirmation are FTIR, and less frequently NMR or XPS (see Table 3). Despite the popularity of this modification approach, only a few studies provided quantitative data on the effectiveness of cellulose modification. Long et al. reported the esterification of cellulose by 70–90% formic acid solution at 70–90 °C for 1–5 h [137]. According to HPLC analysis, the variation of conditions resulted in the binding from 1.7 to 15.8% formyl groups. The maximum amount of formyl groups was introduced during esterification of cellulose at 90 °C for 5 h. At this content of formyl groups, the crystallinity index of modified cellulose fibers was close to that of unmodified fibers. Shojaeiarani et al. modified CNC with valeric acid in DMF in presence of DMAP at 25 °C for 4 h [138]. The reported substitution degree under these conditions was 10% (determined by elemental analysis).

Table 4. Selected studies on covalent modification of cellulose micro- and nanomaterials via esterification and transesterification for the preparation of composites with aliphatic polyesters.

Type of Cellulose	Modifying Agent(s)	Filler Content, (wt%)	Matrix Aliphatic Polyester	Processing/ Design ofComposites	Characterization Methods	Refs.
CNF and CNC	Acetic acid	1	PLA	Casting/Films	Crystallinity, optical, barrier and mechanical properties	[139]
CNC	Acetic acid	3	PLA	Casting/Films	FTIR, XPS, rheological and mechanical tests, TEM, AFM	[140]
CNC	Acetic acid	3	PCL	Casting/Films	Crystallinity, morphology and mechanical properties	[141]
CMC	Acetic acid	0.25–0.75	PHB	Casting/Films	TD-NMR, XRD, WAXD, TGA, DSC, molecular dynamics	[142]
CMF	Butanoic acid	30	PCL/PCL-g-MAGMA	Melt blending/ Films	FTIR, SEM, XRD, DSC, TGA, mechanical tests	[135]
CNC	Valeric acid	1 or 3	PLA	Extrusion + Molding/ Films	FTIR, TEM, SEM, TGA, DMA, mechanical tests	[138]
CNF	Dodecanoic acid	0.05–1.3	PLA (+PEG as plasticizer)	Melt Spinning/ Fibers	Optical microscopy, SEM, TEM, DSC, mechanical tests	[143]
CNF	Oleic Acid	4, 8 or 12	PLA	Casting/Films	FTIR, SEM, XRD, TGA, DSC, WVP and mechanical tests	[136]
CMC	Methacrylic acid	3 or 10	PLA	Extrusion + Molding/ Films	FTIR, SEM, TGA, DSC, flame retardant and mechanical tests	[144]
CNF	Resin acids (from rosin)	2–10	PLA/ Chitosan	Casting/Films	Elemental analysis, TEM, SEM, mechanical and antimicrobial tests, XPS	[145]
CMC	Palmitic acid (from olive oil)	0.1–2	PLA	Casting/Films	FTIR, XRD, WVP, mechanical, UV and biodegradation tests, TGA	[146]
CNC	Benzoic acid	15	PLA	Casting/Films	TEM, SEM, TGA, DMA and tensile tests	[147]
CMF	Formic acid	1	PLA	Casting/Films	FTIR, SEM, XRD, WVP, moisture adsorption, light transmittance and tensile tests	[137]
CNC	Hexanoic ordodecanoic acid	2 or 7	PLLA, PDLLA	Extrusion and melt spinning/Fibers	SEM, DSC, mechanical tests	[134]
Cellulose fibers	Vinyl laurate	5–30	PLA	Melt blending/ Films	ATR-IR, XPS, XRD, DMA, SEM, DSC, TGA, wettability, rheological and tensile tests	[148]
CNF	Triglycerides of Canola oil	1, 3 or 5	PLA + PBS	Extrusion and Molding/Dumbbells	FTIR, SEM, DSC, TGA, tensile and flexural tests	[42]

Methods: TD-NMR: time-domain nucleic magnetic resonance; UV: ultraviolet spectroscopy (transparency); ATR-IR: attenuated total reflection infrared spectroscopy; for other abbreviations see footnote to Table 3.

Transesterification is based on the displacement of an alcohol from an ester by another hydroxyl-containing compound under acid or basic catalysis. In comparison with esterification this approach is less-used for cellulose modification. The modification of cellulose (nano)fibers with lauryl [148] and fatty-acid residues using this method have been recently reported [42]. The modification of cellulose fiber with lauryl moieties was testified by ATR-IR and XPS.

As one could expect, modification of CNC changes the properties of its dispersion and the final composite materials. Modification of cellulose with small hydrophobic molecules was found to improve the rheological properties by suppressing hydrogen bonds abundant in pure cellulose dispersion [148]. As an example, Figure 6 shows images of contact angle changes and SEM images of before and after CMF modification using vinyl laurate.

In total, esterification with carboxylic acids is a quite simple process that does not require the utilization of expensive reagents. However, the reversible nature of the reaction demands a thorough selection of reaction conditions to achieve the sufficient degree of modification.

Figure 6. Images of the neat CMF (**a**) and lauryl-CMF (**d**), SEM images of the neat CMF (**b**) and lauryl-CMF (**e**) and contact angles of the neat CMF (**c**) and lauryl-CMF (**f**). Reproduced from [148] with permission of Elsevier.

b. Acylation with Anhydrides of Carboxylic Acids and Acyl Chlorides

Another way to form an ester bond between cellulose and the modifying agent is to use anhydrides of carboxylic acids or acyl chlorides. Both are highly reactive toward nucleophiles and can acylate a number of functional groups of biomacromolecules (Figure 5). In these cases, the carboxyl is activated and the acylation of hydroxylic groups is facilitated compared to the use of carboxylic acids.

If the anhydride is formed from monocarboxylic acids (e.g., acetic anhydride), acylation requires a deprotonating agent (e.g., pyridine) and is accompanied by the release of one acid molecule as a byproduct. In turn, reactions with dicarboxylic acid anhydrides (e.g., succinic or maleic anhydride) undergo nucleophilic ring opening to form an acylated product containing a newly formed carboxylate group under elevated temperatures. In turn, acyl chlorides have a greater reactivity in comparison with anhydrides. Despite the high reactivity of acyl chlorides compared to other carboxylic acid derivatives (anhydrides, esters, amides), hydroxyl acylation using them can be further promoted by the addition of organic bases such as pyridine. The latter acts as a catalyst by forming an active intermediate with the carbonyl group of acyl chloride. The summary of the application of anhydrides of carboxylic acids and acyl chlorides for cellulose modification is presented in Table 5.

Table 5. Summary on covalent modification of cellulose micro- and nanomaterials via acylation with anhydrides of carboxylic acids and acyl chlorides for the preparation of composites with aliphatic polyesters.

Type of Cellulose	Modifying Agent(s)	Filler Content (wt%)	Matrix Aliphatic Polyester	Processing/ Design of Composites	Characterization Methods	Refs.
CNF	Acetic anhydride	1 or 2	PCL/ Gelatin	Electrospinning/ Nanofibrous scaffolds	FTIR, SEM, WAXS, DSC, biodegradation, conductivity and mechanical tests	[149]
CNF	Acetic anhydride	5	PLA	Extrusion/Strands	FTIR, XRD, TGA, SEM, DMA, tensile and wettability tests	[150]
CNF	Acetic anhydride	0.2–3	PLA	Casting/Films	FTIR, UV, SEM, DSC, mechanical tests	[151]
CNF	Acetic, propionic or butyric anhydride	2	PLA	Casting/Films	ATR-IR, SEM, DSC, wettability, transmittance, transparency, and mechanical tests	[152]
CMF	Acetic anhydride	1–20	PLA/ PLA-EGMA	Casting/Films	FTIR, XRD, TGA, optical microscopy, wettability and mechanical tests	[153]
Cellulose fibers	Acetic anhydride	20, 30 or 40	PLA	Extrusion + molding/ Films	FTIR, TGA, kinetics study	[154]
CNC	Succinic anhydride	1, 2 or 3	PLA	Extrusion + molding/Films	FTIR, TEM, SEM, DSC and DMA	[155]
CNF	Maleic anhydride	5–10	PLA	Casting/Films	FTIR, XRD, SEM, TEM, TGA, mechanical tests	[156]
CNC	Maleic anhydride	1, 3 or 5	PLA	Casting/Films	FTIR, XPS, FE-SEM, DMA and tensile tests	[157]
CNC	Maleic anhydride and furan methylamine	1	PCL/TPU	Extrusion/Filaments	FTIR, shape memory, self-healing, conductivity study, molecular dynamics simulations	[158]
Mixture of celluloses and lignin fibers	Maleic anhydride or APTES	5	PLA	Extrusion + molding/ Films	FTIR, SEM, EDX, wettability and mechanical tests	[159,160]
CMC	Butyryl or lauroyl chlorides	0–9	PLA/BS CMC	Extrusion/ Pellets and Films	NMR, FTIR, SEM, TGA, and mechanical tests	[161]
CNF	Stearoyl chloride	30	PLA	Melt blending/ Blends	DSC, hardness, rheological, wettability and mechanical tests	[162]
CNC	Dodecanoyl chloride or APTES	0.5, 1 or 2	PLA	Extrusion + molding/ Films	ATR-IR, AFM, SEM, XRD, wettability and mechanical tests	[163]
CNC	Palmitoyl chloride	0.5 or 1	PHBV	Melt blending	SEM, HSPOM, TGA, DSC, rheological and mechanical tests	[164]
Lignincellulose	Benzoyl chloride	1–5	PLA	Casting/Films	FTIR, DSC, XRD, SEM, DMTA, rheological and tensile tests	[165]

Methods: HSPOM: hot-stage polarized optical microscope WAXS: wide-angle X-ray scattering; EDX: energy-dispersive X-ray spectroscopy. Abbreviations: APTES: (3-aminopropyl)triethoxysilane; EGMA: ethylene–glycidyl methacrylate copolymer; TPU: thermoplastic polyurethane; BS CMC: banana leaf sheath cellulose microcrystals; for other abbreviations see footnote to Tables 3 and 4.

Under proper conditions, these reactions provide good modification efficiency. For instance, Jamaluddin et al. studied a modification of CNF (in DMF) with acetic, propionic, and butyric anhydrides within 4 h to investigate the reaction rate [152]. Using EDX analysis, the degree of substitution with acetic and propionic anhydride was found to be quite close (around 30%), while the butyric anhydride demonstrated lower modification efficacy (~24%). A 20% acetylation degree of CNF (determined by FTIR) when modified with acetic anhydride in DMF in the presence of pyridine at 105 °C for 20 h was found by Zepic et al. [151]. The average diameter of pure CNF was 43 nm, whereas for acetylated CNF it was 55 nm. In addition, pure CNF was found not to disperse in chloroform, forming large aggregates due to hydrogen bonding. At the same time, acetylated CNF effectively dispersed after 2 min of sonication.

Sojoudiasli et al. reported the modification of CNC with myristoyl chloride in 1,4-dioxane in presence of 1-methylimidazole as a catalyst during 7 h [166]. By applying such conditions, the authors managed to achieve a 31% hydrophobic carboxylic acid coverage of the CNC surface. At the same time, Zhou et al. prepared the cellulose nanocrystals modified with maleic anhydride for reinforcement of the PCL-based composites. Using XPS analysis, the authors detected only about 10% carboxylate groups in the modified CNC samples after the modification, which proceeded in DMF at 120 °C for 20 h under stirring [157].

As several studies have shown, the modification of the cellulose surface with small hydrophobic molecules affects also the crystallinity of micro- and nanomaterials. For instance, Szefer et al. studied the effect of CNC modification with succinic anhydride on the crystallinity fraction content [155]. It was found that the modified CNC showed a higher content of the crystallinity fraction (79.4%) compared to the neat CNC (74.9%). The authors supposed that the increase in crystallinity could be advantageous for the reinforcement of PLA using modified CNC as a filler. Jamaluddin et al. determined the degree of crystallinity (by DSC method) for pure PLA and its composites with neat and hydrophobized CNF. For all composites, the degree of crystallinity was lower than that of pure PLA (39.6%) [152]. However, while this parameter was in the range of 28.2–33.1% for composites with modified CNC, it was only 16.2% for the composite with pure CNC. The reason for such a result was the incompatibility between the neat CNF filler and the PLA matrix, that, in turn, promotes the formation of more aggregates than with the modified CNF. At the same time, the length of the modifying agent, namely acetic, propionic, or butyric anhydride, had no significant effect on the degree of crystallinity. The effect of using anhydrides with different chain lengths on CNF morphology is illustrated in Figure 7.

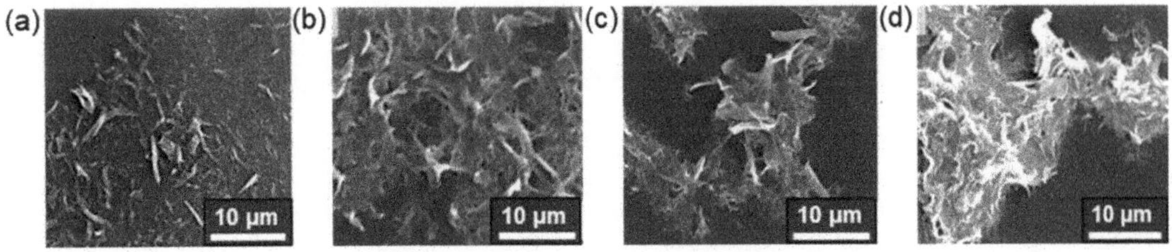

Figure 7. SEM images of CNF before (**a**) and after modification with acetic (**b**), propionic (**c**), and butyric (**d**) anhydrides (reaction time was 4 h). Reproduced from Supplementary Materials of [152], published under the terms of the Creative Commons CC BY license.

Like cellulose materials prepared by chemical modification by (trans)esterification, cellulose hydrophobized by acylation with anhydrides and acyl chlorides also shows good compatibility with hydrophobic matrices. Studying the ultrathin sections of composite materials by TEM, Fumagalli et al. indicated the efficient dispersion of CNC and CMF hydrophobized with the use of myristoyl chloride in comparison with neat cellulose materials [167]. In particular, efficient dispersion of modified CNC in a hydrophobic matrix was achieved with a volume fraction of dry nanocellulose filler up to 20 wt%. A similar result was also observed by Bin et al. [153]. While neat CMF was only able to be introduced into PLA in an amount of 3 wt%, in the case of acetylated CMF, the amount of filler was increased up to 20 wt%. The improved interfacial compatibility has a positive effect on the mechanical properties of the composites (see Section 5.2).

In comparison to esterification with carboxylic acids, acylation with anhydrides of carboxylic acids and acyl chlorides can provide higher degrees of surface modification. In general, ester bonds formed between the modifier and cellulose are suitable for the preparing composites with aliphatic polyesters using any processing technique. At the same time, the sensitivity of the ester bonds to hydrolysis can play a positive role in the degradation of composite materials, e.g., in the case of packaging ones.

4.2.2. Silyl Ethers Formation

Silyl ethers represent chemical bonds consisting of Si covalently bonded to an alkoxy group (-C-O-Si-) due to the reaction between organosilanes and hydroxyl groups (Figure 5). Silanization is an inexpensive and effective method for modifying organic and inorganic surfaces enriched with hydroxyls. Unlike previous modifications, silanization of cellulose materials is used both as an independent functionalization of cellulose and as an intermediate one to introduce the functionality needed for further modification. A wide variety of different silane agents are now available, representing mainly functional ethyl/propyl tri(methoxy/ethoxy)silane agents [168]. Generally, alkoxysilanes have a low reactivity to hydroxylic groups under ambient temperature conditions. In comparison, trimethoxysilanes are more reactive than triethoxysilanes. While trimethoxysilanes react slowly with OH-groups at room temperature, triethoxysilanes are practically unreactive without prior hydrolysis. At the same time, under the right conditions both are sufficiently reactive and can modify OH-bearing surfaces without prior hydrolysis. In particular, the addition of a basic or acidic catalyst increases the reaction rate by increasing the hydrogen-bonding capability of surface hydroxyls and facilitates the reaction at room temperatures.

Selected studies on the cellulose modification by silanization are listed in Table 6. As seen, (3-aminopropyl)triethoxysilane (APTES), (3-methacryloyloxypropyl)trimethoxysilane (MPTMS), and (3-glycidoxypropyl)triethoxysilane (GPTES) are among the most-used silane agents. The effect of variation in silane agent amount on the properties of cellulose micro- and nanomaterials and their composites with aliphatic polyesters has been studied by several research groups [169–173].

Table 6. Selected studies on covalent modification of cellulose micro- and nanomaterials via silanization for the preparation of composites with aliphatic polyesters.

Type of Cellulose	Modifying Agent(s)	Filler Content (wt%)	Matrix Aliphatic Polyester	Processing/ Design of Composites	Characterization Methods	Refs.
CMC	APTES	3	PLA/PP	Extrusion + Molding/Films	FTIR, SEM, DSC, rheological, DMA and tensile tests	[174]
CMC	APTES	0–25	PLA	Automated coating/Films	FTIR, XRD, SEM, TGA, degradation study and mechanical tests	[171]
CNF	APTES	9.5 or 17	PCL	Electro-spinning and compression molding/Films	SEM, XPS, DSC, DMA and tensile tests	[175]
CNC	APTES	2.5	PLA	Compression molding/Sheets	FTIR, SEM, DMTA	[176]
Cellulose fibers	APTES	6, 8 or 10	PLA-co-glycerol	Impregnation of filler into resin/Slides	TGA, DSC, SEM, flexural, wettability, water adsorption, conductivity, DMTA and mechanical tests, element and resonance analysis	[169]
Cellulose fibers	APTES	30	PLA	Blending	FTIR, SEM, DSC, HDT, mechanical tests	[172]
CNC	CETMS	0.5 or 1	PLA	Hot pressing/Films	FTIR, FE-SEM, WAXS, mechanical tests	[177]
CNC	MTMS	2.5	PLA	Casting/Films	FTIR, SEM-EDS, TEM, TGA, DSC, mechanical tests	[178]
CMF	APTES, DMS and TMS	1	PLA	Extrusion + molding/blends	NMR, FTIR, DSC, TGA, and mechanical tests	[179]
CNF	MPTMS	0.25–2	PLA	Casting; Melt blending/Films	NMR, FTIR, SEM, AFM, TGA, and mechanical tests; XPS, DSC	[170]
CNW	MPTMS, APTES, VTMS, MTMS	2.5; 3	PLA	Casting/Films	FTIR, DSC, SEM, and mechanical tests	[173,180]
CNC/DPF	MPTMS and PEG-6000	N/A	PLA	Hot molding/Strips	FTIR, SEM, TGA, DSC, water adsorption, degradation and mechanical tests	[181]
CNF	VTMS, APTES and GPTES	5	PLA	Extrusion + Molding/Films	FTIR, TGA, AFM, SEM, and mechanical tests	[182]
Cellulose fibers	GPTES	30	PLA + PP	Hot molding/Films	FTIR, SEM, XRD, TGA, DMA	[183]
Lignincellulose fibers	MPTMS vs. acetic anhydride	30	PLA	Extrusion/Strands	FTIR, TGA, SEM, GPC, TGA, mechanical tests	[184]

Methods: HDT: heat deflection temperature; for other abbreviations see footnote to Table 3. Abbreviations: DPF: digital printing wastepaper fiber; APTES: γ-aminopropyltriethoxysilane; CETMS: 2-(carbomethoxy)ethyltrimethoxysilane; MTMS: (3-mercaptopropyl) trimethoxysilane (A-189); DMS: N-(2-aminoethyl)-3-aminopropyltrimethoxysilane; TMS: N^1-(3-trimethoxysilyl propyl) diethylenetriamine; MPTMS: 3-methacryloxypropyltrimethoxysilane; VTMS: vinyltrimethoxysilane; VTES: vinyltriethoxysilane; GPTES: glycidyloxypropyl triethoxysilane.

For example, Qu et al. modified CNF with MPTMS at room temperature within 1 h. The determined degrees of substitution were 2.05, 3.84, 5.90, and 6.84% for initial amounts of MPTMS equal to 0.5, 1.0, 1.5, and 2.0 vol% relative to 1 wt% CNF suspension in ethanol. The best mechanical properties were detected for composite prepared by the casting of the PLA solution containing CNF modified by MPTMS (1 vol%): the highest tensile strength and elongation at break increased by 42.3% and 28.2%, respectively, compared to pure PLA. In turn, Li at al. reported the modification of CMC with APTES in 90% ethanol in the presence of a catalytic amount of glacial acetic acid (pH 4–5) at room temperature [171]. The amount of APTES varied from 0.5 to 4.4 wt% relative to CMC. As in the previous case, the tensile properties depended on the amount of silane agent used for modification. The maximum values of tensile strength and elongation at break were found for the PLA composites reinforced with CMC modified with 3 wt% APTES. At the same time, according to XRD analysis, the crystallinity index was decreased during modification with APTES by approximately 5% in comparison to neat CMC. Figure 8 demonstrates images of cellulose fiber, clearly indicating the change in surface morphology after alkaline etching and modification with APTES [172].

Figure 8. SEM images of untreated cellulose fibers (**a**), treated by 10 wt% NaOH solution (**b**) and treated by 10 wt% NaOH solution and 2 wt% of APTES (**c**) (reproduced from [172] with permission of John Wiley & Sons).

Araujo et al. used three different silane agents, namely APTES, GPTES, and VTMS to modify CNF [182]. The modification was carried out in a mixture of ethanol–water (90:10 v/v) at 60 °C for 24 h. The silanization of CNF with organosilanes was proven using NMR spectroscopy. However, quantitative data on the effectiveness of modification with this or that silane were not presented. A comparison of the mechanical properties of PLA filled with CNF modified by different silanes showed a slight increase in Young's modulus for VTMS and GPTES modification and a more evident increase in stress at break for APTES modification. An analogous study, but also with the use of MTMS for modification of cellulose nanowhiskers, was performed by Ma et al. [173]. Silanization was fulfilled in a water–alcohol solution (water/alcohol = 80/20) under acidic catalysis (pH 4 adjusted with acetic acid) at concentrations of silane agents equal to 1, 8, and 16 wt% and 50 °C for 30 min. Regardless of the silane agent, the best tensile properties were established for PLA films produced by solution casting and filled with CNW modified with 8 wt% silane. The elongation at break increased significantly from 12.4% for untreated CNW to approximately 214, 255, and 210% for VTMS, MPTMS, and MTMS treatments, respectively, while APTES demonstrated a relatively weak improvement in strength (111 %). A similar effect was recently observed for composites produced by compression molding and based on PCL containing CNF modified by APTES: cellulose modification had virtually no effect on tensile strength and elongation at break, and only a slight improvement was found for Young's modulus [175]. Virtually no improvement in mechanical properties was observed in the work of Zhang et al. who prepared composites by hot molding from PLA, digital printing waste paper fibers (DPF), and CNC modified with APTES [181]. Most likely, such a dramatic difference in the properties of composites fabricated with the use of silanized cellulose micro- and nanomaterials is strongly related not only to the properties of different silanes, but also to the effectiveness of the cellulose modification, which, unfortunately, is not given sufficient attention in the recent studies.

4.2.3. Urethane-Bond Formation

Urethane or carbamate bonds are formed by the reaction of isocyanate groups with compounds containing active hydrogen atoms, such as in hydroxyl groups (Figure 5). Isocyanates are highly reactive compounds, showing the highest reaction rate at alkaline pH values (e.g., pH 8.5). The main drawback of isocyanates is their high sensitivity to hydrolysis, since moisture rapidly decomposes them. Thus, many reaction protocols recommend performing the reaction with isocyanates in an organic medium (e.g., DMSO, toluene, etc.) [168].

Modification of cellulose with isocyanates is not as popular as acylation for forming an ester bond, and nor is silanization. Nevertheless, there are several recent papers describing such modification for preparation of the aliphatic ester/cellulose composites (Table 7). Among the isocyanates used are aliphatic mono- and bifunctional compounds such as *n*-octadecyl isocyanate and isophorone diisocyanate, and aromatic compounds such as toluene-2,4-diisocyanate.

Table 7. Summary on covalent modification of cellulose micro- and nanomaterials with alkyl/aryl isocyanates for the preparation of composites with aliphatic polyesters.

Type of Cellulose	Modifying Agent(s)	Filler Content (wt%)	Matrix Aliphatic Polyester	Processing/ Design of Composites	Characterization Methods	Refs.
CNF and CMF	*n*-Octadecyl isocyanate	3–12	PCL	Casting/Films	TEM, SEM, DSC, DMA and tensile tests	[185]
CNC	*n*-Octadecyl isocyanate	5 or 10	PCL	Casting/Films	FTIR, XPS, SEM, TGA, rheological and wettability tests	[29]
CMC	Toluene-2,4-diisocyanate	1–5	PLA	Extrusion + Molding/ Films	Elemental analysis, FTIR, SEM, wettability and mechanical tests	[32]
CNC	Toluene-2,4-diisocyanate	1–5; 1–9	PLA	Casting/Films	ATR-FTIR, UVis, TEM, AFM, TGA, tensile tests; NMR, SEM, XRD, DSC	[186,187]
CNC	Isophorone diisocyanate	1 or 5	PLA	Casting/Films	Elemental analysis, NMR, XRD, SEM, wettability and mechanical tests	[188]
Holocellulose powder	4,4′-Methylenebis (phenyl isocyanate)	5–30	PBS	Hot pressing/Sheets	FTIR, SEM, wettability, water adsorption, degradation and mechanical tests	[189]

Methods: for abbreviations see footnote to Table 3.

The modification of cellulose micro- and nanomaterials with *n*-octadecyl isocyanate is reported to be performed in toluene at 90–110 °C for 30 min [29,185]. Recently, Ogunsona et al. studied the CNC modification with isophorone diisocyanate in toluene, toluene/DMSO mixture and DMSO at 105 °C for 1 h [188]. According to the elemental analysis, a higher coupling efficiency was achieved using toluene/DMSO = 10/90 (11%) or pure DMSO (10%). Figure 9 demonstrates the change in the CNC wettability as a result of the surface hydrophobization [188].

Olonisakin et al. investigated the substitution degree in the CMC modification reaction with toluene-2,4-diisocyante [32]. The reaction was performed in THF at 75 °C for various times (from 1 to 24 h). The determined substitution degree ranged from 11 to 16% and reached 14% after 10 h. A change in the hydrophilic–hydrophobic properties of CNC after modification with toluene-2,4-diisocyanate [186] is shown in Figure 10.

In sum, the reactivity of isocyanates is excellent and the reaction with hydroxyls run quite fast. Compared to other modifications, the carbamates, formed by modifying cellulose with isocyanate-bearing molecules, are extremely stable bonds. Thus, the probability of surface-modifier leakage during further manipulations with the modified filler is minimized.

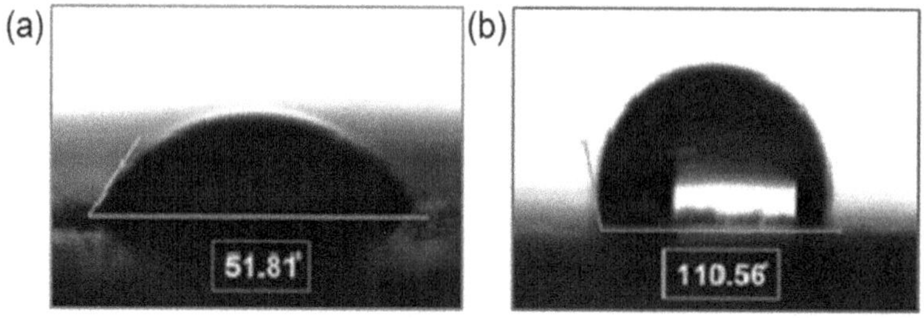

Figure 9. Contact angles of neat CNC (**a**) and CNC modified with isophorone diisocyanate (**b**) (reproduced from [188] with permission of Elsevier).

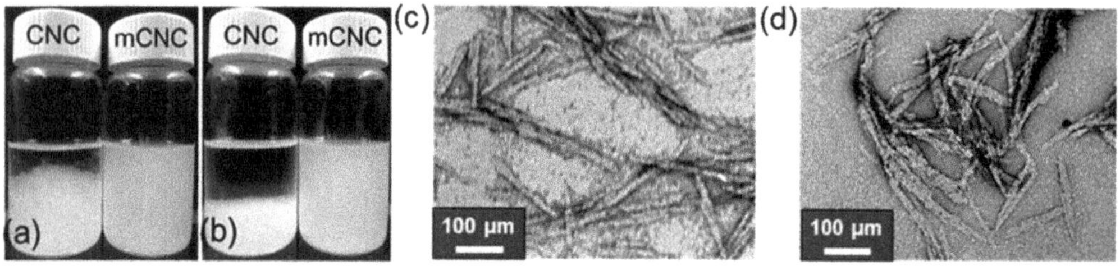

Figure 10. Images of neat CNC and modified CNC (mCNC) in chloroform after 15 min sonication (**a**) and after 15 min sonication and standing (**b**); TEM images of neat (**c**) and modified (**d**) CNC (reproduced from [186] with permission of Royal Society of Chemistry).

4.2.4. Other Modifications

Some other ways to modify the surface of cellulose micro- and nanomaterials with low-molecular compounds [190–195] are also reported. For instance, the activation of carboxyl groups (introduced by TEMPO-mediated oxidation of cellulose hydroxyls) allowed the further attachment of octadecylamine to form an amide bond between cellulose and a modifier [191]. The synthesis of hexyl-CNF can be carried out by two-step method, which includes the reaction of CNF hydroxyls with mono-chloroacetic acid to form carboxymethylated CNF followed by substitution of carboxymethyl-moiety with hexyl one under acidic catalysis as described by Eyholzer et al. [193]. Given the high reactivity of hexahydroxyl N-containing heterocyclic compounds, Yin et al. used 2,4,6-trichloro-1,3,5-triazine as a crosslinking agent for grafting dodecylamine onto the surface of CNC [194].

In order to provide conductive properties to cellulose-based materials, the surface of cellulose can be modified by metals [196–198]. In particular, Sundar et al. reported the modification of CMC with Fe (II) ions by the reaction of iron hydroxide obtained in situ with carboxylic groups generated in CMC via oxidation of surface hydroxyls [196]. Another approach to introduce Fe (II)-ion cellulose materials was proposed by Hassan et al. [197]. They used 4′-chloro-2.2′:6,2″-terpyridine to introduce the terpyridine-chelating group into the surface of CNC. Introduced terpyridine groups are capable of binding Fe (II) ions during further treatment with $FeSO_4$. To prepare the electroactive material, Ummartyotin et al. developed a protocol for the functionalization of bacterial cellulose with Sr ions [198]. For this, the cellulose suspension was treated with strontium chloride at an elevated temperature for 3 h to react with cellulose hydroxyls and finally to form Sr-O bonds.

4.3. Covalent Modification with Polymers

Recently, the functionalization of cellulose with polymers has become a new tool for modifying the properties of cellulose as a filler for aliphatic polyesters in order to improve the mechanical, thermal, or biological properties of composites. There are two techniques that can be used for such modification, namely polymer grafting from the surface of cellulose micro- or nanomaterials by in situ polymerization (the grafting "from" method), or covalent immobilization of presynthesized polymers on the surface of cellulose materials (the grafting "to" method). Summarized results on cellulose modification using both grafting "from" and grafting "to" approaches are discussed below.

4.3.1. Grafting "from"

The presence of a large number of hydroxyl groups on the surface of cellulose materials makes it extremely attractive for surface-initiated ring-opening polymerization for cyclic monomers of aliphatic hydroxy acids, e.g., lactide, glycolide, and lactones. Indeed, the first works devoted to the grafting of polymers from the cellulose materials focused on the ROP of ε-caprolactone [67,199] and L-lactide [199] initiated by surface hydroxyls (Figure 11). The application of surface-initiated ROP makes it possible to obtain cellulose grafted with aliphatic polyesters, which are ideal candidates for increasing the compatibility of the filler with the aliphatic polyester matrix. Thus, it is not surprising that by now the modification of cellulose micro- and nanomaterials of PLA and PCL, followed by the production of composites and the evaluation of their properties, has been studied and discussed by many authors (Table 8).

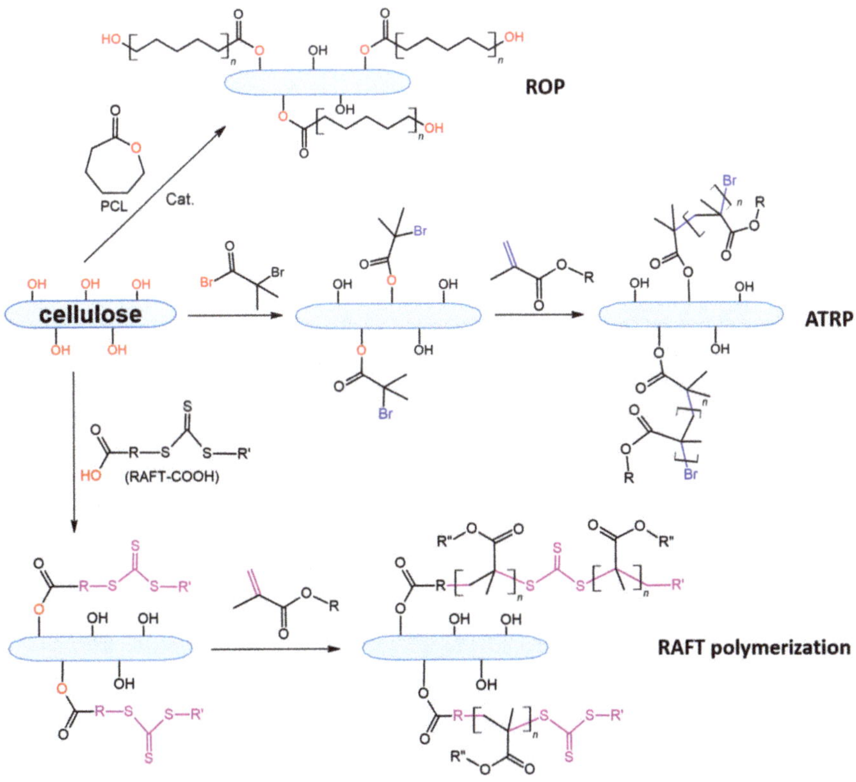

Figure 11. Examples of cellulose modification by the grafting "from" technique (the details and corresponding references are presented in Table 8).

Table 8. Summary on covalent modification of cellulose micro- and nanomaterials by grafting "from" technique.

Type of Cellulose	Grafted Polymer	Cellulose Premodification/Polymerization Technique	Filler Content (wt%)	Matrix Aliphatic Polyester	Processing/Design	Characterization Methods	Refs.
CNC	PLA	−/ROP	1 or 5	PLA	Extrusion + Casting/Films	Elemental analysis, FTIR, SEM, TEM, XRD, TGA	[200]
CNC	PLLA, PDLA	−/ROP	1–10	PLA	Casting/Films	NMR, FTIR, XRD, TEM, SEM, POM, DSC, mechanical tests	[201]
CNW	PLLA	−/ROP	2, 4 or 8	PLA	Molding	FTIR, AFM, DSC, DMTA	[202]
CNC	PLLA	−/ROP	5, 10 or 20	PLA, PLGA, PHB	Casting/Films	NMR, FTIR, AFM, mechanical tests	[203,204]
CNC	PLLA, PDLA	−/ROP	5 or 15	PLLA	Casting/Films	NMR, FTIR, WAXD, POM, DSC, rheological tests	[205]
CMC	OLLA	−/ROP	10, 30 or 50	PLA	Hot pressing/Films	NMR, FTIR, SEM, DSC, mechanical tests	[82]
CNF	PLA	−/ROP	2	PLA	Extrusion/Filaments	FTIR, SEM, WVP, oxygen permeability, water adsorption and mechanical tests	[206]
CNC	PLLA, PDLA	−/ROP	0.2, 0.5 or 1	PLA	Casting/Sheets	NMR, FTIR, GPC, TEM, SEM, DSC, TGA, rheology study	[207]
CNC	PLLA	−/ROP	5	PCL	Casting/Films	FTIR, TGA, DLS, POM, mechanical and MTT-tests	[208]
CNC	PLA	−/ROP	2	PHB	Melt mixing	NMR, FTIR, DSC, XRD, TEM, XPS, SAXS	[209]
CNF	PCL	−/ROP	10	PCL	Molding/Films	FTIR, SEM, mechanical tests	[210]
CNW	PCL	−/ROP	1 or 5	PBSA	Casting/Films	FTIR, XRD, SEM, TGA, DSC	[211]
CNC	PCL	−/ROP	0.5 or 1	PHVB	Melt Blending	FTIR, FE-SEM, DSC, HSPOM, rheological and mechanical tests	[164]
CNW	PCL	−/ROP	–	–	–	SEC, XPS, FTIR, contact angles measurements	[212]
CNW	PCL	−/ROP	0–40	PCL	Casting/Films	FTIR, TOF-SIMS, WAXS, XPS, TEM, TGA, DMA, DSC, wettability and tensile tests	[67]
CNC	PLA	APTES + 3,5-diaminobenzoic acid/ROP	1–7	PLA	Electrospining/Nanofibers film	FTIR, XPS, SEM, TGA, tensile and shape memory tests	[213]
CNC	PBS	−/Polycondensation of 1,4-butanediol and succinic anhydride	0.5, 1 or 2	PLA/PBS	Compression molding/Sheets	NMR, FTIR, XPC, SEC, elemental analysis, SEM, DSC, TGA, WAXD, DMA,	[214]
Cellulose fibers	PBA, PEHA or PMMA	Adsorption of monomers/radical polymerization	50	PLA	Hot pressing/Sheets	Optical microscopy, SEM, DMA, fungal growth test	[215]
CNC	PBMA	α-bromoisobutyryl bromide/ATRP	0.5, 1 or 3	PCL	Extrusion + hot pressing/Sheets	FTIR, AFM, SEM, wettability, mechanical tests	[216]
CNF	PMMA	Oxidation with TEMPO + GPTMS/radical polymerization	1, 2 or 3	PLA	Compression molding/Sheets	FTIR, NMR, TEM, SEM, TGA, DSC, wettability and transparency study, mechanical tests	[217]
CMF	PMMA	VTES, MPTMS	2	PHB	Molding/Films	FTIR, TGA, DSC, POM, SEM, DMA and tensile tests	[218]

Methods: TOF-SIMS: time of flight secondary ion mass spectrometry; SEC: size-exclusion chromatography; SAXS: small-angle X-ray scattering; for other abbreviations see footnote to Table 3. Abbreviations: ROP: ring-opening polymerization; ATRP: atom-transfer radical polymerization; OLLA: oligo(lactic acid); PBSA: poly[(butylene succinate)-co-adipate]; PBA: poly(butyl acrylate); PEHA: poly(ethylhexyl acrylate); PMMA: poly(methyl methacrylate); PBMA: poly (butyl methacrylate); GPTMS: 3-glycidoxypropyl trimethoxysilane.

The main feature of ROP in the grafting from the cellulose surface is the heterogeneous nature of this process. A typical procedure of such grafting involves the utilization of stannous (II) octoate as a catalyst (0.2–2.0 wt% with respect to the monomer) and performing the reaction in toluene at 80–130 °C for 18–24 h [67,199,202,205,208]. Chai et al. reported the PLLA and PDLA synthesis in toluene at 170 °C for 8 h [201]. In all cases, the success of the grafting was confirmed by FTIR and in some studies by solid-state NMR and XPS (Table 7). Measurement of the contact angles for the cellulose substrate grafted with PLLA and PCL showed the increase in this parameter after modification. The grafting of short chains of PCL and PLLA resulted in a contact angle equal to 95 and 107 °, while the long chains of the same polymers provided contact angles of 99 and 112 °, respectively [199].

Recently, Chuensangjun et al. reported a chemo-enzymatic preparation of CNC grafted with PLA [219]. The developed procedure included (1) the oxidation of CNC surface with TEMPO, (2) ROP of L-lactide catalyzed with stannous (II) octoate, and finally (3) additional esterification of the CNC surface hydroxyls with the oligomer of lactic acid obtained by lipase-catalyzed ROP. Depending on the reaction conditions (temperature and reaction time), the percentage of grafting ranged from 3.4 to 59.6%. Optimal conditions for ROP included a two-temperature protocol: 15 min at 140 °C and 8 h at 100 °C, followed by incubation of CNC-g-PLA with the enzymatically produced oligomer of lactic acid for an additional 16 h at 100 °C. Under these conditions, the percentage of grafting was maximal along with the high crystallinity of CNC-g-PLA sample (>76%).

The use of stannous (II) octoate as a catalyst does not satisfy the requirements of green chemistry and may remain in the polymer, requiring further purification steps for possible sensitive applications, such as biomedical applications. In this regard, the development of novel grafting methods that exclude the use of stannous (II) octoate is in high demand. One such method was recently proposed by Yoo et al. who used zinc acetate dihydrate to catalyze the surface-initiated polycondensation of D,L-lactic acid taken as aqueous syrup (85 wt%) containing CNC (5 wt%) at 180 °C [220]. In addition, the authors replaced some of the lactic acid with dodecanoic, palmitic, or stearic acids for common polycondensation in the presence of dibutyl tin dilaurate catalyst at 190 °C and 100 mmHg for 30 min, and then at 35 °C and 10 mmHg until the viscous solution was obtained. Using NMR spectroscopy, the degree of polymerization was calculated. The average DP for PLA grafting was 6, while for copolycondensation with palmitic acid it reached 8.

Labet et al. offered the use of citric acid as an alternative to metal catalysis in the production of cellulose grafted with PCL [212]. The reaction conditions were optimized with respect to ratio between ε-caprolactone, citric acid, and cellulose-surface hydroxyls as well as temperature and polymerization time. According to the XPS analysis, the most successful grafting occurred under the following conditions: [ε-caprolactone]:[citric acid]:[surface OH] = 660:10:1, 150 °C and 2 h. Under optimized conditions, the amount of grafted PCL corresponded to 58 wt%.

In addition to PLA and PCL, the successful grafting of PBS from CNC via polycondensation of 1,4-butanediol and succinic acid with cellulose surface hydroxyls was recently reported by Zhang et al. [214]. The reaction was carried out at 220 °C for 4 h at normal pressure in the presence of titanium butoxide (0.1 wt% of the reactants). SEC analysis of PBS dissociated from CNC allowed the determination of the PBS molecular weight. The highest determined M_n was 23,700.

Previously considered studies included the direct grafting of aliphatic polyesters from the surface hydroxyls of cellulose materials. To increase the amount of bound polyester, Peng et al. premodified cellulose with APTES and then with 3,5-diaminobenzoic acid [213]. This approach made it possible to obtain a corona with multiplied amino groups. Thereafter, ROP of D,L-lactide was carried in DMSO in the presence of stannous caprylate as a catalyst at 130 °C for 16 h. As a result, a hyperbranched corona of PLA was produced on the surface of CNC, as speculated by the authors. CNC modification with PLA was confirmed by FTIR spectroscopy. However, the authors did not provide any quantitative data to support the idea that this approach was superior in comparison to PLA grafting initiated by CNC surface hydroxyls. As an example, a change in the morphology of the CMC surface due to the grafting L-lactide acid oligomers is illustrated in Figure 12 [82].

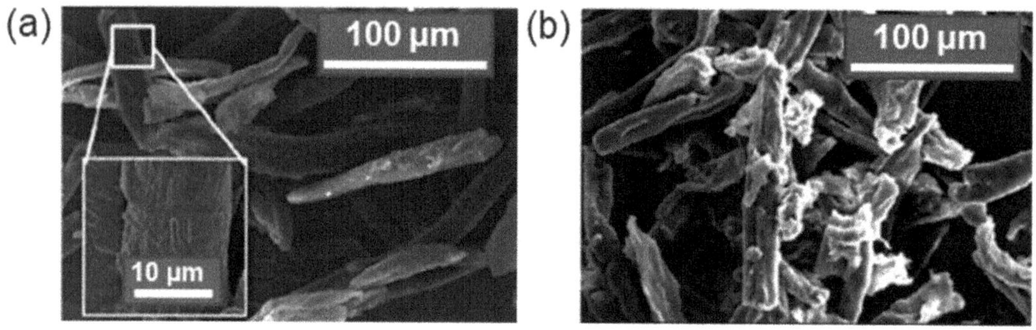

Figure 12. SEM images of CMC before (**a**) and after grafting of oligo(L-lactic acid) (**b**). Reproduced from [82] with permission of Royal Society of Chemistry.

Besides aliphatic polyesters, several research groups used radical polymerization as a tool to functionalize the cellulose surface with various poly(meth)acrylates. The realization of such an approach required the cellulose with a small molecule capable of further polymerization. Following this idea, the adsorption of monomers on the cellulose surface, its silanization with vinyl-, methacryloyl-, or glycidoxy-containing silanes, or treatment with α-bromoisobutyryl bromide were implemented. The adsorption of butyl acrylate (BA), 2-ethylhydroxy acrylate (HEA), and methylmethacrylate (MMA) was the oldest approach proposed for modifying the cellulose fibers with poly(meth)acrylates [215]. Despite the simplicity of this approach, possible polymer leakage from the fiber surface limits the application of such a technique. However, recently, a similar approach was used to modify agave cellulose fibers covalently with PMMA [221]. A decrease in crystallinity from 80% to 68% was reported after PMMA grafting with an efficiency of 55%.

Popa et al. described the use of two silane agents containing vinyl (VTES) or methacryloyl (MPTMS) moieties, for modification of CMF for further radical polymerization of MMA [218]. In both cases, the PMMA grafting was evidenced by FTIR. At the same time, no direct data on the grafting efficiency was reported. Both samples blended with PHB had very similar thermal properties but different tensile strengths. Composites of PHB with CMF grafted through silanization with MPTMS showed 30% better tensile strength, 60% better elongation at break, and 15% better Young's modulus than composites with CMF pretreated with VTES. This fact indirectly indicates a better grafting of PMMA when MPTMS is used to premodify cellulose. The results seem to be expected, since MMA, being a methacrylate type monomer, polymerizes better with methacrylate-type silane because of the similar double-bond activity in these monomers.

An original approach to grafting a methacrylate-type polymer from the CNC surface by controlled radical polymerization was proposed by Boujemaoui et al. [216]. A procedure for pretreatment of CNC with α-bromisobutyryl bromide to obtain CNC-Br was developed. The latter was used for the surface-initiated ATRP of BMA (Figure 11). The amount of PBMA grafted from CNC was found to be 4 or 28% for low (degree of polymerization 110) and high (degree of polymerization 487) molecular weights, respectively.

Another example of the controlled radical polymerization on the surface of cellulose was reported by Aubin et al. [222,223]. In this case, the functionalization of CNC with poly(N-isopropylacrylamide) (PNIPAM) and N,N′-dimethylaminoethyl methacrylate (DMAEMA) by RAFT polymerization was employed (Figure 11). Before polymerization, the CNC surface was modified with a chain-transfer agent (4-cyano-4-[(dodecylsulfanylthio carbonyl)sulfanyl]pentanoic acid) (CTA) under carbodiimide activation and in presence of DMAP as a catalyst in DMF at 50 °C for 48 h. Three different grafting densities, namely 0.006, 0.09 and 0.46 CTA/nm^2, were achieved by varying the initial CTA concentration [222]. The copolymer grafting was carried out in DMF at 70 °C for 72 h using AIBN as an initiator by varying the DMAEMA concentration from 0 to 20 mol%. According to NMR analysis, the molecular weights of the grafted copolymer were close to the theoretical ones, confirming the controlled nature of polymerization. The NIPAM conversion ranged from 80 to 96%, while the degree of polymerization varied from 330 to 1900 depending on the ratio of reactants [222].

Thus, the examples overviewed in this subsection illustrate a diversity of techniques, such as ring-opening polymerization, and free-radical and controlled radical polymerizations, which can be used to modify the surface of cellulose micro- and nanomaterials by a grafting "from" technique. Biodegradable (PLA, PCL, PBS), nondegradable hydrophilic (PHEA) and hydrophobic (PBA, PMMA, PBMA), as well as thermoresponsible (PNIPAM) polymers can be successfully grafted.

4.3.2. Grafting "to"

In contrast to the previous approach, grafting "to" is used to modify the surface of cellulose materials with presynthesized polymers. Figure 13 illustrates the reported pathways to modify cellulose materials with polymers via a grafting "to" technique.

One of the first described methods of grafting "to" is the modification of cellulose with PLA via its intermediate functionalization sequentially with phenyl isocyanate and toluene diisicyanate in a mixture of anhydrous methylene chloride and anhydrous toluene under reflux for 4 h. The resulting intermediate was reacted with CMC or bleached kraft softwood pulps during 72 h for modification [224]. Recently, a similar approach was used to modify CNC with PLLA. The reaction also involved two steps, namely activation of the polyester in DMSO at 60 °C for 3 h followed by reaction with CNC hydroxyls in DMSO at 120 °C for 12 h [207]. In both cases, the modification was evidenced by FTIR and NMR [207] or XPS [224], but neither study contains quantitative data on the effectiveness of grafting.

Besides aliphatic polyesters, poly(amino acids) can be grafted onto the cellulose surface. Recently, Averianov et al. reported on the modification of CNC with hydrophobized poly(glutamic acid) (PGlu) via two-step approach [225]. It was based on (1) the partial oxidation of cellulose vicinal diols to aldehyde groups and (2) their reaction with terminal amino groups of PGlu. The authors compared the grafting of PGlu of two molecular weights (10,400 and 2100) and found a better modification in the case of PGlu with a lower molecular weight. In this case, 90 wt% of PGlu taken for the reaction was bound to CNC.

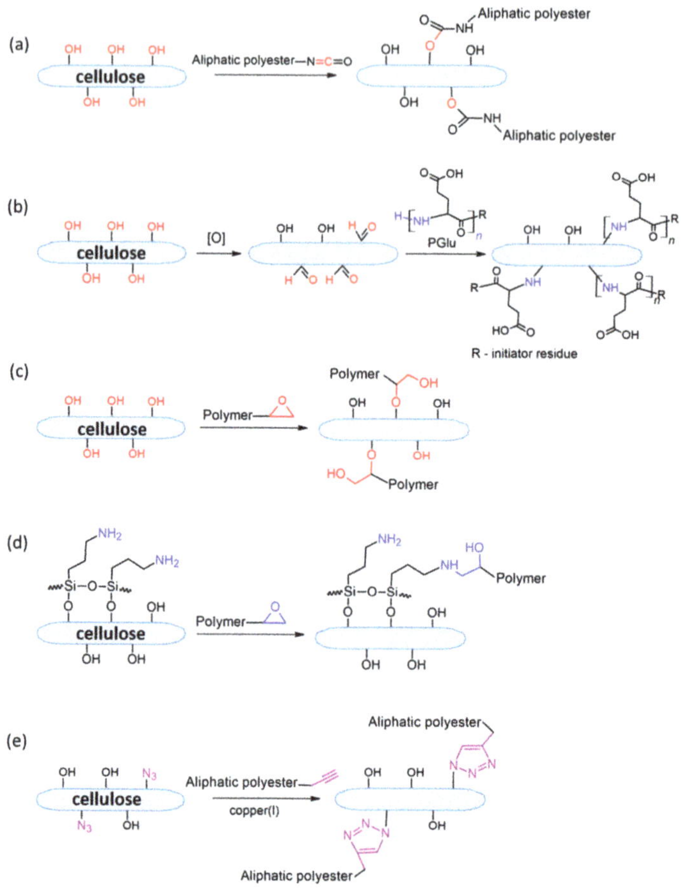

Figure 13. Possible ways of modifying the cellulose surface with macromolecular compounds by grafting "to" (Table 9): (**a**) attachment of polymer containing isocyanate group, (**b**) interaction of polymer amines with aldehyde groups of preoxidized cellulose, (**c**,**d**) modification of cellulose with epoxy-bearing polymers, (**e**) modification via "click"-reaction.

Hong et al. reported the direct modification of CMC with poly(ethylene-co-glycidyl methacrylate) (PEGMA) by the reaction of epoxy groups of the polymer with cellulose hydroxyls [226]. The reaction was carried out in xylene at 120 °C for 3 h under acid catalysis with *p*-toluenesulfonic acid. The success of the modification was proven by both XPS and FTIR spectroscopy. Premodification of CNC with APTES allowed further successful grafting of epoxy-bearing PEG. The modification of cellulose hydroxyls with epoxy group of allyl glycidyl ether, the further epoxidation of a double bond, and the attachment of PVA through its hydroxyls was reported by Virtanen et al. [190]. The reaction between CNC-APTES- and epoxy-PEG was run in water at 65 °C for 6.5 h. According to the elemental analysis (XPS), the grafting efficiency of PEG to CNC was calculated to be 43.36% [227].

In addition, click chemistry may be also an option if grafting of presynthesized polymers is of interest. Krouit at al. reported the modification of cellulose fibers with 10-undecynoic acid by esterification and used azide-PCL to perform a Cu(I)-catalyzed heterogenous click reaction in THF at room temperature for 48 h [228]. The azidation was confirmed by FTIR, XPS, and NMR spectroscopy. The yield of the azidation reaction was over 90%. Another means of click-chemistry performance was recently proposed by

Mincheva et al. [229]. In this case, the authors used azidized CNC and propyl-bearing PLLA for azide-alkyne cycloaddition in THF at 50 °C under copper(I) catalysis. The amount of PLLA grafted onto the CNC surface was calculated to be 12 wt%.

Table 9. Summary on covalent modification of cellulose micro- and nanomaterials by grafting "to" technique.

Type of Cellulose	Grafted Polymer	Cellulose Pre-modification/ Polymerization Technique	Filler Content (wt%)	Matrix Aliphatic Polyester	Processing/ Design	Characterization Methods	Refs.
CNC	PGlu	Amination/ROP	5	PLLA	Casting/Films	NMR, DLS, TGA, mechanical tests	[225]
CNC	PGlu	Amination/ROP	5, 10 or 15	PLLA, PDLLA, PCL	Casting/Films	OTM, ORM, SEM, POM, mechanical tests, MTT-test, in vivo study, histology	[230,231]
CNC	PLA	Toluene diisocyanate	0.2, 0.5 or 1	PLA	Casting/ Sheets	NMR, FTIR, GPC, TEM, SEM, DSC, TGA, rheology study	[207]
CNC	Propargyl-containing PLA/PBS	Tionyl chloride followed with sodium azide	–	–	–	SEC, MALDI, ATR-IR, XPS, NMR, TGA, SEM	[229]
Cellulose fibers	N_3-PCL	10-undecynoic acid	–	–	–	FTIR, NMR, XPS, elemental analysis	[228]
CNC	Epoxy-PEG	APTES	1–5	PLA	Hot pressing	FTIR, TEM, XPS, XRD, SEM, POM, TGA, DSC, wettability and mechanical tests	[227]

Methods: OTM: optical transmitted microscopy; ORM: optical reflected microscopy; MALDI: matrix-assisted laser desorption/ionization mass spectrometry; for other abbreviations see footnote to Tables 3, 4 and 8. Abbreviations: PGlu: poly(glutamic acid).

Compared to the grafting "from", the positive side of the grafting "to" method is the use of presynthesized polymers, which can be synthesized by the methods of controlled polymerization with a narrow molecular weight distribution. In this case, immobilization of narrowly distributed polymers ensures uniform attachment, which ultimately provides a homogeneously modified cellulose filler. In turn, grafting "from" in some cases does not allow for controlling the polymer molecular weight and dispersity. At the same time, it represents a simpler in situ approach that requires less time and lower amounts of reagents.

In general, the functionalization of the cellulose surface with polymers provides a more considerable effect in terms of its hydrophobization and demonstrates more detectable improvements in the properties of composites (see Section 5). Moreover, both the properties of the filler and the composite material can be adjusted by varying the properties of the polymers used for modification.

4.4. Modification with Particles

The modification of cellulose micro- and nanomaterials is a relatively novel approach for influencing cellulose properties. In the current literature, one can find publications on the modification of the cellulose surface by inorganic nanoparticles (e.g., hydroxyapatite [232,233], silver [234,235], zinc-oxide nanoparticles [236]), and organic particles (e.g., latex) [237,238], as well as hybrid ones (e.g., organo-montmorillonite [239] and polydopamine-hydroxyapatite [240]).

In the last decade, interest in the development of biocomposites based on aliphatic polyesters containing cellulose micro- and nanomaterials modified with hydroxyapatite as a filler has attracted much attention. Such biocomposites are considered as scaffolds for bone regeneration. Aliphatic polyesters are hydrophobic and provide low cell adhesion and proliferation on their surface. In turn, introduction of cellulose can provide surface hydrophilicity and improve cell attractiveness. Hydroxyapatite (HA) is responsible for further scaffold biomineralization. Both cellulose particles/whiskers/fibers and hydroxyapatite can improve the mechanical properties of the scaffold. There are several ways to modify cellulose with hydroxyapatite. For example, Lu et al. reported the formation of hydroxyapatite on the surface of CNC in water from calcium-nitrate tetrahydrate and diammonium hydrogen phosphate at 70 ° C for 2 h and then additionally for 48 h at room temperature [232]. TEM analysis the resulting dispersion revealed an increase in the diameter of the cellulose nanocrystals due to adsorption of hydroxyapatite particles. Sridevi et al. separately prepared the yttrium substitute nano-hydroxyapatite and then used it to modify cellulose from rice husk, which was prefunctionalized with citric acid. The mixture of components was stirred for 24 h, ultrasonicated for 1 h, and dried at 80 °C to obtain dry composite powder [233]. The modification was evidenced by FTIR spectroscopy, TEM, and EDX analysis. The modification had almost no effect on the degree of crystallinity of the modified particles.

Li et al. used phosphorylated cellulose nanofibers to induce the growth of the hydroxyapatite particles on the CNF surface [241]. For this, phosphorylated CNF was incubated in the solution containing calcium chloride, sodium chloride, potassium chloride, magnesium chloride, and disodium hydrogen phosphate at 37 °C over 28 days. SEM analysis of treated CNF allowed the detection of only few spherical particles on the CNF surface after 7 days of incubation. More and bigger particles were found after 14 days, while after 21 and 28 days the particle sizes had increased considerably. The size and morphology of the HA particles varied greatly: spherical and cylindrical HA particles were detected on the CNF surface.

Silver nanoparticles are another of the most popular inorganic systems considered for the modification of cellulose materials. Silver nanoparticles are known to have antimicrobial properties, and their introduction into biomedical materials and food packaging is widely studied [242,243]. In the case of cellulose modification with the silver nanoparticles, the latter are obtained by the reduction of silver from silver nitrate. For instance, to modify CNW with Ag nanoparticles, Hasan et al. initially oxidized cellulose hydroxyls to carboxyl groups by TEMPO [244]. The formed carboxyl groups captured Ag ions from solution and were then reduced with sodium tetraborate during 1 h at 25 °C in the presence of 1wt% of CNC. The yield of silver in the product was 1.77 wt%. Different contrasts of CNW and Ag nanoparticles in TEM allowed their noncomplicated visualization and evaluation of the size of individual components. It was found that the CNWs with a length of about 200 nm and a diameter of 20–30 nm were covered by a large number of silver nanoparticles with the average size of about 5 nm. The analogous modification was carried by Lertprapaporn et al., who modified CMC with Ag nanoparticles but without the preoxidation of cellulose by TEMPO [245]. In this case, the yield of silver was reported to be 1.28%. The change in CMC morphology as a result of the modification with Ag nanoparticles is shown in Figure 14.

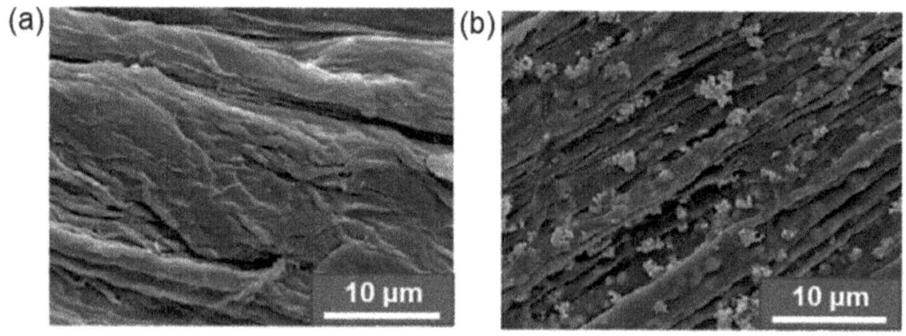

Figure 14. SEM images of neat CMC (**a**) and CMC modified with Ag nanoparticles (**b**). Reproduced from [245] with permission of Elsevier.

Recently, the electrostatic modification of CNF with various positively charged latex nanoparticles formed from amphiphilic copolymers has been reported [237,238]. Such an approach is also one of the possible ways to change the hydrophobicity of the cellulose surface in order to improve the interfacial adhesion to hydrophobic matrices. For this, poly(dimethylaminoethyl methcrylate)-based latexes with diameters from 79 to 146 nm were used to modify negatively charged cellulose surfaces prepared by CNF oxidation be TEMPO. The adsorption of the latex nanoparticles onto the surface of the CNF was confirmed by quartz-crystal microbalance with dissipation and atomic force microscopy. It was established that the total adsorbed mass increased from 10.5 to 44.4 mg/m^2 with increased particle size from 42 to 96 nm. In addition, the increase in contact angle (up to 94°) and surface roughness (AFM) also testified the latex adsorption [238].

5. Aliphatic Polyesters/Cellulose Composites

5.1. Preparation of Aliphatic Polyesters/Cellulose Composites

The effective dispersion of cellulose in the polymer matrix is a key task for producing the homogeneous biocomposites based on hydrophobic aliphatic polyesters and cellulose micro- or nanomaterials [39,45]. Hot blending (direct and continuous melt mixing) or mixing an aliphatic polyester solution with dry/suspended cellulose is generally used to obtain composite blends. In terms of producing aliphatic polyester/cellulose composites on an industrial scale, hot blending is of most interest [39]. Various kinds of equipment, such as an internal mixer or a single/two-screw extruder, were used to optimize this approach with respect to composite components [2,39]. In spite of the applied shear/elongational forces during this type of mixing, the hydrophilicity and strong interchain interactions of cellulose, as well as high temperatures of processing, can lead to heterogeneous material as well as thermal degradation of cellulose or aliphatic polyester [39,79,90,92]. In this regard, various methods have been proposed to improve the distribution of cellulose in the polymer matrix: the main ones are the modification of cellulose, polyesters, and the use of additives (surfactants, compatibilizers) [2,39]. An alternative method of obtaining a blend that avoids thermal degradation of the components is the approach of dispersing cellulose in an aliphatic polyester solution, followed by evaporation of the solvent to obtain the material. The disadvantages of this method include a limited set of suitable solvents due to the possibility of removing them without the use of high temperatures, the solubility of aliphatic polyesters, and increased aggregation of cellulose chains/particles [39]. Subsequent application of such techniques as melt compounding (extrusion, melt spinning, compression molding, injection molding, 3D printing), solution casting, electrospinning, etc. to the resulting blends allows the fabrication of aliphatic polyester/cellulose composites of various shapes [2,39,65]. Types of techniques for the fabrication of composite materials based on cellulose and aliphatic polyesters are summarized in Figure 15.

Extrusion can be implemented both in the laboratory and on an industrial scale, and allows for the pushing of the melt through the extruder die to obtain composite material of different geometries, often in the form of pellets and filament [2,65]. The melt-spinning technique is based on the formation of composite fibers using the melt-filament technique from melt-extruded pellets. The composite sheets can be produced by the compression-molding technique, which is based on the pressing of blends or layered components under certain temperatures and pressures. It allows for the introduction of a large quantity of cellulose materials with high efficiency (up to 70%) [246]. Injection molding is the injection of pressurized samples into a special preheated mold followed by curing. The temperature and holding time can be varied and require optimization.

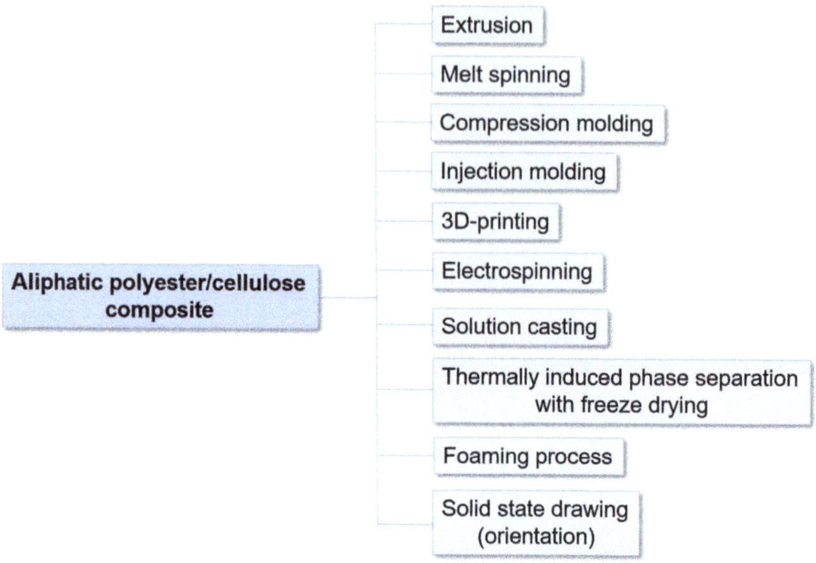

Figure 15. Scheme for different techniques of composite preparation.

3D-printing techniques make it possible to produce products of different—including complex—shapes with high accuracy and reproducibility. The greatest potential for obtaining polymer cellulose-based composites from the melt is the fused-deposition-modeling (FDM) type of 3D printing, which consists in nozzle-deposition-based extrusion. This method allows the production of three-dimensional biodegradable materials for a wide variety of fields (food packaging, construction and automobile industry, and biomedicine) [2,4,65].

Using the solution-casting method, aliphatic polyester/cellulose composites can be formed as films. This method involves casting a suspension of cellulose in an aliphatic polyester solution onto a flat substrate, followed by evaporation of the solvent in the first step in air at room temperature, and then sometimes using vacuum drying and/or slight heating (up to 60 °C) [39,173,231]. The main obstacle of this method is the impossibility of producing composites on an industrial scale. Furthermore, composites from a filler suspension in polymer solution can be fabricated as supermacroporous products using the method of thermally induced phase separation followed by freeze drying [2,4].

Electrospinning (also referred to as Electrostatic fiber spinning [51]) is applicable to both polymer melts and solutions and allows the formation of composite polyester/cellulose fibers from 100 nm to several micrometers thick under electrostatic forces [39,247]. The fiber mats obtained with this technique are characterized by a large surface area and are of great interest for tissue engineering, drug delivery, cosmetics, sensors, conducting nanofibers, etc. [247]. The electrospinning method is economical and performed under ambient conditions. The application of solid-state drawing against aliphatic polyester/cellulose-based composite films using a tensile tester with the ability to heat and thermostat the sample has also been reported. This approach led to an efficient orientation of the filler and polymer, and as a consequence, to the formation of an organized molecular structure with improved mechanical and thermal properties by increasing the crystallinity of the material [2].

In addition, foam composites based on aliphatic esters and cellulose micro-/nanomaterials can be produced, but the implementation of the foaming process is difficult at low melt strength [39]. Recently, hybrid approaches to obtain composites based on aliphatic polyesters and cellulose by combining different types of mixing ("wet" and "dry") and molding have been increasingly utilized [2,39]. This is primarily due to the tendency of cellulose to aggregate, which makes it difficult to disperse homogenously in the polymer matrix. The use of predried cellulose, which is prone to irreversible aggregation, makes it especially difficult to obtain a homogeneous composite mixture. Thus, attempts are made to avoid the pulp-drying stage [51]. Typically, cellulose and polymer-based masterbatches are resorted to (casting from solution, in situ polymerization, centrifugation), followed by dilution with aliphatic polyester by melt blending. There have also been reports on the introduction of nanocellulose suspension into the extruder (liquid-assisted feeding) [39]. A combination of solvent casting and extrusion has been reported to prevent thermal degradation and achieve good dispersibility of the cellulose material [2].

Processing type has a great influence on the properties of the obtained composites. When choosing one or the other approach to obtain aliphatic polyester/cellulose composite, the characteristics and properties (thermal resistance, dispersibility, solubility) of both cellulose and polymer matrix used, as well as the desired final design of the material, should be considered [39,45,51].

5.2. *Effect of Cellulose Modification on Mechanical Properties of Composites*

The production of biocomposites from aliphatic polyesters and cellulose micro- and nanomaterials has attracted much attention due to the prospects of using such materials for biomedicine, food packaging, and other technical biodegradable materials [105,248–251]. However, the high hydrophilicity of cellulose, and in turn, the high hydrophobicity of aliphatic polyesters makes it difficult to obtain homogeneous composites with good mechanical characteristics. A decrease in the tensile properties of composites using neat cellulose has been observed by several research groups [231,252–254]. For instance, Liu et al. reported a decrease in tensile strength of PLA (60 MPa) after its filling with neat cellulose fibers [252]. The decrease from 52 to 42 MPa was observed when the filler content was increased from 3 to 15 wt%. Moreover, the tensile strength depended on the 3D-printing method. The strongest composite was obtained when parallel printing was performed.

Table 10 illustrates a summary of the tensile properties for aliphatic ester/cellulose composites for selected papers. In the case of adsorption and covalent modification of cellulose materials with small molecules, the mechanical properties differ markedly in different studies. Only a slight improvement was observed for composites obtained with cellulose micro- and nanomaterials modified by adsorption [114,120,123,125]. A decrease in tensile modulus, tensile strength, and yield stress was observed when comparing PLA-based composites prepared by electrospinning and filled with neat CNC and CNC modified by adsorption with ethoxylated nonylphenol phosphate ester [122]. In this case, the low effect of modification on the mechanical properties of composites can be partially explained by the possible desorption of small molecules from the surface of cellulose during the production of the composites.

Table 10. Selected studies on the tensile properties of aliphatic polyester/cellulose composites.

Aliphatic Polyester	Filler	Filler Content (wt%)	Processing	Mean Tensile Modulus (GPa)		Mean Tensile Strength (MPa)		Mean Elongation at Break (%)		Refs.
				Polyester	Composite	Polyester	Composite	Polyester	Composite	
PLA	Acetylated CNF	10	Solution casting	1.08	2.37	28.3	44.1	29.9	30.1	[151]
PLA	Acetylated CNC	3	Solution casting	1.8	1.8	57	52	3.3	4.2	[140]
PLA/PBS (70/30)	CNF-fatty acids	5	Moulding	1.5	2.0	34	47	N/A	N/A	[42]
PLA	CNF-oleate	8	Solution casting	0.58	1.0	10	18.5	7	7.5	[136]
PLA	CNF-propionate	4	Solution casting	1.24	1.74	46.1	53	1.54	1.31	[152]
PLA	Silanized CNF	2	Solution casting	1.78	1.82	52.5	54.7	7.3	5.3	[177]
PLA	CNC with adsorbed PVA	1	Solution casting	1.61	1.82	47.9	45.3	3.4	12.3	[126]
PCL	CNC with adsorbed PVP	5	Solution casting	0.18	0.29	20.6	10.4	903	15	[127,131]
PLA	CNC-g-PDLA	2	Solution casting	2.50	3.25	60	80	7.5	4.5	[201]
PCL	CNC-g-PLLA	5	Solution casting	0.32	0.51	25	13	830	25	[208]
PLGA	CNC-g-PLLA	5	Solution casting	1.4	1.2	40	29	4	5	[203]
PCL	CNC-g-PCL	20	Solution casting	0.23	0.48	21	18	640	30	[67]
PLA	CNC-g-PLA	5	Electro-spinning	N/A	N/A	4.7	13	14.5	32.5	[213]
PLA	CNC-g-APTES-PEG	2	Hot pressing	N/A	N/A	25	56	1.9	3.9	[227]

Xu et al. revealed a stronger interfacial adhesion between PLA and CNC covalently modified by acetic acid [140]. As a result, the composite films prepared by solution casting showed higher strength and Young's modulus than ones with neat CNC by about 20%. The improvement in tensile strength and Young's modulus by 38 and 71%, respectively, for PLA filled with hydrophobized MFC due to its transesterification with vinyl-laurate groups (compression-molded composites) was also observed by Li et al. [148]. At the same time, modification of CNF with fatty acids by transesterification and production of composites with PLA-PBS by combination of extrusion and molding showed no significant effect on the tensile strength and Young's modulus when 1 and 3 wt% filler were used [42]. Moreover, the parameters obtained were very close to those obtained for unmodified cellulose. Only the composite containing a 5 wt% filler showed an 25% increase in tensile strength. An improvement in tensile properties of not more than 13% for composites prepared by solution casting and based on PLA filled with bacterial cellulose fibers modified with citric acid was found by Ramirez et al. [132]. Practically no effect on the mechanical properties of PLA-based composites (also prepared by solution casting) was observed when CMC was modified with palmitic-acid residues from olive oil [146]. The modification of cellulose fibers with formic acid and variation of the content of formyl groups from 1.7 to 15.8% provided an increase in tensile strength of the PLA-based composites produced by solution casting [137]. Even at the lowest formyl group content on the cellulose surface, the tensile strength and Young's modulus increased by more than 200%. At the same time, these parameters for a PLA composite filled with 1 wt% cellulose fiber containing 15.8% formyl groups increased by more than 300%. Such different results can be related to the properties of both the matrix polymer and the filler. For the latter, the source of the cellulose, the effectiveness of its modification, and the content in the matrix can lead to differences in composite properties. As illustration, the morphology of PLA composites filled with neat CMC and CMC modified with toluene-2,4-diisocyanate is presented in Figure 16.

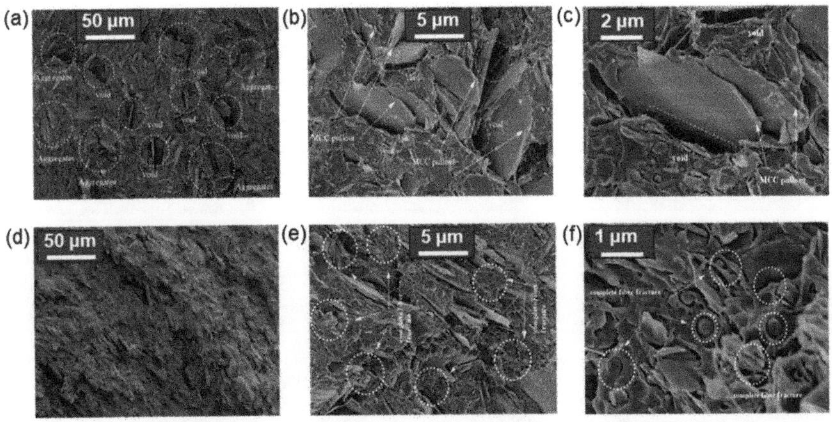

Figure 16. Distribution of neat and modified CMC (3 wt%) in the PLA matrix (SEM): (**a**) surface of PLA/CMC; (**b**,**c**) fractured surface of PLA/CMC at different magnifications; (**d**) surface of PLA/modified CMC; (**e**,**f**) fractured surface of PLA/modified CMC at different magnifications. The CMC modification with toluene-2,4-diisocyanate was carried out for 24 h to reach the maximum functionalization. Reproduced from [32] with permission of American Chemical Society.

Gwon et al. compared three different commercial PLAs (4032D, 3001D, and 2003D, Ingeo, Natureworks LLC, Blair, NE, USA) for preparing composites with CNC modified with toluene-2,4-diisocyanate [187]. In all cases, the use of modified CNC showed higher tensile-strength values than the use of the unmodified filler. The best mechanical properties were observed for composites prepared by solution casting and based on PLA with higher molecular weight and crystallinity (4032D), while composites based on PLA with low molecular weight and high crystallinity (3001D) or low crystallinity and high molecular weight (2003D) showed reduced tensile properties.

Recently, Voronova et al. investigated the effect of the content of CNC with adsorbed PVP in the PCL matrix on the tensile properties of composite films prepared by solution casting [127]. Increasing the filler content from 5 to 15 wt% resulted in a twofold increase in Young's modulus, but at the same time a threefold decrease in tensile strength. A similar tendency was also observed for the composite films fabricated by solution casting when CNC was grafted with poly(glutamic acid) and used as a filler to PCL and PLA [230,231]. The effect on changes in composite morphology and material homogeneity depending on the used filler is demonstrated in Figure 17.

The effect of cellulose-surface grafting with aliphatic polyesters for enhancing the filler compatibility with PLA, PCL, or PLGA matrices have been studied in several papers [67, 201,208,213]. Two main trends can be found for composite films prepared by solution casting: an increase in tensile modulus with a simultaneous decrease in elongation at break and vice versa. For example, Chai et al. varied the content of CNC grafted with PDLA in the PLA matrix from 1 to 10 wt % and found that the best mechanical properties were observed with a filler content of 2 wt% [201]. In this case, an increase in Young's modulus and tensile strength of about 30% was established, while elongation at break was reduced by 34%. A similar trend, but with a sharper decrease in elongation at break, was also observed by Averianov et al. for PCL-based composites filled with CNC-g-PLLA [208] and by Habibi et al. for PCL-based composites filled with CNC-g-PCL [67]. In turn, in the case of PLGA composites filled with CNC-g-PLLA, there was a 17–50% decrease in Young's modulus and a 20–50% increase in elongation at break, depending on the filler content [203]. In the case of using CNC grafted with PLLA as a filler to PHB, a significant reduction in the brittleness of the matrix polymer was observed (Figure 18).

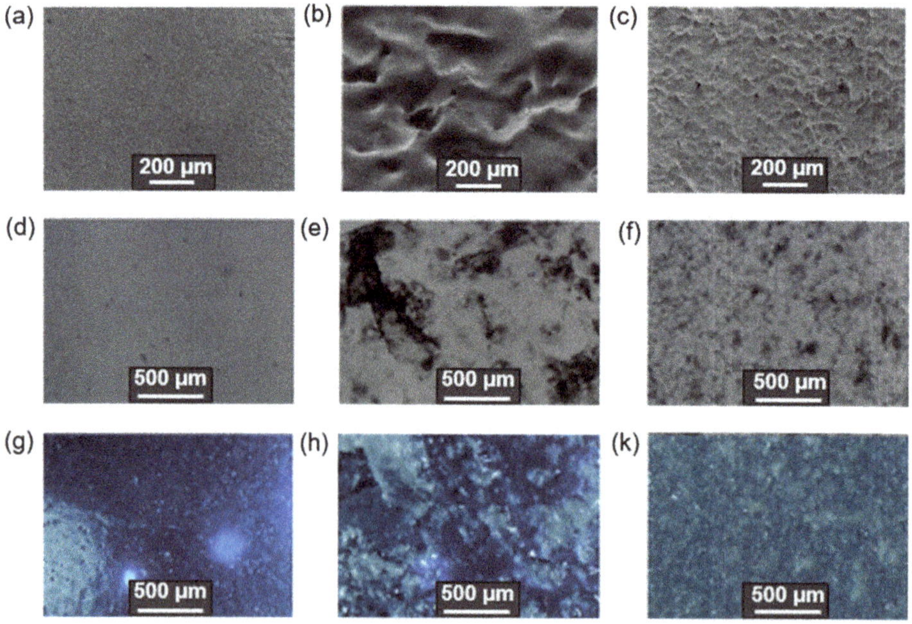

Figure 17. Images obtained by SEM (**a–c**), optical (gray) (**d–f**) and reflected (blue) (**g–k**) microscopy for pure PLLA (**a**,**d**,**g**), and its composites with neat CNC (15 wt%) (**b**,**e**,**h**), and CNC modified with poly(glutamic acid) (15 wt%) (**c**,**f**,**k**) (reproduced from [230] under the terms of the Creative Commons CC BY license).

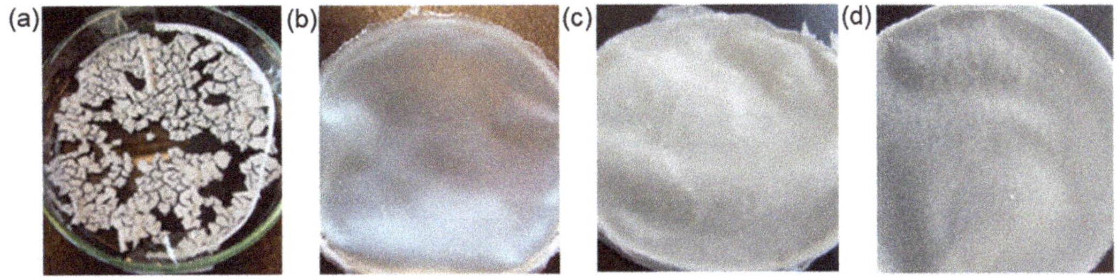

Figure 18. Images of PHB composite films: (**a**) filled with 5 wt% of neat CNC, (**b**) filled with 5 wt% of CNC with adsorbed surfactant, (**c**) filled with 20 wt% of CNC with adsorbed surfactant, and (**d**) filled with 20 wt% of CNC grafted with poly(lactic acid). Reproduced from [203] with permission of Elsevier.

Thus, in general, hydrophobization of the cellulose surface improves the compatibility of modified cellulose with aliphatic polyesters, which contributes to enhanced mechanical properties. However, the approach to cellulose modification, its efficiency, filler content, composite fabrication technique, as well as aliphatic polyester characteristics affect the properties of the resulting composite material.

5.3. Effect of Cellulose Modification on Thermal Properties and Crystallization

The thermal stability of composites is one of the key properties when considering the thermoplastic-processing techniques. In general, the aliphatic polyesters filled with unmodified cellulose micro- and nanomaterials or modified with small molecules exhibit reduced thermal stability [142,171,255]. For example, Hong et al. observed the decrease in thermal stability (TGA) of composite fabricated by injection molding in comparison with pure PLA [255]. In particular, a decrease in onset temperature from 345.2 °C for PLA to 306.8 °C for the PLA filled with the 30 wt% of silanized bagasse fibers (40–50%cellulose, 25–35% hemicellulose, and 15–35% lignin) was detected. In turn, the temperature of a maximum mass loss for pure PLA and its composite with silanized fibers was reduced from 363.7 to 331.4 °C. A similar trend was established by Li et al. who detected a reduction in the onset temperature from 344.2 °C for pure PLA to 246.4 °C for its composite prepared by solution casting with the use of silanized CMC as a filler and acetyl tributyl citrate as a plasticizer [171]. By preparing composites based on PHB with acetylated CMC, Ribero et al. demonstrated that a slight improvement in thermal stability can be observed if the filler content does not exceed 0.5 wt%. In turn, the onset and endset temperatures decreased at 0.75 wt% of the filler [142]. Kasa et al. varied the content of the neat and acetylated CNC from 1 to 7 wt% in the PLA matrix [31]. The highest degradation temperature (325 °C) was detected for composite obtained by the casting of PLA solution containing 1 wt% of acetylated CNC. The degradation temperatures for PLA/neat CNC composite (1 wt%) and PLA were lower by about 20 and 35 °C, respectively. Increasing the filler content to 3 wt% contributed to a temperature decrease of 5 °C and then remained unchanged when the filler content was increased to 7 wt%.

In contrast to the modification with small molecules, modification of cellulose by grafting with aliphatic polyester can improve its thermostability. Recently, Simao et al. compared the thermostability of unmodified cellulose nanowhiskers with these grafted with PCL and found that the maximum mass loss was observed at about 300 °C for the neat CNW and 370 °C for the CNW-g-PCL [211]. As a result, the onset temperature of degradation of the PCL/PBSA (30/70) composites produced by electrospinning and containing 1 and 5 wt% of CNW-g-PCL remained at the same level as for pure aliphatic polyesters (295 and 294 °C, respectively). The maximum degradation temperature was reduced after filling PCL/PBSA with CNW-g-PCL, but not as dramatically as for modification with small molecules. These temperatures were 441, 435, and 438 for nonfilled polyester blend and filled with 1 and 5 wt% CNW-g-PCL, respectively. The same tendency was observed for PLA filled with CNC-g-PLA (composites fabricated by electrospinning). A neat PLA started to degrade at 285 °C and degraded until 390 °C [213]. With the addition of CNC-g-PLA (3–7 wt%), the initial degradation temperature was elevated to 300 °C and the final degradation temperature reached 400 °C.

DSC analysis of pure aliphatic polyesters and their composites with cellulose micro- and nanomaterials allowed the conclusion that the glass-transition temperature (T_g) and melting temperature are reduced for composites with neat cellulose or modified with small molecules. For example, Way et al. observed a decrease in glass-transition temperature from 60.4 °C for pure PLA to 56.5, 56.6, and 54.8 °C for its composite with neat, silanized, and acetylated lignincellulose fibers (30 wt%), respectively [184]. At the same time, the melting temperatures were varied within 1 °C.

Simao et al. studied the thermal properties of PCL/PBSA mixture and its composite with CNW-g-PCL produced by electrospinning [211]. They found that the melting temperature for the 50/50 blend of two aliphatic polyesters was 64 °C. The introduction of 1 and 5 wt% of cellulose nanowhiskers grafted with PCL led to a decrease in melting temperature to 63 and 60 °C. At the same, an increase in crystallization temperature from 24 °C for polymer mixture to 29 and 35 °C for the 1 and 5 wt% composites was detected. In turn, a different trend in the glass-transition temperature was observed for PLA composites with CNC-g-PLA. In this, case T_g increased with increasing filler content from 54.9 for pure PLA to 57.1 °C for composite containing 7 wt% of the filler. Similar to the grafting of cellulose with polymers, the modification with fatty acids also favored the improvement of properties for PLA-based composites fabricated by solution casting [136]. For example, modifying CMF with oleic acid and using it as a filler (12 wt%) in PLA provided an increase in the melting temperature from 125.5 to 159.2 °C.

Fang et al. found that during the melt-crystallization process, CNC-g-PLLA provided better nucleation and less restriction to chain mobility than CNC-g-PDLA at undercooled conditions [205]. In particular, no melt crystallization was detected in neat PLLA during the cooling process, while the melting-crystallization temperatures were 92.3 and 97.8 °C for 5 and 15 wt% composites produced by injection molding, respectively. In turn, no melt crystallization was observed when PLLA was filled with 5 wt% of CNC-g-PDLA, whereas increasing the filler content to 15 wt% revealed a melt-crystallization peak at 92.5 °C. Almasi et al. compared the PLA crystallization capability with its composites with modified CMC prepared by solution casting. A broad crystallization peak was observed by DSC at 57.5 °C [136]. In turn, the composites containing 4 and 8 wt% of CMC modified with oleic acid demonstrated the enhanced crystallizability (about 75 °C). At the same time, Chai et al. revealed that CNC-g-PDLA increased the crystallization ability of the PLLA matrix more than CNC-g-PLLA (Figure 19) [201].

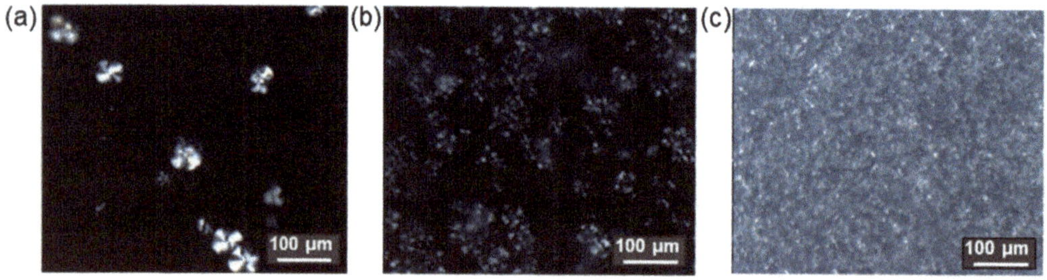

Figure 19. Polarized optical microscopy of PLLA crystallization at 120 °C after heating to 185 °C and hold for 4 min: PLLA (**a**), PLLA with 10 wt% CNC-g-PLLA (**b**) and PLLA with 10 wt% CNC-g-PDLA (**c**) (reproduced from [201] with permission of Elsevier).

5.4. Effect of Cellulose Modification on Composite Degradation

Degradation of materials fabricated from aliphatic polyesters occurs heterogeneously as a result of hydrolysis of ester bonds under the action of water and can occur both in presence or absence of enzymes [80,84,86,98]. Degradation of all aliphatic polyesters follows the same principle. Usually, the degradation starts from a nonenzymatic decrease in molecular weight, and then the enzymatically catalyzed hydrolysis in the case of degradation in the body is joined. The biodegradation of aliphatic polyesters is catalyzed by enzymes with esterase activity [86,256,257]. Some enzymes, such as proteinase K, for example, can cleave the core molecule itself, not just its oligomeric products preproduced during hydrolysis [8]. In the environment, the biodegradation is provided by the metabolism by certain microorganisms to CO_2 and H_2O [85]. In all cases, initially the hydrolytic cleavage of ester bonds occurs in the amorphous sites and then in the crystalline zones [8,85]. Thus, the amorphous PLA is degraded much faster than other semicrystalline aliphatic polyesters.

The rate of degradation is determined by molecular weight, crystallinity, and degradation conditions [89,98]. Polymers with higher molecular weight take a longer time to biodegrade because more ester bonds need to be cleaved to form water-soluble fragments [98]. The degradation rate of aliphatic polyesters is affected by such factors as temperature, the ability of water to penetrate the polymer matrix, material thickness and morphology, hydrolysis-product removal, pH, presence of catalysts, UV, moisture, etc. [5,8,84–86,93]. Aliphatic polyesters have been shown to dissociate very rapidly in strongly acidic and strongly basic environments. Under composting conditions, when the medium temperature can reach 70 °C, the degradation rate can increase significantly [8]. UV exposure also accelerates polymer degradation [85].

As a main product formed during aliphatic-ester degradation, a corresponding carboxylic acid is released, namely lactic acid, glycolic acid, caproic acid, 3-hydroxybutyric acid, succinic acid, etc. In turn, the release of carboxylic acid(s) affects the pH of the surrounding area towards acidification. In vivo, such local acidification favors the appearance of inflammatory reactions in tissues. Therefore, the faster the material destruction rate and the worse the tissue can neutralize this effect, the more inflammation can develop in the implant area [98].

The degradation rate for aliphatic polyesters of similar molecular weight and under similar conditions can be described by the following range: PGA > PLGA $\geq$ PDLLA > PLLA > PHB > PCL > PBS (Table 2). PGA is characterized by a very rapid biodegradation. For instance, in vitro degradation of commercial PGA-based sutures (Dexon®) approved by the FDA in 1969 resulted in a 42% decrease in polymer weight and loss of mechanical properties in 28 days [86]. PLGA and PLA are considered as polymers with moderate degradation rates while PCL and PBS are slowly degradable aliphatic polyesters [80,90,258]. In vitro experiments in various media (composting, burial in soil, seawater, presence of lipase enzyme or activated sludge) demonstrated that PBS biodegradation was significantly lower than for PCL and PHBV but higher than for petrochemical plastics [91]. It is known that microorganisms such as *Fusarium solani* can contribute to the degradation of PBS.

The effect of the cellulose as filler to aliphatic polyesters on the degradation rates of biocomposites have been studied in several papers [171,234,237,251,259,260]. An investigation of enzymatically catalyzed (proteinase K) degradation of PLA composites with neat CNC and CNC modified by adsorption with acid phosphate ester of ethoxylated nonylphenol (surfactant Beycostat A B09) showed a lower weight loss for composites prepared with modified CNC via solution-casting technique [259]. After three days of incubation of the films, the weight loss for pure PLA and PLA filled with neat CNC approached 100%. At the same time, the PLA composites with modified CNC under the same conditions lost only about 32% of their weight within a week. Furthermore, increasing the filler content in PLA from 1 to 3 wt% significantly slowed down the weight loss of the material. While the material containing 1 wt% filler completely degraded within 9 days, the composite filled with 3 wt% modified CNC demonstrated an 85% weight loss within 21 days. Examination of these composites by SEM revealed a change in the morphologies of PLA and PLA/CNC composites after just 2 h of incubation with enzyme. Surface erosion with holes and channels was observed. In turn, PLAs filled with modified CNC retained their topography even after 24 h of incubation in the medium containing proteinase K. Analysis of the polymer crystallinity supported the tendency known for the degradation of pure aliphatic polyesters: in the presence of the enzyme, amorphous regions were destroyed first in respect to crystalline ones. The higher values of crystallinity degrees were detected during the different degradation times.

Vilela et al. investigated the enzymatic degradation of composites produced by injection molding of PCL with neat CNF and CNF electrostatically covered with polymethcarylate-based latexes in presence of lipase over 2.5 months [237]. The highest weight loss was detected for the PCL/CNF composite, while the lowest one was found for pure PCL. The PCL composite containing CNF modified with latexes showed an intermediate weight loss.

A study of the degradation of PLA films and their composites fabricated by solution casting with the use of unmodified CNW and CNW modified by adsorption of PEG monooleate, PEG-300 or PEG-1000 in garden soil confirmed the effect of microorganisms on the substrates [260]. The authors found that CNW-surface modification with PEGs accelerated film disintegration after incubation for 90, 120, and 150 days in soil at 29 °C and 28% soil moisture. The highest disintegration rate was detected when PEG-1000 was used as a modifier. A similar soil-burial experiment was performed by Li et al. for PLA composite films filled with neat CMC and CMC silanized by APTES and prepared by solution casting [171]. It was found that adding CMC contributed to the higher degradation rate of the films. During the degradation process, the films partly lost their transparency, and the color became yellow and the surface wrinkled. After 60 days of incubation, 98%, 96%, and 88% of the original mass were maintained by PLA, PLA/CMC-APTES, and PLA/CMC, respectively. After 90 days, the most pronounced degradation was found for the PLA/CMC composite (~26% weight loss), while the least degradation was found for the pure PLA material (around 7% weight loss). The covalent modification of CMC with APTES favored a slower degradation rate (20% weight loss) compared to the unmodified material, but it was still evident compared to pure PLA.

Thus, the introduction of the unmodified natural fibers into the aliphatic polyesters accelerated their degradation in vitro [80,91,237,251,259]. At the same time, in the case of cellulose modification, different trends were observed that may be connected with the nature of the modifier and the method of modification (adsorption or covalent binding).

5.5. Effect of Cellulose Modification on Biological Properties

Being biocompatible and hydrophilic, cellulose is widely considered not only for production of degradable packaging [261] and technical materials [158,245,262], but also for the development of biomedical materials [263]. The biocompatibility of cellulose materials in vivo and in vitro is supported by several studies [230,231,249]. For example, Codreanu et al. evaluated the viability and cell-proliferation potential for the composites prepared by melt mixing of PHB with bacterial cellulose using the mouse preadipocyte (3T3-L1) cell line [249]. Pure PHB was used as a control material. The MTT assay performed after 24 h and 5 days of incubation revealed no difference between composites containing 1 and 2 wt% of cellulose filler and pure PHB. The macroporous scaffolds produced from pure PHB and its composites with bacterial cellulose were tested in vivo for bone regeneration over 4 and 20 weeks. The enhanced osteogenic differentiation was estimated for the composite materials compared to pure PHB. The best regeneration potential was detected for PHB composite containing 2 wt% of bacterial cellulose.

Taking into account that modification of cellulose micro- and nanomaterials can change surface properties, biological testing of modified cellulose should be performed to confirm the biocompatibility of a potential biomaterial. The proliferation of osteosarcoma cell lines (MG-63) during 48 h at the surface of PLA electrospun-fiber composites containing cellulose nanofibers modified with hydroxyapatite (HAP) particles was recently reported [264]. The highest cell growth was observed when using the PLA/CNF-HAP composite with a 70/30 component ratio (Figure 20).

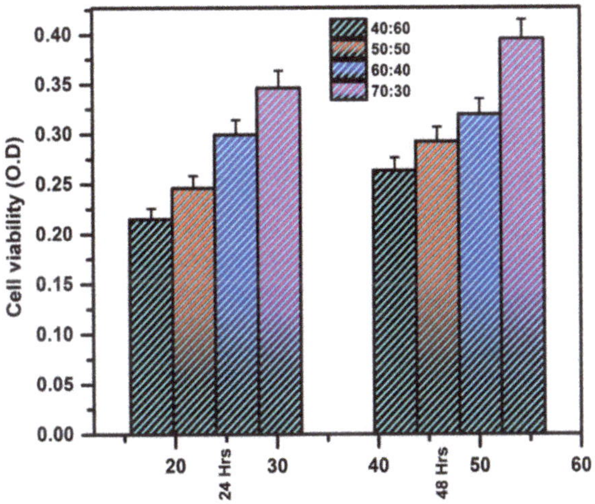

Figure 20. Cell-proliferation study (MG-63 cell line) for the PLA/CNF-HAP electrospun composites of different compositions reproduced from [264] with permission of John Wiley & Sons).

The evaluation of cell viability on the surface of PLLA, PDLLA, or PCL composite films produced by solution casting and containing 5, 10, or 15 wt% of neat CNC or CNC modified with poly(glutamic acid) revealed the similar viability of rabbit mesenchymal stem cells after 24 h [230,231]. The biocompatibility of the composites was assessed after a 1-month in vivo subcutaneous biocompatibility test in rats. All composites demonstrated higher compatibility than pure PDLLA or PCL, which can be related to the partial surface hydrophilization due to the addition of CNC. However, the composites filled with CNC-PGlu caused the formation of a thinner fibrous capsule and less inflammation. This fact can be explained by the lower roughness of the composite due to better distribution of the modified CNC in the polyester matrix. The PCL-and PLLA-based composites, compared to PDLLA ones, demonstrated less inflammation due to slower hydrolysis and less acidification. Moreover, modification of CNC with PGlu improved the mineralization of the composites (Figure 21), which makes these materials promising for the use as scaffolds for bone-tissue regeneration.

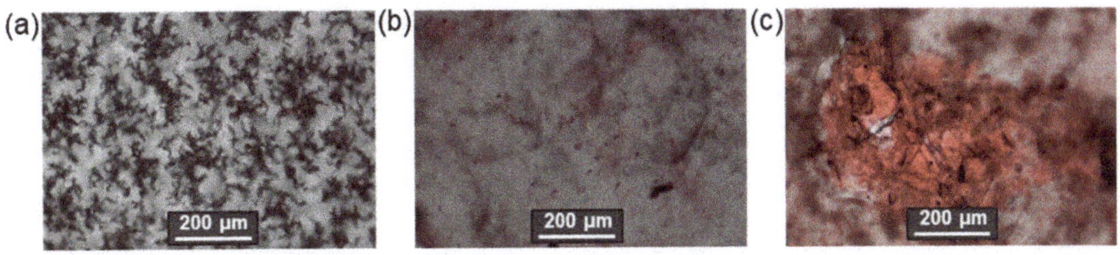

Figure 21. In vitro mineralization study by optical microscopy of pure PCL (**a**) and its composites with 15 wt% neat CNC (**b**) and 15 wt% CNC modified with PGlu (**c**). Staining was performed with alizarin red S. Red color indicates the presence of calcium deposits (reproduced from [224] under the terms of the Creative Commons CC BY license).

Recently, the hemolytic activity of CMC modified with poly(2-hydroxyethyl methacrylate), poly(glycidyl methacrylate) (PGMA), and poly(lauryl methacrylate-co-methyl methacrylate) was evaluated by Rabbi et al. [265]. Incubation of modified CMC in human blood was performed for 1–6 h at 37 °C, varying the concentration from 0.25 to 1.00 mg/mL. In all cases, the degree of hemolysis did not exceed 1%, despite the hydrophobization of the cellulose surface with polymethacrylates. At the same time, CMC-PGMA showed twice as much hemolytic activity (about 1%) as other CMC types.

Fortunati et al. studied the antibacterial activities of composites fabricated by solution casting from PLA and neat CNC or CNC covered with surfactant Beycostat A B09 with or without Ag nanoparticles against *S. Aureus* and *E. coli* [121,234]. All composites showed an antibacterial activity higher than pure PLA, but as it was expected, composites containing Ag nanoparticles demonstrated more pronounced antibacterial activity.

From the presented few works that have reported on the study of the biological properties of modified cellulose, it is obvious that targeted modification of cellulose with appropriate components provides an improvement of the properties of interest.

6. Conclusions and Future Outlook

In this review, the different approaches to the modification of the surface of cellulose micro- and nanomaterials have been summarized and discussed. In general, the surface hydrophobization positively influences the interfacial compatibility between cellulose and aliphatic polyesters. It in turn contributes to the enhancement of mechanical and thermal properties, and improves barrier properties and stability towards degradation in comparison with composites with neat cellulose. However, a degree of these improvements depends on the nature of the modifying agent, the method of modification, the efficiency of modification, and the filler content in the matrix of aliphatic polyester. Obviously, the covalent modification of the cellulose provides more stable linking. The optimal content of the filler is in the range from 1 to 10 wt%. Polymers and fatty acids provide higher impact on the properties of the final composite materials than smaller and more hydrophilic molecules. Furthermore, the characteristics of cellulose, its source and premodification treatment also play an important role since they determine the crystallinity of the sample, surface functionality, degree of purity, and dimensions.

Currently, one of the most-used methods for composite preparation is solution casting. However, this method is not suitable for large-scale production, and also has such drawbacks as retention of solvent traces and possible aggregation and precipitation of filler during solvent evaporation. Hot pressing, melt blending through extrusion, and injection or blow molding are more technologically advanced and better-controlled techniques that have industrial application prospects. Such parameters as crystallization, orientation, dispersion, and distribution of cellulose fillers in the melt-polymer matrix should be controlled when fabricating the composite products.

In general, both neat and modified cellulose-containing composites of aliphatic polyesters demonstrate high biocompatibility in vitro and in vivo. Given that many aliphatic polyesters as well as cellulose are approved for biomedical applications, the introduction of such composites into biomedical practice is very likely in the future.

Author Contributions: Conceptualization, E.K.-V.; formal analysis, M.S. and E.K.-V.; investigation, M.S. and E.K.-V.; visualization, M.S.; writing—original draft preparation, M.S. and E.K.-V.; writing—review and editing, E.K.-V. All authors have read and agreed to the published version of the manuscript.

Funding: This research received no external funding.

Institutional Review Board Statement: Not applicable.

Informed Consent Statement: Not applicable.

Data Availability Statement: Not applicable.

Conflicts of Interest: The authors declare no conflict of interest.

References

1. Sengupta, S.; Manna, S.; Roy, U.; Das, P. Manufacturing of biodegradable Poly Lactic Acid (PLA): Green alternatives to petroleum derived plastics. In *Encyclopedia of Renewable and Sustainable Materials*; Elsevier: Oxford, UK, 2020; pp. 561–569.
2. Mokhena, T.C.; Sefadi, J.S.; Sadiku, E.R.; John, M.J.; Mochane, M.J.; Mtibe, A. Processing of Thermoplastic PLA/Cellulose Nanomaterials Composites. *Polymers* **2018**, *10*, 1363. [CrossRef] [PubMed]
3. Lim, L.-T.; Auras, R.; Rubino, M. Processing technologies for poly(lactic acid). *Prog. Polym. Sci.* **2008**, *33*, 820–852. [CrossRef]
4. Stepanova, M.; Eremin, A.; Averianov, I.; Gofman, I.; Lavrentieva, A.; Korzhikov-Vlakh, V.; Korzhikova-Vlakh, E. Comparison of Supermacroporous Polyester Matrices Fabricated by Thermally Induced Phase Separation and 3D Printing Techniques. *Key Eng. Mater.* **2019**, *822*, 277–283. [CrossRef]
5. Elsawy, M.A.; Kim, K.-H.; Park, J.-W.; Deep, A. Hydrolytic degradation of polylactic acid (PLA) and its composites. *Renew. Sustain. Energy Rev.* **2017**, *79*, 1346–1352. [CrossRef]
6. Liu, G.; Zhang, X.; Wang, D. Tailoring Crystallization: Towards High-Performance Poly(lactic acid). *Adv. Mater.* **2014**, *26*, 6905–6911. [CrossRef]
7. Garlotta, D. A Literature Review of Poly(Lactic Acid). *J. Polym. Environ.* **2001**, *9*, 63–84. [CrossRef]
8. Rydz, J.; Sikorska, W.; Kyulavska, M.; Christova, D. Polyester-Based (Bio)degradable Polymers as Environmentally Friendly Materials for Sustainable Development. *Int. J. Mol. Sci.* **2014**, *16*, 564–596. [CrossRef]
9. Manavitehrani, I.; Fathi, A.; Badr, H.; Daly, S.; Shirazi, A.N.; Dehghani, F. Biomedical applications of biodegradable polyesters. *Polymers* **2016**, *8*, 20. [CrossRef]
10. Seyednejad, H.; Ghassemi, A.H.; Van Nostrum, C.F.; Vermonden, T.; Hennink, W.E. Functional aliphatic polyesters for biomedical and pharmaceutical applications. *J. Control. Release* **2011**, *152*, 168–176. [CrossRef]
11. Zhao, H. Enzymatic ring-opening polymerization (ROP) of polylactones: Roles of non-aqueous solvents. *J. Chem. Technol. Biotechnol.* **2018**, *93*, 9–19. [CrossRef]
12. Siracusa, V.; Rocculi, P.; Romani, S.; Rosa, M.D. Biodegradable polymers for food packaging: A review. *Trends Food Sci. Technol.* **2008**, *19*, 634–643. [CrossRef]
13. Teixeira, S.; Morawa Eblagon, K.; Miranda, F.; Fernando, M.; Pereira, R.; Figueiredo, J.L.; Morin, J.-F.; Savi, P. Towards Controlled Degradation of Poly(lactic) Acid in Technical Applications. *C* **2021**, *7*, 42. [CrossRef]
14. Murariu, M.; Dubois, P. PLA composites: From production to properties. *Adv. Drug Deliv. Rev.* **2016**, *107*, 17–46. [CrossRef] [PubMed]
15. Vakharia, V.S.; Kuentz, L.; Salem, A.; Halbig, M.C.; Salem, J.A.; Singh, M. Additive Manufacturing and Characterization of Metal Particulate Reinforced Polylactic Acid (PLA) Polymer Composites. *Polymers* **2021**, *13*, 3545. [CrossRef] [PubMed]
16. Alam, F.; Verma, P.; Mohammad, W.; Teo, J.; Varadarajan, K.M.; Kumar, S. Architected poly(lactic acid)/poly(ε-caprolactone)/halloysite nanotube composite scaffolds enabled by 3D printing for biomedical applications. *J. Mater. Sci.* **2021**, *56*, 14070–14083. [CrossRef]
17. Chiesa, E.; Dorati, R.; Pisani, S.; Bruni, G.; Rizzi, L.G.; Conti, B.; Modena, T.; Genta, I. Graphene Nanoplatelets for the Development of Reinforced PLA–PCL Electrospun Fibers as the Next-Generation of Biomedical Mats. *Polymers* **2020**, *12*, 1390. [CrossRef]
18. Stepanova, M.; Solomakha, O.; Rabchinskii, M.; Averianov, I.; Gofman, I.; Nashchekina, Y.; Antonov, G.; Smirnov, A.; Ber, B.; Nashchekin, A.; et al. Aminated Graphene-Graft-Oligo(Glutamic Acid) /Poly(ε-Caprolactone) Composites: Preparation, Characterization and Biological Evaluation. *Polymers* **2021**, *13*, 2628. [CrossRef] [PubMed]
19. Liu, Z.; Lei, Q.; Xing, S. Mechanical characteristics of wood, ceramic, metal and carbon fiber-based PLA composites fabricated by FDM. *J. Mater. Res. Technol.* **2019**, *8*, 3741–3751. [CrossRef]
20. Leonés, A.; Salaris, V.; Mujica-Garcia, A.; Arrieta, M.P.; Lopez, D.; Lieblich, M.; Kenny, J.M.; Peponi, L. PLA Electrospun Fibers Reinforced with Organic and Inorganic Nanoparticles: A Comparative Study. *Molecules* **2021**, *26*, 4925. [CrossRef]
21. Sheng, K.; Zhang, S.; Qian, S.; Fontanillo Lopez, C.A. High-toughness PLA/Bamboo cellulose nanowhiskers bionanocomposite strengthened with silylated ultrafine bamboo-char. *Compos. Part B Eng.* **2019**, *165*, 174–182. [CrossRef]
22. Kaščak, J.; Gašpár, Š.; Paško, J.; Husár, J.; Knapčíková, L. Polylactic Acid and Its Cellulose Based Composite as a Significant Tool for the Production of Optimized Models Modified for Additive Manufacturing. *Sustainability* **2021**, *13*, 1256. [CrossRef]
23. Julkapli, N.M.; Bagheri, S. Progress on nanocrystalline cellulose biocomposites. *React. Funct. Polym.* **2017**, *112*, 9–21. [CrossRef]
24. He, W.; Wu, J.; Xu, J.; Mosselhy, D.A.; Zheng, Y.; Yang, S. Bacterial Cellulose: Functional Modification and Wound Healing Applications. *Adv. Wound Care* **2021**, *10*, 623–640. [CrossRef] [PubMed]
25. Teramoto, Y. Recent Advances in Multi-Scale Experimental Analysis to Assess the Role of Compatibilizers in Cellulosic Filler-Reinforced Plastic Composites. *J. Compos. Sci.* **2021**, *5*, 138. [CrossRef]
26. Kargarzadeh, H.; Ioelovich, M.; Ahmad, I.; Thomas, S.; Dufresne, A. Methods for Extraction of Nanocellulose from Various Sources. In *Handbook of Nanocellulose and Cellulose Nanocomposites*; Wiley-VCH Verlag GmbH & Co. KGaA: Weinheim, Germany, 2017; pp. 1–49.
27. Kargarzadeh, H.; Mariano, M.; Gopakumar, D.; Ahmad, I.; Thomas, S.; Dufresne, A.; Huang, J.; Lin, N. Advances in cellulose nanomaterials. *Cellulose* **2018**, *25*, 2151–2189. [CrossRef]
28. Yang, X.; Biswas, S.K.; Han, J.; Tanpichai, S.; Li, M.; Chen, C.; Zhu, S.; Das, A.K.; Yano, H. Surface and Interface Engineering for Nanocellulosic Advanced Materials. *Adv. Mater.* **2021**, *33*, 2002264. [CrossRef]
29. Celebi, H.; Ilgar, M.; Seyhan, A.T. Evaluation of the effect of isocyanate modification on the thermal and rheological properties of poly(ε-caprolactone)/cellulose composites. *Polym. Bull.* **2021**. [CrossRef]

30. Eyley, S.; Thielemans, W. Surface modification of cellulose nanocrystals. *Nanoscale* **2014**, *6*, 7764–7779. [CrossRef]
31. Kasa, S.N.; Omar, M.F.; Abdullah, M.M.A.B.; Ismail, I.N.; Ting, S.S.; Vac, S.C.; Vizureanu, P. Effect of Unmodified and Modified Nanocrystalline Cellulose Reinforced Polylactic Acid (PLA) Polymer Prepared by Solvent Casting Method Morphology, mechanical and thermal properties. *Mater. Plast.* **2017**, *54*, 91–97. [CrossRef]
32. Olonisakin, K.; Li, R.; Zhang, X.-X.; Xiao, F.; Gao, J.; Yang, W. Effect of TDI-Assisted Hydrophobic Surface Modification of Microcrystalline Cellulose on the Tensile Fracture of MCC/PLA Composite, and Estimation of the Degree of Substitution by Linear Regression. *Langmuir* **2021**, *37*, 793–801. [CrossRef]
33. Lv, S.; Gu, J.; Tan, H.; Zhang, Y. Modification of wood flour/PLA composites by reactive extrusion with maleic anhydride. *J. Appl. Polym. Sci.* **2016**, *133*, 43295. [CrossRef]
34. Fox, D.M.; Novy, M.; Brown, K.; Zammarano, M.; Harris, R.H.; Murariu, M.; McCarthy, E.D.; Seppala, J.E.; Gilman, J.W. Flame retarded poly(lactic acid) using POSS-modified cellulose. 2. Effects of intumescing flame retardant formulations on polymer degradation and composite physical properties. *Polym. Degrad. Stab.* **2014**, *106*, 54–62. [CrossRef]
35. Tao, Y.; Liu, M.; Han, W.; Li, P. Waste office paper filled polylactic acid composite filaments for 3D printing. *Compos. Part B Eng.* **2021**, *221*, 108998. [CrossRef]
36. Fox, D.M.; Temburni, S.; Novy, M.; Flynn, L.; Zammarano, M.; Kim, Y.S.; Gilman, J.W.; Davis, R.D. Thermal and Burning Properties of Poly(lactic acid) Composites Using Cellulose-Based Intumescing Flame Retardants. In *Fire and Polymers VI: New Advances in Flame Retardant Chemistry and Science*; American Chemical Society: Washington, DC, USA, 2012; pp. 223–234.
37. Rodríguez Soto, K.X. Laminated composites reinforced with chemically modified sheets-stalk of Musa Cavendish. *Rev. Mex. Ing. Química* **2019**, *18*, 749–758. [CrossRef]
38. Goriparthi, B.K.; Suman, K.N.S.; Mohan Rao, N. Effect of fiber surface treatments on mechanical and abrasive wear performance of polylactide/jute composites. *Compos. Part A Appl. Sci. Manuf.* **2012**, *43*, 1800–1808. [CrossRef]
39. Vatansever, E.; Arslan, D.; Nofar, M. Polylactide cellulose-based nanocomposites. *Int. J. Biol. Macromol.* **2019**, *137*, 912–938. [CrossRef]
40. Saba, N.; Jawaid, M.; Al-Othman, O. An Overview on Polylactic Acid, its Cellulosic Composites and Applications. *Curr. Org. Synth.* **2017**, *14*, 156–170. [CrossRef]
41. Kian, L.K.; Saba, N.; Jawaid, M.; Sultan, M.T.H. A review on processing techniques of bast fibers nanocellulose and its polylactic acid (PLA) nanocomposites. *Int. J. Biol. Macromol.* **2019**, *121*, 1314–1328. [CrossRef]
42. John, M.J.; Dyanti, N.; Mokhena, T.; Agbakoba, V.; Sithole, B. Design and Development of Cellulosic Bionanocomposites from Forestry Waste Residues for 3D Printing Applications. *Materials* **2021**, *14*, 3462. [CrossRef]
43. Getme, A.S.; Patel, B. A Review: Bio-fiber's as reinforcement in composites of polylactic acid (PLA). *Mater. Today Proc.* **2020**, *26*, 2116–2122. [CrossRef]
44. Zhou, L.; Ke, K.; Yang, M.-B.; Yang, W. Recent progress on chemical modification of cellulose for high mechanical-performance Poly(lactic acid)/Cellulose composite: A review. *Compos. Commun.* **2021**, *23*, 100548. [CrossRef]
45. Ferreira, F.V.; Dufresne, A.; Pinheiro, I.F.; Souza, D.H.S.; Gouveia, R.F.; Mei, L.H.I.; Lona, L.M.F. How do cellulose nanocrystals affect the overall properties of biodegradable polymer nanocomposites: A comprehensive review. *Eur. Polym. J.* **2018**, *108*, 274–285. [CrossRef]
46. Seddiqi, H.; Oliaei, E.; Honarkar, H.; Jin, J.; Geonzon, L.C.; Bacabac, R.G.; Klein-Nulend, J. Cellulose and its derivatives: Towards biomedical applications. *Cellulose* **2021**, *28*, 1893–1931. [CrossRef]
47. Bogolitsyn, K.; Parshina, A.; Aleshina, L. Structural features of brown algae cellulose. *Cellulose* **2020**, *27*, 9787–9800. [CrossRef]
48. Zanchetta, E.; Damergi, E.; Patel, B.; Borgmeyer, T.; Pick, H.; Pulgarin, A.; Ludwig, C. Algal cellulose, production and potential use in plastics: Challenges and opportunities. *Algal Res.* **2021**, *56*, 102288. [CrossRef]
49. Heinze, T. Cellulose: Structure and Properties. In *Cellulose Chemistry and Properties: Fibers, Nanocelluloses and Advanced Materials*; Springer: Berlin/Heidelberg, Germany, 2015; pp. 1–52.
50. Liu, Y. Recent Progress in Fourier Transform Infrared (FTIR) Spectroscopy Study of Compositional, Structural and Physical Attributes of Developmental Cotton Fibers. *Materials* **2013**, *6*, 299–313. [CrossRef]
51. Habibi, Y.; Lucia, L.A.; Rojas, O.J. Cellulose Nanocrystals: Chemistry, Self-Assembly, and Applications. *Chem. Rev.* **2010**, *110*, 3479–3500. [CrossRef]
52. Brown, R.M.; Saxena, I.; Kudlicka, K. Cellulose biosynthesis in higher plants. *Trends Plant Sci.* **1996**, *1*, 149–156.
53. Azizi Samir, M.A.S.; Alloin, F.; Dufresne, A. Review of Recent Research into Cellulosic Whiskers, Their Properties and Their Application in Nanocomposite Field. *Biomacromolecules* **2005**, *6*, 612–626. [CrossRef]
54. Klemm, D.; Heublein, B.; Fink, H.-P.; Bohn, A. Cellulose: Fascinating Biopolymer and Sustainable Raw Material. *Angew. Chemie Int. Ed.* **2005**, *44*, 3358–3393. [CrossRef]
55. Lavoine, N.; Desloges, I.; Dufresne, A.; Bras, J. Microfibrillated cellulose—Its barrier properties and applications in cellulosic materials: A review. *Carbohydr. Polym.* **2012**, *90*, 735–764. [CrossRef] [PubMed]
56. Czaja, W.K.; Young, D.J.; Kawecki, M.; Brown, R.M. The Future Prospects of Microbial Cellulose in Biomedical Applications. *Biomacromolecules* **2007**, *8*, 1–12. [CrossRef]
57. Kondo, T. The relationship between intramolecular hydrogen bonds and certain physical properties of regioselectively substituted cellulose derivatives. *J. Polym. Sci. Part B Polym. Phys.* **1997**, *35*, 717–723. [CrossRef]

58. Siqueira, G.; Bras, J.; Dufresne, A. Cellulosic Bionanocomposites: A Review of Preparation, Properties and Applications. *Polymers* **2010**, *2*, 728–765. [CrossRef]
59. Wada, M.; Ike, M.; Tokuyasu, K. Enzymatic hydrolysis of cellulose I is greatly accelerated via its conversion to the cellulose II hydrate form. *Polym. Degrad. Stab.* **2010**, *95*, 543–548. [CrossRef]
60. Krässig, H.A. (Ed.) *Cellulose, Structure, Accessibility and Reactivity*; Gordon and Breach Publishers: Philadelphia, PA, USA, 1993.
61. Mariano, M.; El Kissi, N.; Dufresne, A. Cellulose nanocrystals and related nanocomposites: Review of some properties and challenges. *J. Polym. Sci. Part B Polym. Phys.* **2014**, *52*, 791–806. [CrossRef]
62. Guambo, M.P.R.; Spencer, L.; Vispo, N.S.; Vizuete, K.; Debut, A.; Whitehead, D.C.; Santos-Oliveira, R.; Alexis, F. Natural Cellulose Fibers for Surgical Suture Applications. *Polymers* **2020**, *12*, 3042. [CrossRef]
63. Anglès, M.N.; Dufresne, A. Plasticized Starch/Tunicin Whiskers Nanocomposites. 1. Structural Analysis. *Macromolecules* **2000**, *33*, 8344–8353. [CrossRef]
64. Agoda-Tandjawa, G.; Durand, S.; Berot, S.; Blassel, C.; Gaillard, C.; Garnier, C.; Doublier, J.-L. Rheological characterization of microfibrillated cellulose suspensions after freezing. *Carbohydr. Polym.* **2010**, *80*, 677–686. [CrossRef]
65. Murphy, C.A.; Collins, M.N. Microcrystalline cellulose reinforced polylactic acid biocomposite filaments for 3D printing. *Polym. Compos.* **2018**, *39*, 1311–1320. [CrossRef]
66. Saito, T.; Kuramae, R.; Wohlert, J.; Berglund, L.A.; Isogai, A. An Ultrastrong Nanofibrillar Biomaterial: The Strength of Single Cellulose Nanofibrils Revealed via Sonication-Induced Fragmentation. *Biomacromolecules* **2013**, *14*, 248–253. [CrossRef] [PubMed]
67. Habibi, Y.; Goffin, A.-L.; Schiltz, N.; Duquesne, E.; Dubois, P.; Dufresne, A. Bionanocomposites based on poly(ε-caprolactone)-grafted cellulose nanocrystals by ring-opening polymerization. *J. Mater. Chem.* **2008**, *18*, 5002–5010. [CrossRef]
68. Ebeling, T.; Paillet, M.; Borsali, R.; Diat, O.; Dufresne, A.; Cavaillé, A.J.; Chanzy, H. Shear-Induced Orientation Phenomena in Suspensions of Cellulose Microcrystals, Revealed by Small Angle X-ray Scattering. *Langmuir* **1999**, *15*, 6123–6126. [CrossRef]
69. Li, H.; Cao, Z.; Wu, D.; Tao, G.; Zhong, W.; Zhu, H.; Qiu, P.; Liu, C. Crystallisation, mechanical properties and rheological behaviour of PLA composites reinforced by surface modified microcrystalline cellulose. *Plast. Rubber Compos.* **2016**, *45*, 181–187. [CrossRef]
70. Trache, D.; Hussin, M.H.; Hui Chuin, C.T.; Sabar, S.; Fazita, M.R.N.; Taiwo, O.F.A.; Hassan, T.M.; Haafiz, M.K.M. Microcrystalline cellulose: Isolation, characterization and bio-composites application—A review. *Int. J. Biol. Macromol.* **2016**, *93*, 789–804. [CrossRef] [PubMed]
71. Eichhorn, S.J.; Baillie, C.A.; Zafeiropoulos, N.; Mwaikambo, L.Y.; Ansell, M.P.; Dufresne, A.; Entwistle, K.M.; Herrera-Franco, P.J.; Escamilla, G.C.; Groom, L.; et al. Review: Current international research into cellulosic fibres and composites. *J. Mater. Sci.* **2001**, *36*, 2107–2131. [CrossRef]
72. Tashiro, K.; Kobayashi, M. Theoretical evaluation of three-dimensional elastic constants of native and regenerated celluloses: Role of hydrogen bonds. *Polymer* **1991**, *32*, 1516–1526. [CrossRef]
73. Lahiji, R.R.; Xu, X.; Reifenberger, R.; Raman, A.; Rudie, A.; Moon, R.J. Atomic Force Microscopy Characterization of Cellulose Nanocrystals. *Langmuir* **2010**, *26*, 4480–4488. [CrossRef] [PubMed]
74. Iwamoto, S.; Kai, W.; Isogai, A.; Iwata, T. Elastic Modulus of Single Cellulose Microfibrils from Tunicate Measured by Atomic Force Microscopy. *Biomacromolecules* **2009**, *10*, 2571–2576. [CrossRef] [PubMed]
75. Walther, A.; Timonen, J.V.I.; Díez, I.; Laukkanen, A.; Ikkala, O. Multifunctional High-Performance Biofibers Based on Wet-Extrusion of Renewable Native Cellulose Nanofibrils. *Adv. Mater.* **2011**, *23*, 2924–2928. [CrossRef] [PubMed]
76. Ioelovich, M. Nanoparticles of Amorphous Cellulose and Their Properties. *Am. J. Nanosci. Nanotechnol.* **2013**, *1*, 41. [CrossRef]
77. Frenot, A.; Henriksson, M.W.; Walkenström, P. Electrospinning of cellulose-based nanofibers. *J. Appl. Polym. Sci.* **2007**, *103*, 1473–1482. [CrossRef]
78. Håkansson, K.M.O.; Fall, A.B.; Lundell, F.; Yu, S.; Krywka, C.; Roth, S.V.; Santoro, G.; Kvick, M.; Prahl Wittberg, L.; Wågberg, L.; et al. Hydrodynamic alignment and assembly of nanofibrils resulting in strong cellulose filaments. *Nat. Commun.* **2014**, *5*, 4018. [CrossRef] [PubMed]
79. McAdam, B.; Brennan Fournet, M.; McDonald, P.; Mojicevic, M. Production of Polyhydroxybutyrate (PHB) and Factors Impacting Its Chemical and Mechanical Characteristics. *Polymers* **2020**, *12*, 2908. [CrossRef] [PubMed]
80. Rafiqah, S.A.; Khalina, A.; Harmaen, A.S.; Tawakkal, I.A.; Zaman, K.; Asim, M.; Nurrazi, M.N.; Lee, C.H. A Review on Properties and Application of Bio-Based Poly(Butylene Succinate). *Polymers* **2021**, *13*, 1436. [CrossRef]
81. Lee, B.K.; Yun, Y.; Park, K. PLA micro- and nano-particles. *Adv. Drug Deliv. Rev.* **2016**, *107*, 176–191. [CrossRef] [PubMed]
82. Xiao, L.; Mai, Y.; He, F.; Yu, L.; Zhang, L.; Tang, H.; Yang, G. Bio-based green composites with high performance from poly(lactic acid) and surface-modified microcrystalline cellulose. *J. Mater. Chem.* **2012**, *22*, 15732–15739. [CrossRef]
83. Dwivedi, R.; Kumar, S.; Pandey, R.; Mahajan, A.; Nandana, D.; Katti, D.S.; Mehrotra, D. Polycaprolactone as biomaterial for bone scaffolds: Review of literature. *J. Oral Biol. Craniofacial Res.* **2020**, *10*, 381–388. [CrossRef]
84. Lasprilla, A.J.R.; Martinez, G.A.R.; Lunelli, B.H.; Jardini, A.L.; Filho, R.M. Poly-lactic acid synthesis for application in biomedical devices—A review. *Biotechnol. Adv.* **2012**, *30*, 321–328. [CrossRef]
85. Auras, R.; Harte, B.; Selke, S. An Overview of Polylactides as Packaging Materials. *Macromol. Biosci.* **2004**, *4*, 835–864. [CrossRef]
86. Budak, K.; Sogut, O.; Aydemir Sezer, U. A review on synthesis and biomedical applications of polyglycolic acid. *J. Polym. Res.* **2020**, *27*, 208. [CrossRef]

87. Anjana; Raturi, G.; Shree, S.; Sharma, A.; Panesar, P.S.; Goswami, S. Recent approaches for enhanced production of microbial polyhydroxybutyrate: Preparation of biocomposites and applications. *Int. J. Biol. Macromol.* **2021**, *182*, 1650–1669. [CrossRef] [PubMed]
88. Sudesh, K.; Abe, H.; Doi, Y. Synthesis, structure and properties of polyhydroxyalkanoates: Biological polyesters. *Prog. Polym. Sci.* **2000**, *25*, 1503–1555. [CrossRef]
89. Labet, M.; Thielemans, W. Synthesis of polycaprolactone: A review. *Chem. Soc. Rev.* **2009**, *38*, 3484. [CrossRef]
90. Bartnikowski, M.; Dargaville, T.R.; Ivanovski, S.; Hutmacher, D.W. Degradation mechanisms of polycaprolactone in the context of chemistry, geometry and environment. *Prog. Polym. Sci.* **2019**, *96*, 1–20. [CrossRef]
91. Mochane, M.J.; Magagula, S.I.; Sefadi, J.S.; Mokhena, T.C. A Review on Green Composites Based on Natural Fiber-Reinforced Polybutylene Succinate (PBS). *Polymers* **2021**, *13*, 1200. [CrossRef]
92. Yamane, K.; Sato, H.; Ichikawa, Y.; Sunagawa, K.; Shigaki, Y. Development of an industrial production technology for high-molecular-weight polyglycolic acid. *Polym. J.* **2014**, *46*, 769–775. [CrossRef]
93. Low, Y.J.; Andriyana, A.; Ang, B.C.; Zainal Abidin, N.I. Bioresorbable and degradable behaviors of PGA: Current state and future prospects. *Polym. Eng. Sci.* **2020**, *60*, 2657–2675. [CrossRef]
94. Fortelny, I.; Ujcic, A.; Fambri, L.; Slouf, M. Phase Structure, Compatibility, and Toughness of PLA/PCL Blends: A Review. *Front. Mater.* **2019**, *6*, 206. [CrossRef]
95. Garin, M.; Tighzert, L.; Vroman, I.; Marinkovic, S.; Estrine, B. The influence of molar mass on rheological and dilute solution properties of poly(butylene succinate). *J. Appl. Polym. Sci.* **2014**, *131*, 40887. [CrossRef]
96. Ikada, Y.; Tsuji, H. Biodegradable polyesters for medical and ecological applications. *Macromol. Rapid Commun.* **2000**, *21*, 117–132. [CrossRef]
97. Kliem, S.; Kreutzbruck, M.; Bonten, C. Review on the Biological Degradation of Polymers in Various Environments. *Materials* **2020**, *13*, 4586. [CrossRef] [PubMed]
98. Woodruff, M.A.; Hutmacher, D.W. The return of a forgotten polymer—Polycaprolactone in the 21st century. *Prog. Polym. Sci.* **2010**, *35*, 1217–1256. [CrossRef]
99. Puchalski, M.; Szparaga, G.; Biela, T.; Gutowska, A.; Sztajnowski, S.; Krucińska, I. Molecular and Supramolecular Changes in Polybutylene Succinate (PBS) and Polybutylene Succinate Adipate (PBSA) Copolymer during Degradation in Various Environmental Conditions. *Polymers* **2018**, *10*, 251. [CrossRef] [PubMed]
100. Tserki, V.; Matzinos, P.; Pavlidou, E.; Vachliotis, D.; Panayiotou, C. Biodegradable aliphatic polyesters. Part I. Properties and biodegradation of poly(butylene succinate-co-butylene adipate). *Polym. Degrad. Stab.* **2006**, *91*, 367–376. [CrossRef]
101. Ikada, Y.; Jamshidi, K.; Tsuji, H.; Hyon, S.H. Stereocomplex formation between enantiomeric poly(lactides). *Macromolecules* **1987**, *20*, 904–906. [CrossRef]
102. Saeidlou, S.; Huneault, M.A.; Li, H.; Park, C.B. Poly(lactic acid) crystallization. *Prog. Polym. Sci.* **2012**, *37*, 1657–1677. [CrossRef]
103. Pérez-Fonseca, A.A.; Robledo-Ortíz, J.R.; González-Núñez, R.; Rodrigue, D. Effect of thermal annealing on the mechanical and thermal properties of polylactic acid-cellulosic fiber biocomposites. *J. Appl. Polym. Sci.* **2016**, *133*, 43750. [CrossRef]
104. Gupta, A.; Simmons, W.; Schueneman, G.T.; Hylton, D.; Mintz, E.A. Rheological and Thermo-Mechanical Properties of Poly(lactic acid)/Lignin-Coated Cellulose Nanocrystal Composites. *ACS Sustain. Chem. Eng.* **2017**, *5*, 1711–1720. [CrossRef]
105. Byun, Y.; Rodriguez, K.; Han, J.H.; Kim, Y.T. Improved thermal stability of polylactic acid (PLA) composite film via PLA–β-cyclodextrin-inclusion complex systems. *Int. J. Biol. Macromol.* **2015**, *81*, 591–598. [CrossRef]
106. Zhou, H.; Lawrence, J.G.; Bhaduri, S.B. Fabrication aspects of PLA-CaP/PLGA-CaP composites for orthopedic applications: A review. *Acta Biomater.* **2012**, *8*, 1999–2016. [CrossRef] [PubMed]
107. Casciaro, B.; D'Angelo, I.; Zhang, X.; Loffredo, M.R.; Conte, G.; Cappiello, F.; Quaglia, F.; Di, Y.P.P.; Ungaro, F.; Mangoni, M.L. Poly(lactide- co-glycolide) Nanoparticles for Prolonged Therapeutic Efficacy of Esculentin-1a-Derived Antimicrobial Peptides against Pseudomonas aeruginosa Lung Infection: In Vitro and in Vivo Studies. *Biomacromolecules* **2019**, *20*, 1876–1888. [CrossRef] [PubMed]
108. Alibolandi, M.; Alabdollah, F.; Sadeghi, F.; Mohammadi, M.; Abnous, K.; Ramezani, M.; Hadizadeh, F. Dextran-b-poly (lactide-co-glycolide) polymersome for oral delivery of insulin: In vitro and in vivo evaluation. *J. Control. Release* **2016**, *227*, 58–70. [CrossRef]
109. Abe, H.; Matsubara, I.; Doi, Y.; Hori, Y.; Yamaguchi, A. Physical Properties and Enzymic Degradability of Poly(3-hydroxybutyrate) Stereoisomers with Different Stereoregularities. *Macromolecules* **1994**, *27*, 6018–6025. [CrossRef]
110. Yeo, J.C.C.; Muiruri, J.K.; Thitsartarn, W.; Li, Z.; He, C. Recent advances in the development of biodegradable PHB-based toughening materials: Approaches, advantages and applications. *Mater. Sci. Eng. C* **2018**, *92*, 1092–1116. [CrossRef] [PubMed]
111. Lenz, R.W.; Marchessault, R.H. Bacterial Polyesters: Biosynthesis, Biodegradable Plastics and Biotechnology. *Biomacromolecules* **2005**, *6*, 1–8. [CrossRef]
112. Mitomo, H.; Hsieh, W.-C.; Nishiwaki, K.; Kasuya, K.; Doi, Y. Poly(3-hydroxybutyrate-co-4-hydroxybutyrate) produced by Comamonas acidovorans. *Polymer* **2001**, *42*, 3455–3461. [CrossRef]
113. Pitt, C.G.; Chasalow, F.I.; Hibionada, Y.M.; Klimas, D.M.; Schindler, A. Aliphatic polyesters. I. The degradation of poly(ϵ-caprolactone) in vivo. *J. Appl. Polym. Sci.* **1981**, *26*, 3779–3787. [CrossRef]
114. Bondeson, D.; Oksman, K. Dispersion and characteristics of surfactant modified cellulose whiskers nanocomposites. *Compos. Interfaces* **2007**, *14*, 617–630. [CrossRef]

115. Li, C.; Sun, C.; Wang, C.; Tan, H.; Xie, Y.; Zhang, Y. Cellulose nanocrystal reinforced poly(lactic acid) nanocomposites prepared by a solution precipitation approach. *Cellulose* **2020**, *27*, 7489–7502. [CrossRef]
116. Bayart, M.; Adjallé, K.; Diop, A.; Ovlaque, P.; Barnabé, S.; Robert, M.; Elkoun, S. PLA/flax fiber bio-composites: Effect of polyphenol-based surface treatment on interfacial adhesion and durability. *Compos. Interfaces* **2021**, *28*, 287–308. [CrossRef]
117. Góis, G.S.; Nepomuceno, N.C.; França, C.H.A.; Almeida, Y.M.B.; Hernandéz, E.P.; Oliveira, J.E.; Oliveira, M.P.; Medeiros, E.S.; Santos, A.S.F. Influence of morphology and dispersion stability of CNC modified with ethylene oxide derivatives on mechanical properties of PLA-based nanocomposites. *Polym. Compos.* **2019**, *40*, E399–E408. [CrossRef]
118. Pal, N.; Banerjee, S.; Roy, P.; Pal, K. Melt-blending of unmodified and modified cellulose nanocrystals with reduced graphene oxide into PLA matrix for biomedical application. *Polym. Adv. Technol.* **2019**, *30*, 3049–3060. [CrossRef]
119. Arrieta, M.P.; Fortunati, E.; Dominici, F.; Rayón, E.; López, J.; Kenny, J.M. Multifunctional PLA–PHB/cellulose nanocrystal films: Processing, structural and thermal properties. *Carbohydr. Polym.* **2014**, *107*, 16–24. [CrossRef] [PubMed]
120. Pracella, M.; Mura, C.; Galli, G. Polyhydroxyalkanoate Nanocomposites with Cellulose Nanocrystals as Biodegradable Coating and Packaging Materials. *ACS Appl. Nano Mater.* **2021**, *4*, 260–270. [CrossRef]
121. Fortunati, E.; Armentano, I.; Zhou, Q.; Puglia, D.; Terenzi, A.; Berglund, L.A.; Kenny, J.M. Microstructure and nonisothermal cold crystallization of PLA composites based on silver nanoparticles and nanocrystalline cellulose. *Polym. Degrad. Stab.* **2012**, *97*, 2027–2036. [CrossRef]
122. Cacciotti, I.; Fortunati, E.; Puglia, D.; Kenny, J.M.; Nanni, F. Effect of silver nanoparticles and cellulose nanocrystals on electrospun poly(lactic) acid mats: Morphology, thermal properties and mechanical behavior. *Carbohydr. Polym.* **2014**, *103*, 22–31. [CrossRef]
123. Qin, S.; Hu, Y.; Tian, X.; Tian, Y.; Liu, W.; Zhao, L. Modification of cellulose nanocrystals by self-assembly nucleation agents to improve poly(L-lactide) nanocomposite' properties. *Cellulose* **2020**, *27*, 4337–4353. [CrossRef]
124. Lin, J.; Yang, Z.; Hu, X.; Hong, G.; Zhang, S.; Song, W. The Effect of Alkali Treatment on Properties of Dopamine Modification of Bamboo Fiber/Polylactic Acid Composites. *Polymers* **2018**, *10*, 403. [CrossRef]
125. Gregor-Svetec, D.; Leskovšek, M.; Leskovar, B.; Stanković Elesini, U.; Vrabič-Brodnjak, U. Analysis of PLA Composite Filaments Reinforced with Lignin and Polymerised-Lignin-Treated NFC. *Polymers* **2021**, *13*, 2174. [CrossRef]
126. Alvarado, N.; Romero, J.; Torres, A.; López de Dicastillo, C.; Rojas, A.; Galotto, M.J.; Guarda, A. Supercritical impregnation of thymol in poly(lactic acid) filled with electrospun poly(vinyl alcohol)-cellulose nanocrystals nanofibers: Development an active food packaging material. *J. Food Eng.* **2018**, *217*, 1–10. [CrossRef]
127. Voronova, M.I.; Gurina, D.L.; Surov, O.V.; Zakharov, A.G. Interactions in solvent–polycaprolactone–cellulose nanocrystals–polyvinyl pyrrolidone system: Experiment and molecular dynamics simulation. *J. Mol. Liq.* **2021**, *341*, 117409. [CrossRef]
128. Pal, N.; Banerjee, S.; Roy, P.; Pal, K. Reduced graphene oxide and PEG-grafted TEMPO-oxidized cellulose nanocrystal reinforced poly-lactic acid nanocomposite film for biomedical application. *Mater Sci Eng C* **2019**, *104*, 109956. [CrossRef]
129. Barbosa, R.F.S.; Souza, A.G.; Rosa, D.S. Acetylated cellulose nanostructures as reinforcement materials for PBAT nanocomposites. *Polym. Compos.* **2020**, *41*, 2841–2854. [CrossRef]
130. Suryanegara, L.; Nugraha, R.A.; Achmadi, S.S. Improvement of thermal and mechanical properties of composite based on polylactic acid and microfibrillated cellulose through chemical modification. *IOP Conf. Ser. Mater. Sci. Eng.* **2017**, *223*, 012032. [CrossRef]
131. Abdulkhani, A.; Hosseinzadeh, J.; Ashori, A.; Dadashi, S.; Takzare, Z. Preparation and characterization of modified cellulose nanofibers reinforced polylactic acid nanocomposite. *Polym. Test.* **2014**, *35*, 73–79. [CrossRef]
132. Ávila Ramírez, J.A.; Cerrutti, P.; Bernal, C.; Errea, M.I.; Foresti, M.L. Nanocomposites Based on Poly(lactic acid) and Bacterial Cellulose Acetylated by an α-Hydroxyacid Catalyzed Route. *J. Polym. Environ.* **2019**, *27*, 510–520. [CrossRef]
133. Ling, Z.; Liu, W.; Ren, Y.; Chen, H.; Huang, C.; Lai, C.; Yong, Q. Bioinspired manufacturing of oriented polysaccharides scaffolds for strong, optical haze and anti-UV/bacterial membranes. *Carbohydr. Polym.* **2021**, *270*, 118328. [CrossRef]
134. Blaker, J.J.; Lee, K.-Y.; Walters, M.; Drouet, M.; Bismarck, A. Aligned unidirectional PLA/bacterial cellulose nanocomposite fibre reinforced PDLLA composites. *React. Funct. Polym.* **2014**, *85*, 185–192. [CrossRef]
135. Haque, M.M.-U.; Errico, M.E.; Gentile, G.; Avella, M.; Pracella, M. Functionalization and Compatibilization of Poly(ε-caprolactone) Composites with Cellulose Microfibres: Morphology, Thermal and Mechanical Properties. *Macromol. Mater. Eng.* **2012**, *297*, 985–993. [CrossRef]
136. Almasi, H.; Ghanbarzadeh, B.; Dehghannya, J.; Entezami, A.A.; Asl, A.K. Novel nanocomposites based on fatty acid modified cellulose nanofibers/poly(lactic acid): Morphological and physical properties. *Food Packag. Shelf Life* **2015**, *5*, 21–31. [CrossRef]
137. Long, S.; Zhong, L.; Lin, X.; Chang, X.; Wu, F.; Wu, R.; Xie, F. Preparation of formyl cellulose and its enhancement effect on the mechanical and barrier properties of polylactic acid films. *Int. J. Biol. Macromol.* **2021**, *172*, 82–92. [CrossRef] [PubMed]
138. Shojaeiarani, J.; Bajwa, D.S.; Hartman, K. Esterified cellulose nanocrystals as reinforcement in poly(lactic acid) nanocomposites. *Cellulose* **2019**, *26*, 2349–2362. [CrossRef]
139. Trifol, J.; Plackett, D.; Sillard, C.; Hassager, O.; Daugaard, A.E.; Bras, J.; Szabo, P. A comparison of partially acetylated nanocellulose, nanocrystalline cellulose, and nanoclay as fillers for high-performance polylactide nanocomposites. *J. Appl. Polym. Sci.* **2016**, *133*, 58–66. [CrossRef]
140. Xu, C.; Chen, J.; Wu, D.; Chen, Y.; Lv, Q.; Wang, M. Polylactide/acetylated nanocrystalline cellulose composites prepared by a continuous route: A phase interface-property relation study. *Carbohydr. Polym.* **2016**, *146*, 58–66. [CrossRef]

141. Xu, C.; Wu, D.; Lv, Q.; Yan, L. Crystallization Temperature as the Probe To Detect Polymer–Filler Compatibility in the Poly(ε-caprolactone) Composites with Acetylated Cellulose Nanocrystal. *J. Phys. Chem. C* **2017**, *121*, 18615–18624. [CrossRef]
142. Ribeiro, F.A. dos S.V.; Cavalcante, M. de P.; Tavares, M.I.B.; Melo, A.R.A. Effect of modified microcrystalline cellulose on poly(3-hydroxybutyrate) molecular dynamics by proton relaxometry. *Polym. Polym. Compos.* **2021**, *29*, 553–560.
143. Clarkson, C.M.; El Awad Azrak, S.M.; Chowdhury, R.; Shuvo, S.N.; Snyder, J.; Schueneman, G.; Ortalan, V.; Youngblood, J.P. Melt Spinning of Cellulose Nanofibril/Polylactic Acid (CNF/PLA) Composite Fibers For High Stiffness. *ACS Appl. Polym. Mater.* **2019**, *1*, 160–168. [CrossRef]
144. Zhu, T.; Guo, J.; Fei, B.; Feng, Z.; Gu, X.; Li, H.; Sun, J.; Zhang, S. Preparation of methacrylic acid modified microcrystalline cellulose and their applications in polylactic acid: Flame retardancy, mechanical properties, thermal stability and crystallization behavior. *Cellulose* **2020**, *27*, 2309–2323. [CrossRef]
145. Niu, X.; Liu, Y.; Song, Y.; Han, J.; Pan, H. Rosin modified cellulose nanofiber as a reinforcing and co-antimicrobial agents in polylactic acid /chitosan composite film for food packaging. *Carbohydr. Polym.* **2018**, *183*, 102–109. [CrossRef]
146. Kale, R.D.; Gorade, V.G. Preparation of acylated microcrystalline cellulose using olive oil and its reinforcing effect on poly(lactic acid) films for packaging application. *J. Polym. Res.* **2018**, *25*, 81. [CrossRef]
147. Shojaeiarani, J.; Bajwa, D.S.; Stark, N.M. Green esterification: A new approach to improve thermal and mechanical properties of poly(lactic acid) composites reinforced by cellulose nanocrystals. *J. Appl. Polym. Sci.* **2018**, *135*, 46468. [CrossRef]
148. Li, K.; Mcgrady, D.; Zhao, X.; Ker, D.; Tekinalp, H.; He, X.; Qu, J.; Aytug, T.; Cakmak, E.; Phipps, J.; et al. Surface-modified and oven-dried microfibrillated cellulose reinforced biocomposites: Cellulose network enabled high performance. *Carbohydr. Polym.* **2021**, *256*, 1755302. [CrossRef] [PubMed]
149. Moazzami Goudarzi, Z.; Behzad, T.; Ghasemi-Mobarakeh, L.; Kharaziha, M. An investigation into influence of acetylated cellulose nanofibers on properties of PCL/Gelatin electrospun nanofibrous scaffold for soft tissue engineering. *Polymer* **2021**, *213*, 123313. [CrossRef]
150. Jonoobi, M.; Mathew, A.P.; Abdi, M.M.; Makinejad, M.D.; Oksman, K. A Comparison of Modified and Unmodified Cellulose Nanofiber Reinforced Polylactic Acid (PLA) Prepared by Twin Screw Extrusion. *J. Polym. Environ.* **2012**, *20*, 991–997. [CrossRef]
151. Žepič, V.; Poljanšek, I.; Oven, P.; Čop, M. COST-FP1105: Properties of PLA films reinforced with unmodified and acetylated freeze dried nanofibrillated cellulose. *Holzforschung* **2016**, *70*, 1125–1134. [CrossRef]
152. Jamaluddin, N.; Hsu, Y.-I.; Asoh, T.-A.; Uyama, H. Effects of Acid-Anhydride-Modified Cellulose Nanofiber on Poly(Lactic Acid) Composite Films. *Nanomaterials* **2021**, *11*, 753. [CrossRef]
153. Bin, Y.; Yang, B.; Wang, H. The effect of a small amount of modified microfibrillated cellulose and ethylene–glycidyl methacrylate copolymer on the crystallization behaviors and mechanical properties of polylactic acid. *Polym. Bull.* **2018**, *75*, 3377–3394. [CrossRef]
154. Oza, S.; Ning, H.; Ferguson, I.; Lu, N. Effect of surface treatment on thermal stability of the hemp-PLA composites: Correlation of activation energy with thermal degradation. *Compos. Part B Eng.* **2014**, *67*, 227–232. [CrossRef]
155. Szefer, E.; Leszczyńska, A.; Hebda, E.; Pielichowski, K. The Application of Cellulose Nanocrystals Modified with Succinic Anhydride under the Microwave Irradiation for Preparation of Polylactic Acid Nanocomposites. *J. Renew. Mater.* **2021**, *9*, 1127–1142. [CrossRef]
156. Pandey, J.K.; Lee, C.S.; Ahn, S.-H. Preparation and properties of bio-nanoreinforced composites from biodegradable polymer matrix and cellulose whiskers. *J. Appl. Polym. Sci.* **2010**, *115*, 2493–2501. [CrossRef]
157. Zhou, L.; He, H.; Li, M.; Huang, S.; Mei, C.; Wu, Q. Enhancing mechanical properties of poly(lactic acid) through its in-situ crosslinking with maleic anhydride-modified cellulose nanocrystals from cottonseed hulls. *Ind. Crops Prod.* **2018**, *112*, 449–459. [CrossRef]
158. Bi, H.; Ye, G.; Sun, H.; Ren, Z.; Gu, T.; Xu, M. Mechanically robust, shape memory, self-healing and 3D printable thermoreversible cross-linked polymer composites toward conductive and biomimetic skin devices applications. *Addit. Manuf.* **2022**, *49*, 102487. [CrossRef]
159. Xia, X.; Liu, W.; Zhou, L.; Liu, H.; He, S.; Zhu, C. Study on flax fiber toughened poly (lactic acid) composites. *J. Appl. Polym. Sci.* **2015**, *132*, 42573. [CrossRef]
160. Xia, X.; Liu, W.; Zhou, L.; Hua, Z.; Liu, H.; He, S. Modification of flax fiber surface and its compatibilization in polylactic acid/flax composites. *Iran. Polym. J.* **2016**, *25*, 25–35. [CrossRef]
161. Suchaiya, V.; Aht-Ong, D. Microwave-Assisted Modification of Cellulose as a Compatibilizer for Pla and Mcc Biocomposite Film: Effects of Side Chain Length and Content on Mechanical and Thermal Properties. *Polym. Polym. Compos.* **2014**, *22*, 613–624. [CrossRef]
162. Spiridon, I.; Darie, R.N.; Kangas, H. Influence of fiber modifications on PLA/fiber composites. Behavior to accelerated weathering. *Compos. Part B Eng.* **2016**, *92*, 19–27. [CrossRef]
163. Robles, E.; Urruzola, I.; Labidi, J.; Serrano, L. Surface-modified nano-cellulose as reinforcement in poly(lactic acid) to conform new composites. *Ind. Crops Prod.* **2015**, *71*, 44–53. [CrossRef]
164. Chen, J.; Yang, R.; Ou, J.; Tang, C.; Xiang, M.; Wu, D.; Tang, J.; Tam, K.C. Functionalized cellulose nanocrystals as the performance regulators of poly(β-hydroxybutyrate-co-valerate) biocomposites. *Carbohydr. Polym.* **2020**, *242*, 116399. [CrossRef]
165. Ghorbani Chaboki, M.; Mohammadi-Rovshandeh, J.; Hemmati, F. Poly(lactic acid)/thermoplasticized rice straw biocomposites: Effects of benzylated lignocellulosic filler and nanoclay. *Iran. Polym. J.* **2019**, *28*, 777–788. [CrossRef]

166. Sojoudiasli, H.; Heuzey, M.-C.; Carreau, P.J.; Riedl, B. Rheological behavior of suspensions of modified and unmodified cellulose nanocrystals in dimethyl sulfoxide. *Rheol. Acta* **2017**, *56*, 673–682. [CrossRef]
167. Fumagalli, M.; Berriot, J.; de Gaudemaris, B.; Veyland, A.; Putaux, J.-L.; Molina-Boisseau, S.; Heux, L. Rubber materials from elastomers and nanocellulose powders: Filler dispersion and mechanical reinforcement. *Soft Matter* **2018**, *14*, 2638–2648. [CrossRef] [PubMed]
168. Hermanson, G. *Bioconjugate Techniques*, 2nd ed.; Academic Press: Cambridge, MA, USA, 2008.
169. Ramamoorthy, S.K.; Bakare, F.; Herrmann, R.; Skrifvars, M. Performance of biocomposites from surface modified regenerated cellulose fibers and lactic acid thermoset bioresin. *Cellulose* **2015**, *22*, 2507–2528. [CrossRef]
170. Qu, P.; Zhou, Y.; Zhang, X.; Yao, S.; Zhang, L. Surface modification of cellulose nanofibrils for poly(lactic acid) composite application. *J. Appl. Polym. Sci.* **2012**, *125*, 3084–3091. [CrossRef]
171. Li, X.; Deng, L.; Li, Y.; Li, K. Preparation of Microcrystalline Cellulose from Bagasse Bleached Pulp Reinforced Polylactic Acid Composite Films. *Sugar Tech.* **2020**, *22*, 1138–1147. [CrossRef]
172. Luo, H.; Zhang, C.; Xiong, G.; Wan, Y. Effects of alkali and alkali/silane treatments of corn fibers on mechanical and thermal properties of its composites with polylactic acid. *Polym. Compos.* **2016**, *37*, 3499–3507. [CrossRef]
173. Ma, Y.; Qian, S.; Hu, L.; Qian, J.; Fontanillo Lopez, C.A.; Xu, L. Mechanical, thermal, and morphological properties of PLA biocomposites toughened with silylated bamboo cellulose nanowhiskers. *Polym. Compos.* **2019**, *40*, 3012–3019. [CrossRef]
174. Lee, J.S.; Ryu, Y.S.; Kim, I.; Kim, S.H. Effect of interface affinity on the performance of a composite of microcrystalline cellulose and polypropylene/polylactide blends. *Polym. Int.* **2019**, *68*, 1402–1410. [CrossRef]
175. Inukai, S.; Kurokawa, N.; Hotta, A. Mechanical properties of poly(ε-caprolactone) composites with electrospun cellulose nanofibers surface modified by 3-aminopropyltriethoxysilane. *J. Appl. Polym. Sci.* **2020**, *137*, 48599. [CrossRef]
176. Frone, A.N.; Berlioz, S.; Chailan, J.-F.; Panaitescu, D.M.; Donescu, D. Cellulose fiber-reinforced polylactic acid. *Polym. Compos.* **2011**, *32*, 976–985. [CrossRef]
177. Montes, S.; Azcune, I.; Cabañero, G.; Grande, H.-J.; Odriozola, I.; Labidi, J. Functionalization of Cellulose Nanocrystals in Choline Lactate Ionic Liquid. *Materials* **2016**, *9*, 499. [CrossRef] [PubMed]
178. Qian, S.; Sheng, K. PLA toughened by bamboo cellulose nanowhiskers: Role of silane compatibilization on the PLA bionanocomposite properties. *Compos. Sci. Technol.* **2017**, *148*, 59–69. [CrossRef]
179. Threepopnatkul, P.; Sittattrakul, A.; Supawititpattana, K.; Jittiarpon, P.; Raksawat, P.; Kulsetthanchalee, C. Effect of bacterial cellulose on properties of poly(lactic acid). *Mater. Today Proc.* **2017**, *4*, 6605–6614. [CrossRef]
180. Raquez, J.-M.; Murena, Y.; Goffin, A.-L.; Habibi, Y.; Ruelle, B.; DeBuyl, F.; Dubois, P. Surface-modification of cellulose nanowhiskers and their use as nanoreinforcers into polylactide: A sustainably-integrated approach. *Compos. Sci. Technol.* **2012**, *72*, 544–549. [CrossRef]
181. Zhang, X.; Di, J.; Li, J.; Li, S.; Duan, J.; Lv, J.; Zhu, X.; Xu, L.; Chang, X. Effects of different interfacial modifiers on the properties of digital printing waste paper fiber/nanocrystalline cellulose/poly(lactic acid) composites. *Polym. Eng. Sci.* **2022**, *62*, 781–792. [CrossRef]
182. Araújo, R.S.; Ferreira, L.C.; Rezende, C.C.; Marques, M.F.V.; Errico, M.E.; Avolio, R.; Avella, M.; Gentile, G.; Russo, P. Poly(lactic acid)/Cellulose Composites Obtained from Modified Cotton Fibers by Successive Acid Hydrolysis. *J. Polym. Environ.* **2018**, *26*, 3149–3158. [CrossRef]
183. Kabache, F.; Nekkaa, S.; Guessoum, M. Alkali and epoxy-silane surface modified pine cone flour reinforced polypropylene/poly (lactic acid) blend: Viscoelastic and morphological characterization. *J. Adhes. Sci. Technol.* **2021**, 1–24. [CrossRef]
184. Way, C.; Dean, K.; Wu, D.Y.; Palombo, E.A. Polylactic Acid Composites Utilising Sequential Surface Treatments of Lignocellulose: Chemistry, Morphology and Properties. *J. Polym. Environ.* **2011**, *19*, 849–862. [CrossRef]
185. Siqueira, G.; Bras, J.; Dufresne, A. Cellulose Whiskers versus Microfibrils: Influence of the Nature of the Nanoparticle and its Surface Functionalization on the Thermal and Mechanical Properties of Nanocomposites. *Biomacromolecules* **2009**, *10*, 425–432. [CrossRef]
186. Gwon, J.-G.; Cho, H.-J.; Chun, S.-J.; Lee, S.; Wu, Q.; Lee, S.-Y. Physiochemical, optical and mechanical properties of poly(lactic acid) nanocomposites filled with toluene diisocyanate grafted cellulose nanocrystals. *RSC Adv.* **2016**, *6*, 9438–9445. [CrossRef]
187. Gwon, J.-G.; Cho, H.-J.; Chun, S.-J.; Lee, S.; Wu, Q.; Li, M.-C.; Lee, S.-Y. Mechanical and thermal properties of toluene diisocyanate-modified cellulose nanocrystal nanocomposites using semi-crystalline poly(lactic acid) as a base matrix. *RSC Adv.* **2016**, *6*, 73879–73886. [CrossRef]
188. Ogunsona, E.O.; Panchal, P.; Mekonnen, T.H. Surface grafting of acrylonitrile butadiene rubber onto cellulose nanocrystals for nanocomposite applications. *Compos. Sci. Technol.* **2019**, *184*, 107884. [CrossRef]
189. Huang, L.; Zhao, H.; Xu, H.; An, S.; Li, C.; Huang, C.; Wang, S.; Liu, Y.; Chen, J. Study of 4,4′-Methylene Diisocyanate Phenyl Ester-Modified Cassava Residues/Polybutylene Succinate Biodegradable Composites: Preparation and Performance Research. *Processes* **2019**, *7*, 588. [CrossRef]
190. Virtanen, S.; Vuoti, S.; Heikkinen, H.; Lahtinen, P. High strength modified nanofibrillated cellulose-polyvinyl alcohol films. *Cellulose* **2014**, *21*, 3561–3571. [CrossRef]
191. Soman, S.; Chacko, A.S.; Prasad, V.S. Semi-interpenetrating network composites of poly(lactic acid) with cis-9-octadecenylamine modified cellulose-nanofibers from Areca catechu husk. *Compos. Sci. Technol.* **2017**, *141*, 65–73. [CrossRef]

192. Silva, F.; Gracia, N.; McDonagh, B.H.; Domingues, F.C.; Nerín, C.; Chinga-Carrasco, G. Antimicrobial activity of biocomposite films containing cellulose nanofibrils and ethyl lauroyl arginate. *J. Mater. Sci.* **2019**, *54*, 12159–12170. [CrossRef]
193. Eyholzer, C.; Tingaut, P.; Zimmermann, T.; Oksman, K. Dispersion and Reinforcing Potential of Carboxymethylated Nanofibrillated Cellulose Powders Modified with 1-Hexanol in Extruded Poly(Lactic Acid) (PLA) Composites. *J. Polym. Environ.* **2012**, *20*, 1052–1062. [CrossRef]
194. Yin, Y.; Zhao, L.; Jiang, X.; Wang, H.; Gao, W. Cellulose nanocrystals modified with a triazine derivative and their reinforcement of poly(lactic acid)-based bionanocomposites. *Cellulose* **2018**, *25*, 2965–2976. [CrossRef]
195. Hu, Z.; Berry, R.M.; Pelton, R.; Cranston, E.D. One-Pot Water-Based Hydrophobic Surface Modification of Cellulose Nanocrystals Using Plant Polyphenols. *ACS Sustain. Chem. Eng.* **2017**, *5*, 5018–5026. [CrossRef]
196. Sundar, S.; Sain, M.; Oksman, K. Thermal characterization and electrical properties of Fe-modified cellulose long fibers and micro crystalline cellulose. *J. Therm. Anal. Calorim.* **2011**, *104*, 841–847. [CrossRef]
197. Hassan, M.L.; Fadel, S.M.; Ward, A.A.; Moorefield, C.M.; Newkome, G.R. Electrical properties of Fe II -terpyridine-Modified cellulose nanocrystals and polycaprolactone/Fe II -CTP nanocomposites. *Polym. Compos.* **2016**, *37*, 2734–2743. [CrossRef]
198. Ummartyotin, S.; Thiangtham, S.; Manuspiya, H. Strontium-Modified Bacterial Cellulose and a Polyvinylidene Fluoride Composite as an Electroactive Material. *For. Prod. J.* **2017**, *67*, 288–296. [CrossRef]
199. Lönnberg, H.; Zhou, Q.; Brumer, H.; Teeri, T.T.; Malmström, E.; Hult, A. Grafting of cellulose fibers with poly(ε-caprolactone) and poly(L-lactic acid) via ring-opening polymerization. *Biomacromolecules* **2006**, *7*, 2178–2185. [CrossRef] [PubMed]
200. Bitinis, N.; Verdejo, R.; Bras, J.; Fortunati, E.; Kenny, J.M.; Torre, L.; López-Manchado, M.A. Poly(lactic acid)/natural rubber/cellulose nanocrystal bionanocomposites Part I. Processing and morphology. *Carbohydr. Polym.* **2013**, *96*, 611–620. [CrossRef] [PubMed]
201. Chai, H.; Chang, Y.; Zhang, Y.; Chen, Z.; Zhong, Y.; Zhang, L.; Sui, X.; Xu, H.; Mao, Z. The fabrication of polylactide/cellulose nanocomposites with enhanced crystallization and mechanical properties. *Int. J. Biol. Macromol.* **2020**, *155*, 1578–1588. [CrossRef]
202. Goffin, A.-L.; Raquez, J.-M.; Duquesne, E.; Siqueira, G.; Habibi, Y.; Dufresne, A.; Dubois, P. From Interfacial Ring-Opening Polymerization to Melt Processing of Cellulose Nanowhisker-Filled Polylactide-Based Nanocomposites. *Biomacromolecules* **2011**, *12*, 2456–2465. [CrossRef]
203. Gårdebjer, S.; Bergstrand, A.; Idström, A.; Börstell, C.; Naana, S.; Nordstierna, L.; Larsson, A. Solid-state NMR to quantify surface coverage and chain length of lactic acid modified cellulose nanocrystals, used as fillers in biodegradable composites. *Compos. Sci. Technol.* **2015**, *107*, 1–9. [CrossRef]
204. Gårdebjer, S.; Bergstrand, A.; Larsson, A. A mechanistic approach to explain the relation between increased dispersion of surface modified cellulose nanocrystals and final porosity in biodegradable films. *Eur. Polym. J.* **2014**, *57*, 160–168. [CrossRef]
205. Fang, H.-G.; Yang, K.-J.; Xie, Q.-Z.; Chen, X.; Wu, S.-L.; Ding, Y.-S. Influence of Interfacial Enantiomeric Grafting on Melt Rheology and Crystallization of Polylactide/Cellulose Nanocrystals Composites. *Chinese J. Polym. Sci.* **2022**, *40*, 93–106. [CrossRef]
206. Lafia-Araga, R.A.; Sabo, R.; Nabinejad, O.; Matuana, L.; Stark, N. Influence of Lactic Acid Surface Modification of Cellulose Nanofibrils on the Properties of Cellulose Nanofibril Films and Cellulose Nanofibril–Poly(lactic acid) Composites. *Biomolecules* **2021**, *11*, 1346. [CrossRef]
207. Jiang, Y.; Zhang, Y.; Cao, M.; Li, J.; Wu, M.; Zhang, H.; Zheng, S.; Liu, H.; Yang, M. Combining 'grafting to' and 'grafting from' to synthesize comb-like NCC-g-PLA as a macromolecular modifying agent of PLA. *Nanotechnology* **2021**, *32*, 385601. [CrossRef] [PubMed]
208. Stepanova, M.; Averianov, I.; Gofman, I.; Solomakha, O.; Nashchekina, Y.; Korzhikov-Vlakh, V.; Korzhikova-Vlakh, E. Poly(ε-caprolactone)-based biocomposites reinforced with nanocrystalline cellulose grafted with poly(L-lactic acid). *IOP Conf. Ser. Mater. Sci. Eng.* **2019**, *500*, 012021. [CrossRef]
209. Chen, J.; Wu, D.; Tam, K.C.; Pan, K.; Zheng, Z. Effect of surface modification of cellulose nanocrystal on nonisothermal crystallization of poly(β-hydroxybutyrate) composites. *Carbohydr. Polym.* **2017**, *157*, 1821–1829. [CrossRef] [PubMed]
210. Ichimura, H.; Kurokawa, N.; Hotta, A. Enhancement of the mechanical property of poly(ε-caprolactone) composites with surface-modified cellulose nanofibers fabricated via electrospinning. *MRS Adv.* **2019**, *4*, 385–391. [CrossRef]
211. Simão, J.A.; Bellani, C.F.; Branciforti, M.C. Thermal properties and crystallinity of PCL/PBSA/cellulose nanocrystals grafted with PCL chains. *J. Appl. Polym. Sci.* **2017**, *134*, 44493. [CrossRef]
212. Labet, M.; Thielemans, W. Citric acid as a benign alternative to metal catalysts for the production of cellulose-grafted-polycaprolactone copolymers. *Polym. Chem.* **2012**, *3*, 679. [CrossRef]
213. Peng, Q.; Cheng, J.; Lu, S.; Li, Y. Electrospun hyperbranched polylactic acid–modified cellulose nanocrystals/polylactic acid for shape memory membranes with high mechanical properties. *Polym. Adv. Technol.* **2020**, *31*, 15–24. [CrossRef]
214. Zhang, X.; Zhang, Y. Reinforcement effect of poly(butylene succinate) (PBS)-grafted cellulose nanocrystal on toughened PBS/polylactic acid blends. *Carbohydr. Polym.* **2016**, *140*, 374–382. [CrossRef]
215. Shanks, R.A.; Hodzic, A.; Ridderhof, D. Composites of poly(lactic acid) with flax fibers modified by interstitial polymerization. *J. Appl. Polym. Sci.* **2006**, *101*, 3620–3629. [CrossRef]
216. Boujemaoui, A.; Cobo Sanchez, C.; Engström, J.; Bruce, C.; Fogelström, L.; Carlmark, A.; Malmström, E. Polycaprolactone Nanocomposites Reinforced with Cellulose Nanocrystals Surface-Modified via Covalent Grafting or Physisorption: A Comparative Study. *ACS Appl. Mater. Interfaces* **2017**, *9*, 35305–35318. [CrossRef]

217. Shih, Y.-F.; Chou, M.-Y.; Chang, W.-C.; Lian, H.-Y.; Chen, C.-M. Completely biodegradable composites reinforced by the cellulose nanofibers of pineapple leaves modified by eco-friendly methods. *J. Polym. Res.* **2017**, *24*, 209. [CrossRef]
218. Popa, M.S.; Frone, A.N.; Radu, I.C.; Stanescu, P.O.; Truşcă, R.; Răditoiu, V.; Nicolae, C.A.; Gabor, A.R.; Panaitescu, D.M. Microfibrillated Cellulose Grafted with Metacrylic Acid as a Modifier in Poly(3-hydroxybutyrate). *Polymers* **2021**, *13*, 3970. [CrossRef] [PubMed]
219. Chuensangjun, C.; Kitaoka, T.; Chisti, Y.; Sirisansaneeyakul, S. Chemo-enzymatic preparation and characterization of cellulose nanofibers-graft-poly(lactic acid)s. *Eur. Polym. J.* **2019**, *114*, 308–318. [CrossRef]
220. Yoo, Y.; Youngblood, J.P. Green One-Pot Synthesis of Surface Hydrophobized Cellulose Nanocrystals in Aqueous Medium. *ACS Sustain. Chem. Eng.* **2016**, *4*, 3927–3938. [CrossRef]
221. Rosli, N.A.; Ahmad, I.; Anuar, F.H.; Abdullah, I. Application of polymethylmethacrylate-grafted cellulose as reinforcement for compatibilised polylactic acid/natural rubber blends. *Carbohydr. Polym.* **2019**, *213*, 50–58. [CrossRef]
222. Thérien-Aubin, H.; Wang, Y.; Nothdurft, K.; Prince, E.; Cho, S.; Kumacheva, E. Temperature-Responsive Nanofibrillar Hydrogels for Cell Encapsulation. *Biomacromolecules* **2016**, *17*, 3244–3251. [CrossRef]
223. Li, Y.; Khuu, N.; Gevorkian, A.; Sarjinsky, S.; Therien-Aubin, H.; Wang, Y.; Cho, S.; Kumacheva, E. Supramolecular Nanofibrillar Thermoreversible Hydrogel for Growth and Release of Cancer Spheroids. *Angew. Chemie Int. Ed.* **2017**, *56*, 6083–6087. [CrossRef]
224. Paquet, O.; Krouit, M.; Bras, J.; Thielemans, W.; Belgacem, M.N. Surface modification of cellulose by PCL grafts. *Acta Mater.* **2010**, *58*, 792–801. [CrossRef]
225. Averianov, I.V.; Stepanova, M.A.; Gofman, I.V.; Nikolaeva, A.l.; Korzhikov-Vlakh, V.A.; Karttunen, M.; Korzhikova-Vlakh, E.G. Chemical modification of nanocrystalline cellulose for enchanced interfacial compatibility with poly(lactic acid). *Mendeleev Commun.* **2019**, *29*, 220–222. [CrossRef]
226. Hong, S.-H.; Hwang, S.-H. Enhancing the mechanical performance of surface-modified microcrystalline cellulose reinforced high-density polyethylene composites. *Mater. Today Commun.* **2021**, *27*, 102426. [CrossRef]
227. Yin, Y.; Ma, J.; Tian, X.; Jiang, X.; Wang, H.; Gao, W. Cellulose nanocrystals functionalized with amino-silane and epoxy-poly(ethylene glycol) for reinforcement and flexibilization of poly(lactic acid): Material preparation and compatibility mechanism. *Cellulose* **2018**, *25*, 6447–6463. [CrossRef]
228. Krouit, M.; Bras, J.; Belgacem, M.N. Cellulose surface grafting with polycaprolactone by heterogeneous click-chemistry. *Eur. Polym. J.* **2008**, *44*, 4074–4081. [CrossRef]
229. Mincheva, R.; Jasmani, L.; Josse, T.; Paint, Y.; Raquez, J.-M.; Gerbaux, P.; Eyley, S.; Thielemans, W.; Dubois, P. Binary Mixed Homopolymer Brushes Tethered to Cellulose Nanocrystals: A Step Towards Compatibilized Polyester Blends. *Biomacromolecules* **2016**, *17*, 3048–3059. [CrossRef] [PubMed]
230. Stepanova, M.; Averianov, I.; Solomakha, O.; Zabolotnykh, N.; Gofman, I.; Serdobintsev, M.; Vinogradova, T.; Korzhikov-Vlakh, V.; Korzhikova-Vlakh, E. Composite biomaterials based on poly(L-lactic acid) and functionalized cellulose nanocrystals. *J. Renew. Mater.* **2020**, *8*, 383–395. [CrossRef]
231. Stepanova, M.; Averianov, I.; Serdobintsev, M.; Gofman, I.; Blum, N.; Semenova, N.; Nashchekina, Y.; Vinogradova, T.; Korzhikov-Vlakh, V.; Karttunen, M.; et al. PGlu-Modified Nanocrystalline Cellulose Improves Mechanical Properties, Biocompatibility, and Mineralization of Polyester-Based Composites. *Materials* **2019**, *12*, 3435. [CrossRef]
232. Lu, J.; Sun, C.; Yang, K.; Wang, K.; Jiang, Y.; Tusiime, R.; Yang, Y.; Fan, F.; Sun, Z.; Liu, Y.; et al. Properties of Polylactic Acid Reinforced by Hydroxyapatite Modified Nanocellulose. *Polymers* **2019**, *11*, 1009. [CrossRef]
233. Sridevi, S.; Sutha, S.; Kavitha, L.; Gopi, D. Valorization of biowaste derived nanophase yttrium substituted hydroxyapatite/citrate cellulose/ opuntia mucilage biocomposite: A template assisted synthesis for potential biomedical applications. *Mater. Chem. Phys.* **2021**, *273*, 125144. [CrossRef]
234. Fortunati, E.; Rinaldi, S.; Peltzer, M.; Bloise, N.; Visai, L.; Armentano, I.; Jiménez, A.; Latterini, L.; Kenny, J.M. Nano-biocomposite films with modified cellulose nanocrystals and synthesized silver nanoparticles. *Carbohydr. Polym.* **2014**, *101*, 1122–1133. [CrossRef]
235. Fortunati, E.; Armentano, I.; Zhou, Q.; Iannoni, A.; Saino, E.; Visai, L.; Berglund, L.A.; Kenny, J.M. Multifunctional bionanocomposite films of poly(lactic acid), cellulose nanocrystals and silver nanoparticles. *Carbohydr. Polym.* **2012**, *87*, 1596–1605. [CrossRef]
236. Reis, R.S.; Souza, D. de H.S.; Marques, M. de F.V.; da Luz, F.S.; Monteiro, S.N. Novel bionanocomposite of polycaprolactone reinforced with steam-exploded microfibrillated cellulose modified with ZnO. *J. Mater. Res. Technol.* **2021**, *13*, 1324–1335. [CrossRef]
237. Vilela, C.; Engström, J.; Valente, B.F.A.; Jawerth, M.; Carlmark, A.; Freire, C.S.R. Exploiting poly(ε-caprolactone) and cellulose nanofibrils modified with latex nanoparticles for the development of biodegradable nanocomposites. *Polym. Compos.* **2019**, *40*, 1342–1353. [CrossRef]
238. Engström, J.; Hatton, F.L.; Wågberg, L.; D'Agosto, F.; Lansalot, M.; Malmström, E.; Carlmark, A. Soft and rigid core latex nanoparticles prepared by RAFT-mediated surfactant-free emulsion polymerization for cellulose modification—A comparative study. *Polym. Chem.* **2017**, *8*, 1061–1073. [CrossRef]
239. Liu, R.; Liu, M.; Hu, S.; Huang, A.; Ma, E. Comparison of six WPCs made of organo-montmorillonite-modified fibers of four trees, moso bamboo and wheat straw and poly(lactic acid) (PLA). *Holzforschung* **2018**, *72*, 735–744. [CrossRef]

240. Gao, F.; Zeng, D.; Liu, H.; Qin, R.; Zhang, J.; Chen, Y.; Wang, W.; Peng, C.; Li, M.; Li, Q.; et al. Porous cellulose microspheres coated in one step with a polydopamine suspension of hydroxyapatite for bone tissue engineering. *Cellulose* **2022**, *29*, 1955–1967. [CrossRef]
241. Li, K.; Wang, J.; Liu, X.; Xiong, X.; Liu, H. Biomimetic growth of hydroxyapatite on phosphorylated electrospun cellulose nanofibers. *Carbohydr. Polym.* **2012**, *90*, 1573–1581. [CrossRef]
242. Szymańska-Chargot, M.; Chylińska, M.; Pieczywek, P.M.; Walkiewicz, A.; Pertile, G.; Frąc, M.; Cieślak, K.J.; Zdunek, A. Evaluation of Nanocomposite Made of Polylactic Acid and Nanocellulose from Carrot Pomace Modified with Silver Nanoparticles. *Polymers* **2020**, *12*, 812. [CrossRef]
243. Carbone, M.; Donia, D.T.; Sabbatella, G.; Antiochia, R. Silver nanoparticles in polymeric matrices for fresh food packaging. *J. King Saud Univ. Sci.* **2016**, *28*, 273–279. [CrossRef]
244. Hasan, A.; Waibhaw, G.; Saxena, V.; Pandey, L.M. Nano-biocomposite scaffolds of chitosan, carboxymethyl cellulose and silver nanoparticle modified cellulose nanowhiskers for bone tissue engineering applications. *Int. J. Biol. Macromol.* **2018**, *111*, 923–934. [CrossRef]
245. Lertprapaporn, T.; Manuspiya, H.; Laobuthee, A. Dielectric improvement from novel polymeric hybrid films derived by polylactic acid/nanosilver coated microcrystalline cellulose. *Mater. Today Proc.* **2018**, *5*, 9326–9335. [CrossRef]
246. Nakagaito, A.N.; Fujimura, A.; Sakai, T.; Hama, Y.; Yano, H. Production of microfibrillated cellulose (MFC)-reinforced polylactic acid (PLA) nanocomposites from sheets obtained by a papermaking-like process. *Compos. Sci. Technol.* **2009**, *69*, 1293–1297. [CrossRef]
247. JOHN, M.; THOMAS, S. Biofibres and biocomposites. *Carbohydr. Polym.* **2008**, *71*, 343–364. [CrossRef]
248. Hongwei Ma; Chang Whan Joo Structure and mechanical properties of jute—Polylactic acid biodegradable composites. *J. Compos. Mater.* **2011**, *45*, 1451–1460. [CrossRef]
249. Codreanu, A.; Balta, C.; Herman, H.; Cotoraci, C.; Mihali, C.V.; Zurbau, N.; Zaharia, C.; Rapa, M.; Stanescu, P.; Radu, I.-C.; et al. Bacterial Cellulose-Modified Polyhydroxyalkanoates Scaffolds Promotes Bone Formation in Critical Size Calvarial Defects in Mice. *Materials* **2020**, *13*, 1433. [CrossRef] [PubMed]
250. Braun, B.; Dorgan, J.R.; Knauss, D.M. Reactively Compatibilized Cellulosic Polylactide Microcomposites. *J. Polym. Environ.* **2006**, *14*, 49–58. [CrossRef]
251. Fortunati, E.; Puglia, D.; Kenny, J.M.; Minhaz-Ul Haque, M.; Pracella, M. Effect of ethylene-co-vinyl acetate-glycidylmethacrylate and cellulose microfibers on the thermal, rheological and biodegradation properties of poly(lactic acid) based systems. *Polym. Degrad. Stab.* **2013**, *98*, 2742–2751. [CrossRef]
252. Liu, H.; He, H.; Peng, X.; Huang, B.; Li, J. Three-dimensional printing of poly(lactic acid) bio-based composites with sugarcane bagasse fiber: Effect of printing orientation on tensile performance. *Polym. Adv. Technol.* **2019**, *30*, 910–922. [CrossRef]
253. Wei, L.; Agarwal, U.P.; Hirth, K.C.; Matuana, L.M.; Sabo, R.C.; Stark, N.M. Chemical modification of nanocellulose with canola oil fatty acid methyl ester. *Carbohydr. Polym.* **2017**, *169*, 108–116. [CrossRef]
254. Chen, P.-Y.; Lian, H.-Y.; Shih, Y.-F.; Chen-Wei, S.-M. Chemically Functionalized Plant Fibers and Carbon Nanotubes for High Compatibility and Reinforcement in Polylactic Acid (PLA) Composite. *J. Polym. Environ.* **2018**, *26*, 1962–1968. [CrossRef]
255. Hong, H.; Xiao, R.; Guo, Q.; Liu, H.; Zhang, H. Quantitively Characterizing the Chemical Composition of Tailored Bagasse Fiber and Its Effect on the Thermal and Mechanical Properties of Polylactic Acid-Based Composites. *Polymers* **2019**, *11*, 1567. [CrossRef]
256. Volokitina, M.V.; Korzhikov-Vlakh, V.A.; Tennikova, T.B.; Korzhikova-Vlakh, E.G. Macroporous monoliths for biodegradation study of polymer particles considered as drug delivery systems. *J. Pharm. Biomed. Anal.* **2017**, *145*, 169–177. [CrossRef]
257. Azevedo, H.S.; Reis, R.L. Understanding the Enzymatic Degradation of Biodegradable Polymers and Strategies to Control Their Degradation Rate. In *Biodegradable Systems in Tissue Engineering and Regenerative Medicine*; Reis, R.L., Roman, J.S., Eds.; CRC Press: Boca Raton, FL, USA, 2004; pp. 177–201.
258. Ambrosio, G.; Faglia, G.; Tagliabue, S.; Baratto, C. Study of the Degradation of Biobased Plastic after Stress Tests in Water. *Coatings* **2021**, *11*, 1330. [CrossRef]
259. Luzi, F.; Fortunati, E.; Puglia, D.; Petrucci, R.; Kenny, J.M.; Torre, L. Study of disintegrability in compost and enzymatic degradation of PLA and PLA nanocomposites reinforced with cellulose nanocrystals extracted from Posidonia Oceanica. *Polym. Degrad. Stab.* **2015**, *121*, 105–115. [CrossRef]
260. Gois, G.; Da, S.; de Andrade, M.F.; Garcia, S.M.S.; Vinhas, G.M.; Santos, A.S.F.; Medeiros, E.S.; Oliveira, J.E.; Almeida, Y.M.B. de Soil Biodegradation of PLA/CNW Nanocomposites Modified with Ethylene Oxide Derivatives. *Mater. Res.* **2018**, *20*, 899–904. [CrossRef]
261. Abdul Khalil, H.P.S.; Bhat, A.H.; Ireana Yusra, A.F. Green composites from sustainable cellulose nanofibrils: A review. *Carbohydr. Polym.* **2012**, *87*, 963–979. [CrossRef]
262. Yang, W.; Zhao, X.; Fortunati, E.; Dominici, F.; Kenny, J.M.; Puglia, D.; Wang, D.-Y. Effect of Cellulose Nanocrystals on Fire, Thermal and Mechanical Behavior of N,N′-Diallyl-phenylphosphoricdiamide Modified Poly(lactic acid). *J. Renew. Mater.* **2017**, *5*, 423–434. [CrossRef]
263. Lin, N.; Dufresne, A. Nanocellulose in biomedicine: Current status and future prospect. *Eur. Polym. J.* **2014**, *59*, 302–325. [CrossRef]

264. Mary Stella, S.; Vijayalakshmi, U. Influence of chemically modified Luffa on the preparation of nanofiber and its biological evaluation for biomedical applications. *J. Biomed. Mater. Res. Part A* **2019**, *107*, 610–620. [CrossRef] [PubMed]
265. Rabbi, M.A.; Rahman, M.M.; Minami, H.; Rahman, M.A.; Hoque, S.M.; Ahmad, H. Biocomposites of synthetic polymer modified microcrystalline jute cellulose particles and their hemolytic behavior. *Cellulose* **2019**, *26*, 8713–8727. [CrossRef]

polymers MDPI

Review

Recent Progress of Rice Husk Reinforced Polymer Composites: A Review

Mohamed Azlan Suhot *, Mohamad Zaki Hassan, Sa'ardin Abdul Aziz and Mohd Yusof Md Daud

Razak Faculty of Technology and Informatics, Universiti Teknologi Malaysia, Jalan Sultan Yahya Petra, Kuala Lumpur 54100, Malaysia; mzaki.kl@utm.my (M.Z.H.); saa.kl@utm.my (S.A.A.); yusof.kl@utm.my (M.Y.M.D.)
* Correspondence: azlans.kl@utm.my

Abstract: Recently, because of the rising population, carbon overloading, and environmental distress, human beings have needed to increase awareness and responsibility for the reduction of agricultural waste. The utilization of agricultural waste as a filler material in reinforced polymers is a fascinating discovery. This review paper attempts to study the physical, mechanical, and thermal behavior of rice husk (RH) as a fiber for reinforcing various synthetic polymers, based on recent studies, conducted between 2017 and 2021. It also highlights that advanced modification techniques could further improve the performance of composites by tailoring the physical and chemical substances of the fiber or matrix. The thermal properties, including flame-retardance and thermal behavior, are also discussed. The characteristics of the fiber–matrix interaction between RH and the polymer matrix provide essential insights into the future-ready applications of this agricultural waste fiber. The way forward in researching RH polymer composites is finally reviewed.

Keywords: natural fiber; rice husk; biocomposites; physicomechanical; thermal behavior

Citation: Suhot, M.A.; Hassan, M.Z.; Aziz, S.A.; Md Daud, M.Y. Recent Progress of Rice Husk Reinforced Polymer Composites: A Review. *Polymers* **2021**, *13*, 2391. https://doi.org/10.3390/polym13152391

Academic Editor: Evgenia G. Korzhikova-Vlakh

Received: 20 June 2021
Accepted: 14 July 2021
Published: 21 July 2021

1. Introduction

Disposing of waste is problematic, and the most significant challenge today is to find novel ways to utilize these residues. Between 2003 and 2013, Cherubin et al. [1] reported that residues such as rejected crops in the form of leaf litter, straws, sawdust, forest waste, leaves, weeds, and other by-products surged by approximately 33%, as a percentage of the total product. Moreover, agriculture waste reached more than 5 billion Mg in 2013, whereby 47% of the leftover residues were from the Asian continent, followed by America (29%), Europe (16%), Africa (6%), and Oceania (2%) [1]. Landfilling is the primary option to treat this waste; however, issues related to air quality, global availability of land, and greenhouse gas releases, such as methane and leachate, have come to the fore [2–4]. Other options are to develop new processing techniques for a higher utilization rate of residues [5].

Malaysia is surrounded by the Straits of Malacca and the South China Sea, and enjoys typical tropical weather, with proximity to water, which gives this country a quite humid, hot [6], and rainy climate throughout the year, with a temperature range from a mild 20 to 30 °C [7]. It is located near the equator and blessed with natural resources, such as crop biomass [8–10], hydro, and solar power. In the 1990s, roughly 4.2 million tonnes of vegetable residue and 2.3 million tonnes of livestock waste were produced in Peninsular Malaysia [11]. Table 1 shows the production of agricultural waste in Malaysia in 2007, which was generated from the production of palm oil, rice, rubber, coconut, sugar cane waste, forest products, and municipal waste [12]. Malaysia is an ethnically heterogeneous country and has experienced a radical growth in population. Currently, it is focused on a self-sufficiency policy in rice and paddy production, and this is the country's primary staple food and food crop. Increasing demand for rice production has significantly increased the waste from rice husk and straw.

Table 1. Crop waste production in Malaysia in 2007 [12].

Types	Quantity (kt)	Source	Source (kt)
Agricultural waste			
Oil palm fronds	46,837	Oil palm FFB	81.92
EPFB	18,022		
Oil palm fibers	11,059		
Oil palm shells	4506		
Oil palm trunks	10,827		
Paddy straw	880	Replanting paddy	2375
RH	484		
Banana residues	265	Banana	530
Sugarcane bagasse	234	Sugarcane	730
Coconut husk	171	Coconut	505
Pineapple waste	48	Pineapples for factories	69
Forest residues			
Logging residues	2649	Logs	2649
Plywood residues	2492	Plywood	2492
Sawmill residues	1.16	Sawn timber	1418
Municipal solid waste			
Organic waste	4653	MSW	6744

Rice (*Oryza sativa* L. genus) is the primary source of daily food intake and has become the world's second most important cereal crop sector due to the demand of billions of human beings. In 2019, approximately 756 million metric tons of rice were produced globally, and 90% of the total output came from Asia [13]. In Malaysia, about 700,000 hectares of paddy are planted on the extensive agricultural land, yielding more than 800,000 tonnes of rice husk (RH) and stalk waste annually [14]. These wastes should never be burned, due to various reasons, such as the ashes, harmful gases, and fumes that contribute to air pollution [15].

Typically, the RH can be used as biochar, extracted silica, or husk itself. In general, RH is a hull to protect seeds or grains. It is formed from rigid materials, is water-insoluble, and is abrasive, with a high level of cellulose–silica structures. The exterior of the hulls consists of silica covered with a cuticle, with a small amount of silica content at the innermost epidermis.

Recently, several attempts have been made to utilize these waste materials in composite structures. The study of RH as a filler has been of interest to researchers since the 1970s. This paper presents a compressive review of the physical, mechanical, and thermal durability of RH composites between 2017 and 2021, and it details the knowledge gaps that need to be filled in the respective research areas. Furthermore, it discusses the potential of RH composites to be used in photonics, construction materials, and automotive and furniture applications, based on their strength and thermal characteristics.

2. Tensile Strength of RH Composites

The exploitation of RH residues in biocomposites offers multiple advantages, for example, reducing the relative amount of constituents derived from synthetic polymers, such as resin polymers and some additives. The tensile strength is mainly used to evaluate the strength behavior of a composite material. The behavior of composites is dependent upon the filler type, matrix material, concentration, size, dispersion, and the adhesion between the filler and the matrix material. Various studies have been conducted on the variation of tensile properties of RH-reinforced composites at different filler loadings using different types of matrix materials as polymer matrices, as tabulated in Table 2. However, they can be classified into four groups: matrix modification, filler treatment/modification, and hybridization.

Table 2. Reported studies of the mechanical properties of RH reinforced polymers.

Matrix	Parametric Study	Tensile Strength (MPa)	Flexural (MPa)	Impact Strength (kJ/m^2)	References
HDPE	Matrix modification	22.5 ± 0.5	49.6 ± 1.2		Abdulkarem et al. [16]
PP	Filler loading	19.7	39.2		Zafar et al. [17]
HDPE	Filler modification	20	53.7	13	Zhang et al. [18]
HDPE	Hybrid	15.8	25.7	15.2	Zhang et al. [19]
Natural rubber	Filler modification	21.3 ± 0.7			Xue et al. [20]
Epoxy	Filler modification	120			Fernandes et al. [21]
Epoxidized natural rubber	Filler modification	18.5 ± 0.5			Pongdong et al. [22]
HDPE/PET	Matrix modification	22.2 ± 0.1	48 ± 2	3 ± 0.1	Chen et al. [23]
Epoxidized natural rubber	Matrix modification	35	45	22	Raghu et al. [24]
rHDPE	Filler treatment	18.37			Rajendran et al. [25]
Epoxy	Filler treatment	46 ± 1	87 ± 2	2.7 ± 0.1	Bisht et al. [26]
rABS/PP	Matrix modification	21 ± 1			Santiago et al. [27]
HDPE	Filler treatment	26.3 ± 0.50			Zhang et al. [28]
rPP	Filler treatment	28 ± 0.25		3.0 ± 0.5	Moreno et al. [29]
TPS	Filler treatment	2.43 ± 0.25			Boonsuk et al. [30]
PU	Filler treatment	0.25 ± 0.11			Olcay et al. [31]
Epoxy	Hybrid	43			Shubbar [32]
PP	Hybrid	40 ± 2			Awang et al. [33]
Epoxy	Hybrid	30 ± 2	25 ± 2		Kumar et al. [34]
PP	Filler loading	33.2 ± 0.5	39.8 ± 0.3		Hidalgo-Salazar et al. [35]
Corn starch	Hybrid	10.7	19.6		Singh et al. [36]
HDPE	Matrix modification		30 ± 2		Sun et al. [37]
PLA	Matrix modification		5254 ± 25		Běhálek et al. [38]
PP	Hybrid	15.6 ± 0.25	37.6 ± 1.88		Guna et al. [39]
PVC	Matrix modification	51.9 ± 2.54		74.9 ± 5.81	Petchwattana et al. [40]
PVC	Hybrid			5.5 ± 0.80	Jiang et al. [41]
Cassava starch	Filler modification	3.3 ± 0.5			Kargarzadeh et al. [42]
Corn starch	Matrix modification	14.3 ± 1.13			Yap et al. [43]
PLA and PBAT	Matrix modification		10.0 ± 1.0		Spada et al. [44]

Using RH as a reinforcement has offered significant enhancements to the tensile properties of composites, as reported by Abdulkareem et al. [16]. They clarified that the Young's modulus of RH/waste polystyrene (PS) composite increased with increasing the RH content; up to 40 wt % compared to pure PS. A similar improvement in strength was also discovered by Zafar et al. [17] when studying an RH reinforced polypropylene (PP) matrix composite. The maximum tensile strength was achieved at 5 wt % of RH loading, with the size of the RH filler being 355–500 micron. In contrast, this contradicted the findings mentioned by Zhang et al. [18], where increasing the RH filler to 70 wt % decreased the tensile properties due to fiber agglomeration in the matrix. Zhang et al. [19] also studied the tensile strength of RH in a high-density polyethene (HDPE) matrix for different RH loadings, and the best level of tensile strength was attained at 40 wt % loadings. This was the result of the uniform distribution of RH in the matrix, making the matrix tightly wrap the RH, and thus improving the interface bonding. In another study, a unique type of RH, called hydrochar, was reinforced with polylactic acid (PLA) as the matrix, and it was observed that the tensile modulus improved from 2.63 GPa in virgin PLA to 4.24 GPa after blending with hydrochar. Xue et al. [20] used the ball milling technique to enhance the filler–matrix interaction by refining the particle size. They found that the tensile strength increased 44% compared to unmilled RH.

While studying the performance between unfilled and filled epoxidized natural rubber (ENR) with RH ash, it was shown that the filled ENR provided a higher tensile strength than the unfilled ENR. A study of the differences in mechanical properties between RH ash filler and high-purity silica in an epoxy matrix composite by Fernandes et al. [21] found that similar characteristics were observed. They claimed that RH ash could replace silica with little loss of desirable properties. Pongdong et al. [22] indicated a similar conclusion, whereby they found that RH ash filler exhibited a similar reinforcement compared to conventional siliceous earth for epoxidized natural rubber matrix composites.

In order to reduce the level of alkalinity of the pore water in a synthetic polymer, matrix modification has been promoted. This method can typically enhance the durability of the fiber–matrix interaction by using cementitious materials. The tensile strength of the RH reinforced hybrid recycled HDPE/polyethylene terephthalate (PET) composites was optimum at 70 wt % of filler loading, as reported by Chen et al. [23]. In a similar study, Raghu et al. [24] used maleic anhydride grafted polypropylene (MAPP) and m-isopropenyl α–α-dimethylbenzyl-isocyanate grafted polypropylene (m-TMI-g-PP) as coupling agents. They found that the tensile properties of the RH/PP composites were better than the control samples. They observed that at 50 wt % RH loading, the tensile strength increased by 52% as compared with another type of filler, which were encouraging results.

Several researchers evaluated the improved tensile strength properties between RH and the matrix resin using surface modification techniques such as esterification, silane treatment, fiber mercerization, or fiber surface modification. For example, Rajendran et al. [25] treated RH with ultraviolet-ozonolysis and found that the treated RH composites improved the tensile strength by 5% compared to the composite with untreated RH. Bisht et al. [26] used a mercerization treatment on RH flour and studied the effect on the tensile strength of RH/epoxy composites. The tensile strength of the composite improved by 36% with treated sodium hydroxide (NaOH) solution, by up to 8%. Santiago et al. [27] compared the tensile strength of RH powder in a recycled acrylonitrile butadiene rubber/PP hybrid matrix between a silane treatment and anhydride (AC) treatment of the fillers. Again, the AC treatment exhibited better tensile strength compared to the silane treatment.

Zhang et al. [28] analyzed an extracted RH biochar reinforced HDPE composite at different pyrolysis temperatures using injection moulding. The best tensile properties of the composites were obtained in the temperature range of 500–600 °C, due to their outstanding physical interlocking structures. A similar pyrolysis of RH work was conducted by Moreno et al. [29]. It was shown that the increased RH content in the PP matrix led to a proportional decrease in the tensile strength. However, the decrease in tensile strength was less significant for the pyrolysis composites, as verified by the fracture surface.

In addition, Boonsuk et al. [30] mentioned that an alkaline treatment with 11% *w/v* of NaOH removed the hemicellulose layer of RH and offered an outstanding tensile strength improvement, by a factor of 220%, compared to the neat thermoplastic starch. It improved the matrix-filler load transfer capabilities due to the loss of hemicellulose and the rougher outer surfaces after alkaline treatment. By contrast, the combination between untreated and 5% RH loading in a flexible polyurethane (PU) was found to have the best tensile performance of the composites [31]. The treated RH with 10% *w/v* NaOH adversely affected the surface of the filler and decreased the tensile behavior.

Some researchers fabricated hybrid RH composites by combining two or more different types of fillers within a common matrix. For example, Shubbar [32] evaluated the tensile properties of RH combined with fumed silica nanopowder in an epoxy matrix. The tensile properties increased by 50%, just by adding 5 wt % RH, compared to the sample with pure resin. Furthermore, Awang et al. [33] evaluated RH combined with titanium oxide (TiO_2) and zirconium oxide (ZnO) in the PP matrix and proposed that the addition of TiO_2 gave a higher tensile strength and Young's modulus compared to the addition of ZnO. From the scanning electron microscope (SEM) images, they concluded that this higher tensile strength was due to a better interaction between the matrix and the RH particles. Additionally, Kumar et al. [34] assessed a combination of RH/bauhinia-vahilii-weight/sisal filler with epoxy as the matrix and concluded that the addition of RH improved the tensile strength by 34.42% compared to not using the RH filler loading.

The application of RH as a filler in polymer matrix composites increased the tensile strength in all the research that was reviewed in this paper. RH could replace silica and other fillers; however, some of the research showed that the tensile strength increment had a maximum point after a certain amount of RH loading. Some researchers conducted additional studies on the improvement of the interface properties of the filler–matrix, either by surface modification of RH or to the matrix formulation. It is believed that this is the way forward for increasing the usage of RH in polymer matrix composites.

3. Flexural Strength of RH Composites

In order to characterise the bending properties of the composite material, the most classical test used to characterize this behaviour is the flexural test (three or four points). A study by Zhang et al. [18] reported that the bending strength of a RH biochar/HDPE composite reached 53.7 MPa, which was far beyond wood–plastic composites. It was indicated that the biochar behaved as a rigid grain and locked the movement of a particle in the polymer chains. Hidalgo-Salazar et al. [35] analyzed a RH-reinforced PP composite and recorded an increase of 75% in flexural strength for the RH/PP composite compared with neat PP. They attributed the increase in bending properties to the stiffening effect of RH in the PP matrix. Singh et al. [36] also measured the flexural strength of a fully recycled RH-reinforced corn starch matrix composite and mentioned that the maximum flexural strength was 19.60 MPa for a RH/corn starch composite with 15 wt % RH content.

Flexural modulus is a material characteristic that is significantly influenced by the morphology and crystallinity of polymers. In particular, the heterogeneous structure of the surface layers is important for high values of flexural modulus. Using a compatibilizer, Chen et al. [23] used an ethylene-glycidyl methacrylate (E-GMA) copolymer as a compatibilizer between recycled HDPE and recycled PET, and maleic anhydride polyethene (MAPE) as a coupling agent between the filler and matrix. They reported an increase in flexural strength of 62% with the increase of RH concentration in the polymer blends of recycled HDPE and recycled PET. It was discovered that the use of a compatibilizer increased the strength of the RH composite with the matrix blend. The coupling agent also improved the flexural strength of the RH/PP composites, and an increase of 46% was reported by Raghu et al. [24]. Moreover, when comparing the effect of silane coupling and compatibilizer MAPE on interfacial adhesion properties in RH/HDPE composites, Sun et al. [37] found that the bending strength and flexural strength were improved by 11.5% and 40.7%, respectively. It was observed that the flexural modulus increased with the increase in RH and the technical cellulose fiber amount. It was obvious that the flexural modulus reached higher values at higher quantities of cellulose fibers (20–30 mass%). Furthermore, there was no positive effect on the flexural modulus with a variety of plasma surface treatments of technical cellulose fibers or grafted maleic anhydride (PLA-g-MAH/PLA/30CeF). The smallest effect on the flexural modulus was noted for ozone-treated fillers [38].

Kumar et al. [34] reported an increase of 33% in the flexural strength for RH/bauhinia-vahilii-weight/sisal epoxy composites compared to unfilled composites at all filler loadings. The effects of hybridized RH with groundnut shell (GNS) reinforced with PP were obtained by Guna et al. [39]. The maximum flexural strength of the hybrid composites was obtained with a 20/60/20 GNS/RH/PP ratio, which was 40% higher than the non-hybrid composites. This could suggest that a higher loading of small fillers was inclined to extensive delamination, and the misalignment of the filler in the matrix thus decreased the strength properties.

4. Impact Strength of RH Composites

Singh [36] reported that the impact energy of RH/corn starch composites increased with the increase of the amount of RH content. The impact strength reached 0.362 J for composites with 15 wt % RH content.

The mercerization of fibers improved the impact strength, and Bisht et al. [26] reported that the impact strength of RH flour–epoxy composites were highest at 8% NaOH concentration. The reason for the increase of the impact strength was due to the mercerisation treatment, which improved the adhesion between the matrix and fiber by way of removing the voids on the surface of the untreated RHs. Surface modification by silane treatment of a PVC matrix in RH–PVC composites also increased the impact strength to 44%, as reported by Petchwattan et al. [40].

The use of coupling agents, as studied by Raghu [24], showed that the impact strength of RH–PP composites decreased with increasing filler loadings. Jiang et al. [41] explored the possibility of reinforcing RH–PVC composites with basalt fibers (BF) and found a noticeably increase in impact strength, whereby the BF acted as a reinforcing agent and strengthened the mobility of the matrix chains. Additionally, the aspect ratio of BF was higher than RH, thus the shift of the stress from the matrix to the fiber was more effective.

5. Water Diffusion Behavior of RH Composites

The water diffusion behavior of fiber-reinforced composites is dependent on the relative mobility of penetrants between the water molecules and polymer parts. In general, this obeys Fick's diffusion theory, and three classes of diffusion can be determined [45,46]. The measurement of the kinetic diffusion mechanism was evaluated based on Fick's theory and the fitting of experimental values, as follows:

$$log\,log\left(\frac{M_t}{M_\infty}\right) = log\,log\,k + n\,log\,log\,t \tag{1}$$

where M_t and M_∞ are the water absorption at time t and the saturation point, respectively. k and n are constants.

The diffusion mechanism is reflected in the value of n. When the rate of diffusion of the infiltrate is less than the polymer part, Case I of the Fickian diffusion mechanism is obtained. For this case, the value of $n = 0.5$, where the saturated condition corresponding to a time is rapidly gained and conserved inside the composite [47]. However, when $n = 1.0$, this indicates that the diffusion activity is faster than the relaxation process [48]. The mechanism is distinguished by the progressive barrier between the bulging outer part and the inner glassy part of the synthetic polymer. In Case II, an equilibrium penetration diffusion is reached at a constant velocity. The non-Fickian is justified at a $0.5 < n < 1.0$ diffusion mechanism and does not obey the Fickian laws. At this condition, Melo et al. [49] used a Langmuir-type model to closely interpret the physical phenomenon of water absorption relaxation of natural fibre composites. In some cases, when n is larger than 1, it is known as Super Case II kinetics [50]; however, when $n < 0.5$, this can be classified as 'Less Fickian' behaviour.

Table 3 summarises the water absorption kinetics of an RH-reinforced synthetic polymer. Chen and Ahmad [51] reported that the water absorption and swelling showed a linear increase with the increase of RH content. The higher water absorption and swelling with higher RH fiber content were due to the hydrophilicity of RH. This finding agreed with the finding of Abdulkareem et al. [16], where it was observed that the percentage of water absorbed increased with the addition of RH. Abdulkareem et al. [16] attributed the increase in water absorption to the pores and gaps in the RH structure. A different mechanism was observed in the epoxy matrix by Shubbar [32], whereby it was reported that due to the swelling of the composite as a result of water absorption, the epoxy matrix cracked, which in turn generated a capillary effect and caused further water absorption.

Table 3. Reported studies of moisture absorption kinetics of RH-reinforced polymers.

Matrix	Parametric Study	M_∞(%)	Thickness Swelling (%)	Diffusion Coefficients ($D \times 10^{-5}$ mm^2/s)	References
rHDPE	Hybrid	1.8–4.0	4.8–6.8		Chen and Ahmad [51]
HDPE	Filler content	12.0–13.0	8.5–10.0		Sheykh et al. [52]
PE	Filler treatment	3.0–7.0			Nabinejad et al. 2017
Cassava starch	Hybrid	0.5–2.7			Huner [53]
PVC	Coupling agent	0.4–2.4			Saidi et al. [54]
PE	Hybrid	2.5–13.0			Mohamed et al. [55]
Corn starch	Filler content	5.1–11.9			Battegazzore et al. [56]
PVC	Coupling agent	4.2–6.3			Petchwattana et al. [40]
Epoxy	Hybrid	1.2–2.4			Shubbar [32]
rHDPE	Filler treatment	1.7–4.0			Rajendran et al. [25]
rPE	Filler content			2.8–1.6	Abdulkareem et al. [16]
rHDPE/rPET	Hybrid	3.0–9.5	4.0–8.8		Chen et al. [23]
Epoxy	Filler content	0.06–0.17			Fernandes et al. [21]
PLA	Filler treatment	2.5–3.5			Prappuddivongs et al. [57]
PLA/PLB	Blending effect	0.8–5.2			Akindoyo et al. [58]
HDPE	Filler content	0.12–0.28			Daramola [59]
Epoxy	Filler content	0.08–0.13			Hamid et al. [60]
Soy bean	Coating	4.0–11.0		3.2–15.9	Chalapud et al. [61]
ABS	Hybrid	6.9–9.1		1.1 1.4	Norhasnan et al. [62]
rHDPE	Filler treatment	3.7–26.6	0.63		Chen et al. [63]

The RH was found to be better in terms of its water absorption properties when it was compared with other types of fillers. Muthuraj et al. [64] found that composites containing RH showed lower water absorption compared to other types of fillers, such as wheat husk, wood fibers, and textile waste. This observation was explained by the higher hydrophobicity of RH compared to other fillers. Yusuf et al. [65] compared composites containing RH with composites containing bamboo stem fiber. They found that composites with RH were better in terms of their lower water absorption and swelling thickness due to the lower affinity of RH to water. Sheykh et al. [52] compared RH and bagasse ash in an HDPE composite. The RH–HDPE composite was found to have lower water absorption and thickness swelling properties. This was due to the lower accessible -OH group on the surface of RH compared to bagasse fibers. Mohamed et al. [55] compared the water absorption properties of different contents of hybrid kenaf–RH in a polypropylene composite. Similarly, other researchers found that a higher RH content exhibited lower water absorption properties. This is because kenaf has larger voids and has more hydroxyl groups that can interact with water.

Antunes et al. [66] studied the ability of RH panels (to be used as wall panels) to absorb and desorb moisture using the moisture buffer value test. They found that the higher RH content panels had a better ability for absorbing and desorbing moisture compared to panels with a lower content of RH, which makes them excellent for high humidity applications.

Akindoyo et al. [58] presented that all composite structures massively absorbed more water than neat PLA due to a natural fiber composite, which contained a higher abundance hydroxyl groups and easily interacted with water molecules. The increase in water absorption was higher in the reinforced blends, which could be credited to the water uptake properties of natural fibers. In general, all the composites conformed with Fickian's law, where there was an initial rapid water uptake before reaching a saturation plateau region, with further increases in the soaking period. The effect of nano-silica particles extracted from RH on the water absorption characteristics was evaluated by Daramola [59]. An enhancement of the moisture absorption resistance of a nano silica-reinforced HDPE composite was observed at a lower particle weight fraction. However, increasing the filler loading resulted in an increase in the void content, interfacial bonding, and exposure surface between the filler and blend. Similar work was also reported by Hamid et al. [60].

In contrast, they reported that the silica concentration had no significant effect on the water moisture kinetic. Additionally, a nano-silica crystalline composite offered a higher water resistance than a nano-silica amorphous coupon. Both composites had a more hydrophobic resistance compared to epoxy resin. Furthermore, Norhasnan et al. [62] evaluated a hybridized RH/coco peat reinforced ABS, which showed reduced water-resistance biocomposite structures. Figure 1 shows the moisture absorption behaviour of the RH/CP reinforced ABS, and a maximum water kinetic behaviour for 20 wt % of coco peat composite composition was found, due to the higher hydrophilicity of the coco peat particle.

Fiber surface treatment and matrix modification improves the water absorption properties of RH composites. This was confirmed by several kinds of research that used NaOH and silane treatments on RH and coupling agents on the matrix. Huner [53] used 10% NaOH, while Nabinejad et al. [67] used 5% NaOH. Both found that NaOH decreased the water absorption of the RH–PP composite. NaOH treatment caused the surface of the RH to be polar. The same result was also produced by silane treatment of RH. Water absorption decreased by up to 38%, as reported by Petchwattana et al. [40], due to the silane reacting with free OH groups and due to the elimination of voids. Huner [53] reported that the tendency for reaction was lower than NaOH, causing the water absorption rate for silane treated composite to be lower. The use of MAPP decreased the water absorption rate due to the decrease of micro gaps in the interface, as a result of enhanced bonding between the filler and matrix [53]. The comparison between NaOH treatment and UV/O_2 treatment by Rajendran Royan et al. [25] showed that the NaOH-treated RH exhibited higher water absorption properties. The reason for this was due to the dry treatment with UV/O_2, where the RH was not soaked in any liquid and as a result, there was no fiber swelling that could give access to water in the reactive region. Saidi et al. [54] used a titanate coupling agent for a RH–PVC composite. Titanate coupling improved the interfacial adhesion between the RH and PVC matrix, preventing the diffusion of water molecules. As a result, the water absorption was reduced by 26%.

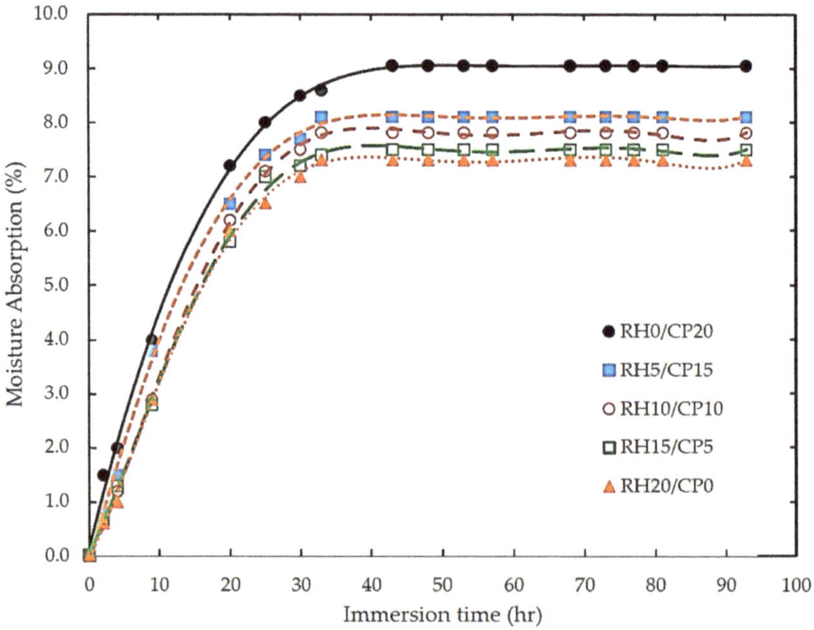

Figure 1. Typical of water absorption plots of RH and cocopeat ABS polymer blend composites.

Researchers have also used hydrophobic materials to increase the water resistance of the composite structure. As an example, Chalapud et al. [61] used a tung oil in RH that was adhesively bonded by a soy protein composite to improve the moisture resistance of particleboard. The impregnation of the composite panel with tung oil decreased its moisture absorption capacity and reduced the volume of voids, as obtained by surface microscopy. Since the oil was hydrophobic, a longer time period was required to reach saturation conditions, and this reduced the water kinetic mechanism, diffusing inside the hydroxyl groups of the RH and matrix to form hydrogen bonding. Moreover, a gamma radiation post-treatment was also employed by Chen et al. [63] and proved that the irradiation process increased the moisture kinetics and swelling effect on the composites. However, increasing the filler content after post-treatment also increased the moisture absorption, due to a huge quantity of carbonyl and hydroxyl groups in the composite, which allowed more molecule water to be diffused via the capillary effect.

6. Thermal Properties of RH Composites

The thermal stability studies using differential scanning calorimetry and thermogravimetry analysis conducted by researchers and reported in this review are summarised in Table 4. Many researchers have studied the thermal stability of polymer composites that were reinforced with RH fillers. Some researchers found a positive effect while others reported a negative effect on the thermal stability of RH in polymer composites. For example, the addition of RH ash to LDPE increased the onset degradation temperature and peak melting temperature, as reported by Zulkiple and Romli [68]. This effect also increased with the increasing percentage of filler loading. They explained that the filler acted as an insulator that could delay thermal degradation. A similar effect was also reported by Hidalgo-Salazar and Salinas [35], and Majeed et al. [69] with the addition of RH to PP. Hidalgo-Salazar and Salinas [35] explained that RH contributed to the increase in thermal stability of a PP composite due to the nucleation of spherulites, which increased the crystallinity of the PP matrix. A positive effect on thermal stability can also be seen in the increase of residual weight of a composite with the increase of RH content, attributed to the increase in thermal degradation of the composite due to the lower thermal degradation of RH compared to PP [69]. RH based silica, which is silica obtained from RH, provided a positive effect on the thermal stability of composites to which it was added. This was reported by many researchers, such as Krishnadevi et al. [70] and Tipachan et al. [71]. Functionalized RH ash, for example RH functionalized with 3-aminopropyltrimethoxysilane (3-APTMS) [70], improved the thermal stability of cardanol based benzoxazine composite. This was due to the formation of a complex network structure of silica reinforcement. The addition of silica nanoparticles from RH [71] improved the thermal stability of the PLA composite due to the thermal stability of the silica.

Many other researchers reported the negative effect of the incorporation of RH in polymer composites. In PP matrix composites, the addition of RH decreased the decomposition temperature due to RH being less thermally stable than PP. This was reported by Arjmandi et al. [72]. When studying natural rubber filled with unmodified RH ash, Zeng et al. [73] reported that the thermal stability was decreased due to the presence of pore structures with a high specific surface energy unevenly dispersed, and which when heated caused poor thermal stability. Similarly, the addition of RH to PLA [74] and PBAT–PLA hybrid composite [64] lowered the thermal degradation temperature, due to the lower thermal degradation nature of RH.

Due to the negative effect of RH on the thermal stability of polymer composites, many researchers have investigated various methods to improve the thermal stability, for example by adding flame retardants, by increasing the fiber–matrix interaction, and by hybridization with other fillers. Flame retardants have been successfully used to increase the thermal stability of RH composites. Ammonium polyphosphate (APP) [72] helped to increase the maximum degradation temperature, due to the formation of phosphorus and charred layers that shielded the PP matrix from further degradation. The addition of clay [69,75,76] thermally stabilized the RH composites, because the exfoliation of clay can delay the decomposition process of composites. The silicate layers of clay produced a twisted path for the diffusion of volatiles, which in turn improved the thermal stability of the composite. The use of halogen-free flame retardants [77] increased the thermal stability by generating thermally stable substances such as aluminum oxide, silicon dioxide, and phosphorus-containing compounds. Increasing the fiber–matrix interaction increased the thermal stability of RH composites [78]. For example, the increase of thermal stability of an epoxy–RH composite was attributed to an improvement of interfacial bonding due to the modification of RH using 3-glycidoxypropyltrimethoxy silane (GPS) [79]. The addition of TiO_2 also increased the thermal stability of PP–RH composites. Awang et al. [33] attributed this to the enhancement of the PP matrix and RH interaction. Another method of increasing the thermal stability of RH composites is hybridization. For example, a combination of RH and walnut shell [36] raised the thermal stability of a corn starch matrix composite due to the thermal stability of cellulose and lignin. The addition of volcanic ash, RH, and treated solid waste [80] was proven to improve the thermal stability of PP.

Table 4. Reported studies of the thermal stability of RH-reinforced polymers.

Matrix	Parametric Study	T_g, °C	5% Weight Loss, °C	10% Weight Loss, °C	T_m, °C	Char Yield at (600–900 °C), %	References
Cardanol based polybenzoxazine	Filler content	145	319	370	470	47	Krishnadevi et al. [70]
PLA	Matrix modification		335	521		6.1	Tipachan et al. [71]
Organic paraffin	Filler content	51.5			341	7.46	Lai et al. [79]
PLA	Filler treatment				472	3	Chen and Ahmad [51]
PP	Filler content				449.5	3	Arjmandi et al. [72]
Epoxy	Filler treatment	140			156.5		Fernandes et al. [21]
PMMA	Filler content	129			453		Oleiwi et al. [78]
LDPE	Filler comparison				112.9		Zulkipli and Romli [68]
Natural rubber	Coupling agent				447.2		Zeng et al. [73]
PBAT/PLA	Filler content					40	Muthuraj et al. [64]
PP	Filler content				166		Hidalgo and Salinas [35]
rHDPE	Filler content				131		Guo et al. [81]
Corn starch	Filler content		139	166		3.5	Singh et al. [36]
PP	Filler content		356.3	422.4	489.8	9.2	Awang et al. [33]
PP	Filler content				403	22.8	Almiron et al. [80]
PP	Filler content				350	30	Das et al. [82]
PET	Flame retardant effect		234			1.3	Phan et al. [77]
PVC	Clay addition				455	11	Dutta and Maji [75]
PP	Clay addition			311	162.9		Majeed et al. [69]

7. Flame Retardance of RH Composites

The addition of RH improved the flame retardant properties of polymer composites. The reason for the improvement of flame retardant properties is mainly the existence of char layers that act as barriers to further degradation. Krishnadevi et al. [70] reported that 20% functionalized RH ash added to cardanol based benzoxazine matrix produced the highest value of limiting oxygen index (LOI) analysis, of 36%. The reason was that the Si-O-Si phase promoted the formation of intumescent char, which could enhance the flame retardance. Other researchers added flame retardants to RH composites to further enhance flame retardancy. Attia and Saleh [83] reported a reduction of peak release heat of up to 33% during a flammability test of styrene-butadiene rubber containing RH silica nanoparticles using a cone calorimeter instrument. This improvement in flammability properties was attributed to the existence of a char layer from the combustion of molokhia extract and that was coated on the RH. The formation of an insulative char layer was also the reason why the addition of clay improved the flame retardant properties of a PVC–RH composite studied by Dutta and Maji [76]. Vu et al. [84] and Vu and Bach [85] studied RH-based silica added to epoxy with the addition of phosphorus-jointed epoxidized oil (DOPO-J-ESO). They found that thick char layers were formed by the combination of a hydroxyl group in the silica and the phosphaphenanthrene group of the DOPO-J-ESO, which improved the flame retardance of the composite. Similarly, the addition of diammonium hydrogen phosphate (DAP) to a polyurethane–RH composite [86] enhanced the fire resistance, due to the existence of residual char when DAP decomposed. This residual char acted as a flame inhibitor and dehydration agent, which further accelerated the formation of char. An increase of char residue was reported by Wu et al. [87] with the incorporation of 30 wt% melamine phosphate modified lignin (MAP-lignin) to 5 wt% RHA in poly(3-hydroxybutyrate-co-4-hydroxybutyrate) (P34HB) matrix composite. The increment of up to 24.3% char residue helped enhance the combustion behavior of the composite. Another type of flame retardant that has been studied is magnesium phytate (Mg-Phyt), which was added to the epoxy–RH composite studied by Xu et al. [88]. The addition of 5% Mg-Phyt to 5% RHA produced a compact silica-rich char layer containing Si-P and P-O-C structures that were thermally stable. Figure 2 represents the flame retardant mechanism evaluation of RH composite samples according to ASTM D 2863 [89].

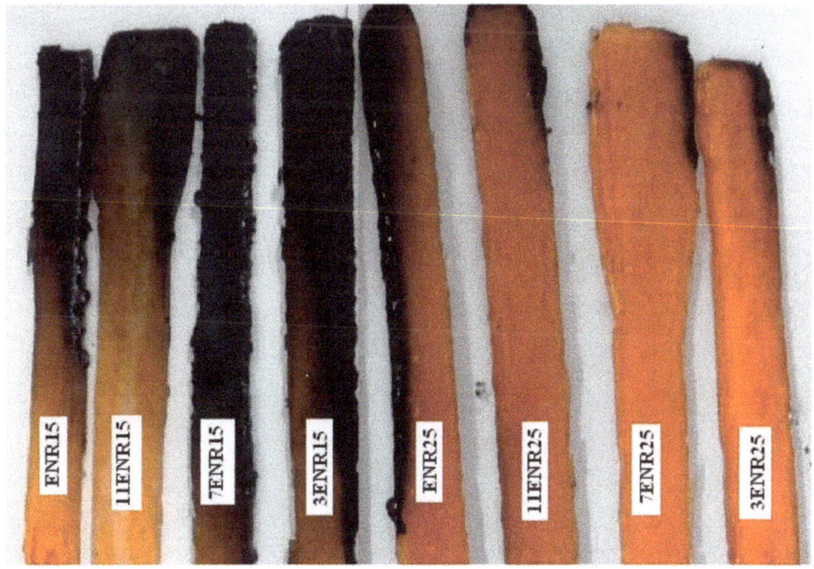

Figure 2. Burn test on RH biocomposites, according to ASTM D 2863 [89].

8. RH Composites Applications

RH itself has several benefits, for example, it can be turned into fuel using the pyrolysis process [90]. The highest energy content of RH can be harnessed by this gasification technique, earning substantial combustion products to reduce the use of fossil fuels. RH also contains high levels of lignin and cellulose, which are useful in activated carbon capture [91–93]. Wang et al. [91] modified the adsorbent X zeolite from RH for carbon dioxide (CO_2) capture, prepared using the consolidation of thermal and chemical activation. These modified zeolites improved the mechanism of heat and gas uptake during adsorption and desorption, and they were applicable for long-term capture and separation of CO_2 from an industrial exhaust gas. Moreover, RH is an advanced material with a high composition of silica that can be used as polishing and cleaning agent [94]. Furthermore, Okoya et al. [95] proposed employing RH biochar as a catalyst in conventional water treatment for removing chlorpyrifos from pesticide polluted water.

Rice husk as a biomass can be converted into various types of carbon products through the process of pyrolysis. One of the applications for this derived carbon is electronic wave absorption, as detailed by Chen et al. [96]. In addition, rice husk has a high silica content and, significantly, offers unique physical and chemical properties, which can improve the thermal and acoustic features of structures. The utilisation of RH in several applications in the construction industry has been discussed in the literature, such as in engineering cementitious composites (ECC) to produce lighter cement [97] or flexible mortar [98]; in resilient mats [99], where it was used to attenuate the impact of sound in buildings; and in rubber-modified asphalt concrete [100,101]. Zhang et al. [97] suggested that replacing fly ash with RH accelerated the hydration process, pozzolanic reaction, tensile strength, and refined pore distribution in lighter ECC. Nano silica fumes extracted from RH were used as an additive mineral to improve the physical and mechanical properties of flexible mortar composites. Significant improvements in resistance to thermal damage, higher strength, and lower permeability compared to the control cenosphere-based lightweight mortar were observed [97]. Furthermore, composite boards made from a combination of RH and recycled rubber were examined by Marques et al. [99]. These panels were tested as the top layer of a floating insulation floor system. They suggested these composite boards mitigated vibrations, improved impact sound insulation, and performed better as floor coatings. In the short-term ageing at high-temperature performance of an asphalt binder modified with a combination of crumb rubber and RH [100,101] indicated that adding crumb rubber and RH enhanced the ageing resistance as compared with the base asphalt binder. In addition, RH can be used as kiln bricks for plastering mortar to control the thermal properties of buildings [102]. De Silva et al. [103] suggested that a combination of cement, sand, and waste RH increased the strength, durability, and thermal performances of the brick. This mortar was used as a refractory material in a furnace application and retained its physicochemical properties at a temperature of more than 1000 °C [103,104].

The application of RH as biodegradable composites that to be used for interior and exterior parts in the automotive industries such as boot/spare tire lining [105] and brake pads [106] was widely examined. Currently, Volkswagen Group motors use a RH reinforced polymer to fabricate the tailgate, double load floor of the trunk, and covering of the roof for the SEAT León car [105]. The panel specimens were successfully tested to support up to 100 kilos of concentrated load and thermally examined in a climatic chamber, to measure the resistance to heat, cold, and humidity. Furthermore, RH was promoted to be used as a friction material, which has been dominated by the asbestos material. This asbestos-based brake pad was banned due to the carcinogenic and hazardous effects on the human body [106]. The husk filler could reduce the rate of wear, improve the friction behavior, and gave a harsh surface roughness, which resembled a commercial brake pad [107].

High-value applications, such as the use of RH in the manufacturing of silica gels, silicon chips, and ingredients for lithium-ion have been discovered. In semiconductor applications, zinc oxide (ZnO) was utilized due to its better ultraviolet absorbing properties compared to other semiconductor materials. However, it tends to agglomerate, which led to poor degradation. A combination of ZnO and RH as a composite material improved photocatalytic activity under ultraviolet (UV) irradiation [108,109]. Furthermore, RH was also successfully synthesized as a microporous activated carbon through carbonizing and activating it with zinc chloride ($ZnCl_2$) for a lithium–sulfur battery application. The composite offered a capacity of 426 $mAhg^{-1}$ at a 2C rate, and was suitable for energy storage devices for electric power [110]. Moreover, Suwanprateeb et al. [111] suggested that RH-reinforced epoxy could be used as an embedding material in electrical and electronic applications. Chen et al. [112] successfully produced a low-cost synthesis of SiC whiskers using a combination of RH and graphene. This whisker displayed excellent chemical reaction thermodynamics and outstanding performance for degrading rhodamine B.

Several studies reported the application of RH fibers in furniture and household appliances, which could support the concept of reducing agriculture waste [113]. Nicolao et al. [13] evaluated hybridized hemp and RH fibers for fabricating medium-density particleboard. They mentioned that this material combination can be considered for applications in building and furniture. Pratheep et al. [114] also verified that high mechanical properties can be obtained with smaller size RH utilization, such as wood powder composite resulting in the reduction of voids and cavities. This wood–plastic composite could be eligible as a decking and flooring fixture. Sadik et al. [115] also proved that the effect of nano-silica extracted from RH in wood plastic composite applications improved the water absorption and thickness swelling of composites. In addition, the thermal stability was slightly enhanced compared with the neat composite. The utilization of RH fibers in composites for food packaging has received a lot of attention from researchers [16,17]. Gupta et al. [116] synthesized a carboxymethyl cellulose (CMC) element from RH fiber to fabricate a biodegradable film. To improve the tensile strength and elongation of the film, the CMC was blended with glycerol and citric acid. Datta et al. [117] also extracted RH fiber to produce crystalline starch phthalate. These thin films offered outstanding mechanical (tensile, tear, stiffness), optical (haze, transmittance), and biodegradation properties, which were suitable for biodegradable food packaging applications. Figure 3 shows the RH fiber that was used in advanced applications.

Figure 3. Application of RH biocomposites: (**a**) medium density board [56], (**b**) brake pad [106], (**c**) car trunk [105], (**d**) polymer film [118], and (**e**) refractory mortar [104].

9. A Way Forward: RH Composites

The advantages of RH, including its biodegradability, abundant availability, and low-cost, make this fiber a potential replacement for synthetic polymers. However, the incompatibility between the hydrophilic matrix and hydrophobic fiber reduces the mechanical properties of RH composites. Due to this, chemical and surface treatment techniques to enhance fiber–matrix interfacial adhesion have increased the fabrication cost of these structures. The adoption of optimization techniques such as the Taguchi and response surface method [119–121] has great potential and could reduce the cost of experimental setups and the final product itself. In addition, there is also great potential to develop the finite element method software package to compare the mechanical properties of RH composites from experimental works.

The recyclability of waste composite products has been given tremendous attention lately. Many attempts have been made to recycle RH composite waste via manufacturing the product using thermoplastic matrices. The usage of natural resins, including PLA, thermoplastic starch, and corn/soy starch, would help to reduce the environmental impact [122]. These discoveries have great potential to help preserve natural resources in large amounts.

In general, silica is the major constituent of RH ash, which is suitable to be used in thermal insulation, electrical devices, and photonic applications. Inhomogeneous nanosilica, through burning RH in a pure oxygen atmosphere, and costly extraction methods are the major drawbacks of these techniques [123]. New and low cost routes for producing silica, such as by suspending RH in chemical solutions and then precipitating it in an acidic condition for extracting a silica gel, have great potential for further investigation.

Moreover, the degradation rate of RH biocomposite products needs to be examined for the ageing, ultraviolet deterioration, and humidity effects. There is also great potential for evaluating the kinetic degradation reaction rate of RH composite products affected by ubiquitous microorganisms, microbials, and isolated bacteria [124].

10. Conclusions

To conclude, recent works on RH-reinforced polymer composites between 2017 and 2021, mainly on physical, mechanical, and thermal behavior have been discussed. In addition, an insightful overview of the future-ready applications has also been given. As highlighted in this review, RH polymer composites have mainly focused on the interface interaction problem between RH and the polymer matrix. Several researchers improved the surface modification of the fiber and matrixes by esterification, silane treatment, and fiber mercerization. Another method is a thorough hybridization with other types of fillers or a mixed blend of the matrices. Both techniques have significantly improved tensile, flexural, and impact strength properties.

Research on moisture kinetics is also related to the fiber surface treatment and matrix modification of composites, due to the high hydrophilicity of RH. Many of the treatments produced improvements of the water absorption properties and swelling effect of RH composites. The thermal stability and flame retardant properties were enhanced by adding flame retardants, increasing the fiber–matrix interaction, and by hybridization. Incorporating flame particles in composites significantly increased the maximum degradation temperature and LOI, however, it reduced the peak release heat.

Potential applications for RH polymer composites have been briefly described. RH in itself has been used as a fuel, for activated carbon capture, polishing particles, cleaning agents, and catalysts for water treatments. As a filler, RH is used in many fields, such as construction, automotive, rechargeable batteries, semiconductors, electricals, electronics, furniture, household appliances, food packaging, and biodegradable films.

The vast potential applications of RH fillers have been described, and we believe that cost reduction should be emphasized in terms of research and product development. We also suggest that the degradation of performance from an environmental and biological organism perspective is an interesting subject to be investigated.

Author Contributions: Conceptualization, M.A.S. and M.Z.H.; writing—original draft preparation, M.A.S. and M.Z.H.; writing—review and editing, S.A.A.; funding acquisition, M.Y.M.D. All authors have read and agreed to the published version of the manuscript.

Funding: This research was funded by Universiti Teknologi Malaysia with UTM Transdiscipli-nary Research Grant Scheme Q.K130000.3556.05G83, Ministry of Higher Education (MOHE) with Fundamental Research Grant (FRGS) Scheme R.K130000.7856.5F404, Research University Grant (RUG) UTMFR Scheme Q.K.130000.2656.21H13, Ministry of Higher Education (MOHE) with Fundamental Research Grant (FRGS) Scheme FRGS/1/2020/TK0/UTM/02/59 and UTM Encouragement Research Q.K130000.2656.18J24 that provide financial supports throughout this research project.

Institutional Review Board Statement: Not applicable.

Informed Consent Statement: Not applicable.

Data Availability Statement: The data presented in this study are available on request from the corresponding author.

Conflicts of Interest: The authors declare no conflict of interest.

References

1. Cherubin, M.R.; Oliveira, D.M.D.S.; Feigl, B.J.; Pimentel, L.G.; Lisboa, I.P.; Gmach, M.R.; Varanda, L.L.; Morais, M.C.; Satiro, L.S.; Popin, G.V. Crop residue harvest for bioenergy production and its implications on soil functioning and plant growth: A review. *Sci. Agric.* **2018**, *75*, 255–272. [CrossRef]
2. Jittin, V.; Bahurudeen, A.; Ajinkya, S. Utilisation of rice husk ash for cleaner production of different construction products. *J. Clean. Prod.* **2020**, *263*, 121578. [CrossRef]
3. Beidaghy Dizaji, H.; Zeng, T.; Hartmann, I.; Enke, D.; Schliermann, T.; Lenz, V.; Bidabadi, M. Generation of high quality biogenic silica by combustion of rice husk and rice straw combined with pre-and post-treatment strategies—A review. *Appl. Sci.* **2019**, *9*, 1083. [CrossRef]
4. Nguyen, H.; Moghadam, M.J.; Moayedi, H. Agricultural wastes preparation, management, and applications in civil engineering: A review. *J. Mater. Cycles Waste Manag.* **2019**, *21*, 1039–1051. [CrossRef]
5. Lou, Z.; Wang, Q.; Sun, W.; Zhao, Y.; Wang, X.; Liu, X.; Li, Y. Bamboo flattening technique: A literature and patent review. *Eur. J. Wood Wood Prod.* **2021**, 1–14.
6. Mohamed Rashidi, A.H.; Jamal, M.H.; Hassan, M.Z.; Mohd Sendek, S.S.; Mohd Sopie, S.L.; Abd Hamid, M.R. Coastal structures as beach erosion control and sea level rise adaptation in malaysia: A review. *Water* **2021**, *13*, 1741. [CrossRef]
7. Tan, M.L.; Samat, N.; Chan, N.W.; Lee, A.J.; Li, C. Analysis of precipitation and temperature extremes over the Muda River Basin, Malaysia. *Water* **2019**, *11*, 283. [CrossRef]
8. Roslan, S.; Rasid, Z.; Hassan, M. The natural fiber composites based on bamboo fibers: A review. *ARPN J. Eng. Appl. Sci.* **2015**, *10*, 6279–6288.
9. Hassan, M.Z.; Sapuan, S.; A Rasid, Z.; Nor, A.F.M.; Dolah, R.; Md Daud, M.Y. Impact damage resistance and post-impact tolerance of optimum banana-pseudo-stem-fiber-reinforced epoxy sandwich structures. *Appl. Sci.* **2020**, *10*, 684. [CrossRef]
10. Ibrahim, M.I.; Hassan, M.Z.; Dolah, R.; Yusoff, M.Z.M.; Salit, M.S. Tensile behaviour for mercerization of single kenaf fiber. *Malays. J. Fundam. Appl. Sci.* **2018**, *14*, 437–439. [CrossRef]
11. Mekhilef, S.; Saidur, R.; Safari, A.; Mustaffa, W. Biomass energy in Malaysia: Current state and prospects. *Renew. Sustain. Energy Rev.* **2011**, *15*, 3360–3370. [CrossRef]
12. Goh, C.S.; Tan, K.T.; Lee, K.T.; Bhatia, S. Bio-ethanol from lignocellulose: Status, perspectives and challenges in Malaysia. *Bioresour. Technol.* **2010**, *101*, 4834–4841. [CrossRef] [PubMed]
13. FAOSTAT. Available online: http://www.fao.org/faostat (accessed on 15 July 2021).
14. Manickam, T.; Cornelissen, G.; Bachmann, R.T.; Ibrahim, I.Z.; Mulder, J.; Hale, S.E. Biochar application in Malaysian sandy and acid sulfate soils: Soil amelioration effects and improved crop production over two cropping seasons. *Sustainability* **2015**, *7*, 16756–16770. [CrossRef]
15. Athira, G.; Bahurudeen, A.; Appari, S. Sustainable alternatives to carbon intensive paddy field burning in India: A framework for cleaner production in agriculture, energy, and construction industries. *J. Clean. Prod.* **2019**, *236*, 117598. [CrossRef]
16. Abdulkareem, S.A.; Adeniyi, A.G. Tensile and water absorbing properties of natural fibre reinforced plastic composites from waste polystyrene and rice husk. *J. Eng. Res. Dev.* **2018**, *1*, 199–204.
17. Zafar, M.F.; Siddiqui, M.A. Raw natural fiber reinforced polystyrene composites: Effect of fiber size and loading. *Mater. Today Proc.* **2018**, *5*, 5908–5917. [CrossRef]
18. Zhang, Q.; Yi, W.; Li, Z.; Wang, L.; Cai, H. Mechanical Properties of Rice Husk Biochar Reinforced High Density Polyethylene Composites. *Polymers* **2018**, *10*, 286. [CrossRef]
19. Zhang, Q.; Li, Y.; Cai, H.; Lin, X.; Yi, W.; Zhang, J. Properties comparison of high density polyethylene composites filled with three kinds of shell fibers. *Results Phys.* **2019**, *12*, 1542–1546. [CrossRef]
20. Xue, B.; Wang, X.; Sui, J.; Xu, D.; Zhu, Y.; Liu, X. A facile ball milling method to produce sustainable pyrolytic rice husk bio-filler for reinforcement of rubber mechanical property. *Ind. Crops Prod.* **2019**, *141*, 111791. [CrossRef]
21. Fernandes, I.J.; Santos, R.V.; Santos, E.C.A.d.; Rocha, T.L.A.C.; Domingues, N.S., Jr.; Moraes, C.A.M. Replacement of commercial silica by rice husk ash in epoxy composites: A comparative analysis. *Mater. Res.* **2018**, *21*. [CrossRef]
22. Pongdong, W.; Kummerlöwe, C.; Vennemann, N.; Thitithammawong, A.; Nakason, C. A comparative investigation of rice husk ash and siliceous earth as reinforcing fillers in dynamically cured blends of epoxidized natural rubber (ENR) and thermoplastic polyurethane (TPU). *J. Polym. Environ.* **2018**, *26*, 1145–1159. [CrossRef]
23. Chen, R.S.; Ahmad, S.; Gan, S. Rice husk bio-filler reinforced polymer blends of recycled HDPE/PET: Three-dimensional stability under water immersion and mechanical performance. *Polym. Compos.* **2018**, *39*, 2695–2704. [CrossRef]
24. Raghu, N.; Kale, A.; Chauhan, S.; Aggarwal, P.K. Rice husk reinforced polypropylene composites: Mechanical, morphological and thermal properties. *J. Indian Acad. Wood Sci.* **2018**, *15*, 96–104. [CrossRef]
25. Rajendran Royan, N.R.; Sulong, A.B.; Yuhana, N.Y.; Chen, R.S.; Ab Ghani, M.H.; Ahmad, S. UV/O3 treatment as a surface modification of rice husk towards preparation of novel biocomposites. *PLoS ONE* **2018**, *13*, e0197345. [CrossRef]

26. Bisht, N.; Gope, P. Effect of alkali treatment on mechanical properties of rice husk flour reinforced epoxy bio-Composite. *Mater. Today Proc.* **2018**, *5*, 24330–24338. [CrossRef]
27. Santiagoo, R.; Ismail, H.; Suharty, N. Comparison of processing and mechanical properties of polypropylene/recycled acrylonitrile butadiene rubber/rice husk powder composites modified with silane and acetic anhydride compound. In *Natural Fibre Reinforced Vinyl Ester and Vinyl Polymer Composites*; Elsevier: Amsterdam, The Netherlands, 2018; pp. 333–347.
28. Zhang, Q.; Zhang, D.; Lu, W.; Khan, M.U.; Xu, H.; Yi, W.; Lei, H.; Huo, E.; Qian, M.; Zhao, Y. Production of high-density polyethylene biocomposites from rice husk biochar: Effects of varying pyrolysis temperature. *Sci. Total Environ.* **2020**, *738*, 139910. [CrossRef]
29. Moreno, D.D.P.; de Camargo, R.V.; dos Santos Luiz, D.; Branco, L.T.P.; Grillo, C.C.; Saron, C. Composites of recycled polypropylene from cotton swab waste with pyrolyzed rice husk. *J. Polym. Environ.* **2021**, *29*, 350–362. [CrossRef]
30. Boonsuk, P.; Sukolrat, A.; Bourkaew, S.; Kaewtatip, K.; Chantarak, S.; Kelarakis, A.; Chaibundit, C. Structure-properties relationships in alkaline treated rice husk reinforced thermoplastic cassava starch biocomposites. *Int. J. Biol. Macromol.* **2021**, *167*, 130–140. [CrossRef] [PubMed]
31. Olcay, H.; Kocak, E.D. Rice plant waste reinforced polyurethane composites for use as the acoustic absorption material. *Appl. Acoust.* **2021**, *173*, 107733. [CrossRef]
32. Shubbar, S.D.A. Experimental investigation of rice husk particles as filler in hybrid composites. *J. Univ. Babylon Eng. Sci.* **2018**, *26*, 307–315.
33. Awang, M.; Mohd, W.W. Comparative studies of Titanium Dioxide and Zinc Oxide as a potential filler in Polypropylene reinforced rice husk composite. In Proceedings of the IOP Conference Series: Materials Science and Engineering, Pekan, Malayzia, 1–2 March 2018; p. 012046.
34. Kumar, S.; Mer, K.K.S.; Gangil, B.; Patel, V.K. Synergy of rice-husk filler on physico-mechanical and tribological properties of hybrid Bauhinia-vahlii/sisal fiber reinforced epoxy composites. *J. Mater. Res. Technol.* **2019**, *8*, 2070–2082. [CrossRef]
35. Hidalgo-Salazar, M.A.; Salinas, E. Mechanical, thermal, viscoelastic performance and product application of PP-rice husk Colombian biocomposites. *Compos. Part. B Eng.* **2019**, *176*, 107135. [CrossRef]
36. Singh, T.; Gangil, B.; Patnaik, A.; Biswas, D.; Fekete, G. Agriculture waste reinforced corn starch-based biocomposites: Effect of rice husk/walnut shell on physicomechanical, biodegradable and thermal properties. *Mater. Res. Express* **2019**, *6*, 045702. [CrossRef]
37. Sun, J.; Pang, Y.; Yang, Y.; Zhao, J.; Xia, R.; Li, Y.; Liu, Y.; Guo, H. Improvement of rice husk/HDPE bio-composites interfacial properties by silane coupling agent and compatibilizer complementary modification. *Polymers* **2019**, *11*, 1928. [CrossRef] [PubMed]
38. Běhálek, L.; Borůvka, M.; Brdlík, P.; Habr, J.; Lenfeld, P.; Kroisová, D.; Veselka, F.; Novák, J. Thermal properties and non-isothermal crystallization kinetics of biocomposites based on poly (lactic acid), rice husks and cellulose fibres. *J. Therm. Anal. Calorim.* **2020**, *142*, 629–649. [CrossRef]
39. Guna, V.; Ilangovan, M.; Rather, M.H.; Giridharan, B.; Prajwal, B.; Krishna, K.V.; Venkatesh, K.; Reddy, N. Groundnut shell/rice husk agro-waste reinforced polypropylene hybrid biocomposites. *J. Build. Eng.* **2020**, *27*, 100991. [CrossRef]
40. Petchwattana, N.; Sanetuntikul, J. Static and Dynamic Mechanical Properties of Poly (vinyl chloride) and Waste Rice Husk Ash Composites Compatibilized with γ-aminopropyltrimethoxysilane. *Silicon* **2018**, *10*, 287–292. [CrossRef]
41. Jiang, L.; Fu, J.; Liu, L.; Du, P. Wear and thermal behavior of basalt fiber reinforced rice husk/polyvinyl chloride composites. *J. Appl. Polym. Sci.* **2021**, *138*, 50094. [CrossRef]
42. Kargarzadeh, H.; Johar, N.; Ahmad, I. Starch biocomposite film reinforced by multiscale rice husk fiber. *Compos. Sci. Technol.* **2017**, *151*, 147–155. [CrossRef]
43. Yap, S.Y.; Sreekantan, S.; Hassan, M.; Sudesh, K.; Ong, M.T. Characterization and Biodegradability of Rice Husk-Filled Polymer Composites. *Polymers* **2021**, *13*, 104. [CrossRef]
44. Spada, J.C.; Seibert, S.F.; Tessaro, I.C. Impact of PLA Poly (Lactic Acid) and PBAT Poly (butylene adipate-co-terephthalate) coating on the properties of composites with high content of rice husk. *J. Polym. Environ.* **2021**, *29*, 1324–1331. [CrossRef]
45. Deo, C.; Acharya, S. Effect of moisture absorption on mechanical properties of chopped natural fiber reinforced epoxy composite. *J. Reinf. Plast. Compos.* **2010**, *29*, 2513–2521. [CrossRef]
46. Shakeri, A.; Ghasemian, A. Water absorption and thickness swelling behavior of polypropylene reinforced with hybrid recycled newspaper and glass fiber. *Appl. Compos. Mater.* **2010**, *17*, 183–193. [CrossRef]
47. Miao, J.; Tsige, M.; Taylor, P.L. Generalized model for the diffusion of solvents in glassy polymers: From fickian to super case II. *J. Chem. Phys.* **2017**, *147*, 044904. [CrossRef] [PubMed]
48. Siriwardena, S.; Ismail, H.; Ishiaku, U. A comparison of the mechanical properties and water absorption behavior of white rice husk ash and silica filled polypropylene composites. *J. Reinf. Plast. Compos.* **2003**, *22*, 1645–1666. [CrossRef]
49. Melo, R.Q.; Lia Fook, M.V.; de Lima, A.G. Non-Fickian Moisture Transport in Vegetable-Fiber-Reinforced Polymer Composites Using a Langmuir-Type Model. *Polymers* **2020**, *12*, 2503. [CrossRef]
50. Lee, S.-H.; Ohkita, T.; Kitagawa, K. Eco-composite from poly (lactic acid) and bamboo fiber. *Walter de Gruyter* **2004**, *58*, 529–536. [CrossRef]
51. Chen, R.S.; Ahmad, S. Mechanical performance and flame retardancy of rice husk/organoclay-reinforced blend of recycled plastics. *Mater. Chem. Phys.* **2017**, *198*, 57–65. [CrossRef]

52. Sheykh, M.J.; Tarmian, A.; Doosthoseini, K. Wear resistance and friction coefficient of nano-SiO_2 and ash-filled HDPE/lignocellulosic fiber composites. *Polym. Bull.* **2017**, *74*, 4537–4547. [CrossRef]
53. Huner, U. Effect of chemical treatment and maleic anhydride grafted polypropylene coupling agent on rice husk and rice husk reinforced composite. *Mater. Express* **2017**, *7*, 134–144. [CrossRef]
54. Saidi, M.A.; Ahmad, M.; Arjmandi, R.; Hassan, A.; Rahmat, A.R. The effect of titanate coupling agent on water absorption and mechanical properties of rice husk filled poly (vinyl chloride) composites. In *Natural Fibre Reinforced Vinyl Ester and Vinyl Polymer Composites*; Elsevier: Amsterdam, The Netherlands, 2018; pp. 197–210.
55. Mohamed, R.; Mohamed, M.M.F.; Norizan, M.N.; Mohamed, R.R.R. Physical and morphological properties of filled calcium carbonate/kenaf fibre/rice husk polypropylene hybrid composite. In Proceedings of the AIP Conference Proceedings, Selangor, Malaysia, 21–22 October 2017; p. 050003.
56. Battegazzore, D.; Alongi, J.; Frache, A.; Wågberg, L.; Carosio, F. Layer by Layer-functionalized rice husk particles: A novel and sustainable solution for particleboard production. *Mater. Today Commun.* **2017**, *13*, 92–101. [CrossRef]
57. Prapruddivongs, C.; Apichartsitporn, M.; Wongpreedee, T. Effect of silica resources on the biodegradation behavior of poly (lactic acid) and chemical crosslinked poly (lactic acid) composites. *Polym. Test.* **2018**, *71*, 87–94. [CrossRef]
58. Akindoyo, J.O.; Husney, N.A.A.b.; Ismail, N.H.; Mariatti, M. Structure and performance of poly (lactic acid)/poly (butylene succinate-co-L-lactate) blend reinforced with rice husk and coconut shell filler. *Polym. Polym. Compos.* **2020**, 0967391120954047. [CrossRef]
59. Daramola, O.O. Flesural properties and water absorotion characteristics of high density polyethylene-based siliceous composites. *Acta Tech. Corviniensis-Bull. Eng.* **2019**, *12*, 57–62.
60. Hamid, N.H.; Hisan, W.S.I.W.B.; Abdullah, U.H.; Azim, A.A.A.; Tahir, P.M. Mechanical properties and moisture absorption of epoxy composites mixed with amorphous and crystalline silica from rice husk. *BioResources* **2019**, *14*, 7363–7374.
61. Chalapud, M.C.; Herdt, M.; Nicolao, E.S.; Ruseckaite, R.A.; Ciannamea, E.M.; Stefani, P.M. Biobased particleboards based on rice husk and soy proteins: Effect of the impregnation with tung oil on the physical and mechanical behavior. *Constr. Build. Mater.* **2020**, *230*, 116996. [CrossRef]
62. Norhasnan, N.H.A.; Hassan, M.Z.; Nor, A.F.M.; Zaki, S.; Dolah, R.; Jamaludin, K.R.; Aziz, S.A.A. Physicomechanical properties of rice husk/coco peat reinforced acrylonitrile butadiene styrene blend composites. *Polymers* **2021**, *13*, 1171. [CrossRef]
63. Chen, R.S.; Ab Ghani, M.H.; Ahmad, S.; Mou'ad, A.T.; Gan, S. Tensile, thermal degradation and water diffusion behaviour of gamma-radiation induced recycled polymer blend/rice husk composites: Experimental and statistical analysis. *Compos. Sci. Technol.* **2021**, *207*, 108748. [CrossRef]
64. Muthuraj, R.; Lacoste, C.; Lacroix, P.; Bergeret, A. Sustainable thermal insulation biocomposites from rice husk, wheat husk, wood fibers and textile waste fibers: Elaboration and performances evaluation. *Ind. Crops Prod.* **2019**, *135*, 238–245. [CrossRef]
65. Yusuf, T.A. Characterization of selected properties of composites of waste paper with untreated bamboo stem fibre and rice husk. *Acta Polytech.* **2017**, *57*, 295–303. [CrossRef]
66. Antunes, A.; Faria, P.; Silva, V.; Brás, A. Rice husk-earth based composites: A novel bio-based panel for buildings refurbishment. *Constr. Build. Mater.* **2019**, *221*, 99–108. [CrossRef]
67. Nabinejad, O.; Debnath, S.; Ying, T.J.; Liew, W.Y.; Davies, I.J. Influence of alkali treatment and nanoclay content on the properties of rice husk filled polyester composites. *Mater. Sci. Forum* **2017**, *882*, 89–100. [CrossRef]
68. Zulkipli, A.H.; Romli, A.Z. Thermal characterization of low density polyethylene (LDPE)/rice husk ash and rice husk ash derived silica aerogel composites. In Proceedings of the AIP Conference Proceedings, Selangor, Malaysia, 21–22 October 2017; p. 030011.
69. Majeed, K.; Ahmed, A.; Abu Bakar, M.S.; Indra Mahlia, T.M.; Saba, N.; Hassan, A.; Jawaid, M.; Hussain, M.; Iqbal, J.; Ali, Z. Mechanical and thermal properties of montmorillonite-reinforced polypropylene/rice husk hybrid nanocomposites. *Polymers* **2019**, *11*, 1557. [CrossRef]
70. Krishnadevi, K.; Devaraju, S.; Sriharshitha, S.; Alagar, M.; Priya, Y.K. Environmentally sustainable rice husk ash reinforced cardanol based polybenzoxazine bio-composites for insulation applications. *Polym. Bull.* **2020**, *77*, 2501–2520. [CrossRef]
71. Tipachan, C.; Gupta, R.K.; Agarwal, S.; Kajorncheappunngam, S. Flame retardant properties and thermal stability of polylactic acid filled with layered double hydroxide and rice husk ash silica. *J. Polym. Environ.* **2020**, *28*, 948–961. [CrossRef]
72. Arjmandi, R.; Ismail, A.; Hassan, A.; Bakar, A.A. Effects of ammonium polyphosphate content on mechanical, thermal and flammability properties of kenaf/polypropylene and rice husk/polypropylene composites. *Constr. Build. Mater.* **2017**, *152*, 484–493. [CrossRef]
73. Zeng, Z.; Li, Y.; Yu, H.; Wang, Q. Study on Modified Rice Husk Ash and Its Filling Natural Rubber Composite. *DEStech Trans. Eng. Technol. Res.* **2018**. [CrossRef]
74. Nizamuddin, S.; Jadhav, A.; Qureshi, S.S.; Baloch, H.A.; Siddiqui, M.; Mubarak, N.; Griffin, G.; Madapusi, S.; Tanksale, A.; Ahamed, M.I. Synthesis and characterization of polylactide/rice husk hydrochar composite. *Sci. Rep.* **2019**, *9*, 1–11.
75. Dutta, N.; Maji, T.K. Valorization of waste rice husk by preparing nanocomposite with polyvinyl chloride and montmorillonite clay. *J. Thermoplast. Compos. Mater.* **2019**, *34*, 801–806. [CrossRef]
76. Dutta, N.; Maji, T.K. Synergic effect of montmorillonite and microcrystalline cellulose on the physicochemical properties of rice husk/PVC composite. *SN Appl. Sci.* **2020**, *2*, 1–10. [CrossRef]

77. Phan, H.T.; Nguyen, B.T.; Pham, L.H.; Pham, C.T.; Do, T.V.V.; Hoang, C.N.; Nguyen, N.N.; Kim, J.; Hoang, D. Excellent fireproof characteristics and high thermal stability of rice husk-filled polyurethane with halogen-free flame retardant. *Polymers* **2019**, *11*, 1587. [CrossRef]
78. Oleiwi, J.K.; Hamad, Q.A.; Rahman, H.J.A. Study thermal behavior of heat cure poly (methyl methacrylate) reinforced by bamboo and rice husk powders for denture applications. *Al-Qadisiyah J. Eng. Sci.* **2018**, *11*, 417–425. [CrossRef]
79. Lai, S.M.; Han, J.L.; Yu, Y.F. Properties of rice husk/epoxy composites under different interfacial treatments. *Polym. Compos.* **2017**, *38*, 1992–2000. [CrossRef]
80. Almirón, J.; Roudet, F.; Duquesne, S. Influence of volcanic ash, rice husk ash, and solid residue of catalytic pyrolysis on the flame-retardant properties of polypropylene composites. *J. Fire Sci.* **2019**, *37*, 434–451. [CrossRef]
81. Guo, Y.; Zhu, S.; Chen, Y.; Li, D. Thermal properties of wood-plastic composites with different compositions. *Materials* **2019**, *12*, 881. [CrossRef]
82. Das, O.; Hedenqvist, M.S.; Prakash, C.; Lin, R.J. Nanoindentation and flammability characterisation of five rice husk biomasses for biocomposites applications. *Compos. Part A Appl. Sci. Manuf.* **2019**, *125*, 105566. [CrossRef]
83. Attia, N.F.; Saleh, B.K. Novel synthesis of renewable and green flame-retardant, antibacterial and reinforcement material for styrene–butadiene rubber nanocomposites. *J. Therm. Anal. Calorim.* **2020**, *139*, 1817–1827. [CrossRef]
84. Vu, C.M.; Nguyen, V.-H.; Bach, Q.-V. Phosphorous-jointed epoxidized soybean oil and rice husk-based silica as the novel additives for improvement mechanical and flame retardant of epoxy resin. *J. Fire Sci.* **2020**, *38*, 3–27. [CrossRef]
85. Vu, C.M.; Bach, Q.-V. Effects of DOPO-grafted epoxidized soybean oil on fracture toughness and flame retardant of epoxy resin/rice husk silica hybrid. *Macromol. Res.* **2020**, *28*, 826–834. [CrossRef]
86. Son, T.; Nguyen, H.T.; Phan, H.T.; Pham, C.T.; Nguyen, B.T.; Pham, L.H.; Do, T.V.V.; Vothi, H.; Kim, J.; Hoang, D.Q. Phosphorus/phosphorus-nitrogen flame retardants applied to polyurethane/rice husk eco-composites: Thermal behavior, flame retardancy, and physico-mechanical properties. *Polym. Bull.* **2021**, *78*, 2727–2743. [CrossRef]
87. Wu, W.; He, H.; Liu, T.; Wei, R.; Cao, X.; Sun, Q.; Venkatesh, S.; Yuen, R.K.; Roy, V.A.; Li, R.K. Synergetic enhancement on flame retardancy by melamine phosphate modified lignin in rice husk ash filled P34HB biocomposites. *Compos. Sci. Technol.* **2018**, *168*, 246–254. [CrossRef]
88. Xu, Y.; Li, J.; Shen, R.; Wang, Z.; Hu, P.; Wang, Q. Experimental study on the synergistic flame retardant effect of bio-based magnesium phytate and rice husk ash on epoxy resins. *J. Therm. Anal. Calorim.* **2021**, 1–12. [CrossRef]
89. Kavitha, D.; Murugavel, S.C.; Thenmozhi, S. Flame retarding cardanol based novolac-epoxy/rice husk composites. *Mater. Chem. Phys.* **2021**, *263*, 124225. [CrossRef]
90. Bakari, R.; Kivevele, T.; Huang, X.; Jande, Y.A. Simulation and optimisation of the pyrolysis of rice husk: Preliminary assessment for gasification applications. *J. Anal. Appl. Pyrolysis* **2020**, *150*, 104891. [CrossRef]
91. Wang, Y.; Jia, H.; Chen, P.; Fang, X.; Du, T. Synthesis of La and Ce modified X zeolite from rice husk ash for carbon dioxide capture. *J. Mater. Res. Technol.* **2020**, *9*, 4368–4378. [CrossRef]
92. He, S.; Chen, G.; Xiao, H.; Shi, G.; Ruan, C.; Ma, Y.; Dai, H.; Yuan, B.; Chen, X.; Yang, X. Facile preparation of N-doped activated carbon produced from rice husk for CO_2 capture. *J. Colloid Interface Sci.* **2021**, *582*, 90–101. [CrossRef] [PubMed]
93. Panda, D.; Saini, C.; Kumar, E.A.; Singh, S.K. In situ casting of rice husk ash in metal organic frameworks induces enhanced CO_2 capture performance. *Sci. Rep.* **2020**, *10*, 1–12. [CrossRef]
94. Tiwari, S.; Pradhan, M. Effect of rice husk ash on properties of aluminium alloys: A review. *Mater. Today Proc.* **2017**, *4*, 486–495. [CrossRef]
95. Okoya, A.A.; Adegbaju, O.S.; Akinola, O.E.; Akinyele, A.B.; Amuda, O.S. Comparative assessment of the efficiency of rice husk biochar and conventional water treatment method to remove chlorpyrifos from pesticide polluted water. *Curr. J. Appl. Sci. Technol.* **2020**, *39*, 1–11. [CrossRef]
96. Chen, Z.; Wu, Z.; Su, J.; Li, J.; Gao, B.; Fu, J.; Zhang, X.; Huo, K.; Chu, P.K. Large-scale and low-cost synthesis of in situ generated SiC/C nano-composites from rice husks for advanced electromagnetic wave absorption applications. *Surf. Coat. Technol.* **2021**, *406*, 126641. [CrossRef]
97. Zhang, Z.; Yang, F.; Liu, J.-C.; Wang, S. Eco-friendly high strength, high ductility engineered cementitious composites (ECC) with substitution of fly ash by rice husk ash. *Cem. Concr. Res.* **2020**, *137*, 106200. [CrossRef]
98. Gupta, S.; Kua, H.W. Application of rice husk biochar as filler in cenosphere modified mortar: Preparation, characterization and performance under elevated temperature. *Constr. Build. Mater.* **2020**, *253*, 119083. [CrossRef]
99. Marques, B.; António, J.; Almeida, J.; Tadeu, A.; de Brito, J.; Dias, S.; Pedro, F.; Sena, J.D. Vibro-acoustic behaviour of polymer-based composite materials produced with rice husk and recycled rubber granules. *Constr. Build. Mater.* **2020**, *264*, 120221. [CrossRef]
100. Abdelmagid, A.A.; Feng, C.P. Laboratory evaluation of the effects of short-term aging on high temperature performance of asphalt binder modified with crumb rubber and rice husk ash. *Pet. Sci. Technol.* **2019**, *37*, 1557–1565. [CrossRef]
101. Abdelmagid, A.A.; Pei Feng, C. Evaluating the effect of rice-husk ash and crumb-rubber powder on the high-temperature performance of asphalt binder. *J. Mater. Civ. Eng.* **2019**, *31*, 04019296. [CrossRef]
102. Bodie, A.R.; Micciche, A.C.; Atungulu, G.G.; Rothrock, M.J., Jr.; Ricke, S.C. Current trends of rice milling byproducts for agricultural applications and alternative food production systems. *Front. Sustain. Food Syst.* **2019**, *3*, 47. [CrossRef]

103. De Silva, G.S.; Vishvalingam, S.; Etampawala, T. Effect of waste rice husk ash from rice husk fuelled brick kilns on strength, durability and thermal performances of mortar. *Constr. Build. Mater.* **2021**, *268*, 121794. [CrossRef]
104. Thang, N.H. Novel Porous Refractory Synthesized from Diatomaceous Earth and Rice Husk Ash. *J. Polym. Compos.* **2020**, *8*, 128–137.
105. SEAT Uses Recycled Rice Husks to Reduce Plastics in New Leon. Available online: https://blog.swanswaygarages.com/seat-uses-recycled-rice-husks-to-reduce-plastics-in-new-leon. (accessed on 16 July 2021).
106. Primaningtyas, W.; Sakura, R.; Syafi'i, I.; Adhyaksa, A. Asbestos-free Brake Pad Using Composite Polymer Strengthened with Rice Husk Powder. In Proceedings of the IOP Conference Series: Materials Science and Engineering, Surabaya, Indonesia, 29 September 2019; p. 012015.
107. Paramasivam, K.; Jayaraj, J.J.; Ramar, K.; Subramani, Y.; Ajithkumar, K.; Kabilan, N. Evaluation of natural fibers for the production of automotive brake pads replacement for asbestos brake pad. In Proceedings of the AIP Conference Proceedings, Chennai, India, 7–9 August 2020; p. 040005.
108. Eddy, D.R.; Noviyanti, A.R.; Solihudin, S.; Ishmayana, S.; Tjokronegoro, R.-A. Rice Husk for Photocatalytic Composite Material Fabrication. In *Visible-Light Photocatal. Carbon-Based Mater*; IntechOpen: London, UK, 2018; pp. 19–28.
109. Chen, H.; Zhao, L.; Wang, G.; He, X.; Wang, X.; Fang, W.; Du, X. Direct growth of ZnO nanorods on biogenic hierarchical rice husk SiO_2 and their application to dye degradation. *Clean Technol. Environ. Policy* **2017**, *19*, 1335–1345. [CrossRef]
110. Vu, D.-L.; Seo, J.-S.; Lee, H.-Y.; Lee, J.-W. Activated carbon with hierarchical micro–mesoporous structure obtained from rice husk and its application for lithium–sulfur batteries. *RSC Adv.* **2017**, *7*, 4144–4151. [CrossRef]
111. Suwanprateeb, J.; Hatthapanit, K. Rice-husk-ash-based silica as a filler for embedding composites in electronic devices. *J. Appl. Polym. Sci.* **2002**, *86*, 3013–3020. [CrossRef]
112. Chen, J.-P.; Song, G.; Liu, Z.; Kong, Q.-Q.; Zhang, S.-C.; Chen, C.-M. Preparation of SiC whiskers using graphene and rice husk ash and its photocatalytic property. *J. Alloys Compd.* **2020**, *833*, 155072. [CrossRef]
113. Nicolao, E.; Leiva, P.; Chalapud, M.; Ruseckaite, R.A.; Ciannamea, E.M.; Stefani, P.M. Flexural and tensile properties of biobased rice husk-jute-soybean protein particleboards. *J. Build. Eng.* **2020**, *30*, 101261. [CrossRef]
114. Pratheep, V.; Priyanka, E.; Prasad, P.H. Characterization and Analysis of Natural Fibre-Rice Husk with Wood Plastic Composites. In Proceedings of the IOP Conference Series: Materials Science and Engineering, Tamil Nady, India, 12–13 April 2019; p. 012066.
115. Sadik, W.A.A.; El Demerdash, A.G.M.; Abbas, R.; Bedir, A. Effect of Nanosilica and Nanoclay on the Mechanical, Physical, and Morphological Properties of Recycled Linear Low Density Polyethylene/Rice Husk Composites. *J. Polym. Environ.* **2021**, *29*, 1600–1615. [CrossRef]
116. Gupta, H.; Kumar, H.; Kumar, M.; Gehlaut, A.K.; Gaur, A.; Sachan, S.; Park, J.-W. Synthesis of biodegradable films obtained from rice husk and sugarcane bagasse to be used as food packaging material. *Environ. Eng. Res.* **2020**, *25*, 506–514. [CrossRef]
117. Datta, D.; Halder, G. Blending of phthalated starch and surface functionalized rice husk extracted nanosilica with LDPE towards developing an efficient packaging substitute. *Environ. Sci. Pollut. Res.* **2020**, *27*, 1533–1557. [CrossRef]
118. Datta, D.; Samanta, S.; Halder, G. Surface functionalization of extracted nanosilica from rice husk for augmenting mechanical and optical properties of synthesized LDPE-Starch biodegradable film. *Polym. Test.* **2019**, *77*, 105878. [CrossRef]
119. Hassan, M.Z.; Sapuan, S.; Roslan, S.A.; Sarip, S. Optimization of tensile behavior of banana pseudo-stem (*Musa acuminate*) fiber reinforced epoxy composites using response surface methodology. *J. Mater. Res. Technol.* **2019**, *8*, 3517–3528. [CrossRef]
120. Hassan, M.Z.; Roslan, S.A.; Sapuan, S.; Rasid, Z.A.; Mohd Nor, A.F.; Md Daud, M.Y.; Dolah, R.; Mohamed Yusoff, M.Z. Mercerization optimization of bamboo (*Bambusa vulgaris*) fiber-reinforced epoxy composite structures using a box–behnken design. *Polymers* **2020**, *12*, 1367. [CrossRef]
121. Nor, A.F.M.; Hassan, M.Z.; Rasid, Z.A.; Sarip, S.; Daud, M.Y.M. Optimization on Tensile Properties of Kenaf/Multi-walled CNT Hybrid Composites with Box-Behnken Design. *Appl. Compos. Mater.* **2021**, *28*, 607–632. [CrossRef]
122. Bisht, N.; Gope, P.C.; Rani, N. Rice husk as a fibre in composites: A review. *J. Mech. Behav. Mater.* **2020**, *29*, 147–162. [CrossRef]
123. Motlagh, E.K.; Asasian-Kolur, N.; Sharifian, S.; Pirbazari, A.E. Sustainable rice straw conversion into activated carbon and nano-silica using carbonization-extraction process. *Biomass Bioenergy* **2021**, *144*, 105917. [CrossRef]
124. Adebajo, S.; Akintokun, P.; Ojo, A.; Akintokun, A.; Badmos, O. Recovery of Biosurfactant Using Different Extraction Solvent by Rhizospheric Bacteria Isolated from Rice-husk and Poultry Waste Biochar Amended Soil. *Egypt. J. Basic Appl. Sci.* **2020**, *7*, 252–266. [CrossRef]

polymers

MDPI

Review

Advocating Electrically Conductive Scaffolds with Low Immunogenicity for Biomedical Applications: A Review

Dania Adila Ahmad Ruzaidi [1], Mohd Muzamir Mahat [1,*], Saiful Arifin Shafiee [2], Zarif Mohamed Sofian [3], Awis Sukarni Mohmad Sabere [4], Rosmamuhamadani Ramli [1], Hazwanee Osman [5], Hairul Hisham Hamzah [6], Zaidah Zainal Ariffin [1,*] and Kishor Kumar Sadasivuni [7,*]

1 Faculty of Applied Sciences, Universiti Teknologi MARA, Shah Alam 40450, Malaysia; dniahmad1998@gmail.com (D.A.A.R.); rosma614@uitm.edu.my (R.R.)
2 Kulliyyah of Science, International Islamic University Malaysia, Bandar Indera Mahkota, Kuantan 25200, Malaysia; sabs@iium.edu.my
3 Department of Pharmaceutical Technology, Faculty of Pharmacy, Universiti Malaya, Kuala Lumpur 50603, Malaysia; ms_zarif@um.edu.my
4 Kulliyyah of Pharmacy, International Islamic University Malaysia, Bandar Indera Mahkota, Kuantan 25200, Malaysia; awissabere@iium.edu.my
5 Centre of Foundation Studies UiTM, Universiti Teknologi MARA (UiTM), Cawangan Selangor, Kampus Dengkil, Dengkil 43800, Malaysia; hazwanee@uitm.edu.my
6 School of Chemical Sciences, Universiti Sains Malaysia (USM), Gelugor 11800, Malaysia; hishamhamzah@usm.my
7 Center for Advanced Materials, Qatar University, Doha P.O. Box 2713, Qatar
* Correspondence: mmuzamir@uitm.edu.my (M.M.M.); drzaidah@uitm.edu.my (Z.Z.A.); Kishorkumars@qu.edu.qa (K.K.S.)

Citation: Ahmad Ruzaidi, D.A.; Mahat, M.M.; Shafiee, S.A.; Mohamed Sofian, Z.; Mohmad Sabere, A.S.; Ramli, R.; Osman, H.; Hamzah, H.H.; Zainal Ariffin, Z.; Sadasivuni, K.K. Advocating Electrically Conductive Scaffolds with Low Immunogenicity for Biomedical Applications: A Review. *Polymers* **2021**, *13*, 3395. https://doi.org/10.3390/polym13193395

Academic Editor: Evgenia G. Korzhikova-Vlakh

Received: 1 September 2021
Accepted: 27 September 2021
Published: 2 October 2021

Abstract: Scaffolds support and promote the formation of new functional tissues through cellular interactions with living cells. Various types of scaffolds have found their way into biomedical science, particularly in tissue engineering. Scaffolds with a superior tissue regenerative capacity must be biocompatible and biodegradable, and must possess excellent functionality and bioactivity. The different polymers that are used in fabricating scaffolds can influence these parameters. Polysaccharide-based polymers, such as collagen and chitosan, exhibit exceptional biocompatibility and biodegradability, while the degradability of synthetic polymers can be improved using chemical modifications. However, these modifications require multiple steps of chemical reactions to be carried out, which could potentially compromise the end product's biosafety. At present, conducting polymers, such as poly(3,4-ethylenedioxythiophene) poly(4-styrenesulfonate) (PEDOT: PSS), polyaniline, and polypyrrole, are often incorporated into matrix scaffolds to produce electrically conductive scaffold composites. However, this will reduce the biodegradability rate of scaffolds and, therefore, agitate their biocompatibility. This article discusses the current trends in fabricating electrically conductive scaffolds, and provides some insight regarding how their immunogenicity performance can be interlinked with their physical and biodegradability properties.

Keywords: PEDOT: PSS; conducting polymer; conductive scaffolds; degradation rate; biocompatibility; fabrication of scaffolds; biomedical application; tissue engineering

1. Introduction

Advances in tissue engineering (TE) promise novel techniques to accelerate the recovery of damaged tissues by overcoming autologous, allogeneic, and xenogeneic tissue repair [1]. The three essential elements of TE are cells, scaffolds, and growth factors [2]. The success rate of TE is dependent on the ability of porous 3D scaffolds to mimic the function of the extracellular matrix (ECM) of a specific tissue. Scaffolds should also provide a compatible environment for the regeneration of tissues and the transplantation of organs [1,3]. Implantable scaffolds can meet the required criteria of biocompatibility, porosity, cell viability, and mechanical properties [4] by ensuring: (1) a greater control over the scaffold's surface topography, surface wettability and surface charge [5]; (2) an optimised scaffold's mass transportability to facilitate nutrient exchange and waste removal; (3) optimised biocompatibility to minimise the risk of toxicity [6–8]; (4) the presence of biological cues for cell fate to induce tissue regeneration [9–11]; (5) similarity in the rate of scaffold biodegradation and tissue growth [12,13]. The continuous development of biocompatible and biomimetic scaffolds is paramount to realise its clinical applications in improving the patient's care and quality of life [8,14].

Initially, scaffolds only act as a support system for cells to attach and proliferate. Due to this limited functionality, scaffold-based tissue engineering is primarily focused on improving tissue recovery by electrical stimuli through the application of conductive polymers (CPs). CPs exhibit electrical conductivity due to a reduction in their neutral state [15,16], and because of the presence of conjugated double bonds along the backbone. Dopant ions were commonly added into the CP's chemical structure to neutralise the unstable backbone of the polymer in its oxidised state by donating or accepting electrons [17]. Poly (3,4-ethylene dioxythiophene): poly (4-styrene sulfonate) (PEDOT: PSS) is an example of a biocompatible CP that is commonly used to produce conductive scaffolds. In another study focusing on CPs, the electrical deterioration of scaffolds could be prevented by immobilising the dopant in a polyaniline (PANI) conductive scaffold patch. This was achieved as a result of the strong chelation bonding between a phytic acid dopant and chitosan [18]. In addition, several studies have demonstrated that conductive scaffolds with polypyrrole (PPy) can promote the regeneration of nerves, bones, muscles, and cardiac cells through electrical stimulation [19]. Figure 1 shows that the electrical conduction mechanism from conductive scaffolds enables cellular signalling and function in tissues to replicate normal electrophysiology; therefore, it causes electroactive cells to align in a specific direction, as well as to migrate and proliferate. Despite the advantages of using electroactive polymers in TE, the poor biodegradability of CPs is a barrier to realising their true potential [20–23].

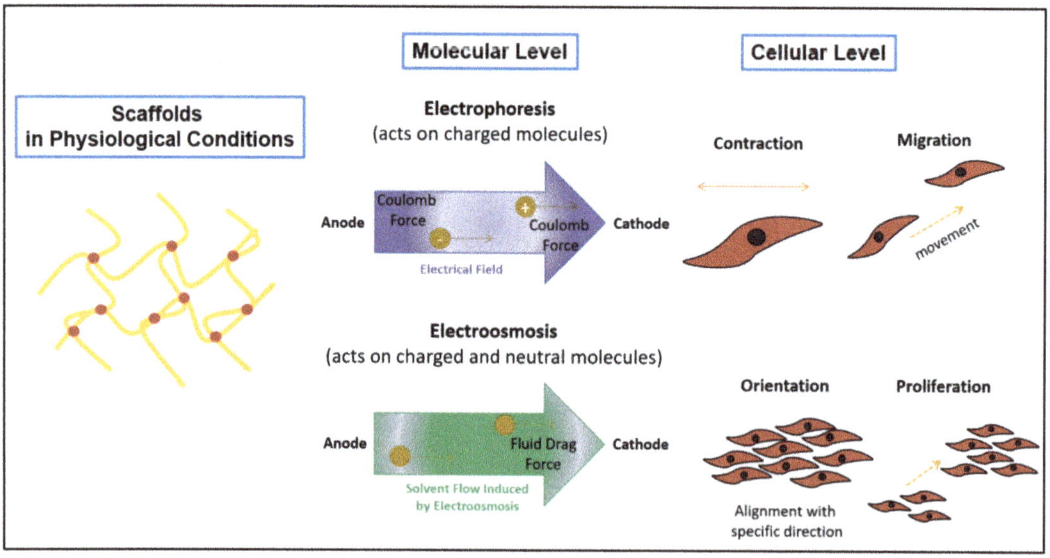

Figure 1. Illustration of the electrical conduction stimulation from conductive scaffolds, which enables cellular signalling and function in tissues.

The subject of biodegradability demands great attention in the pursuit of fabricating a conductive scaffold. The biodegradability of scaffolds is desirable, as it allows the materials to naturally dissipate, either via absorption or elimination in physiological conditions [24]. According to previous studies, conductive scaffolds were incorporated with natural compounds to make them biodegradable. Collagen and chitosan are examples of natural compounds that can physiologically degrade, without leaving behind any toxic residues. The application of non-toxic materials in scaffolds can minimise the adverse responses of the immune systems. However, studies on the immunogenicity performances of conductive and biodegradable scaffolds are scarce [5,12,14,25,26].

It is vital to assess the efficiency of conductive scaffolds, in terms of their biodegradability and electrical conductivity. Introducing biomaterials into the human body, theoretically, triggers an immune response. However, a mild impact can be expected from a conductive-implantable scaffold with a stable biodegradation rate, compared to the scaffolds with an unstable biodegradation rate. Regardless, the by-products of the scaffold biodegradation process should be non-toxic to the host. Toxic by-products cause excessive dead cells and damaging tissues, thus compromising the functions of organs. Therefore, the use of safer biodegradable materials with non-toxic by-products is desirable for the growth and proliferation of the cell. Figure 2 illustrates that, with the use of suitable materials, we can ideally alter the immunogenicity of scaffolds.

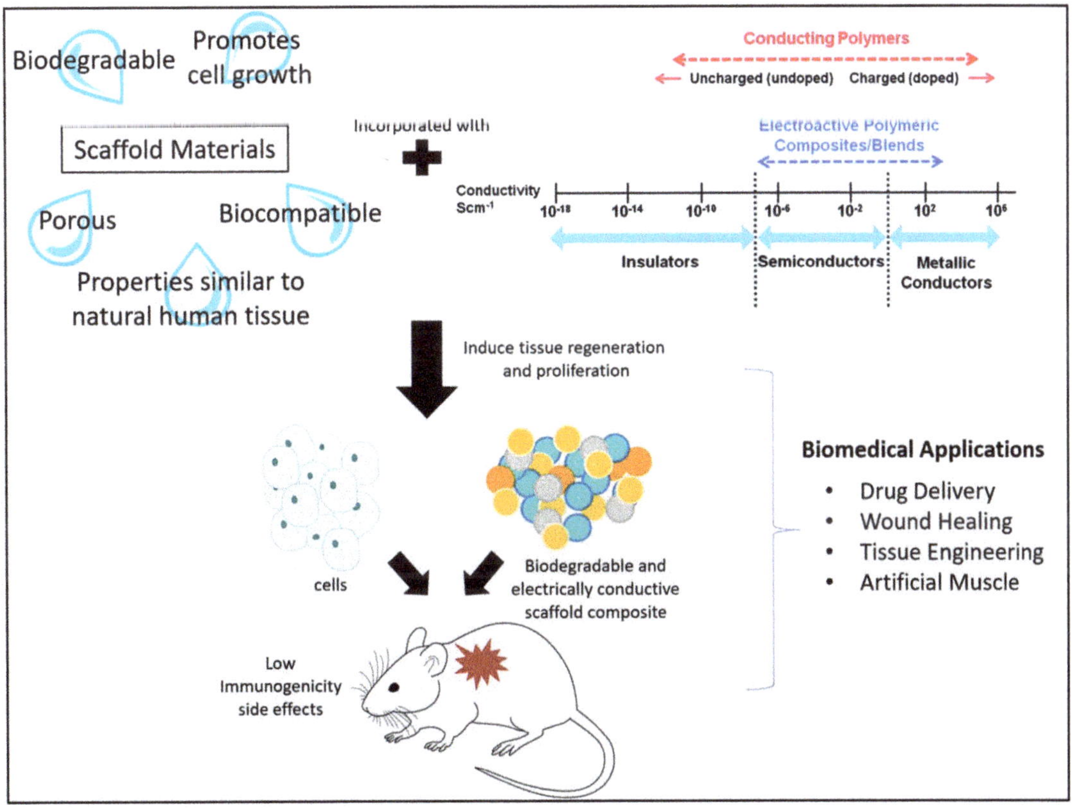

Figure 2. Illustration of the use of biodegradable and electrically conductive materials to promote cell growth with low immunogenicity.

Although organ transplantation is a proven method to address organ complications, this method is severely limited by the difficulties in securing replacement organs from donors [11]. Scaffold-based TE could be the key to this conundrum. This review article will discuss the recent fabrication techniques that can be employed to fabricate scaffolds. The contribution of the crosslinking process in regulating the degradation rate of scaffolds will be presented. A review of scaffolds' physical properties, mainly regarding their mechanical and electrical conductivity performance, will also be discussed, according to their type of CP substituents. In addition, the biodegradation trends of conductive scaffolds in phosphate-buffered saline (PBS) will be reviewed, together with a juxtaposition with human immunogenic responses. Another feature of this field of research is that its outcomes can spur innovations in drug delivery processes [27]. In the latter part of this review, we outline several of the current developments of biodegradable natural materials and synthetic polymeric materials for various biomedical applications, including tissue engineering, wound healing, and drug delivery.

2. Techniques in Fabricating Electrically Conductive Scaffolds

The fabrication and the design of conductive scaffolds influence their performance in physiological conditions. Therefore, the required properties for a specific tissue should be identified before a scaffold is produced. The selection of materials for the scaffolds is essential to promote a safe environment for cell proliferation. Scaffolds can be fabricated from natural polymers, such as silk, collagen, keratin, cellulose, and chitin. Synthetic polymers, such as polylactic acid (PLA), polyglycolide acid (PGA), and polyhydroxyalkanoates (PHA) can also act as the base material of scaffolds. Figure 3 and Table 1 list the common techniques that are used to fabricate conductive scaffolds [17,24,28–32]. Additionally, Figure 4 shows that distinct scaffold fabrication techniques yield distinct forms and structures of scaffolds, including interconnected porous, hydrogel, and nanofibrous mat [18,25,29,30,33].

At present, electrospinning and lyophilisation are two standard techniques that are used to fabricate porous scaffolds [34,35]. The electrospinning method is used to obtain the nanofiber. Nanofibrous structured scaffolds can resemble ECM tissues on the nanoscopic scale [14]. In order to produce an electrically conductive nanofibrous scaffold, CPs are usually mixed into the scaffold matrix using physical blending or through coating techniques. The combination of CPs with spinnable polymers was found to facilitate the electrospinning process, while encouraging a micro- or nano-fibre structure. A composite-conductive scaffold can be realised by adding a layer of CPs onto electrospun fibres, during the combined process of electrospinning and spin coating [25,36–38]. However, a porous scaffold that is structurally similar to foams and sponges is more stable than a nanofibrous-structured scaffold created from electrospinning. The distribution of pores in a scaffold can be rearranged randomly or by following an organised pattern. The choice of pore distribution can be accomplished by manipulating the solvent and phase separating conditions during the scaffold fabrication process. The mechanical mismatch between the host tissue and the scaffolds can be minimised by applying the best scaffold fabrication technique for the target tissue. Minimising the discrepancies in mechanical compatibility is essential to encourage the host tissue's acceptance of foreign scaffolds.

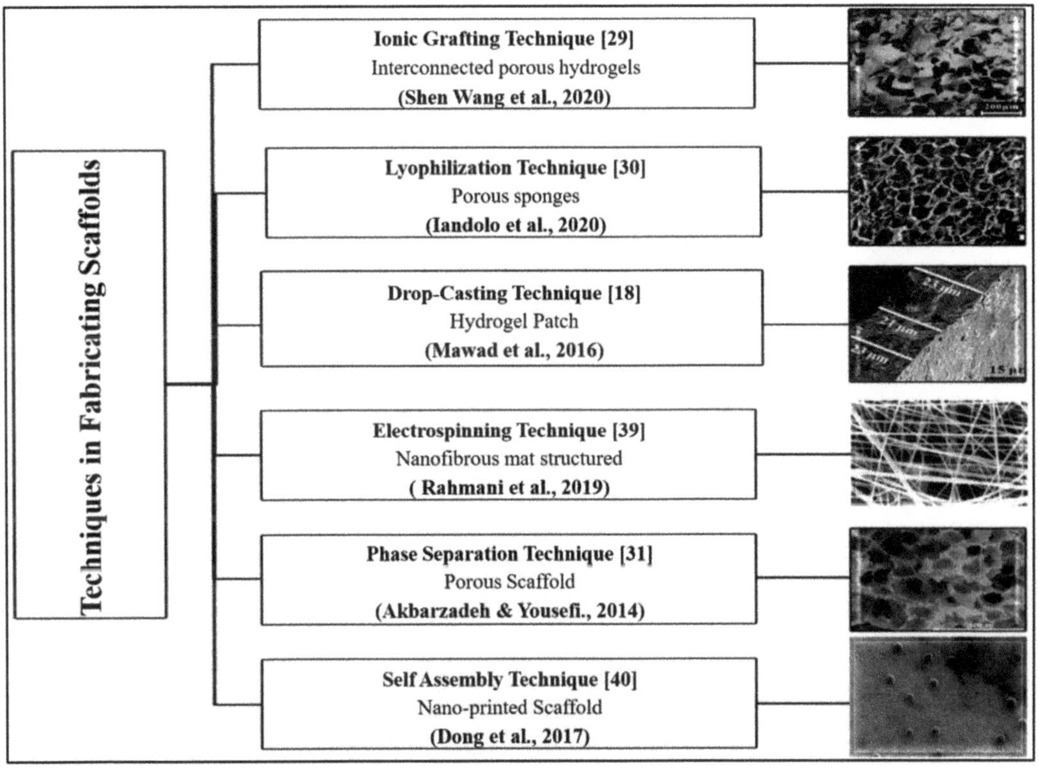

Figure 3. Techniques in fabricating scaffolds and their respective scanning electron microscope images [18,29–31,39,40].

A three-dimensional (3D) composite with a high concentration of PEDOT: PSS-gelatine-bioactive glass scaffold was fabricated with the freeze-drying method to promote cell response, attachment, and viability. These advantages increase the number of functional human mesenchymal stem cells (hMSC) [41]. Shen Wang et al. (2020) fabricated interconnected porous hydrogel scaffolds with an ionic grafting technique. In this technique, the conductive PPy polymer was grafted onto methacrylic anhydride gelatine in the presence of ferric ions. The reversible ionic interactions between ferric ions, the gelatine, and PPy mixtures afforded the hydrogels with self-healing abilities, injectable capabilities, and high electrical conductivity [29]. In addition, the incorporation of gelatine in scaffold composites introduced injectable capabilities, which is a valuable feature for nerve tissue regeneration. Although the biodegradation processes did reduce the electrical conductivity of the scaffolds, the scaffolds still met the minimum requirement of the electrical stimulation characteristic for application in neural TE [42,43]. A strong understanding of the interaction between cells and conductive materials is indispensable in researching smart biomaterials for their application in TE. Recently, electrically conductive hydrogels have been fabricated and designed using a 3D printing method, which includes fused deposition modelling (FDM), direct ink writing (DIW), inkjet printing, and stereolithography (SLA) methods [44]. For example, conductive polymer carbon nanocomposites as the main thermoplastic filament materials for the FDM method have been employed as emerging electrochemical sensing devices [45]. However, the main drawback of 3D printing methods is the low printing resolution that is produced at the end of the fabrication process.

Mawad et al. (2016) reported that the electrical stability of the PANI scaffold patch could be improved by immobilising the dopant, phytic acid, together with a PANI on the surface of a biocompatible chitosan film [18]. The PANI patch scaffold was prepared by drop-casting chitosan solution onto a glass slide, followed by drop-casting a mixture of aniline, phytic acid, and ammonium persulfate onto the chitosan film. The ability of the scaffold patch to sustain its conductive state under physiological conditions can instigate an electronic interface between the biomaterials and the implantation site. A dopant is typically introduced during the synthesis of CPs, to produce favourable electrical characteristics. Dimethyl sulfoxide (DMSO), ethylene glycol (EG), sulfuric acid (H_2SO_4), phytic acid, and *p*-Toluenesulfonic acid (*p*TSA) are commonly employed as dopants [46–54]. In their research of crosslinked porous 3D hydrogel–PEDOT CP scaffolds, Mawad et al. (2016) mixed all of the required chemicals, followed by a filtration process [55]. It was found that the electroactive region of the PEDOT hydrogel network was covalently bonded to a hydrophilic polymer, and could be further altered to fabricate a covalently linked polymeric network.

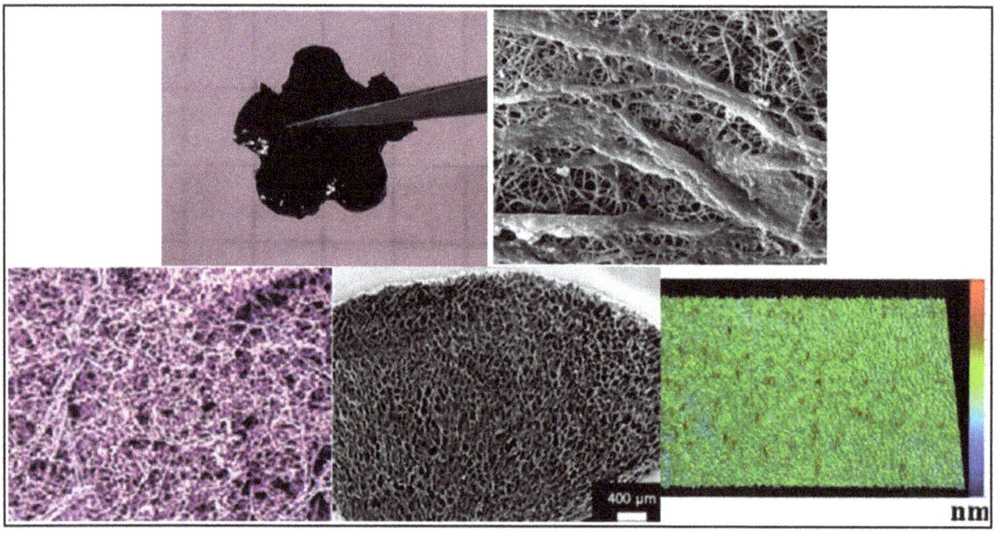

Figure 4. Images of scaffold structures: porous hydrogel, nanofibrous, porous hydrogel scaffold, porous sponge scaffold, and hydrogel patch scaffold [18,29,30,34,39].

Recently, the antibacterial, mechanical, and physical properties of a collagen-chitosan sponge scaffold, constructed from aquatic sources, were studied [56]. The researcher applied the lyophilisation technique (−50 °C, 0.5 Mbar vacuum pressure) to acquire a highly porous-structured scaffold. Furthermore, they added an alginate compound to the scaffold to improve its mechanical properties. The sponge scaffold's porosity was in the range of 88–95%, with the addition of biopolymers, such as alginate and chitosan, onto the collagen [56]. Moreover, injectable-electroconductive hydrogels have the potential to improve cell survival, which could be translated into a novel treatment protocol [29,42,57], while minimising the need for invasive surgery. Additionally, the incorporation of electro-conductive nanomaterials in hydrogels may influence their bulk electrical properties and topography, which can also affect the retention and biology of living cells. For instance, the application of CPs, such as PEDOT: PSS, in hydrogel scaffolds promotes the rhythmic beating in a neonatal rat cellular matrix [34,58]. The use of PEDOT in chitosan-based gels also resulted in an excellent regenerative capacity [43].

Table 1. Summary of scaffold fabrication techniques.

Fabrication Method	Scaffold Structure	Major Findings	Reference
Ionic grafting technique	Interconnected porous-hydrogels	Formation of conductive and self-healable hydrogels.	[29]
Lyophilization technique	Sponge scaffold	The mechanical properties of collagen sponges were improved with the addition of alginate. Future research can confirm its potency in the healing of skin ulcers.	[33]
Drop-casting technique	Hydrogel-patch scaffold	Stable and conductive scaffold patch.	[18]
Chemical mixing—a filtered technique	Hydrogel scaffold	Electroactive hydrogels with advantageous characteristics: covalently crosslinked porous 3D scaffolds with notable swelling ratio, excellent mechanical properties, electroactivity in physiological conditions and cell proliferation and differentiation.	[56]
Lyophilization technique	Porous-sponge scaffold	Scaffold biomimicry was enhanced with the addition of collagen. Collagen increases electrochemical impedance responses. Controlling scaffold's mechanical properties is highly beneficial for understanding the factors influencing cell behaviour in 3D scaffold structures.	[30]
Electrospinning technique	Nanofibrous mat-structured scaffold	The application of PEDOT: PSS with special electrical and mechanical properties as a scaffold is recommended for cardiac TE.	[25]

Iandolo et al. (2020) studied biomimetic scaffolds that supported neural crest-derived stem cell osteogenic differentiation [30]. The highly porous scaffolds were prepared by combining PEDOT: PSS with collagen type I, using the ice-templated technique. Collagen type I and PEDOT: PSS were blended in an ultrasonic bath to ensure the mixture's homogeneity. Next, the mixture was poured into a specific mould with the desired shape, followed by freeze-drying and thermal treatment. (3-glycidyloxypropyl) trimethoxysilane (GOPS) was used as the crosslinker to enhance the mechanical properties of the scaffolds. Heat treatment was performed to reticulate the crosslink [30]. The outcome of this study was a porous scaffold with the desired geometry. The ice templating process was necessary to create highly porous scaffolds that were compatible with cell infiltration and proliferation.

Another technique that is used to fabricate conductive scaffolds is electrospinning. Abedi et al. (2019) studied the fabrication and characterisation of conductive nanofibrous chitosan/PEDOT: PSS for cardiac TE. PEDOT: PSS was paired with chitosan due to its chemical and thermal stability. The combination promoted the scaffold's electrical properties, mirroring the myocardium ECM. A double nozzle electrospinning apparatus was used to form a fibrous mat structure of a conductive scaffold. The scaffold was then crosslinked using the glutaraldehyde solution vapor technique [25]. They discovered that the addition of PEDOT: PSS into chitosan/polyvinyl alcohol (PVA) increased the scaffold's conductivity. The electrospinning and crosslinking of the fibrous mat scaffold increased its mechanical strength. The presence of hydrogenic bonds, between OH groups in PVA and chitosan, and SO^{-3} groups in PSS in the CPs' dispersion, also contributed to its higher mechanical strength. Interestingly, the fibrous mat of the chitosan/PVA/PEDOT: PSS scaffold promoted the attachment and proliferation of cells.

When fabricating and designing a suitable and safe TE product, scaffolds should be able to biodegrade and should support cellular growth in vitro and in vivo. However, other aspects, such as attachment, migration, and cell proliferation, must also be taken into consideration. These aspects are highly dependent on the scaffold's surface properties and its interaction with transmembrane proteins. Low-risk tissue regeneration processes can be achieved with advanced scaffold technology.

3. The Crosslinking Process in Fabricating Conductive-Polymeric Scaffolds

During the fabrication of polymeric scaffolds, the crosslinking process is adopted to chemically bind the molecules within the scaffold's structure. This, in turn, significantly influences their mechanical properties. For example, the crosslinking process upon a nanofibrous-structured scaffold causes the fibres to attach and stack between each other, resulting in better resistance performance to the application of shearing forces. The crosslinker can be divided into three types: chemical crosslinker, biophysical crosslinker, and enzymatic crosslinker. Adding crosslinkers can enhance their mechanical properties; however, this can affect their physical condition, causing the scaffolds to become contracted or shrunk. However, in the presence of dehydrothermal (DHT), crosslinking can prevent scaffold contraction, thus making scaffolds more stable for a long period of time. Meanwhile, crosslinking with ultraviolet (UV) light can enhance the scaffold's stability in an aqueous environment, and does not cause any physical changes to the scaffolds [58–60]. This may be due to the fact that the UV light crosslinking process does not take place chemically, and interacts with the compound that is present in the scaffolds. UV crosslinking hardened the liquid polymeric material to be more stable and to possess a rigid shape, with no heat exposure exerted on the material [61]. In addition, crosslinking can also avoid the premature dissolution of the scaffold at the normal body temperature [62]. Figure 5 shows an illustration of the mechanism of the scaffold's crosslinking processes. Extensive research has been performed to develop biomedical scaffolds that meet the criteria of cost-effective fabrication, ease of customisation, and safe application in clinical settings, particularly in the areas of infection treatment and drug delivery [38,63].

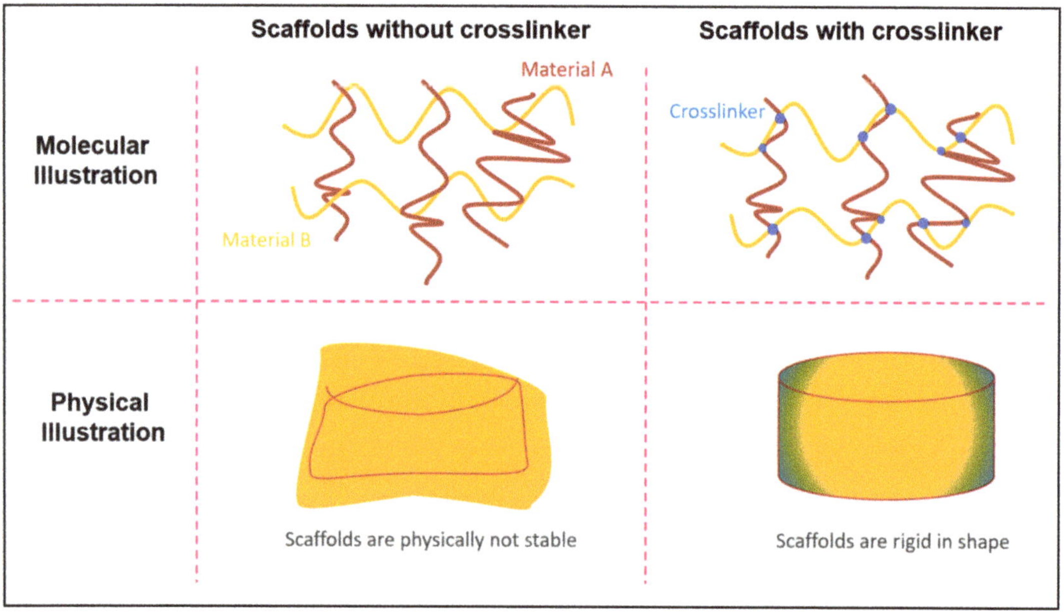

Figure 5. Illustration of the enhancement of a scaffold's mechanical properties by employing the crosslinking processes.

The intermolecular crosslinking of collagen-based scaffolds, either through chemical or physical methods, can modify their mechanical properties [12,56]. However, some well-known crosslinking agents, such as glutaraldehyde (GTA), glyoxal, and glycol diglycidyl ether, are toxic [64,65]. According to Hua et al. (2020), sodium tripolyphosphate (TPP) is a safe crosslinker for collagen and chitosan, with excellent biocompatibility [66]. A comparative study of crosslinkers reported that genipin (GP) and TPP had better biocompatibility, compared to GTA, towards the collagen/chitosan scaffold. Nonetheless, TPP was recommended as the most suitable crosslinking agent for myocardial TE [67]. In their study, collagen and chitosan were dissolved in a 1% (v/v) acetic acid solution at room temperature. After fabrication was carried out using the lyophilisation method (−20 °C for 48 h), a 1:1 ratio of a lyophilised collagen/chitosan scaffold was crosslinked. In the case of collagen, crosslinking could be achieved through the formation of an amide, from the activation of the carboxylic group with the amine. In comparison, 1-(3-dimethyl aminopropyl)-3-ethyl carbodiimide hydrochloride and N-hydroxysuccinimide were employed as crosslinkers to strengthen the covalent attachment between the carbonyl group of gelatine and the amino group of chitosan in the porous PEDOT/Chitosan/Gel scaffolds [41].

Interestingly, a study focusing on poly(vinyl) alcohol-gellan gum-based nanofiber proved that a scaffold's degradation rate could be controlled by various crosslinking agents [38]. In addition, for electroconductive hydrogels (EH), there are several types of suitable crosslinking processes, such as physical crosslinking, covalent crosslinking, and supramolecular crosslinking. However, each of these crosslinking techniques may affect the biodegradability of the EH scaffold. The density and concentration of the crosslinker can alter the properties of the scaffolds, including their water content, mesh size, porosity, diffusivity, and mechanical characteristics [60,68–70]. Overall, the use of crosslinkers could physically change the properties of scaffolds, which can affect the biodegradability of the scaffolds. This idea warrants further investigation, focusing on the execution of scaffolds' optimum degradation rates by varying the concentration of any crosslinker.

4. Physical Properties of Conductive-Polymeric Scaffolds

No single polymeric system can be considered the ideal biomaterial for all medical applications, due to the complexity of the human body and the scope of applications that polymeric biomaterials are currently utilized for. Thus, a mixture of natural and synthetic polymers can overcome the limitations of a monocomponent system [12]. Theoretically, a composite scaffold that is composed of more than two sub-materials exhibits a greater characteristic value, due to the combination of materials that hold some desired properties. Due to the various needs of scaffolds, composite materials with excellent properties are commonplace in TE [71]. For example, Lari et al. (2016) fabricated nanohydroxyapatite/chitosan (nHAp/CS) composite scaffolds that were embedded with PEDOT: PSS through a lyophilization technique. They found that nHAp and PEDOT: PSS were homogeneously dispersed in the chitosan matrix. The CS/nHAp/PEDOT: PSS scaffolds exhibited a high cell attachment rate, due to their surface roughness. The electrical conductivity that was recorded for the CS/nHAp/PEDOT: PSS sample was 9.72 ± 0.78 μS/m. They also found that both the compressive modulus and the yield strength of the chitosan increased to 0.5–1.0 MPa and 5–5.5 MPa, respectively, with the addition of nHAp and PEDOT: PSS. Unfortunately, the CS/nHAp/PEDOT: PSS had poor biodegradability, due to the strong bonding between PEDOT: PSS and the rest of the substrate chains [26]. Meanwhile, Iandolo et al. (2020) prepared highly porous biomimetic scaffolds by combining the PEDOT: PSS with collagen type I (the most abundant protein in bone) for inactive support. However, aggregation took place when collagen was a part of the mixture, due to the interaction between the positively charged protein chains and the negatively charged polystyrene sulfonic acid groups, which caused the restructuring of the CPs. Unfortunately, this process contributed to the de-doping of the conductive PEDOT segments.

Abedi et al. (2019) demonstrated that the addition of PEDOT: PSS in chitosan scaffolds could improve their biocompatibility, cell viability, and mechanical and electrical properties. In their study, the electrical conductivity, elongation at break (%), ultimate strength, and toughness of CS/PVA/PEDOT: PSS scaffolds were recorded at 7.63×10^{-3} S/m, 5.6 ± 0.3%, 18.78 ± 0.95 MPa, and 48.87×10^{6} J m^{-3}, respectively. They also discovered that the attachment length of the cells increased as the nanofiber diameter was decreased to 40 nm.

Further shrinkage of the fibre diameter to less than 40 nm affected the attachment of cells. However, the addition of PEDOT: PSS can be toxic, depending on the weight composition. The addition of more than 0.6 wt.% PEDOT: PSS in gelatine scaffolds causes toxicity in stem cells; this means that the gelatine amount must be lower than 0.6 wt.% in a single scaffold composition [25]. Overall, previous studies established that PEDOT: PSS, which is favourable for cell proliferation, can be the predecessor for a highly biomimetic, electroactive scaffold for stem cell expansion and differentiation [30].

It is essential to establish a link between materials degradation in vitro and mechano-morphological characteristics, since biodegradation can negatively affect the mechanical and structural integrity of the scaffolds. Biodegradation is manifested by material loss in grafted scaffold structures because of ageing in the PBS medium. The biodegradation rate of a scaffold can be determined by the calculation of weight loss in PBS conditions. A higher percentage of mass loss in the scaffolds can be attributed to the high surface area and porosity of scaffolds [72]. For example, Zhang et al. (2008) investigated the in vitro biodegradation of an electrospun tubular protein scaffold, by immersing the scaffold in a PBS condition (pH = 7.3) for various timeframes. This was then followed by determining the scaffolds' weight loss percentage. Ideally, the degradation rate of the scaffold should be similar to the rate of new tissue formation. The mechanical and electrical properties of various scaffolds are summarised in Table 2. In addition, the conductivity of the fabricated scaffold should refer to the sheet conductivity of the material, which is different from the commonly evaluated electrochemical impedance responses of the coated electrodes.

Table 2. Mechanical and electrical properties of conductive-based scaffold materials.

Scaffold Material	Conductivity (S/m)	Mechanical Strength (MPa)	Major Findings	References
CS/nHAp/PEDOT sponges	$9.72 \pm 0.78 \times 10^{-6}$	5.0–5.5	PEDOT: PSS/nHAp/CS is a promising scaffold, due to its porosity, microstructure, conductivity, and cell response.	[26]
Chitosan/aniline Patch	-	6.73–1.14	A better understanding of the role of conductive materials in electro-responsive tissues in ex vivo and in vivo models can be achieved by applying bioelectronic devices onto the biotic–abiotic interface.	[18]
CS/PEDOT: PSS nanofibrous	$(1.5 \times 10^{-3}$–$7.63 \times 10^{-3})$	13.07 ± 1.09–18.78 ± 0.95	Scaffolds with PEDOT: PSS showed greater cell support without any cell toxicity. The smaller fibre diameter of the fibrous mat structure can aid cell attachment.	[25]
8% PEDOT-HA/Cs/Gel hydrogel	(3.16×10^{-3})	$47.3 \pm 0.3 \times 10^{-3}$ (compressive)	An 8% PEDOT-HA/Cs/Gel hydrated scaffold, with the compressive modulus of $47.3 \pm 0.3 \times 10^{-3}$ MPa, is a viable candidate for brain tissue in nerve TE.	[9]
Conductive PEDOT layer assembled Cs/Gel	(6.51×10^{-3})—6th week (1.82×10^{-3})—8th week	-	Although the conductivity of scaffolds depreciated with progressing biodegradation, they still met the electrical conductivity requirements for electrical stimulation in neural TE application.	[10]

5. Biodegradation Mechanisms of Polymeric-Based Scaffolds

Recent advances in biodegradable biomaterial synthesis have been directed toward developing and synthesizing polymers with properties that are tailored for specific biomedical applications. Understanding the biodegradation mechanism of scaffolds in the physiological environment is necessary to optimise their functionality [73]. There are two types of biodegradation process: surface degradation and bulk degradation. As illustrated in Figure 6, the type of biodegradation depends upon the diffusivity of water inside the matrix, the degradation rate of the polymer's functional groups, and the size of the matrix. Polymeric scaffolds that experience surface degradation will preserve their bulky structures, even when their overall size is reduced. The bulk degradation of polymeric material will demolish the scaffold's internal structure and reduce its molecular mass [74].

Biodegradable scaffold materials will experience a gradual breakdown that is dependent on biological, chemical, and biophysical aspects and factors. The four types of polymeric in vivo degradation mechanisms are hydrolytic, oxidation, stimuli-associated, and enzymatic [6,27,75,76], as summarized in Figure 7. The biodegradation of polymeric biomaterials such as scaffolds involves a hydrolysis process that is initiated by water molecules and the disintegration of sensitive bonds in the polymer by enzymes. These events cause the erosion of the polymer. The biodegradation rate depends upon the physiological environment and the intrinsic properties of the scaffolds, as follows: (1) the chemical structure, (2) the presence of hydrolytically unstable bonds, (3) the level of hydrophilicity and hydrophobicity, (4) the crystalline morphology, (5) the glass transition temperature, (6) the copolymer ratio, (7) molecular weight, (8) tacticity, (9) loading direction, (10) pH, and (11) the treatment processes that are involved during scaffold fabrication, such as crosslinking process [69,77].

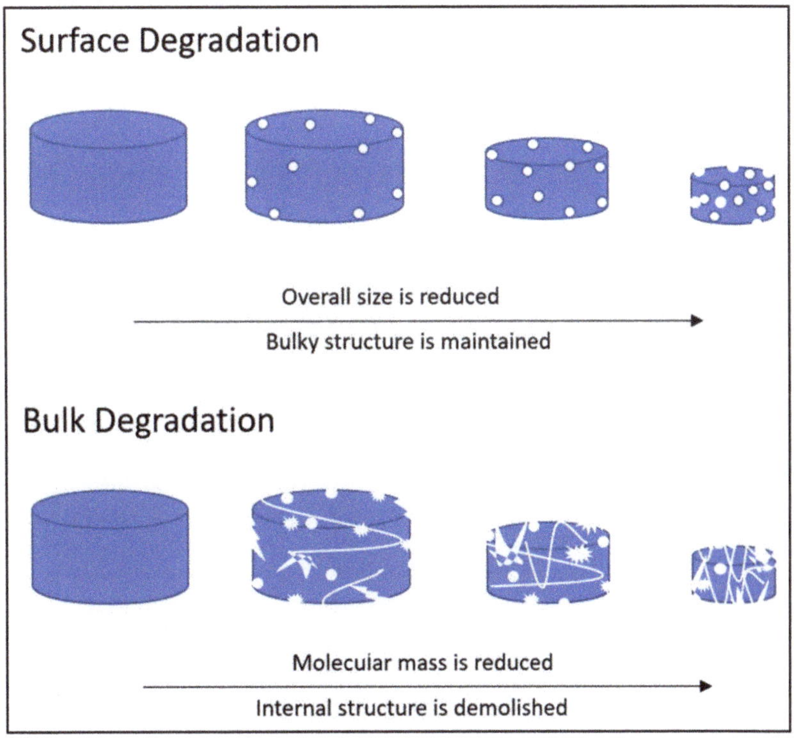

Figure 6. Illustrations of the surface and bulk degradation process.

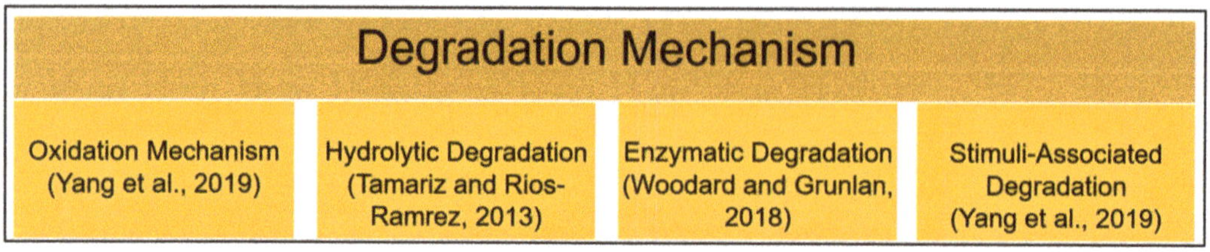

Figure 7. Different types of degradation mechanisms in physiological condition.

The hydrolytic degradation of polymers entails the collapse of chemical bonds in the polymer backbone by water molecules. Acids, bases, or salts catalyse the degradation to form oligomers and monomers. This form of degradation significantly reduces the molecular weight of the polymer. A previous study reported that the addition of PPy causes the polymer chains in scaffolds to aggregate and resist water diffusion, which slows down the degradation rate. This is due to the increasing number of hydrophobic bonds that are present after the blending of PPy with the PCL-CS mixture [78]. Figure 8 is an example that illustrates the hydrolytic degradation mechanism of PLA scaffolds. Mild hydrolysis results in the slight degradation of the polymer's surface, revealing the surface carboxyl and hydroxyl groups. Carboxyl may be present in several forms, including the carboxylate ion, carboxylic acid, and carboxyl salt [79].

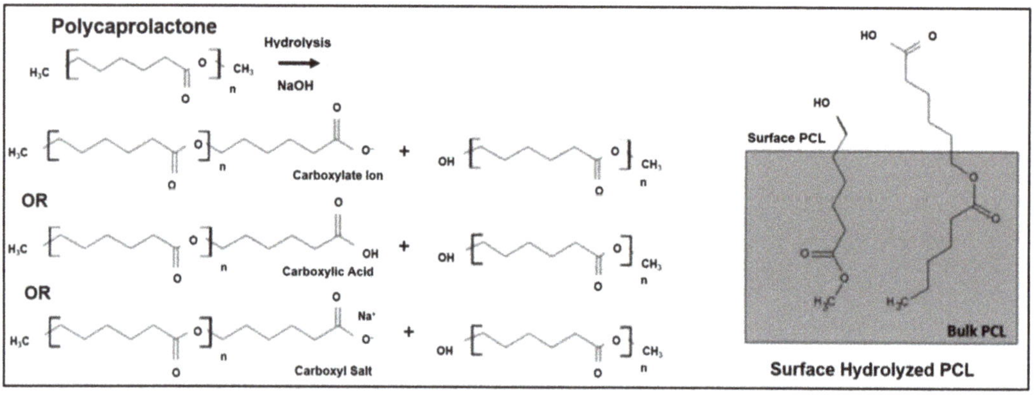

Figure 8. The hydrolytic degradation mechanism of PCL scaffolds [79].

The degradation mechanism process can also be mediated by biological agents, such as enzymes, that partake in tissue remodelling. Polymeric scaffolds are also vulnerable to oxidation mechanisms. When scaffolds are exposed to body fluids and tissues, the host's immune cells will initiate inflammatory responses. This situation can cause the release of highly reactive oxygenic molecules, such as hydrogen peroxide (H_2O_2), superoxide (O_2-), nitric oxide (NO), and hypochlorous acid (HOCl). These molecules accelerate polymer chain scission and the degradation of scaffolds. Figure 9 shows an example of the oxidative degradation of poly (urethane) derivatives by hydrogen peroxide compound, including (A) poly(ether urethanes), (B) poly(carbonate urethanes), and (C) aromatic polyurethanes, which produce glycol ether radicals as by-products.

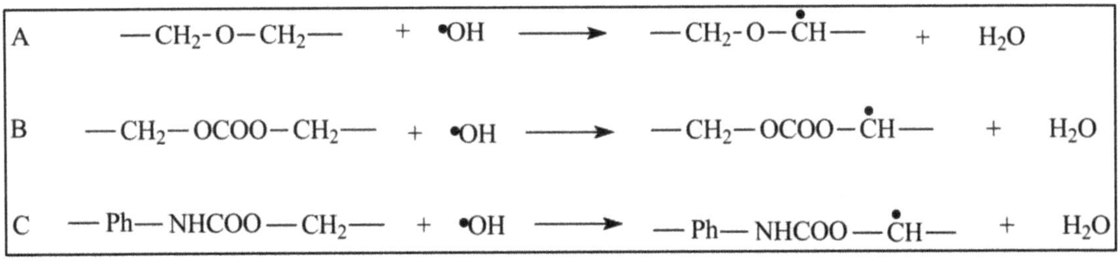

Figure 9. A mechanism of oxidative degradation by H_2O_2 in poly(ether urethanes) (**A**), poly(carbonate urethanes) (**B**), and aromatic polyurethanes (**C**) [80].

The degradation mechanism process can also be mediated by biological agents, such as enzymes, that partake in tissue remodelling [73]. Figure 10 shows an example of the enzymatic degradation mechanism of methyl methacrylate-poly 3-(trimethoxysilyl) propyl methacrylate (MMA-TMSPMA) star polymers that are synthesised with the arms of three different architectures (random, inner, and outer), crosslinked with a dimethacryloyl peptide (MaCh-peptide) core, and cleaved with collagenase activity [81]. The enzymatic mechanism often proceeds, concurrently, with hydrolytic degradation. The presence of hydrolases enzymes, such as proteases, esterases, glycosidases, and phosphatases, catalyses the hydrolysis reaction of biomaterial disintegration [27]. The interaction between the enzymes and the polymeric chains begins with the diffusion of specific enzymes on the polymer's solid surface. This diffusion is followed by the enzyme–substrate complex formation, whereby the substrate causes a conformational or shape change of the enzyme–substrate complex. Catalysis of the hydrolysis reaction occurs, and the soluble by-products diffuse into body fluids [6].

Additionally, Figure 11 shows examples of the stimuli that are associated with degradation mechanisms; a pH-sensitive drug-gold nanoparticle system for tumour chemotherapy, and surface-enhanced Raman scattering (SERS) imaging [82]. This concept can be applied to scaffold degradation mechanisms with the use of doxorubicin drugs as fillers for a specific treatment. According to Yang et al. (2019), a stimuli-associated degradation mechanism that encourages scaffold swelling normally manifests in sol-gel degradation behaviour. In this case, the degradation process occurs by allowing the scaffold network structure to be cleaved by external triggers: pH-responsive, light-responsive and redox-responsive [8]. A pH-responsive smart hydrogel offers targeted and controlled release behaviour to wounds, while its network architecture remains intact, with slower degradation in normal tissues. A photo-responsive hydrogel goes through light-mediated degradation, while redox-responsive hydrogels react to internal and external oxidative and reductive stimuli.

The scaffold degradation rate should be on par with the tissue ingrowth to maximise healing or to deliver healable drugs. Generally, the degradation of polymers in physiological conditions is caused by a molecular chain scission that is initiated by hydrolysis (anhydride, ortho-ester, ester, urea, urethane/carbonate, and amide bonds) or enzyme-catalysed hydrolysis. A number of degradable polymeric scaffolds contain labile bonds that tend to hydrolyse. Additionally, these bonds are too stable under physiological conditions. Thus, they require an enzymatic catalyst to encourage degradation [70]. There are several non-invasive techniques to monitor in vivo scaffold degradation. Electron paramagnetic resonance (EPR) is an efficient and accurate technique to investigate radical and oxidative stresses [83]. Ultrasound elasticity imaging (UEI) can be used to characterise the structural, functional, and compositional changes of biodegradable scaffolds via phase-sensitive speckle tracking [84,85]. Several non-invasive and non-destructive techniques to investigate parameters such as a scaffold's pH value, distribution, and cell viability are: (i) confocal laser scanning microscopy (CLSM); (ii) nuclear magnetic resonance (NMR); (iii) optical coherence microscopy (OCM); (iv) optical coherence tomography (OCT). OCT can be used in tandem with various light sources, such as near-infrared fluorescence (NIR) [86,87]. Zhang et al. (2020) innovated a multifunctional hydrogel system with tetraphenylethene (TPE), that has similar traits to aggregation-induced emission (AIE) nanoparticles, to monitor the degradation of hydrogel scaffolds in physiological conditions [88,89].

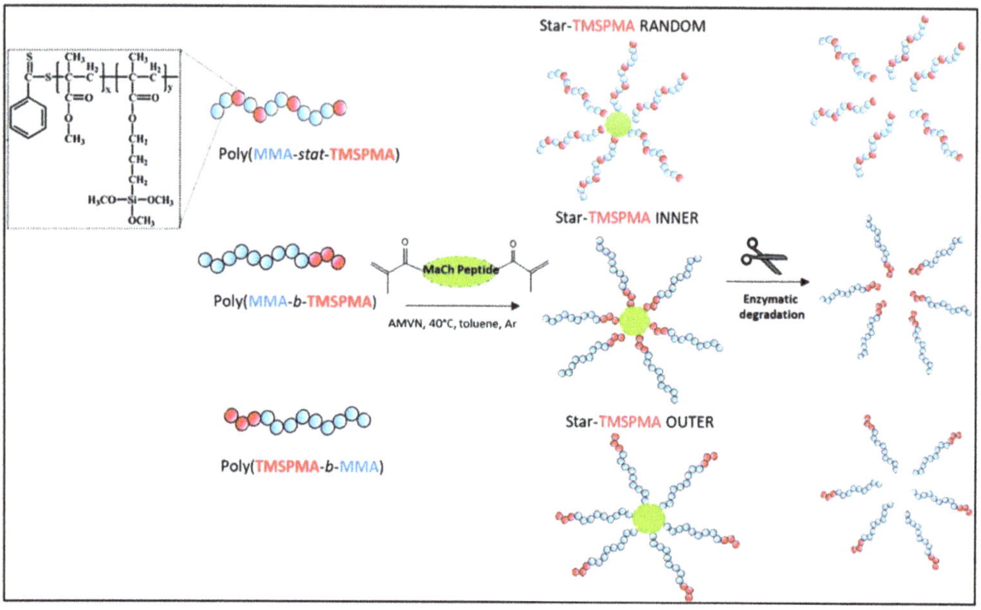

Figure 10. Illustration of the enzymatic degradation process [81].

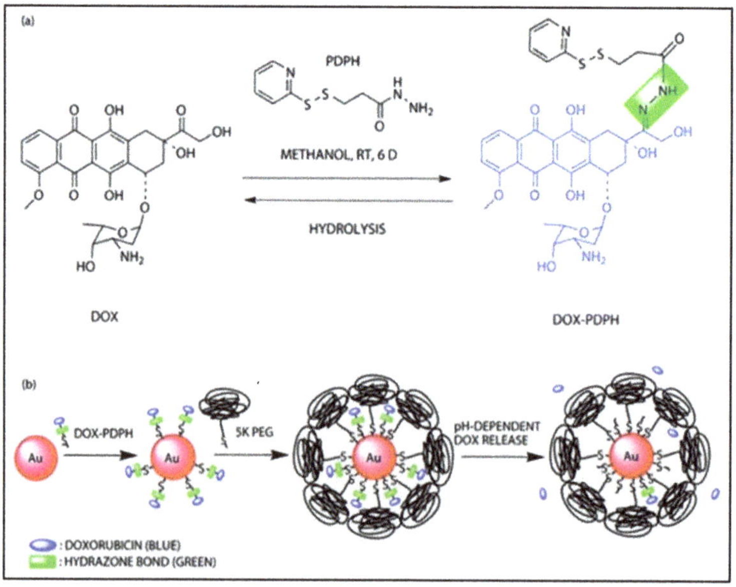

Figure 11. A pH-sensitive drug-gold nanoparticle system: (**a**) chemical synthesis of the doxorubicin–hydrazone linker conjugate (dox–PDPH); (**b**) schematic illustration of the synthesis of the multifunctional drug delivery system and its pH-dependent doxorubicin release [82].

6. Immunogenic Effects on the Biodegradation Behaviour of Scaffolds

Immunogenicity is a biological response to the presence of non-compatible foreign substances or living organisms in the body. This reaction can cause complications for one's health. Therefore, biocompatible scaffolds are imperative to circumvent immunogenic effects. The by-products of degradation, such as monomers, oligomers, and polymer fragments, should be biocompatible with the human body, and should be able to pass through filtering organs without causing any complications. Scaffold materials with a low degradation rate might lead to transplant failure, due to negative immune responses. Woodard and Grunlan (2018) [75] underlined the scaffold's dimensions as a factor that influences degradation. This idea was also presented in another study [74].

Figure 12 illustrates how different biodegradation rate trends are capable of tuning the immunogenicity of scaffolds. A poor degradation rate of a non-compatible scaffold will inhibit tissue regeneration, since the scaffold will continuously trigger immune reactions. However, incompatible scaffolds with high degradation rates could compromise tissue regeneration rates, due to the absence of cellular support. Altering the material degradation rate seems to be a promising means of optimising the scaffold's biocompatibility. Natural polymers often exhibit better biocompatibility and biodegradability. However, the degradation of synthetic polymers can be tailored accordingly. Acidic by-products, which are hydrolytically produced from polyester scaffold degradation, may also cause physiological inflammatory responses [75]. In some cases, a low pH environment can unnecessarily accelerate the scission of scaffolds. Therefore, biodegradable polyester for physiological purposes can be optimised by introducing basic salts into the polyester, such as calcium carbonate, sodium bicarbonate, and calcium hydroxyapatite [75]. The use of biodegradable natural polymers, such as collagen, was suggested, according to the presence of collagenase as a physiological enzyme [90]. The presence of natural polymers in scaffolds only triggers mild biological responses, and this can prevent severe immunogenicity side effects.

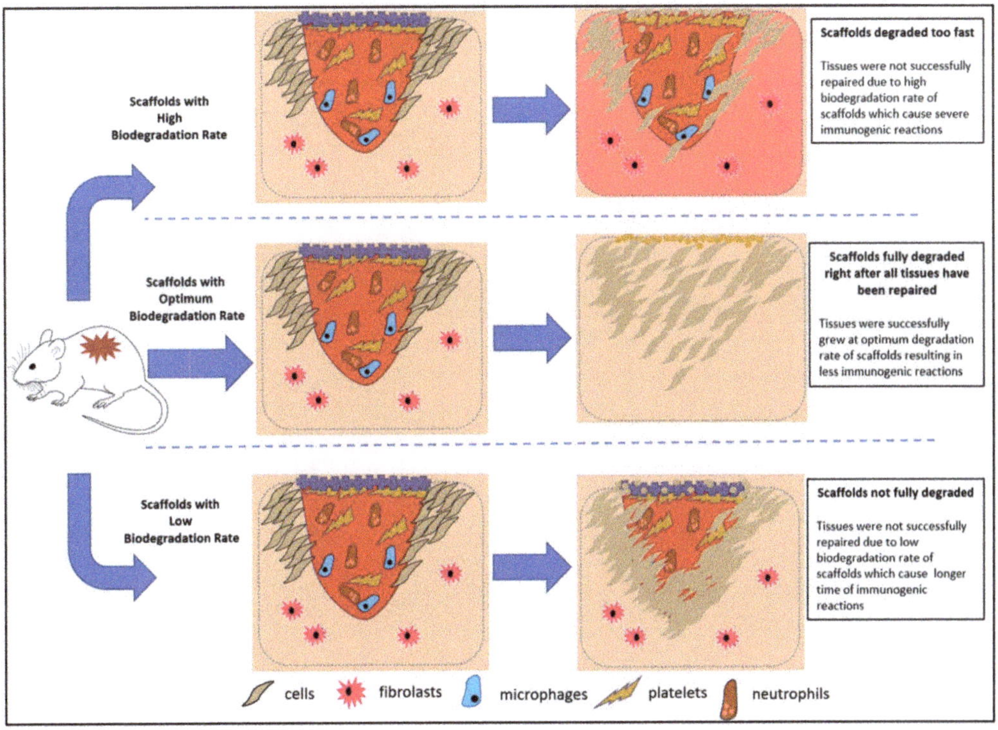

Figure 12. Illustration of the different immunogenic responses towards the specific biodegradation trends of scaffolds.

7. Common Conductive and Biodegradable Scaffolds

7.1. PEDOT-Based Scaffolds

7.1.1. Characteristics of PEDOT-Based Scaffolds

To utilize CPs for developing and synthesizing scaffolds with properties that are tailored for use in tissue engineering, the suitable conductive and hybrid systems of biocompatible scaffolds must be discussed. Extensive research is ongoing to optimise the stability of scaffolds, by incorporating composite materials to overcome the problems with tissues [18,38,58,68,91]. The scaffold materials must fulfil the criteria of convenient sterilisation, biocompatibility, and non-toxicity [92]. The seamless electrical communication among cells, and the optimised growth of cells, can be achieved by exploiting scaffolds with sufficient electrical conductivity. Recently, the focus of scaffold-based TE is on the enhancement of bone healing via electrical stimuli with CPs [40]. The addition of modified electroactive oligomers can control the biodegradability of CPs. These grafted copolymers are connected via degradable ester linkages, and are highly sought after in biomedical applications that employ pyrrole, aniline, or thiophene groups [23]. Unfortunately, electroactive aniline-based oligomers are toxic to human bodies. In response to this observation, Mawad et al. (2016) proposed 3,4-ethylene dioxythiophene (EDOT)-based oligomers as an alternative to replace aniline-based oligomers, to minimise the biomaterial's toxicity [54].

PEDOT can be chemically tuned to alter the mechanical and electrical properties of scaffolds. This tuning can promote the covalent attachment between biomolecules and scaffolds to become biocompatible [93]. In addition, conductive scaffolds are conductive for the differentiation and proliferation of electrically stimulated responsive cells [28]. PEDOT: PSS is a common biocompatible CP behind conductive scaffolds. Figure 13 shows the chemical structure of PEDOT: PSS. This copolymer has a moderate band gap and excellent stability [25]. For example, the addition of PEDOT: PSS in a chitosan-based electrospun scaffold not only enhances the scaffold's mechanical and electrical conductivity, but also improves its biocompatibility and cell viability [24]. The findings regarding the biodegradability of various scaffolds including conductive scaffolds from previous research are summarised in Table 3. Predominant crosslinking in scaffold fabrication (as discussed in the Crosslinking Process section in Fabricating Conductive-Polymeric Scaffolds (Section 3)), increases the mechanical properties and stability of the scaffold at the possible expense of the scaffold's biodegradability.

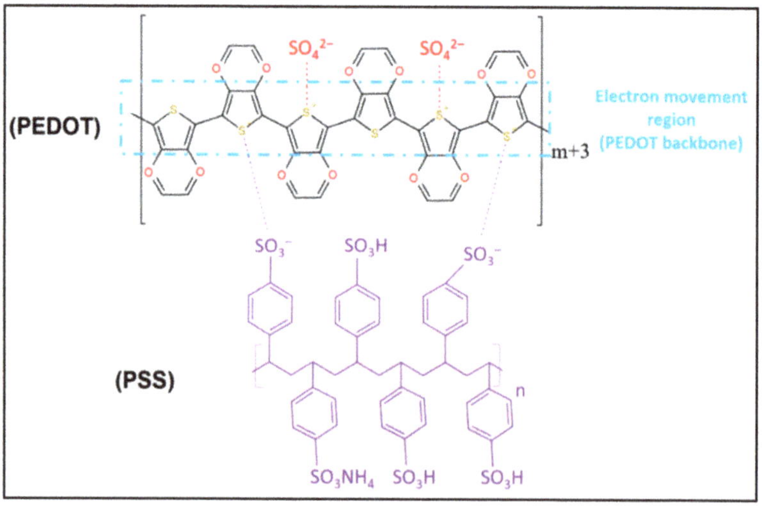

Figure 13. The chemical structure of PEDOT: PSS with the presence of sulfonate ions from an oxidant [94].

7.1.2. Biodegradable Trends of Various Conductive PEDOT-Based Scaffold Composites

The research focusing on scaffolds for biomedical applications has made significant strides with the advent of new technologies. The mechanical and electrical properties of scaffolds will affect the scaffold's biodegradation rate and biocompatibility [72,95–97]. Wang et al. (2017) experimented with varying weight percentages of PEDOT-HA nanoparticles in the fabrication of hyaluronic acid-doped PEDOT/chitosan/gelation porous conductive scaffolds [9]. They reported that the in vitro biodegradation of the scaffold had an inverse relationship with the weight percentage of PEDOT-HA. The scaffold displayed high biodegradability when there was a minute amount of PEDOT-HA. A large volume of PEDOT-HA enhanced the stability and biodegradation resistance of the scaffold. The addition of hydrophobic PEDOT likely reduced the hydrophilicity of the scaffolds [9].

Table 3. Biodegradation index of various composite scaffolds.

Scaffold Composition	Scaffold's Biodegradability Claims	Reference
PEDOT: PSS Chitosan Hydroxyapatite Nanoparticles	• CS and nHAp/CS scaffolds showed approximately 30% weight loss after 1 month. • PEDOT: PSS/nHAp/CS scaffold degraded with 10% weight loss. • Incorporation of PEDOT: PSS enhanced the stability of scaffold in PBS.	[26]
PEDOT Hyaluronic acid Chitosan Gelatine	• CS/Gel scaffold showed 86% weight loss after 8 weeks of incubation in lysozyme enzymatic solution. • 10% PEDOT-HA/CS/Gel scaffold experienced 43% weight loss after 8 weeks of incubation. • The presence of PEDOT and hyaluronic acid enhanced the stability and biodegradation resistance of scaffolds, due to low water permeability in hydrophobic PEDOT.	[9]
PEDOT Chitosan Gelatine Hyaluronic acid	• Biodegradation of 0.5 wt.% PEDOT/CS/Gel scaffold was 72.55 ± 3.79% at week 8 in an enzymatic solution. • Biodegradation of PEDOT/Cs/Gel scaffold was 59.97 ± 3.22%, while it was 47.15 ± 2.17% for PEDOT-HA/CS gel scaffold. • The addition of PEDOT lowers the degradation rate of the scaffold. • PEDOT nanoparticles mediated cell–surface interactions and enhanced cell's growth and proliferation.	[41]
Chitosan PEDOT Gelatine	• The conductivity of hydrated and dehydrated PEDOT/CS/Gel scaffold gradually decreased over the degradation period. • Chitosan and gelatine gradually degraded in an enzymatic solution. • The conductivity value of conductive scaffolds still met the required electrical value for neural tissue engineering application after 8 weeks of incubation time.	[10]
Gelatine Sodium Alginate PEDOT: PSS	• Scaffolds showed 30% and 70% weight loss after 7 and 30 days of incubation in DMEM: F12 shaker incubator, respectively. • There was no significant difference between scaffolds with or without a conductive-polymer.	[98]
Chitosan k-carrageenan N-isopropyl acrylamide (NIPAM)	• Chitosan/k-carrageenan with 0 µL, 200 µL and 400 µL of gold (Au) recorded weight loss of 27.41%, 36.52%, and 40.12% in PBS solution at week 8, respectively. • The addition of gold nanoparticles increased the weight loss due to decreasing the molecular weight of chitosan through the catalytic activity of Au. • Hydrolysis, dissolution, and enzymatic cleavage are the types of degradation mechanisms involved.	[44]

Table 3. *Cont.*

Scaffold Composition	Scaffold's Biodegradability Claims	Reference
Gellan PVA	• PG-NFs scaffold was stable and can maintain its structure in PBS solution for up to 14 days of incubation. • The initial weight loss is due to non-crosslinked gellan and PVA molecules.	[38]
Polycaprolactone (PCL) Polypyrrole-block- poly(caprolactone) (PPy-b-PCL)	• PCL and PCL/PPy 2% scaffolds had degradation rates of 37.28% and 55.8%, respectively. • The degradation rate of the scaffold in the incubator shaker increased with an increasing PPy-b-PCL concentration.	[68]
Hydroxyapatite Collagen I Chitosan Glutaraldehyde	• Ha-Col-CS-GTA recorded 10% and 16% weight loss on day 1 and 21, respectively, after incubation for the biodegradation test using AF-MSCs. • Scaffold without a crosslinker GTA showcased a higher degradation rate of 39% and 55% after day 1 and 21 of incubation time, respectively.	[13].
Polycaprolactone polypyrrole	• The maximum weight loss of the PCL/chitosan mat was 20% in PBS solution after 14 days of incubation. • An amount of 7.5% PPy in the PCL/chitosan mat scaffold showed 12% weight loss after incubation time. • Adding PPy into the PCL/chitosan fibres slowed the weight loss and degradation rate.	[78].
Collagen Chitosan GTA Genipin TPP	• The highest degradation rate (55.0 ± 3.78%) at day 21 and (62.0 ± 4.23%) 28 were recorded on the TPP crosslinked scaffold. • The GTA solution crosslinked scaffold had the slowest degradation rate (17.7 ± 1.57% and 26.5 ± 2.98%) at day 21 and 28, respectively. • The GTA vapour crosslink degradation rate was relatively higher (3.4 ± 2.85%) on day 28. • The higher the crosslinking degree, the lower the degradation rate. • GP and TPP were only able to crosslink 78.38 ± 8.20% and 143.27 ± 4.03%, respectively (crosslinking degree between the crosslinker and collagen/chitosan scaffolds).	[67].

Wang et al. (2017) also researched PEDOT nanoparticles/chitosan/gelatinous porous scaffolds. They claimed that the presence of PEDOT nanoparticles significantly reduced the degradation rate of the scaffolds. Interestingly, the presence of PEDOT nanoparticles in the scaffold increased the cell viability. This is due to the genial interactions between the scaffold and the cell surface, which encourage cell proliferation and growth [40]. Another contribution from Wang et al. (2018) was the study of PEDOT/chitosan/gelatinous scaffolds for neural cells. The electrical conductivity of hydrated and dehydrated PEDOT/chitosan/gel scaffolds gradually diminished over time [10]. This observation could be attributed to the disintegration of the PEDOT layers on the surface of the chitosan/gelatinous matrix. The dissipation of PEDOT layers occurs when chitosan and gelatine are gradually degraded in the presence of the enzymes that are supposed to stimulate physiological conditions. Although the electrical conductivity of the scaffold reduced over time, as summarised in Table 3, it still met the electrical conductivity requirements for electrical stimulation in neural TE application.

Another study demonstrated that the addition of conductive PEDOT: PSS in nHAp/chitosan composite scaffolds using the lyophilisation method reduced the scaffold's biodegradability rate in the PBS solution. Although the scaffold with PEDOT: PSS had low biodegradability, its mechanical properties were consistent [26]. Lari et al. (2020) attempted to decrease the wettability, while dialling up the mechanical properties, of PEDOT: PSS/nHA/CS biocomposite, by integrating polycaprolactone (PCL) into the scaffold. PCL was selected due to its biodegradability and ease of blending with chitosan [5]. In another study, a PCL-CS-PPy conductive biocomposite nanofibrous scaffold is also a subject of interest in TE. The nanofibrous-structured scaffolds with sufficient biodegradability can be fashioned using electrospinning. It is worth mentioning that the mechanical properties decrease in tandem with the weight percentage of PPy. When the weight percentage of PPy used was reduced, it was easier for the scaffold to disintegrate, exhibiting the non-polymeric scaffold's typical mechanical behaviour [68].

In addition, Abedi et al. (2019) fabricated a conductive nanofibrous chitosan/PEDOT: PSS scaffold using the electrospinning method [25]. They reported that the addition of PEDOT: PSS in the scaffold may support cell growth without any toxic effects. Nonetheless, the biodegradability of the scaffold has yet to be explored. It was reported that the use of PEDOT: PSS in fabricating the scaffold did not change the biodegradability of the scaffold, due to the presence of alginate [98]. In addition, they also stated that an increase in PEDOT: PSS concentration in gelatine-alginate scaffolds could increase cell proliferation, although they did not report the cause. A takeaway from these studies is that the addition of CPs in the polymer matrix is a prerequisite for acquiring electrically conductive scaffolds. Adjusting the weight percentage of the CPs in the matrix composite of the scaffolds will affect their biodegradability. Therefore, further studies focusing on the relationship between the addition of CPs and the superior biodegradability of conductive scaffolds are warranted.

7.2. Collagen-Based Scaffolds

The addition of collagen or silk fibroin in scaffolds as biodegradable substituents can enhance their biodegradability, due to the nature of their proteins. Bioactive molecules, such as collagen, chitosan, and hydroxyapatite (HAp) were described as compatible, non-toxic, non-carcinogenic, non-immunogenic, and soluble in physiological conditions [13]. Due to these properties, the application of a collagen scaffold is prevalent in the field of damaged tissue regeneration [12,13]. Unfortunately, collagen is vulnerable to rapid degradation in body fluid or cell culture media [99]. The breakdown of collagen fibres depends upon the proteolytic action of collagenases, which are part of the large family of matrix metalloproteinases. For type I collagen, the cleavage site is specific, generating three-quarter and one-quarter length fragments. These fragments are further degraded by their matrix proteinases, as illustrated in Figure 14. Therefore, it is mandatory to blend collagen with other materials to augment the mechanical properties of collagen-based scaffolds.

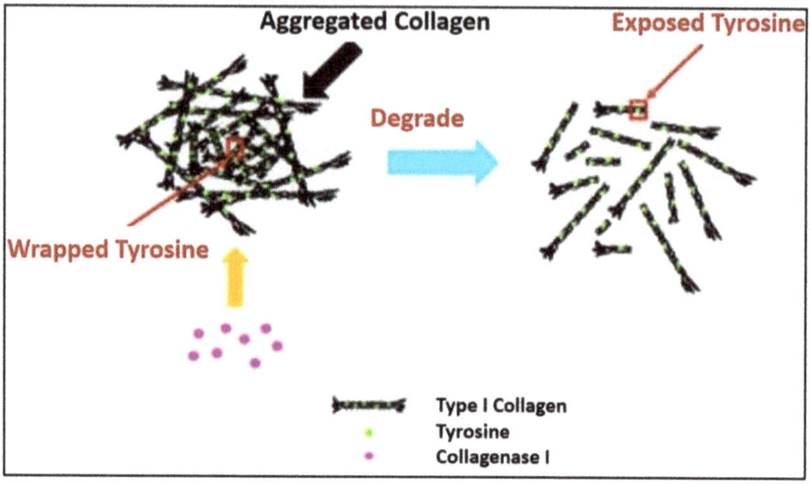

Figure 14. Process of collagen breakdown with the presence of a physiologic collagenase enzyme [100].

7.3. Chitosan-Based Scaffolds

Another biodegradable polymer, chitosan, has become relevant in TE, due to its features and properties of low toxicity, non-immunogenic, and biodegradability, that are similar to the native ECM [5,9,14]. Chitosan degrades the body through physical and chemical degradation. The former entails swelling, cracking, and dissolution, while chemical degradation results from depolymerisation, oxidation, and hydrolysis [78]. Chitosan behaves as a hydrophilic cation, due to the electronegativity of its amino groups. The deacetylation degree (DA) of chitosan generally influences its polarity, pH, ionic strength and, ultimately, its water-soluble behaviour. Chitosan usually degrades at a pH below 6 [76]. The breakdown of chitosan through the pH degradation mechanism is illustrated in Figure 15. Chitosan is a polysaccharide with a cationic nature and displays outstanding properties, such as biocompatibility, hydrophilicity, and anti-thrombogenicity. Furthermore, chitosan can be combined with various polymer materials or drugs, using the appropriate preparation techniques [4,10,14,18,65,101–103]. Nonetheless, their mechanical and electrical attributes are unique, depending on their fabrication technique.

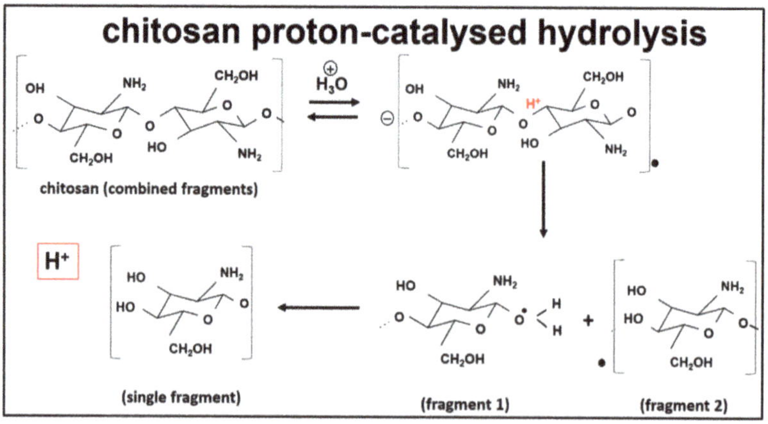

Figure 15. The breakdown of chitosan at pH 3 [76].

A study demonstrated that the addition of PEDOT: PSS into chitosan/PVA altered the mechanical properties and conductivity of the scaffold [24]. Conductive polymeric scaffolds with low degradation rates can be turned into composite scaffolds with high degradation rates by mixing them with biodegradable materials. Modifying the CPs backbone by adding typical enzymatically cleavable or hydrolysable linkages between biodegradable materials and CPs backbone also offers a scaffold with a higher biodegradability rate [104]. Table 4 shows a summary of the biodegradation trend of collagen, chitosan, and PEDOT: PSS-based scaffolds.

8. Current Developments of Polymeric Materials for Biomedical Applications

CPs hold favourable characteristics, such as electronic–ionic hybrid conductivity, mechanical softness, permeable porosity, and versatile chemical modification. This means they are recommended for a wide range of biomedical applications, including biosensors, chemical sensors, drug delivery systems, artificial muscles, and neural interfaces [105]. Additionally, CPs are utilized in the application of artificial muscles, due to their electrochemical deformation properties. The magnitude of the CPs' strain depends upon their number of anions. Briefly, by applying a positive voltage with a suitable electrolyte, the polymer becomes oxidized, and the material loses electrons from the polymer. Then, a pair of anions are formed in the electrolyte. These anions cause the expansion of the polymer. The CP's contraction mechanism (reduction reaction) is similar to the expansion mechanism (oxidation reaction) [106]. In addition, CPs can also be used to replace alkaline metal as biosensors, such as for non-enzymatic glucose sensors, hydrogen peroxide (H_2O_2) sensors, and dissolved oxygen sensors [107]. Recently, CPs like PANI was reported to have antibacterial properties through a disruption process against the native surface charge of bacterial cells [108]. This finding successfully proved that PANI can be utilized in producing antibacterial medical appliances. However, the use of CPs for skin biosensors was doubted, due to their biocompatibility issues (including inflammation and serious disorders) when in contact with living human physiology [109]. Hence, the utilisation of CP biomaterials in biomedical applications warrants more research in the near future, since CPs hold several drawbacks alongside their advantageous properties. Table 4 shows the current development of biodegradable natural and synthetic polymeric materials for various biomedical applications, including tissue engineering, temporary implants, wound healing, and drug delivery.

Table 4. Polymeric materials for biomedical applications.

Specific Application	Polymer Type	Material	Reference
Drug delivery	Synthetic Biopolymer	poly(lactic-*co*-glycolic) acid (PLGA)	[110]
	Biodegradable natural polymer	chitosan	[111]
	Biodegradable natural polymer	kefiran	[112]
	Synthetic biopolymer	Poly (ethylene) glycol (PEG)	[113]
Tissue Engineering	Biodegradable natural polymer	gelatine	[94]
	Synthetic biopolymer	Polyurethane and modified polyurethane	[114]
	Biodegradable natural polymer	collagen	[115]
	Synthetic biopolymer	polyester derivatives, such as poly ϵ-caprolactone (PCL), polylactic acid (PLA), and polyglycolic acid (PGA)	[116]
Temporary Implants	Synthetic degradable polymer	shape memory polymers (with shape memory effects)	[117]
	Synthetic biopolymer	poly(lactic) acid (PLA)	[118]
Wound Healing	Biodegradable natural polymer	fibrinogen, hyaluronic acid, cellulose	[119]
	Synthetic polymer	polyvinyl alcohol	[120]
	Biodegradable natural polymer	protein derivatives, such as silk; collagen bacterial polyester, such as bacterial cellulose	[121]

9. Conclusions

Fabrication techniques for conductive and biodegradable scaffolds could affect the physical and mechanical properties of the scaffold. Injectable scaffolds have the potential to be the most suitable low-risk method for TE-related medical treatments. Optimising the scaffold's properties to resemble the properties of human ECM tissue is an ideal goal to realise a scaffold's potential. This will lower the property gap between the scaffold and the tissues at the implantation site. Identical biodegradation and tissue regeneration rates are pivotal to ensure optimised healing. The by-products of scaffolds' degradation should be non-toxic and biocompatible to inhibit immunogenic effects, particularly at the implantable site. In addition to the scaffold's degradation rate, a proof of concept regarding biodegradation should be studied in vivo and in vitro. The biodegradability of CP-based scaffolds should be prioritised, without neglecting their electrical conductivity behaviour in future studies. The uses of CPs in biomedical applications are very broad, and each different CP has its own potential and speciality. Research, at the basic level, must be carried out, together with a focus on the varied aspects of study environment (in vitro and in vivo studies). This is because human bodies are very complex. Extensive studies focusing on the biocompatibility and immunogenicity of an electrically conductive scaffold composite must be conducted to obtain a clearer view of the use of CPs for tissue engineering applications.

Funding: This work was carried by the NPRP grant # NPRP11S-1221-170116 from the Qatar National Research Fund (a member of Qatar Foundation). The statements made herein are solely the responsibility of the authors.

Data Availability Statement: Not applicable.

Acknowledgments: The authors wish to thank Norzawani Buang M Yassin, of Imperial College London, United Kingdom, for the useful discussion.

Conflicts of Interest: The authors declare no conflict of interest.

References

1. Sultana, N.; Chang, H.C.; Jefferson, S.; Daniels, D.E. Application of conductive poly (3,4-ethylenedioxythiophene):poly (styrene-sulfonate) (PEDOT:PSS) polymers in potential biomedical engineering. *J. Pharm. Investig.* **2020**, *50*, 437–444. [CrossRef]
2. Kaviani, A.; Zebarjad, S.M.; Javadpour, S.; Ayatollahi, M.; Bazargan-Lari, R. Fabrication and characterization of low-cost freeze-gelated chitosan/collagen/hydroxyapatite hydrogel nanocomposite scaffold. *Int. J. Polym. Anal. Charact.* **2019**, *24*, 191–203. [CrossRef]
3. O'Brien, F.J. Biomaterials & scaffolds for tissue engineering. *Mater. Today* **2011**, *14*, 88–95. [CrossRef]
4. Roducts, P.; Matica, A.; Ostafe, V. Biodegradability of chitosan based products. *New Front. Chem.* **2017**, *26*, 75–86.
5. Lari, A.; Sultana, N.; Soon, C.F. Biocomposites conductive scaffold based on PEDOT: PSS/nHA/Chitosan/PCL: Fabrication and characterization. *Malays. J. Fundam. Appl. Sci.* **2019**, *15*, 146–149. [CrossRef]
6. Tamariz, E.; Rios-Ramrez, A. Biodegradation of medical purpose polymeric materials and their impact on biocompatibility. *Biodegrad.-Life Sci.* **2013**, 3–30.
7. Flaig, I.; Radenković, M.; Najman, S.; Pröhl, A.; Jung, O.; Barbeck, M. in vivo analysis of the biocompatibility and immune response of jellyfish collagen scaffolds and its suitability for bone regeneration. *Int. J. Mol. Sci.* **2020**, *21*, 4518. [CrossRef]
8. Yang, D.; Xiao, J.; Wang, B.; Li, L.; Kong, X.; Liao, J. The immune reaction and degradation fate of scaffold in cartilage/bone tissue engineering. *Mater. Sci. Eng. C* **2019**, *104*, 109927. [CrossRef]
9. Wang, S.; Guan, S.; Zhu, Z.; Li, W.; Liu, T.; Ma, X. Hyaluronic acid doped- poly (3,4-ethylene dioxythiophene)/chitosan/gelatin (PEDOT-HA/Cs/Gel) porous conductive scaffold for nerve regeneration. *Mater. Sci. Eng. C* **2017**, *71*, 308–316. [CrossRef] [PubMed]
10. Wang, S.; Guan, S.; Li, W.; Ge, D.; Xu, J.; Sun, C.; Liu, T.; Ma, X. 3D culture of neural stem cells within conductive PEDOT layer-assembled chitosan/gelatin scaffolds for neural tissue engineering. *Mater. Sci. Eng. C* **2018**, *93*, 890–901. [CrossRef]
11. Alaribe, F.N.; Manoto, S.L.; Motaung, S.C.K.M. Scaffolds from biomaterials: advantages and limitations in bone and tissue engineering. *Biologia* **2016**, *71*, 353–366. [CrossRef]
12. Dong, C.; Lv, Y. Application of collagen scaffold in tissue engineering: Recent advances and new perspectives. *Polymers* **2016**, *8*, 42. [CrossRef] [PubMed]
13. Rahman, S.; Rana, M.; Spitzhorn, L.-S.; Akhtar, N.; Hasan, Z.; Choudhury, N.; Fehm, T.; Czernuszka, J.T.; Adjaye, J.; Asaduzzaman, S.M. Fabrication of biocompatible porous scaffolds based on hydroxyapatite/collagen/chitosan composite for restoration of defected maxillofacial mandible bone. *Prog. Biomater.* **2019**, *8*, 137–154. [CrossRef] [PubMed]
14. Zhang, T.; Jin, L.; Fang, Y.; Lin, F.; Sun, W.; Xiong, Z. Fabrication of biomimetic scaffolds with oriented porous morphology for cardiac tissue engineering. *J. Biomater. Tissue Eng.* **2014**, *4*, 1030–1039. [CrossRef]

15. Plieth, W. Intrinsically conducting polymers. *Electrochem. Mater. Sci.* **2008**, 323–363. [CrossRef]
16. Peng, H.; Sun, X.; Weng, W.; Fang, X. Flexible Electronic Devices Based on Polymers. *Polymer Mater. Energy Electron. Appl.* **2017**, 325–354. [CrossRef]
17. Omar, S.N.I.; Ariffin, Z.Z.; Zakaria, A.; Safian, M.F.; Halim, M.I.A.; Ramli, R.; Sofian, Z.M.; Zulkifli, M.F.; Aizamddin, M.F.; Mahat, M.M. Electrically conductive fabric coated with polyaniline: Physicochemical characterisation and antibacterial assessment. *Emergent Mater.* **2020**, *3*, 469–477. [CrossRef]
18. Mawad, D.; Mansfield, C.; Lauto, A.; Perbellini, F.; Nelson, G.W.; Tonkin, J.; Bello, S.O.; Carrad, D.J.; Micolich, A.P.; Mahat, M.M.; et al. A conducting polymer with enhanced electronic stability applied in cardiac models. *Sci. Adv.* **2016**, *2*, e1601007. [CrossRef]
19. Liang, Y.; Goh, J.C.-H. Polypyrrole-incorporated conducting constructs for tissue engineering applications: A review. *Bioelectricity* **2020**, *2*, 101–119. [CrossRef]
20. Palza, H.; Zapata, P.A.; Angulo-Pineda, C. Electroactive smart polymers for biomedical applications. *Materials* **2019**, *12*, 277. [CrossRef]
21. Jadoun, S.; Riaz, U.; Budhiraja, V. Biodegradable conducting polymeric materials for biomedical applications: A review. *Med. Devices Sens.* **2021**, *4*, e10141. [CrossRef]
22. Dubey, N.; Kushwaha, C.S.; Shukla, S.K. A review on electrically conducting polymer bionanocomposites for biomedical and other applications. *Int. J. Polym. Mater. Polym. Biomater.* **2019**, *69*, 709–727. [CrossRef]
23. Prasopthum, A.; Deng, Z.; Khan, I.M.; Yin, Z.; Guo, B.; Yang, J. Biomaterials science three dimensional printed degradable and conductive polymer scaffolds promote chondrogenic differentiation. *Biomater. Sci.* **2020**, *8*, 4287–4298. [CrossRef]
24. Da Silva, A.C.; Córdoba de Torresi, S.I. Advances in conducting, biodegradable and biocompatible copolymers for biomedical applications. *Front. Mater.* **2019**, *6*, 1–9. [CrossRef]
25. Abedi, A.; Hasanzadeh, M.; Tayebi, L. Conductive nanofibrous Chitosan/PEDOT: PSS tissue engineering scaffolds. *Mater. Chem. Phys.* **2019**, *237*, 121882. [CrossRef]
26. Lari, A.; Sun, T.; Sultana, N. PEDOT: PSS-containing nanohydroxyapatite/chitosan conductive bionanocomposite scaffold: Fabrication and evaluation. *J. Nanomater.* **2016**, *2016*, 1–12. [CrossRef]
27. Lončarević, A.; Ivanković, M.; Rogina, A. Lysozyme-induced degradation of chitosan: The characterisation of degraded chitosan scaffolds. *J. Tissue Repair Regen.* **2017**, *1*, 12–22. [CrossRef]
28. Dhandayuthapani, B.; Yoshida, Y.; Maekawa, T.; Kumar, D.S. Polymeric scaffolds in tissue engineering application: A review. *Int. J. Polym. Sci.* **2011**, *2011*, 1–19. [CrossRef]
29. Wang, S.; Lei, J.; Yi, X.; Yuan, L.; Ge, L.; Li, D.; Mu, C. Fabrication of polypyrrole-grafted gelatin-based hydrogel with conductive, self-healing, and injectable properties. *ACS Appl. Polym. Mater.* **2020**, *2*, 3016–3023. [CrossRef]
30. Iandolo, D.; Sheard, J.; Levy, G.K.; Pitsalidis, C.; Tan, E.; Dennis, A.; Kim, J.-S.; Markaki, A.E.; Widera, D.; Owens, R.M.; et al. Biomimetic and electroactive 3D scaffolds for human neural crest-derived stem cell expansion and osteogenic differentiation. *MRS Commun.* **2020**, *10*, 179–187. [CrossRef]
31. Akbarzadeh, R.; Yousefi, A.-M. Effects of processing parameters in thermally induced phase separation technique on porous architecture of scaffolds for bone tissue engineering. *J. Biomed. Mater. Res. Part B Appl. Biomater.* **2014**, *102*, 1304–1315. [CrossRef]
32. Wade, R.J.; Burdick, J.A. Engineering ECM signals into biomaterials. *Mater. Today* **2012**, *15*, 454–459. [CrossRef]
33. Roshanbinfar, K.; Vogt, L.; Greber, B.; Diecke, S.; Boccaccini, A.R.; Scheibel, T.; Engel, F.B. Electroconductive biohybrid hydrogel for enhanced maturation and beating properties of engineered cardiac tissues. *Adv. Funct. Mater.* **2018**, *28*, 1803951. [CrossRef]
34. Conoscenti, G.; La Carrubba, V.; Brucato, V. A versatile technique to produce porous polymeric scaffolds: The thermally induced phase separation (TIPS) method. *Arch. Chem. Res.* **2017**, *1*, 10–20. [CrossRef]
35. Lee, J.K.; Link, J.M.; Hu, J.C.Y.; Athanasiou, K.A. The self-assembling process and applications in tissue engineering. *Cold Spring Harb. Perspect. Med.* **2017**, *7*, a025668. [CrossRef]
36. Wang, L.; Wu, Y.; Hu, T.; Guo, B.; Ma, P.X. Electrospun conductive nanofibrous scaffolds for engineering cardiac tissue and 3D bioactuators. *Acta Biomater.* **2017**, *59*, 68–81. [CrossRef]
37. Aadil, K.R.; Nathani, A.; Sharma, C.S.; Lenka, N.; Gupta, P. Investigation of poly (vinyl) alcohol-gellan gum based nanofiber as scaffolds for tissue engineering applications. *J. Drug Deliv. Sci. Technol.* **2019**, *54*, 101276. [CrossRef]
38. Tseghai, G.B.; Mengistie, D.A.; Malengier, B.; Fante, K.A.; Van Langenhove, L. PEDOT: PSS-based conductive textiles and their applications. *Sensors* **2020**, *20*, 1881. [CrossRef]
39. Rahmani, A.; Nadri, S.; Kazemi, H.S.; Mortazavi, Y.; Sojoodi, M. Conductive electrospun scaffolds with electrical stimulation for neural differentiation of conjunctiva mesenchymal stem cells. *Artif. Organs* **2019**, *43*, 780–790. [CrossRef]
40. Dong, S.; Han, L.; Du, C.; Wang, X.; Li, L.; Wei, Y. 3D printing of aniline tetramer-grafted-polyethylenimine and pluronic F127 composites for electroactive scaffolds. *Macromol. Rapid Commun.* **2017**, *38*, 1600551. [CrossRef] [PubMed]
41. Fani, N.; Hajinasrollah, M.; Vostikolaee, M.A.; Eslaminejad, M.B.; Mashhadiabbas, F.; Tongas, N.; Rasoulianboroujeni, M.; Yadegari, A.; Ede, K.; Tahriri, M.; et al. Influence of conductive PEDOT: PSS in a hard tissue scaffold: In vitro and in vivo study. *J. Bioact. Compat. Polym.* **2019**, *34*, 436–441. [CrossRef]
42. Wang, S.; Sun, C.; Guan, S.; Li, W.; Xu, J.; Ge, D.; Zhuang, M.; Liu, T.; Ma, X. Chitosan/gelatin porous scaffolds assembled with conductive poly (3, 4-ethylenedioxythiophene) nanoparticles for neural tissue engineering. *J. Mater. Chem. B* **2017**, *5*, 4774–4788. [CrossRef] [PubMed]

43. Pourjavadi, A.; Doroudian, M.; Ahadpour, A.; Azari, S. Injectable chitosan/κ—carrageenan hydrogel designed with au nanoparticles: A conductive scaffold for tissue engineering demands. *Int. J. Biol. Macromol.* **2019**, *126*, 310–317. [CrossRef] [PubMed]
44. Athukorala, S.; Tran, T.; Balu, R.; Truong, V.; Chapman, J.; Dutta, N.; Choudhury, N. 3D printable electrically conductive hydrogel scaffolds for biomedical applications: A review. *Polymers* **2021**, *13*, 474. [CrossRef] [PubMed]
45. Omar, M.H.; Razak K., A.; Ab Wahab, M.N.; Hamzah, H.H. Recent progress of conductive 3D printed electrodes based upon polymers/carbon nanomaterials using a fused deposition modelling (FDM) method as emerging electrochemical sensing devices. *RSC Adv.* **2021**, *11*, 16557–16571. [CrossRef]
46. Lu, B.; Yuk, H.; Lin, S.; Jian, N.; Qu, K.; Xu, J.; Zhao, X. Pure PEDOT: PSS hydrogels. *Nat. Commun.* **2019**, *10*, 1–10. [CrossRef]
47. Gao, Q.; Wang, M.; Kang, X.; Zhu, C.; Ge, M. Continuous wet-spinning of flexible and water-stable conductive PEDOT: PSS/PVA composite fibres for wearable sensors. *Compos. Commun.* **2020**, *17*, 134–140. [CrossRef]
48. Zhang, L.; Yang, K.; Chen, R.; Zhou, Y.; Chen, S.; Zheng, Y.; Li, M.; Xu, C.; Tang, X.; Zang, Z.; et al. The role of mineral acid doping of PEDOT: PSS and its application in organic photovoltaics. *Adv. Electron. Mater.* **2020**, *6*, 1900648. [CrossRef]
49. Pathak, C.S.; Singh, J.P.; Singh, R. Effect of dimethyl sulfoxide on the electrical properties of PEDOT: PSS/n-Si heterojunction diodes. *Curr. Appl. Phys.* **2015**, *15*, 528–534. [CrossRef]
50. Ouyang, L.; Musumeci, C.; Jafari, M.J.; Ederth, T.; Inganäs, O. Imaging the phase separation between PEDOT and polyelectrolytes during processing of highly conductive PEDOT: PSS films. *ACS Appl. Mater. Interfaces* **2015**, *7*, 19764–19773. [CrossRef]
51. Sun, Y.; Yang, S.; Du, P.; Yan, F.; Qu, J.; Zhu, Z.; Zuo, J.; Zhang, C. Investigate the effects of EG doping PEDOT/PSS on transmission and anti-reflection properties using terahertz pulsed spectroscopy. *Opt. Express* **2017**, *25*, 1723–1731. [CrossRef] [PubMed]
52. Kim, T.; Park, S.; Seo, J.; Lee, C.W.; Kim, J.-M. Highly conductive PEDOT: PSS with enhanced chemical stability. *Org. Electron.* **2019**, *74*, 77–81. [CrossRef]
53. Li, S.; Tao, Y.; Maryum, P.; Wang, Q.; Zhu, J.; Min, F.; Cheng, H.; Zhao, S.; Wang, C. Bifunctional polyaniline electroconductive hydrogels with applications in supercapacitor and wearable strain sensors. *J. Biomater. Sci. Polym. Ed.* **2020**, *31*, 938–953. [CrossRef]
54. Omar, S.; Ariffin, Z.; Akhir, R.; Shri, D.; Halim, M.; Safian, M.; Azman, H.; Ramli, R.; Mahat, M. Polyaniline (PANI) fabric doped p-toluene sulfonic acid (pTSA) with anti-infection properties. *Mater. Today Proc.* **2019**, *16*, 1994–2002. [CrossRef]
55. Mawad, D.; Artzy-Schnirman, A.; Tonkin, J.; Ramos, J.; Inal, S.; Mahat, M.M.; Darwish, N.; Zwi-Dantsis, L.; Malliaras, G.; Gooding, J.J.; et al. Electroconductive hydrogel based on functional poly (Ethylenedioxy Thiophene). *Chem. Mater.* **2016**, *28*, 6080–6088. [CrossRef]
56. Valenzuela-Rojo, R.D.; López-Cervantes, J.; Sánchez-Machado, D.I.; Escárcega-Galaz, A.A.; Macias, M.D.R.M. Antibacterial, mechanical and physical properties of collagen - chitosan sponges from aquatic source. *Sustain. Chem. Pharm.* **2020**, *15*, 100218. [CrossRef]
57. Zhang, C.; Hsieh, M.-H.; Wu, S.-Y.; Li, S.-H.; Wu, J.; Liu, S.-M.; Wei, H.-J.; Weisel, R.D.; Sung, H.-W.; Li, R.-K. A self-doping conductive polymer hydrogel that can restore electrical impulse propagation at myocardial infarct to prevent cardiac arrhythmia and preserve ventricular function. *Biomaterials* **2020**, *231*, 119672. [CrossRef]
58. Rowland, C.R.; Lennon, D.P.; Caplan, A.; Guilak, F. The effects of crosslinking of scaffolds engineered from cartilage ECM on the chondrogenic differentiation of MSCs. *Biomaterials* **2013**, *34*, 5802–5812. [CrossRef]
59. Davidenko, N.; Bax, D.V.; Schuster, C.F.; Farndale, R.W.; Hamaia, S.W.; Best, S.M.; Cameron, R.E. Optimisation of UV irradiation as a binding site conserving method for crosslinking collagen-based scaffolds. *J. Mater. Sci. Mater. Med.* **2016**, *27*, 1–17. [CrossRef]
60. Haugh, M.G.; Murphy, C.M.; McKiernan, R.C.; Altenbuchner, C.; O'Brien, F.J. Crosslinking and mechanical properties significantly influence cell attachment, proliferation, and migration within collagen glycosaminoglycan scaffolds. *Tissue Eng.-Part A* **2011**, *17*, 1201–1208. [CrossRef]
61. Ramli, J.; Hadi, A.S.; Jeefferie, A.R.; Mahat, M.M. A preliminary study on the effects of photoinitiator and UV curing exposure time to the mechanical and physical properties of the epoxy and vinyl ester fibreglass laminated composites. *Int. J. Eng. Technol.* **2010**, *1*, 14–20.
62. Campiglio, C.E.; Negrini, N.C.; Farè, S.; Draghi, L. Cross-linking strategies for electrospun gelatin scaffolds. *Materials* **2019**, *12*, 2476. [CrossRef]
63. Saidin, S.; Jumat, M.A.; Amin, N.A.A.M.; Al-Hammadi, A.S.S. Organic and inorganic antibacterial approaches in combating bacterial infection for biomedical application. *Mater. Sci. Eng. C* **2020**, *118*, 111382. [CrossRef]
64. Rizeq, B.R.; Younes, N.N.; Rasool, K.; Nasrallah, G.K. Synthesis, bioapplications, and toxicity evaluation of chitosan-based nanoparticles. *Int. J. Mol. Sci.* **2019**, *20*, 5776. [CrossRef] [PubMed]
65. Islam, N.; Wang, H.; Maqbool, F.; Ferro, V. In vitro enzymatic digestibility of glutaraldehyde-crosslinked chitosan nanoparticles in lysozyme solution and their applicability in pulmonary drug delivery. *Molecules* **2019**, *24*, 1271. [CrossRef]
66. Hua, Y.; Ma, C.; Wei, T.; Zhang, L.; Shen, J. Collagen/chitosan complexes: Preparation, antioxidant activity, tyrosinase inhibition activity, and melanin synthesis. *Int. J. Mol. Sci.* **2020**, *21*, 313. [CrossRef] [PubMed]
67. Fang, Y.; Zhang, T.; Song, Y.; Sun, W. Assessment of various crosslinking agents on collagen/chitosan scaffolds for myocardial tissue engineering. *Biomed. Mater.* **2020**, *15*, 045003. [CrossRef]
68. VijayaVenkataRaman, S.; Kannan, S.; Cao, T.; Fuh, J.Y.H.; Sriram, G.; Lu, W.F. 3D-printed PCL/PPy conductive scaffolds as three-dimensional porous nerve guide conduits (NGCs) for peripheral nerve injury repair. *Front. Bioeng. Biotechnol.* **2019**, *7*, 266. [CrossRef] [PubMed]

69. Söntjens, S.; Nettles, D.L.; Carnahan, M.A.; Setton, L.A.; Grinstaff, M. Biodendrimer-based hydrogel scaffolds for cartilage tissue repair. *Biomacromolecules* **2006**, *7*, 310–316. [CrossRef]
70. Lu, T.; Hu, H.; Li, Y.; Jiang, Q.; Su, J.; Lin, H.; Xiao, Y.; Zhu, X.; Zhang, X. Bioactive scaffolds based on collagen filaments with tunable physico-chemical and biological features. *Soft Matter* **2020**, *16*, 4540–4548. [CrossRef] [PubMed]
71. Qu, H.; Fu, H.; Han, Z.; Sun, Y. Biomaterials for bone tissue engineering scaffolds: a review. *RSC Adv.* **2019**, *9*, 26252–26262. [CrossRef]
72. Zhang, X.; Thomas, V.; Vohra, Y.K. In vitro biodegradation of designed tubular scaffolds of electrospun protein/polyglyconate blend fibers. *J. Biomed. Mater. Res. Part B Appl. Biomater.* **2008**, *89*, 135–147. [CrossRef]
73. Zhang, F.; King, M.W. Biodegradable polymers as the pivotal player in the design of tissue engineering scaffolds. *Adv. Health Mater.* **2020**, *9*, e1901358. [CrossRef] [PubMed]
74. Schedl, L.; von Burkersroda, F.; Gopferich, A. Why degradable polymers undergo surface erosion or bulk erosion. *Biomaterials* **2002**, *23*, 4221–4231. [CrossRef]
75. Woodard, L.N.; Grunlan, M.A. Hydrolytic degradation and erosion of polyester biomaterials. *ACS Macro Lett.* **2018**, *7*, 976–982. [CrossRef]
76. Guarino, V.; Caputo, T.; Altobelli, R.; Ambrosio, L. Degradation properties and metabolic activity of alginate and chitosan polyelectrolytes for drug delivery and tissue engineering applications. *AIMS Mater. Sci.* **2015**, *2*, 497–502. [CrossRef]
77. Ding, Y.; Invernale, M.A.; Sotzing, G.A. Conductivity trends of PEDOT-PSS impregnated fabric and the effect of conductivity on electrochromic textile. *ACS Appl. Mater. Interfaces* **2010**, *2*, 1588–1593. [CrossRef]
78. Talebi, A.; Labbaf, S.; Karimzadeh, F. Polycaprolactone-chitosan-polypyrrole conductive biocomposite nanofibrous scaffold for biomedical applications. *Polym. Compos.* **2019**, *41*, 645–652. [CrossRef]
79. Sarkar, S.; Baker, B.A.; Chen, D.; Pine, P.S.; McDaniel, J.H.; Salit, M.L.; Losert, W.; Simon, G.G.; Dunkers, J. Roles of nanofiber scaffold structure and chemistry in directing human bone marrow stromal cell response. *Adv. Tissue Eng. Regen. Med.* **2016**, *1*, 00003. [CrossRef]
80. Cauich-Rodríguez, J.V.; Chan-Chan, L.H.; Hernandez-Sánchez, F.; Cervantes-Uc, J.M. Degradation of polyurethanes for cardiovascular applications. *Adv. Biomater. Sci. Biomed. Appl.* **2013**, 51–82. [CrossRef]
81. Volsi, A.L.; Tallia, F.; Iqbal, H.; Georgiou, T.K.; Jones, J.R. Enzyme degradable star polymethacrylate/silica hybrid inks for 3D printing of tissue scaffolds. *Mater. Adv.* **2020**, *1*, 3189–3199. [CrossRef]
82. Tang, H.; Zhao, W.; Yu, J.; Li, Y.; Zhao, C. Recent development of pH-responsive polymers for cancer nanomedicine. *Molecules* **2019**, *24*, 4. [CrossRef]
83. Zech, J.; Mader, M.; Gündel, D.; Metz, H.; Odparlik, A.; Agarwal, S.; Mader, K.; Greiner, A. Noninvasive characterization (EPR, μCT, NMR) of 3D PLA electrospun fibre sponges for controlled drug delivery. *Int. J. Pharm. X* **2020**, *2*, 100055.
84. Hamedani, Y.; Teixeira, R.B.; Karbasiafshar, C.; Wipf, P.; Bhowmick, S.; Abid, M.R. Delivery of a mitochondria-targeted antioxidant from biocompatible, polymeric nanofibrous scaffolds. *FEBS Open Bio* **2020**, *11*, 35–47. [CrossRef] [PubMed]
85. Talacua, H.; Söntjens, S.H.M.; Thakkar, S.H.; Brizard, A.M.A.; van Herwerden, L.A.; Vink, A.; van Almen, G.C.; Dankers, P.Y.W.; Bouten, C.V.C.; Budde, R.P.J.; et al. Imaging the in vivo degradation of tissue engineering implants by use of supramolecular radiopaque biomaterials. *Macromol. Biosci.* **2020**, *20*, 2000024. [CrossRef] [PubMed]
86. Leng, X.; Liu, B.; Su, B.; Liang, M.; Shi, L.; Li, S.; Qu, S.; Fu, X.; Liu, Y.; Yao, M.; et al. In situ ultrasound imaging of silk hydrogel degradation and neovascularization. *J. Tissue Eng. Regen. Med.* **2015**, *11*, 822–830. [CrossRef] [PubMed]
87. Kim, S.H.; Park, J.H.; Kwon, J.S.; Cho, J.G.; Park, K.G.; Park, C.H.; Yoo, J.J.; Atala, A.; Choi, H.S.; Kim, M.S.; et al. NIR fluorescence for monitoring in vivo scaffold degradation along with stem cell tracking in bone tissue engineering. *Biomaterials* **2020**, *258*, 120267. [CrossRef]
88. Gil, C.J.; Tomov, M.L.; Theus, A.S.; Cetnar, A.; Mahmoudi, M.; Serpooshan, V. In vivo tracking of tissue-engineered constructs. *Micromachines* **2019**, *10*, 474. [CrossRef]
89. Zhang, M.; Wang, Z.; Huang, P.; Jiang, G.; Xu, C.; Zhang, W.; Guo, R.; Li, W.; Zhang, X. Real-time and noninvasive tracking of injectable hydrogel degradation using functionalized AIE nanoparticles. *Nanophotonics* **2020**, *9*, 2063–2075. [CrossRef]
90. Agrawal, C.M.; Athanasiou, K.A. Technique to control pH in vicinity of biodegrading PLA-PGA implants. *J. Biomed. Mater. Res.* **1997**, *38*, 105–114. [CrossRef]
91. Meng, Z.; He, J.; Li, J.; Su, Y.; Li, D. Melt-based, solvent-free additive manufacturing of biodegradable polymeric scaffolds with designer microstructures for tailored mechanical/biological properties and clinical applications. *Virtual Phys. Prototyp.* **2020**, *15*, 417–444. [CrossRef]
92. Szymczyk-Ziółkowska, P.; Łabowska, M.B.; Detyna, J.; Michalak, I.; Gruber, P. A review of fabrication polymer scaffolds for biomedical applications using additive manufacturing techniques. *Biocybern. Biomed. Eng.* **2020**, *40*, 624–638. [CrossRef]
93. Kayser, L.V.; Lipomi, D.J. Stretchable conductive polymers and composites based on PEDOT and PEDOT: PSS. *Adv. Mater.* **2019**, *31*, e1806133. [CrossRef]
94. Ruzaidi, D.A.A.; Mahat, M.M.; Sofian, Z.M.; Hashim, N.A.N.; Osman, H.; Nawawi, M.A.; Ramli, R.; Jantan, K.A.; Aizamddin, M.F.; Azman, H.H.; et al. Synthesis and Characterization of Porous, Electro-Conductive Chitosan–Gelatin–Agar-Based PEDOT: PSS Scaffolds for Potential Use in Tissue Engineering. *Polymers* **2021**, *13*, 2901. [CrossRef]
95. Feng, P.; Wu, P.; Gao, C.; Yang, Y.; Guo, W.; Yang, W.; Shuai, C. A multimaterial scaffold with tunable properties: Toward bone tissue repair. *Adv. Sci.* **2018**, *5*, 1700817. [CrossRef] [PubMed]

96. Sarvari, R.; Akbari-Alanjaraghi, M.; Massoumi, B.; Beygi-Khosrowshahi, Y.; Agbolaghi, S. Conductive and biodegradable scaffolds based on a five-arm and functionalized star-like polyaniline–polycaprolactone copolymer with a d -glucose core. *N. J. Chem.* **2017**, *41*, 6371–6384. [CrossRef]
97. Ghosh, S.; Haldar, S.; Gupta, S.; Bisht, A.; Chauhan, S.; Kumar, V.; Roy, P.; Lahiri, D. Anisotropically conductive biodegradable scaffold with coaxially aligned carbon nanotubes for directional regeneration of peripheral nerves. *ACS Appl. Bio Mater.* **2020**, *3*, 5796–5812. [CrossRef]
98. Mahmoudinezhad, M.H.; Karkhaneh, A.; Jadidi, K. Effect of PEDOT: PSS in tissue engineering composite scaffold on improvement and maintenance of endothelial cell function. *J. Biosci.* **2018**, *43*, 307–319. [CrossRef] [PubMed]
99. Peng, L.; Xiang, R.C.; Jia, W.W.; Dong, X.X.; Wang, G. Preparation and evaluation of porous chitosan/collagen scaffolds for periodontal tissue engineering. *J. Bioact. Compat. Polym.* **2006**, *21*, 207–220. [CrossRef]
100. Shen, Y.; Zhu, D.; Lu, W.; Liu, B.; Li, Y.; Cao, S. The characteristics of intrinsic fluorescence of type I collagen influenced by collagenase I. *Appl. Sci.* **2018**, *8*, 1947. [CrossRef]
101. Talebi, A.; Labbaf, S.; Karimzadeh, F. A conductive film of chitosan- polycaprolactone-polypyrrole with potential in heart patch application. *Polym. Test.* **2019**, *75*, 254–261. [CrossRef]
102. Hardiansyah, A.; Tanadi, H.; Yang, M.-C.; Liu, T.-Y. Electrospinning and antibacterial activity of chitosan-blended poly (lactic acid) nanofibers. *J. Polym. Res.* **2015**, *22*, 1–10. [CrossRef]
103. Xu, T.; Yang, H.; Yang, D.; Yu, Z.Z. Polylactic acid nanofiber scaffold decorated with chitosan island like topography for bone tissue engineering. *ACS Appl. Mater. Interfaces* **2017**, *9*, 21094–21104. [CrossRef]
104. Boehler, C.; Aqrawe, Z.; Asplund, M. Applications of PEDOT in bioelectronic medicine. *Bioelectron. Med.* **2019**, *2*, 89–99. [CrossRef]
105. Lee, S.; Ozlu, B.; Eom, T.; Martin, D.C.; Shim, B.S. Electrically conducting polymers for bio-interfacing electronics: From neural and cardiac interfaces to bone and artificial tissue biomaterials. *Biosens. Bioelectron.* **2020**, *170*, 112620. [CrossRef] [PubMed]
106. Namsheer, K.; Rout, C.S. Conducting polymers: A comprehensive review on recent advances in synthesis, properties and applications. *RSC Adv.* **2021**, *11*, 5659–5697. [CrossRef]
107. Shafiee, S.A.; Perry, S.C.; Hamzah, H.H.; Mahat, M.M.; Al-Lolage, F.A.; Ramli, M.Z. Recent advances on metal nitride materials as emerging electrochemical sensors: A mini review. *Electrochem. Commun.* **2020**, *120*, 106828. [CrossRef]
108. Mahat, M.M.; Sabere, A.S.M.; Azizi, J.; Amdan, N.A.N. Potential Applications of Conducting Polymers to Reduce Secondary Bacterial Infections among COVID-19 Patients. A Review. *Emergent Mater.* **2021**, 1–14. [CrossRef]
109. Ramanavicius, S.; Ramanavicius, A. Conducting polymers in the design of biosensors and biofuel cells. *Polymers* **2020**, *13*, 49. [CrossRef]
110. Lim, H.; Ha, S.; Bae, M.; Yoon, S.H. A highly robust approach to fabricate the mass-customizable mould of sharp-tipped biodegradable polymer microneedles for drug delivery. *Int. J. Pharm.* **2021**, *600*, 120475. [CrossRef]
111. Ahmad, S.; Abbasi, A.; Manzoor, K.; Mangla, D.; Aggarwal, S.; Ikram, S. Chitosan-based bionanocomposites in drug delivery. In *Bionanocomposites in Tissue Engineering and Regenerative Medicine*; Woodhead Publishing Series in Biomaterials; Woodhead Publishing: Sawston, UK, 2021; pp. 187–203.
112. Tan, K.-X.; Chamundeswari, V.N.; Loo, S.C.J. Prospects of kefiran as a food-derived biopolymer for agri-food and biomedical applications. *RSC Adv.* **2020**, *10*, 25339–25351. [CrossRef]
113. Englert, C.; Brendel, J.C.; Majdanski, T.C.; Yildirim, T.; Schubert, S.; Gottschaldt, M.; Windhab, N.; Schubert, U.S. Pharmapolymers in the 21st century: Synthetic polymers in drug delivery applications. *Prog. Polym. Sci.* **2018**, *87*, 107–164. [CrossRef]
114. Naureen, B.; Haseeb, A.; Basirun, W.; Muhamad, F. Recent advances in tissue engineering scaffolds based on polyurethane and modified polyurethane. *Mater. Sci. Eng. C* **2020**, *118*, 111228. [CrossRef]
115. Perez-Puyana, V.; Jiménez-Rosado, M.; Romero, A.; Guerrero, A. Fabrication and Characterization of Hydrogels Based on Gelatinised Collagen with Potential Application in Tissue Engineering. *Polymers* **2020**, *12*, 1146. [CrossRef]
116. Asadi, N.; Del Bakhshayesh, A.R.; Davaran, S.; Akbarzadeh, A. Common biocompatible polymeric materials for tissue engineering and regenerative medicine. *Mater. Chem. Phys.* **2019**, *242*, 122528. [CrossRef]
117. Holman, H.; Kavarana, M.N.; Rajab, T.K. Smart materials in cardiovascular implants: Shape memory alloys and shape memory polymers. *Artif. Organs* **2020**, *45*, 454–463. [CrossRef] [PubMed]
118. Da Silva, D.; Kaduri, M.; Poley, M.; Adir, O.; Krinsky, N.; Shainsky-Roitman, J.; Schroeder, A. Biocompatibility, biodegradation and excretion of polylactic acid (PLA) in medical implants and theranostic systems. *Chem. Eng. J.* **2018**, *340*, 9–14. [CrossRef]
119. Bombin, A.D.J.; Dunne, N.J.; McCarthy, H.O. Electrospinning of natural polymers for the production of nanofibres for wound healing applications. *Mater. Sci. Eng. C* **2020**, *114*, 110994. [CrossRef] [PubMed]
120. Shende, P.; Gupta, H. Formulation and comparative characterization of nanoparticles of curcumin using natural, synthetic and semi-synthetic polymers for wound healing. *Life Sci.* **2020**, *253*, 117588. [CrossRef]
121. Puertas-Bartolomé, M.; Mora-Boza, A.; García-Fernández, L. Emerging biofabrication techniques: A review on natural polymers for biomedical applications. *Polymers* **2021**, *13*, 1209. [CrossRef] [PubMed]

polymers MDPI

Review

Waste Natural Polymers as Potential Fillers for Biodegradable Latex-Based Composites: A Review

D. N. Syuhada and A. R. Azura *

School of Materials and Mineral Resources Engineering, Engineering Campus, Universiti Sains Malaysia, Nibong Tebal 14300, Penang, Malaysia; nsyuhadadzulkafly@gmail.com
* Correspondence: srazura@usm.my; Tel.: +60-45996111

Abstract: In recent years, biodegradable composites have become important in various fields because of the increasing awareness of the global environment. Waste natural polymers have received much attention as renewable, biodegradable, non-toxic and low-cost filler in polymer composites. In order to exploit the high potential for residual natural loading in latex composites, different types of surface modification techniques have been applied. This review discusses the preparation and characterization of the modified waste natural fillers for latex-based composites. The potency of the waste natural filler for the latex-based composites was explored with a focus on the mechanical, thermal, biodegradability and filler–latex interaction. This review also offers an update on the possible application of the waste natural filler towards the biodegradability of the latex-based composites for a more sustainable future.

Keywords: extractions; cellulose; crosslink mechanism; biodegradable; waste materials; natural filler; latex; composite; mechanical properties; surface modifications

Citation: Syuhada, D.N.; Azura, A.R. Waste Natural Polymers as Potential Fillers for Biodegradable Latex-Based Composites: A Review. *Polymers* **2021**, *13*, 3600. https://doi.org/10.3390/polym13203600

Academic Editor: Evgenia G. Korzhikova-Vlakh

Received: 20 September 2021
Accepted: 13 October 2021
Published: 19 October 2021

1. Introduction

Municipal solid waste consists of everyday items such as bottles, food scraps, paints, food wrappers, packaging and newspapers are used and thrown away every day. Some of these materials are biodegradable, while others are not. Because non-biodegradable items do not decompose over time, they are frequently thrown in landfills or incinerated, creating a serious environmental concern. This is one of the most important concerns affecting society today, as the concentration of non-biodegradable plastics continue to rise at an alarming rate [1]. Conversion of industrial waste into valuable product has raised the interest in the research area and various effort is taken towards the waste to wealth concept all around the world.

Rubber or elastomer is a polymeric substance with flexibility and extensibility qualities. When a force is applied, the molecules stretch out in the direction of the applied force; when the force is released, the molecules abruptly return to their initial state. Natural rubber (NR) is a biodegradable and environmentally friendly elastomer with exceptional elasticity and outstanding characteristics under cyclical stresses [2]. NR latex is obtained from the Hevea Brasiliensis tree. NR latex is a colloidal dispersion of small polymeric particles (3 to 5 μm) covered by a layer of proteins and lipids [3]. Table 1 shows the component in the colloidal dispersion of NR latex.

NR latex is useful for a wide range of products due to the ability of the rubber particles to coalesce and form a coherent polymer layer impenetrable to air and water. The vulcanization process in the latex transforms the materials into a highly versatile raw material for the manufacture of a variety of rubber goods via dipping, molding, casting, or spreading operations. However, the film cast from NR latex is soft and tacky; thus, the latex must be prevulcanized using a sulfur, peroxide or radiation technique to give a higher grade output [4]. To achieve the elastic qualities of the rubber, the latex was treated with many chemicals that act as preservatives, anticoagulants, vulcanizing agents and antioxidants [3]. NR latex products outperform synthetic counterparts due to their excellent strength properties along with low modulus and high wet gel strength, elasticity, resilience, heat dissipation and abrasion resistance characteristics which cannot easily be mimicked by synthetic polymers.

Table 1. Component in colloidal dispersion of NR latex [3].

Components	Percentage (*w*/*w*, %)
Rubber polymer (cis 1,4 polyisoprene)	25–35
Proteins	1.0–1.8
Carbohydrates	0.4–1.1
Neutral lipids	0.5–0.6
Inorganic components	0.4–0.6
Amino acids, amides, etc.	0.4
Water	50–70

Synthetic latex is primarily produced using emulsion polymerization and can be synthesised from a variety of monomers, including acrylates, styrene, vinyl acetate and butadiene [5]. The properties of synthetic rubber such as high temperature resistance, good resistance to abrasion, high strength is depending on the composition of the copolymers. Both NR and synthetic rubber are different in microstructure but contain isoprene as the main chain [3]. Even though NR latex products can biodegrade in the soil, the process is slow due to the large molecular weight of NR latex macromolecules with additional chemical crosslinking that formed through the vulcanization process. These cross-linkages are stable under ambient conditions and can withstand the soil environment for a prolonged period of time. Another factor that may conduce to the delayed degradation rate is the use of additives such as antioxidants in vulcanized NR latex, which is intended to limit microbial growth upon disposal [6].

Currently, the majority of rubber waste is burned or disposed of in landfills. Microbes play a role in the biodegradation process by releasing extracellular enzymes that cleave polymer chains into tiny molecules, which are then absorbed into the microorganism's cell to be used as a source of carbon and energy. CO_2, H_2O and other metabolic products are released at the end of the process and can be utilized by other living creatures. Therefore, biodegradation is another way of degrading rubber waste that has the potential to resolve environmental problems. Microbial breakdown of polyisoprene rubber, on the other hand, is a very sluggish process that can take months or even years [7]. The addition of compounding ingredients such as reinforcing fillers, accelerators and sulfur increased the resistance of the latex towards microbial degradation caused by the decreased of access for microorganism to the rubber matrix [8].

Biodegradation is an environmentally benign alternative to conventional disposal methods in which microorganisms biologically break down complex organic chemicals in commercial items into cell biomass and less complex molecules, as well as water and either carbon dioxide or methane. For microbes to use polymers as a nutrition source, polymers must first be oxidized through an abiotic process, such as exposure to ultraviolet (UV) irradiation, heat and/or chemicals in the environment. Some microbes, such as rubber degraders, can commence the oxidation process on their own, with secreted enzymes causing polymer biofragmentation followed by bioassimilation of small cleavage pieces [3].

The global demand for natural source NR rubber gloves has been growing in recent years. Due to an increased in NR glove consumption, the formation of a significant volume of NR latex-based solid waste is unavoidable. The use of chemical agents such as fillers, antioxidants, accelerators, vulcanizing agents and inhibitors during the manufacturing process have a significant impact on the chemistry of NR latex, decreasing its inherent degradability. An excessive amount of solid waste has been generated due to increased consumption of NR latex products and their low degradation, which has negatively impacted the sustainability of the environment. The development of biodegradable NR latex composites has been widely investigated in order to reduce the environmental issues caused by rubber waste disposal [6,9–12].

Natural polymer composites made from filler have received much attention in recent years due to the benefits of natural filler and environmental concerns. In order to reduce the greenhouse effect, the composite material based on renewable and biodegradable materials is progressively replacing the synthetic fiber reinforced composite. The strength of a natural fiber composite is determined not only by the interfacial strength, but also by the fiber's basic strength. The fiber quality is determined by the crop's location and climate, as well as the fiber age, plant species, transit mode, storage time and inventory condition. Chemical treatments of natural fibers are used to change the physical, mechanical and thermal properties of composites to address this problem. Chemical treatments may reduce the hydrophilicity of plant fibers, clean the surface of fibers, improve the roughness of fibers and reduce the moisture content of fibers, resulting in better reinforcement–matrix interactions [13].

The growth of agricultural industries also leads to an increase in the production of biomass from the agricultural sector. The agro waste production is generated in ongoing basis and produced in large quantities. Excessive agricultural waste creates a disposal problem through waste management problems, pollution problems and also becomes a waste of primary resources. In order to deal with this problem, recycling and reusing this waste is more important. Agricultural waste is used as feedstock, biochemicals, biomaterials and also as fuel to produce heat and electricity [14]. However, due to poor fuel properties, the usage of agro waste as fuel is still minimal. Agro waste also can be beneficial if it is used properly. There are several research has been done on the agro waste as an alternative filler to synthetic filler in order to obtain renewable and environmentally friendly products. The agro waste can be generated from various sources such as date palm waste [15,16], cocoa pod husk [17], cotton [18–20], coconut [21–23], rubber wood and kenaf bast [24]. Basically, from agro waste, it contains three main constituents which is cellulose, hemicellulose and lignin.

Cellulose is the most abundant polymer produced in nature and microorganism. The major conventional resources of cellulose can be obtained from cotton and wood. The cellulose and its derivatives have been utilized for years in the industrial applications such as textile, paper and medical due to its unique properties, biodegradable and come from renewable resources [25,26]. Cellulose is polysaccharides with a chemical formula of $(C_6H_{10}O_5)_n$ composed of repeating units of β-D-glucopyranose with covalent link between the OH group of C4 and C1 carbon atoms [27]. Properties of cellulose is depending on its degree of polymerization which depends on the source of the cellulose. However, due to the high number of hydroxyl group on the glucose ring along the skeleton; thus, there is hydrogen binding between the chains. The crystallization of the chain and the presence of the two regions in cellulose, which is crystalline and amorphous, resulting the general properties of cellulose such as high strength, stiffness, durability and biocompatibility. The hydroxyl group in the cellulose gives the hypdrophilicity, chirality and biodegradability to the cellulose; in addition, these hydroxyl groups allow the chemical modification of the cellulose [28]. However, the characteristics of cellulose are highly dependent on the plant sources and the extraction process [29,30]. There are few types of cellulose that are extracted from natural resources, which are microcrystalline cellulose (MCC), cellulose nanowhiskers (CNW) and cellulose nanofibers (CNF). MCC has the properties of biodegradable, insoluble and biopolymer [31]. CNW is a needle-like cellulose, having at least 1 dimension equal or less than 100 nm and highly crystalline. CNF can be divided into two categories which are nanofibrillated cellulose and microfibrillated cellulose [29]. CNF is a long, flexible nano-string consist of alternation crystalline and amorphous. CNF has a large surface to volume ratio, high mechanical properties and ability to form a highly porous mesh [32].

The purpose of this review is to provide a summary of recent studies on the effect of incorporating waste natural filler into latex-based composites. This review also covers the extraction method and properties of waste natural filler, surface modification of waste natural filler and the preparation of latex-based composites. This review paper also discusses the potential applications of waste natural filler.

2. Type of Waste Natural Filler and the Extraction Methods

Various methods are used to extract the natural cellulose from plant. The most commonly used method is an acid hydrolysis method and alkaline method. The aim of these processes is to completely remove the amorphous region, for example, lignin and hemicellulose, to obtain the cellulose crystalline region which may affect the properties of the cellulose [18].

The technique involves in the extraction of cellulose from waste natural filler, including the chemical method, mechanical method and a combination of mechanical and chemical method. Chemical treatment usually involves the use of alkaline [33–38], acid hydrolyzed [39–42] and bleaching of the natural filler. Alkaline treatment ruptures the intramolecular crossed bond between lignin and hemicellulose. This process increases the filler porosity and allows the access of the alkali to the lignin which allow intensive dissolution of lignin structure [43]. The commonly use alkaline treatment for the natural filler is sodium hypochlorite [33–36] and sodium hydroxide (NaOH) [37,38]. Since the alkaline treatment mostly remove the lignin structure, this treatment is usually used as pre-treatment in the isolation of cellulose. The usage of the acid hydrolysis technique is the most common technique found to isolate nanocellulose after the pre-treatment process. The usage of acid includes hydrochloric acid [39], oxalic acid [44], formic acid [42] and sulfuric acid [41]. The drawback of using acid treatment is a large number of effluents generated during washing the filler, corrosion hazard of the acid, low thermal stability for the sulfuric acid and generate a low yield of cellulose due to degradation of cellulose [32].

A part of chemical treatment, mechanical treatment has also been used as a method to isolate nanocellulose. The example of the mechanical methods to produce nanocellulose are ball milling [37], ultrasonication [45] and homogenization [46]. High pressure and high intensity ultrasound are found to be effective mechanical treatment to produce cellulose. High-pressure homogenization generates a variety of disruptive forces that can cause the structural organization of cellulosic materials to partially dissolve, including cavitation, turbulence and shear effects [46]. Meanwhile, high-intensity ultrasonic treatments may also degrade cellulosic materials by producing strong cavitation stresses [45]. The combined technique of using mechanical and chemical treatment is called as mechano-chemical treatment. The mechanochemical technique is applied by Song et al. (2018) [40] in order to use a mild acid to isolate the cellulose. The usage of mild acid aims to reduce the degradation of cellulose, which impact the yield of the cellulose. However, mild acid needs more reaction time in order to isolate the cellulose; hence, it is assisted with mechanical forces in order to reduce the reaction time.

In order to determine the crystallinity and the crystallite size of the natural filler, x-ray diffraction (XRD) is generally used. XRD analysis is used to compare the effect of the treatment towards the improvement of the crystallinity of the natural filler, which helps to comprehend the final properties of the composites. The crystallinity index measures the difference of intensity of the lattice's diffraction with the intensity diffraction of the amorphous. Meanwhile, the crystallite size can be measured by following Scherrer's equation [47]. Most studies showed that the crystallinity of waste natural filler subjected to the treatment or extraction will be improved. The amorphous peak of the waste natural filler will be reduced, while the crystalline peak will be increased. Thambiraj and Ravi Shankaran [18] mentioned that the plane reflection of the original industrial waste cotton is wide and round, while after treatment, the curve becomes more sharper and narrower, which indicated the removal of lignin and hemicellulose, thus resulting the increase in crystallinity. The increase in crystallinity was related to the removal of the amorphous constituent and the arrangement of the crystalline region into more ordered structure [17].

Another important characterization of the modified waste natural filler is the morphology of the waste natural filler. The morphology of the filler is usually determined by scanning electron microscopy (SEM) and transmission electron microscopy (TEM). Using this test, the structure of the filler can be obtained, including the estimated length and width of the filler. The structure and the size of the modified waste natural filler depends on the sources of the waste natural filler and the treatment process. The structure of the treated waste natural filler will turn into a needle-like shape [24], rod-like shape [17,18,20,48], skeletal rod-like microfibril structure [15] and also acicular structure [49], as shown in Figure 1. Basically, the size of the treated waste natural filler may be reduced after treatment, resulting in the separation of individual fibers due to the removal of the cementing materials, which are lignin and hemicellulose [21]. Table 2 summarized the results from different sources of natural waste filler and the extraction method used.

Table 2. Properties of cellulose from different sources and extraction method.

Sources	Treatment	Structure	Size	Crystallinity, %	Reference
Date palm biomass waste	Acid hydrolysis and bleaching	Skeletal rod-like macro fibril structure,	180 μm	Rachis: 53.71 Leaflet: 50.66 Fiber: 52.43	[15]
Cocoa pod husk	Acid hydrolysis	Rod-like morphology,	10–60 nm 41–155 nm	67.60	[17]
Industrial waste cotton	Acid hydrolysis	Rod-like structure,	10 nm 190 nm	N/A	[18]
Soy hull	Acid hydrolysis	Long crystal,	49 ± 1.1 nm 503 ± 155 nm	80.4	[50]
Rubberwood (RW), Kenaf bast (KB)	Acid hydrolysis	Needle shape	Rubberwood: 33.54 nm Kenaf bast: 5.14 nm	RW:74.34 KB: 73.19	[24]
Jute fiber	Steam explosion and low acid hydrolysis	Nanocrystal	50 nm	82.22	[51]
Coconut spathe	Acid hydrolysis	Nanofiber	30–60 nm	N/A	[21]
Shrimp shell waste	Acid hydrolysis	Rod-like structure	25–32 nm 400 nm	86	[48]
Cotton, cotton stalk pulp	H_2SO_4 hydrolysis	Rod shape nanofiber	Cotton: 460 nm, 25 nm Cotton stalk: 100–850 nm, 25–100 nm	91, 86.3	[20]
Hemp fiber	H_2SO_4 hydrolysis	Rod shape structure	(160 ± 20) nm (4.5 ± 1) nm	N/A	[52]
Wood	Ethanol, peroxide and ultrasonic	Rod-like shape	1–9 nm, l < 500 nm	84.37	[53]
Barley straw and husk	NaOH and sulfuric acid	Acicular structure	L: (270 ± 40) nm W:(15 ± 5) nm	N/A	[49]
Grain straw	H_2SO_4 hydrolysis, homogenizer and ultrasonic	Rod shape	L:120–800 nm W:10–25 nm	28.6–41.4	[54]

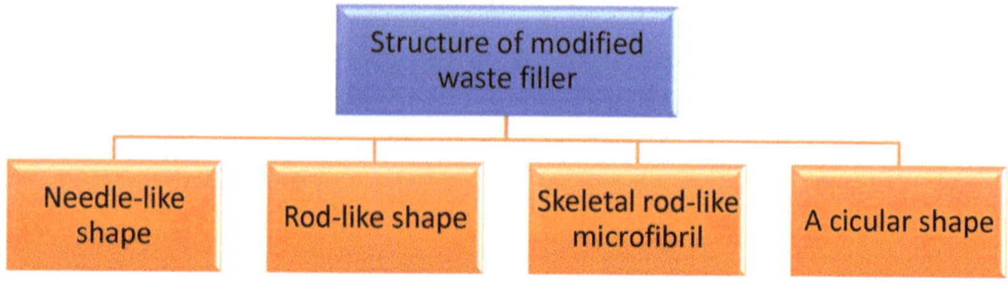

Figure 1. The different structure of modified waste/raw natural filler.

However, in comparison to synthetic fibers, these natural cellulosic fibers have poor mechanical capabilities and their compatibility with the composite matrix should be improved. Furthermore, most hydroxyl and other polar groups increase the moisture absorption of natural fibers, making it more difficult to make composites with hydrophobic polymer matrices). As a result, natural fibers should be surface-treated to increase thermodynamic miscibility and interface bonding strength before being used as reinforcement materials in polymer composites. In addition, surface modification is a required step that is rapidly becoming a major research area, particularly in the field of natural fiber modification. Surface modification methods such as silane treatment, radiation and discharge treatments, benzoylation treatment, peroxide treatment, use of maleate coupling agents, alkali treatment or mercerization and acetylation treatment have all been widely used so far and have the ability to improve the properties of natural fibers and bio-composites [55]. Figure 2 summarize the surface modification technique for cellulose-based fillers.

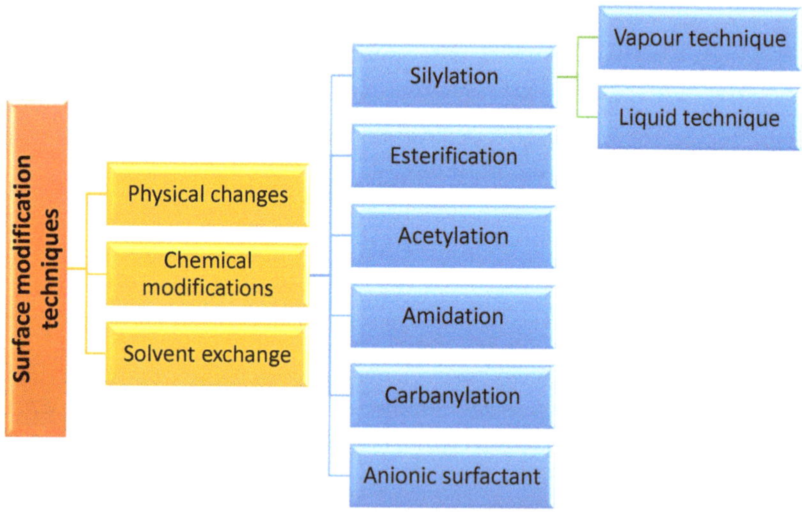

Figure 2. Summary on the surface modification technique for cellulose-based fillers.

One disadvantage of cellulose nanocrystal (CNC), which is isolated and extracted from natural resources via hydrochloric acid (HCl) hydrolysis, is that it tends to aggregate, making it difficult to disperse in common solvents, which might disrupt processing. Dispersibility is generally determined by the surface functioning, aspect ratio and solvent's capacity to disrupt the hydrogen bond interaction. CNC dispersion is commonly aided by the addition of negatively charged groups such as sulphate ester and carboxylic acid [20]. Due to the presence of the hydroxyl group of the NC surface, it has a significant agglomeration tendency, resulting in bigger particles with poor dispersion. To solve this problem, surface modification can boost the hydrophobicity and electrostatic stability of the NCs.

Solvent exchange, physical changes and chemical modifications are the three basic types of NCs surface modification.

The most common modifications are physical and chemical, which are based on the addition of polymers or surfactants to the surface of the NCs. The chemical modification is the most common, characterized by the covalent attachment of a polymer grafting onto nanocellulose surface. This approach is effective at changing the polarity of nanoparticles, which offers a wide range of applications. Silylation, esterification, amidation, acetylation and carbanylation are the most commonly used technique in chemical modifications. Physical modification is a straightforward method that avoids the formation of covalent bonds, while ensuring cellulose stability. These techniques change the fibers' dispersion and increase their hydrophobicity. Moreover, after nanocellulose isolation, most modification activities are carried out in aqueous solution, which requires more energy and time. In situ modification has been investigated by a few authors, while most modification is still conducted in aqueous solution to achieve a stable suspension [37].

CNC extracted from miscanthus gigantus (MxG) were modified by functionalized the CNC extracted with hydrochloric acid with 2,2,6,6-tetramethylpiperidinyloxy (TEMPO) oxidation to improve dispersibility of the CNC. The schematic reaction of TEMPO oxidation is shown in Figure 3. The hydrolysis of cellulose with sulfuric acid (H_2SO_4) also led to the formation of negatively charged sulphate ester groups on the surface of CNC. From these two reactions, it found that the crystallinity of the CNC also increased. In the dispersibility test, CNC modified with TEMPO oxidization improved dispersibility by being able to be held up to 7 days in water. Similar observations are also found on CNC hydrolyzed with (H_2SO_4) with the more viscous sample. Thermal degradation of the carboxylic treated MxG showed lower degradation temperature with higher weight residual loss, compared to both acid-hydrolyzed samples (H_2SO_4 and HCl). From the TEM results, the acid hydrolysis sample of the MxG shows aggregate, which makes it difficult to determine the dimension. The dimension of treated CNC is easier to measure since it can be dispersed in solution. It was found that the CNC has a ribbon-like shaped cross-section aspect ratio of the TEMPO-CNC and sulfuric acid-CNC is 69 and 63, respectively [39].

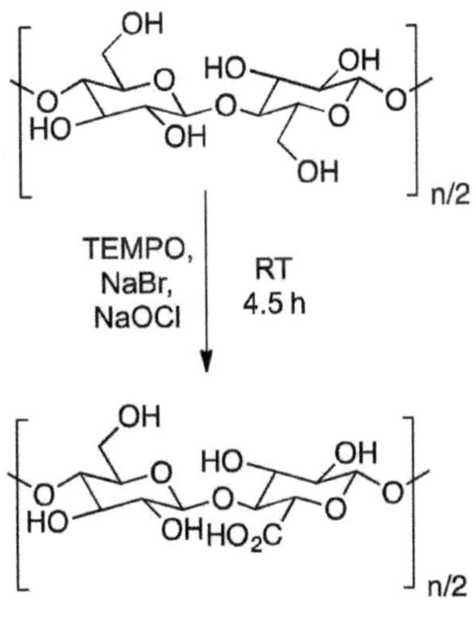

Figure 3. TEMPO oxidation of MxG-CNC to MxG-CNC-COOH.

Another option for surface modification by chemical method is the use of anionic surfactant. The modification is conducted during the isolation process, which known as in situ ball milling modification. The raw material, which is soybean straw (SS), was first pre-treated with sodium hydroxide and bleaching process to remove non-cellulosic components.

The surface modification takes place by using a mechanical method which is using ball milling with the addition of surfactants, which is condensed sodium alkyl naphthalene sulfonate (Surfom WG 8168). Fourier transform infrared (FTIR) results show the changes in the spectral changes of the cellulose side groups, which are attributed to the active surfactant groups. These changes indicate physical interactions between the compounds. Chemical bonds change the functional groups present on the material surface, while the physical interactions between the materials are the result of an electrical approximation between them, with no substitution of functional groups. The minor spectrum fluctuations are caused by conformational changes in the molecules as a result of their approximation, which alters the chain's vibration modes.

X-ray photon spectroscopy (XPS) results show that the decrease in carbon and oxygen, together with the presence of Na and S, confirms the presence of surfactant on the cellulose surface. The milling time influenced the size distribution and as the longer milling time decreased the nanoparticle sizes (9 and 12 h). The in-situ modification generated the electrostatic interactions between the NC and the surfactant, which decreased the NC dimensions. From the TEM image, it can be observed that the modification process affects the morphology of the NC, where unmodified NCs showed a nanofibrillar shape with a diameter of ~400 nm, while in-situ modified NCs showed smaller particle sizes, nanocrystals shape and diameters between 100 and 150 nm. This indicates the efficiency of surfactant stabilization and electrostatic repulsions. It also suggests that the surfactant alters the milling process, making it more efficient. The surface-modified NC showed a similar crystalline structure than the unmodified NC and good thermal stability, expanding the applicability of nanocellulose in areas such as emulsifiers, coatings or fillers in polymer nanocomposites [37].

Esterification is the chemical modification technique that adjusts the hydrophobic/hydrophilic balance in cellulose, which improves its solubility, augmenting dispersion and interfacial adhesion. Surface modification based on the esterification method is explained in the project conducted by Singh et al. [47]; based on their study, the surface modification of cellulose isolated from wheat straw was conducted by using indirect esterification methods using propionic anhydride. The reaction mechanism of surface modification of nanofibrils is a two-stage process, an indirect method where the first involves the esterification of the primary and secondary hydroxyl groups of cellulosic surface present in MFCs followed by the isolation of surface-modified NFCs using mechanical disintegration.

The usage of propionic acid is found to have better dimensional stability in the resultant fiber and composites. The indirect method limits the reaction materials on the exposed surface of nanofibrils without causing any destruction in the fiber's inner surface. Micro-fibrillated cellulose was soaked in a reaction medium of toluene, pyridine and H_2SO_4. Where toluene acts to restrict the swelling of the fiber and does not allow the reactants to enter the bulk site of fiber, pyridine is used to increase number of accessible reactive hydroxyl sites and enhance acetylation rate and H_2SO_4 as a catalyst. The presence of new bands on the modified MFC confirms the formation of ester binds with the free hydroxyl group present on the exposed surface fibers. The influence of propionylation on the morphology of modified MFCs was demonstrated by SEM examination. The surface of the steam exploded and the hydrolyzed MFCs of wheat straw are cleaner and smoother, but the surface of propionylated MFCs is rougher. Propynated groups on the surface are responsible for the roughness. From the TEM image, it was found that the high shear action of the homogenizer at a set time interval caused MFCs to disintegrate into fine NFCs and showed that the diameter of the nanofibrils in all samples was less than 100 nm. The tendency for treated samples to agglomerate is relatively low, showing that surface modification confers good hydrophobicity.

As a result of the indirect approach of propionylation, the hydrophobic character of the fiber surface significantly improved. With the increasing reaction time and temperature, the contact angle of the water drop rose. The rate of diffusion of the reactants and the rate of the reaction determined the course of the esterification reaction on the cellulose surface. The enhanced hydrophobicity of the nanofibrils has an impact on the thermal properties. At 100 °C, there was a decrease in the percentage weight loss after the change, which is linked to the evaporation of adsorbed water. The changed nanofibrils are less stable at higher temperatures, as observed by thermogravimetric analysis (TGA) and derivative thermogravimetric (DTG) plots. This phenomenon could be owing to the introduction of ester bonds into the cellulose structure or it could be related to a modest drop in crystallinity in the samples obtained after esterification.

In the study conducted by Barbosa et al. [35], microcrystalline cellulose (MCC) first underwent acid hydrolyzed with different type of acid which are sulfuric acid and hydrochloric acid. The sample was then acetylated using acetic anhydride in a homogeneous system and by using sulfuric acid as catalyst. From the result, the sulfuric acid that hydrolyzed sample showed smaller particle size compared to the hydrochloric acid sample. However, the thermal stability is lower for the sulfuric sample. This happens due to the presence of the sulphate ester group. From the morphology test, it was found that, after acetylation, the agglomeration is improved by observing the uniform size that obtained. After modification with acetic anhydride, the zeta potential of the sample was found, showing changes in the surface charge. The high negative charges results show the electrostatic stabilization of the sulfuric sample due to the sulphate ester group, which makes it stable for a longer time. The introduction of the acetyl group reduces the electrostatic stabilization due to the acetyl group, which diminishes the free surface charges of OH group and acts as a steric stabilization mechanism. This is supported by the FTIR, XPS and NMR analysis; the results show that the replacement of the OH group by acetyl group and has a high degree of substitution. The thermal stability of the sulfuric acid sample after acetylation also improved.

Other chemical modifications were conducted by using silylation techniques. The silylation technique, also known as silane grafting, is where the cellulose was introduced with hydrophobic alkyl groups such as chlorosilane. In the work by Khanjanzadeh et al. [56], a silylation technique was reported which modified CNC using the direct method and without uses of hazardous solvent. In their work, CNC was directly mixed with 3-aminopropyltriethoxysilane (APTES) solution at pH 4 and stirred. The sample was then centrifuged to collect the precipitate. In their result, the APTES-treated (ATR) sample shows the presence of N–H bonding which shows that the functional group is successfully added in the CNC. This is supported by the EDX and XPS results, which shows the presence of Si and N, which are attributed to APTES. The XPS detailed the increases of C, N and SI and reduction in O group due to the attenuation of oxygen from the cellulose surface, where C increased due to the presence of the propyl group from APTES. Nuclear magnetic resonance (NMR) results also show the presence of a new peak representing the resonance of an aminopropyl group of silanes coupling agent grafted into CNC. The crystallinity of the CNC showed no effect after modification, indicating that the crystalline structure of CNC was preserved and the atomic force microscope (AFM) shows that the modification does not impose any changes of the particle size and shape. While TGA results showed an increase in thermal stability after modification which, due to the bonding between APTES and CNS and the residue, are increasing due to siloxy moieties.

Jin et al. [57] also proposed the use 3-aminopropyltriethoxysilane (KH-550). Their work also used NCC, which is isolated using sulfuric acid and using the direct method for the surface modification of NCC. They proposed that, under acidic condition, the KH-550 form -Si-OH, which then formed hydrogen bonds with the hydroxyl group of the NCC or formed an ether bond through the dehydration or condensation process. The proposed alkylation mechanism is shown in Figure 4. The SEM result showed that SNCC has the image of the clubbed fibril network with a fuzzy boundary. The FTIR shows that the SNCC has a sharper intensity compared to NCC, which is due to the formation of hydrogen bonds. Other than that, the 1000–1100 band become broader and more complex, presumably due to the presence of the C-Si-O and Si-O-Si bonds. The thermal stability of the NCC also improved after treatment, suggesting that this is due to the hydrogen bonding of the NCC and the silane. Char residue also increased due to the presence of the alkoxysilane group. However, the crystallinity of the NCC after treatment was reduced. This is one of its more important properties, since the crystallinity may impact the final results of the composite. The reduced SNCC result is due to NCC being grafted with amorphous silane.

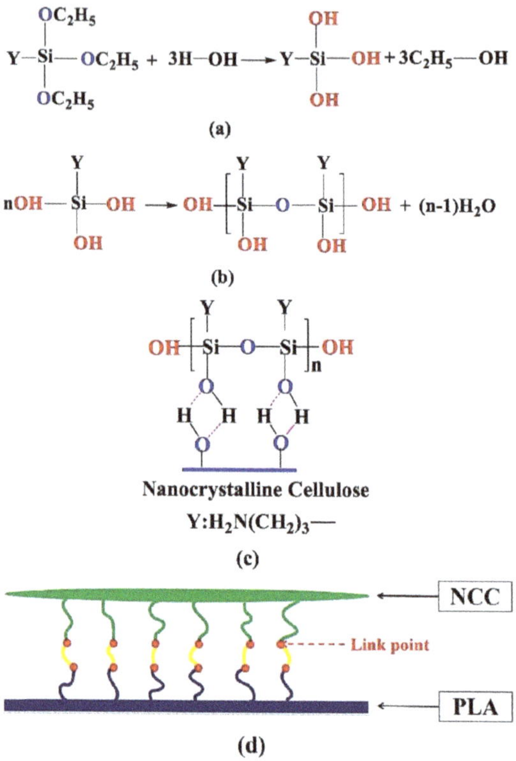

Figure 4. Proposed alkylation mechanism of NCC (**a**) Hydrolyzed of KH-550, (**b**) Oligomers formation, (**c**) Formation of hydrogen bond on surface of NCC (**d**) Coupling of KH-550 and NCC [57].

A comparison is made between two different methods of silylation using vapor technique and liquid technique [58]. The vapor technique is conducted by introducing the cellulose fiber with hexamethyldisilazane (HDMS) under the chemical vapor condition, while the liquid technique is conducted by soaking the CF in HDMS liquid. During silylation, the hydroxy group is converted to trimethylsiloxy group, which imparts hydrophobic properties. The FTIR shows the presence of a new peak at 1254 and 845, which, due to Si-C stretching, indicates that chemical moieties with trimethylsilyl appear on the surface of CF. Under SEM morphology, after modification, the surface of the fiber becomes smoother, especially the vapor techniques. The EDX shows the presence of the Si group on the modified CF, with vapor showing a higher value compared to the liquid. This was supported by the XPS result, with similar findings. XPS also shows a reduction in O/C ratio, which indicates a higher degree of silylation in the vapor method. The degree of surface substitution is higher in vapor than in the liquid treatment, which shows more hydrophobic wetting behavior and can be proved by the water contact angle result. Vapor treatment has higher water contact angle value compared to liquid treatment. This can be concluded that vapor technique has a higher functionalization degree of cellulose fiber surface which resulting more hydrophobic wetting behavior.

Cichosz et al. [59] investigated the effect of hybrid surface modification by combining the solvent exchange method and chemical modification and observed the effect of drying and not drying the cellulose before grafting. The grafting of cellulose was conducted by incorporating with maleic anhydride (MA). All the cellulose fibers show a decrease in moisture content after modifications. From the gathered data, it can be noted that the employment of ethanol highly contributed to the lowering of the water absorption. The cellulose, which was not dried before being grafted with MA, is more prone to grafting compared to a dried sample. All the samples that were treated with MA had lower thermal stability. The obtained results show that the cellulose fiber has a sufficient hydrophobilization level, suggesting that it is a promising method to modify the cellulose fiber.

3. Preparation and Characterization of Latex-Based Waste Natural Filler Composite

Metroxylan sagu pith waste (MSPW) is utilized as a new biodegradable filler in NR latex films. MSPW is a residue from the starch extraction process. It contains a high number of sago starch granules and high cellulose: up to 60%. The obtained results from incorporating MSPW with NR latex showed a reduction in tensile strength due to lignin, which hinder the formation of chains between rubber particles and creates distance between NR latex colloidal particles. A decreasing trend was also noted at the elongation at the break results and the modulus, due to the interparticle integration effect, which decreases the ability of the NR latex films to retain their stiffness. However, the tear strength increased with the addition of MSPW, due to the ability of the particle to deviate cracks, indicating a good cementing mechanism in NR latex films [12].

The presence of the glucopyranosyl group of starch, which consists of three hydroxyl groups, tends to form a strong hydrogen bond which results in filler agglomeration when incorporated with the XNBR latex. To solve this problem, amino-functional starch was produced by grafting the acrylonitrile monomer on the starch. This modification also helps increase the homogeneity of the filler dispersion in latex compound. The addition of the ANS, however, reduces the mechanical properties of the ANS/XNBR composites. This occurs due to the reduction in crosslink density between the rubber molecules, which was hindered by the formation of the ANS rich region. As biodegradation continues, further reductions in the mechanical properties are recorded due to the microbial activity in the ANS rich region, which were spotted in the SEM and optical micrograph. The formation of micro-voids was also detected in the XNBR latex films during biodegradation process. It was proved by the FTIR analysis that the reduction in starch aliphatic chain at peak 1607 cm^{-1} and stretching of N-H bond at peak 3302 cm^{-1} can be observed [11].

Another study was carried out using the acid hydrolyzed method in order to improve the mechanical properties of latex-sago starch composite, [60]. This method was able to induce the formation of the sulphate ester group on the surface of starch, which can increase the interaction between the rubber and starch. The XNBR latex mixed the compounding ingredients and acid hydrolyzed sago starch and formed into films using the coagulant dipping technique. AHSS-latex composite showed that the swelling percentage and mechanical properties strength improved compared to native sago starch, due to the increase of surface activity from smaller particles and the presence of sulphate ester, which creates the C-O-C bond, hence improving the rubber–filler interactions. The mass loss in AHSS-XNBR latex composite shows the highest value due to the loss of the amorphous region after acid hydrolysis, which led to microorganism attack on rubber and glycosidic chains.

Un-crosslinked NBR latex, which was added with nanocellulose from cotton seed linter pulp (CN), shows an increase in its viscosity, indicating the strong reaction between the CN and NBR chains. The foamed rubber produced has a smaller size, which provides better mechanical properties. Increasing the content of CN showed an improvement in tensile strength, which increased from 3.72 MPa to 6.54 MPa. The increment of tensile strength can be justified due to good interaction between CN and NBR matrix. This is supported through the SEM image shown in Figure 5; here, it can be observed that the CN was embedded in NBR matrix, which indicates a good interfacial adhesion. CN functioned as a physical crosslinking point with NBR chains and created hydrogen bonding and, hence, enhancing the crosslinked network of the composites. The possible scheme of the crosslink network between the NBR and CN is shown in Figure 6. CN also serves as a tensile stress carrier by transferring stress and prevent cracking and flaws from enlarging [61].

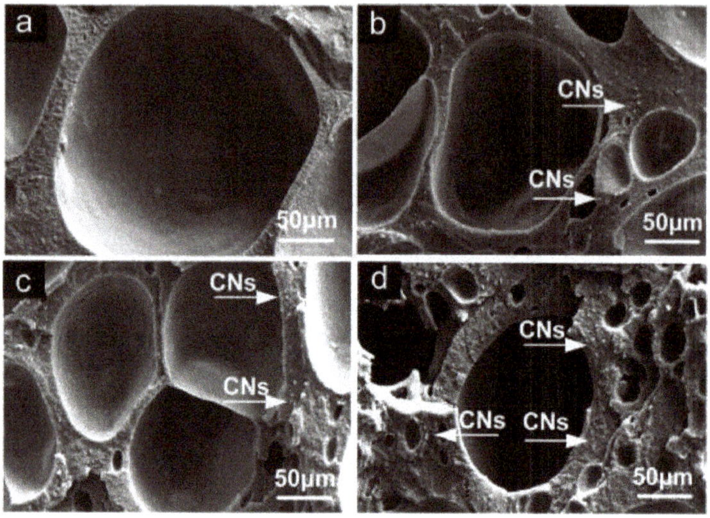

Figure 5. SEM images fractured surface of NBR/CNs nanocomposites (**a**) 3phr CNs, (**b**) 5 phr CNs, (**c**) 10 phr CNs and (**d**) 15 phr CNs [61].

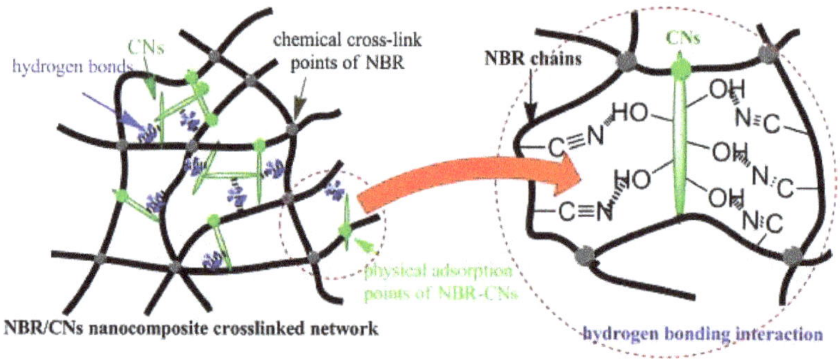

Figure 6. Possible scheme of crosslink network between NBR and CNs [61].

The incorporation of the nanocellulose fiber (CNF) from coconut spathe with NR latex was studied by Gopalakrishnan et al. [21]. The SEM result of the nanocellulose isolation showed that the diameter of nanocellulose is 40–50 nm; this is supported by the TEM results, where the diameter is around 30–60 nm. The optimum cure time of the NC/NR composite decreased with the increase of CNF content. Tensile, tear, modulus at 300 and hardness increase with increase of CNF, which indicates the high reinforcement of an NR matrix by well dispersed CNF. However, elongation at the break decreased with the increase of CNF, due to strong interaction between the rubber and filler, which increases its ability to resist the deformation of the NR chains. On the fracture surface, CNF was found to be embedded in the matrix, showing good compatibility between fiber and matrix with the presence of bonding agents. There were no voids or pull-out crack observed, which indicates good interfacial adhesion. The thermal properties of the maximum degradation remain the same, as suggested: that the thermal stability was not affected by the addition of CNF. The DMA modulus is directly proportional to the elastic nature of the composite. The modulus increased with the increase in CNF loading at low strain level, due to hydrodynamic reinforcement and the interactions of the filler–filler and filler–polymer. Solvent resistance towards toluene uptake decreased with higher loading CNF; this is due to the increase in the interfacial reaction between the NR chains and cellulose in the presence of the bonding agent, which increases the crosslinking point, hence preventing penetration of the solvent into rubber matrix.

Kenaf bast and kenaf core are used to determine their effect toward incorporation with NR latex. Kenaf/NR composite was prepared using the Dunlop method [62]. From the results, it can be observed that the tensile strength reduced with the increase in filler for both types of filler used. This happened due to kenaf core form agglomeration and led to a higher stress concentration, while kenaf bast has a larger pore size and is less interconnected with the matrix. Elongation at the break was also found to decrease with the increase in loading, due to the increase in the rigidity of the foam. The high loading fillers resulted in an NR latex foam with a higher stiffness, where the kenaf aggregates and stack acts as a filler network. Kenaf bast/NR foam has a higher stiffness than kenaf core due to its high cellulose content. The highly crystalline compact structure promotes higher stiffness. From the SEM image, the kenaf core has an irregular and particulate shape, and a tendency to agglomerate, making it less interconnected with rubber; meanwhile, bast has a fibrous shape and a higher length to cause the physical interaction between the filler and rubber. As incorporated with NR, the kenaf core forms an agglomeration, which forms a weak interaction between rubber–filler. Meanwhile bast is in a fibrous shape, which tends to cause a strong adhesion with rubber.

In the study on the effect of oxidization and hydrolyzation of corn starch, it can be noted that the particle size of the cornstarch was reduced by hydrolysis under alkali conditions at elevated temperature and high shearing force [63]. This can be explained by both the amide bond in protein and the ether bond in starch, which can be hydrolyzed under acid and base conditions in order to break up the polymer into smaller segment. The treatment shows that the CF has a size reduction of 33 times. The decrease in particle size helps improve the tensile strength, modulus and toughness, but there is a reduction in the elongation at the break. This occurs due to the increase in the filler–matrix contact area and an increase in the number of particles for filler network formation. Since corn flour has a greater modulus than NR, the rigidity of the connected article network penetrating throughout the NR matrix contributes to an increase in the modulus. The agglomeration effect can be explained through the Kraus plots in Figure 7. As the filler content of hydrolyzed CF composite was increased to 20%, it caused agglomeration, which led to the reduction of the restriction of polymer motion and, hence, increased the swelling ratio.

The hydrolysis of protein in the corn flour to generate negative charges was insufficient to cause particle–particle repulsion and reduce particle agglomeration at higher filler concentrations. The oxidation of corn flour converts hydroxyl functional groups in cornstarch to carboxylic acid functional groups, which are negatively charged under alkali conditions. The particles repel each other and prevent the formation of a particle network because of the increase in the negative charges on the surface of the corn flour particles. The composites filled with CF of low-level oxidation in the Kraus plot have a negative slope of the swelling behavior, which indicates that charge repulsion between particles prevented its agglomeration at the higher filler fractions. Lower crosslink density caused by high levels of oxidation resulted in poor tensile properties. As there are more COOH functional groups on the CF particle surface, the sulfur crosslink is retarded. A similar process uses a traditional retarder containing carboxylic acid groups, which interferes with the accelerator's activity and reduces the cure time. The oxidized CF has a significantly lower viscosity, greater surface area and lower reinforcement effect than the hydrolyzed CF. Consequently, the dispersion viscosity was reduced [63].

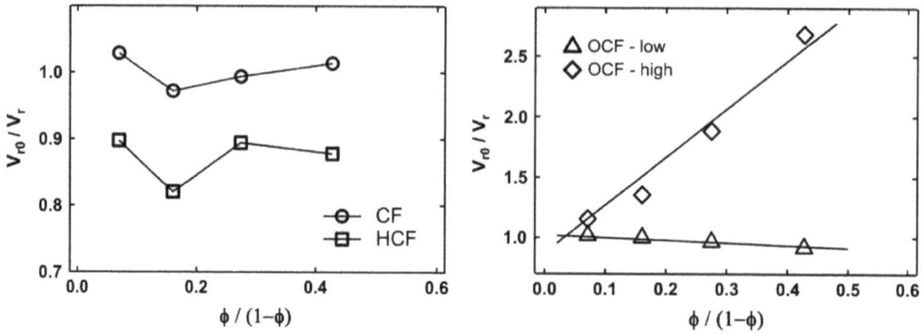

Figure 7. Kraus plot for corn flour (CF), hydrolyzed corn flour (HCF), oxidation corn flour-low oxidation level (OCF-low) and oxidation corn flour-high oxidation level (OCF-high) [63].

In another work by Ab Rahman et al. [6], NR latex was incorporated with sago starch (SS) to observe the impact of dispersing agent towards the properties and biodegradability of NR latex films. The dispersing agents that were incorporated with sago starch were sodium alginate (SA), carboxymethyl cellulose (CMC) and anchoid (A). From the result, the particle size of SA/SS dispersion shows the smallest particle size and has the most stable result in zeta potential, which indicate good compatibility with sago starch. As the dispersion is incorporated with NR latex, the swelling percentage of NR/SA/SS decreases, compared to anchoid; this due to the increase in the rubber–filler interaction between the fillers. High crosslink density for NR/SA/SS was observed due to the increase in the interaction between sago and alginate to form a better filler–rubber interaction. The molecular entanglement in the rubber–filler can overcome the disadvantage of poor adhesion between rubber and sago starch. The presence of physical entanglement restricts the polymer chain mobility and reduces the penetration of the solvent into the NR latex film, which results in decreased of the swelling index.

The tensile strength of the unfilled is the highest; however, SS, which uses anchoid and SA as dispersing agents, has a tensile strength that meets the ASTM requirements. NR/CMC/SS has lowest tensile strength due to bigger particle size, which acts as stress concentration factor and, thus, reduces its ability to withstand further elongation. Larger and incompatible filler particles diminish or eliminate the ability to strain-induce crystallization, in which the particles physically disrupt the development of the crystalline lattice. The tear strength also showed unfilled NR film as the highest. NR/SA/SS has a high tear strength, compared to NR/CMC/SS; due to its smaller particle size and the presence of filler, it could resist crack growth during tear deformation. NR/SA/SS has a high tensile modulus due to SA, meaning that it can make a good stabilizer in the NR latex films, which enables the filler able to disperse well in the latex and reduce the mobility polymer chain mobility and, hence, increase the rigidity. From the visual analysis of degradation, it can be clearly observed that NR/SA/SS film shows more signs of biodegradation, with the formation of more yellow marks on the latex film; this indicates that the SA/SS disperses well in NR latex films and, hence, increases the rate of degradation [6].

NR latex was added to corn derivative bio-filler in order to study its impact on the properties, including biodegradation. The addition of corn starch and corn flour generally reduce the tensile strength of the film. From these three corn derivatives, corn grain (GC) has the optimum filer composition and highest compatibility with NR latex, which results in better physical and mechanical properties, which are low density, moderate hardness, elongation at break and higher tensile strength. At higher CG loading, the SEM confirmed the occurrence of biodegradation with the observation of mycelia forming microbial colonies at the composite surface. In addition, FTIR analysis proved that there are increases of aldehyde and ketone groups during soil burial, which indicate the progression of biodegradation. Another sign of degradation was also the detection of C-O-C and C-O bonds of primary alcohols, reducing of the carbon double bond and widening the epoxide group peak. Although higher CG loadings are better for achieving a higher degree of biodegradability, mechanical characteristics are drastically reduced. CG loadings greater than 20 phr of tensile retention are reduced, implying the rapid loss of initial tensile strength under ageing circumstances at larger biofiller loadings. As a result, the most ideal CG loading in NR latex films was determined to be 20 phr for the manufacture of biodegradable NR gloves, in line with ASTM D3578, which reported a biodegradability of roughly 50% after 15 weeks of soil burial [64].

4. Potential Applications

Cellulose can be widely applied in industry. Previously, cellulose was mostly used in the textile, paper and fine chemical industries; however, recently, cellulose has gained more interest in other applications, including the food, agriculture, pharmaceutical, medical, environmental and energy industries [30].

The incorporation of the cellulose nanofiber (CNF) with graphene (GNS) and NR latex creates a composite that is suitable for flexible display, sensor, biomedical and energy storage applications. The incorporation of CNF in GNS helps to distribute GNS and form a multi-layer crosslinked network structure. Except the increase in the maximum stress and elongation at the break of the composite, the CNF also helps improve the conductivity of the composites. The addition of CNF reduces the percolation threshold of the composite. The addition of CNF lowered the composite's percolation threshold. Due to the high aspect ratio of the filler and its strong interfacial bonding, the multi-layered three-dimensional conductive network increased the electrical conductivity and overlaps easily in order to form a conductive network [65].

Hydrogel is hydrophilic polymer which consists of the 3-dimensional network. It has the capability to retain large amounts of water or biological fluids without being dissolved. Thanks to their features, they have recently gained the interest of numerous researchers in a variety of sectors, including drug delivery, wastewater treatment and agriculture. Most of the hydrogels on the market are synthesized from synthetic polymers, due to their excellent physical, chemical and mechanical properties. However, synthetic hydrogels have a high manufacturing cost and are non-renewable, difficult to decompose and not environmentally friendly. To solve these limitations, the hydrogels derived from natural polymers are very appealing solutions due to their good biodegradability, environmentally friendly nature and low production cost. Tanan et al. [66] conducted a study on the hydrogel of cassava starch-g-polyacrylic acid/natural rubber/polyvinyl alcohol (PVA) blend. The hydrogel was synthesized by aqueous solution polymerization. The semi-interpreting network is introduced by grafting cassava starch with acrylic acid and then mixing this with the NR/PVA blend; next, casting is conducted to obtain the film. The grafting of cassava starch results in the growth of the polyacrylic chain, which is then semi-interpenetrated with a PVANR solution to enhance the mechanical strength and flexibility of final products. From their research, it was found that the hydrogel exhibited an excellent water retention capacity, able to be reusable and have a high level of biodegradation. This property is a good indicator that it has a potential application in agriculture and horticulture. The improvement in its mechanical strength and thermal stability can be related to the H-bonding interaction between the OH group of PVA and the hydrophilic group on the network of grafted cassava.

Chitosan incorporated with epoxidized NR latex (ENR) is currently one of the most promising composites, which benefits the agricultural field. The bio-composite is used as a carrier in the slow release of fertilizer. The development of the slow-release copper compound is based on the entrapment of chitosan in ENR. This method will allow a matrix to be steadily altered with water-soluble polymer and released copper as a micronutrient in the fertilizer. In the slow released fertilizer, it is important to determine the effects of absorption, desorption and degradability. From the results, it can be observed that, as the chitosan loading increased, the absorption rate also increased due to the increase in the amino group, which creates an adsorption-active site. In addition to that, by its combination with ENR latex, chitosan altered its 3-dimensional structure and enhanced the contact with copper ion. The presence of the void from the SEM image also increases the adsorption site. The desorption rate of the bio-composite also depends on the chitosan loading. From kinetic study, it has been found that the desorption rate also depends on the chitosan content, where, with the increasing of chitosan content, the desorption rate also increased. The incremental increase in chitosan content increases the porosity of the rubber. As the chitosan swells, it allows the ions to diffuse through the pore formed and, hence, increases the release rate of the ions. For the biodegradation study, it was found that the increase of chitosan increases the biodegradation rate of the bio-composite. This result is a good indicator that once the release of copper ion through diffusion is stopped, the release of ions will be followed by the biodegradation method [67].

Junior et al. [68] studied the effect of bleached paper kraft and latex incorporated with cement. The stress–strain curve of the composite explains that the composite has a strain-softening behavior after the cracking of the matrix. This explained that the force is being transferred from the matrix to the fiber and thus, enables the composite to have a ductile behavior. The addition of cellulose to the composite has the biggest impact on the flexural result of the composite. The fiber, which is well distributed in the composite, helps to control the crack propagation and absorbing fractional energy in the post-cracking. While latex content assists in the reduction in the void of composites, it also helps by increasing the adhesion and tenacity of the composite by reducing the detachment of the interface between the fiber and the matrix.

Waterborne latexes are widely used in industrial application, such as adhesives, coatings and paintings. The most important requirement of waterborne latexes is the appearance after the application. The appearance depends on some properties, including rheology, substrate geometry and process parameters. The most common defect of waterborne latexes was the formation of sag. Jiang et al. [69] suggest that the addition of cellulose to acrylic latex improved the sag resistance of the latex and is thus beneficial for the waterborne applications. Cellulose nanocrystal has been used as a rheology modifier, in which it can help improve the viscosity. From the research results, acrylic latex with cellulose provides a gel-like behavior, where it shows that, as the concentration increases, the elastic and viscous moduli also increase. The increase in gel strength is due to stronger network formation. The suspension shows shear thinning behavior for viscosity. In the application of waterborne latexes, the viscosity of the latex must be low enough for atomization during spraying and increased in viscosity to prevent sagging. Creep properties determined the sag resistance of the composites. The addition of cellulose shows lower creep compliance under constant stress and high recovery of the strain. With the increase of cellulose content, the creep compliance at the end of constant stress clearly decreased, which occurred due to the network formation of cellulose and particle–particle interaction. A slight increase in the tensile strength and tensile modulus and reduction in the tensile strain were observed from the acrylic latex cellulose film, which occur due to the percolation phenomena. The improvement in thermal stability is explained by the high char formation, which acts as layers against further thermal degradation. The addition of nano size cellulose and it being well dispersed in latex does not disturb the transmittance properties of the latex, where it still provides high transparency to the latex films.

Eslami et al. [70] discuss the effect of modified nanocellulose, which is suitable for the dipped latex product. In their work, cellulose nanocrystal was modified using the grafting method by adding polylactic acid (PLA). The grafting approach suggests that the partial substitution of the OH group with PLA and, hence, CNC will have a dual nature, with mutual dispersibility in aqueous media. In this work, CNC and modified CNC (mCNC) were added into polychloroprene (CR) latex composite. Both CNC and mCNC-CR latex composites show an increase tensile strength and tensile modulus. A slightly increased tensile strength and modulus of the mCNC-CR composite contributed to better interaction between the PLA graft and the CR. From the TEM result, the CNC-CR composite shows the polygonal structure of the CNC, which is believed to occur due to the CNC being trapped in between coalescing particle and does not diffuse in CR. This structure is considered a partially structured network and a slight agglomeration, which results in early percolation. The polygonal structure is reduced in the mCNC-CR composite, explaining that mCNC is well dispersed and able to interact with CR and, thus, is in good agreement with the tensile strength of the composites. However, the incorporation of CNC and mCNC in CR caused increased water permeability, with mCNC-CR increasing until it reached the percolation limit of 3wt%. With good tensile and tear properties, these composites have the potential for wider application in dipped goods, such as medical gloves, membranes, catheters and balloons.

5. Conclusions

Natural filler-based composites have been widely used in order to preserve nature. Owing to their good mechanical properties, such as high strength, stiffness, durability and cellulose, they have caught the attention of many researchers aiming to fully utilize the potential of this natural resource. However, in comparison to synthetic fibers, these natural cellulosic fibers have poor mechanical properties, due to poor compatibility with the composite matrix. They are hydrophilic, which means that they disperse in the presence of hydroxyl, and the polar groups make the cellulose hydrophilic, which tends to form a strong network and agglomerates, making it difficult to uniformly dispersed in the latex matrix. As a result, natural fibers should be surface treated to increase the thermodynamic miscibility and interface bonding strength before being used as reinforcement materials in polymer composites. Through surface modification, the hydrophobicity of the cellulose can be improved and helps improve the dispersibility of the cellulose in the latex matrix. After surface modification, the mechanical and thermal properties of the composite were improved, and at the same time, provide biodegradability to the final products. The improvement in the dispersibility and properties of waste/raw natural filler in latex-based composite provides a wide range of potential applications in industries including biodegradable gloves, flexible display, sensor, medical application, agriculture and the construction industry.

Author Contributions: D.N.S. involved in writing the manuscript and A.R.A. involved in reviewing and editing. All authors have read and agreed to the published version of the manuscript.

Funding: This research was funded by the Ministry of Higher Education Malaysia under the Fundamental Research Grant Scheme (grant no: FRGS/1/2020/TK0/USM/01/2).

Institutional Review Board Statement: Not applicable.

Informed Consent Statement: Not applicable.

Data Availability Statement: The data presented in this study are available on request from the corresponding author.

Acknowledgments: The authors gratefully acknowledge the support provided by School of Materials and Mineral Resources, Universiti Sains Malaysia.

Conflicts of Interest: The authors declare no conflict of interest.

References

1. Dugvekar, M.; Dixit, S. High density polyethylene composites reinforced by jute fibers and rice stalk dust: A mechanical study. *Mater. Today Proc.* **2021**. [CrossRef]
2. Abdulrahman, S.; Hainin, M.R.; Satar, M.K.I.M.; Hassan, N.A.; Al Saffar, Z.H. Review on the potentials of natural rubber in bitumen modification. In *IOP Conference Series: Earth and Environmental Science*; IOP Publishing Ltd.: Chiyoda, Tokyo, 2020; Volume 476, p. 012067. [CrossRef]
3. Basik, A.; Sanglier, J.-J.; Yeo, C.; Sudesh, K. Microbial Degradation of Rubber: Actinobacteria. *Polymers* **2021**, *13*, 1989. [CrossRef] [PubMed]
4. Sanguansap, K.; Suteewong, T.; Saendee, P.; Buranabunya, U.; Tangboriboonrat, P. Composite natural rubber based latex particles: A novel approach. *Polymer* **2005**, *46*, 1373–1378. [CrossRef]

5. Keddie, J.; Routh, A.F. *Fundamentals of Latex Film Formation: Processes and Properties. Routh Google Books*; Springer Science & Business Media: Berlin, Germany, 2010. Available online: https://books.google.com.my/books?hl=en&lr=&id=ONqr3cBXRWEC&oi=fnd&pg=PR1&dq=what+is+latex&ots=pc8szKklf3&sig=av5i3zswKKbbXJ5BBthPeZ12gjw&redir_esc=y#v=onepage&q=what%20is%20latex&f=false (accessed on 7 October 2021).
6. Rahman, M.F.A.; Norfaizal, N.S.; Azura, A.R. The Influence of Sago Starch Dispersion on Mechanical Properties of Biodegradable Natural Rubber Latex Films. *Mater. Today Proc.* **2019**, *17*, 1040–1046. [CrossRef]
7. Nawong, C.; Umsakul, K.; Sermwittayawong, N. Rubber gloves biodegradation by a consortium, mixed culture and pure culture isolated from soil samples. *Braz. J. Microbiol.* **2018**, *49*, 481–488. [CrossRef]
8. Abraham, E.; Cherian, B.M.; Elbi, P.A.; Pothen, L.A.; Thomas, S. Recent Advances in the Recycling of Rubber Waste. In *Recent Developments in Polymer Recycling*; Fainleib, A., Grigoryeva, O., Eds.; Transworld Research Network: Trivandrum, India, 2011; pp. 47–100. Available online: https://www.researchgate.net/publication/282235405 (accessed on 10 October 2021).
9. Riyajan, S.A. Biodegradable Green Glove Containing Ascorbic Acid from Maleated Epoxidized Natural Rubber/Poly(vinyl alcohol) Blend: Preparation and Physical Properties. *J. Polym. Environ.* **2021**, 1–10. [CrossRef]
10. Misman, M.A.; Azura, A.R. Overview on the potential of biodegradable natural Rubber Latex gloves for commercialization. *Adv. Mater. Res.* **2014**, *844*, 486–489. [CrossRef]
11. Misman, M.A.; Azura, A.R. The Mechanical Properties Enhancement of Biodegradable Aminofunctional Starch/Carboxylate Nitrile Butadiene Rubber 2 Latex Films. *Preprints* **2018**, 2018040324. [CrossRef]
12. Azura, A.R.; Adzami, N.S.; Tajarudin, H.A. Utilization of metroxylan sagu pith waste as biodegradable filler for Natural Rubber (NR) latex films. *Solid State Phenom.* **2017**, *264*, 198–201. [CrossRef]
13. Wei, D.W.; Wei, H.; Gauthier, A.C.; Song, J.; Jin, Y.; Xiao, H. Superhydrophobic modification of cellulose and cotton textiles: Methodologies and applications. *J. Bioresour. Bioprod.* **2020**, *5*, 1–15. [CrossRef]
14. Horvat, I.; Dović, D. Combustion of agricultural biomass—Issues and solutions. *Trans. Famena* **2018**, *42*, 75–86. [CrossRef]
15. Galiwango, E.; Rahman, N.S.A.; Al-Marzouqi, A.H.; Abu-Omar, M.M.; Khaleel, A.A. Isolation and characterization of cellulose and α-cellulose from date palm biomass waste. *Heliyon* **2019**, *5*, e02937. [CrossRef]
16. Alothman, O.Y.; Kian, L.K.; Saba, N.; Jawaid, M.; Khiari, R. Cellulose nanocrystal extracted from date palm fibre: Morphological, structural and thermal properties. *Ind. Crop. Prod.* **2021**, *159*, 113075. [CrossRef]
17. Akinjokun, A.I.; Petrik, L.F.; Ogunfowokan, A.O.; Ajao, J.; Ojumu, T.V. Isolation and characterization of nanocrystalline cellulose from cocoa pod husk (CPH) biomass wastes. *Heliyon* **2021**, *7*, e06680. [CrossRef] [PubMed]
18. Thambiraj, S.; Shankaran, D.R. Preparation and physicochemical characterization of cellulose nanocrystals from industrial waste cotton. *Appl. Surf. Sci.* **2017**, *412*, 405–416. [CrossRef]
19. Pandi, N.; Sonawane, S.H.; Kishore, K.A. Synthesis of cellulose nanocrystals (CNCs) from cotton using ultrasound-assisted acid hydrolysis. *Ultrason. Sonochemistry* **2021**, *70*, 105353. [CrossRef] [PubMed]
20. Shamskar, K.R.; Heidari, H.; Rashidi, A. Preparation and evaluation of nanocrystalline cellulose aerogels from raw cotton and cotton stalk. *Ind. Crop. Prod.* **2016**, *93*, 203–211. [CrossRef]
21. Gopalakrishnan, J. Reinforcement of Natural Rubber Using Cellulose Nanofibres Isolated from Coconut Spathe. 2018. Available online: https://www.materialstoday.com/proceedings (accessed on 8 September 2021).
22. Farah, N.H.; Salmah, H.; Marliza, M. Effect of Butyl Methacrylate on Properties of Regenerated Cellulose Coconut Shell Biocomposite Films. *Procedia Chem.* **2016**, *19*, 335–339. [CrossRef]
23. Taksitta, K.; Sujarit, P.; Ratanawimarnwong, N.; Donpudsa, S.; Songsrirote, K. Development of tannin-immobilized cellulose fiber extracted from coconut husk and the application as a biosorbent to remove heavy metal ions. *Environ. Nanotechnol. Monit. Manag.* **2020**, *14*, 100389. [CrossRef]
24. Tuerxun, D.; Pulingam, T.; Nordin, N.I.; Chen, Y.W.; Kamaldin, J.B.; Julkapli, N.B.M.; Lee, H.V.; Leo, B.F.; Johan, M.R.B. Synthesis, characterization and cytotoxicity studies of nanocrystalline cellulose from the production waste of rubber-wood and kenaf-bast fibers. *Eur. Polym. J.* **2019**, *116*, 352–360. [CrossRef]
25. Gabriel, T.; Belete, A.; Syrowatka, F.; Neubert, R.H.H.; Gebre-Mariam, T. Extraction and characterization of celluloses from various plant byproducts. *Int. J. Biol. Macromol.* **2020**, *158*, 1248–1258. [CrossRef]
26. Farahhanim, W.; Fathilah, W.; Othaman, R. Electrospun cellulose fibres and applications. *Sains Malays.* **2019**, *48*, 1459–1472. [CrossRef]
27. Menon, M.P.; Selvakumar, R.; Suresh, P.; Ramakrishna, S. Extraction and Modification of Cellulose Nanofibers Derived from Biomass for Environmental Application. 2017. Available online: www.webometrics.info/en/node/58 (accessed on 18 September 2021).
28. Shaghaleh, H.; Xu, X.; Wang, S. Current progress in production of biopolymeric materials based on cellulose, cellulose nanofibers, and cellulose derivatives. *RSC Adv.* **2018**, *8*, 825–842. [CrossRef]
29. Kumar, A.; Negi, Y.S.; Choudhary, V.; Bhardwaj, N.K. Characterization of Cellulose Nanocrystals Produced by Acid-Hydrolysis from Sugarcane Bagasse as Agro-Waste. *J. Mater. Phys. Chem.* **2020**, *2*, 1–8. [CrossRef]
30. Wu, C.; McClements, D.J.; He, M.; Zheng, L.; Tian, T.; Teng, F.; Li, Y. Preparation and characterization of okara nanocellulose fabricated using sonication or high-pressure homogenization treatments. *Carbohydr. Polym.* **2021**, *255*, 117364. [CrossRef] [PubMed]

31. Abdullah, M.A.; Nazir, M.S.; Raza, M.R.; Wahjoedi, B.A.; Yussof, A.W. Autoclave and ultra-sonication treatments of oil palm empty fruit bunch fibers for cellulose extraction and its polypropylene composite properties. *J. Clean. Prod.* **2016**, *126*, 686–697. [CrossRef]
32. Zhao, J.; Zhang, W.; Zhang, X.; Zhang, X.; Lu, C.; Deng, Y. Extraction of cellulose nanofibrils from dry softwood pulp using high shear homogenization. *Carbohydr. Polym.* **2013**, *97*, 695–702. [CrossRef]
33. Pujiasih, S.; Kurnia; Masykur, A.; Kusumaningsih, T.; Saputra, O.A. Silylation and characterization of microcrystalline cellulose isolated from indonesian native oil palm empty fruit bunch. *Carbohydr. Polym.* **2018**, *184*, 74–81. [CrossRef]
34. Prabhu, L.; Krishnaraj, V.; Sathish, S.; Ramesh, M. Mechanical, Chemical and Acoustical Behavior of Sisal-Tea Waste-Glass Fiber Reinforced Epoxy Based Hybrid Polymer Composites. *Mater. Today Proc.* **2019**, *16*, 653–660. [CrossRef]
35. Barbosa, R.F.S.; Souza, A.G.; Ferreira, F.F.; Rosa, D.S. Isolation and acetylation of cellulose nanostructures with a homogeneous system. *Carbohydr. Polym.* **2019**, *218*, 208–217. [CrossRef]
36. Ganguly, P.; Sengupta, S.; Das, P.; Bhowal, A. Valorization of food waste: Extraction of cellulose, lignin and their application in energy use and water treatment. *Fuel* **2020**, *280*, 118581. [CrossRef]
37. Souza, A.G.; Santos, D.F.; Ferreira, R.R.; Pinto, V.Z.; Rosa, D.S. Innovative process for obtaining modified nanocellulose from soybean straw. *Int. J. Biol. Macromol.* **2020**, *165*, 1803–1812. [CrossRef]
38. Khan, M.N.; Rehman, N.; Sharif, A.; Ahmed, E.; Farooqi, Z.H.; Din, M.I. Environmentally benign extraction of cellulose from dunchi fiber for nanocellulose fabrication. *Int. J. Biol. Macromol.* **2020**, *153*, 72–78. [CrossRef]
39. Cudjoe, E.; Hunsen, M.; Xue, Z.; Way, A.E.; Barrios, E.; Olson, R.A.; Hore, M.J.; Rowan, S.J. Miscanthus Giganteus: A commercially viable sustainable source of cellulose nanocrystals. *Carbohydr. Polym.* **2017**, *155*, 230–241. [CrossRef]
40. Song, K.; Ji, Y.; Wang, L.; Wei, Y.; Yu, Z. A green and environmental benign method to extract cellulose nanocrystal by ball mill assisted solid acid hydrolysis. *J. Clean. Prod.* **2018**, *196*, 1169–1175. [CrossRef]
41. Landry, V.; Alemdar, A.; Blanchet, P. Nanocrystalline cellulose: Morphological, physical, and mechanical properties. *For. Prod. J.* **2011**, *61*, 104–112. [CrossRef]
42. Teow, Y.H.; Amirudin, S.N.; Ho, K.C. Sustainable approach to the synthesis of cellulose membrane from oil palm empty fruit bunch for dye wastewater treatment. *J. Water Process. Eng.* **2020**, *34*, 101182. [CrossRef]
43. Candido, R.G.; Godoy, G.G.; Gonçalves, A. Characterization and application of cellulose acetate synthesized from sugarcane bagasse. *Carbohydr. Polym.* **2017**, *167*, 280–289. [CrossRef] [PubMed]
44. Xie, H.; Zou, Z.; Du, H.; Zhang, X.; Wang, X.; Yang, X.; Wang, H.; Li, G.; Li, L.; Si, C. Preparation of thermally stable and surface-functionalized cellulose nanocrystals via mixed H2SO4/Oxalic acid hydrolysis. *Carbohydr. Polym.* **2019**, *223*, 115116. [CrossRef] [PubMed]
45. Wang, Q.; Yan, X.; Chang, Y.; Ren, L.; Zhou, J. Fabrication and characterization of chitin nanofibers through esterification and ultrasound treatment. *Carbohydr. Polym.* **2018**, *180*, 81–87. [CrossRef] [PubMed]
46. Ilyas, R.A.; Sapuan, S.M.; Ishak, M.R.; Zainudin, E.S. Sugar palm nanofibrillated cellulose (*Arenga pinnata (Wurmb.) Merr*): Effect of cycles on their yield, physic-chemical, morphological and thermal behavior. *Int. J. Biol. Macromol.* **2019**, *123*, 379–388. [CrossRef]
47. Singh, M.; Kaushik, A.; Ahuja, D. Surface functionalization of nanofibrillated cellulose extracted from wheat straw: Effect of process parameters. *Carbohydr. Polym.* **2016**, *150*, 48–56. [CrossRef] [PubMed]
48. Cd, M.D.; Joseph, R.; Begum, P.S.; Raghunandanan, A.; Vackkachan, N.T.; Padmanabhan, D.; Formela, K. Chitin nanowhiskers from shrimp shell waste as green filler in acrylonitrile-butadiene rubber: Processing and performance properties. *Carbohydr. Polym.* **2020**, *245*, 116505. [CrossRef]
49. Fortunati, E.; Benincasa, P.; Balestra, G.M.; Luzi, F.; Mazzaglia, A.; Del Buono, D.; Puglia, D.; Torre, L. Revalorization of barley straw and husk as precursors for cellulose nanocrystals extraction and their effect on PVA_CH nanocomposites. *Ind. Crop. Prod.* **2016**, *92*, 201–217. [CrossRef]
50. Neto, W.P.F.; Mariano, M.; da Silva, I.S.V.; Silvério, H.A.; Putaux, J.L.; Otaguro, H.; Pasquini, D.; Dufresne, A. Mechanical properties of natural rubber nanocomposites reinforced with high aspect ratio cellulose nanocrystals isolated from soy hulls. *Carbohydr. Polym.* **2016**, *153*, 143–152. [CrossRef]
51. Thomas, M.G.; Abraham, E.; Jyotishkumar, P.; Maria, H.J.; Pothen, L.A.; Thomas, S. Nanocelluloses from jute fibers and their nanocomposites with natural rubber: Preparation and characterization. *Int. J. Biol. Macromol.* **2015**, *81*, 768–777. [CrossRef] [PubMed]
52. Luzi, F.; Fortunati, E.; Jiménez, A.; Puglia, D.; Pezzolla, D.; Gigliotti, G.; Kenny, J.M.; Chiralt, A.; Torre, L. Production and characterization of PLA_PBS biodegradable blends reinforced with cellulose nanocrystals extracted from hemp fibres. *Ind. Crop. Prod.* **2016**, *93*, 276–289. [CrossRef]
53. Li, Y.; Liu, Y.; Chen, W.; Wang, Q.; Liu, Y.; Li, J.; Yu, H. Facile extraction of cellulose nanocrystals from wood using ethanol and peroxide solvothermal pretreatment followed by ultrasonic nanofibrillation. *Green Chem.* **2016**, *18*, 1010–1018. [CrossRef]
54. Oun, A.A.; Rhim, J.W. Isolation of cellulose nanocrystals from grain straws and their use for the preparation of carboxymethyl cellulose-based nanocomposite films. *Carbohydr. Polym.* **2016**, *150*, 187–200. [CrossRef]
55. Liu, Y.; Lv, X.; Bao, J.; Xie, J.; Tang, X.; Che, J.; Ma, Y.; Tong, J. Characterization of silane treated and untreated natural cellulosic fibre from corn stalk waste as potential reinforcement in polymer composites. *Carbohydr. Polym.* **2019**, *218*, 179–187. [CrossRef]

56. Khanjanzadeh, H.; Behrooz, R.; Bahramifar, N.; Gindl-Altmutter, W.; Bacher, M.; Edler, M.; Griesser, T. Surface chemical functionalization of cellulose nanocrystals by 3-aminopropyltriethoxysilane. *Int. J. Biol. Macromol.* **2018**, *106*, 1288–1296. [CrossRef]
57. Jin, K.; Tang, Y.; Zhu, X.; Zhou, Y. Polylactic acid based biocomposite films reinforced with silanized nanocrystalline cellulose. *Int. J. Biol. Macromol.* **2020**, *162*, 1109–1117. [CrossRef]
58. Jankauskaitė, V.; Balčiūnaitienė, A.; Alexandrova, R.; Buškuvienė, N.; Žukienė, K. Effect of Cellulose Microfiber Silylation Procedures on the Properties and Antibacterial Activity of Polydimethylsiloxane. *Coatings* **2020**, *10*, 567. [CrossRef]
59. Cichosz, S.; Masek, A. Cellulose Fibers Hydrophobization via a Hybrid Chemical Modification. *Polymers* **2019**, *11*, 1174. [CrossRef] [PubMed]
60. Daud, S.; You, Y.S.; Azura, A.R. The Effect of Acid Hydrolyzed Sago Starch on Mechanical Properties of Natural Rubber and Carboxylated Nitrile Butadiene Rubber Latex. 2019. Available online: https://www.materialstoday.com/proceedings2214-7853 (accessed on 18 September 2021).
61. Chen, Y.; Zhang, Y.; Xu, C.; Cao, X. Cellulose nanocrystals reinforced foamed nitrile rubber nanocomposites. *Carbohydr. Polym.* **2015**, *130*, 149–154. [CrossRef] [PubMed]
62. Kudori, S.N.I.; Ismail, H.; Khimi, S.R. Tensile and Morphological Properties on Kenaf Core or Bast Filled Natural Rubber Latex Foam (NRLF). *Mater. Today Proceeding* **2019**, *17*, 609–615. [CrossRef]
63. Jong, L. Particle size and particle-particle interactions on tensile properties and reinforcement of corn flour particles in natural rubber. *Eur. Polym. J.* **2016**, *74*, 136–147. [CrossRef]
64. Jayathilaka, L.P.I.; Ariyadasa, T.U.; Egodage, S.M. Development of biodegradable natural rubber latex composites by employing corn derivative bio-fillers. *J. Appl. Polym. Sci.* **2020**, *137*. [CrossRef]
65. Xiong, X.Q.; Bao, Y.L.; Liu, H.; Zhu, Q.; Lu, R.; Miyakoshi, T. Study on mechanical and electrical properties of cellulose nanofibrils/graphene-modified natural rubber. *Mater. Chem. Phys.* **2019**, *223*, 535–541. [CrossRef]
66. Tanan, W.; Panichpakdee, J.; Saengsuwan, S. Novel biodegradable hydrogel based on natural polymers: Synthesis, characterization, swelling/reswelling and biodegradability. *Eur. Polym. J.* **2019**, *112*, 678–687. [CrossRef]
67. Raju, G.; Haris, M.R.H.M.; Azura, A.R.; Eid, A.M.A.M. Chitosan Epoxidized Natural Rubber Biocomposites for Sorption and Biodegradability Studies. *ACS Omega* **2020**, *5*, 28760–28766. [CrossRef]
68. Júnior, P.; Machado, C.; Alberto, R.; Ferreira, R.; Motta, L.A.D.; Pasquini, D. Characterization and properties of cementitious composites with cellulose fiber, silica fume and latex. *Constr. Build. Mater.* **2020**, *257*, 119602. [CrossRef]
69. Jiang, Y.; Zhang, Y.; Ding, L.; Wang, B.; Chen, Z.; Zhong, Y.; Zhang, L.; Mao, Z.; Xu, H.; Sui, X. Sag control of waterborne acrylic latex with regenerated nanocellulose suspension. *Prog. Org. Coat.* **2018**, *123*, 146–152. [CrossRef]
70. Eslami, H.; Tzoganakis, C.; Mekonnen, T.H. Constructing pristine and modified cellulose nanocrystals based cured polychloroprene nanocomposite films for dipped goods application. *Composites Part C Open Access* **2020**, *1*, 100009. [CrossRef]

polymers

MDPI

Article

Improving the Processability and Performance of Micronized Fiber-Reinforced Green Composites through the Use of Biobased Additives*

Bruno F. A. Valente [1], Armando J. D. Silvestre [1], Carlos Pascoal Neto [2], Carla Vilela [1] and Carmen S. R. Freire [1,*]

[1] CICECO–Aveiro Institute of Materials, Department of Chemistry, University of Aveiro, 3810-193 Aveiro, Portugal

[2] RAIZ, Research Institute of Forest and Paper (The Navigator Company), Rua José Estevão, Eixo, 3800-783 Aveiro, Portugal

* Correspondence: cfreire@ua.pt

Abstract: Green composites made of bioplastics reinforced with natural fibers have gained considerable attention over recent years. However, the use of natural fibers in composites usually compromise some key properties, such as the impact strength and the processability of the final materials. In the present study, two distinct additives, namely an epoxidized linseed oil (ELO) and a sugar-based surfactant, viz. GlucoPure® Sense (GPS), were tested in composite formulations of poly(lactic acid) (PLA) or poly(hydroxybutyrate) (PHB) reinforced with micronized pulp fibers. Both additives showed a plasticizing effect, which led to a decrease in the Young's and flexural moduli and strengths. At the same time, the elongation and flexural strain at break were considerably improved on some formulations. The melt flow rate was also remarkably improved with the incorporation of the additives. In the PHB-based composites, an increment of 230% was observed upon incorporation of 7.5 wt.% ELO and, in composites based on PLA, an increase of around 155% was achieved with the introduction of 2.5 wt.% GPS. ELO also increased the impact strength to a maximum of 29 kJ m^{-2}, in formulations with PLA. For most composites, a faster degradation rate was observed on the formulations with the additives, reaching, in the case of PHB composites with GPS, a noteworthy weight loss over 75% under burial testing in compost medium at room temperature.

Keywords: green composites; poly(lactic acid); poly(hydroxybutyrate); cellulose; micronized fibers; epoxidized linseed oil; sugar-based surfactant; mechanical properties; thermal properties; biodegradability

Citation: Valente, B.F.A.; Silvestre, A.J.D.; Neto, C.P.; Vilela, C.; Freire, C.S.R. Improving the Processability and Performance of Micronized Fiber-Reinforced Green Composites through the Use of Biobased Additives. *Polymers* **2022**, *14*, 3451. https://doi.org/10.3390/polym14173451

Academic Editor: Evgenia G. Korzhikova-Vlakh

Received: 8 July 2022
Accepted: 12 August 2022
Published: 24 August 2022

1. Introduction

The global polymeric composites market is still dominated, in terms of value and volume, by composites based on petroleum-based matrices reinforced with glass fibers [1]. Nonetheless, natural fibers are becoming increasingly used as substitutes for glass fibers in composite materials due to their high availability, lower production cost, renewability and biodegradability [2]. Additionally, the manufacturing and processing of the composites is less hazardous when using natural fibers and causes less wear and tear on the machinery [2–4]. With a compound annual growth rate (CAGR) of 14.2% between 2020 and 2028, the natural fibers reinforced composites market is forecasted to reach over USD 60 billion, which clearly validates the establishment of natural fibers in the composites industry [5]. However, the sustainability of these so-called biocomposites is still not satisfactory, as the most common thermoplastic matrices, such as polypropylene (PP) or polyethylene (PE), are derived from non-renewable resources [6]. Additionally, the separation of the individual components of the composite is difficult, which makes them unsuitable for recycling, and disposal on landfills or incineration procedures are not sustainable options due to their negative long term impact on the environment [7–9].

To fabricate fully sustainable and environmentally friendly composites, all the components must have an entirely green path from the origin to its disposal. Poly(lactic acid) (PLA) and poly(hydroxybutyrate) (PHB) are good candidates to replace the conventional thermoplastic matrices given their biobased origins and biodegradability [10,11]. PLA, which can be produced by polycondensation of lactic acid or ring opening polymerization of a lactide intermediate, is a linear polyester with good mechanical properties, nontoxicity, stability to ultraviolet light and easily dyeable [12]. PHB is a polyester that is bio-synthetized by a wide variety of microorganisms (e.g., *Ralstonia eutropha*, *Aeromonas hydrophila*), belonging to the poly(hydroxyalkanoate) family, and with properties comparable to those of PP [11]. Despite their relatively high costs, both polymers are among the few bioplastics produced at a commercial level and can be easily processed through the mainstream technologies, such as injection molding, blow molding, thermoforming, and extrusion, among others, which facilitates their industrial processing and usage [11,12]. However, and although both polymers are promising for use as matrices in composites, their brittleness and low impact resistance still constitute a challenge [13,14].

Additionally, the combination of thermoplastic polymers (regardless of their biobased or fossil nature) with natural fibers, even though environmentally advantageous, still has some limitations. For instance, the hydrophilicity and high aspect ratio of the fibers makes them poorly compatible with the hydrophobic thermoplastic matrices and causes the formation of bundles or aggregates of fibers [15,16]. Likewise, the low impact strength, high water uptake and dimensional instability are also serious concerns for their long term application [6]. Mechanical treatments, such as milling processes or micronization, have been proposed to overcome the agglomeration issue by improving the dispersion of the fibers on the matrices through the decrease in the fiber size to the micrometer range [16–18]. Chemical pre-treatments of the fibers, i.e., alkali treatments [6] and surface modifications, such as acetylation, benzoylation or silylation, generally improve the compatibility between the composites constituents due to the hydrophobization of the fibers surface [6]. Additives, such as coupling agents, plasticizers, impact modifiers and lubricants, among others, may also be used to tailor the properties of composites, by improving the processability and minimizing some of the drawbacks mentioned before [13,19]. From an industrial perspective, the use of additives still provides an additional advantage as they are simpler to use and do not require multiple steps and time consuming processes, as in the case of the chemical modifications of the cellulose fibers [13].

Plasticizers are one of the most common class of additives used in thermoplastic and composite industries, being primarily applied to decrease the rigidity and brittleness of the thermoplastic matrices and to lower their melting and glass transition temperatures, thus improving their flexibility and processability [20–22]. Plasticization may also influence a variety of other properties, such as density, viscosity, resistance to biological degradation, hardness, resistance to fracture, and degree of crystallinity, among others [22]. Several families of compounds, including polyols (e.g., glycerol [23] and poly(ethylene glycol) (PEG) [24]), triacetin [25], citrate esters (e.g., triethyl citrate (TEC) [26] or acetyl tributyl citrate (ATC) [27]), and epoxidized vegetable oils [19,28], have been investigated as plasticizers in composite formulations. Sugar-based surfactants, e.g., sophorolipids, which are highly desirable given their green character, have also been tested with some pure biobased thermoplastic matrices, such as PLA, PHB and polycaprolactone (PCL) [29], but not in composites of these thermoplastics and cellulose fibers. Alongside plasticizers, coupling agents are also extensively used additives in the composite industry. Such compounds act as a bridge between the matrices and the fibers, thus promoting better interfacial adhesion between them [13]. Maleated coupling agents [30], diisocyanates [31], silanes [32] and even functionalized vegetable oils [33] have been targets of research. Within the catalogue of additives, functionalized vegetable oils are particularly attractive given their sustainability, commercial availability, non-toxicity and relatively low cost [34]. More specifically, epoxidized linseed [19] and soybean [14] oils have been highlighted as additives in green composites of PLA acting simultaneously as coupling agents and plasticizers [14,19,33].

Despite recent studies on the use of additives, more specifically epoxidized oils, on green composites based on PLA or PLA/PHB blends and different plant fibers [28,35–37], they have never been used in composites reinforced with micronized cellulose fibers and the examination of crucial properties, such as the melt flow rate (MFR), water uptake, flexural and impact properties, is scarce or non-existent. Moreover, to the best of our knowledge, the effect of surfactants and epoxidized oils as additives in green composites based on PHB has not been evaluated yet. Therefore, in the present study, a sugar-based surfactant and epoxidized linseed oil were individually evaluated as additives in composites with two distinct biobased thermoplastic matrices, namely PLA and PHB, reinforced with micronized bleached eucalyptus kraft pulp (BEKP). The interfacial morphology and the tensile, flexural and impact mechanical performance alongside with the melt flow rate, thermal properties, water uptake capacity and burial behavior in compost medium were thoroughly investigated.

2. Materials and Methods

2.1. Materials

Poly(hydroxybutyrate) (PHB), grade P226, with a melt flow rate of 10 g·10 min^{-1} (180 °C, 5 kg), density of 1.25 g·cm^{-3} and number-average molecular weight of 22,200 ± 4500 [38], was supplied by Biomer (Schwalbach, Germany). Poly(lactic acid) (PLA), grade 3100HP, with a melt flow rate of 24 g·10 min^{-1} (210 °C, 2.16 kg) and molecular weight of 148 kDa [39] was supplied by NatureWorks (Plymouth, MN, USA). The mechanically treated (micronized) cellulose pulp fibers (average length and width of 332 μm and 12.5 μm, respectively) were obtained through a micronization procedure from eucalyptus bleached kraft pulp and were kindly provided by a Portuguese pulp mill. The epoxidized linseed oil (ELO), composed of stearic (3–5%), palmitic (5–7%), oleic (18–26%), linoleic (14–20%) and linolenic (51–56%) acids, was acquired from Traquisa (Barcelona, Spain). ELO has an oxirane oxygen minimum of 8%, iodine value under 5%, density of 1.1 g·cm^{-3} and viscosity between 800 and 1200 cP. The nonionic surfactant GlucoPure® Sense (GPS), composed of sunflower oil methylglucamide (52%), glycerin (5%), water (10%) and propyleneglycol (33%), was obtained from Clariant (Barcelona, Spain). Compost medium Nutrimais Pulverulento, obtained by selective composting of lignocellulosic residues from forest exploration and from food wastes, was acquired from Nutrimais (Gondomar, Portugal). The specifications of the compost medium are as follows: moisture = 29.23 ± 1.41%, water-holding capacity = 170.33 ± 12.34%, organic matter = 57.66 ± 11.5%, pH = 8.9 ± 0.6, and elemental composition = 32.03 ± 2.66% (total carbon (C)), 2.41 ± 0.48% (total nitrogen (N)), 1.96 ± 0.39 (total potassium (K)), 0.71 ± 0.14% (total magnesium (Mg)), 14.50 ± 2.90% (total calcium (Ca)), and 1.33 ± 0.27% (total phosphorous (P)).

2.2. Compounding and Processing of the Composites

Reference composites without additives were obtained by melt mixing the thermoplastic polymers with the micronized cellulose fibers for a fixed fiber load of 40 wt.%. For the formulations with additives, the fibers were pre-mixed with ELO or GPS, followed by the melt mixing with the corresponding matrices. The additives were incorporated in percentages ranging from 2.5 to 7.5 wt.% relative to the total weight of the composites. The formulations were mixed at 180 °C for PLA and 170 °C for PHB, for 15 min at 50 rpm in a Brabender W 30 EHT Plastograph EC mixer (Brabender, Duisburg, Germany) with a total volume capacity of 30 cm^3. The rectangular and dog-bone shaped test specimens for the characterization assays were produced through injection molding in a Thermo Scientific Haake Minijet II (Thermo Scientific, Waltham, MA, USA). PLA composites were injected at 195 °C with the mold temperature between 100 °C and 125 °C and PHB composites were injected at 185 °C with the mold at 65 °C. The injection pressure was set at 800 bar during 15 s with a post-injection pressure of 200 bar for another 5 s.

2.3. Characterization Techniques

The density of the composites was calculated by dividing the test specimens' weights by their volumes. Rectangular specimens (80 × 10 × 4 mm^3) with a 3.2 cm^3 volume were used for density calculations using at least five replicates.

Scanning electron microscopy (SEM) images of the fracture zones of the composites (after the tensile tests) were acquired on a FE-SEM Hitachi SU70 microscope (Hitachi High-Technologies Corporation, Tokyo, Japan) operated at 15.0 kV. Prior to the analysis, the samples were coated with a carbon film.

The determination of the Young's modulus, tensile strength and elongation at break were performed according to ISO-527-2 (bar type 5A). The tests of at least six dog bone-like specimens with test dimensions of 30 × 4.25 × 2.10 mm^3 were conducted at a crosshead velocity of 5 mm·min^{-1} using a 10 kN static load cell. The flexural properties of the composites were determined following the three-point loading model according to ISO 178. Five rectangular specimens (80 × 10 × 4 mm^3) were tested with a 500 N load cell at a velocity of 5 mm·min^{-1}. The span length between supports was set at 64 mm. The strain at break was calculated as the ratio between the extension at break and the maximum deflection (20 mm). Both the tensile and flexural tests were conducted on a universal testing machine Instron 5966 (Instron Corporation, Norwood, MA, USA). The unnotched Charpy (edgewise) impact strength of ten specimens with dimensions of 80 × 10 × 4 mm^3 was acquired on a Ray Ran Universal Pendulum impact system (Ray-Ran Test Equipment Ltd., Nuneaton, UK). The equipment was operating a pendulum of 4 J and the support span was set at 62 mm, according to ISO 179/1eU.

The melt flow rate of the composites was evaluated on a Melt Flow Indexer Devenport (MFR-9) (Ametek, Denmark), according to ASTM D1238. The temperature selected for analysis of the PHB-based composites was 175 °C and for PLA-based composites was 190 °C. At least five cut-offs for each sample were weighted and the melt flow rate (MFR) was calculated as follows:

$$MFR\left(g\cdot 10\ min^{-1}\right) = \frac{600 \times m}{t} \quad (1)$$

where m is the average mass of the cut-offs, in grams, and t is the cut-off time interval, in seconds.

Thermogravimetric analysis (TGA) was carried out on a SETSYS Setaram TGA analyzer (SETARAM Instrumentation, Lyon, France) equipped with a platinum cell. Approximately 10 mg of each sample was heated from room temperature to 800 °C at a rate of 10 °C·min^{-1} under a nitrogen flow.

The differential scanning calorimetry (DSC) thermograms of the matrices and composites were obtained on a Perkin Elmer Diamon DSC unit (PerkinElmer, Waltham, MA, USA). The samples were placed in aluminum capsules and heated from −40 to 220 °C, hold for 5 min at 220 °C to eliminate the thermal history, cool back to −40 °C and reheated from −40 to 220 °C at a heating rate of 10 °C·min^{-1}. The results were recorded on the second heating cycle.

To determine the water uptake capacity, three specimens of each sample (60 × 10 × 1 mm^3) were immersed in distilled water, at room temperature, over a period of 31 days. Then, the specimens were periodically removed from the water, wiped with tissue paper and weighed to determine the weight of the wet specimens. The water uptake (%) at time t was calculated as represented in Equation (2):

$$Water\ uptake\ (\%) = \frac{(W_t - W_0)}{W_0} \times 100 \quad (2)$$

where W_0 is the specimen's initial weight and W_t is the weight of the specimens after immersion for a t time period.

Burial tests of the neat matrices and composites (with and without additives) were performed by measuring the weight loss of the materials after burying them in compost

medium. Prior to the tests, composite specimens ($60 \times 10 \times 1$ mm^3) were preconditioned by drying them at 40 °C in an oven for a period of 48 h and the initial mass was recorded (W_0). Then, the specimens were buried in compost in a 1 L container, at a distance of 1 cm from the bottom. Water was added to the containers to adjust the water content to 60 wt.% of the water-holding capacity. The assays were conducted at room temperature (18–25 °C) and the humidity of the compost was constantly adjusted by adding water every couple of days. At the end of 15, 30, 60, 90, 120, 150 and 180 days, three specimens from each sample were retrieved, carefully washed to remove any compost residues and dried at 40 °C for 48 h. Finally, the mass of the specimens was recorded (W_t) to calculate the weight loss according to Equation (3):

$$Weight\ loss\ (\%) = \frac{W_0 - W_t}{W_0} \times 100 \quad (3)$$

Statistical analysis of all mechanical properties data was performed using the analysis of variance (ANOVA) and Tukey's mean comparison test (OriginPro 9.6.5, OriginLab Corporation, MA, USA) with the statistical significance established at $p < 0.05$.

3. Results and Discussion

Fully biobased composites composed of PHB or PLA, reinforced with cellulose micronized fibers, were manufactured by melt-mixing and processed by injection molding (Figure 1). Two distinct biobased commercial additives, viz. epoxidized linseed oil (ELO) and a sugar-based surfactant GlucoPure® Sense (GPS), were chosen given their sustainability, non-toxicity, and commercial availability. GPS was selected given that the main constituent is an amphiphilic molecule, potentially enabling interactions with the hydrophilic cellulose and the hydrophobic matrix [40]. Moreover, GPS is also constituted by propylene glycol and glycerin, which are well-known plasticizers. ELO is a recognized plasticizer with acid scavenger abilities, mainly used for poly(vinyl chloride) (PVC), which has also proved efficient in plasticizing green composites and improving the interfacial adhesion between matrices and reinforcements. A fiber load of 40 wt.% was chosen based on the best performance of the PHB and PLA-based composites reported in our previous study [18], as well as by taking into consideration the reinforcement percentage of fibers on commercial petroleum-based composites [41]. The effect of the incorporation of both additives separately, on the properties and performance of the composites, was evaluated.

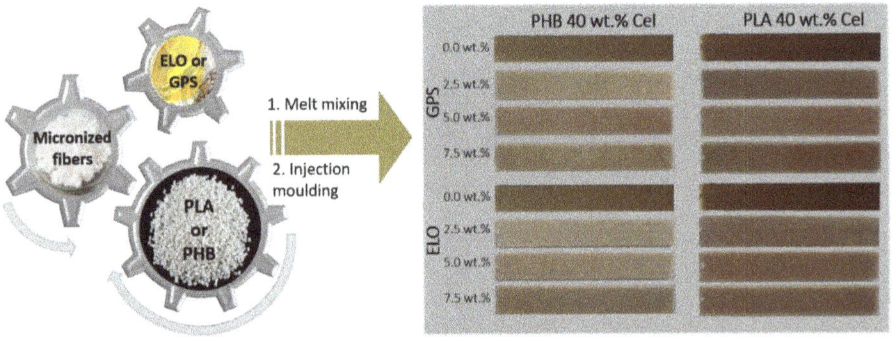

Figure 1. Schematic illustration of the experimental procedure and photographs of the injection-molded specimens of poly(hydroxybutyrate) (PHB) and poly(lactic acid) (PLA)-based composites reinforced with cellulose micronized fibers with different percentages of GPS and ELO additives.

According to Figure 1, the reference materials, i.e., the composites without additives, are darker than the corresponding materials with either ELO or GPS. This darker color of the reference materials can be attributed to some thermal degradation of cellulose

during processing at high temperatures [42]. However, this effect is softened upon the incorporation of ELO and GPS. In fact, Immonen et al. [19] also observed a similar color change on composites made of PLA reinforced with 40 wt.% bleached softwood kraft pulp (BSKP) in the presence of ELO. The lightning was attributed to the plasticizing effect of ELO that reduces the viscosity of the composite material and the friction during processing. From Figure 1, it is also perceptible that composites based on PHB with higher amounts of GPS (7.5 wt.%) have brighter areas on the surface, unlike the ones with 7.5 wt.% ELO. That can be related to an eventual lack of compatibility between the hydrophobic PHB and GPS, which is clearly more hydrophilic than ELO.

The composites, with or without additives, display Fourier Transform Infrared-Attenuated Total Reflection (FTIR-ATR) spectra similar to those of the corresponding thermoplastic polymers (Figure S1), which can be related with the high percentage of the polymers and the relatively low loads of the additives in the composite formulations. Additionally, all the composites have similar density values as listed in Table 1. In fact, and given the relatively low amounts of additives, the average density of the composites is only marginally affected. Even so, a slight decrease trend on the density of the composites can be noted as the percentage of additive increases. That is due to the lower density of the additives (1.06 $g{\cdot}cm^{-3}$ for both ELO [43] and GPS [44]) in comparison with the thermoplastic matrices (1.25 $g{\cdot}cm^{-3}$ for PHB [45] and 1.24 $g{\cdot}cm^{-3}$ for PLA [46]) and cellulose (1.5–1.6 $g{\cdot}cm^{-3}$) [47]. Thus, the raise in the percentage of a lower density material leads to the slight decrease in the density of the overall composite material.

Table 1. Density of the PHB- and PLA-based composites reinforced with cellulose micronized fibers without additives and with different percentages of GPS and ELO additives.

Sample	Density ($g{\cdot}cm^{-3}$)	Sample	Density ($g{\cdot}cm^{-3}$)
PHB_Cel	1.34 ± 0.01	PLA_Cel	1.38 ± 0.00
PHB_Cel 2.5 wt.% GPS	1.34 ± 0.02	PLA_Cel 2.5 wt.% GPS	1.36 ± 0.02
PHB_Cel 5.0 wt.% GPS	1.33 ± 0.02	PLA_Cel 5.0 wt.% GPS	1.37 ± 0.00
PHB_Cel 7.5 wt.% GPS	1.30 ± 0.03	PLA_Cel 7.5 wt.% GPS	1.37 ± 0.01
PHB_Cel 2.5 wt.% ELO	1.34 ± 0.01	PLA_Cel 2.5 wt.% ELO	1.37 ± 0.00
PHB_Cel 5.0 wt.% ELO	1.33 ± 0.02	PLA_Cel 5.0 wt.% ELO	1.36 ± 0.00
PHB_Cel 7.5 wt.% ELO	1.30 ± 0.02	PLA_Cel 7.5 wt.% ELO	1.35 ± 0.00

3.1. Morphological Characterization

Scanning electronic microscopy was used to evaluate the morphology of the composites with different additive contents. The SEM micrographs of the fracture zones after tensile testing are displayed in Figure 2. From the analysis of these micrographs, it can be observed that, despite some fiber pull-outs and voids, the compatibility between the fibers and matrices on the composites without additives is relatively good, particularly in the composites based on PLA. This observation is in line with our previous work, where different grades of PLA and PHB matrices were melt compounded with four micronized pulp fibers with distinct aspect ratios [18].

Concerning the composites loaded with the additives, no significant differences were observed when the sugar-based surfactant GPS was added to both PLA and PHB composites, independently of the added content. On the other hand, and regarding the composites with ELO, mainly the ones based on PLA, the interface between fibers and matrix is sometimes indistinguishable, which is indicative of a superior compatibility and interfacial adhesion. That points out that besides acting as a plasticizer, ELO can enhance even further the interaction between the matrix and the fibers. Despite some authors reporting that no major differences could be observed on the SEM micrographs upon incorporation of ELO in green composites of PLA reinforced with birch kraft pulp fibers [33,48], which was attributed to the reasonable interfacial adhesion already observed on the composites without the additive, some other studies revealed similar results to the ones obtained in the present work. For instance, composites of PLA with hazelnut shell flour (HSF) containing

7.5 wt.% ELO showed remarkably different fracture micrographs revealing improved interfacial adhesion [49]. It has been proposed in a few studies that the epoxy groups of the epoxidized oils can react, particularly with cellulose, but also with PLA. More specifically, the ring opening of the epoxy groups by the hydroxyl groups of cellulose, leading to the formation of a covalent ether bond between the epoxidized oil and the cellulosic fibers and the reaction between the epoxy functional groups with the hydroxyl and carboxyl end-groups of PLA, which leads, in the last case, to an ester bond [14,19,49].

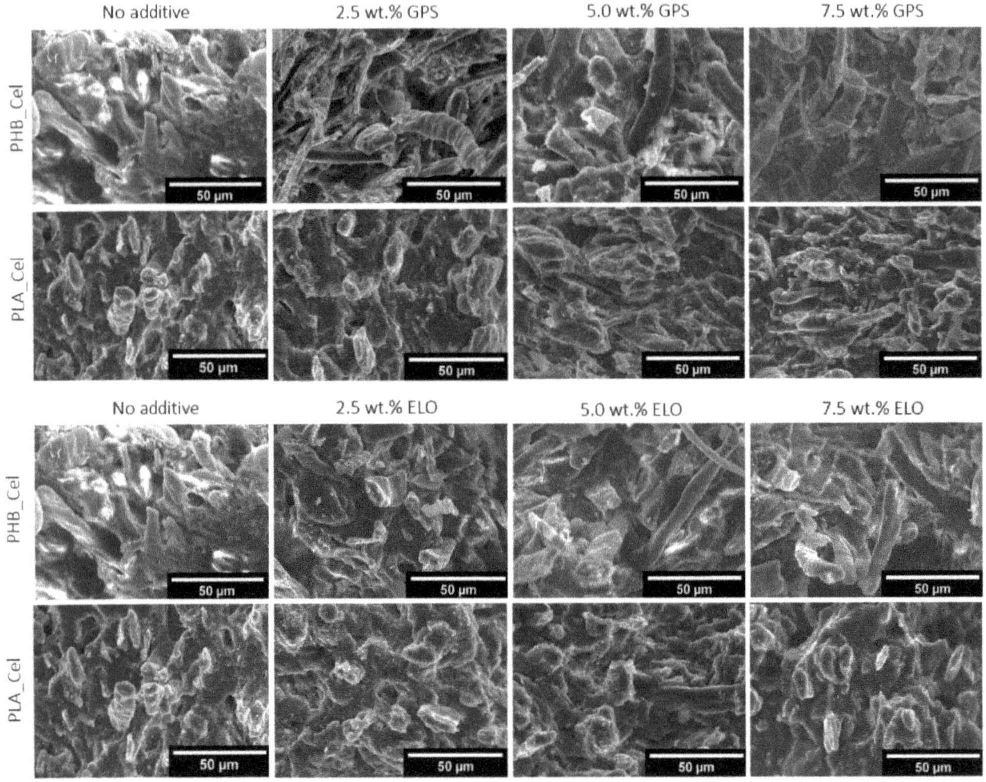

Figure 2. SEM micrographs of the fractured surfaces of PHB- and PLA-based composites without additives and with different contents of GPS and ELO additives.

3.2. Mechanical Properties

3.2.1. Tensile Properties

The tensile performance of the composites with and without additives is presented in Figure 3. The Young's modulus and tensile strength decreased progressively with the augmentation of the content of the GPS and ELO additives. Moreover, this decline is more pronounced for the PHB-based composites where, for example, the Young's modulus and tensile strength decreased from 3.83 ± 0.02 GPa and 35.9 ± 1.7 MPa to 2.02 ± 0.06 GPa and 13.6 ± 1.3 MPa, respectively, when 7.5 wt.% of GPS was added. Concerning the elongation at break, while for the PHB-based composites with ELO, the changes were not statistically different, for composites with PLA, this parameter more than doubled (1.62 ± 0.29 to 3.67 ± 0.30%) for higher loads (7.5 wt.%). On the other hand, the GPS caused a decrease in the elongation at break of all the composites, regardless of the thermoplastic matrix.

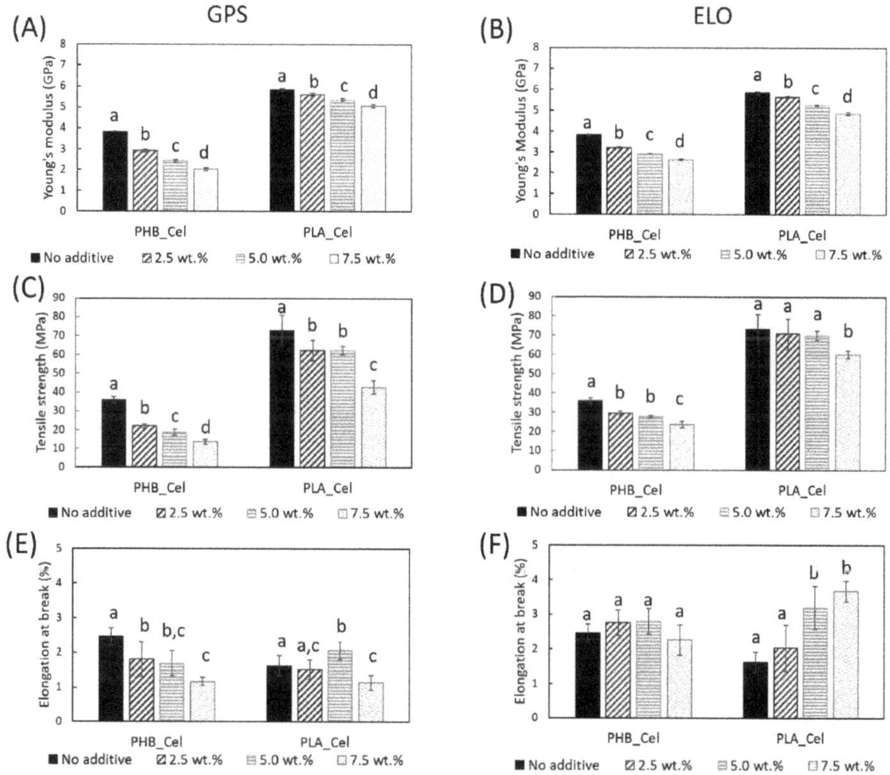

Figure 3. Tensile mechanical properties of the PHB- and PLA-based composites without additives and with GPS (**A**,**C**,**E**) and ELO (**B**,**D**,**F**) additives. Different letters (a,b,c,d) indicate statistically significant differences ($p < 0.05$).

It is known that the incorporation of plasticizers, such as PEG or citrate esters, usually results in the decrease in the Young's modulus and tensile strength, and in the increase in the elongation at break, either for PLA [50] or PHB [51,52] matrices or composites. As for the ELO, literature shows contradictory results for the tensile behavior of PLA-based composites. Taking for instance the example of the HSF-reinforced PLA composites, the addition of ELO led to a decrease in the Young's modulus and tensile strength and a correspondent increase in the elongation at break [49]. On the contrary, for the PLA composites reinforced with bleached hardwood kraft pulp (BHKP), the addition of ELO increased all three parameters [33]. In another study about PLA composites reinforced with 40 wt.% BSKP, the Young's modulus and tensile strength increased by 5% and 7%, respectively, for a load of 5 wt.% of ELO (relative to fiber mass). On its turn, the elongation at break was reduced by 12.5%. However, for a higher load, i.e., 12 wt.% of ELO, the composites had inferior Young's modulus and tensile strength, and higher elongation at break than the reference material, which also corroborates the plasticizing effect of ELO [19]. To the best of our knowledge, ELO was never used in PHB-based composites. However, on a neat matrix of PHB, small amounts of ELO worked as plasticizers, decreasing the tensile strength and modulus, and increasing the elongation at break [53], thus acting in accordance to the present results.

3.2.2. Flexural Properties

The flexural properties agree with the tensile data, which means that, for the most part, the incorporation of both additives diminishes the flexural modulus and flexural strength, and that the ELO improves the flexural strain at break of the composites (Figure 4). The flexural modulus of the PHB-based composites containing 7.5 wt.% of additives decreased to less than half of the initial value (from 5.1 ± 0.2 GPa to 2.4 ± 0.1 for GPS and 2.5 ± 0.1 GPa for ELO), while the reduction for the PLA-based composites is much less noticeable. For instance, there are no statistical differences between the flexural modulus of PLA_Cel without ELO and with 2.5 wt.% ELO and between the PLA_Cel with 5.0 wt.% or 7.5 wt.%. Concerning the strain at break, with the incorporation of 7.5 wt.% of ELO, a raise of 39% and 115% was observed for PHB and PLA-based composites, respectively. This increase in the flexibility of the composites is also due to the plasticizing effect of ELO, which decreases the intermolecular forces between the polymeric chains, thus reducing the brittleness of the materials, consequently increasing their flexibility, ductility and extensibility [49,54]. Similar trends on the flexural behavior have been reported by Balart et al. [49] for composites of PLA reinforced with HSF and using ELO as plasticizer. On the other hand, the flexibility of the composites with GPS, represented by the strain at break, decreased with the incorporation of the sugar-based surfactant. This behavior is certainly related to the lack of compatibility of this hydrophilic additive with the thermoplastic matrices, as previously observed on the SEM images, particularly for composites of PHB.

3.2.3. Impact Properties

The impact resistance is one of the mechanical parameters that is often seriously compromised in green composites, either due to the brittleness of the matrices or its poor interfacial adhesion with the cellulosic fibers. Therefore, the understanding of the effect of both additives on the Charpy impact strength of the composites is of utmost importance (Figure 5). As observed from the results, the incorporation of ELO, at a 5 wt.% load, raises the impact strength of the composites by 36.7% and 135.7% for PHB and PLA-based composites, respectively. These outstanding improvements, especially in composites with PLA, are the result of the plasticizing effect of ELO combined with the enhanced interfacial adhesion induced by this additive. Both the plasticizer effect of ELO, which reduces the brittleness of ternary mixtures, and the enhanced interfacial adhesion between the polymeric matrices and the reinforcing fibers contribute to averting the initiation and propagation of cracks [54]. A plateau on the impact resistance of the materials is achieved for 5 wt.% of ELO. A further increase in the additive content does not cause any significant upgrade on the impact resistance, most likely because after that threshold, the ELO molecules start to interact more with themselves than with cellulosic fibers or even with the PHB or PLA [55]. It is worth mentioning that the use of ELO in the present work brings the values of the impact strength of these green composites close to those of biocomposites based on PP and PE currently available on the market (33 to 42 kJ·m^{-2}) [41,56].

Regarding the influence of the sugar-based surfactant GPS, this additive did not contribute to the improvement of the impact strength of the composite materials. Instead, the impact strength decreased for higher loads of the additive, mainly for PHB-based composites. The lack of compatibility between the hydrophilic constituents of GPS and the hydrophobic matrices, previously mentioned, are probably the main cause for the loss of the resistance to impact. In fact, these results are also in agreement with the outcomes of the tensile and flexural tests previously described.

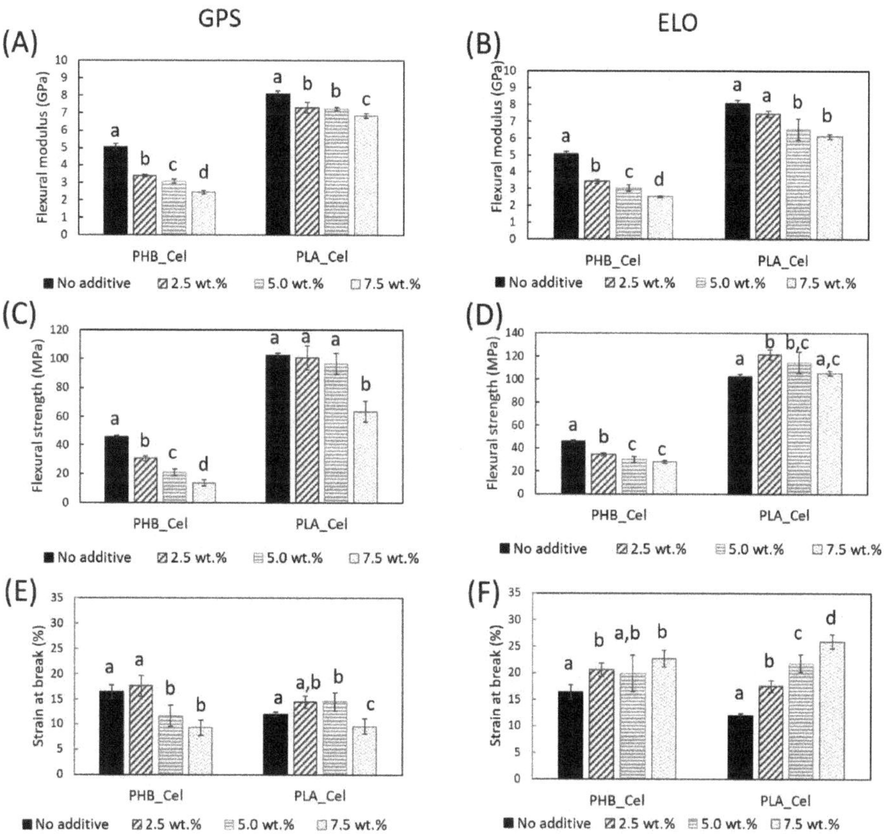

Figure 4. Flexural mechanical properties of the PHB- and PLA-based composites without additives and with GPS (**A**,**C**,**E**) and ELO (**B**,**D**,**F**) additives. Different letters (a,b,c,d) indicate statistically significant differences ($p < 0.05$).

Several other works have previously reported improvements on the impact properties of green composites with PLA using various epoxidized oils, such as epoxidized soybean (ESO) [55], linseed (ELO) [19], palm (EPO) [57] and jatropha (EJO) oils [36]. Yet, the improvements were modest in comparison with the results obtained in the present study. For instance, an ELO load of 8 wt.% relative to the weight of the BSKP fibers yielded a 37% increase in the impact resistance, raising the Charpy unnotched impact strength to only 18 kJ·m^{-2} [19]. In a similar attempt to induce better mechanical properties, the utilization of ELO in composites with BHKP increased the impact strength by 25.9% [33]. The use of low aspect ratio-micronized fibers in the present study certainly contributed to the better performance of the materials when compared with the aforementioned studies. As disclosed in a previous work, the use of fibers with smaller aspect ratios reduces the formation of defects and consequent initiation and propagation of cracks [18].

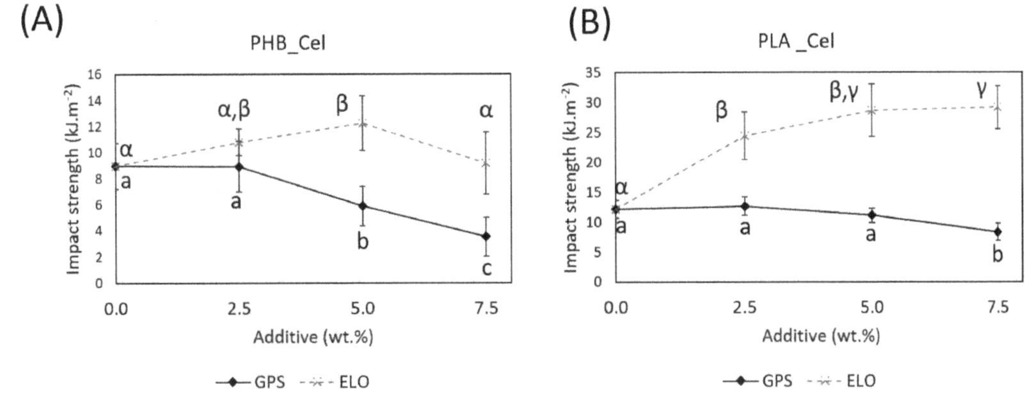

Figure 5. Impact strength of the PHB- (**A**) and PLA-based composites (**B**) without additives and with different loads of GPS and ELO additives (the lines are for visual guidance only). Different letters (a,b,c) or (α,β,γ) indicate statistically significant differences ($p < 0.05$).

Concerning the PHB-based composite materials, and as far as our literature search could verify, no data are available regarding the use of any sugar based surfactants or epoxidized oils as additives. However, on a neat PHB matrix, ELO proved useful for increasing the impact strength [53]. Moreover, plasticizers, such as soybean oil (SO), ESO or TEC, have been tested in films made of a PHB copolymer, more specifically poly(3-hydroxybutyrate-*co*–3-hydroxyvalerate) (PHBV). Both ESO and TEC plasticizers promoted considerable improvements on the impact strength of the blends [58].

3.3. Melt Flow Rate

The melt flow rate, expressed in grams of composite extruded through a nozzle in ten minutes, and measured at a constant temperature when a standard weight is applied, is an indirect measurement of melt viscosity and molecular weight of the materials and an indication of their processability [59,60]. The variation of the MFR of composites as a function of the load of additives is illustrated in Figure 6. With the additive GPS, none of the composites based on PHB flowed at all, at any of the percentages. The lack of flowability of these composites is certainly related with the migration of the additive. In fact, the clear separation of a liquid with the same color and consistency of GPS observed during the MFR assays provides evidence of phase separation. As a matter of fact, the diffusion and leaching of the plasticizer from the bulk material to the surroundings is one of the main disadvantages reported for the use of this class of additives, which can seriously compromise the performance of the plasticized composite materials [61]. Several characteristics of the plasticizer, such as the type, molecular weight, branching degree and polarity, can affect its migration [20,22]. The compatibility of the plasticizer with the thermoplastic polymer and some conditions, such as the temperature, can also have an effect on this behavior [20,22]. Thus, considering that low molecular weight plasticizers tend to have higher migration rates and the high temperatures also favor the migration [62], it is not surprising that GPS, given its formulation and the temperature at which the MFR assay is conducted, is highly prone to migration. Moreover, the lack of compatibility between GPS and PHB, as noticed before, certainly contributes to its diffusion from the bulk material to the surroundings. As for the PLA-based composites, at low percentages of GPS (2.5 wt.%), the migration phenomenon did not take place and a raise on the MFR from 2.71 to 6.93 $g{\cdot}10\ min^{-1}$ could be observed. With increasing amounts of this additive (higher than 5.0 wt.%), the melt flow rate started to drop and the separation of the additive from the composite also became clear.

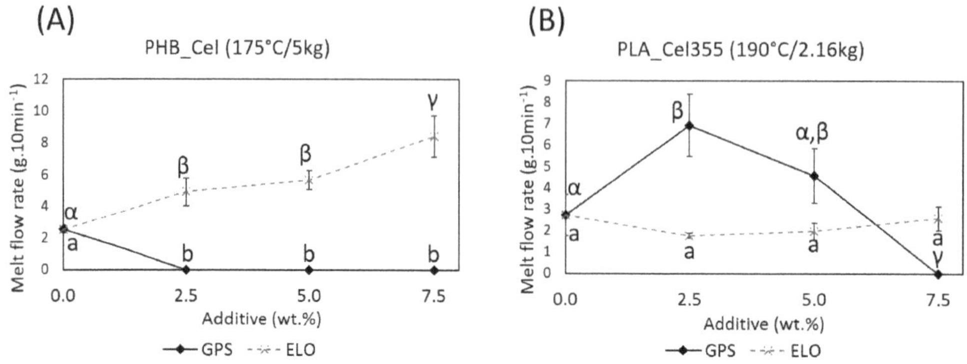

Figure 6. Melt flow rate of the PHB- (**A**) and PLA-based composites (**B**), without additives and with different loads of GPS and ELO additives (the lines are for visual guidance only). Different letters (a,b) or (α,β,γ) indicate statistically significant differences ($p < 0.05$).

On the contrary, ELO has a low migration tendency [63]. The melt flow rate of PHB-based composites with 7.5 wt.% ELO experiences a remarkable increase of 230%, from 2.55 to 8.42 g·10 min^{-1}. Such improvement is undoubtedly related to the plasticizing effect of ELO in PHB based composites, which reflects the inferences made from the mechanical assays. Aside from the increase in the molecular mobility of the polyester chains, the plasticizer reduces the fiber/fiber and fiber/matrix friction, which contributes to the improvement of the melt flowability [24]. For the composites with PLA, the addition of ELO did not change the MFR of the composites. As depicted in the SEM micrographs (Figure 2), ELO promotes the increase in the interfacial adhesion of the fibers with PLA, which prevents the melt flow rate from increasing. Immonen et al. [19] also claimed that a decrease was expected in the apparent viscosity of the PLA/BSKP composites with ELO, if the epoxidized oil was working solely as a plasticizer. However, due to the improved compatibilization of the fibers with PLA caused by the addition of ELO, an increase was noted instead.

Although of extreme importance, the influence of the addition of eco-friendly plasticizers in the melt flow rate of green composites has not been the subject of intensive research. Previous studies reported that composites of bamboo reinforced PLA plasticized with PEG had better flowability than the corresponding composites without additives [24] and that ESO could be used as plasticizer for neat PLA [64]. However, to the best of our knowledge, the melt flow rate of green composites with epoxidized oils or sugar-based surfactants as additives was not yet reported. The present work demonstrates that ELO can improve the processability of PHB-based composites, and GPS, at low amounts, can also contribute to a better flowability of PLA-based composite materials.

3.4. Thermal Properties

Differential scanning calorimetry was performed to evaluate the impact of the additives on the onset and peak melting temperature (T_m) of the composites. The results portrayed in Figure S2 and summarized in Table 2 show that, unlike PLA-based composites where only one melting peak can be observed between 160.4–169.3 °C, the PHB ones have two melting peaks (135.6–149.2 °C and 152.1–163.0 °C). The existence of double or multiple peaks is not uncommon for PHB and has also been reported elsewhere [38]. That behavior can be related with the partial melting and recrystallization and remelting (mrr) process, to the melting of crystals with different lamellar thicknesses or to the melting of different crystalline structures [65].

Table 2. DSC and TGA data of the PHB- and PLA-based composites without additives and with different loads of GPS and ELO additives.

Samples	DSC				TGA	
	$Onset_1$ (°C)	Tm_1 (°C)	$Onset_2$ (°C)	Tm_2 (°C)	T_{max1} (°C)	T_{max2} (°C)
PHB_Cel	142.7	148.2	154.2	162.4	276.2	351.2
PHB_Cel 2.5 wt.% GPS	142.1	147.5	153.5	162.2	281.4	338.1
PHB_Cel 5.0 wt.% GPS	136.9	141.9	149.0	157.2	275.4	335.9
PHB_Cel 7.5 wt.% GPS	126.3	135.6	141.7	152.1	269.2	336.5
PHB_Cel 2.5 wt.% ELO	144.0	149.2	155.9	163.0	283.1	346.9
PHB_Cel 5.0 wt.% ELO	141.4	146.8	153.1	160.9	283.0	352.5
PHB_Cel 7.5 wt.% ELO	140.6	146.0	152.6	160.9	276.4	349.7
PLA_Cel	155.7	165.0	-	-	336.2	-
PLA_Cel 2.5 wt.% GPS	159.0	166.8	-	-	338.3	-
PLA_Cel 5.0 wt.% GPS	150.1	168.1	-	-	331.8	-
PLA_Cel 7.5 wt.% GPS	150.7	160.4	-	-	332.9	-
PLA_Cel 2.5 wt.% ELO	160.3	169.3	-	-	341.7	-
PLA_Cel 5.0 wt.% ELO	158.9	168.9	-	-	339.4	-
PLA_Cel 7.5 wt.% ELO	152.2	166.9	-	-	341.5	-

The second heating scans also reveal that, for the PHB-based composites, the incorporation of increasing amounts of GPS gradually lower both the onset and peak melting temperatures of the composites. For instance, the addition of 7.5 wt.% of GPS leads the melting temperatures' peak to drop by 12.6 °C and 10.3 °C. On the other hand, in composites with PLA, the peak melting temperature is increased for lower loads of GPS and decreased 4.6 °C for 7.5 wt.%. With ELO, a reduction on the melting temperature was only observed for composites with PHB, mainly at higher concentrations. All the aforementioned decreases are related to the plasticizing action of the additives, which reduces the intermolecular interactions, consequently lowering the melting point [27]. Similar outcomes on the lowering of the melting point as a cause of plasticization have also been reported, for example, for PHB [53] and PLA [66] matrices with epoxidized vegetable oils and also in green composites of PHB plasticized with ATC [27]. For the case of the composites with PLA and ELO, the melting point shifts to higher temperatures with the incorporation of this additive. For instance, 5 wt.% ELO raises the melting point from 165.0 °C to 168.9 °C. It is suggested that the increased interaction between the components of the ternary mixture with the incorporation of ELO is responsible for the increase in the melting peak. These results can once more be linked to the enhanced interfacial morphology discussed above. Surprisingly, in previous works about the addition of epoxidized oils in green composites of PLA, the authors reported lower melting temperatures on the composites with the epoxidized oil, which was mainly associated to its plasticizing effect [14,19,33].

The thermal stability of the composites, accessed through TGA, is also presented in Table 2 and Figure S3. The observation of two degradation steps in the composites with PHB and only one in the composites with PLA was already reported in a previous work [18]. For composites with PHB, the first step, which corresponds to the degradation of the thermoplastic polymer, shifts to lower temperatures when the GPS load is over 5.0 wt.%. Additionally, for the second step, which is related with the degradation of the cellulosic fibers, the incorporation of GPS promotes a decrease in the maximum degradation temperature from 351.2 °C to 336.5 °C. The low compatibility of GPS with the cellulosic fibers and with PHB, as seen on the test specimens on Figure 1, may play an important role on the decrease in the thermal stability of the PHB-based composites with GPS. For the PLA-based composite materials, the maximum degradation temperature also decreased from 336.2 °C to 332.9 °C with the addition of 7.5 wt.% GPS. With regards to ELO, for composites with PHB, a 5 wt.% load gives the best thermal stability with maximum degradation temperatures of 283.0 °C and 352.5 °C. For PLA-based materials, the maximum degradation temperature is increased up to 5.5 °C. This increase in the thermal stability has been attributed in other works to the improved interaction between the ELO with the polymer chains, leading to the formation of a barrier on the surface which restricts the permeability of volatile compounds to the exterior [49,53]. Similar to the action of ELO in the present study, the incorporation of other epoxidized vegetable oils, such as EPO [54] and epoxidized palm and soybean oil (ESPO) [67], have also shown improved thermal stability, particularly with PLA.

3.5. Water Uptake Capacity

It is well known that hydrophobic matrices have almost negligible water uptakes. For instance, the maximum water absorption of the neat PHB and PLA used in the present work is under 1.3 ± 0.1%, after 30 days of immersion [18]. On the other hand, cellulose is hydrophilic and, when added for the manufacturing of composite materials, may cause some level of swelling and dimensional instability [68,69]. Moreover, the stress generated by the swelling of the fibers may lead to the occurrence of cracks, seriously compromising the mechanical performance of the composites [70]. In Figure 7, it is perceptible that composites with 40 wt.% of cellulose fibers absorbed water quickly in an early stage, reaching a plateau at the end of approximately 5 days. With the addition of the additives, the water uptake of the composites increased, for both PHB- and PLA-based materials. For example, the water uptake raised from 9.7 % (no additive) to a maximum of 12.6% (GPS) and 9.9% (ELO), for composites based on PHB, and from 6.6 % (no additive) to 9.8 % (GPS) and 8.1% (ELO), for composites based on PLA. Such increments in water uptake, explicitly more expressive for GPS, must be related to its plasticizing effect and hydrophilic character: the plasticizer molecules weaken the intermolecular interactions between the polymeric chains, increasing the ductility of the matrices, ultimately allowing fibers to absorb more water and to swell more easily. The swollen fibers may lead to the formation of cracks, which further contributes to the increase in the water uptake [71].

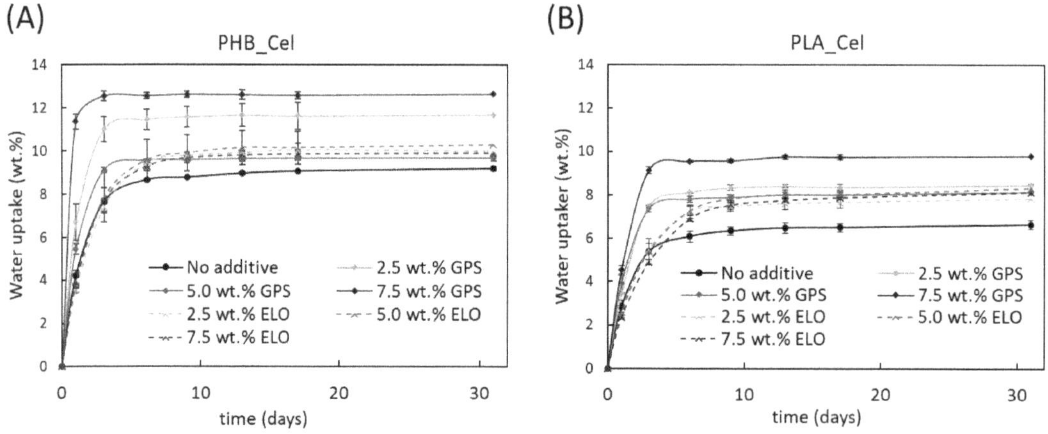

Figure 7. Water uptake capacity of the PHB- (**A**) and PLA-based composites (**B**) without additives and with different loads of GPS and ELO additives (the lines are for visual guidance only).

Also visible on the results presented in Figure 7 is that, while for composites with GPS, higher percentages of the additive resulted in higher amounts of water absorbed by the composite, the water uptake capacity remained similar for composites with increasing amounts of ELO. That can be explained by taking into consideration that GPS has in its composition considerable amounts of hydrophilic molecules (glycerin and propylene glycol), which certainly contribute to higher water absorption. Furthermore, the sugar portion of the surfactant may also contribute to this outcome. Thus, as the amount of GPS is increased in the composites, and with it the content of hydrophilic moieties, the affinity for water gradually increases. On the contrary, ELO is a more hydrophobic molecule which, in addition, and as mentioned before, seems to improve the interfacial adhesion, thus reducing the number of gaps in the interfacial region [49,70]. In fact, Balart et al. [71] investigated the water uptake of PLA composites with HSF in the presence of ELO, and also concluded that ELO contributed to the increase in the water uptake and that the load of this additive was not a differentiating factor [71].

3.6. Degradation in Compost Medium

The evaluation of the degradation of the composite materials in environmental conditions is becoming increasingly important. In reality, the alleged biodegradability of PLA and PHB is one of the main motivations behind their use in green composites with natural fibers and one of the leading advantages over traditional PP- or PE-based composites [6]. Since several conditions, such as the temperature and moisture of the soil/compost medium, the characteristics of the polymers, namely chemical composition, crystallinity, molecular weight and the presence of additives, have an influence on the biodegradation of the composites [72], it is also key to test the effect of the additives used in this work on the biodegradation profile of the obtained composites. Moreover, and considering that additives influence the properties of the composites, it is also expected that they interfere with the resistance to biological degradation [20]. Thus, the weight loss of the composites with the additives and the reference materials was evaluated under controlled conditions, in compost medium, and the visual representation of the data is illustrated in Figure 8.

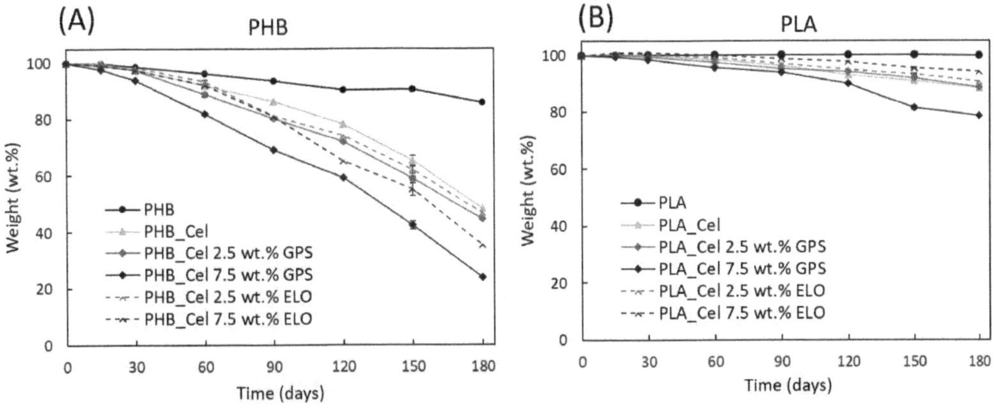

Figure 8. Weight loss as a function of time for the PHB- (**A**) and PLA-based composites (**B**) without additives and with different loads of GPS and ELO additives. The standard deviations (not shown) are under 7.6% for PHB-based materials and under 3.9% for PLA-based materials, and the lines are for visual guidance only.

First, it is evident that the neat matrices have different degradation rates. While the pristine PHB lost 14.2% of its initial weight, the mass of pure PLA specimens was virtually unchanged over six months. That can be explained by taking into consideration that PHB and PLA have different biodegradation mechanisms [10,73]. If on the one hand, PHB biodegradation is solely enzymatic, on the other hand, the biodegradation of PLA first starts with non-enzymatic hydrolysis, followed by enzymatic degradation [10,73]. The non-enzymatic hydrolysis begins with water absorption that allows for a random cleavage of the ester bonds of the polymer, resulting in lower molecular weight oligomers and lactic acid. Such hydrolysis occurs preferably at temperatures above the PLA glass transition temperature (55–62 °C) [10,74]. Given that the present compost burial experiments were conducted at room temperature (18–25 °C), that explains why PLA did not show any signs of degradation. Similar results were obtained by Zhang et al. [75] who also demonstrated that PHB could in fact be degraded in compost, while PLA did not show any weight loss at room temperature. In yet another study, in a one-year test run, injection molded tensile bars of PLA did not show any signs of degradation at 25 °C both on compost and soil. However, at temperatures near thermophilic conditions (50 °C), weight losses of near 45% were recorded after only 4 weeks [76].

As cellulosic fibers are easily degraded under environmental conditions [77], their incorporation into the biobased matrices promoted a considerable increase in the weight loss of the composites. A closer look at the test specimens in Figure 9 and the degradation profiles displayed in Figure 8 reveals that the composites with PHB have higher weight losses than the ones with PLA, which is obviously related to the degradation of the matrices, or in the case of PLA, of its absence. As PLA showed no degradation at these specific conditions, the weight loss of the composites is only attributed to the degradation of cellulose fibers, while for PHB-based composites, the weight loss is due to the combination of the degradation of cellulose and PHB. It can also be observed that the degradation rate speeds up with time, which is a consequence of surface erosion that facilitates the diffusion of water and extracellular enzymes from the microorganisms, making the polymers more easily accessible for degradation [72].

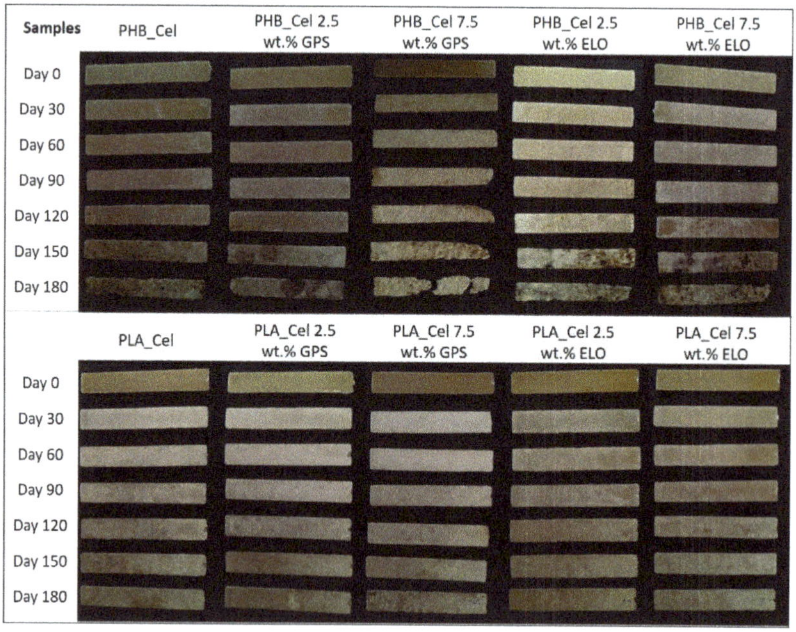

Figure 9. Digital photographs of the test specimens prior and after burial tests in compost medium for 30, 60, 90, 12 and 180 days.

With the exception of PLA-based composites with ELO, the composites with additives showed higher weight losses than the reference materials without additives. For instance, the incorporation of 7.5 wt.% of GPS increases the weight loss from 53.4% to 76.2% in composites with PHB and from 11.6% to 21.4% in composites with PLA. Such results can be correlated with the water uptake capacity (Figure 7), which, as discussed before, continuously raised with the load of GPS. Higher water absorption promotes the swelling of the fibers and increases the availability of bulk material to enzymatic activity, thus resulting in higher degradation rates [77]. Another possible contributing factor is the leaching of the hydrophilic components of GPS, leading to an increase in internal surface area. A similar conclusion on the leaching of hydrophilic plasticizers and its effect on the biodegradation was disclosed on PHB films plasticized with triethyl citrate [78]. For the case of the composites with ELO, this additive still raises the weight loss in the mixtures with PHB, mainly because of its plasticizing effect and consequent increase in the water uptake. However, although the water uptake is still higher in PLA composites with ELO than in the reference materials, the weight loss is inferior in the compost burial experiments, which can be associated with the improved interfacial adhesion between fibers and PLA promoted by ELO, as previously discussed. Since the biodegradation of composites mainly begins around the interface between the matrix and the reinforcements, an improvement on the interfacial adhesion decelerates the hydrolysis mechanism of the composites [77,79]. Other authors reported a similar outcome with the use of ELO on PLA-based composites. Despite the slight increase in the water uptake of the material with the incorporation of ELO, the disintegration rate in composting conditions was found to be lower, which was credited to the coupling effect of ELO [49,71].

4. Conclusions

Two eco-friendly and commercially available additives, namely an epoxidized linseed oil and the sugar-based surfactant GlucoPure® Sense, were successfully incorporated on green composites of PLA and PHB reinforced with cellulose micronized fibers. GPS, a mixture of a sugar-based surfactant with propylene glycol and glycerin, mostly functioned as a plasticizer, making the composite materials more ductile, and consequently decreasing their tensile and flexural properties. This additive improved the composite degradation rates on compost medium and, in certain amounts, was beneficial to improve the melt flow rate in composites with PLA. Nevertheless, this additive is not suitable for PHB-based materials since it lacks compatibility with this thermoplastic polymer, thus compromising its processability. Regarding ELO, it performed differently on PLA and PHB-based composites. If from one side, on the composites with PHB, ELO mainly caused a drastic improvement on the melt flow rate, on the other side, in composites with PLA, it also enhanced the interfacial adhesion, which contributed to an outstanding increase in the impact resistance of the green composites. Both additives led to an increase in the water uptake and, for most cases, also accelerated the weight loss on compost under environmental conditions.

Overall, ELO and GPS additives proved to be viable as functional additives to improve the performance and processability of green composites. Properties, such as the impact strength and melt flow rate, may be tuned using different amounts of such additives. Moreover, their use is of simple implementation on industrial procedures with the additional advantage of being entirely environmentally friendly.

Supplementary Materials: The following supporting information can be downloaded at: https://www.mdpi.com/article/10.3390/polym14173451/s1, Figure S1: FTIR-ATR spectra of PHB, PHB_Cel and PHB_Cel 7.5 wt.% ELO (A) and PLA, PLA_Cel and PLA_Cel 7.5 wt.% ELO (B), Figure S2: Differential scanning calorimetry curves of PHB- (A,C) and PLA-based composites (B,D) without and with different loads of GPS and ELO additives, Figure S3: Thermogravimetric and derivative curves of PHB- (A) PLA-based composites (B) without and with different loads of GPS and ELO additives.

Author Contributions: Conceptualization, C.P.N. and C.S.R.F.; investigation, B.F.A.V. and C.V.; resources, A.J.D.S., C.P.N. and C.S.R.F.; data curation, B.F.A.V.; writing—original draft preparation, B.F.A.V.; writing—review and editing, A.J.D.S., C.P.N., C.V. and C.S.R.F.; supervision, C.P.N., C.V. and C.S.R.F.; funding acquisition, A.J.D.S., C.P.N. and C.S.R.F. All authors have read and agreed to the published version of the manuscript.

Funding: This work was carried out under the Project InPacTus—Innovative products and technologies from Eucalyptus, Project N.° 21874 (POCI-01-0247-FEDER-021874) funded by Portugal 2020 through the European Regional Development Fund (ERDF) in the frame of COMPETE 2020 n°246/AXIS II/2017, and project CICECO—Aveiro Institute of Materials, UIDB/50011/2020, UIDP/50011/2020 & LA/P/0006/2020, financed by national funds through the FCT/MCTES (PIDDAC).

FCT is also acknowledged for the research contracts under Scientific Employment Stimulus to C.V. (CEECIND/00263/2018 and 2021.01571.CEECIND) and C.S.R.F. (CEECIND/00464/2017).

Conflicts of Interest: The authors declare no conflict of interest.

References

1. Composites Market by Fiber Type (Glass Fiber Composites, Carbon Fiber Composites, Natural Fiber Composites), Resin Type (Thermoset Composites, Thermoplastic Composites), Manufacturing Process, End-Use Industry and Region–Global Forecast to 2026. Available online: https://www.marketsandmarkets.com/Market-Reports/composite-market-200051282.html (accessed on 20 January 2021).
2. Jariwala, H.; Jain, P. A review on mechanical behavior of natural fiber reinforced polymer composites and its applications. *J. Reinf. Plast. Compos.* **2019**, *38*, 441–453. [CrossRef]
3. Pickering, K.L.; Efendy, M.G.A.; Le, T.M. A review of recent developments in natural fibre composites and their mechanical performance. *Compos. Part A Appl. Sci. Manuf.* **2016**, *83*, 98–112. [CrossRef]
4. Dhakal, H.N. Mechanical performance of PC-based biocomposites. In *Biocomposites: Design and Mechanical Performance*; Mirsa, M., Pandey, J., Mohanty, A., Eds.; Woodhead Publishing: Sawston, UK, 2015; pp. 303–317; ISBN 9781782423942.
5. Verified Market Research. Global Biocomposites Market Size by Fiber (Wooden Fiber and Non-wood Fiber), by End-Use Industry (Building & Construction, Transportation, Consumer Goods), by Geographic Scope and Forecast. Available online: https://www.verifiedmarketresearch.com/product/biocomposite-market/ (accessed on 27 October 2021).
6. Gholampour, A.; Ozbakkaloglu, T. A review of natural fiber composites: Properties, modification and processing techniques, characterization, applications. *J. Mater. Sci.* **2020**, *55*, 829–892. [CrossRef]
7. La Mantia, F.P.; Morreale, M. Green composites: A brief review. *Compos. Part A Appl. Sci. Manuf.* **2011**, *42*, 579–588. [CrossRef]
8. Ho, M.-P.; Wang, H.; Lee, J.-H.; Ho, C.-K.; Lau, K.-T.; Leng, J.; Hui, D. Critical factors on manufacturing processes of natural fibre composites. *Compos. Part B Eng.* **2012**, *43*, 3549–3562. [CrossRef]
9. Thompson, R.C.; Moore, C.J.; vom Saal, F.S.; Swan, S.H. Plastics, the environment and human health: Current consensus and future trends. *Philos. Trans. R. Soc. B Biol. Sci.* **2009**, *364*, 2153–2166. [CrossRef] [PubMed]
10. Karamanlioglu, M.; Preziosi, R.; Robson, G.D. Abiotic and biotic environmental degradation of the bioplastic polymer poly(lactic acid): A review. *Polym. Degrad. Stab.* **2017**, *137*, 122–130. [CrossRef]
11. Dietrich, K.; Dumont, M.-J.; Del Rio, L.F.; Orsat, V. Producing PHAs in the bioeconomy—Towards a sustainable bioplastic. *Sustain. Prod. Consum.* **2017**, *9*, 58–70. [CrossRef]
12. Henton, D.E.; Gruber, P.; Lunt, J.; Randall, J. Polylactic acid technology. In *Natural Fibers, Biopolymers, and Biocomposites*; CRC Press: Boca Raton, FL, USA, 2005; pp. 527–577; ISBN 9780203508206.
13. Ghaffar, S.H.; Madyan, O.A.; Fan, M.; Corker, J. The Influence of Additives on the Interfacial Bonding Mechanisms between Natural Fibre and Biopolymer Composites. *Macromol. Res.* **2018**, *26*, 851–863. [CrossRef]
14. Chen, T.; Wu, Y.; Qiu, J.; Fei, M.; Qiu, R.; Liu, W. Interfacial compatibilization via in-situ polymerization of epoxidized soybean oil for bamboo fibers reinforced poly(lactic acid) biocomposites. *Compos. Part A Appl. Sci. Manuf.* **2020**, *138*, 106066. [CrossRef]
15. Amiandamhen, S.O.; Meincken, M.; Tyhoda, L. Natural Fibre Modification and Its Influence on Fibre-matrix Interfacial Properties in Biocomposite Materials. *Fibers Polym.* **2020**, *21*, 677–689. [CrossRef]
16. Qiang, T.; Wang, J.; Wolcott, M.P. Facile Fabrication of 100% Bio-Based and Degradable Ternary Cellulose/PHBV/PLA Composites. *Materials* **2018**, *11*, 330. [CrossRef]
17. Qiang, T.; Wang, J.; Wolcott, M.P. Facile Preparation of Cellulose/Polylactide Composite Materials with Tunable Mechanical Properties. *Polym. Plast. Technol. Eng.* **2017**, *57*, 1288–1295. [CrossRef]
18. Valente, B.F.A.; Silvestre, A.J.D.; Neto, C.P.; Vilela, C.; Freire, C.S.R. Effect of the Micronization of Pulp Fibers on the Properties of Green Composites. *Molecules* **2021**, *26*, 5594. [CrossRef]
19. Immonen, K.; Anttila, U.; Wikström, L. Coupling of PLA and bleached softwood kraft pulp (BSKP) for enhanced properties of biocomposites. *J. Thermoplast. Compos. Mater.* **2018**, *32*, 328–341. [CrossRef]
20. Menčík, P.; Přikryl, R.; Stehnová, I.; Melčová, V.; Kontárová, S.; Figalla, S.; Alexy, P.; Bočkaj, J. Effect of Selected Commercial Plasticizers on Mechanical, Thermal, and Morphological Properties of Poly(3-hydroxybutyrate)/Poly(lactic acid)/Plasticizer Biodegradable Blends for Three-Dimensional (3D) Print. *Materials* **2018**, *11*, 1893. [CrossRef]
21. Kontárová, S.; Přikryl, R.; Melčová, V.; Menčík, P.; Horálek, M.; Figalla, S.; Plavec, R.; Feranc, J.; Sadílek, J.; Pospíšilová, A. Printability, Mechanical and Thermal Properties of Poly(3-Hydroxybutyrate)-Poly(Lactic Acid)-Plasticizer Blends for Three-Dimensional (3D) Printing. *Materials* **2020**, *13*, 4736. [CrossRef]
22. Wypych, G. *Handbook of Plasticizers*, 3rd ed.; Wypych, G., Ed.; ChemTec Publishing: Toronto, ON, Canada, 2017; ISBN 978-1-5231-0906-7.
23. Xie, G.; Zhang, Y.; Lin, W. Plasticizer Combinations and Performance of Wood Flour–Poly(Lactic Acid) 3D Printing Filaments. *BioResources* **2017**, *12*, 6736–6748. [CrossRef]
24. Long, H.; Wu, Z.; Dong, Q.; Shen, Y.; Zhou, W.; Luo, Y.; Zhang, C.; Dong, X. Effect of polyethylene glycol on mechanical properties of bamboo fiber-reinforced polylactic acid composites. *J. Appl. Polym. Sci.* **2019**, *136*, 47709. [CrossRef]
25. Kumar, R.; Ofosu, O.; Anandjiwala, R.D. Macadamia Nutshell Powder Filled Poly Lactic Acid Composites with Triacetin as a Plasticizer. *J. Biobased Mater. Bioenergy* **2013**, *7*, 541–548. [CrossRef]

26. Scalioni, L.V.; Gutiérrez, M.C.; Felisberti, M.I. Green composites of poly(3-hydroxybutyrate) and curaua fibers: Morphology and physical, thermal, and mechanical properties. *J. Appl. Polym. Sci.* **2016**, *134*, 44676. [CrossRef]
27. Panaitescu, D.M.; Nicolae, C.A.; Gabor, A.R.; Trusca, R. Thermal and mechanical properties of poly(3-hydroxybutyrate) reinforced with cellulose fibers from wood waste. *Ind. Crop. Prod.* **2020**, *145*, 112071. [CrossRef]
28. Nguyen, V.K.; Nguyen, T.T.; Thi, T.H.P.; Pham, T.T. Effects of Pulp Fiber and Epoxidized Tung Oil Content on the Properties of Biocomposites Based on Polylactic Acid. *J. Compos. Sci.* **2020**, *4*, 56. [CrossRef]
29. Solaiman, D.K.; Ashby, R.D.; Zerkowski, J.A.; Krishnama, A.; Vasanthan, N. Control-release of antimicrobial sophorolipid employing different biopolymer matrices. *Biocatal. Agric. Biotechnol.* **2015**, *4*, 342–348. [CrossRef]
30. Yatigala, N.S.; Bajwa, D.S.; Bajwa, S.G. Compatibilization improves physico-mechanical properties of biodegradable biobased polymer composites. *Compos. Part A Appl. Sci. Manuf.* **2018**, *107*, 315–325. [CrossRef]
31. Yu, T.; Hu, C.; Chen, X.; Li, Y. Effect of diisocyanates as compatibilizer on the properties of ramie/poly(lactic acid) (PLA) composites. *Compos. Part A Appl. Sci. Manuf.* **2015**, *76*, 20–27. [CrossRef]
32. Georgiopoulos, P.; Kontou, E.; Georgousis, G. Effect of silane treatment loading on the flexural properties of PLA/flax unidirectional composites. *Compos. Commun.* **2018**, *10*, 6–10. [CrossRef]
33. Peltola, H.; Immonen, K.; Johansson, L.; Virkajärvi, J.; Sandquist, D. Influence of pulp bleaching and compatibilizer selection on performance of pulp fiber reinforced PLA biocomposites. *J. Appl. Polym. Sci.* **2019**, *136*, 47955. [CrossRef]
34. Samarth, N.B.; Mahanwar, P.A. Modified Vegetable Oil Based Additives as a Future Polymeric Material—Review. *Open J. Org. Polym. Mater.* **2015**, *5*, 1–22. [CrossRef]
35. Orue, A.; Eceiza, A.; Arbelaiz, A. The effect of sisal fiber surface treatments, plasticizer addition and annealing process on the crystallization and the thermo-mechanical properties of poly(lactic acid) composites. *Ind. Crop. Prod.* **2018**, *118*, 321–333. [CrossRef]
36. Kamarudin, S.H.; Abdullah, L.C.; Aung, M.M.; Ratnam, C.T. Mechanical and physical properties of Kenaf-reinforced Poly(lactic acid) plasticized with epoxidized Jatropha Oil. *BioResources* **2019**, *14*, 9001–9020.
37. Lopera-Valle, A.; Caputo, J.V.; Leão, R.; Sauvageau, D.; Luz, S.M.; Elias, A. Influence of Epoxidized Canola Oil (eCO) and Cellulose Nanocrystals (CNCs) on the Mechanical and Thermal Properties of Polyhydroxybutyrate (PHB)—Poly(lactic acid) (PLA) Blends. *Polymers* **2019**, *11*, 933. [CrossRef]
38. Srubar, W.; Wright, Z.; Tsui, A.; Michel, A.; Billington, S.; Frank, C. Characterizing the effects of ambient aging on the mechanical and physical properties of two commercially available bacterial thermoplastics. *Polym. Degrad. Stab.* **2012**, *97*, 1922–1929. [CrossRef]
39. Mysiukiewicz, O.; Barczewski, M.; Skórczewska, K.; Matykiewicz, D. Correlation between Processing Parameters and Degradation of Different Polylactide Grades during Twin-Screw Extrusion. *Polymers* **2020**, *12*, 1333. [CrossRef]
40. Kumar, R.; Yakubu, M.K.; Anandjiwala, R.D. Flax fibre reinforced polylactic acid composites with amphiphilic additives. *Plast. Rubber Compos.* **2010**, *39*, 437–444. [CrossRef]
41. UPM Formi. Available online: https://www.upmformi.com/biocomposite-products/materials/ (accessed on 4 January 2022).
42. Ozyhar, T.; Baradel, F.; Zoppe, J. Effect of functional mineral additive on processability and material properties of wood-fiber reinforced poly(lactic acid) (PLA) composites. *Compos. Part A Appl. Sci. Manuf.* **2020**, *132*, 105827. [CrossRef]
43. Traquisa. Traquisa Transformaciones Químico Industriales. Available online: https://www.traquisa.com/es/productos/ (accessed on 20 December 2021).
44. Clariant. GlucoPure Sense Sugar-Based Co-Surfactant. Available online: https://www.clariant.com/en/Solutions/Products/2017/01/04/15/29/GlucoPure-Sense# (accessed on 20 December 2021).
45. Biomer Biopolyesters. Available online: www.biomer.de (accessed on 23 April 2021).
46. NatureWorks. 3 Series for Injection Molding. Available online: https://www.natureworksllc.com/Products/3-series-for-injection-molding (accessed on 22 April 2021).
47. Dufresne, A. Nanocellulose: A new ageless bionanomaterial. *Mater. Today* **2013**, *16*, 220–227. [CrossRef]
48. Paunonen, S.; Berthold, F.; Immonen, K. Poly(lactic acid)/pulp fiber composites: The effect of fiber surface modification and hydrothermal aging on viscoelastic and strength properties. *J. Appl. Polym. Sci.* **2020**, *137*, 49617. [CrossRef]
49. Balart, J.F.; Fombuena, V.; Fenollar, O.; Boronat, T.; Sánchez-Nacher, L. Processing and characterization of high environmental efficiency composites based on PLA and hazelnut shell flour (HSF) with biobased plasticizers derived from epoxidized linseed oil (ELO). *Compos. Part B Eng.* **2016**, *86*, 168–177. [CrossRef]
50. Harte, I.; Birkinshaw, C.; Jones, E.; Kennedy, J.; de Barra, E. The effect of citrate ester plasticizers on the thermal and mechanical properties of poly(DL-lactide). *J. Appl. Polym. Sci.* **2012**, *127*, 1997–2003. [CrossRef]
51. Requena, R.; Jiménez, A.; Vargas, M.; Chiralt, A. Effect of plasticizers on thermal and physical properties of compression-moulded poly[(3-hydroxybutyrate)-co-(3-hydroxyvalerate)] films. *Polym. Test.* **2016**, *56*, 45–53. [CrossRef]
52. Seggiani, M.; Cinelli, P.; Verstichel, S.; Puccini, M.; Vitolo, S.; Anguillesi, I.; Lazzeri, A. Development of fibres-reinforced biodegradable composites. *Chem. Eng. Trans.* **2015**, *43*, 1813–1818. [CrossRef]
53. Garcia-Garcia, D.; Ferri, J.M.; Montanes, N.; Lopez-Martinez, J.; Balart, R. Plasticization effects of epoxidized vegetable oils on mechanical properties of poly(3-hydroxybutyrate). *Polym. Int.* **2016**, *65*, 1157–1164. [CrossRef]

54. Silverajah, V.S.G.; Ibrahim, N.A.; Yunus, W.M.Z.W.; Hassan, H.A.; Woei, C.B. A Comparative Study on the Mechanical, Thermal and Morphological Characterization of Poly(lactic acid)/Epoxidized Palm Oil Blend. *Int. J. Mol. Sci.* **2012**, *13*, 5878–5898. [CrossRef]
55. Mahmud, S.; Long, Y.; Wang, J.; Dai, J.; Zhang, R.; Zhu, J. Waste Cellulose Fibers Reinforced Polylactide Toughened by Direct Blending of Epoxidized Soybean Oil. *Fibers Polym.* **2020**, *21*, 2949–2961. [CrossRef]
56. SAPPI Symbio. Available online: https://www.sappi.com/symbio (accessed on 22 April 2021).
57. Rashid, S.M.S.A.; Anuar, H.; Apandi, S.N.E.M.; Buys, Y.F.; Hasan, N.A. Effects of plasticizer on mechanical properties of durian skin fiber reinforced polylactic acid biocomposite. In *AIP Conference Proceedings, Proceedings of the International Conference on X-rays and Related Techniques in Research and Industry 2018, Kota Bharu, Malaysia, 18–19 August 2018*; AIP Publishing: Long Island, NY, USA, 2019; Volume 2068, p. 020007.
58. Choi, J.S.; Park, W.H. Effect of biodegradable plasticizers on thermal and mechanical properties of poly(3-hydroxybutyrate). *Polym. Test.* **2004**, *23*, 455–460. [CrossRef]
59. Singh, R.; Kumar, R.; Hashmi, M.S.J. Friction Welding of Dissimilar Plastic-Based Material by Metal Powder Reinforcement. *Ref. Modul. Mater. Sci. Mater. Eng.* **2017**, *101*, 77. [CrossRef]
60. Zhou, Q.; Xanthos, M. Nanoclay and crystallinity effects on the hydrolytic degradation of polylactides. *Polym. Degrad. Stab.* **2008**, *93*, 1450–1459. [CrossRef]
61. Bialecka-Florjanczyk, E.; Florjanczyk, Z. Solubility of Plasticizers, Polymers and Environmental Pollution. In *Thermodynamics, Solubility and Environmental Issues*; Letcher, T., Ed.; Elsevier: Amsterdam, The Netherlands, 2007; pp. 397–408; ISBN 9780444527073.
62. Sun, X.S. Plastics derived from starch and poly (lactic acids). In *Bio-Based Polymers and Composites*; Wool, R., Sun, X.S., Eds.; Elsevier: Amsterdam, The Netherlands, 2005; pp. 369–410; ISBN 9780080454344.
63. Vieira, M.G.A.; da Silva, M.A.; Dos Santos, L.O.; Beppu, M.M. Natural-based plasticizers and biopolymer films: A review. *Eur. Polym. J.* **2011**, *47*, 254–263. [CrossRef]
64. Xu, Y.; You, M.; Qu, J. Melt rheology of poly (lactic acid) plasticized by epoxidized soybean oil. *Wuhan Univ. J. Nat. Sci.* **2009**, *14*, 349–354. [CrossRef]
65. Gunaratne, L.; Shanks, R.; Amarasinghe, G. Thermal history effects on crystallisation and melting of poly(3-hydroxybutyrate). *Thermochim. Acta* **2004**, *423*, 127–135. [CrossRef]
66. Ali, F.; Chang, Y.-W.; Kang, S.C.; Yoon, J.Y. Thermal, mechanical and rheological properties of poly (lactic acid)/epoxidized soybean oil blends. *Polym. Bull.* **2008**, *62*, 91–98. [CrossRef]
67. Chieng, B.W.; Ibrahim, N.A.; Then, Y.Y.; Loo, Y.Y. Epoxidized Vegetable Oils Plasticized Poly(lactic acid) Biocomposites: Mechanical, Thermal and Morphology Properties. *Molecules* **2014**, *19*, 16024–16038. [CrossRef]
68. Ishak, Z.A.M.; Yow, B.N.; Ng, B.L.; Khalil, H.P.S.A.; Rozman, H.D. Hygrothermal aging and tensile behavior of injection-molded rice husk-filled polypropylene composites. *J. Appl. Polym. Sci.* **2001**, *81*, 742–753. [CrossRef]
69. Lin, Q.; Zhou, X.; Dai, G. Effect of hydrothermal environment on moisture absorption and mechanical properties of wood flour-filled polypropylene composites. *J. Appl. Polym. Sci.* **2002**, *85*, 2824–2832. [CrossRef]
70. Espert, A.; Vilaplana, F.; Karlsson, S. Comparison of water absorption in natural cellulosic fibres from wood and one-year crops in polypropylene composites and its influence on their mechanical properties. *Compos. Part A Appl. Sci. Manuf.* **2004**, *35*, 1267–1276. [CrossRef]
71. Balart, J.F.; Montanes, N.; Fombuena, V.; Boronat, T.; Sánchez-Nacher, L. Disintegration in Compost Conditions and Water Uptake of Green Composites from Poly(Lactic Acid) and Hazelnut Shell Flour. *J. Polym. Environ.* **2018**, *26*, 701–715. [CrossRef]
72. Iglesias-Montes, M.L.; Soccio, M.; Luzi, F.; Puglia, D.; Gazzano, M.; Lotti, N.; Manfredi, L.B.; Cyras, V.P. Evaluation of the factors affecting the disintegration under a composting process of poly(Lactic acid)/poly(3-hydroxybutyrate) (pla/phb) blends. *Polymers* **2021**, *13*, 3171. [CrossRef]
73. Arrieta, M.P.; Samper, M.D.; Aldas, M.; López, J. On the Use of PLA-PHB Blends for Sustainable Food Packaging Applications. *Materials* **2017**, *10*, 1008. [CrossRef]
74. Siracusa, V.; Rocculi, P.; Romani, S.; Rosa, M.D. Biodegradable polymers for food packaging: A review. *Trends Food Sci. Technol.* **2008**, *19*, 634–643. [CrossRef]
75. Zhang, M.; Thomas, N.L. Blending polylactic acid with polyhydroxybutyrate: The effect on thermal, mechanical, and biodegradation properties. *Adv. Polym. Technol.* **2011**, *30*, 67–79. [CrossRef]
76. Karamanlioglu, M.; Robson, G.D. The influence of biotic and abiotic factors on the rate of degradation of poly(lactic) acid (PLA) coupons buried in compost and soil. *Polym. Degrad. Stab.* **2013**, *98*, 2063–2071. [CrossRef]
77. Meereboer, K.W.; Misra, M.; Mohanty, A.K. Review of recent advances in the biodegradability of polyhydroxyalkanoate (PHA) bioplastics and their composites. *Green Chem.* **2020**, *22*, 5519–5558. [CrossRef]
78. Freier, T.; Kunze, C.; Nischan, C.; Kramer, S.; Sternberg, K.; Saß, M.; Hopt, U.T.; Schmitz, K.-P. In vitro and in vivo degradation studies for development of a biodegradable patch based on poly(3-hydroxybutyrate). *Biomaterials* **2002**, *23*, 2649–2657. [CrossRef]
79. Pérez-Fonseca, A.A.; Herrera-Carmona, V.S.; Gonzalez-García, Y.; del Campo, A.S.M.; González-López, M.E.; Ramírez-Arreola, D.E.; Robledo-Ortíz, J.R. Influence of the blending method over the thermal and mechanical properties of biodegradable polylactic acid/polyhydroxybutyrate blends and their wood biocomposites. *Polym. Adv. Technol.* **2021**, *32*, 3483–3494. [CrossRef]

polymers

Article

Polymerizable Choline- and Imidazolium-Based Ionic Liquids Reinforced with Bacterial Cellulose for 3D-Printing

Michael A. Smirnov [1,2,*], Veronika S. Fedotova [1,2], Maria P. Sokolova [1], Alexandra L. Nikolaeva [1], Vladimir Yu. Elokhovsky [1] and Mikko Karttunen [1,3,4,5,*]

[1] Institute of Macromolecular Compounds, Russian Academy of Sciences, V.O. Bolshoi pr. 31, 199004 St. Petersburg, Russia; fedotova.veronicka2016@yandex.ru (V.S.F.); pmarip@mail.ru (M.P.S.); alexandra.l.nikolaeva@gmail.com (A.L.N.); Vladimir_Elokovskiy@hq.macro.ru (V.Y.E.)

[2] Institute of Chemistry, Saint Petersburg State University, Universitetsky pr. 26, Peterhof, 198504 St. Petersburg, Russia

[3] Department of Chemistry, The University of Western Ontario, 1151 Richmond Street, London, ON N6A 5B7, Canada

[4] Department of Physics and Astronomy, The University of Western Ontario, 1151 Richmond Street, London, ON N6A 5B7, Canada

[5] The Centre of Advanced Materials and Biomaterials Research, The University of Western Ontario, 1151 Richmond Street, London, ON N6A 5B7, Canada

* Correspondence: Smirnov_Michael@mail.ru (M.A.S.); mkarttu@uwo.ca (M.K.)

Abstract: In this work, a novel approach is demonstrated for 3D-printing of bacterial cellulose (BC) reinforced UV-curable ion gels using two-component solvents based on 1-butyl-3-methylimidazolium chloride or choline chloride combined with acrylic acid. Preservation of cellulose's crystalline and nanofibrous structure is demonstrated using wide-angle X-ray diffraction (WAXD) and atomic force microscopy (AFM). Rheological measurements reveal that cholinium-based systems, in comparison with imidazolium-based ones, are characterised with lower viscosity at low shear rates and improved stability against phase separation at high shear rates. Grafting of poly(acrylic acid) onto the surfaces of cellulose nanofibers during UV-induced polymerization of acrylic acid results in higher elongation at break for choline chloride-based compositions: 175% in comparison with 94% for imidazolium-based systems as well as enhanced mechanical properties in compression mode. As a result, cholinium-based BC ion gels containing acrylic acid can be considered as more suitable for 3D-printing of objects with improved mechanical properties due to increased dispersion stability and filler/matrix interaction.

Keywords: bacterial cellulose; ionic liquids; 3D-printing; mechanical properties; ion gels

Citation: Smirnov, M.A.; Fedotova, V.S.; Sokolova, M.P.; Nikolaeva, A.L.; Elokhovsky, V.Y.; Karttunen, M. Polymerizable Choline- and Imidazolium-Based Ionic Liquids Reinforced with Bacterial Cellulose for 3D-Printing. *Polymers* **2021**, *13*, 3044. https://doi.org/10.3390/polym13183044

Academic Editor: Agnieszka Tercjak

Received: 10 August 2021
Accepted: 6 September 2021
Published: 9 September 2021

1. Introduction

Manufacturing of products with complex shapes is a problem that has both practical and fundamental material-related challenges. The rise of 3D-printing based on biopolymer-based materials offers, however, a viable and versatile method for natural polymers [1–4]. Cellulose—the most abundant natural polymer—is a prospective candidate for the next generation of sustainable and biocompatible materials. Its crystalline parts demonstrate extremely high mechanical properties [5], which makes cellulose nanomaterials prospective reinforcing fillers [6–10]. Highly crystalline cellulose organized in nanofibrils can be produced by bacterial synthesis and is commonly referred to as bacterial cellulose (BC).

Due to the unique structure and chemical purity of BC, it has been studied as a prospective material for preparation of tough and strong fibers [11], scaffolds for cartilage tissue engineering [12], components for artificial cartilages [13] and as a good insulator composite material [14]. What complicates processing of cellulose is that its strong network of hydrogen bonds (hydrogen bonding in the context of cellulose in contact with membranes has been characterised e.g., in Refs. [15,16]) makes it insoluble in common solvents and non-meltable at temperatures lower than its decomposition temperature.

Ionic liquids (ILs) have melting points below 100 °C and demonstrate such attractive properties as high thermal [17] and chemical stability, low vapor pressure, structural diversity and miscibility with many other solvents [18–21]. ILs offer a potential solution to alleviate many of the problems and they have been applied in cellulose and biomass processing [22–24]. For example, they have been used for cellulose extraction from corn stover [25], in fractionation of lignocellulosic biomass [26], for the preparation of cellulose-based ionic porous materials [27] and conversion of cellulose into 5-hydroxymethylfurfural [28]. Moreover, cellulose-based composites with ILs can be obtained by radical polymerization, where imidazolium and choline acrylic derivatives act as monomers [29,30].

Over recent years, rheological properties of cellulose solutions in ILs have been studied intensely. This is particularly the case for imidazolium-based ILs. Viscosity and viscoelasticity are among the main properties of compositions used for 3D-printing of cellulose-based materials [31]. Continuous flow through the nozzles is achieved due to shear thinning, while shape preservation of the material after extrusion is provided by high storage modulus [32–34]. The details of viscous flow of cellulose solutions in ILs with concentrations from diluted up to concentrated regimes are attracting increasing attention and have been studied using 1-allyl-3-methylimidazolium chloride [35], 1-ethyl-3-methylimidazolium acetate [36–38] and 1-butyl-3-methylimidazolium chloride [36]. However, the use of imidazolium ILs is limited by their increased viscosity [39,40], which makes the use cellulose solutions in IL for 3D-printing difficult. This problem can be resolved by the addition of co-solvents [41,42]. For example, rheological properties of IL-cellulose formulations used for 3D-printing have been modified by the addition of dimethyl sulfoxide (DMSO) [43]. An important benefit of application of IL-based solvents is their ability to prevent the collapse of 3D-printed materials—a common problem in the case of water-based inks [31]. However, investigations regarding the application of ILs for 3D-printing of cellulose-based materials remain few. Among the disadvantages of ILs that limit their practical applications is the high cost [44].

A very promising cheaper and "green" subclass of ILs, deep eutectic solvents (DES), has been attracting tremendous interest over the recent years [45–47]. DES can be easily prepared from two components (hydrogen bond donor (HBD) and acceptor (HBA)) with melting temperature significantly below the melting temperatures of the individual components. Additionally, DES can be composed of components of natural origin, for example choline chloride, citric acid or glucose [48,49]. This makes them interesting for increasing sustainability in chemical technology. Due their capability to form hydrogen bond networks, DES have demonstrated outstanding ability to interact with polysaccharides [22] and they are being intensely studied as prospective plasticizers for chitosan [50–53], starch [54] and as media for treatment of chitin [55–57] or cellulose [58–65]. Additionally, using unsaturated acids such as acrylic or methacrylic acids as hydrogen bond donors [66] provides DES the ability to polymerize in a new green way for preparation of functional materials [67].

The main approaches for 3D-printing of materials with cellulose are fused deposition modeling [68], digital light processing [69] and robocasting [70] with or without subsequent UV-curing. Among the UV-curable components of materials for 3D-printing, acrylates have attracted the most attention because acrylic acid and its derivatives are capable of rapid photopolymerization [71,72]. For example, the use of water dispersions of cellulose nanocrystals mixed with 2-hydroxyethyl methacrylate [73], acrylic acid (AA) [71] and a mixture of AA with 3-dimethyl (methacryloyloxyethyl) ammonium propanesulfonate [74] for 3D-printing of cellulose-reinforced hydrogels has been reported. At the same time, composite ion gels based on cellulose and ILs are also considered as attractive materials with a wide range of prospective applications such as strain sensors [75], components for flexible electronic devices [76], preparation of a bactericidal hydrogel dressings [77] and air filters that have a great potential in air purification [78]. In approaches based on

UV-curing, DES and ILs containing photo-polymerizable components can be proposed as a promising medium for preparation of cellulose-based viscous gels suitable for 3D-printing. However, DES-containing acrylic acid for direct 3D-printing of cellulose reinforced ion gel has been utilized only in the recent work of Lai and Yu [75], in which the ink was prepared using commercial cellulose nanocrystals, and DES based on choline chloride (ChCl) and ethylene glycol mixed with acrylic acid as UV-polymerizable monomer and $Al_2(SO_4)_3$ as a cross-linker; ethylene glycol was used as the main HBD in DES. To the best of our knowledge, rheological properties and 3D-printing of cellulose dispersion in DES containing acrylic acid as the only HBD has not been reported yet. In order to fill this gap, we have studied two cases of polymerizable compositions for the preparation of cellulose nanofibers (CNF)-reinforced ion gels. The first one is based on IL 1-butyl-3-methylimidazolium chloride (ImCl) mixed with AA and the second one is a mixture of ChCl as a "greener" alternative for imidazolium-based salts and AA. Based on the method we have proposed for the preparation of CNF gel from BC using DES based on ChCl [79], here we suggest that mixtures based on organic chloride salt and acrylic monomer can be directly used as a media for the preparation of a CNF ion gel from BC. Furthermore, we also suggest that they are directly applicable for 3D-printing with subsequent UV-curing.

2. Materials and Methods

2.1. Materials

Lyophilized culture of *Acetobacter xylinum* was purchased form All-Russian collection of industrial microorganisms (National Bioresource Center, GosNIIgenetika, Moscow, Russia) and cultivated as described further. 1-butyl-3-methylimidazolium chloride (Sigma-Aldrich, Buchs SG, Switzerland, CAS 79917-90-1, purity > 98%) and choline chloride (Glentham Life Sciences Ltd., Corsham, UK, CAS 67-48-1, purity > 99%) were dried under vacuum at 60 °C for at least 24 h before use. N,N′-methylene-bis-acrylamide (Sigma-Aldrich, St. Louis, MO, USA, CAS 110-26-9, purity > 99.5%), acrylic acid (Sigma-Aldrich, CAS 67-48-1, purity > 99%) and 2-hydroxy-2-methylpropiophenone (Sigma-Aldrich, Milan, Italy, CAS 7473-98-5, purity > 97%) were used as received. Peptone and D-mannitol (CAS 69-65-8) were obtained from LenReaktiv (Saint Petersburg, Russia), yeast extract from Research Center for Pharmacotherapy (Saint Petersburg, Russia) and NaOH (CAS 1310-73-2) from NevaReaktiv (Saint Petersburg, Russia).

2.2. Methods

2.2.1. Bacterial Synthesis of Cellulose

The activation of lyophilized *A. xylinum* was carried out by introducing a seed culture medium. A single *A. xylinum* colony was transferred to 400 mL of seed culture medium and cultivated statically for 3 months at 27 °C (Figure 1). Then, 200 mL of the cell suspension was introduced into a bottle with 1 L of culture medium containing 0.3 wt% of peptone, 0.5 wt% of yeast extract and 2.5 wt% of D-mannitol. The medium was incubated statically at 27 °C for 15–20 days in an incubator (BINDER, Tuttlingen, Germany) until the formation of a BC membrane with a thickness of 1 cm. After that, the BC membrane was purified by boiling in 0.5 wt% NaOH solution for 30 days, and then thoroughly washed with distilled water until pH reached 7. After that, the membrane was mechanically ground, and the obtained suspension was diluted in order to prepare 0.5 wt% concentration of cellulose. The obtained mass was lyophilized with a Scientz-10ND lyophilic dryer (Scientz, Ningbo, China) and then ground into small pieces with 1–3 mm average size.

Figure 1. BC growing (**a**) a purified BC hydrogel (**b**) and BC after shredding and freeze drying (**c**).

2.2.2. Preparation of Dispersions and Regeneration of Cellulose

The solvents were prepared by mixing of ChCl or ImCl with AA in the mole ratio 1:3. For this purpose, 3.161 g of ChCl or 3.199 g of ImCl were mixed with 4.894 and 3.960 g of AA, respectively. The mixtures were stirred at 70 °C until a clear liquid formed. After that, lyophilized bacterial cellulose was added and the viscous mixture was stirred for 20 min at 20 °C and relative humidity of 40–45% until complete wetting of cellulose with the solvent. After that, the mixture was homogenized by multiple pushing of the mass through the die with 0.58 mm diameter. These systems (Figure 2) were used for rheological measurements and regeneration of cellulose with excess of water. The regenerated cellulose was precipitated with centrifugation for 20 min at 4000 rpm using Microprocessor Centrifuge (Digisystem Laboratory Instruments Inc., New Taipei City, Taiwan). In order to remove residual components of IL, the regenerated cellulose was redispersed in distilled water and centrifuged again; this was repeated 5 times.

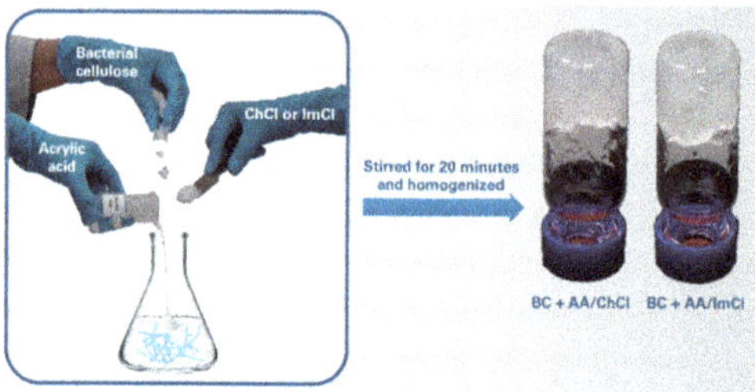

Figure 2. Preparation scheme and images of BC ion gels in AA/ChCl and AA/ImCl mixtures.

2.2.3. Microscopic Investigation

Atomic force microscopy (AFM) studies were performed using a SPM-9700HT scanning probe microscope (Shimadzu, Kyoto, Japan). The AFM images were taken in air at room temperature. The instrument operated in the force dynamic mode, and an NSG30_SS Silicon tip (tip curvature radius 2 nm) was employed. Images with 512 × 512 points were obtained.

2.2.4. Wide-Angle X-ray Diffraction Study

Crystallinity of the samples was studied by wide-angle X-ray diffraction (WAXD) using a D8 Discover diffractometer (Bruker, Rheinstetten, Germany) equipped with a CuKa radiation source (λ = 1.54 Å). The WAXD patterns were studied in the 2θ range of 5° to 40°.

2.2.5. Fourier Transform Infrared Spectroscopy

Attenuated total reflection (ATR)-FTIR spectroscopy study was performed on the IRAffinity-1S spectrometer (Shimadzu, Kyoto, Japan) with 100 scans at resolution 2 cm^{-1} from 3750 to 700 cm^{-1}.

2.2.6. Rheological Studies

Kinematic viscosities of the binary mixtures AA/ChCl and AA/ImCl were determined at 20 °C using a capillary viscometer with inner diameters of 0.99 and 0.73 mm (according to GOST 10028-81), respectively. Before the experiment, the viscometer constant was calibrated using a solution of water in glycerol (water content 20.85 and 30.45 wt%). Temperature control after filling the viscometer was carried out using LT-400 thermostat (LOIP, Moscow, Russia). The time of sample expiration through the capillary was recorded by a stopwatch. Kinematic viscosities were calculated by multiplying the expiration time by the viscometer constant and converted to dynamic viscosity using the density of systems.

Rheological studies were performed with the aid of a Physica MCR 301 rheometer (Anton Paar, Graz, Austria) equipped with cone-plate (20 mm diameter, 2° cone angle) geometry in shear and dynamic regimes at 20 °C.

2.2.7. Polymerization in Bulk and 3D-Printing

The UV-curable compositions were prepared by the addition of 48.9 and 39.6 mg of bifunctional cross-linker (N,N′-methylene-bis-acrylamide), and 48.9 and 39.6 mg of photoinitiator (2-hydroxy-2-methylpropiophenone) were added, respectively, to the CNF dispersion based on AA/ChCl (8.137 g) and AA/ImCl (7.232 g). The amounts of cross-linker and initiator were chosen as 1 wt% from the mass of the polymerizable monomer (AA). The prepared compositions were used for UV-induced polymerization in polyethylene tubes for obtaining bulk material in the form of cylinders, or 3D-printed in the form of strips (Figure 3) in order to test the mechanical properties of filaments prepared via extrusion. For this purpose, the 3D-printer 3D BioScaffolder BS3.2 (GeSIM, Radeberg, Saxony, Germany) equipped with a pneumatic syringe with a 0.58 mm diameter nozzle was applied. UV-curing in all cases was performed using UV-lampOmniCure S1500 (Lumen Dynamics, Mississauga, Ontario, Canada) with light power 0.15 W cm^{-2}. All samples (including polymerized in bulk and 3D-printed) were irradiated 7 times with 15 s and 20 s pauses between irradiations.

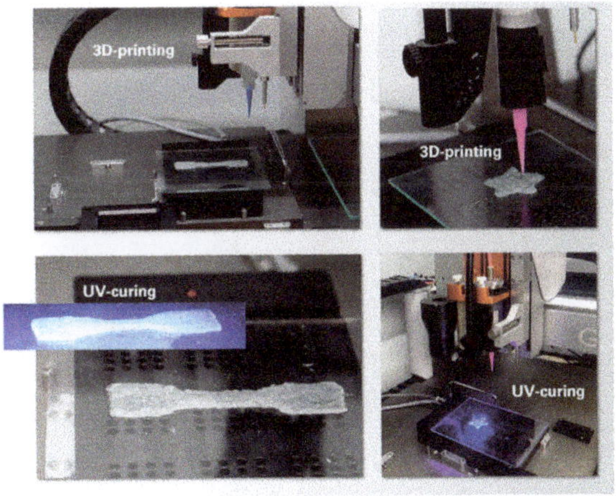

Figure 3. Examples of 3D-printed structures prepared using compositions based on AA/ChCl (dog-bone-shaped) and AA/ImCl (star).

2.2.8. Measurements of Mechanical Properties

An AG-100kNX Plus setup (Shimadzu, Kyoto, Japan) was used to study the mechanical characteristics. The measurements were performed in two testing modes. In the first one, the strip samples (with cross-section 1 × 5 mm^2) were stretched at a rate of 50 mm/min, the gauge length being 15 mm. Experiments using the strip samples were performed according to the ASTM D882 requirements. Young's modulus E, the break stress σ_b, and the ultimate deformation ε_b were determined. The specimens were printed by aligning 4 filaments co-linearly with the direction of extension with subsequent UV-curing. The second testing mode consisted of a single-shot uniaxial compression of block cylinders at a rate of 2 mm/min up to 80% deformation. The so-called current compression moduli [13] $E_{10-15\%} = \Delta\sigma/\Delta\varepsilon$ were determined in the deformation range of 10–15%, as well as stress values corresponding to 50 and 80% compression of the samples ($\sigma_{50\%}$ and $\sigma_{80\%}$), respectively. The latter were fabricated as follows: Photopolymerization of the compositions was conducted in the tube with the inner diameter of 9 mm by applying UV-irradiation. The obtained cylinder was cut into small ones with heights of about 5 mm. All the mechanical tests were performed at room temperature.

The design of materials preparation and investigation is summarized in Figure 4.

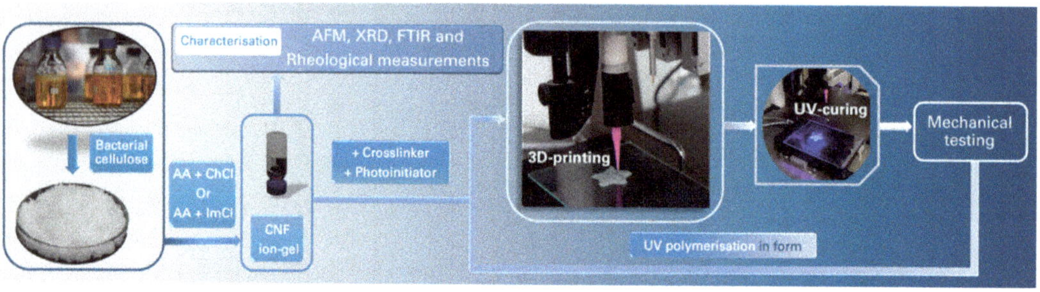

Figure 4. Block scheme demonstrating the preparation of materials and their characterization.

3. Results

3.1. Microscopy Investigation

To evaluate the effects of AA containing ionic liquids on the morphology of BC, an AFM study was conducted. It was performed in two cases: (1) Only treatment with IL and (2) treatment with subsequent polymerization of AA. The height maps for the initial BC (Figure 5a), BC regenerated before polymerization from AA/ImCl (Figure 5b) and AA/ChCl (Figure 5c) demonstrate an approximately similar fibrous structure.

The profiles demonstrating the changing of fibrous diameter after different treatments are given in Figure S1. It is seen that BC nanofibers regenerated before polymerisation are thicker than in the initial BC sample: The diameter can be estimated to be 78–95 nm and 55–75 nm in samples regenerated from AA/ChCl and AA/ImCl, respectively, in comparison with 40–70 nm in the initial BC. The thickening effect is more pronounced in the case of the AA/ImCl solvent. This can be due to swelling of the cellulose fiber surfaces as a result of possible introduction of ImCl components between the cellulose macromolecules. This idea is supported by the earlier results [60,80] that demonstrate the swelling of cellulose nanofibers in the ChCl/imidazole mixture and 1-ethyl-3-methylimidazolium acetate, respectively. Nanofiber morphologies regenerated after AA polymerization from AA/ImCl and AA/ChCl are given in Figure 5d,e, respectively. The profiles (Figure S1) demonstrate that polymerisation of AA leads to an increase in nanofiber diameter up to 86–100 nm and 85–140 nm for AA/ImCl and AA/ChCl, respectively; the increase in thickness during polymerisation is more pronounced with the AA/ChCl solvent. Thickening during polymerisation can be attributed to possible grafting of poly(acrylic acid) to the cellulose surface, which was confirmed by FTIR measurements and will be discussed later in Section 3.3.

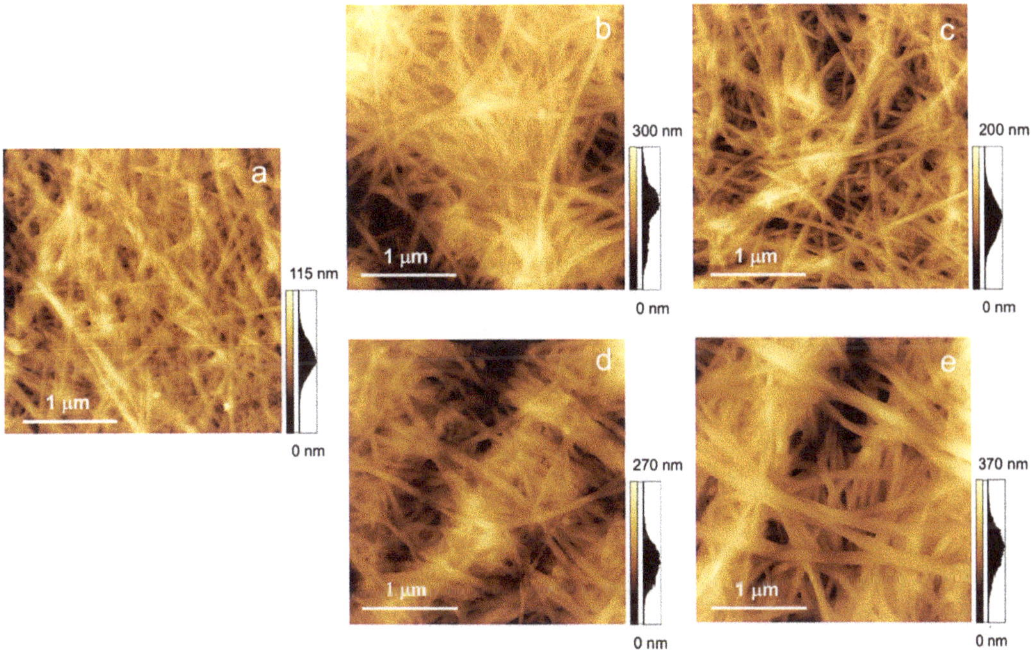

Figure 5. Surface topography of the initial BC (**a**), BC after AA/ImCl (**b**), after AA/ChCl (**c**), BC-PAA from AA/ImCl (**d**) and from AA/ChCl (**e**).

3.2. WAXD Study

WAXD patterns of the initial BC, BC after AA/ImCl and after AA/ChCl are presented in Figure 6. Typical diffraction peaks are seen at 2θ = 14.8°, 16.6°, 22.6° and 34.4° corresponding to (1–10), (110), (200) and (004) lattice planes of the cellulose Iβ polymorph. The positions of reflections are preserved after treatment of BC with both types of used ILs. Using the Segal method [81], the degree of crystallinity (χ_c) can be estimated as

$$\chi_c = \frac{I_{200} - I_{am}}{I_{200}} \times 100 \quad (1)$$

where I_{am} and I_{200} are the diffraction intensities at 2θ = 18° and at the maximum of the (200) lattice diffraction plane. Using this equation, the values of χ_c were calculated as 96, 92 and 92% for the initial BC and CNF after treatment with AA/ChCl and AA/ImCl, respectively. Thus, the XRD results confirm the AFM data about preservation of the BC nanofibrilar structure during the treatment with AA/ChCl and AA/ImCl. The redistribution of intensities between the diffraction peaks at 2θ = 14.8° and 16.6° can be connected to the preferential orientations of CNF in dried films along the surface plane.

3.3. Fourier Transform Infrared Spectra

FTIR spectra were measured for the initial BC, for regenerated CNF from ionic liquids and for CNF purified after polymerization of the AA component with water (Figure 7). The results demonstrate that typical absorption bands of BC in the region of 1200–1000 cm^{-1} remain approximately at the same positions for all treated samples. That allows one to conclude that the chemical structure of the initial BC is not significantly affected. Samples regenerated from non-polymerized ILs (AA/ImCl and AA/ChCl) did not demonstrate significant changes in the FTIR spectra. At the same time, CNFs regenerated and purified via multiple repetitions of centrifugation/redispersion after polymerization of ILs (without cross-linker) demonstrate the appearance of intensive bands with maxima at 1704 and

1545 cm^{-1}. These maxima can be attributed to the carboxylic groups in protonated and deprotonated forms, respectively.

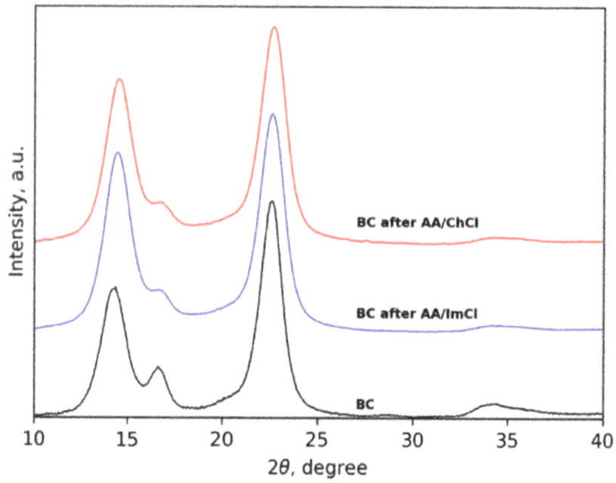

Figure 6. WAXD patterns of initial BC, BC after AA/ImCl and after AA/ChCl.

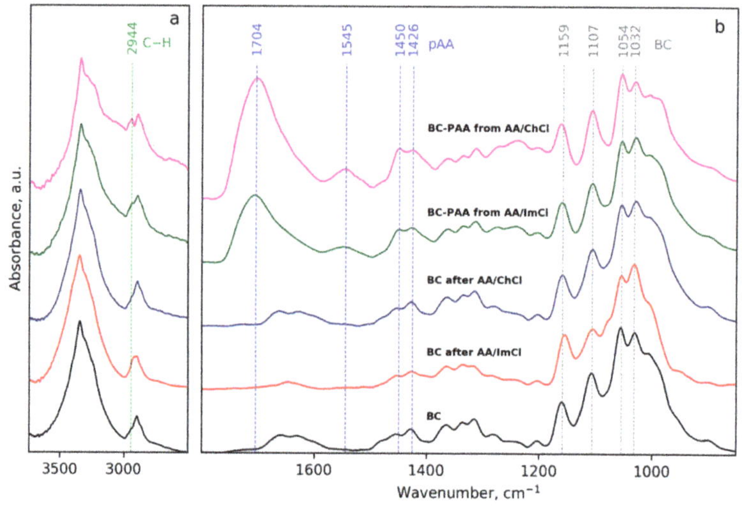

Figure 7. FTIR spectra of initial BC, BC after AA/ImCl and after AA/ChCl, BC-poly(acrylic acid) (BC-PAA) from AA/ImCl and from AA/ChCl in the regions 3700–2500 cm^{-1} (**a**) and 1800–900 cm^{-1} (**b**).

Comparison of intensities of bands near 1054 and 1704 cm^{-1} (Figure 7) allows us to propose that in the case of AA/ChCl, the amount of grafted poly(acrylic acid) is higher than in the case of AA/ImCl. This observation is in agreement with the AFM data in Section 3.1. The appearance of grafted poly(acrylic acid) onto the CNF surface is also supported by the increase in absorption intensities near 2944 and 1450 cm^{-1}, which can be attributed to the vibrations of the C–H bonds in the poly(acrylic acid) main chain [82].

3.4. Rheological Measurements

The dynamic viscosity of AA/ChCl at room temperature is higher than that of AA/ImCl: 56.7 and 24.5 mPa·s^{-1}, respectively. The dependence of shear viscosity on the shear rate for ion gels containing BC (Figure 8) demonstrates shear thinning in all systems and concentrations. In the case of dispersions in AA/Im, no plateau at high shear rates was achieved because of the instability at shear rates higher than 1 s^{-1}. In the case of

ChCl-based systems, the measurements gave stable results up to 63 s^{-1} and 1000 s^{-1} for dispersions with 1 and 0.5 wt% of CNF, respectively. The appearance of a plateau at the high shear rate region with viscosities about 6.3 $Pa{\cdot}s^{-1}$ and 0.05 $Pa{\cdot}s^{-1}$ was observed for dispersions with 1 and 0.5 wt% of CNF, respectively. The flow instabilities are connected to solid-like behavior of gels, as will be demonstrated further.

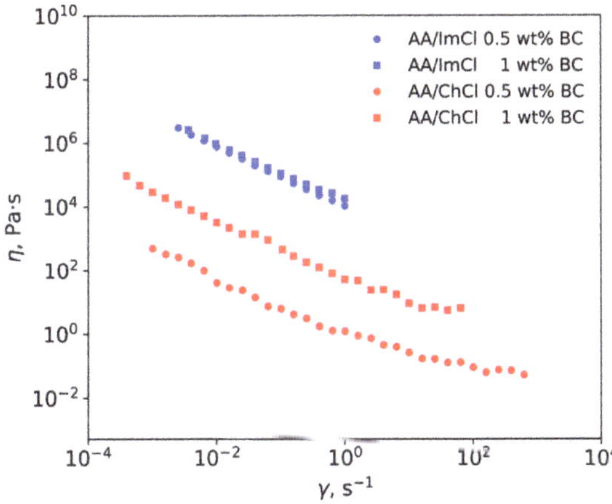

Figure 8. Steady-state viscosity (η) of the different compositions as a function of shear rate (γ) at 20 °C.

CNF gels in AA/Im have higher viscosities in comparison with dispersions in AA/ChCl. The AFM results (Figure 5) demonstrate that CNF regenerated from AA/ImCl have slightly higher diameters in comparison with the AA/ChCl systems. The swelling of the CNF surface layer in imidazolium-based solvents possibly leads to an increasing amount of surface −OH groups being accessible for an interaction with solvent and between the CNFs. This leads to the formation of a stronger hydrogen bond network in the case of CNF gels in AA/ImCl in comparison with the AA/ChCl-based systems.

The results of viscosity measurements in the dynamic regime are given in Figure 9. The higher G′ (filled symbols)values in comparison with G″ (hollow symbols) for all studied systems demonstrate the gel-like behavior of the material with predominantly elastic properties. This implies the formation of a strong CNF network even at the smallest cellulose concentration in dispersions. Taking this into account, it can be proposed that instability in the rheological measurements in shear mode can be connected to pronounced elastic behavior. The lack of viscous properties can result in the disruption of the CNF network at high shear rates with a sharp drop in mechanical properties. The lesser influence of CNF concentration on G′ and G″ in AA/ImCl systems in comparison with AA/ChCl-based systems (difference between Figure 9a,b) is in agreement with results of viscosity measurements in shear mode. This demonstrates that in the case of AA/ImCl, a strong percolation network of CNF is formed even at 0.5 wt% and increasing the concentration up to 1 wt% leads to a two-fold increase of the G′ modulus. In the case of AA/ChCl, similar changes in CNF concentration lead to the growth of G′ by an order of magnitude. In the case of 1 wt% of CNF, the values of G′ and G″ are approximately equal for both studied solvents. This means that the mechanical properties at small deformations are determined only by the mechanics of CNF. At the same time, the viscosities of these systems in shear mode are different (Figure 8) demonstrating easier rearrangement of the CNF network in the case of AA/ChCl.

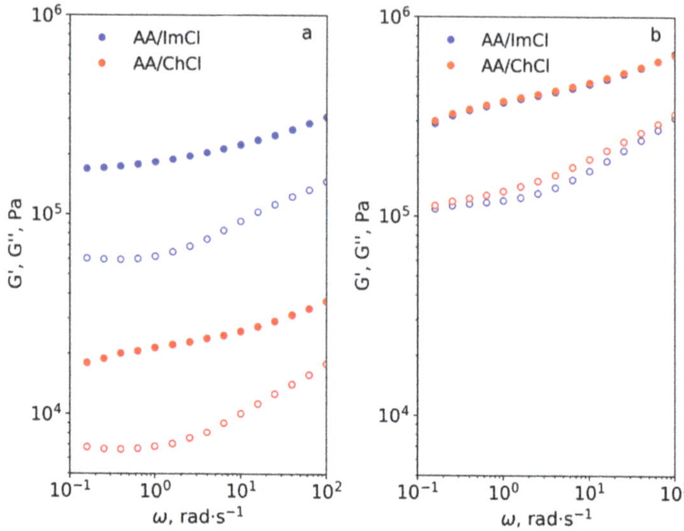

Figure 9. Storage (G′, filled symbols) and loss (G″, hollow symbols) moduli as a function of angular frequency (ω) for dispersions containing 0.5 wt% (**a**) and 1 wt% (**b**) of CNF at 20 °C.

3.5. 3D-Printing and Mechanical Properties

Compositions with 1 wt% of BC nanofiber content were chosen for 3D-printing. Before performing 3D-printing, the dependence of extrusion weight on the extrusion pressure was measured (Figure S2). The minimum pressure needed for extrusion was 20 kPa for both compositions. The extrusion weight for AA/ImCl was higher than in the case of AA/ChCl, which is in agreement with rheological measurements using pure solvents. However, the dependence of extrusion weight on the type of cation is in the opposite direction with steady-state rheological viscosity of ILs containing BC (Figure 8) as will be discussed further.

The mechanical properties measured in compression and extension modes are summarized in Tables 1 and 2, and the corresponding stress–strain curves are given in Figure 10a,b. Figure S3 shows images of initial and stretched ion gel under mechanical testing.

Table 1. Mechanical properties of AA/ChCl- and AA/ImCl-based polymerized ion gels measured in compression mode for samples prepared by bulk polymerization.

Sample	$E_{10–15\%}$, MPa	$\sigma_{50\%}$, MPa	$\sigma_{80\%}$, MPa
AA/ImCl-based ion gel	2.1 ± 0.2	2.7 ± 0.3	45 ± 15
AA/ChCl-based ion gel	6.8 ± 0.9	5.7 ± 0.4	49 ± 6

Table 2. Mechanical properties of AA/ChCl- and AA/ImCl-based polymerized ion gels measured in extension mode for samples prepared via 3D-printing.

Sample	E, MPa	σ_b, MPa	ε_b, %
AA/ImCl-based ion gel	0.51 ± 0.05	0.38 ± 0.05	94 ± 14
AA/ChCl-based ion gel	0.56 ± 0.06	0.56 ± 0.02	175 ± 25

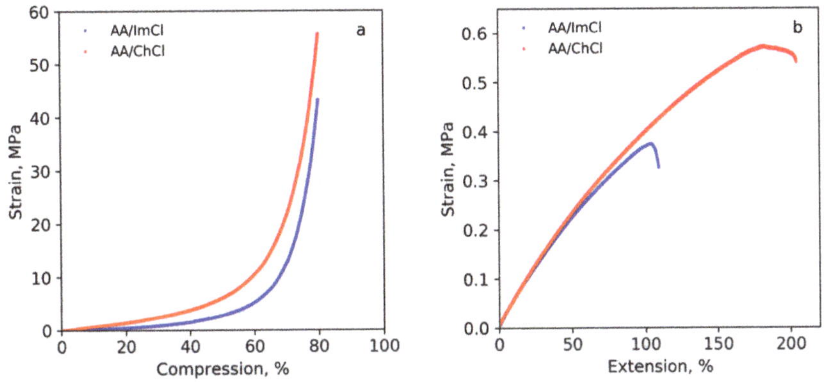

Figure 10. Stress–strain curves of AA/ChCl- and AA/ImCl-based samples prepared by polymerization in bulk and measured in compression mode (**a**) and prepared via 3D-printing and measured in extension mode (**b**).

When analyzing mechanical properties of polymerized samples in compression mode, one should take into account that because of the nonlinearity of the stress–strain curves at small deformations, it is more appropriate to characterize the stiffness of such objects by the so-called current compression modulus $E_{10-15\%} = \Delta\sigma/\Delta\varepsilon$ in the deformation range of 10–15% (Table 1) instead of elastic modulus [13]. Figure 10a shows that samples based on both compositions are able to withstand high compression fairly well, demonstrating no visible signs of destruction even at 80% deformation. A slow increase was observed for deformations up to about 40–50% followed by a steep increase in the compression stress at higher deformations. It is noteworthy that both $E_{10-15\%}$ and $\sigma_{50\%}$ for the AA/ChCl-based sample are substantially higher than those for the AA/ImCl-based one (Table 1). This is probably related to the higher degree of grafting of growing poly(acrylic acid) onto the cellulose nanofiber surfaces during UV-induced radical polymerization as it was shown earlier with AFM and FTIR measurements. This process can lead to stronger interactions between reinforcing filler (cellulose nanofibers) and matrix–crosslinked poly(acrylic acid) network and chloride ions via additional hydrogen bonding.

It is evident from Figure 10b and Table 2 that both types of specimens, namely those fabricated from either AA/ChCl or AA/ImCl via 3D-printing, are quite soft and characterized by an elastic modulus of about 0.5 MPa. Replacing ImCl with ChCl in the polymerizable composition led to an enhancement of mechanical properties, with both strength and ultimate deformation of AA/ChCl-based samples being about 1.5 times higher than those of the AA/ImCl-based ones.

4. Discussion

In this work, viscous nanofibrous ion gels were prepared via dispersion of bacterial cellulose in two types of ionic liquids: (1) Based on 1-butyl-3-methylimidazolium chloride or (2) choline chloride. UV-curing was achieved due to using acrylic acid and a bifunctional cross-linker in the compositions. Structures were characterized by three main methods, AFM, WAXD and FTIR. The WAXD study demonstrates that the crystalline structure of the polymer remained unchanged. Additionally, the absence of noticeable scattering from amorphous parts demonstrates a high degree of crystallinity of both the initial and treated materials.

Together, the AFM and WAXD results prove that cellulose's crystalline and fibrillar structures remain preserved under treatment with the given solvents. More pronounced swelling of nanofibers in the presence of 1-butyl-3-methylimidazolium chloride was observed, demonstrating the increased possibility of 1-butyl-3-methylimidazolium chloride to interact with cellulose. FTIR results further demonstrated the grafting of poly(acrylic acid) onto the surface of cellulose nanofibers in the course of UV-induced radical polymerization.

This observation suggests that the CNF surface becomes modified during photo-initiated AA polymerization in choline- or imidazolium-based Ils [82]. The appearance of strong adsorption near 1700 cm^{-1} has been reported in the literature as evidence for grafting of PAA to starch [83], while those at 1557 and 1451 cm^{-1} are evidence of grafting of poly(sodium acrylate) onto carboxymethylcellulose [84]. Such a process can occur due to the formation of different types of free radicals on the cellulose molecules under UV-irradiation [85] in the presence of a photosensitizer (2-hydroxy-2-methylpropiophenone). This process is more pronounced for choline-based compositions. Furthermore, the higher viscosity of deep eutectic solvents has been proposed as one of the factors leading to an enhancement of the radical polymerisation rate [86], and this can be the reason for the higher degree of grafting of the AA/ChCl solvent in our case. Additionally, the lower accessibility of the cellulose surface for acrylic acid in the case of the 1-butyl-3-methylimidazolium chloride-based solvent can be speculated, but this needs further evaluation and is beyond the scope of the current study. Partial deprotonation of poly(acrylic acid) can be attributed to the possible capture of cholinium or imidazolium cations onto the surface-grafted acidic polymer brush with the formation of salts between acrylic acid and imidazolium or cholinium cations.

The higher viscosity of pure AA/ChCl in comparison with AA/ImCl can be explained by the following: As discussed above, high viscosity of ILs and DES is connected with the formation of hydrogen bond networks [87]. In the case of ChCl, mixing with AA leads to the formation of an ionic liquid due to the network of hydrogen bonds between the −COOH group of AA and the −OH groups of choline with Cl^- [88,89]. In the case of ImCl, the ionic liquid appears due to the formation of network-like interactions between Im and Cl^-, and in this case, AA can be considered as a co-solvent, which destructs the network [90] thus reducing viscosity [91].

On the rheological side, the obtained dispersions exhibit shear thinning behavior, which makes them amenable for 3D-printing. This effect is due to the orientation of the CNFs in the direction of flow and slip on the rheometer wall since phase separation from the dispersed cellulose results in lubrication of the contact between the gel and the rheometer wall [92]. The dispersant can be depleted from the vicinity of walls by the presence of large particles [93], which in our case can be aggregates of CNF or CNF itself because their length can be longer than 3 μm. It has been reported earlier that hydrogels based on cellulose nanocrystals also exhibit shear thinning behavior [31]. However, it is worth noting that high viscosity at low shear rates (in our work, 2.6×10^6 and 7976 $Pa{\cdot}s^{-1}$ for AA/ImCl and AA/ChCl, respectively) can be achieved with a large content of cellulose nanocrystals (10–25 wt%) [73,75] or by using additional polymers such as poly(ethylene glycol) hexadecyl ether and poly(ethylene glycol) diacrylate [94]. High viscosity at low shear rates is necessary to prevent material leakage from the nozzle during 3D-printing [31]. In our work, the use of CNF allows obtaining dispersions with a suitable viscosity at low shear rates even at 1 wt% of cellulose content. Comparing our results with the reported rheological properties of cellulose solutions in pure ILs, it is worth noting that shear thinning is either not observed, for example, in the case of 1-ethyl-3-methylimidazolium acetate [37] or 1-butyl-3-methylimidazolium chloride [38], or it is observed at high shear rates, as was demonstrated in the case of 1-allyl-3-methylimidazolium chloride [35]. This makes it difficult to use such systems for 3D-printing.

The viscosity of ion gels based on ImCl is higher than based on ChCl; that is, opposite to the viscosity of pure solvents. This observation can potentially be explained as follows: The less viscous solvent containing the more polarizable imidazolium cations can result in the formation of more dispersed cellulose because the more polarizable imidazolium cations are better for cellulose stabilization in solution in comparison with ternary ammonium cations [95,96]. Additionally, the π-electron delocalization of the unsaturated heterocyclic ring makes the cations more active in interacting with cellulose [97]. Thus, a higher number of entanglements between the CNFs will result in higher viscosity. In the case of 1-butyl-3-methylimidazolium chloride, bacterial nanocellulose dispersions demonstrate higher viscosity in comparison with choline-based systems at low shear rates.

However, the higher flow stability at high shear rates for dispersions in choline chloride/acrylic acid solvent makes them more suitable for 3D-printing. The flow instabilities can be discussed on the basis of results of rheological measurements in the dynamic mode that implies the formation of a strong CNF network even at the smallest cellulose concentration in dispersions. Taking this into account, it can be proposed that the instability in the rheological measurements in the shear mode can be connected to pronounced elastic behavior. The lack of viscous properties can result in the disruption of the CNF network at high shear rates with a sharp drop in mechanical properties.

The dependence of extrusion weight on the type of cation correlates with viscosities of pure solvents: Higher rates were observed for AA/ImCl-based composition. This allows us to propose that during extrusion, flow is strongly affected by slippage near the walls of the nozzle. As a result, the lower viscosity of the AA/ImCl solvent leads to higher flow rates for BC dispersion in this solvent in comparison with AA/ChCl as can be seen from Figure S2. The wider error bars for BC in AA/ImCl (Figure S2) demonstrate lower flow stability for this system. This is in agreement with limits for steady-state viscosity measurements discussed earlier.

In our study, structural characterization was complemented by studies of mechanical properties. The measurements of 3D-printed strips in the extension mode showed that choline-based ion gels have higher elongation at break: 175% in comparison with 94% for imidazolium-based systems. Additionally, choline-based ion gels demonstrate a higher compression modulus and strength at 50% of compression. Considering the data on structural and rheological properties of the polymerizable mixtures discussed above, one could infer that higher mechanical properties might be associated with higher stability of ChCl-based dispersions under shear stress. Taking into account the assumption that phase stratification near the wall of the nozzle proceeds more intensely in the case of imidazolium-based system, formation of a more defective CNF network structure in the 3D-printed material based on this cation can be proposed. Additionally, the increased interaction between the filler and the matrix in AA/ChCl-based material can be proposed due to more pronounced grafting of poly(acrylic acid) onto the cellulose surface. This result is very interesting, since it demonstrates the higher properties for a "greener" composition.

The values of Young's modulus reported in this work were 0.51 and 0.56 MPa for the 3D-printed samples based on AA/ImCl and AA/ChCl, respectively. These values are higher than those reported in Ref. [75] for similar systems based on cellulose nanocrystals (22.5 wt%) in mixtures of choline chloride and ethylene glycol containing acrylic acid (about 0.17 MPa). This can be connected to the use of a covalent cross-linker N,N′-methylenebisacrylamide, in our case in comparison with the use of reversible ionic cross-links in the work of Lai and Yu [75]. On the other hand, covalent cross-links provide lower values of elongation at break for samples prepared in this work: 94 and 175%, respectively, for AA/ImCl and AA/ChCl-based ion gels in comparison with 1100% reported in [75]. Because of the high concentration of the polymerizable monomer, the tensile strength in our case was also significantly greater than that in ChCl-urea-glycerol-based ion gels reinforced with 4 wt% of cellulose nanocrystals and obtained via polymerization of acrylamide with MBA as a cross-linker [98]: 55.5 kPa in comparison with 0.56 MPa for the AA/ChCl-based material in our case.

5. Conclusions

To conclude, in this work, we prepared UV-curable bacterial cellulose ion gels using two systems, choline- and imidazolium-based ionic liquids containing acrylic acid. First, and importantly, AFM and WAXD data demonstrate that the nanofibrous and crystalline structures of bacterial cellulose are preserved. In addition, the nanofibers swelled more in the presence of 1-butyl-3-methylimidazolium chloride.

Second, both compositions showed shear thinning behavior, making them amenable for 3D-printing. At the same time, the results from dynamic rheological measurements demonstrate the predominantly elastic behavior for all compositions ($G' > G''$) that can lead to instability in their viscous flow. The dispersions showed, however, higher resistance toward phase separation at high shear rates in choline chloride/acrylic acid solvent, thus making the choline-based composition more suitable for 3D-printing of the two.

Finally, FTIR data suggest more pronounced grafting of poly(acrylic acid) onto the surfaces of bacterial cellulose nanofibers during UV-curing in choline chloride/acrylic acid mixtures. Consequently, this leads to stronger interactions between the matrix and filler and yields higher mechanical properties. As a result, the "greener" choline-based ion gels demonstrate a higher compression modulus and strength at 50% of compression as well as higher elongation at break: 175% in comparison with 94% for imidazolium-based systems. In summary, a novel approach based on using mixtures of ionic liquids with acrylic acid for the preparation of bacterial cellulose-reinforced ion gel with subsequent 3D-printing and UV-curing has been demonstrated.

Supplementary Materials: The following are available online at https://www.mdpi.com/article/10.3390/polym13183044/s1, Figure S1: Profiles extracted from surface topography maps of initial BC (a), BC after AA/ImCl (b) and after AA/ChCl (c), BC-PAA from AA/ImCl (d) and from AA/ChCl (e), Figure S2: Dependences of extrusion weight on extrusion pressure for compositions containing 1 wt% of BC through nozzle with diameter of 0.58 mm. Each point was measured 3–5 times. Figure S3: Images of ion gel stretched in a mechanical testing machine.

Author Contributions: Conceptualization, original draft preparation, methodology M.A.S.; original draft preparation, investigation V.S.F., M.P.S. and A.L.N.; investigation V.Y.E.; supervision, conceptualization, review and editing, M.K. All authors have read and agreed to the published version of the manuscript.

Funding: This research was supported by Russian Science Foundation (project number 21-13-00111).

Institutional Review Board Statement: Not applicable.

Informed Consent Statement: Not applicable.

Data Availability Statement: Data available within the article or its supplementary materials.

Acknowledgments: The experimental work was facilitated by the equipment of the Resource Centre of X-ray Diffraction Studies at St. Petersburg State University.

Conflicts of Interest: Authors declare no conflict of interest.

References

1. Li, N.; Qiao, D.; Zhao, S.; Lin, Q.; Zhang, B.; Xie, F. 3D printing to innovate biopolymer materials for demanding applications: A review. *Mater. Today Chem.* **2021**, *20*, 100459. [CrossRef]
2. Wang, J.; Chiappone, A.; Roppolo, I.; Shao, F.; Fantino, E.; Lorusso, M.; Rentsch, D.; Dietliker, K.; Pirri, C.F.; Grützmacher, H. All-in-One Cellulose Nanocrystals for 3D Printing of Nanocomposite Hydrogels. *Angew. Chemie Int. Ed.* **2018**, *57*, 2353–2356. [CrossRef] [PubMed]
3. Wang, Q.; Sun, J.; Yao, Q.; Ji, C.; Liu, J.; Zhu, Q. 3D printing with cellulose materials. *Cellulose* **2018**, *25*, 4275–4301. [CrossRef]
4. Hausmann, M.K.; Rühs, P.A.; Siqueira, G.; Läuger, J.; Libanori, R.; Zimmermann, T.; Studart, A.R. Dynamics of Cellulose Nanocrystal Alignment during 3D Printing. *ACS Nano* **2018**, *12*, 6926–6937. [CrossRef]
5. Moon, R.J.; Martini, A.; Nairn, J.; Simonsen, J.; Youngblood, J. Cellulose nanomaterials review: Structure, properties and nanocomposites. *Chem. Soc. Rev.* **2011**, *40*, 3941–3994. [CrossRef] [PubMed]
6. Doench, I.; Tran, T.; David, L.; Montembault, A.; Viguier, E.; Gorzelanny, C.; Sudre, G.; Cachon, T.; Louback-Mohamed, M.; Horbelt, N.; et al. Cellulose Nanofiber-Reinforced Chitosan Hydrogel Composites for Intervertebral Disc Tissue Repair. *Biomimetics* **2019**, *4*, 19. [CrossRef] [PubMed]
7. Ghanbari, A.; Tabarsa, T.; Ashori, A.; Shakeri, A.; Mashkour, M. Thermoplastic starch foamed composites reinforced with cellulose nanofibers: Thermal and mechanical properties. *Carbohydr. Polym.* **2018**, *197*, 305–311. [CrossRef]
8. Tarrés, Q.; Oliver-Ortega, H.; Alcalà, M.; Espinach, X.X.; Mutjé, P.; Delgado-Aguilar, M. Research on the strengthening advantages on using cellulose nanofibers as polyvinyl alcohol reinforcement. *Polymers* **2020**, *12*, 974. [CrossRef]
9. Zhang, B.; Huang, C.; Zhao, H.; Wang, J.; Yin, C.; Zhang, L.; Zhao, Y. Effects of cellulose nanocrystals and cellulose nanofibers on the structure and properties of polyhydroxybutyrate nanocomposites. *Polymers* **2019**, *11*, 2063. [CrossRef] [PubMed]
10. Stepanova, M.; Averianov, I.; Serdobintsev, M.; Gofman, I.; Blum, N.; Semenova, N.; Nashchekina, Y.; Vinogradova, T.; Korzhikov-Vlakh, V.; Karttunen, M.; et al. PGlu-modified nanocrystalline cellulose improves mechanical properties, biocompatibility, and mineralization of polyester-based composites. *Materials* **2019**, *12*, 3435. [CrossRef]

11. Yao, J.; Chen, S.; Chen, Y.; Wang, B.; Pei, Q.; Wang, H. Macrofibers with High Mechanical Performance Based on Aligned Bacterial Cellulose Nanofibers. *ACS Appl. Mater. Interfaces* **2017**, *9*, 20330–20339. [CrossRef]
12. Svensson, A.; Nicklasson, E.; Harrah, T.; Panilaitis, B.; Kaplan, D.L.; Brittberg, M.; Gatenholm, P. Bacterial cellulose as a potential scaffold for tissue engineering of cartilage. *Biomaterials* **2005**, *26*, 419–431. [CrossRef]
13. Buyanov, A.L.; Gofman, I.V.; Saprykina, N.N. High-strength cellulose–polyacrylamide hydrogels: Mechanical behavior and structure depending on the type of cellulose. *J. Mech. Behav. Biomed. Mater.* **2019**, *100*, 103385. [CrossRef]
14. Sheykhnazari, S.; Tabarsa, T.; Mashkour, M.; Khazaeian, A.; Ghanbari, A. Multilayer bacterial cellulose/resole nanocomposites: Relationship between structural and electro-thermo-mechanical properties. *Int. J. Biol. Macromol.* **2018**, *120*, 2115–2122. [CrossRef]
15. Gurtovenko, A.A.; Mukhamadiarov, E.I.; Kostritskii, A.Y.; Karttunen, M. Phospholipid-Cellulose Interactions: Insight from Atomistic Computer Simulations for Understanding the Impact of Cellulose-Based Materials on Plasma Membranes. *J. Phys. Chem. B* **2018**, *122*, 9973–9981. [CrossRef]
16. Gurtovenko, A.A.; Karttunen, M. How to control interactions of cellulose-based biomaterials with skin: The role of acidity in the contact area. *Soft Matter* **2021**, *17*, 6507–6518. [CrossRef]
17. Arellano, I.H.J.; Guarino, J.G.; Paredes, F.U.; Arco, S.D. Thermal stability and moisture uptake of 1-alkyl-3-methylimidazolium bromide. *J. Therm. Anal. Calorim.* **2011**, *103*, 725–730. [CrossRef]
18. Kasprzak, D.; Krystkowiak, E.; Stępniak, I.; Galiński, M. Dissolution of cellulose in novel carboxylate-based ionic liquids and dimethyl sulfoxide mixed solvents. *Eur. Polym. J.* **2019**, *113*, 89–97. [CrossRef]
19. Isik, M.; Sardon, H.; Mecerreyes, D. Ionic liquids and cellulose: Dissolution, chemical modification and preparation of new cellulosic materials. *Int. J. Mol. Sci.* **2014**, *15*, 11922–11940. [CrossRef] [PubMed]
20. Swatloski, R.P.; Spear, S.K.; Holbrey, J.D.; Rogers, R.D. Dissolution of cellose with ionic liquids. *J. Am. Chem. Soc.* **2002**, *124*, 4974–4975. [CrossRef] [PubMed]
21. Lv, Y.; Wu, J.; Zhang, J.; Niu, Y.; Liu, C.Y.; He, J.; Zhang, J. Rheological properties of cellulose/ionic liquid/dimethylsulfoxide (DMSO) solutions. *Polymer* **2012**, *53*, 2524–2531. [CrossRef]
22. Morais, E.S.; Da Costa Lopes, A.M.; Freire, M.G.; Freire, C.S.R.; Coutinho, J.A.P.; Silvestre, A.J.D. Use of ionic liquids and deep eutectic solvents in polysaccharides dissolution and extraction processes towards sustainable biomass valorization. *Molecules* **2020**, *25*, 3652. [CrossRef]
23. Hou, Q.; Ju, M.; Li, W.; Liu, L.; Chen, Y.; Yang, Q.; Zhao, H. Pretreatment of lignocellulosic biomass with ionic liquids and ionic liquid-based solvent systems. *Molecules* **2017**, *22*, 490. [CrossRef] [PubMed]
24. Zhang, J.; Wu, J.; Yu, J.; Zhang, X.; He, J.; Zhang, J. Application of ionic liquids for dissolving cellulose and fabricating cellulose-based materials: State of the art and future trends. *Mater. Chem. Front.* **2017**, *1*, 1273–1290. [CrossRef]
25. Glińska, K.; Gitalt, J.; Torrens, E.; Plechkova, N.; Bengoa, C. Extraction of cellulose from corn stover using designed ionic liquids with improved reusing capabilities. *Process Saf. Environ. Prot.* **2021**, *147*, 181–191. [CrossRef]
26. Magalhães Da Silva, S.P.; Da Costa Lopes, A.M.; Roseiro, L.B.; Bogel-Łukasik, R. Novel pre-treatment and fractionation method for lignocellulosic biomass using ionic liquids. *RSC Adv.* **2013**, *3*, 16040–16050. [CrossRef]
27. Prasad, K.; Mine, S.; Kaneko, Y.; Kadokawa, J.I. Preparation of cellulose-based ionic porous material compatibilized with polymeric ionic liquid. *Polym. Bull.* **2010**, *64*, 341–349. [CrossRef]
28. Chiappe, C.; Rodriguez Douton, M.J.; Mezzetta, A.; Pomelli, C.S.; Assanelli, G.; De Angelis, A.R. Recycle and Extraction: Cornerstones for an Efficient Conversion of Cellulose into 5-Hydroxymethylfurfural in Ionic Liquids. *ACS Sustain. Chem. Eng.* **2017**, *5*, 5529–5536. [CrossRef]
29. Murakami, M.; Kaneko, Y.; Kadokawa, J. Preparation of cellulose-polymerized ionic liquid composite by in-situ polymerization of polymerizable ionic liquid in cellulose-dissolving solution. *Carbohydr. Polym.* **2007**, *69*, 378–381. [CrossRef]
30. Isik, M.; Gracia, R.; Kollnus, L.C.; Tomé, L.C.; Marrucho, I.M.; Mecerreyes, D. Cholinium-based poly(ionic liquid)s: Synthesis, characterization, and application as biocompatible ion gels and cellulose coatings. *ACS Macro Lett.* **2013**, *2*, 975–979. [CrossRef]
31. Ma, T.; Lv, L.; Ouyang, C.; Hu, X.; Liao, X.; Song, Y.; Hu, X. Rheological behavior and particle alignment of cellulose nanocrystal and its composite hydrogels during 3D printing. *Carbohydr. Polym.* **2021**, *253*, 117217. [CrossRef]
32. Long, W.J.; Tao, J.L.; Lin, C.; Gu, Y.C.; Mei, L.; Duan, H.B.; Xing, F. Rheology and buildability of sustainable cement-based composites containing micro-crystalline cellulose for 3D-printing. *J. Clean. Prod.* **2019**, *239*, 118054. [CrossRef]
33. Shin, S.; Hyun, J. Rheological properties of cellulose nanofiber hydrogel for high-fidelity 3D printing. *Carbohydr. Polym.* **2021**, *263*, 117976. [CrossRef]
34. Shao, Y.; Chaussy, D.; Grosseau, P.; Beneventi, D. Use of Microfibrillated Cellulose/Lignosulfonate Blends as Carbon Precursors: Impact of Hydrogel Rheology on 3D Printing. *Ind. Eng. Chem. Res.* **2015**, *54*, 10575–10582. [CrossRef]
35. Song, H.; Zhang, J.; Niu, Y.; Wang, Z. Phase transition and rheological behaviors of concentrated cellulose/ionic liquid solutions. *J. Phys. Chem. B* **2010**, *114*, 6006–6013. [CrossRef] [PubMed]
36. Sescousse, R.; Le, K.A.; Ries, M.E.; Budtova, T. Viscosity of cellulose-imidazolium-based ionic liquid solutions. *J. Phys. Chem. B* **2010**, *114*, 7222–7228. [CrossRef]
37. Rudaz, C.; Budtova, T. Rheological and hydrodynamic properties of cellulose acetate/ionic liquid solutions. *Carbohydr. Polym.* **2013**, *92*, 1966–1971. [CrossRef]
38. Gericke, M.; Schlufter, K.; Liebert, T.; Heinze, T.; Budtova, T. Rheological properties of cellulose/ionic liquid solutions: From dilute to concentrated states. *Biomacromolecules* **2009**, *10*, 1188–1194. [CrossRef] [PubMed]

39. Zhang, S.; Sun, N.; He, X.; Lu, X.; Zhang, X. Physical properties of ionic liquids: Database and evaluation. *J. Phys. Chem. Ref. Data* **2006**, *35*, 1475–1517. [CrossRef]
40. Hapiot, P.; Lagrost, C. Electrochemical reactivity in room-temperature ionic liquids. *Chem. Rev.* **2008**, *108*, 2238–2264. [CrossRef]
41. Fendt, S.; Padmanabhan, S.; Blanch, H.W.; Prausnitz, J.M. Viscosities of acetate or chloride-based ionic liquids and some of their mixtures with water or other common solvents. *J. Chem. Eng. Data* **2011**, *56*, 31–34. [CrossRef]
42. Yang, F.; Wang, X.; Tan, H.; Liu, Z. Improvement the viscosity of imidazolium-based ionic liquid using organic solvents for biofuels. *J. Mol. Liq.* **2017**, *248*, 626–633. [CrossRef]
43. Gunasekera, D.H.A.T.; Kuek, S.; Hasanaj, D.; He, Y.; Tuck, C.; Croft, A.K.; Wildman, R.D. Three dimensional ink-jet printing of biomaterials using ionic liquids and co-solvents. *Faraday Discuss.* **2016**, *190*, 509–523. [CrossRef]
44. Amde, M.; Liu, J.-F.; Pang, L. Environmental application, fate, effects, and concerns of ionic liquids: A review. *Environ. Sci. Technol.* **2015**, *49*, 12611–12627. [CrossRef]
45. Smith, E.L.; Abbott, A.P.; Ryder, K.S. Deep Eutectic Solvents (DESs) and Their Applications. *Chem. Rev.* **2014**, *114*, 11060–11082. [CrossRef]
46. Hansen, B.B.; Spittle, S.; Chen, B.; Poe, D.; Zhang, Y.; Klein, J.M.; Horton, A.; Adhikari, L.; Zelovich, T.; Doherty, B.W.; et al. Deep Eutectic Solvents: A Review of Fundamentals and Applications. *Chem. Rev.* **2021**, *121*, 1232–1285. [CrossRef]
47. Gurkan, B.E.; Maginn, E.J.; Pentzer, E.B. Deep eutectic solvents: A new class of versatile liquids. *J. Phys. Chem. B* **2020**, *124*, 11313–11315. [CrossRef]
48. Paiva, A.; Craveiro, R.; Aroso, I.; Martins, M.; Reis, R.L.; Duarte, A.R.C. Natural deep eutectic solvents—Solvents for the 21st century. *ACS Sustain. Chem. Eng.* **2014**, *2*, 1063–1071. [CrossRef]
49. Galvis-Sánchez, A.C.; Castro, M.C.R.; Biernacki, K.; Gonçalves, M.P.; Souza, H.K.S. Natural deep eutectic solvents as green plasticizers for chitosan thermoplastic production with controlled/desired mechanical and barrier properties. *Food Hydrocoll.* **2018**, *82*, 478–489. [CrossRef]
50. Sokolova, M.P.; Smirnov, M.A.; Samarov, A.A.; Bobrova, N.V.; Vorobiov, V.K.; Popova, E.N.; Filippova, E.; Geydt, P.; Lahderanta, E.; Toikka, A.M. Plasticizing of chitosan films with deep eutectic mixture of malonic acid and choline chloride. *Carbohydr. Polym.* **2018**, *197*, 548–557. [CrossRef] [PubMed]
51. Smirnov, M.A.; Nikolaeva, A.L.; Bobrova, N.V.; Vorobiov, V.K.; Smirnov, A.V.; Lahderanta, E.; Sokolova, M.P. Self-healing films based on chitosan containing citric acid/choline chloride deep eutectic solvent. *Polym. Test.* **2021**, *97*, 107156. [CrossRef]
52. Xie, Q.; Zheng, X.; Li, L.; Ma, L.; Zhao, Q.; Chang, S.; You, L. Effect of Curcumin Addition on the Properties of Biodegradable Pectin/Chitosan Films. *Molecules* **2021**, *26*, 2152. [CrossRef] [PubMed]
53. Jakubowska, E.; Gierszewska, M.; Nowaczyk, J.; Olewnik-Kruszkowska, E. The role of a deep eutectic solvent in changes of physicochemical and antioxidative properties of chitosan-based films. *Carbohydr. Polym.* **2021**, *255*, 117527. [CrossRef] [PubMed]
54. Zdanowicz, M. Deep eutectic solvents based on urea, polyols and sugars for starch treatment. *Int. J. Biol. Macromol.* **2021**, *176*, 387–393. [CrossRef]
55. Özel, N.; Elibol, M. A review on the potential uses of deep eutectic solvents in chitin and chitosan related processes. *Carbohydr. Polym.* **2021**, *262*, 117942. [CrossRef]
56. Yuan, Y.; Hong, S.; Lian, H.; Zhang, K.; Liimatainen, H. Comparison of acidic deep eutectic solvents in production of chitin nanocrystals. *Carbohydr. Polym.* **2020**, *236*, 116095. [CrossRef]
57. Mukesh, C.; Mondal, D.; Sharma, M.; Prasad, K. Choline chloride-thiourea, a deep eutectic solvent for the production of chitin nanofibers. *Carbohydr. Polym.* **2014**, *103*, 466–471. [CrossRef]
58. Smirnov, M.A.; Sokolova, M.P.; Tolmachev, D.A.; Vorobiov, V.K.; Kasatkin, I.A.; Smirnov, N.N.; Klaving, A.V.; Bobrova, N.V.; Lukasheva, N.V.; Yakimansky, A.V. Green method for preparation of cellulose nanocrystals using deep eutectic solvent. *Cellulose* **2020**, *27*, 4305–4317. [CrossRef]
59. Jiang, J.; Zhu, Y.; Jiang, F. Sustainable isolation of nanocellulose from cellulose and lignocellulosic feedstocks: Recent progress and perspectives Jungang. *Carbohydr. Polym.* **2021**, 118188. [CrossRef]
60. Sirviö, J.A.; Hyypiö, K.; Asaadi, S.; Junka, K.; Liimatainen, H. High-strength cellulose nanofibers produced: Via swelling pretreatment based on a choline chloride-imidazole deep eutectic solvent. *Green Chem.* **2020**, *22*, 1763–1775. [CrossRef]
61. Sirviö, J.A.; Visanko, M.; Liimatainen, H. Deep eutectic solvent system based on choline chloride-urea as a pre-treatment for nanofibrillation of wood cellulose. *Green Chem.* **2015**, *17*, 3401–3406. [CrossRef]
62. Liu, Y.; Guo, B.; Xia, Q.; Meng, J.; Chen, W.; Liu, S.; Wang, Q.; Liu, Y.; Li, J.; Yu, H. Efficient Cleavage of Strong Hydrogen Bonds in Cotton by Deep Eutectic Solvents and Facile Fabrication of Cellulose Nanocrystals in High Yields. *ACS Sustain. Chem. Eng.* **2017**, *5*, 7623–7631. [CrossRef]
63. Fan, Q.; Jiang, C.; Wang, W.; Bai, L.; Chen, H.; Yang, H.; Wei, D.; Yang, L. Eco-friendly extraction of cellulose nanocrystals from grape pomace and construction of self-healing nanocomposite hydrogels. *Cellulose* **2020**, *27*, 2541–2553. [CrossRef]
64. Li, C.; Huang, C.; Zhao, Y.; Zheng, C.; Su, H.; Zhang, L.; Luo, W.; Zhao, H.; Wang, S.; Huang, L.J. Effect of choline-based deep eutectic solvent pretreatment on the structure of cellulose and lignin in Bagasse. *Processes* **2021**, *9*, 384. [CrossRef]
65. Jablonsky, M.; Haz, A.; Majova, V. Assessing the opportunities for applying deep eutectic solvents for fractionation of beech wood and wheat straw. *Cellulose* **2019**, *26*, 7675–7684. [CrossRef]
66. Fazende, K.F.; Phachansitthi, M.; Mota-Morales, J.D.; Pojman, J.A. Frontal Polymerization of Deep Eutectic Solvents Composed of Acrylic and Methacrylic Acids. *J. Polym. Sci. Part A Polym. Chem.* **2017**, *55*, 4046–4050. [CrossRef]

67. Mota-Morales, J.D.; Sánchez-Leija, R.J.; Carranza, A.; Pojman, J.A.; del Monte, F.; Luna-Bárcenas, G. Free-radical polymerizations of and in deep eutectic solvents: Green synthesis of functional materials. *Prog. Polym. Sci.* **2018**, *78*, 139–153. [CrossRef]
68. Yue, C.; Li, M.; Liu, Y.; Fang, Y.; Song, Y.; Xu, M.; Li, J. Three-dimensional Printing of Cellulose Nanofibers Reinforced PHB/PCL/Fe3O4 Magneto-responsive Shape Memory Polymer Composites with Excellent Mechanical Properties. *Addit. Manuf.* **2021**, 102146. [CrossRef]
69. Li, V.C.F.; Kuang, X.; Mulyadi, A.; Hamel, C.M.; Deng, Y.; Qi, H.J. 3D printed cellulose nanocrystal composites through digital light processing. *Cellulose* **2019**, *26*, 3973–3985. [CrossRef]
70. Li, V.C.F.; Dunn, C.K.; Zhang, Z.; Deng, Y.; Qi, H.J. Direct Ink Write (DIW) 3D Printed Cellulose Nanocrystal Aerogel Structures. *Sci. Rep.* **2017**, *7*, 8018. [CrossRef] [PubMed]
71. Kam, D.; Braner, A.; Abouzglo, A.; Larush, L.; Chiappone, A.; Shoseyov, O.; Magdassi, S. 3D Printing of Cellulose Nanocrystal-Loaded Hydrogels through Rapid Fixation by Photopolymerization. *Langmuir* **2021**, *37*, 6451–6458. [CrossRef]
72. Melilli, G.; Carmagnola, I.; Tonda-Turo, C.; Pirri, F.; Ciardelli, G.; Sangermano, M.; Hakkarainen, M.; Chiappone, A. DLP 3D printing meets lignocellulosic biopolymers: Carboxymethyl cellulose inks for 3D biocompatible hydrogels. *Polymers* **2020**, *12*, 12081655. [CrossRef] [PubMed]
73. Siqueira, G.; Kokkinis, D.; Libanori, R.; Hausmann, M.K.; Gladman, A.S.; Neels, A.; Tingaut, P.; Zimmermann, T.; Lewis, J.A.; Studart, A.R. Cellulose Nanocrystal Inks for 3D Printing of Textured Cellular Architectures. *Adv. Funct. Mater.* **2017**, *27*, 1604619. [CrossRef]
74. Lai, P.C.; Yu, S.S. Cationic cellulose nanocrystals-based nanocomposite hydrogels: Achieving 3d printable capacitive sensors with high transparency and mechanical strength. *Polymers* **2021**, *13*, 688. [CrossRef] [PubMed]
75. Lai, C.W.; Yu, S.S. 3D Printable Strain Sensors from Deep Eutectic Solvents and Cellulose Nanocrystals. *ACS Appl. Mater. Interfaces* **2020**, *12*, 34235–34244. [CrossRef] [PubMed]
76. Ahmed, K.; Kawakami, M.; Khosla, A.; Furukawa, H. Soft, conductive nanocomposites based on ionic liquids/carbon nanotubes for 3D printing of flexible electronic devices. *Polym. J.* **2019**, *51*, 511–521. [CrossRef]
77. Zhang, Y.; Yuan, B.; Zhang, Y.; Cao, Q.; Yang, C.; Li, Y.; Zhou, J. Biomimetic lignin/poly(ionic liquids) composite hydrogel dressing with excellent mechanical strength, self-healing properties, and reusability. *Chem. Eng. J.* **2020**, *400*, 125984. [CrossRef]
78. Zhu, M.; Cao, Q.; Liu, B.; Guo, H.; Wang, X.; Han, Y.; Sun, G.; Li, Y.; Zhou, J. A novel cellulose acetate/poly (ionic liquid) composite air filter. *Cellulose* **2020**, *27*, 3889–3902. [CrossRef]
79. Smirnov, M.A.; Vorobiov, V.K.; Bobrova, N.V.; Smirnov, N.N.; Lyulin, S.V. METHOD OF PRODUCING BACTERIAL CELLULOSE NANOFIBERS RU. In Proceedings of the 15 th International Saint Petersburg Conference of Young Scientists, Saint Petersburg, Russia, 28–31 October 2019; Institute of Macromolecular Compounds of the Russian Academy of Sciences Saint Petersburg: Sant Petersburg, Russia, 2019; pp. 1–8.
80. Xu, J.; Zhang, B.; Lu, X.; Zhou, Y.; Fang, J.; Li, Y.; Zhang, S. Nanoscale Observation of Microfibril Swelling and Dissolution in Ionic Liquids. *ACS Sustain. Chem. Eng.* **2018**, *6*, 909–917. [CrossRef]
81. Segal, L.; Creely, J.J.; Martin, A.E.; Conrad, C.M. An Empirical Method for Estimating the Degree of Crystallinity of Native Cellulose Using the X-Ray Diffractometer. *Text. Res. J.* **1959**, *29*, 786–794. [CrossRef]
82. Dong, J.; Ozaki, Y.; Nakashima, K. Infrared, Raman, and near-infrared spectroscopic evidence for the coexistence of various hydrogen-bond forms in poly(acrylic acid). *Macromolecules* **1997**, *30*, 1111–1117. [CrossRef]
83. Czarnecka, E.; Nowaczyk, J. Semi-Natural superabsorbents based on Starch-g-poly(acrylic acid): Modification, synthesis and application. *Polymers* **2020**, *12*, 1794. [CrossRef] [PubMed]
84. Wei, L.; Chen, C.; Hou, Z.; Wei, H. Poly (acrylic acid sodium) grafted carboxymethyl cellulose as a high performance polymer binder for silicon anode in lithium ion batteries. *Sci. Rep.* **2016**, *6*, 1–8. [CrossRef] [PubMed]
85. Hon, N.S. Formation of Free Radicals in Photoirradiated Cellulose—8. Mechanisms. *J. Polym. Sci. Polym. Chem. Ed.* **1976**, *14*, 2497–2512. [CrossRef]
86. Fazende, K.F.; Gary, D.P.; Mota-Morales, J.D.; Pojman, J.A. Kinetic Studies of Photopolymerization of Monomer-Containing Deep Eutectic Solvents. *Macromol. Chem. Phys.* **2020**, *221*, 1900511. [CrossRef]
87. Okoturo, O.O.; VanderNoot, T.J. Temperature dependence of viscosity for room temperature ionic liquids. *J. Electroanal. Chem.* **2004**, *568*, 167–181. [CrossRef]
88. Hammond, O.S.; Bowron, D.T.; Edler, K.J. Liquid structure of the choline chloride-urea deep eutectic solvent (reline) from neutron diffraction and atomistic modelling. *Green Chem.* **2016**, *18*, 2736–2744. [CrossRef]
89. Gontrani, L.; Bonomo, M.; Plechkova, N.V.; Dini, D.; Caminiti, R. X-Ray structure and ionic conductivity studies of anhydrous and hydrated choline chloride and oxalic acid deep eutectic solvents. *Phys. Chem. Chem. Phys.* **2018**, *20*, 30120–30124. [CrossRef]
90. Stassen, H.K.; Ludwig, R.; Wulf, A.; Dupont, J. Imidazolium salt ion Pairs in solution. *Chem. A Eur. J.* **2015**, *21*, 8324–8335. [CrossRef]
91. Li, W.; Zhang, Z.; Han, B.; Hu, S.; Xie, Y.; Yang, G. Effect of water and organic solvents on-the ionic dissociation of ionic liquids. *J. Phys. Chem. B* **2007**, *111*, 6452–6456. [CrossRef]
92. Czaikoski, A.; da Cunha, R.L.; Menegalli, F.C. Rheological behavior of cellulose nanofibers from cassava peel obtained by combination of chemical and physical processes. *Carbohydr. Polym.* **2020**, *248*, 116744. [CrossRef] [PubMed]
93. Barnes, H.A. A review of the slip (wall depletion) of polymer solutions, emulsions and particle suspensions in viscometers: Its cause, character, and cure. *J. Nonnewton. Fluid Mech.* **1995**, *56*, 221–251. [CrossRef]

94. Reid, M.S.; Kedzior, S.A.; Villalobos, M.; Cranston, E.D. Effect of Ionic Strength and Surface Charge Density on the Kinetics of Cellulose Nanocrystal Thin Film Swelling. *Langmuir* **2017**, *33*, 7403–7411. [CrossRef]
95. Wang, H.; Gurau, G.; Rogers, R.D. Ionic liquid processing of cellulose. *Chem. Soc. Rev.* **2012**, *41*, 1519–1537. [CrossRef] [PubMed]
96. Li, Y.; Wang, J.; Liu, X.; Zhang, S. Towards a molecular understanding of cellulose dissolution in ionic liquids: Anion/cation effect, synergistic mechanism and physicochemical aspects. *Chem. Sci.* **2018**, *9*, 4027–4043. [CrossRef]
97. Li, Y.; Liu, X.; Zhang, Y.; Jiang, K.; Wang, J.; Zhang, S. Why Only Ionic Liquids with Unsaturated Heterocyclic Cations Can Dissolve Cellulose: A Simulation Study. *ACS Sustain. Chem. Eng.* **2017**, *5*, 3417–3428. [CrossRef]
98. Hong, S.; Yuan, Y.; Liu, C.; Chen, W.; Chen, L.; Lian, H.; Liimatainen, H. A stretchable and compressible ion gel based on a deep eutectic solvent applied as a strain sensor and electrolyte for supercapacitors. *J. Mater. Chem. C* **2020**, *8*, 550–560. [CrossRef]

polymers MDPI

Article

Barrier Properties and Hydrophobicity of Biodegradable Poly(lactic acid) Composites Reinforced with Recycled Chinese Spirits Distiller's Grains

Zhi-Jun Chen [1,†], Chi-Hui Tsou [1,2,3,4,*,†], Meng-Lin Tsai [5,†], Jipeng Guo [1,†], Manuel Reyes De Guzman [1], Tao Yang [1], Chen Gao [1], Yan Lei [1], Pei-Wen Gan [1], Shuang Chen [1], Lian-Jie Tu [1], Chang-Lei Qu [1], Ruo-Yao Wang [5] and Chin-San Wu [6,*]

1 Material Corrosion and Protection Key Laboratory of Sichuan Province, School of Materials Science and Engineering, Sichuan University of Science and Engineering, Zigong 643000, China; czj2356068001@163.com (Z.-J.C.); Bb7211260@163.com (J.G.); manuelrdg@yahoo.com (M.R.D.G.); y19130115495@126.com (T.Y.); s19871016gc@sohu.com (C.G.); yanlei224@suse.edu.cn (Y.L.); peiwen@suse.edu.cn (P.-W.G.); cs68811190@163.com (S.C.); tfzhhy@163.com (L.-J.T.); quchangleixuexi@shse.edu.cn (C.-L.Q.)
2 Sichuan Yibin Plastic Packaging Materials Co. Ltd., Yibin 644007, China
3 Sichuan Golden-Elephant Sincerity Chemical Co. Ltd., Meishan 620010, China
4 Sichuan Zhixiangyi Technology Co. Ltd., Chengdu 610051, China
5 Department of Materials Science and Engineering, National Taiwan University of Science and Technology, Taipei 10607, Taiwan; mltsai@gapp.ntust.edu.tw (M.-L.T.); steven5202001@yahoo.com.tw (R.-Y.W.)
6 Department of Applied Cosmetology, Kao Yuan University, Kaohsiung 82101, Taiwan
* Correspondence: mayko0301@hotmail.com (C.-H.T.); t50008@cc.kyu.edu.tw (C.-S.W.)
† Authors contributed equally to this work.

Citation: Chen, Z.-J.; Tsou, C.-H.; Tsai, M.-L.; Guo, J.; De Guzman, M.R.; Yang, T.; Gao, C.; Lei, Y.; Gan, P.-W.; Chen, S.; et al. Barrier Properties and Hydrophobicity of Biodegradable Poly(lactic acid) Composites Reinforced with Recycled Chinese Spirits Distiller's Grains. *Polymers* **2021**, *13*, 2861. https://doi.org/10.3390/polym13172861

Academic Editor: Evgenia G. Korzhikova-Vlakh

Received: 24 July 2021
Accepted: 20 August 2021
Published: 26 August 2021

Abstract: Adding natural biomass to poly(lactic acid) (PLA) as a reinforcing filler is a way to change the properties of PLA. This paper is about preparing PLA/biomass composites by physically melting and blending Chinese Spirits distiller's grains (CSDG) biomass and PLA to optimize the composite performance. Composites of modified PLA (MPLA) with varying amounts of CSDG were also prepared by the melt-mixing method, and unmodified PLA/CSDG composites were used as a control group for comparative analysis. The functional groups of MPLA enhanced the compatibility between the polymer substrate and CSDG. The composite water vapor/oxygen barrier and mechanical properties were studied. It was found that the barrier and mechanical properties of MPLA/CSDG composites were significantly improved. SEM was adopted to examine the tensile section structure of the composites, and the compatibility between the filler and the matrix was analyzed. An appropriate amount of CSDG had a better dispersibility in the matrix, and it further improved the interfacial bonding force, which in turn improved the composite mechanical properties. X-ray diffraction, thermogravimetric analysis, and differential scanning calorimetry were conducted to determine the crystalline properties and to analyze the stability of the composites. It was found that the CSDG content had a significant effect on the crystallinity. Barrier and biodegradation mechanisms were also discussed.

Keywords: poly(lactic acid); Chinese spirits distiller's grains; barrier; mechanical behavior; hydrophobicity; biodegradable properties

1. Introduction

Research and development on environment-friendly polymers and polymer composites based on renewable resources has attracted increasing attention [1,2]. Natural [3,4] and various functional materials used in manufacturing have gradually replaced traditional ones [5,6] because of their clean, easy-to-use, harmless, and environment-friendly features, and their usage can reduce energy consumption [7] and environmental pollution [8,9]. Poly(lactic acid) (PLA) can be naturally metabolized in the human body into carbon dioxide and water. It is called "an environmental recycling material of the 21st century" [10–12]. PLA/natural plant fiber composites have considerable performance and excellent biodegradability; they are known as "green" composites and are currently a hot spot in the field of biodegradable materials.

Cellulose sources are rich and diverse, including crop straw, bagasse, wood, bamboo, and corn cobs, and other agricultural waste [13,14]. Filling PLA with natural biomass fibers to prepare composites reduces the density of the polymer matrix, improves the strength and biodegradability of the composites, and greatly reduces the cost. Application fields include construction and civil engineering, sports and entertainment products, medical equipment, bionic products, and home office supplies, and they have expanded to various fields such as aerospace, national defense, and military industry [15,16]. This expansion not only improves the added value of plant fibers but also makes full use of degradable PLA to replace original nonrenewable petroleum-based plastics. The use of degradable plastics is of great significance to the environment and human development [17].

Zhao et al. [18] carried out surface modification treatment on rice straw fiber (RSF) and on combined modified and unmodified RSF with PLA, and they compared and analyzed different samples. The results showed that modified RSF had obvious improvement in the composite tensile strength. When the RSF content was 30%, the elongation of the modified sample was 60.8% higher than that of the unmodified one. Yaacab et al. [19] used rice straw powder (RSP) as filler and PLA as matrix, and they mixed them with hot pressing, followed by high-pressure cooling, to obtain PLA/RSP biocomposites. The results indicated that the tensile strength of PLA/RSP composites decreased as the RSP content increased, but the tensile modulus and stiffness increased significantly. Increase in RSP reduced the thermal stability and crystallinity.

Way et al. [20] applied some maple wood fiber (WF) as a reinforcing filler to prepare PLA/WF composites by melt blending. They studied the composite mechanical and thermodynamic properties, and they reported that the interfacial compatibility between PLA and WF was enhanced, and the mechanical properties of the composites were improved as a result of the chemical treatment with a silane coupling agent. Yusoff et al. [21] adopted an effective method to extract bamboo fiber and subsequently characterized its mechanical properties. Bamboo, coconut, or kenaf fiber was used as reinforcement to prepare an environment-friendly composite with PLA. The mechanical properties were tested, and the fiber structure was observed using SEM. The results showed that when three kinds of fibers were added, the mechanical properties or strength of the composites were improved. Wassamon et al. [22] combined natural fibers, such as bamboo, vanilla, and coconut, with modified PLA to prepare a new type of environment-friendly functional materials. Their study demonstrated that the rigidity of composite materials increased significantly.

Bourmaud et al. [23] used maleic anhydride-grafted PLA (PLA-g-MAH) as a compatibilizer in PLA/reed fiber composites, and they found that adding the compatibilizer enhanced the reed fiber adhesion to the polymer matrix. The degree of fiber dispersion significantly improved the compatibility between the fiber and PLA. Zhang et al. [24] chose PLA-g-MAH as a compatibilizer for wood flour-reinforced PLA composites, and they observed that when the mass fraction of PLA-g-MAH was 30 wt %, the mechanical properties were significantly improved compared with those of the composites without a compatibilizer.

A category of biomass is a by-product mainly obtained from wine industries, known as distiller's grains, one type of which is referred to as Chinese Spirits distiller's grains (CSDG). The majority of CSDG is burned or buried as waste, which not only wastes resources but also causes environmental pollution. However, biomass can be used as fillers to make biodegradable materials. For example, plant fibers can act as nucleating agents in the crystallization process of different thermoplastic polymers and interfere with their supramolecular structure [25]. However, plant fibers also have disadvantages such as poor water resistance and poor plasticity. CSDG can be an alternative, as it is also an important part of biomass, and it is likely to replace glass fibers, inorganic fibers, and other fillers commonly used in plastics to prepare environment-friendly composites. Therefore, CSDG may be combined with PLA to improve the water resistance of composites. At the same time, it may expand further the application of green plastics or the application fields of natural fibers.

CSDG, an unfermented grain, is a by-product from ethanol fuel industries. At present, there is little research on the application of CSDG in PLA composites. Only Lu et al. [26] prepared biodegradable composites in 2014 by mixing PLA with dried distiller's grains. Only the two materials—PLA and dried distiller's grains—were mixed. Neither PLA nor distiller's grains was modified, and the change in mechanical strength was not mentioned in the study. It was only estimated that the mechanical properties should not be ideal. Moreover, the research was mainly focused on biodegradable and thermal properties, and the content of distiller's grains was only 20%. There is still much room for discussion and further studies. Interestingly, CSDG contains a lot of grains, and it is more complicated than distiller's grains in general. In addition, CSDG has never been reported in studies dealing with PLA. In this present work, CSDG was used as a reinforcing filler, and a composite material was obtained by melt blending it with PLA. A modified form of PLA (MPLA) was also considered to improve the interfacial compatibility between CSDG and PLA. The hydrophobicity, biodegradability, as well as the mechanical, barrier, crystallization, and thermodynamic properties of the composites were characterized and analyzed. The effect of varying amounts of CSDG on the physical and mechanical properties of the composites was investigated to explore the potential application of CSDG in the field of green plastics.

2. Experimental

2.1. Experimental Materials

A 2002D grade of PLA was used in this present study. It was from Natureworks LLC (Minnesota, MN, USA), with a number molecular weight of 100,000 g/mol. The modification method to form MPLA was similar to that in previous studies [27,28], and it involved grafting maleic anhydride to PLA (thus, maleic anhydride-grafted PLA was produced). The grafting rate of MPLA was about 1.2%, and the tensile strength of MPLA was 37.08 ± 0.7 MPa.

2.2. Preparation of Composites

PLA/CSDG and MPLA/CSDG composites were prepared, with the CSDG concentration varied from 10 to 50%. The proportion of components in each material is shown in Table 1. The steps of preparing the composite materials are as follows:

(1) Drying of materials: Before the sample preparation, PLA and CSDG were placed in a vacuum oven at a drying temperature of 85 °C for a drying time of 8 h.

(2) Preparation of PLA/CSDG or MPLA/CSDG composites: A torque rheometer (HAAKE PolyLab OS) was used. The conditions at which it was operated were at a temperature of 180 °C and a speed of 100 rpm. First, PLA or MPLA were added to melt them for 1 min. Then, CSDG powder was added, and its content was varied (10, 20, 30, 40, and 50 wt %). The two materials were blended for 10 min to prepare PLA/CSDG or MPLA/CSDG composites with different concentrations of CSDG.

The prepared composite was hot-pressed using a vulcanizer at a temperature of 180 °C for 10 min. The material was taken out of the vulcanizer to left to cool. A custom-made cutter was used to cut the material and form a dumbbell-shaped sample, which was used for testing and characterization.

Table 1. Composition of PLA/CSDG and MPLA/CSDG composites.

Sample	PLA (%)	CSDG (%)	MPLA (%)
PLA/CSDG10	90	10	0
PLA/CSDG20	80	20	0
PLA/CSDG30	70	30	0
PLA/CSDG40	60	40	0
PLA/CSDG50	50	50	0
MPLA/CSDG10	0	10	90
MPLA/CSDG20	0	20	80
MPLA/CSDG30	0	30	70
MPLA/CSDG40	0	40	60
MPLA/CSDG50	0	50	50

2.3. Fourier Transform Infrared Spectroscopy

The infrared absorption spectrum of a material can be obtained by detecting infrared absorption, also known as molecular vibration. Before the test, samples were dried at 80 °C and were cut into pieces. The spectral analysis of the samples was characterized by Fourier transform infrared (FTIR) (NICOLET 6700, Thermo Scientific, Waltham, MA, USA) spectroscopy to obtain the characteristic peaks of the samples.

For the test parameter setting, the test wavelength range was 4000–500 cm^{-1}.

2.4. Mechanical Properties

A microcomputer-controlled electronic universal testing machine (FBS-10KNW, Xiamen Forbes Testing Equipment Co. Ltd., Xiamen, China) in plastic-film tensile test mode was used to determine the mechanical properties of the PLA/CSDG and MPLA/CSDG composites. Tensile strength referred to the ratio of the maximum load P before the test sample broke to the cross-sectional area of the test sample under a specific test temperature and humidity. A tensile load was applied along the axial direction. Tensile strength is usually expressed as δ_t, and its calculation formula is:

$$\delta_t = P/(b \times d) \tag{1}$$

where P = maximum breaking load, N; b = sample width, mm; d = sample thickness, mm.

The elongation at break was the relative elongation of the test sample when it broke, and was calculated by the following formula:

$$\varepsilon_t = (F - G)/G \times 100\% \tag{2}$$

where G = distance between marking lines of sample, mm; F = distance between marking lines when sample broke, mm.

The characterization parameter setting is as follows: sample in dumbbell shape; 1000 N sensor range; 2 mm/min set pulling speed.

2.5. X-ray Diffraction

X-ray diffraction (XRD) technology is the most important structural test method. The prepared sample was analyzed using XRD (D2 PHASER, Bruker, Germany), and corresponding peaks for PLA/CSDG and MPLA/CSDG composites were obtained.

The test parameter setting is given as follows: working voltage = 40 kV, working current = 30 mA, scanning area = 0.02°/s, and step length = 10–90°.

2.6. Morphology Characterization

Scanning electron microscopy (SEM) was used to observe the fracture surface morphology of tensile samples taken from composite materials. SEM is an excellent technique for obtaining morphological information by scanning the surface with electron beams. It can generate high-resolution 3D surface images to describe the surface structure of samples. The generated SEM images clearly depicted the morphology of PLA/CSDG and MPLA/CSDG composites and the distribution of CSDG in them. Before SEM tests were conducted, samples were attached to a support and sputtered with a coating or layer of gold. Then, SEM (VEGA3SBU, TESCAN, Brno, Czechia) was operated to characterize the cross-sectional morphology of PLA/CSDG and MPLA/CSDG composites.

The SEM test parameter setting is indicated as follows: time of spraying gold on samples, 30 s; working current, 5 mA; scanning or acceleration voltage, 3 kV.

2.7. Thermal Analysis

Differential scanning calorimetry (DSC) or a thermal analyzer (DSC200 F3, NETZSCH, Selb, Germany) operated under nitrogen gas was used to test the crystallization temperature and crystallinity of composite films. The sample to be tested was placed in a crucible, and a reference material was placed in another crucible. Then, the two crucibles were heated at the same rate. The sample had to undergo melting, crystallization, oxidation, and degradation processes, so as to obtain the following parameters: T crystallization, T oxidation, T melting, and T decomposition.

$$X_C = (\frac{\Delta H}{\Delta H_0 \times wt\%}) \times 100\% \tag{3}$$

In Equation (3), X_C = degree of crystallinity, ΔH = test sample heat of fusion, ΔH_0 = heat of fusion of a pure PLA substrate when the crystallinity is 100%, and $wt\%$ = percentage of PLA in the sample. PLA with 100% crystallinity has a theoretical enthalpy of 93.7 J/g.

The test parameter setting is given as follows: heating rate of 25 °C/min (heating was from room temperature to 200 °C); maintaining 200 °C constant for 10 min; reduction of temperature to 2 °C; finally, increase of temperature to 200 °C at a rate of 10 °C/min. The sample crystallization temperature was measured.

2.8. Thermogravimetric Analysis

Thermogravimetric analysis (TGA) is a method used to study the composition and thermal stability of materials. Results were obtained by measuring changes in the sample weight. After data were obtained, DTG (differential TG) could be analyzed to determine the degree of change in the sample weight as the temperature increased (i.e., mass loss rate). A thermal analyzer (STA 409PC, Netzsch Company, Erlangen, Germany) was used, wherein both the test samples and the balance were under nitrogen flow. Subsequently, TGA curves were obtained.

The test parameter setting is given as follows: nitrogen was used as shielding gas with a flow rate of 70 mL/min; starting from room temperature, the temperature was increased at a rate of 10 °C /min, and the heating was stopped when it reached 600 °C. Origin software was used to make a diagram for temperature–mass loss ratio and for temperature–DTG data. At the same time, the following were listed: initial degradation temperature, maximum temperature, and total weight loss rate.

2.9. Oxygen Barrier Properties

The penetration of small gas molecules through defect-free films is a molecular diffusion process. First, small gas molecules would be adsorbed and dissolved on the film. Under the action of concentration gradient, the gas molecules would then diffuse from a high concentration to a low concentration, when the gas concentration increases to a certain level. Finally, they would diffuse out on the other side of the film. A differential pressure gas permeation meter (VAC-V2, Labthink Instrument Co. Ltd., Jinan, China) in proportional mode was used to test PLA/CSDG and MPLA/CSDG composites in terms of their oxygen barrier performance, which is expressed as oxygen permeability. Its unit was cm^3/m^2 d Pa.

The test parameter setting was as follows: temperature of test chamber was set to 25 °C, film thickness = 0.08 mm, test environment contained oxygen + nitrogen, judgment ratio = 10%, and relative humidity was 100%.

2.10. Water Vapor Barrier Properties

A water vapor transmission (WVT) rate test system (W3/060, Labthink Instrument Co. Ltd.; Jinan, China) in standard mode was used to test the water vapor barrier properties of PLA/CSDG and MPLA/CSDG composites. The WVT rate was calculated using the equipment software. The WVT performance includes two meanings: water vapor permeability (WVP) and WVT coefficient. These two meanings are different, but they can both be used to indicate the ability of water vapor to pass through a certain material. The WVT rate indicates the weight of water vapor passing through the material in a certain period of time, under certain temperature and humidity conditions. The WVT coefficient is the standard value of WVT calculated by the system. It is used for comparing different test results and standard values of different samples. The result was expressed as WVP in $g/m^2/d$.

The test parameter setting was as follows: temperature of test chamber, 25 °C; film thickness, 0.08 mm; test interval, 30 min; humidity, 50%.

2.11. Contact Angle Test

A contact angle measuring instrument (JC2000D, Shanghai Zhongchen Digital Technology Equipment Co. Ltd., Shanghai, China) was used to test and characterize PLA/CSDG and MPLA/CSDG composites in terms of their hydrophilic properties. Contact angle is a measure of the degree of wetting a solid with a liquid. If the test value is less than 90°, the surface of the sample is hydrophilic, and the smaller the value, the better the wettability of the liquid to the sample. If the test value is greater than 90°, the sample surface is hydrophobic, and the larger the value, it means that it is not easy for the liquid to wet the sample.

A microsyringe was used to extract 2 μL of distilled water, which was dropped on a sample surface. Contact angles were recorded following a five-point fitting method. For all samples, three measurements were taken, and the average was calculated.

The test parameter setting was as follows: sample thickness, 0.08 mm; test time was 0, 3, and 5 s; repetition for each test, 3 times.

2.12. Water Absorption

Water uptake (WU) was measured according to standard methods [29]. Samples were cut into 2 cm × 2 cm and dried to constant weight in an oven at 105 °C. Before tests were done, samples were weighed and soaked in distilled water for 24 h at room temperature. Then, they were taken out of the water, rid of adhering water drops, and weighed. Water absorption or *WU (g/g)* was calculated using Equation (4). The sample weight before soaking was represented as m_0 (g), and m_f was the sample weight after soaking (g).

$$WU(g/g) = \frac{m_f - m_0}{m_0} = 100\% \quad (4)$$

2.13. Degradation Performance Test

The dimension of samples was 2 cm × 2 cm. They were dried in an oven at 105 °C until constant weights were recorded. Before the degradation test was conducted, samples were weighed and buried in soil at room temperature. The test cycle was 180 days. Samples were retrieved from the soil every 30 days, and they were washed, dried, and weighed. The degradation rate was calculated according to Equation (5). C_0 was the initial sample weight (g), and C_t was the sample weight after the degradation (g).

$$\text{degradation rate } (\%) = \frac{C_0 - C_t}{C_0} \times 100\% \quad (5)$$

3. Results and Discussion

3.1. Fourier Transform Infrared Spectra

Figure 1 represents the infrared spectra of PLA, MPLA, and MPLA/CSDG composites. The infrared characterization of these three materials at around 2993 and 2940 cm^{-1} wavenumbers gave a description of symmetric and asymmetric vibration peaks of −CH− in the PLA structure. Pure PLA and the composite material had an obvious absorption peak at about 2357 cm^{-1}, corresponding to the characteristic absorption peak of carbonyl CO_2, which may be due to the inevitable air exposure during the process of placing the sample into the detector. The wavenumber at 1430 cm^{-1} corresponded to -CH- scissor bending vibration, and that at 1370 cm^{-1} referred to the characteristic absorption peak of $-CH_3-$. Absorption peaks at 1185, 1132, 1092, and 1045 cm^{-1} denoted −C−O− stretching vibration peaks. Due to the presence of hydroxyl at the other end of the PLA molecular structure, there would be a weak absorption peak at 3450 cm^{-1}, which is not evident (Figure 1). The main reason is that the hydroxyl absorption peak intensity of PLA was relatively small. Figure 1 does not show it clearly, but when biomass fillers that contained a large number of hydroxyl groups were added to the MPLA matrix, the peak position of the hydroxyl groups slightly shifted to the right. This implies that the hydroxyl groups on the surface of the biomass fillers could interact with PLA. That is, the molecular segments of MPLA and CSDG may be bound together. On the other hand, the position of the characteristic carbonyl peak in PLA was about 1757 cm^{-1}, and the characteristic peak of carbonyl in the MPLA composite shifted to the right. This also implies that the hydroxyl group on the surface of MPLA could interact with the carbonyl group in PLA. On the basis of the above analysis, along with the research by Wu and Tsou [30] on PLA and rice husk regarding a modification treatment through the use of a coupling agent, the PLA component indicated a behavior similar to the shift of carbonyl characteristic peaks. This interaction could improve the related properties of biomass filler and PLA composite materials.

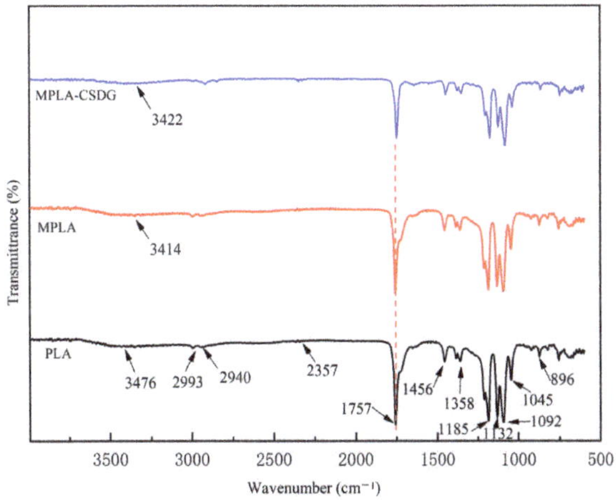

Figure 1. Fourier transform infrared spectra of PLA, MPLA, and MPLA/CSDG.

3.2. Data on Mechanical Properties

Figure 2 and Table 2 provide test results on the influence of different amounts of CSDG on the mechanical properties of PLA/CSDG and MPLA/CSDG composites. Figure 2a depicts tensile strength as a function of the CSDG content, Figure 2b presents data on elongation at break, and Figure 2c plots stress–strain curves. The tensile strength of pure PLA was indicated to be 43.2 MPa. When the content of CSDG was 10%, the tensile strength of the composite material was reduced to 22.15 MPa. At 20% CSDG, the tensile strength increased to 29.34 MPa, which was the highest value. When the amount of added

CSDG reached 50%, the tensile strength was significantly reduced to 18.5 MPa. Compared with the overall strength of pure PLA, that of the PLA/CSDG composites was lower. In the tensile strength data for MPLA/CSDG composites, the tensile strength of the MPLA composite with 10% CSDG was 38.73 MPa. When the CSDG content was increased to 20% and 30%, the tensile strength increased, and the maximum value of 52.68 MPa was reached at the 30% content. However, when 40% CSDG was added, the tensile strength was greatly reduced to 30.2 MPa. Stress–strain curves in Figure 2c indicated a significant difference between PLA and MPLA composites containing 30% CSDG—the tensile strength and elongation at break of the MPLA/CSDG composites were significantly higher than those of the PLA/CSDG composites.

Table 2. Data on mechanical properties of PLA, PLA/CSDG, MPLA, and MPLA/CSDG.

Sample	Tensile Strength (MPa)	Elongation at Break (%)
PLA	43.20 ± 1.9	5.08 ± 0.1
PLA/CSDG10	22.18 ± 2.9	2.43 ± 0.3
PLA/CSDG20	29.33 ± 2.8	2.80 ± 0.2
PLA/CSDG30	23.11 ± 1.9	3.05 ± 0.1
PLA/CSDG40	21.32 ± 2.2	2.60 ± 0.06
PLA/CSDG50	18.47 ± 2.1	2.35 ± 0.07
MPLA	37.08 ± 0.7	2.71 ± 0.3
MPLA/CSDG10	38.72 ± 1.9	2.75 ± 0.08
MPLA/CSDG20	41.13 ± 3.3	2.61 ± 0.4
MPLA/CSDG30	52.65 ± 2.0	6.02 ± 0.1
MPLA/CSDG40	30.30 ± 2.0	4.05 ± 0.5
MPLA/CSDG50	23.73 ± 1.8	2.72 ± 0.3

When CSDG was added to PLA, the elongation at break showed a trend of an initial large decrease, which was followed by a slight increase. The elongation at break reached 3.07% for the sample with 30% CSDG, which was higher than that for the other samples but still lower than that of the pure PLA sample. For the MPLA/CSDG composites, it is apparent that the trend of elongation at break differed much from that for the PLA/CSDG composites. When the content was 30% CSDG, the elongation at break for MPLA/CSDG reached 6.01%, which is the highest among the samples, and it was 1.2 times higher than that for pure PLA.

According to the analysis, the tensile strength and elongation at break of the MPLA/CSDG samples were higher overall than those of PLA/CSDG. These data verify the results from the FTIR analysis, which indicated that the mechanical properties of composite materials became stronger due to the internal interaction. As illustrated by the SEM morphology, the compatibility between CSDG and the PLA matrix became poor when the addition of CSDG was too high, resulting in decreased toughness of the composite material and increased brittleness. In addition, when CSDG increased to a certain level, agglomeration appeared in the film, causing brittle fracture to be more likely to occur [31]. This explanation is consistent with the results obtained by Zhao et al. [32]. In other words, improving the interfacial adhesion is of great significance in enhancing the mechanical properties of composite materials.

3.3. X-ray Diffraction Patterns

With the help of the XRD patterns shown in Figure 3, PLA, MPLA, and two sets of composite materials were analyzed. Figure 3a is the XRD spectra of PLA/CSDG composites, and Figure 3b is the XRD spectra of MPLA/CSDG composites. Different proportions of CSDG in the composites affected the crystallization and the degree of enhancement in the structure and other properties.

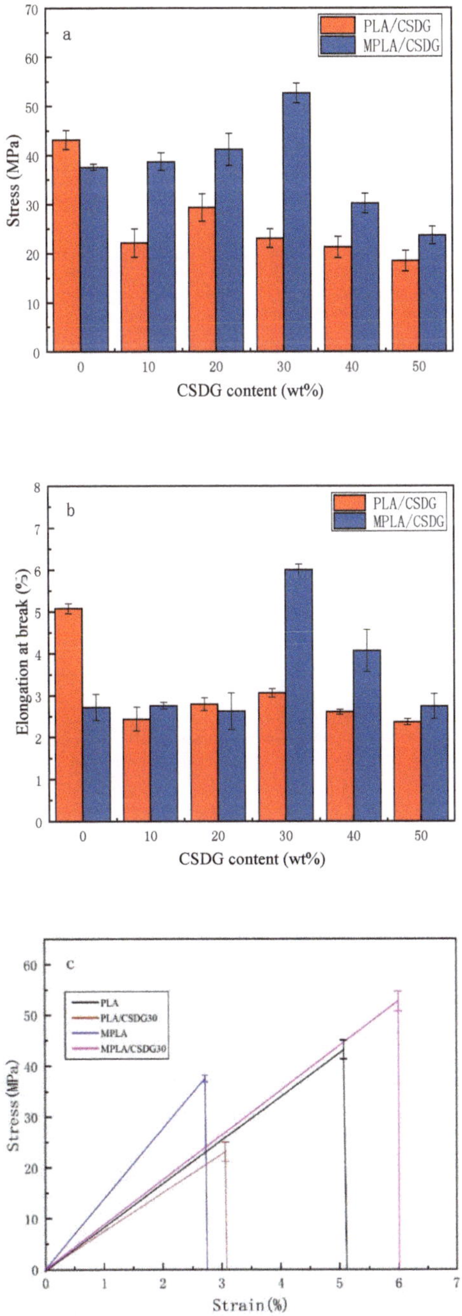

Figure 2. Mechanical properties of PLA/CSDG and MPLA/CSDG composites: (**a**) tensile strength; (**b**) elongation at break; (**c**) stress–strain curves.

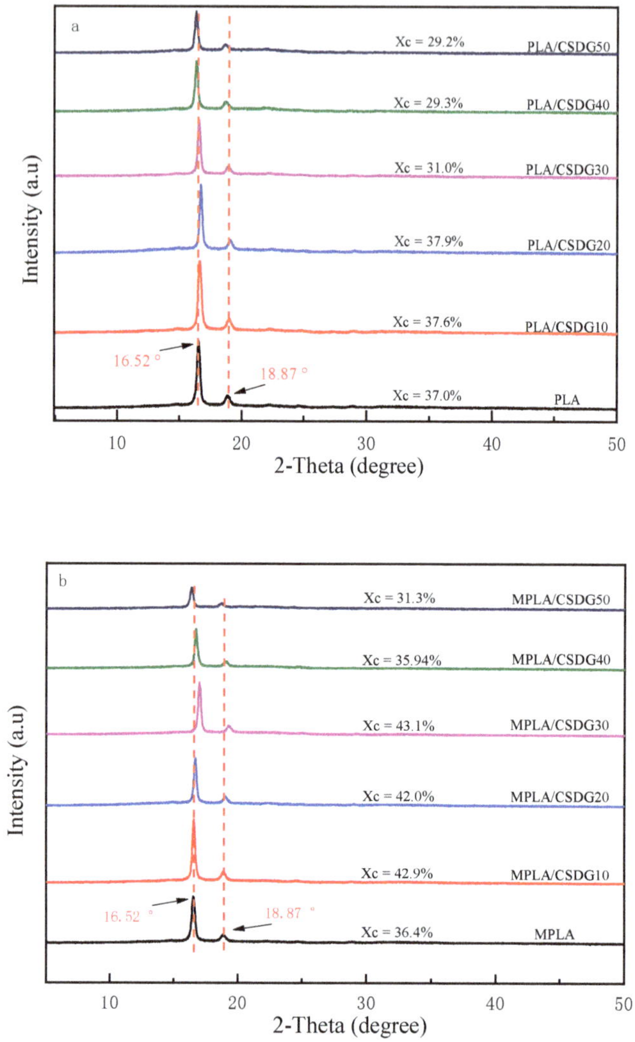

Figure 3. X-ray diffraction patterns: (**a**) PLA/CSDG; (**b**) MPLA/CSDG.

There are obvious sharp diffraction peaks in PLA at $2\theta = 16.52$ and 18.87°, which refer to (200/110) and (203) planes corresponding to typical α crystals of PLA [33,34]. Throughout the spectral curve, PLA/CSDG with 10% and 30% CSDG had almost no shift in peak, indicating that the crystal morphology was basically unchanged [35]. When the content was 20%, PLA/CSDG characteristic peaks shifted slightly to the right. However, when the CSDG content was 40–50%, the characteristic peak gradually shifted to the left. Evident self-aggregation may be the reason for this phenomenon. Similarly, in MPLA/CSDGG samples, the characteristic peak did not shift when the CSDG content was 10%. When the CSDG content was 20–30%, the characteristic peak gradually shifted to the right [36]. When CSDG promoted the completion of crystallization, and as the addition of CSDG was continued, the characteristic peak of MPLA shifted to the left. According to the Bragg equation: $2d\sin\theta = n\lambda$, θ increases when $n\lambda$ does not change, and the value of d decreases, which reduces the interplanar spacing d in a composite material, which may increase the crystallinity and increase the tensile strength. When CSDG was continually added, most CSDG particles occupied the PLA crystal array, so that θ decreased and d increased, resulting in decreased crystallinity. In other words, too many crystal nuclei may hinder the growth of crystals, which may lead to a decline in the degree of crystallization.

3.4. Morphological Images

Generally, when polymers deform under stress, the presence of fillers would produce concentration effects and cause micro-cracks around the polymer. The contact area of fillers with the polymer is large, and when the filler amount is small, micro-cracks would be generated. However, when the filler amount becomes too large, such micro-cracks would turn into macro-cracks, which should lead to poor mechanical properties [37]. Therefore, the shape characteristics of filler, the dispersion in the polymer matrix, and the adhesion between the filler and the matrix all have an important impact on the mechanical properties of composite materials [38–41]. SEM images of the tensile section of PLA/CSDG composites are shown in Figure 4. The surface of pure PLA (Figure 4a) was dense and uniform. It is observed that the fracture surface of PLA/CSDG composites exhibited a rough appearance, and the irregular appearance of CSDG particles interspersed and distributed in the section could be clearly observed. At higher CSDG concentrations (30, 40, 50 wt %), it is observed that the aggregates gradually increased, and it can be established that the dispersibility was better when the filler concentration was lower, and the interfacial effect was relatively good. This means that when the amount of CSDG was higher, poor interfacial adhesion would lead to uneven internal structure of the obtained PLA/CSDG composites, which may lead to deterioration of the mechanical properties [42].

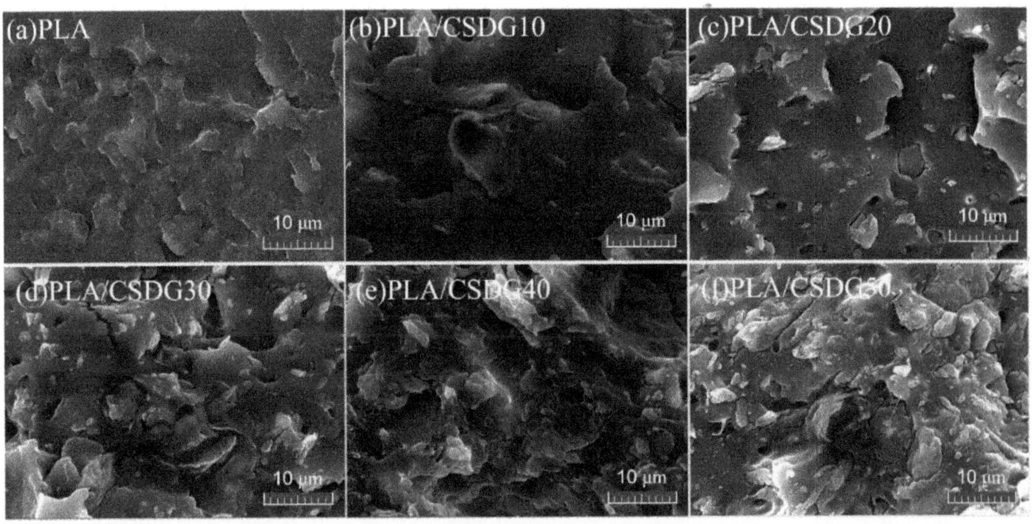

Figure 4. SEM images of PLA and PLA/CSDG composites: (**a**) PLA; (**b**) PLA/CSDG10; (**c**) PLA/CSDG20; (**d**) PLA/CSDG30; (**e**) PLA/CSDG40; (**f**) PLA/CSDG50.

Figure 5 illustrates the tensile section morphology of MPLA/CSDG samples with different amounts of CSDG. Although we can see that there are CSDG particles on the sample surface, MPLA/CSDG showed relatively flat morphology (relative to the rough surface of PLA/CSDG samples). Only a small amount of CSDG aggregates appeared in the matrix, and the distribution was relatively uniform. Nearly no gaps existed between the filler and the matrix, except for the sample with 20 wt % CSDG (Figure 5c). A small number of pores are visible, which may cause the elongation at break to be lower than that of the other samples. The relatively compact structure of MPLA/CSDG samples indicated that a dense structure was formed between MPLA and CSDG. In addition, when the CSDG content was 40 wt % (Figure 5e) and 50 wt % (Figure 5f), significant agglomeration occurred, which led to a decrease in the compatibility between CSDG and the MPLA matrix [43].

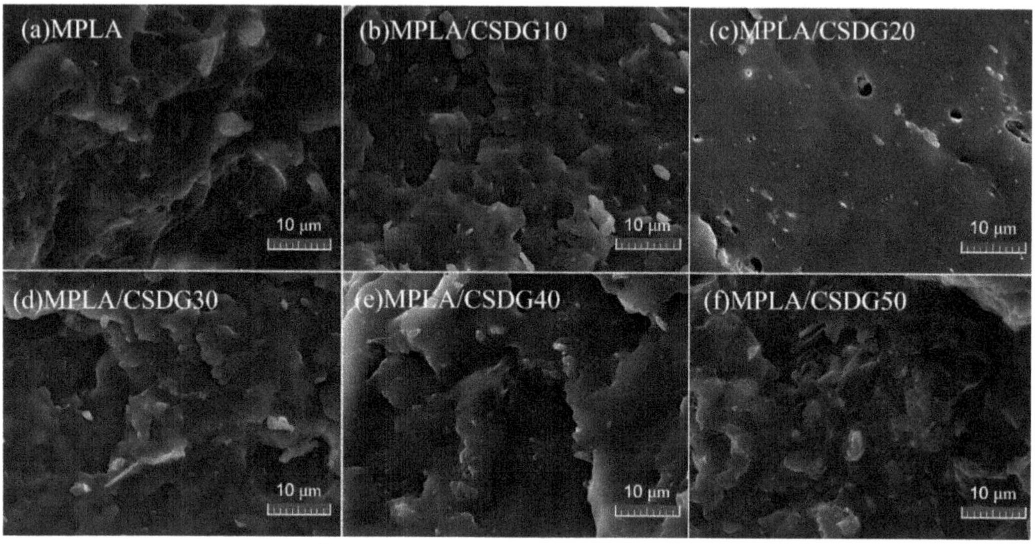

Figure 5. SEM images of MPLA and MPLA/CSDG composites: (**a**) MPLA; (**b**) MPLA/CSDG10; (**c**) MPLA/CSDG20; (**d**) MPLA/CSDG30; (**e**) MPLA/CSDG40; (**f**) MPLA/CSDG50.

3.5. Thermal Stability Analysis

DSC test results are shown in Figure 6, as well as in Table 3. DSC was conducted to determine the glass transition temperature (T_g), melting temperature, recrystallization temperature (T_c), enthalpy of fusion, enthalpy of crystallization, and degree of crystallinity of composite materials. From the cooling curves in Figure 6, PLA and MPLA indicated no crystallization peaks. This is because PLA had very slow crystallization, and the cooling rate of 10 °C/min was too fast, making it too late for PLA to crystallize [44]. With the addition of CSDG, both PLA and MPLA indicated crystallization peaks, which may be attributed to CSDG acting as a nucleating agent for polymers and accelerating the crystallization rate of PLA. From the first heating curves, because the samples were in the initial thermal history, the shape of the curves was irregular, so the sample second heating curves (Figure 7) were considered. In Figure 7, when the CSDG content gradually increased from 0 to 50%, T_g decreased from 68.7 to 58.6 °C, reaching the lowest temperature. The reason for the decrease in T_g may be the presence of many polar groups in CSDG itself. When the number of polar groups in the composite chain exceeded a certain value, the electrostatic repulsion between them exceeded the attractive force, which led to an increase in the distance between molecular chains and a decrease in T_g. At the same time, the overall melting temperature of PLA/CSDG composites during the second heating was slightly lower than that of pure PLA, indicating that the rigidity of the blend was affected to a lesser extent, and the melting temperature was reduced from 170.1 to 168.2 °C (CSDG content increased from 0 to 20%), and it then began to rise (CSDG content, 20–50%). From the analysis, it is believed that when the CSDG content was too much (30%), so that it was not easy for CSDG to disperse in PLA, CSDG produced some subtle aggregation effects, making the crystallization less complete [45,46]. T_c decreased more significantly by 8–17 °C, indicating that CSDG exerted a nucleation effect on PLA [47]. The data for MPLA/CSDG samples showed that T_c and T_g were significantly lower than those for PLA/CSDG. When CSDG was added to PLA, due to the interaction between the hydroxyl groups in the PLA matrix and the filler, the degree of restriction in the movement of molecular chains was increased, hindering the movement of the polymer chains [48]. The existence of more functional groups in MPLA improved the molecular bonding between CSDG and MPLA as well as the uniform dispersion of CSDG in the MPLA matrix. The interaction between the MPLA matrix and the filler improved, free volume in the polymer increased, the degree of hindrance in the movement of molecular chains decreased, and the average chain length between cross-linking points became large, so T_g decreased.

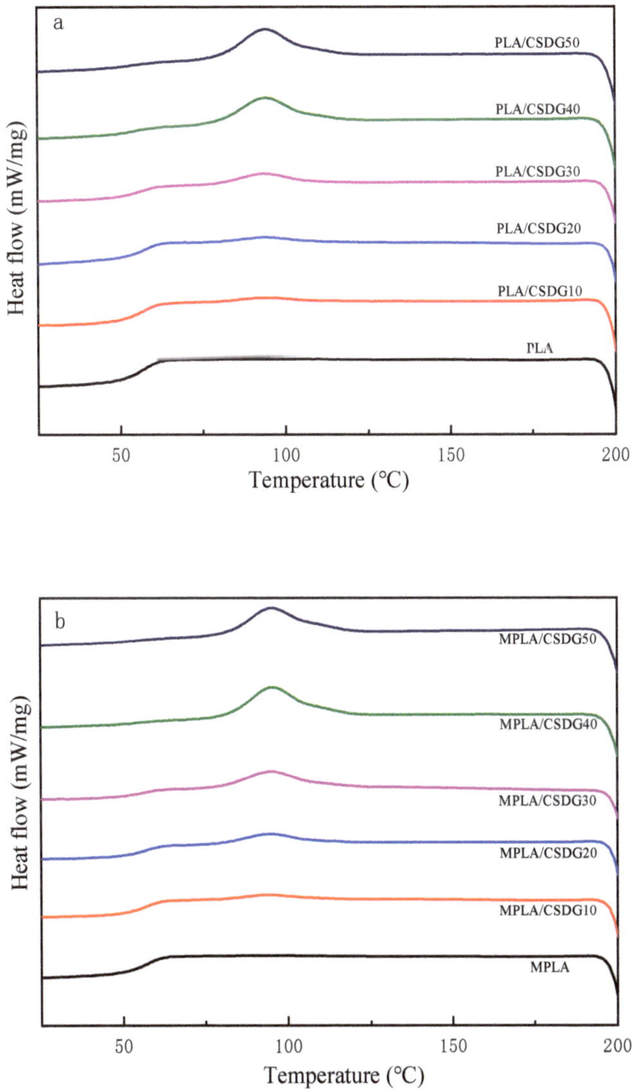

Figure 6. Cooling curves: (**a**) PLA/CSDG; (**b**) MPLA/CSDG.

Table 3. Differential scanning calorimetry data for PLA, PLA/CSDG, MPLA, and MPLA/CSDG.

Sample	Glass Transition Temperature (°C)	Recrystallization Temperature (T_c) (°C)	Enthalpy of Crystallization (J/g)	Melting Temperature (°C)	Melting Enthalpy (J/g)	Crystallinity (%)
PLA	61.7	111.2	29.79	170.1	31.84	37.0
PLA/CSDG10	61.2	102.5	21.67	169.3	32.67	38.74
PLA/CSDG20	61.1	102.8	16.77	168.2	26.12	34.84
PLA/CSDG30	61.1	102.2	11.82	169	24.6	37.51
PLA/CSDG40	59.9	97.9	7.692	169.5	23.1	40.93
PLA/CSDG50	58.6	94.8	3.198	169.7	19.4	41.43
MPLA	48.8	102.3	21.27	165.7	30.3	32.31
MPLA/SDG10	61.1	105.4	21.92	169	32.3	38.33
MPLA/SDG20	60.6	102.7	15.94	168.7	31.2	41.58
MPLA/SDG30	60.5	99.6	8.66	169.5	26.8	43.78
MPLA/SDG40	59.2	94.2	2.87	169.8	24.5	41.54
MPLA/CSDG50	59.5	93.6	1.51	170.1	20.3	41.22

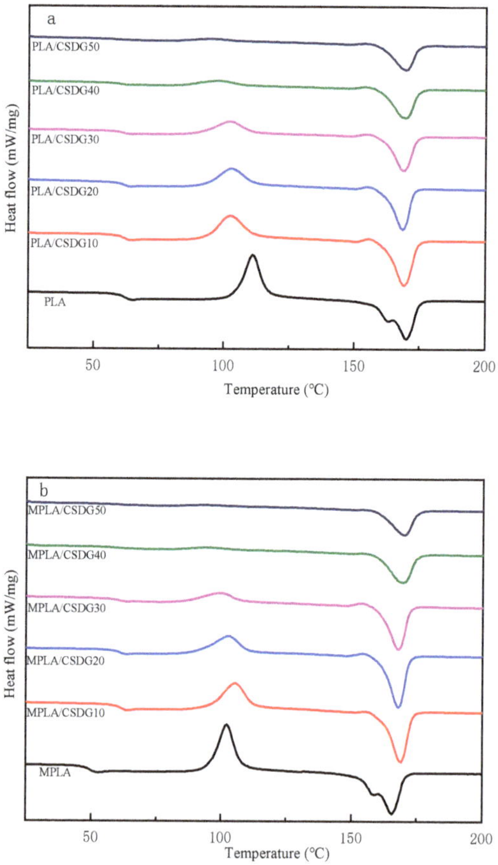

Figure 7. Second heating curves: (**a**) PLA/CSDG; (**b**) MPLA/CSDG.

3.6. Thermogravimetric Analysis

DTG results for PLA/CSDG and MPLA/CSDG composites are shown in Figure 8. Table 4 lists the thermal degradation data. The thermal stability of composite materials was examined through TGA. PLA/CSDG and MPLA/CSDG composites both exhibited three stages of mass loss. First, mass loss started at around 200 °C due to the volatilization of water. The second step involved the loss of PLA at around 350 °C. The mass loss at about 450 °C in the third step was due to the decomposition of composite materials and modifiers. It can be deduced that the addition of CSDG fibers significantly reduced the initial decomposition temperature of the composite materials. A previous researcher analyzed the thermal stability of PLA/ramie composites and found that the addition of ramie fibers reduced the initial decomposition temperature of the composites [49], which is consistent with the finding from this present experiment. The initial degradation temperature of PLA/CSDG decreased with increase in the CSDG content. The temperature at maximum mass loss of PLA/CSDG decreased slightly compared with that of pure PLA. For MPLA/CSDG composites, the temperature at maximum mass loss decreased more significantly, and the initial degradation temperature also decreased substantially. This indicates that unmodified composites maintained better thermal stability, but the thermal stability of both modified and unmodified composites was lower compared with that of pure PLA. The thermal results agree with the formation of aggregates revealed by SEM images. Similar observations were detected for biopolymer matrices filled with coffee grounds [50] and inorganic clay nanoparticles [51]. In general, the clustering of fillers generates a reduction of the polymer thermal stability.

Table 4. Differential thermogravimetric data for PLA, PLA/CSDG, MPLA, and MPLA/CSDG.

Sample	Initial Degradation Temperature (°C)	Temperature at Maximum Mass Loss (°C)	Mass Loss Rate (%)
PLA	311.98 ± 1.7	368.38 ± 0.9	67.33 ± 2.2
PLA/CSDG10	290.79 ± 1.5	360.21 ± 1.3	72.37 ± 3.5
PLA/CSDG20	269.85 ± 3.6	349.86 ± 1.1	60.70 ± 3.3
PLA/CSDG30	265.58 ± 0.9	347.89 ± 1.3	65.18 ± 4.1
PLA/CSDG40	257.58 ± 1.7	345.77 ± 0.8	56.52 ± 2.0
PLA/CSDG50	255.88 ± 1.9	342.99 ± 1.5	57.77 ± 2.9
MPLA	307.83 ± 0.4	368.70 ± 0.5	64.60 ± 1.6
MPLA/CSDG10	286.28 ± 1.4	357.61 ± 1.0	60.33 ± 3.0
MPLA/CSDG20	266.40 ± 1.0	349.25 ± 0.6	57.22 ± 2.1
MPLA/CSDG30	264.09 ± 0.7	345.19 ± 0.5	60.67 ± 2.8
MPLA/CSDG40	254.96 ± 1.8	341.13 ± 0.9	57.30 ± 2.5
MPLA/CSDG50	249.77 ± 0.9	340.52 ± 1.4	55.32 ± 2.9

3.7. Analysis of Oxygen Barrier Performance

Figure 9 describes the effect of different amounts of CSDG on the oxygen permeability in PLA/CSDG and MPLA/CSDG composites. The oxygen permeability in pure PLA was 2.626 $cm^3/m^2 \cdot d \cdot Pa$. With the increase in the CSDG content, the trend in oxygen permeability increased. When the CSDG content was 40%, the oxygen permeability in PLA/CSDG increased to 113.725 $cm^3/m^2 \cdot d \cdot Pa$. In particular, when the CSDG content was 50%, at which excessive self-aggregation of CSDG occurred and CSDG and PLA had poor compatibility with each other, the composite material exhibited too many defects and low mechanical properties. Therefore, we infer that due to the presence of a hydrophilic substance, more voids and micro-cracks were formed, resulting in the decreased oxygen barrier performance of PLA/CSDG. However, the oxygen barrier properties of MPLA/CSDG composites improved. When the CSDG content was 10%, the oxygen permeability in MPLA/CSDG dropped to 2.241 $cm^3/m^2 \cdot d \cdot Pa$, which is slightly lower than that in pure PLA film. However, when the CSDG content was greater than 20%, the oxygen permeability started to increase. At the maximum CSDG content of 50%, the oxygen permeability in MPLA/CSDG was 7.551 $cm^3/m^2 \cdot d \cdot Pa$, which was lower than that in PLA/CSDG (see sub-Figure 9). It is clear that MPLA had a significant effect on enhancing the oxygen barrier performance. Due to the acid anhydride group in MPLA, the composite material was able to form a relatively tight network structure and connections, leading to a significant decrease in oxygen permeability and a significant improvement in oxygen barrier performance.

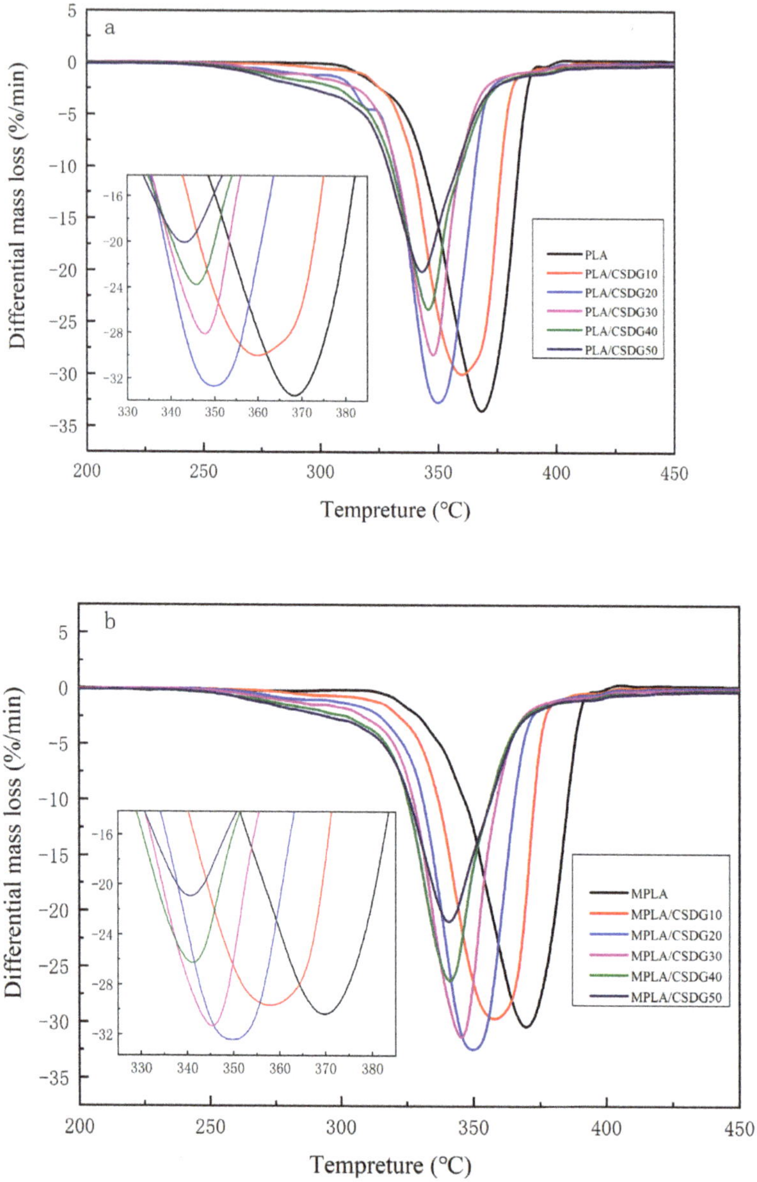

Figure 8. Differential thermogravimetric curves: (**a**) PLA/CSDG; (**b**) MPLA/CSDG.

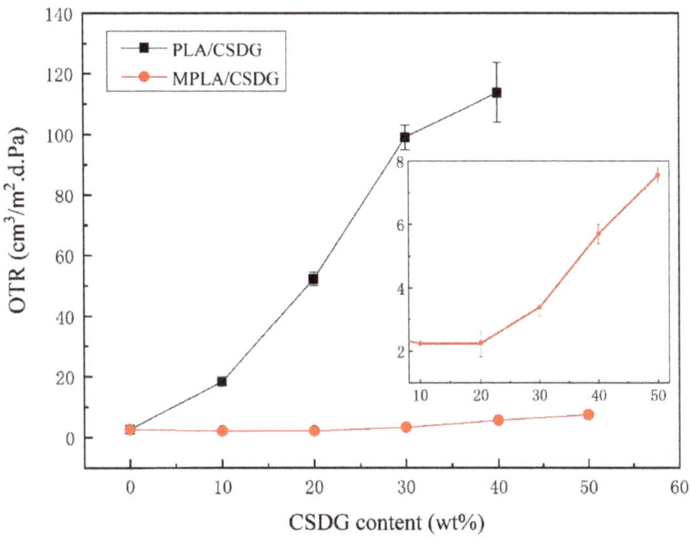

Figure 9. Oxygen barrier properties of PLA/CSDG and MPLA/CSDG composites.

3.8. Analysis of Water Vapor Barrier Properties

Figure 10 plots test results about the influence of different amounts of CSDG on the water vapor permeability in PLA/CSDG and MPLA/CSDG composites. Due to the presence of a large amount of hydrophilic –OH in the molecular chain of PLA, the water-blocking performance of pure PLA film was poor. From the test results, the water vapor permeability in pure PLA was 19.64 $g/m^2/d$. When CSDG was added, the water vapor permeability in PLA/CSDG began to increase. At 50% CSDG, the water vapor permeability in the composite film increased to 250.25 $g/m^2/d$. Therefore, we infer that due to the CSDG hydrophilicity, as indicated from the analysis of electron microscope, excessive CSDG easily led to its agglomeration, and a large number of hydrophilic groups were exposed, thereby reducing the water vapor barrier properties of the composite material [52]. For the case of MPLA, especially when the CSDG content was 10%, the water vapor permeability in MPLA/CSDG dropped to 10.81 $g/m^2/d$. When the added CSDG was 20%, the water vapor permeability in the MPLA/CSDG composite was higher than that in pure PLA film. When CSDG was 50%, the water vapor permeability reached 25.03 $g/m^2/d$. These findings are consistent with the reported barrier properties of PLA-based composites [1]. However, this result for MPLA/CSDG is better than that for the PLA/CSDG composite films. The results showed that the introduction of the acid anhydride group in MPLA improved the adhesion between CSDG and the polymer, significantly improving the water vapor barrier properties of the composite material. However, at the same time, because the CSDG content was too high, agglomeration occurred, resulting in voids in the composite matrix, making the structure loose. This resulted in a decrease in the composite film resistance to water vapor permeability [53,54]. Figure 11 clearly depicts changes in the permeation path of water vapor or oxygen in PLA/CSDG and MPLA/CSDG composites. These findings are consistent with the reported barrier properties of PLA films [55]. However, due to the poor compatibility and poor dispersion of CSDG in the PLA matrix, PLA/CSDG samples failed to form a dense structure and a good barrier effect against water vapor or oxygen. Figure 11b indicates that MPLA/CSDG composites had denser structure, so it was more difficult for water or oxygen to pass through.

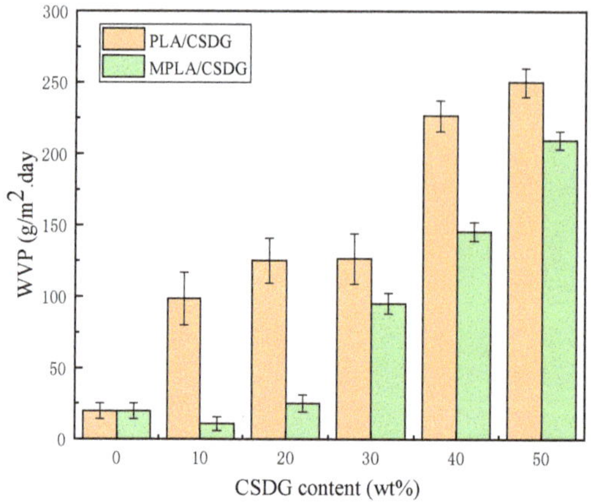

Figure 10. Water vapor barrier properties of PLA/CSDG and MPLA/CSDG composites.

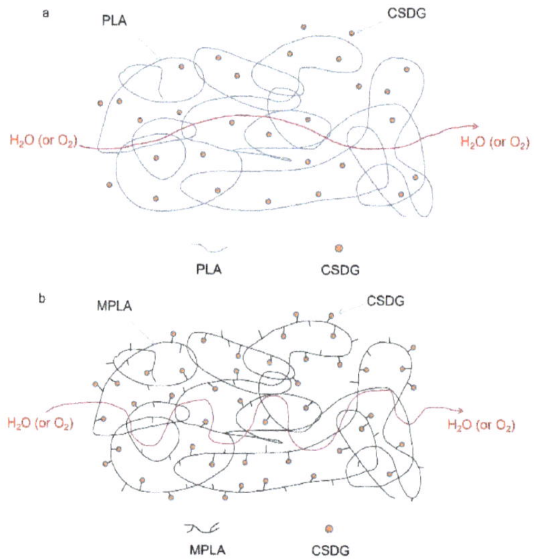

Figure 11. Schematic of molecular permeation path for water vapor or oxygen: (**a**) PLA/CSDG; (**b**) MPLA/CSDG.

3.9. Contact Angle Data

From the data analysis for PLA/CSDG and MPLA/CSDG composites in Figure 12, the contact angle of composites increased significantly. The overall contact angle of MPLA/CSDG was higher than that of unmodified composites. This can be explained by the reaction between the acid anhydride group in MPLA and –OH in CSDG to form a relatively tight network structure, making the material internal connections closer. The binding force at the interface between CSDG and PLA was enhanced, so that the hydrophobicity was also enhanced. With increase in the CSDG content, PLA/CSDG and MPLA/CSDG samples reached the highest contact angle when CSDG was 30% and 20%, respectively. At higher CSDG content, the contact angle began to decrease. The reason is the occurrence of CSDG agglomeration in the PLA matrix, suggesting that a certain maximum content of CSDG would cause the hydrophobicity of PLA to increase. PLA gave an initial contact angle of $73.5 \pm 2.5°$, while PLA/CSDG and MPLA/CSDG samples both showed higher contact angles. The maximum contact angle was close to 90° in the case of PLA/CSDG and 85° for MPLA/CSDG, indicating improvement in the hydrophobicity of PLA.

3.10. Water Absorption Analysis

Test results on water absorption in PLA/CSDG and MPLA/CSDG composite films are indicated in Figure 13. It is clear that the rate of water absorption indicated a rising trend. This is because CSDG contained cellulose, hemicellulose, and lignin—all of them rich in hydroxyl groups—and therefore, materials with CSDG should have high water absorption. This same phenomenon is also reflected in the research by Wen et al. [56] on vinasse and polyethylene. The water absorption in MPLA/CSDG was lower than that in PLA/CSDG samples, and the increase in water absorption tended to be slow, indicating that MPLA/CSDG was more compact, so the water absorption was relatively poor. These results are consistent with the results of contact angle analysis. However, when the filler content was too high, CSDG would agglomerate, and the bond absorption capacity of hydroxyl groups on the surface of CSDG would increase. This may be the main reason for the increase in water absorption in samples with high CSDG content.

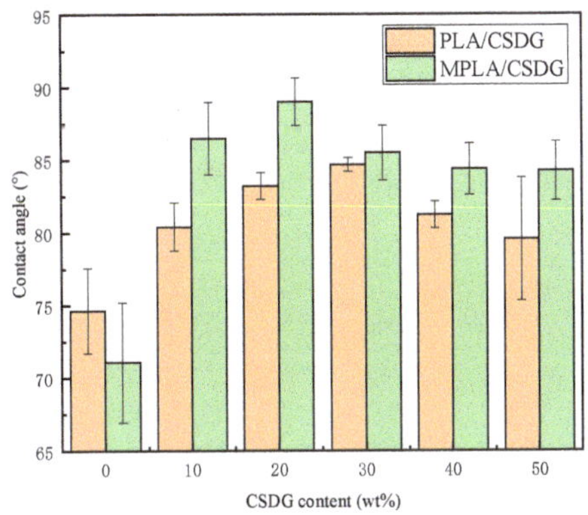

Figure 12. Contact angle of PLA/CSDG and MPLA/CSDG composites.

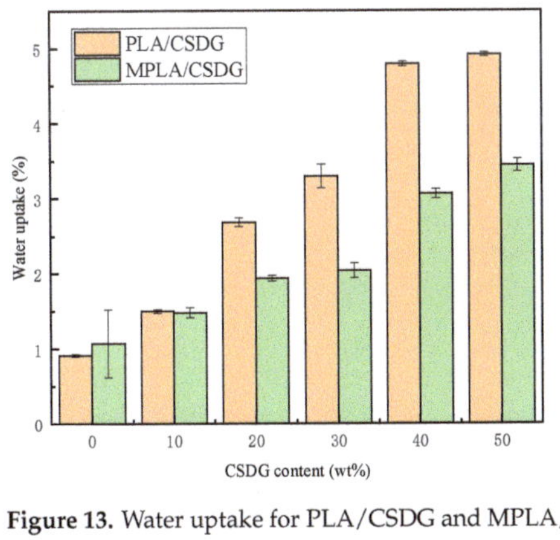

Figure 13. Water uptake for PLA/CSDG and MPLA/CSDG composites.

3.11. Biodegradation Rates

Figure 14 presents a macroanalysis of PLA, MPLA, as well as PLA/CSDG and MPLA/CSDG composites during the process of degradation; changes in the biodegradation rates of different samples with time are shown. For all samples, the degradation could be roughly divided into two stages: initial stage (I) for the first 60 days, and final stage (II) after 60 days. In stage I, the sample biodegradation rate changed little with time. However, in stage II, changes in a short period of time were more observable, indicating that the sample degradation rate was greatly increased. In other words, the samples exhibited a "self-accelerating effect" during the biodegradation process [57]. The soil medium penetrated into the polymer matrix, causing the polymer molecular chains to relax, the chemical bonds to gradually decompose, the molecular weight to decrease, and the material to gradually degrade into oligomers. After 60 days of degradation, there would be more and more free hydroxyl (−OH) and carboxyl (−COOH) groups that accelerated their internal degradation, and further degradation would lead the oligomer to decompose into small molecules, resulting in increased degradation rate for the composite materials in later stages. It was observed that the degradation rate for MPLA/CSDG after 90 days was slightly lower than that for PLA/CSDG. On one hand, the MPLA composite exhibited relatively good internal compatibility. On the other hand, MPLA had reactive acid anhydride groups that would react with the hydroxyl groups on the surface of CSDG and with the carboxyl groups in the polymer matrix to increase the interfacial force between CSDG and MPLA, making the composite material relatively stable. Although the carboxyl group in the PLA matrix could also react with CSDG, the −COOH group was only at the end of the PLA molecular chain, which could not be compared with the large number of anhydride groups in the side chain of MPLA.

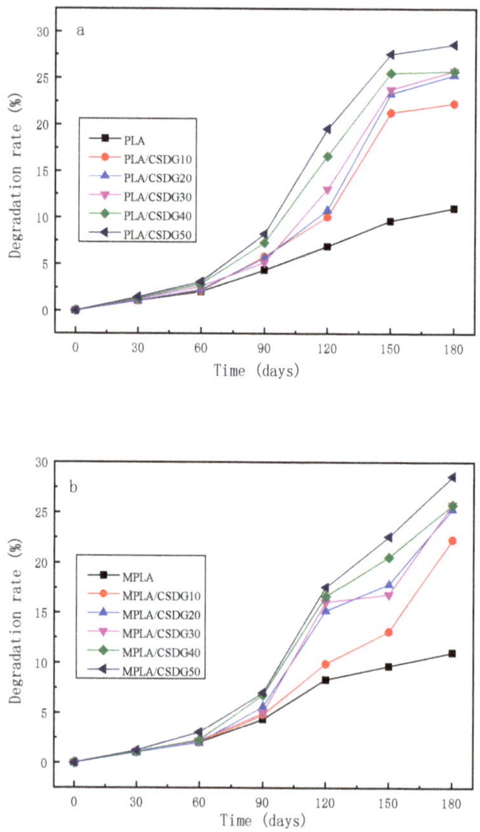

Figure 14. Degradability of (**a**) PLA/CSDG and (**b**) MPLA/CSDG composites.

4. Conclusions

In this study, CSDG was incorporated as a reinforcing filler, and PLA/CSDG composites were prepared by melt blending. To improve the interfacial bonding force between CSDG and PLA, MPLA containing acid anhydride groups was adopted as well. Thus, MPLA/CSDG composites were also fabricated. For the two systems (PLA/CSDG and MPLA/CSDG), mechanical properties, hydrophobicity, gas barrier properties, biocompatibility, and thermodynamic properties were measured and comprehensively analyzed. The analysis revealed that the determining parameters were the dispersion of CSDG and the binding force between PLA and CSDG. Results demonstrated that with increasing CSDG content, the mechanical, barrier, hydrophobic, and thermal degradation properties all had corresponding changes that indicated different trends. The addition of MPLA improved the composite performance. Under the conditions considered in this work, the optimal composition of composites was 20–30% CSDG. In this composition range, MPLA/CSDG composites had good filler dispersion, enhanced hydrophobicity, improved stability, and high mechanical properties. The analysis from biodegradation experiments pointed out that as the degradation progressed, all the samples exhibited a "self-accelerating effect". As the filler content increased, CSDG began to agglomerate, the brittleness of composites increased, and the barrier performance decreased. At the same time, MPLA/CSDG composites delivered stronger performance than PLA/CSDG composites. As CSDG is generally categorized as waste resources, recycling it not only reduces environmental issues but also expands the application of the natural biomass in environment-friendly functional materials.

Author Contributions: Conceptualization, C.-S.W., C.-H.T. and Z.-J.C.; methodology, C.-H.T. and C.-S.W.; software, Z.-J.C., J.G., T.Y., S.C. and R.-Y.W.; validation, C.-S.W. and M.R.D.G.; formal analysis, Z.-J.C., J.G., P.-W.G., Y.L., L.-J.T. and C.-L.Q.; investigation, C.-H.T., J.G., R.-Y.W. and M.-L.T.; resources, M.-L.T., C.-S.W. and C.G.; data curation, J.G., Z.-J.C., T.Y. and C.G.; writing—original draft preparation, C.-H.T. and Z.-J.C.; writing—review and editing, C.-S.W., M.R.D.G. and C.-H.T.; visualization, C.-S.W.; supervision, C.-S.W. and C.-H.T.; project administration, C.-H.T. and C.G.; funding acquisition, C.-S.W., C.-H.T. and C.G. All authors have read and agreed to the published version of the manuscript.

Funding: This research was funded by Sichuan Province Science and Technology Support Program, grant number (2019JDRC0029), by Wuliangye Group Co. Ltd. (CXY2019ZR001) and by the Opening Project of Key Laboratories of Fine Chemicals and Surfactants in Sichuan Provincial Universities (2020JXY04).

Institutional Review Board Statement: Not applicable.

Informed Consent Statement: Not applicable.

Data Availability Statement: The data presented in this study are available on request from the corresponding author.

Acknowledgments: The authors would like to acknowledge the financial support from the following organizations: Wuliangye Group Co. Ltd. (CXY2019ZR001); Sichuan Province Science and Technology Support Program (2019JDRC0029); Zigong City Science and Technology (2017XC16; 2019CXRC01; 2020YGJC13); Opening Project of Material Corrosion and Protection Key Laboratory of Sichuan Province (2017CL03; 2019CL05; 2018CL08; 2018CL07; 2016CL10); Opening Project of Sichuan Province, the Foundation of Introduced Talent of Sichuan University of Science and Engineering (2017RCL31; 2017RCL36; 2017RCL16; 2019RC05; 2019RC07; 2014RC31; 2020RC16); the Opening Project of Key Laboratories of Fine Chemicals and Surfactants in Sichuan Provincial Universities (2020JXY04). Appreciation is also extended to Sichuan Jinxiang Sairui Chemical Co. Ltd.; Apex Nanotek Co. Ltd.; Ratchadapisek Sompote Fund for Postdoctoral Fellowship (Chulalongkorn University).

Conflicts of Interest: The authors declare no conflict of interest.

References

1. Kalia, S.; Kaith, B.S.; Kaur, I. (Eds.) *Cellulose Fibers: Bio-and Nano-Polymer Composites, Green Chemistry and Technology*; Springer Science and Business Media: Berlin, Germany, 2011; Volume 13, pp. 209–217.
2. Guo, J.; Tsou, C.H.; Yu, Y.; Wu, C.S.; Zhang, X.; Chen, Z.; Yang, T.; Ge, F.; Liu, P.; De Guzman, M.R. Conductivity and mechanical properties of carbon black-reinforced poly(lactic acid) (PLA/CB) composites. *Iran. Polym. J.* **2021**. [CrossRef]
3. Wang, M.; Shen, Y.; Jiang, L.; Huang, Y.; Dan, Y. Polylactide materials with ultraviolet filtering function by introducing natural compound. *Polym.-Plast. Technol. Mater.* **2021**, *60*, 1098–1105.
4. Latos-Brozio, M.; Masek, A. Environmentally friendly polymer compositions with natural amber acid. *Int. J. Mol. Sci.* **2021**, *22*, 1556. [CrossRef]
5. Heidari-Asil, S.A.; Zinatloo-Ajabshir, S.; Amiri, O.; Salavati-Niasari, M. Amino acid assisted-synthesis and characterization of magnetically retrievable $ZnCo_2O_4$–Co_3O_4 nanostructures as high activity visible-light-driven photocatalyst. *Int. J. Hydrog. Energy* **2020**, *45*, 22761–22774. [CrossRef]
6. Mousavi-Kamazani, M.; Zinatloo-Ajabshir, S.; Ghodrati, M. One-step sonochemical synthesis of $Zn(OH)_2/ZnV_3O_8$ nanostructures as a potent material in electrochemical hydrogen storage. *J. Mater. Sci. Mater. Electron.* **2020**, *31*, 17332–17338. [CrossRef]
7. Ghodrati, M.; Mousavi-Kamazani, M.; Zinatloo-Ajabshir, S. $Zn_3V_3O_8$ nanostructures: Facile hydrothermal/solvothermal synthesis, characterization, and electrochemical hydrogen storage. *Ceram. Int.* **2020**, *46*, 28894–28902. [CrossRef]
8. Ashrafi, S.; Mousavi-Kamazani, M.; Zinatloo-Ajabshir, S.; Asghari, A. Novel sonochemical synthesis of $Zn_2V_2O_7$ nanostructures for electrochemical hydrogen storage. *Int. J. Hydrog. Energy* **2020**, *45*, 21611–21624. [CrossRef]
9. Zinatloo-Ajabshir, S.; Mousavi-Kamazani, M. Effect of copper on improving the electrochemical storage of hydrogen in CeO_2 nanostructure fabricated by a simple and surfactant-free sonochemical pathway. *Ceram. Int.* **2020**, *46*, 26548–26556. [CrossRef]
10. Ouyang, W.; Huang, Y.; Luo, H.; Wang, D. Poly (lactic acid) blended with cellulolytic enzyme lignin: Mechanical and thermal properties and morphology evaluation. *J. Polym. Environ.* **2012**, *20*, 1–9. [CrossRef]
11. Mina, M.F.; Beg, M.D.H.; Islam, M.R.; Nizam, A.; Alam, A.K.M.M.; Yunus, R.M. Structures and properties of injection-molded biodegradable poly (lactic acid) nanocomposites prepared with untreated and treated multiwalled carbon nanotubes. *Polym. Eng. Sci.* **2014**, *54*, 317–326. [CrossRef]
12. Jamshidian, M.; Tehrany, E.A.; Imran, M.; Akhtar, M.J.; Cleymand, F.; Desobry, S. Structural, mechanical and barrier properties of active PLA–antioxidant films. *J. Food Eng.* **2012**, *110*, 380–389. [CrossRef]
13. Petinakis, E.; Yu, L.; Edward, G.; Dean, K.; Liu, H.; Scully, A.D. Effect of matrix–particle interfacial adhesion on the mechanical properties of poly (lactic acid)/wood-flour micro-composites. *J. Polym. Environ.* **2009**, *17*, 83–94. [CrossRef]
14. Wang, X.; Wang, Y.; Hou, H.; Wang, J.; Hao, C. Ultrasonic method to synthe-size glucan-g-poly(acrylic acid)/sodium lignosulfonate hydrogelsand studies of their adsorption of Cu^{2+} from aqueous solution. *ACS Sustain. Chem. Eng.* **2017**, *5*, 6438–6446. [CrossRef]
15. Azwa, Z.N.; Yousif, B.F.; Manalo, A.C.; Karunasena, W. A review on the degradability of polymeric composites based on natural fibres. *Mater. Des.* **2013**, *47*, 424–442. [CrossRef]
16. Zhou, Y.; Fan, M.; Chen, L. Interface and bonding mechanisms of plant fibre composites: An overview. *Compos. Part B* **2016**, *101*, 31–45. [CrossRef]
17. Lee, J.; Park, S.; Roh, H.G.; Oh, S.; Kim, S.; Kim, M.; Park, J. Preparation and characterization of super absorbent polymers based on starch aldehyde acrylic es and car-boxymethyl cellulose. *Polymers* **2018**, *10*, 605. [CrossRef]
18. Zhao, Y.; Qiu, J.; Feng, H.; Zhang, M.; Lei, L.; Wu, X. Improvement of tensile and thermal properties of poly (lactic acid) composites with admicellar-treated rice straw fiber. *Chem. Eng. J.* **2011**, *173*, 659–666. [CrossRef]
19. Yaacab, N.D.; Ismail, H.; Ting, S.S. Potential Use of Paddy Strawas Filler in Poly Lactic Acid/Paddy Straw Powder Biocomposite: Thermal and Therm alProperties. *Procedia Chem.* **2016**, *19*, 757–762. [CrossRef]
20. Way, C.; Dean, K.; Wu, D.Y.; Palombo, E.A. Palombo Poly(lactic acid) Composites Utilising Sequential Surface Treatments of Lignocellulose: Chemistry, Morphology and Properties. *J. Polym. Environ.* **2011**, *19*, 849–862. [CrossRef]
21. Yusoff, R.B.; Takagi, H.; Nakagaito, A.N. Tensile and flexural properties of poly(lactic acid)-based hybrid green composites reinforced by kenaf, bamboo and coir fibers. *Ind. Crop. Prod.* **2016**, *94*, 562–573. [CrossRef]
22. Sujaritjun, W.; Uawongsuwan, P.; Pivsa-Art, W.; Hamada, H. Mechanic property of surface modified natural fiber reinforced PLA biocomposites. *Energy Procedia* **2013**, *34*, 664–672. [CrossRef]
23. Bourmaud, A.; Pimbert, S. Investigations on Mechanical Properties of Poly(Propylene) and Poly(Lactic Acid)Reinforced by Miscanthus Fibers. *Compos. Part A* **2008**, *39*, 1444–1454. [CrossRef]
24. Zhang, L.; Lv, S.; Sun, C.; Wan, L.; Tan, H.; Zhang, Y. Effect of MAH-g-PLA on the Properties of Wood Fiber/Poly(lactic acid) Composites. *Polymers* **2017**, *9*, 591. [CrossRef] [PubMed]
25. Bertini, F.; Canetti, M.; Cacciamani, A.; Elegir, G.; Orlandi, M.; Zoia, L. Effect of ligno-derivatives on thermal properties and degradation behavior of poly (3-hydroxybutyrate)-based biocomposites. *Polym. Degrad. Stab.* **2012**, *97*, 1979–1987. [CrossRef]
26. Lu, H.; Madbouly, S.A.; Schrader, J.A.; Srinivasan, G.; McCabe, K.G.; Grewell, D.; Graves, W.R. Biodegradation Behavior of Poly(lactic acid) (PLA)/Distiller's Dried Grains with Solubles (DCSDGS) Composites. *ACS Sustain. Chem. Eng.* **2014**, *2*, 2699–2706. [CrossRef]

27. Chi-Hui, T.; Wei-Song, H.; Chin-San, W.; Jui-Chin, C.; Huang, C.; Shih-Hsuan, C.; Chih-Yuan, T.; Wei-Hua, Y.; Shang-Ming, L.; Chih-Kuei, C.; et al. New composition of maleic-anhydride-grafted poly (lactic acid)/rice husk with methylenediphenyl diisocyanate. *Mater. Sci.* **2014**, *20*, 446–451.
28. Tsou, C.Y.; Wu, C.L.; Tsou, C.H.; Chiu, S.H.; Suen, M.C.; Hung, W.S. Biodegradable composition of poly(lactic acid) from renewable wood flour. *Polym. Sci. Ser. B* **2015**, *57*, 473–480. [CrossRef]
29. ASTM D1037-12. *Standard Test Methods for Evaluating Properties of Wood-Base Fiber and Particle Panel Materials*; ASTM International: West Conshohocken, PA, USA, 2012.
30. Wu, C.S.; Tsou, C.H. Fabrication, characterization, and application of biocomposites from poly (lactic acid) with renewable rice husk as reinforcement. *J. Polym. Res.* **2019**, *26*, 44. [CrossRef]
31. Tang, G.; Jiang, Z.G.; Li, X.; Zhang, H.B.; Hong, S.; Yu, Z.Z. Electrically conductive rubbery epoxy/diamine-functionalized graphene nanocomposites with improved mechanical properties. *Compos. Part B* **2014**, *67*, 564–570. [CrossRef]
32. Xin, Z.; Qinghua, Z.; Dajun, C. Enhanced mechanical properties of graphene-based poly(vinylalcohol) composites. *Macromolecules* **2010**, *43*, 2357–2363.
33. Faria, D.L.; Júnior, L.M.; de Almeida Mesquita, R.G.; Júnior, M.G.; Pires, N.J.; Mendes, L.M.; Junior, J.B.G. Production of castor oil-based polyurethane resin composites reinforced with coconut husk fibres. *J. Polym. Res.* **2020**, *27*, 249. [CrossRef]
34. Zheng-Lu, M.; Chi-Hui, T.; You-Li, Y.; De Guzman, M.R.; Chin-San, W.; Chen, G.; Tao, Y.; Zhi-Jun, C.; Zeng, R.; Yu, L.; et al. Thermal Properties and Barrier Performance of Antibacterial High-Density Polyethylene Reinforced with Carboxyl Graphene-Grafted Modified High-Density Polyethylene. *Ind. Eng. Chem.* **2021**. [CrossRef]
35. Pan, P.; Liang, Z.; Zhu, B.; Dong, T.; Inoue, Y. Blending Effects on Polymorphic Crystallization of Poly(L-Lactide). *Macromolecules* **2009**, *42*, 3374–3380. [CrossRef]
36. Yang, T.C.; Wu, T.L.; Hung, K.C.; Chen, Y.L.; Wu, J.H. Mechanical properties and extended creep behavior of bamboo fiber reinforced rec ycled poly (lactic acid) composites using thetime-temperature superposition principle. *Constr. Build. Mater.* **2015**, *93*, 558–563. [CrossRef]
37. Guo, J.; Tsou, C.H.; De Guzman, M.R.; Wu, C.S.; Zhang, X.; Chen, Z.; Wen, Y.-H.; Yang, T.; Zhuang, Y.-J., Ge, F.; et al. Preparation and characterization of bio-based green renewable composites from poly (lactic acid) reinforced with corn stover. *J. Polym. Res.* **2021**, *28*, 199. [CrossRef]
38. Lu, W.; Gramlich, W.M.; Gardner, D.J. Improving the impact strength of Poly(lactic acid) (PLA) in fused layer modeling (FLM). *Polymer* **2017**, *114*, 242–248.
39. Zhou, Y.; Lei, L.; Yang, B.; Li, J.; Ren, J. Preparation of PLA-based nanocomposites modified by nano-CBapulgite with good toughness-strength balance. *Polym. Test.* **2017**, *60*, 78–83. [CrossRef]
40. Qi, Z.; Ye, H.; Xu, J.; Peng, J.; Chen, J.; Guo, B. Synthesis and characterizations of attapulgite reinforced branched poly (butylene succinate) nanocomposites. *Colloids Surf. A* **2013**, *436*, 26–33. [CrossRef]
41. Zhao, L.; Liu, P.; Liang, G.; Gu, A.; Yuan, L.; Guan, Q. The origin of the curing behavior, mechanical and thermal properties of surface functionalized attapulgite/bismaleimide/diallylbisphenol composites. *Appl. Surf. Sci.* **2014**, *288*, 435–443. [CrossRef]
42. Lu, T.; Liu, S.; Jiang, M.; Xu, X.; Wang, Y.; Wang, Z.; Gou, J.; Hui, D.; Zhou, Z. Effects of modifications of bamboo cellulose fibers on the improved mechanical properties of cellulose reinforced poly (lactic acid) composites. *Compos. Part B* **2014**, *62*, 191–197. [CrossRef]
43. Tsou, C.H.; Yao, W.H.; Wu, C.S.; Tsou, C.Y.; Hung, W.S.; Chen, J.C.; De Guzman, M.R. Preparation and characterization of renewable composites from Polylactide and Rice husk for 3D printing applications. *J. Polym. Res.* **2019**, *26*, 227–237. [CrossRef]
44. Ge, F.F.; Tsou, C.H.; Yuan, S.; De Guzman, M.R.; Zeng, C.Y.; Li, J.; Jia, C.F.; Cheng, B.Y.; Yang, P.C.; Gao, C. Barrier Performance and Biodegradability of Antibacterial Poly (butylene adipate-co-terephthalate) Nanocomposites Reinforced with a New MWCNT-ZnO Nanomaterial. *Nanotechnology* **2021**. [CrossRef] [PubMed]
45. Carballeira, P.; Haupert, F. Toughening effects of titanium dioxide nanoparticles on TiO_2/epoxy resin nanocomposites. *Polym. Compos.* **2010**, *31*, 1241–1246. [CrossRef]
46. Aliotta, L.; Cinelli, P.; Coltelli, M.B.; Righetti, M.C.; Gazzano, M.; Lazzeri, A. Effect of nucleating agents on crystallinity and properties of Poly (lactic acid) (PLA). *Eur. Polym. J.* **2017**, *93*, 822–832. [CrossRef]
47. Lee, S.; Hong, J.Y.; Jang, J. The effect of graphene nanofiller on the crystallization behavior and mechanical properties of poly (vinyl alcohol). *Polym. Int.* **2013**, *62*, 901–908. [CrossRef]
48. Dittanet, P.; Pearson, R.A. Effect of silica nanoparticle size on toughening mechanisms of filled epoxy. *Polymer* **2012**, *53*, 1890–1905. [CrossRef]
49. Mukherjee, T.; Kao, N. PLA based biopolymer reinforced with natural fibre: A review. *J. Polym. Environ.* **2011**, *19*, 714. [CrossRef]
50. Cataldo, V.A.; Cavallaro, G.; Lazzara, G.; Milioto, S.; Parisi, F. Coffee grounds as filler for pectin: Green composites with competitive performances dependent on the UV irradiation. *Carbohydr. Polym.* **2017**, *170*, 198–205. [CrossRef]
51. Lisuzzo, L.; Cavallaro, G.; Milioto, S.; Lazzara, G. Effects of halloysite content on the thermo-mechanical performances of composite bioplastics. *Appl. Clay Sci.* **2020**, *185*, 105416. [CrossRef]
52. Yang, Y.H.; Bolling, L.; Priolo, M.A.; Grunlan, J.C. Super gas barrier and selectivity of graphene oxide-polymer multilayer thin films. *Adv. Mater.* **2013**, *25*, 503–508. [CrossRef] [PubMed]
53. Li, X.; Zhang, Y.J.; Tsou, C.H.; Wen, Y.H. A new application of hollow nano-silica added to modified polypropylene to prepare nanocomposite films. *NANO Brief Rep. Rev.* **2021**, *16*, 2150117. [CrossRef]

54. Yao, Y.L.; De Guzman, M.R.; Duan, H.; Gao, C.; Lin, X.; Wen, Y.H.; Du, J.; Lin, L.; Chen, J.C.; Wu, C.S.; et al. Infusing high-density polyethylene with graphene-zinc oxide to produce antibacterial nanocomposites with improved properties. *Chin. J. Polym. Sci.* **2020**, *38*, 898–907. [CrossRef]
55. Ramos, M.; Jiménez, A.; Peltzer, M.; Garrigós, M.C. Development of novel nano-biocomposite antioxidant films based on poly (lactic acid) and thymol for active packaging. *Food Chem.* **2014**, *162*, 149–155. [CrossRef] [PubMed]
56. Wen, Y.; Tsou, C.H.; Gao, C.; Chen, J.C.; Guzman, M. Evaluating distillers grains as bio-fillers for high-density polyethylene. *J. Polym. Res.* **2020**, *27*, 167. [CrossRef]
57. Liu, H.; Song, W.; Chen, F.; Guo, L.; Zhang, J. Interaction of microstructure and interfacial adhesion on impact performance of polylactide (PLA) ternary blends. *Macromolecules* **2011**, *44*, 1513–1522. [CrossRef]

polymers MDPI

Article

Aminated Graphene-Graft-Oligo(Glutamic Acid) /Poly(ε-Caprolactone) Composites: Preparation, Characterization and Biological Evaluation

Mariia Stepanova [1], Olga Solomakha [1], Maxim Rabchinskii [2,*], Ilia Averianov [1], Iosif Gofman [1], Yuliya Nashchekina [3], Grigorii Antonov [2], Aleksey Smirnov [2], Boris Ber [2], Aleksey Nashchekin [2] and Evgenia Korzhikova-Vlakh [1,*]

1 Institute of Macromolecular Compounds, Russian Academy of Sciences, 199004 St. Petersburg, Russia; maristepanova@hq.macro.ru (M.S.); solomanya@bk.ru (O.S.); aaardvark@yandex.ru (I.A.); gofman@imc.macro.ru (I.G.)

2 Ioffe Institute, Politekhnicheskaya st. 26, 194021 St. Petersburg, Russia; antonov@scamt-itmo.ru (G.A.); sab@mail.ioffe.ru (A.S.); boris.ber@mail.ioffe.ru (B.B.); nashchekin@mail.ioffe.ru (A.N.)

3 Institute of Cytology, Russian Academy of Sciences, 194064 St. Petersburg, Russia; ulichka@mail.ru

* Correspondence: rabchinskii@mail.ioffe.ru (M.R.); vlakh@hq.macro.ru (E.K.-V.)

Citation: Stepanova, M.; Solomakha, O.; Rabchinskii, M.; Averianov, I.; Gofman, I.; Nashchekina, Y.; Antonov, G.; Smirnov, A.; Ber, B.; Nashchekin, A.; et al. Aminated Graphene-Graft-Oligo(Glutamic Acid) /Poly(ε-Caprolactone) Composites: Preparation, Characterization and Biological Evaluation. *Polymers* **2021**, *13*, 2628. https://doi.org/10.3390/polym13162628

Academic Editor: Ilaria Armentano

Received: 15 July 2021
Accepted: 4 August 2021
Published: 7 August 2021

Abstract: Biodegradable and biocompatible composites are of great interest as biomedical materials for various regeneration processes such as the regeneration of bones, cartilage and soft tissues. Modification of the filler surface can improve its compatibility with the polymer matrix, and, as a result, the characteristics and properties of composite materials. This work is devoted to the synthesis and modification of aminated graphene with oligomers of glutamic acid and their use for the preparation of composite materials based on poly(ε-caprolactone). Ring-opening polymerization of N-carboxyanhydride of glutamic acid γ-benzyl ester was used to graft oligomers of glutamic acid from the surface of aminated graphene. The success of the modification was confirmed by Fourier-transform infrared and X-ray photoelectron spectroscopy as well as thermogravimetric analysis. In addition, the dispersions of neat and modified aminated graphene were analyzed by dynamic and electrophoretic light scattering to monitor changes in the characteristics due to modification. The poly(ε-caprolactone) films filled with neat and modified aminated graphene were manufactured and carefully characterized for their mechanical and biological properties. Grafting of glutamic acid oligomers from the surface of aminated graphene improved the distribution of the filler in the polymer matrix that, in turn, positively affected the mechanical properties of composite materials in comparison to ones containing the unmodified filler. Moreover, the modification improved the biocompatibility of the filler with human MG-63 osteoblast-like cells.

Keywords: aminated graphene; graphene modification; grafting from; oligomers of glutamic acid; poly-ε-caprolactone; biocompatible polymer composites

1. Introduction

Currently, aliphatic polyesters are the widely used class of biodegradable polymers for the preparation of various biomedical materials [1–3]. The application of poly(lactic acid) (PLA), poly(glycolic acid-co-lactic acid) (PLGA), poly(ε-caprolactone) (PCL) and polyhydroxybutyrate (PHB) in medicine is due to their biocompatibility with human tissues and their ability to biodegrade to nontoxic metabolites [4]. In addition to their applications as drug delivery systems [3] and surgical suture threads [5], they are widely studied as materials for regenerative medicine including scaffolds for bone tissue engineering [6,7]. However, the hydrophobic properties of aliphatic polyesters impair the adhesion of cells while the lack of a sufficient number of the reactive groups limits the possibility of modifying those polymers with bioactive molecules [8,9]. Moreover, the mechanical properties of neat materials based on aliphatic polyesters should be improved so that they can be considered as scaffolds for bone regeneration [10,11].

Among the strategies allowing the regulation of properties of a polymer material, the preparation of biocomposites via the incorporation of biofunctional and/or reinforcing nano- or microparticles in the polymer matrix can be a matter of choice [10,12]. Such particles could improve the mechanical properties [11,13,14] or create the time–spatial distribution of biological ligands [15,16], while the polymer matrix would be degraded in vivo. Moreover, the ability to influence mineralization is one of the key properties of the resulting materials for bone tissue regeneration [16,17]. In this regard, hydroxyapatite [18] or ceramic nanoparticles [19] are often used as fillers both to improve mechanical properties and induce the mineralization of implanted materials.

Recently, it was shown that the application of graphene as a filler can improve the mechanical properties of polymer materials [20,21]. The use of graphene derivatives such as graphene oxide (GO) or reduced graphene oxide (rGO), which contain hydroxylic and carboxylic groups, can also improve the interaction of composites with cells and biomolecules and enhance cell growth, cell differentiation and cell proliferation [22,23]. At the same time, GO and rGO similar to graphene enhances the mechanical properties of polymer materials [24,25]. Recently, Krystyjan et al. and Mohamad et al. prepared the starch/chitosan [26] and chitosan/PLA-based [27] composites containing GO as filler, respectively. In both cases, the addition of GO improved the mechanical properties of polymer materials. Furthermore, the developed polymer composites containing GO showed bacteriostatic activity but were not toxic to human cells. Luo et al. have reported that GO incorporated on electrospun PLGA demonstrated osteoinductive properties when compared to the scaffolds without GO [28].

One of the main drawbacks in preparing of many composites is the aggregation of the fillers in a polymer matrix [29,30]. To overcome this obstacle, many studies have focused on modifying fillers to increase their phase compatibility and homogeneous dispersion state. Surface grafting with oligomers or polymers is a well-known approach to improve the filler distribution in the polymer matrix [31–33]. Recently, we have successfully applied this approach to improve the compatibility of the hydrophobic PLA and PCL with hydrophilic filler (nanocrystalline cellulose) by its grafting with amphiphilic polypeptide [32]. Such modification allowed the enhancement of the composite mechanical characteristics and also favored the matrix mineralization [12,34].

In the case of graphene-based materials, the strong π–π interaction contributes to a pronounced aggregation of graphene and, as a result, its poor distribution in the polymer matrix. Partly, this effect can be overcome by the use of GO or rGO. Furthermore, GO and rGO can give more stable dispersions in water and polar solvents than graphene. In particular, Gu et al. reported the modification of GO with polyamide in ethanol followed by melt spinning [35]. The developed composites provided excellent mechanical properties. Wang et al. proposed a modification of GO with poly(lactic acid) to adjust GO surface properties and enhance phase compatibility with polymer matrix (PLA) [36].

The goal of this study was the development of a technique for the modification of aminated rGO (rGO-Am) with oligomers of glutamic acid (oligo(Glu)) as a moiety with osteoinductive properties [16]. The grafting of oligo(Glu) from the rGO-Am surface was performed by the ring-opening polymerization of N-carboxyanhydride of γ-benzyl ester of glutamic acid initiated by primary amino groups of rGO-Am. The modification of rGO-Am was carefully tested by a number of physicochemical methods. Neat and modified rGO-Am was applied as a filler to prepare PCL-based biocompatible composite films whose mechanical and biological properties were investigated and compared. The biocompatibility of dispersions of the fillers as well as composite films with human MG-63 osteoblast-like cells was studied in order to assess the applicability of the developed composites as potential materials for bone tissue regeneration.

2. Materials and Methods

2.1. Materials

An aqueous dispersion of GO was purchased from Graphene Technologies (Moscow, Russia, www.graphtechrus.com accessed on 30 June 2019). γ-benzyl-L-glutamate (Glu(OBzl)) (>99%), triphosgene (98%) and α-pinene (98%) used for the synthesis of N-carboxyanhydride as well as n-hexylamine (98%) used as an initiator were purchased from Sigma-Aldrich (Darmstadt, Germany). All organic solvents used in this work were purchased from Vecton Ltd. (St. Petersburg, Russia) and distilled before use.

Low-molecular-weight poly(glutamic acid) (PGlu) used for comparison was synthesized in IMC RAS as described earlier [32] and had the following characteristics according to size-exclusion chromatography (SEC): M_w = 10,800, M_n = 7700 and Đ = 1.40. According to ^{1}H NMR, the used PGlu sample contained 15% of residual benzyl groups (PGlu(OBzl)).

Membranes for dialysis with molecular weight cutoff (MWCO) 1000, 3500, 6000−8000 and 12,000–14,000 were purchased from Orange Scientific (OrDialDClean regenerated cellulose dialysis tubing, Anaheim, CA, USA). Vivaspin concentrators used for ultrafiltration were products of Sartorius (Göttingen, Germany).

Human MG-63 osteoblast-like cell line was obtained from the Vertebrate Cell Culture Collection (Institute of Cytology RAS, St. Petersburg, Russia). They were cultured at 37 °C in humidified 5% CO_2 and cultured in EMEM (Lonza, St. Louis, MO, USA) containing 10% fetal bovine serum (FBS; HyClone, Logan, UT, USA), 1% NEAA (FBS; HyClone, Logan, UT, USA) and 1% penicillin/streptomycin (Sigma Aldrich, Darmstadt, Germany). In all, 4~6 passage cells were used for this study. On attaining 80–90% confluency, the cells were trypsinized with trypsin-EDTA (Sigma-Aldrich, Darmstadt, Germany).

2.2. Synthesis of Aminated Graphene

The synthesis of the rGO-Am was carried out according to the previously described method [37]. In brief, GO suspension of 0.05 wt.% concentration was centrifuged at 18,100 g for 20 min (Sigma S-16 centrifuge, Sigma-Aldrich, Darmstadt, Germany), the supernatant was decanted away, and the sediment was transferred into a fluoroplastic cup and rinsed with 46% HBr (Sigma-Aldrich, Darmstadt, Germany). The obtained suspension was further heated at 80 °C for 48 h in air while stirring. Afterward, the synthesized intermediate (brominated graphene) was copiously washed by isopropyl alcohol using a glass filter (of 40 μm of pore size), the sediment was placed into a fluoroplastic cup, and rinsed with a saturated solution of ammonia in isopropyl alcohol. The obtained suspension was stirred for 24 h, resulting in the formation of rGO-Am. To purify the synthesized rGO-Am from the residuals of the reaction mixture, it was washed several times with isopropyl alcohol using a glass filter (of 40 μm of pore size).

2.3. Modification of Aminated Graphene with Oligomers of Glutamic Acid

Two milliliters of distilled and dried ethyl acetate were added to a glass Schlenk tube containing 110 mg of rGO-Am and the mixture was ultrasonicated with an ultrasonic probe UP 50H Hielscher Ultrasonics (Hielscher, Teltow, Germany) at 15–20% power for 1 min. The obtained dispersion was purged with argon for 3 min.

N-carboxyanhydride of γ-benzyl ester of glutamic acid (Glu(OBzl) NCA; monomer) was prepared using a standard protocol as described elsewhere [38,39]. Then, 230 mg of monomer was dissolved in 2 mL of dry ethyl acetate and the solution was purged with argon for 10 min. The resulting solution was added to the rGO-Am dispersion, ultrasonicated with an ultrasonic probe for 1–2 min, sealed, again purged with argon for 10 min, and incubated at 35 °C for 2 days. After this, 5 mL of DMF was added to the reaction mixture. The dispersion was ultrasonicated in the ultrasound bath for 2 min and finally centrifuged at 15,300 g for 15 min. After the removal of the supernatant, a fresh portion of DMF (5 mL) was added to the sediment, and the procedure of redispersion and centrifugation was repeated again. The washing procedure was repeated five times. Finally, the modified rGO-Am was redispersed in 7 mL of methanol, centrifuged as described above, and the sediment was dried in a vacuum at 22 °C for 12 h. The product weight was 103 mg (~30%).

To remove the benzyl protective groups, 2 mL of the DMSO/TFA (1/1, *v/v*) mixture was added to 50 mg of rGO-Am-oligo(Glu(OBzl)). The reaction was carried out in an ice bath (0–5 °C) for 1 h. After this time, the ice bath was removed, 50 mL of DMSO/TFMSA solution (4/1, *v/v*) was added, and the dispersion was stirred for additional 1 h at 22 °C. After that, 6 mL of DMF was added to the resulting dispersion and dialysis against water using a membrane bag with MWCO = 1000 was carried out for 5 days. The purified product was freeze-dried and stored at 4 °C before use. The product weight was 33 mg (~66%).

2.4. Characterization of rGO-Am and Its Derivatives

The survey, C 1s, N 1s and O 1s X-ray photoelectron spectra (XPS) were measured using a Thermo Fisher ESCALAB 250Xi XPS system (Thermo Fisher Scientific, Waltham, MS, USA) with a monochromatic Al Kα X-ray source ($h\nu$ = 1486.68 eV). For all the spectroscopic measurements, the studied materials were deposited onto the silicon wafers by the drop-casting of 20–30 μL of the corresponding isopropyl suspension, of 5×10^{-1} wt.% of concentration with subsequent drying overnight at room temperature. Prior to the measurements, samples were evacuated down to a pressure $P = 10^{-9}$ Torr at least for 20 h to remove all adsorbates. The spectra were calibrated with respect to the Au 4f7/2 line (84.0 eV).

CasaXPS© software (Version 2.3.16Dev52, Casa Software Ltd.; Teignmouth, United Kingdom) was used for the deconvolution and quantification of the acquired C 1s, N 1s and O 1s X-ray photoelectron spectra. All the spectra were fitted with a Shirley background. In the case of C 1s spectra, a set of one asymmetric Doniach-Sunjic function ((DS0.10,100)(GL90)) and six symmetric Gaussian–Lorentzian convoluted functions of 70–30% ratio (GL(30)) was applied. At the same time, three symmetric Gaussian–Lorentzian converged functions of 70–30% ratio (GL(30)) were used for the deconvolution of the N 1s and O 1s spectra. The χ^2 minimization was ensured using the nonlinear least squares routine. Afterward, the C/O ratios and the relative concentration of the carbon atoms in different states were calculated.

Fourier-transform infrared spectroscopy (FTIR) was performed using IRAffinity-1 S Shimadzu (Shimadzu, Kyoto, Japan). The spectra were recorded for 0.2 mg of sample evenly distributed in 20 mg in a KBr tablet.

The hydrodynamic diameter and electrokinetic potential of neat and modified rGO-Am particles were determined by dynamic and electrophoretic light scattering (DLS and ELS, respectively). All measurements were performed in water at concentration of particles equal to 0.5 mg/mL using a Zetasizer Nano-ZS (Malvern Instruments, Malvern, United Kingdom) equipped with a He–Ne laser at 633 nm at scattering angles of 173° and 25°.

The morphology of the studied materials was evaluated with scanning electron microscopy (SEM) using JSM-7001F Jeol microscope (Jeol Ltd., Tokyo, Japan). Optical microscopy in transmitted and reflected light was carried out with the use of the Nikon Eclipse E200 microscope (Nikon Corp., Tokyo, Japan).

Thermogravimetric analysis (TGA) was performed with the use of a DTG-60 Shimadzu (Shimadzu, Kyoto, Japan) in an air atmosphere at a constant heating rate of 5 °C/min. The analysis was performed in the temperature range from 40 to 600 °C using the pre-homogenized samples.

2.5. Synthesis of PCL

Polymerization of ε-caprolactone was carried out in bulk at 130 °C for 20 h. Before polymerization, the weighted portion of the monomer was placed into a Schlenk flask and purged with argon for 10 min. The ratio of monomer to stannous octoate was 3600. After synthesis, the polymer was dissolved in a minimum amount of chloroform and precipitated into cold methanol. The precipitated polymer was dried in a vacuum (ca. 150 Pa). PCL yield was 91%.

The molecular weight characteristics (M_w and M_n) and dispersity ($Đ$) of the polymer obtained were determined by SEC with the use of a Shimadzu HPLC system (Shimadzu, Kyoto, Japan) consisting of a pump LC-10AD VP, system controller SCL-10A VP, and refractometric detector RID-10A (Shimadzu, Canby, OR, USA) supplied with a Rheodyne 725i injection valve (Rohnert Park, CA, USA) and two columns of Agilent PLgel MIXED-D (7.5 × 300 mm, 5 μm) (Agilent, Santa-Clara, CA, USA). The analysis was carried out in THF at a temperature of 40 °C, a flow rate of the mobile phase of 1.0 mL/min. Calculations of molecular weight were fulfilled regarding polystyrene standards with molecular weights in the range of 2000–450,000. Data processing was performed using LC Solution Shimadzu software (version 1.25, Shimadzu, Kyoto, Japan). In addition, the intrinsic viscosity (η) of the synthesized PCL was calculated after measurements of viscosity for a set of solutions with different concentrations in $CHCl_3$ using Ostwald's capillary viscosimeter.

2.6. Manufacturing of Composite Films

The pure and composite films based on PCL were manufactured using the previously developed procedure for the PCL composite films with nanocrystalline cellulose [12]. Briefly, a 5% PCL solution in $CHCl_3$ (6.5 mL) was poured inside the glass cylinder (i.d. = 75 mm) with fixed cellophane bottom. The solution was left for 12 h in the air for chloroform evaporation. After cellophane removal, the obtained PCL-based films were dried in the air thermostat at 45 °C for 3 days. In the case of composite films, unmodified or modified rGO-Am was dispersed in PCL solution in chloroform and shortly ultrasonicated with the ultrasonic probe for 15–20 s at 15–20% power. The amount of a filler was equal to 0.5 and 1.0 wt.%. For neat rGO-Am, 3 wt.% of the filler was used also for comparison. Other manipulations were done as described above for the manufacturing of unfilled PCL films.

2.7. Mechanical Testing

Mechanical properties of the films were studied under uniaxial extension using band-like specimens of 2 mm × 20 mm using the AG-100kNX Plus Shimadzu universal mechanical system (Shimadzu, Kyoto, Japan). The thickness of the tested specimens was 90 ± 10 μm. The extension speed was 10 mm/min. The characteristics of the tested samples were calculated basing on the results of measurements for 7–11 fragments of the material and the results are given as average value ± SD.

2.8. Biological Evaluation

The MTT assay was employed to assess the cell viability and proliferation of the MG-63 on interaction with the material. This is a colorimetric assay measuring the reduction of yellow 3-(4,5-dimethythiazol-2-yl)-2,5-diphenyl tetrazolium bromide (MTT) substrate to an insoluble purple formazan product by mitochondrial succinate dehydrogenase enzyme. The MTT enters the cells where it gets reduced to an insoluble, dark purple colored formazan product. Neat and modified rGO-Am dispersions were evaluated in the concentration range from 4 to 1000 μg/mL using an adhesive 96-well plate (n = 3). In the case of films, the round-shaped specimens with a dimeter of 5 mm were glued by BF-6 medical glue (Tula Pharmaceutical Company, Tula, Russia) to the bottom of the nonadhesion 96-well plate (n = 3). Sterilization of glued films was performed by their exposure under a UV light of wide spectrum for 10 min. Next, 100 μL of medium containing 1×10^4 cells was added to each well. The plates were incubated at 37 °C with 5% CO_2. As a control for dispersions, MG-63 cells were cultured in a culture medium in an adhesive 96-well plate, while pure PCL material was considered as a control for composite films. After 3 days of incubation, MTT (Sigma, St. Louis, MO, USA) solution was added to each well and incubated for 2 h at 37 °C with 5% CO_2. After the incubation time, the above solution was discarded, and the colored formazan crystals formed were solubilized by adding 50 μL of dimethyl sulfoxide. The absorbance was read at 570 nm in a multiwell plate reader (Thermo Fisher Multiscan Labsystems, Waltham, MA, USA). The absorbance values were plotted using MS Excel software.

3. Results and Discussion

3.1. Synthesis of rGO-Am

Figure 1a displays the survey spectra of the GO presented by dominant O 1s and C 1s lines at $h\upsilon$ = 532.5 eV and $h\upsilon$ = 284.7 eV with the absence of other spectral features, verifying the purity of the initial material. Upon the amination, the O 1s line substantially diminished, while the N 1s peak at ca. $h\upsilon$ = 400.1 eV appeared, pointing out the successful introduction of amines by the substitution of the oxygenic moieties in the treated GO. The concentration of nitrogen was estimated to be ca. 3.02 at.%. To further analyze the composition of nitrogen-containing groups embedded upon the applied treatment, high-resolution N 1s spectra were acquired and processed (Figure 1b). Three distinct peaks positioned at $h\upsilon$ = 398.9 eV, $h\upsilon$ = 399.8 eV and $h\upsilon$ = 401.9 eV were discerned, which are related to the presence of pyridines, amines (both primary and secondary), and implemented graphitic nitrogen, respectively [40,41]. As seen, amines are the dominant form of the introduced nitrogen functionalities with their relative content estimated to be up to 88.4%. Accordingly, the concentration of the amines in the synthesized rGO-Am is ca. 2.63 at.%.

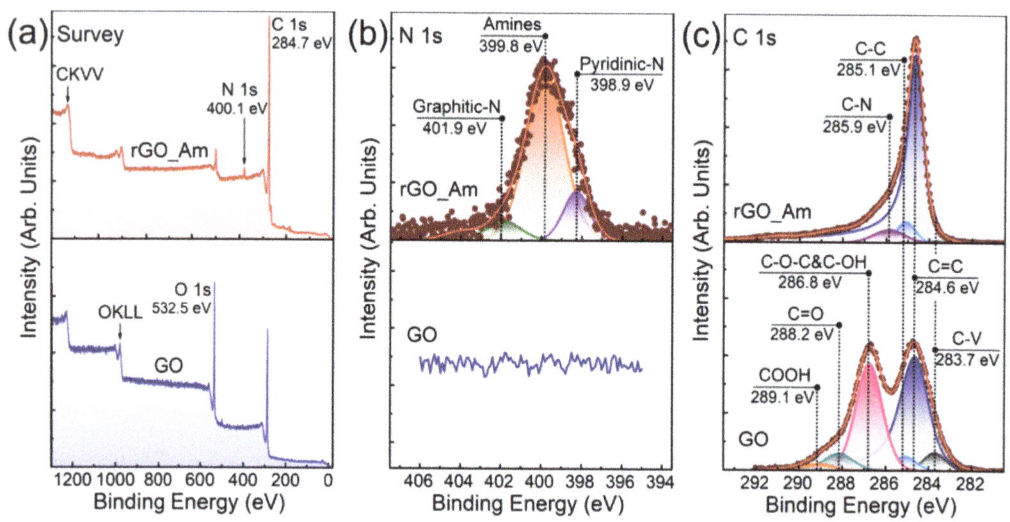

Figure 1. The XPS characterization of initial GO and rGO-Am layers: (**a**) survey spectra, (**b**) high-resolution N 1s spectra and (**c**) high-resolution C 1s spectra.

The elimination of the oxygenic groups upon the consecutive bromination and amination is pointed out by the C 1s spectra of the GO and rGO-Am (Figure 1c). The initial GO is functionalized by the hydroxyls and epoxides at the basal plane of the graphene layer along with the carbonyls and carboxyls at its edges. The presence of these functional groups is signified by the C–O–C&C–OH, C=O and COOH peaks centered at $h\upsilon$ = 286.8 eV, $h\upsilon$ = 288.2 eV and $h\upsilon$ = 289.1 eV, respectively [42,43]. The C–V peak is related to the nonterminated carbon atoms of the vacancy defects and Stone–Wales defects [44]. The relative concentration of the oxygenic groups and carbon atoms in various states is presented in Table 1, whereas the C/O ratio was calculated to be 2.51. After the amination, all the spectral features related to the oxygenic groups were absent with the retention of only a dominant C=C peak at ~284.6 eV, corresponding to the pristine π-conjugated graphene network, and a C–C peak at ~285.1 eV related to nonconjugated C–C bonds [43]. The concentration of the preserved oxygenic groups is less than 3 at.% with the corresponding C/O ratio of rGO-Am of 32.51 (Table 1). In addition, a C–N peak at ~285.9 eV appears due to the formation of carbon–nitrogen bonds upon the introduction of amines and other nitrogen species [37,45]. Given XPS data on the composition of functional groups and the size distribution of the rGO-Am derived from both the measurements using LD technique and the acquired SEM images of arrays of flakes (Supplementary Materials, Section S1, Figure S1), the molar ratio of the presented amines was estimated. The calculated values are ca. 7.1 mmol/g, which corresponds to 5.4×10^9 amine groups per 1 g of rGO-Am. The details on the calculation of the molar ratio are presented in Supplementary Materials, Section S2.

Table 1. Composition of functional groups (at.%), carbon in different states and C/O ratio estimated from the analysis of the deconvoluted C 1s X-ray photoelectron spectra of GO and rGO-Am.

Component	C–V	C=C	C–C	C–OH and C–O–C/C–OH(p)	C=O	COOH	C–N	C/O Ratio
Binding Energy (eV)	283.7	284.6	285.1	286.8 / 286.4	288.2	289.0	285.9	
GO	5.31	52.94	4.49	28.03	6.68	2.55	-	2.51
rGO-Am	<0.1	89.05	4.93	1.5	<0.6	<0.9	3.02	32.51

3.2. Modification of rGO-Am with Oligomers of Glutamic Acid

Modification of rGO-Am was carried out using a "grafting from" technique. One of the common methods for the synthesis of oligomers and polymers of amino acids is the ring-opening polymerization of α-amino acid N-carboxyanhydrides (NCAs) catalyzed with the primary amines [46]. In our case, the primary amino groups of rGO-Am played a role of initiating groups for polymerization of γ-protected glutamic acid NCA to prepare the rGO-Am grafted with oligomers of glutamic acid (oligo(Glu)). The scheme of rGO-Am modification with oligo(Glu) is displayed in Figure 2. At first step, the ring-opening polymerization of Glu(OBzl) NCA was carried out to modify rGO-Am. Then, the Bzl protection was removed from γ-carboxylic groups at acidic conditions. The successful grafting of rGO-Am with oligo(Glu) was testified by FTIR spectroscopy and XPS.

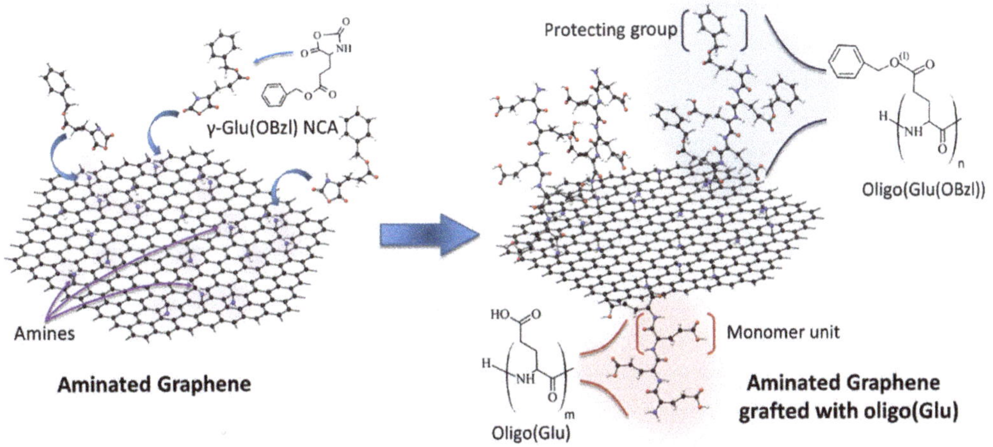

Figure 2. Scheme of rGO-Am modification with oligomers of glutamic acid via ring-opening polymerization of Glu(OBzl) NCA as monomer.

In comparison with neat rGO-Am and protected and unprotected poly(glutamic acid) used as standards, an increase of several characteristic bands corresponding to the groups presented in PGlu was detected in the FTIR spectrum of rGO-Am-oligo(Glu). In particular, an increase in the intensity for the characteristic bands at 1165, 1459, 1561 and 1744 cm^{-1} corresponding to C–N stretching, CH_2 bending, N–C=O stretching (amid II) and C=O stretching vibrations, respectively, was detected after the grafting of oligo(Glu) from rGO-Am (Figure 3).

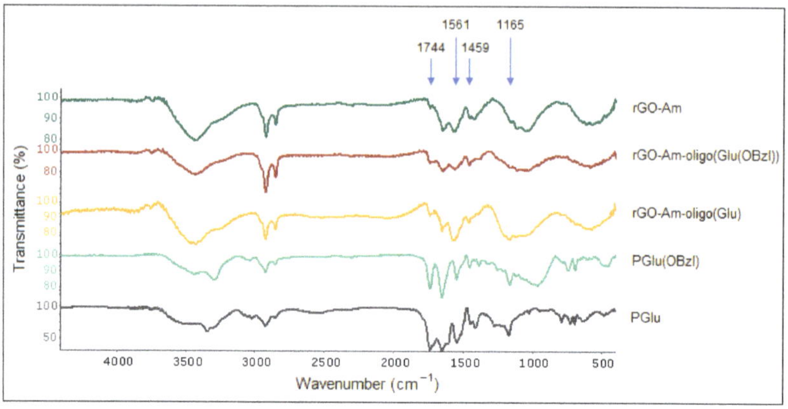

Figure 3. FTIR spectra of neat and modified rGO-Am as well as protected and unprotected poly(glutamic acid) as the standards.

The success of oligo(Glu) grafting from the rGO-Am was also revealed by the corresponding changes in the X-ray photoelectron spectra presented in Figure 4a. In the case of both rGO-Am-oligo(Glu(OBzl)) and rGO-Am-oligo(Glu), the rise of the intensity of the O 1s peak, as well as the N 1s peak, was observed. The quantitative analysis of the acquired survey X-ray photoelectron spectra showed that after the ring-opening polymerization of Glu(OBzl) NCA from rGO-Am, the concentration of the nitrogen increased from 3.02 at.% to ca. 3.52 and 3.47 at.% for protected and unprotected forms, respectively. Given these differences in the nitrogen content and the known chemical structure of a oligo(glutamic acid) (Supplementary Materials, Section S3), the number of oligo(Glu) branches was calculated and estimated to be ca. 2.2×10^9 per 1 g of the obtained material (Supplementary Materials, Section S4). The rate of the functionalization of amines were estimated to be about 4–8 branches and 6–12% of the total number of amines, respectively. It means that the average length of the oligo(Glu) chains is in the range from 4 to 8 monomer units.

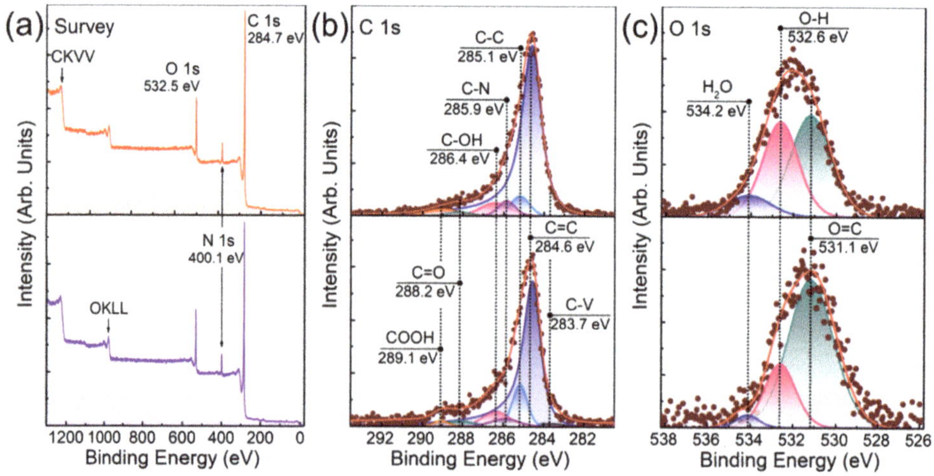

Figure 4. XPS characterization of the rGO-Am grafted with oligo(Glu) (top row) and oligo(Glu(OBzl)) (bottom row): (**a**) survey spectra, (**b**) high-resolution C 1s spectra and (**c**) high-resolution O 1s spectra.

The introduction of oligo(Glu) is further signified by the changes of the C 1s spectra displayed in Figure 4b. As seen, after the synthesis of oligo(Glu) the peaks related to carboxyls and carbonyls centered at hv = 289.1 eV and hv = 288.2 eV, respectively, increased. The appearance of the C–OH peak at hv = 286.2 eV related to phenol groups can be also noted, most probably due to the modification of the graphene layer concurrently with the oligo(Glu) growth. Furthermore, the intensity of the C–C peak at hv = 285.1 eV increased owing to the presence of nonconjugated C–C bonds in the oligo(Glu) backbone. As one might expect, the most prominent rise of the C–C peak was observed in the case of rGO-Am-oligo(Glu(OBzl)), where the relative content of the carbon in the corresponding state was up to 12.12 at.% (Table 2) [43]. In turn, the relative concentration of the nonconjugated C–C bonds detected for rGO-Am-oligo(Glu) was almost two times lower, i.e., 6.77 at.%. Combined with the redistribution of the relative content of the carboxyls and carbonyls from 1.48 and 3.39 at.% in rGO-Am-oligo(Glu(OBzl)) to 2.84 and 1.65 at.% in rGO-Am-oligo(Glu), respectively, this evolution of the C 1s spectra evidences that, in the latter sample, the protective groups were successfully eliminated from at least 45% of the carboxyl groups in the grafted oligo(Glu). This is also illustrated by the comparative analysis of the O 1s X-ray photoelectron spectra displayed in Figure 4c. In both spectra, three peaks are discerned positioned at hv = 531.1 eV, hv = 532.5 eV and hv = 534.2 eV and attributed to double-bonded oxygen, O–H bonded oxygen, and oxygen in the water molecules, respectively [42,43]. As seen, in the case of rGO-Am-oligo(Glu) the peak of the O–H bonded oxygen related to hydroxyl functionality in the deprotected carboxyl group rises with the increase of its relative concentration from 22.39% to 40.48%. Such changes are commonly observed in the case of moving from carbonylated to carboxylated nanocarbon structures [43,47]. Note that the relative content of the adsorbed water also rises from 3.19% in rGO-Am-oligo(Glu(OBzl)) to 10.83% in rGO-Am-oligo(Glu) upon the elimination of the protecting Bzl groups and, thus, a higher concentration of hydrophilic carboxyl groups. As a net result, the XPS data implies both the successful grafting of oligo(Glu) from rGO-Am and the following deprotection of up to 45% of the Bzl groups from the carboxyls presented in oligo(Glu).

Table 2. Composition of functional groups (at.%) and C/O ratio estimated from the analysis of the deconvoluted C 1s X-ray photoelectron spectra.

Component	C–V	C=C	C–C	C–OH and C–O–C/C–OH(p)	C=O	COOH	C–N	C/O Ratio
Binding Energy (eV)	283.7	284.6	285.1	286.8 / 286.4	288.2	289.0	285.9	
rGO-Am-oligo(Glu(OBzl))	1.93	69.55	12.12	8.01	3.39	1.48	3.52	6.87
rGO-Am-oligo(Glu)	<0.1	78.79	6.77	6.48	1.65	2.84	3.47	7.38

3.3. *Characterization of Modified rGO-Am*

It is known that chemical modifications of materials can change their characteristics and properties [48–50]. In order to characterize the modified rGO-Am, such physicochemical methods as DLS and ELS, TGA and DTG were applied for the obtained materials.

According to the DLS and ELS analyses, the aqueous dispersion of neat rGO-Am (pH 7.4) was characterized by the presence of submicron negatively charged particles with a fairly broad size distribution (Table 3). Modification of rGO-Am with protected and unprotected oligomers was accompanied by a decrease in the hydrodynamic diameter of graphene particles due to a reduction of their aggregation. In turn, the modification of rGO-Am with oligo(Glu(OBzl)) was followed by an increase in the surface electrokinetic potential, while, as expected, the deprotection diminished this parameter.

Table 3. Characteristics of the neat and modified rGO-Am particles in aqueous solution (pH 7.4) determined by dynamic and electrophoretic light scattering.

Sample	D_H (nm)	PDI	ζ-potential (mV)
rGO-Am	465 ± 71	0.56	−38.4 ± 7.1
rGO-Am-oligo(Glu(OBzl))	369 ± 66	0.42	−32.1 ± 5.6
rGO-Am-oligo(Glu)	302 ± 49	0.52	−40.0 ± 5.1

Neat and modified rGO-Am as well as PGlu and PGlu(OBzl) were analyzed by TGA. The obtained TGA and DTG curves are presented in Figure 5. Before the beginning of the region where the destruction of the carbon base of rGO-Am starts (410–420 °C), the mass of the neat rGO-Am consistently falls, starting from 120 to 130 °C (Figure 5a). When the intensive thermal destruction process starts (after 400 °C), the sample has already lost ~18% of the initial mass. This may probably be due to the removal of the residual oxygenic functional groups and solvent retained between the layers of rGO-Am at the early stages of sample heating. In the case of PGlu(OBzl), the main thermal degradation proceeds in two stages: the sample loses ~59% and 29% of its mass in the regions of 240–360 and 420–570 °C, respectively. There is also a low-temperature mass loss in the region of 160–240 °C, where a loss of ~4% of mass is noticeable. These three processes are clearly visible on the differential curve of the sample mass change (Figure 5b). The process of thermal destruction of a PGlu sample proceeds in a significantly different way. In this case, three areas of significant mass drop are clearly visible on the TGA and DTG curves: the sample loses ~38% of its mass in the region of 190–320 °C; the mass falls by ~18% in the region of 330–400 °C; and the sample loses the last ~35% of the mass in the region of 430–560 °C. As can be seen when comparing the differential curves for PGlu(OBzl) and PGlu, only the high-temperature mass loss areas (last peaks on the DTG curve, Figure 5b) coincide for the two compared samples. In turn, rGO-Am-oligo(Glu(OBzl)) and rGO-Am-oligo(Glu) were characterized with a more intensive destruction process in the low-temperature region of 200–360 °C (the maximum intensity of this process is 290 °C) in comparison with neat rGO-Am and PGlu/PGlu(OBzl). Since there are no intense mass losses on the TGA and DTG curves of unmodified rGO-Am in this temperature range, it can be stated that the process observed for the modified samples is the cleavage of oligo(Glu)/oligo(Glu(OBzl)) from rGO-Am and their transition to the gas phase, apparently, with the simultaneous destruction of their oligomer chains. The cleavage of the oligomer fragments from rGO-Am is the most likely due to the identity of the thermodestruction process (the coincidence of temperature regions) for samples containing protected and unprotected oligomer chains. The mass losses for both samples were also very close: ~17% for rGO-Am-oligo(Glu) and 14% for rGO-Am-oligo(Glu(OBzl)). Thus, the mass losses in the region of 200–360 °C may be considered as estimated values of the rGO-Am surface grafting with oligomers of glutamic acid: 14–17%.

The analysis of the morphology of the neat and modified rGO-Am platelets did not reveal any obvious surface changes (Figure 6). Both materials demonstrate the corrugated structure with a high number of wrinkles and folds of several μm in scale. This results in the reduction of the π–π^* interlayer stacking between the layers of rGO-Am, resulting in a highly developed surface of the material and irregular porous network structure. This is in contrast to lamellar GO and rGO platelets, having smoothed surface with a low extent of wrinkling and folding. The corrugation of rGO-Am is an intrinsic characteristic of this material related to both the distortion of the graphene layer at nanoscale due to introduced amines and the interaction of amines with the retained oxygenic groups, resulting in folding at a microscale [34]. Notably, it does not depend on the type of solvent used for the deposition varied from the polar ones (isopropyl alcohol) to nonpolar solvents (trichloromethane and tetrachloromethane). This fact is also supported by the absence of alterations of the rGO-Am morphology after its grafting with both PGlu and PGlu(OBzl), modifying its wetting characteristics.

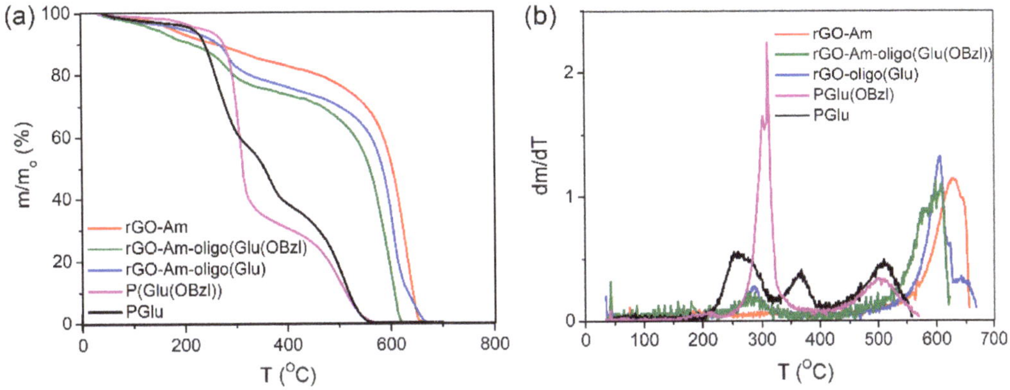

Figure 5. TGA (**a**) and DTG (**b**) curves for the neat and modified rGO-Am as well as protected and unprotected PGlu as standards.

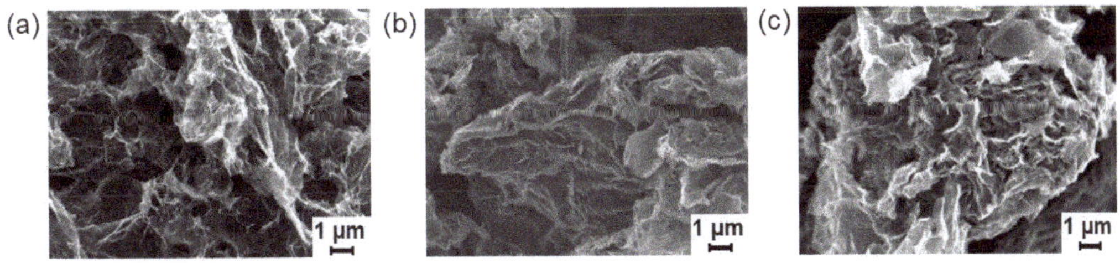

Figure 6. SEM images of rGO-Am platelets on the silicon support (×5000): (**a**) neat rGO-Am, (**b**) rGO-Am-oligo(Glu(OBzl)) and (**c**) rGO-Am-oligo(Glu).

3.4. Manufacturing and Characterization of the Composite Films

In order to prepare the composite materials with rGO-Am and its derivatives, PCL synthesized by ring-opening polymerization of ε-caprolactone was utilized as a polymer matrix in this work. The used PCL had the following characteristics: M_w = 89,000 and Đ = 1.7 (according to SEC), and η = 0.91 dL/g (according to viscosimetric analysis in $CHCl_3$).

The films of pure PCL and its composites with neat and modified rGO-Am were manufactured by casting a polymer solution in $CHCl_3$ on the cellophane substrate as described in the experimental part (Section 2.6). After evaporation of chloroform at room temperature, the films were additionally dried to a constant mass at 45 °C to remove solvent traces.

The PCL-based composite films containing neat and modified rGO-Am were tested at room temperature in the uniaxial stretching mode. Pure PCL films were used as a benchmark. During the tests, the following characteristics of the material were determined: the modulus of elasticity (or Young's modulus) (E), tensile strengths (σ_b) and elongation at break (ε_b).

It is known that the aggregation of graphene and its derivatives due to the strong π–π interaction prevents the homogeneous distribution of these fillers in the polymer matrix. Recently, it has been shown that grafting of rGO with poly (L-lactic acid) increases the compatibility of this filler with a poly(L-lactic acid)-based matrix [36]. Indeed, the introduction of neat rGO-Am into the PCL matrix led to a sharply negative result. In particular, a noticeable decrease in the Young's modulus, strength characteristics and deformation resource of the material was observed (Table 4, Figure 7a).

Table 4. Mechanical properties of PCL and its composites with neat and modified rGO-Am.

Specimen	Filler Content (%)	E (MPa)	σ_b (MPa)	ε_b (%)
PCL	-	396 ± 13	13.9 ± 0.5	25 ± 2
PCL/rGO-Am	0.5	369 ± 43	13.6 ± 0.3	25 ± 3
	1.0	348 ± 32	11.4 ± 0.5	12 ± 1
	3.0	270 ± 28	7.5 ± 0.6	10 ± 1
PCL/rGO-Am-oligo(Glu(OBzl)	0.5	415 ± 20	13.5 ± 0.5	21 ± 2
	1.0	396 ± 32	13.1 ± 0.4	15 ± 2
PCL/rGO-Am-oligo(Glu)	0.5	434 ± 39	14.8 ± 0.4	27 ± 2
	1.0	444 ± 17	14.6 ± 0.4	23 ± 2

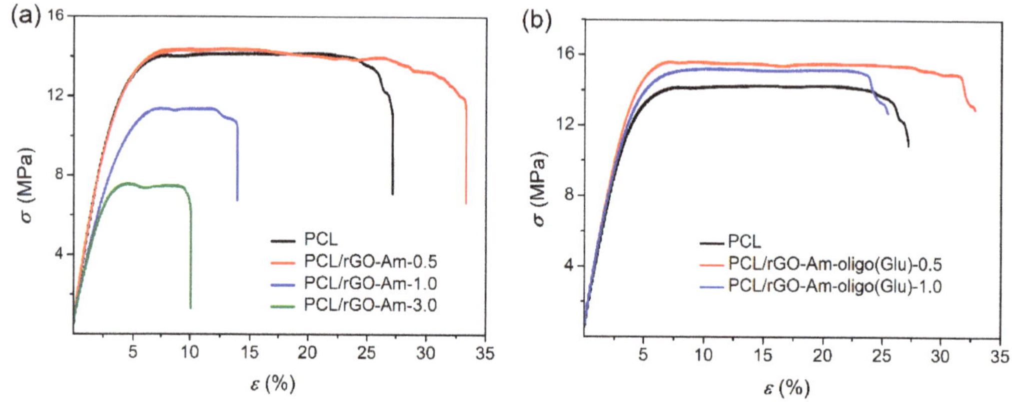

Figure 7. Stress–strain curves for PCL and its composites with neat (**a**) and modifed (**b**) rGO-Am.

Moreover, the higher the content of the filler added to the matrix, the worse mechanical properties were detected. The reason for this negative effect is the aggregation of the filler forming large agglomerates in the films. As a consequence, a pronounced heterogeneity of the material structure and the destruction of samples at the phase boundaries took place. The analysis of films by optical microscopy in transmitted light revealed the evident aggregates of rGO-Am in the PCL matrix at filler contents higher than 0.5 wt.% (Figures 8 and 9). At the same time, the modification of rGO-Am with both Bzl-protected and unprotected oligomers of glutamic acid favored a much more uniform distribution of the filler in the PCL matrix due to enhanced compatibility (Figures 8 and 9; Figure S2). In turn, this fact considerably affected the enhancement of the mechanical properties of the manufactured composite materials (Table 4, Figure 7b). The best mechanical properties were established for the PCL-based composite containing 0.5 wt.% of rGO-Am-oligo(Glu) as a filler. A similar trend has been recently observed by Wang et al. for PLA-based films filled with neat GO and GO grafted with PLA [36]. The authors found that composites containing neat GO had poorer mechanical properties than nonfilled PLA. At the same time, the modification of GO with PLA contributed to the enchantment of mechanical properties.

3.5. In Vitro Biological Evaluation

An in vitro biocompatibility study was carried out with the use of human osteoblast-like cells (MG-63 cell line) for both neat and oligo(Glu)-modified rGO-Am as dispersions and as a part of composite films. Figure 10 illustrates the cell viability in the presence of rGO-Am and rGO-Am-oligo(Glu) taken in the range of concentrations from 4 to 1000 μg/mL. As seen, neat rGO-Am was nontoxic up to the concentration of 500 μg/mL, while rGO-Am-oligo(Glu) was biocompatible in the entire tested concentration range. Thus, the modification of rGO-Am by oligomers of glutamic acid improved the biocompatibility of aminated graphene with living cells.

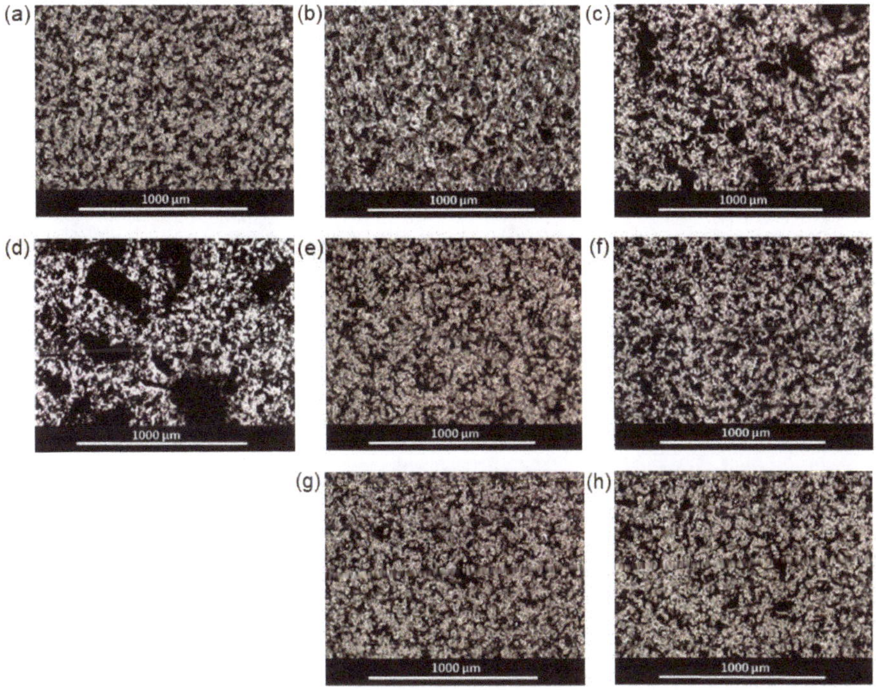

Figure 8. Images of pure PCL and composite films obatined by optical microscopy in transmitted light (×4): (**a**) PCL; (**b**) PCL/rGO-Am-0.5 wt.%; (**c**) PCL/rGO-Am-1.0 wt.%; (**d**) PCL/rGO-Am-3.0 wt.%; (**e**) PCL/rGO-Am-oligo(Glu(OBzl))-0.5 wt.%; (**f**) PCL/rGO-Am-oligo(Glu(OBzl))-1.0 wt.%; (**g**) PCL/rGO-Am-oligo(Glu)-0.5 wt.%; (**h**) PCL/rGO-Am-oligo(Glu)-1.0 wt.%.

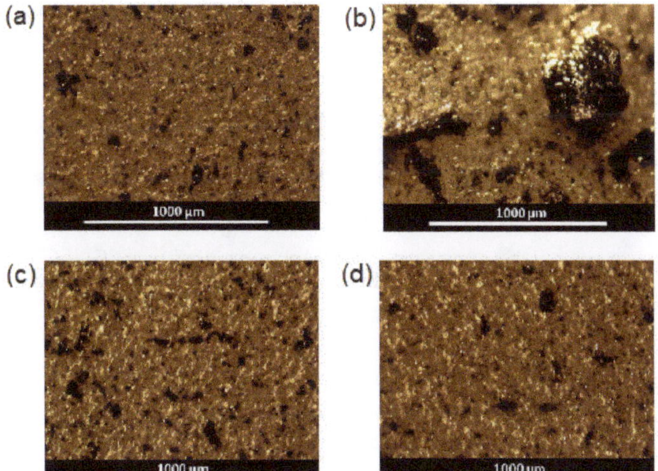

Figure 9. Images of pure PCL and composite films obatined by optical microscopy in reflected light (×4): (**a**) PCL; (**b**) PCL/rGO-Am; (**c**) PCL/rGO-Am-oligo(Glu(OBzl)); (**d**) PCL/rGO-Am-oligo(Glu). The content of the filler in all composites was 1 wt.%.

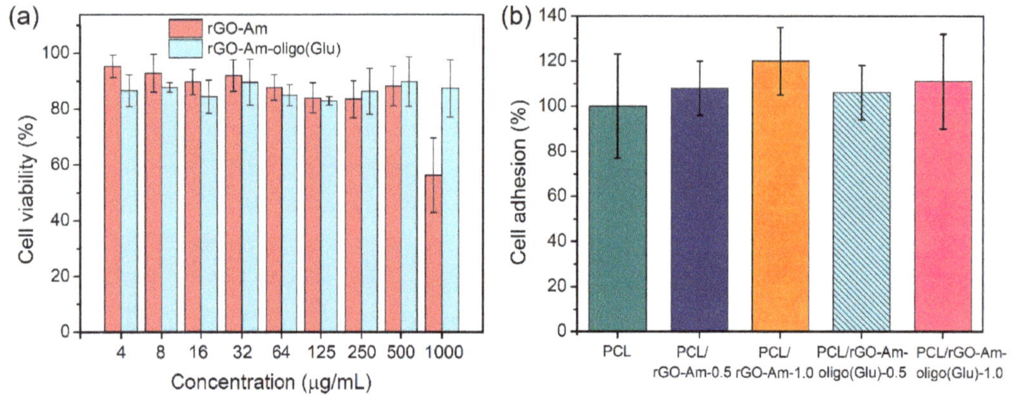

Figure 10. Cytotoxicity study (MTT-test) with MG-63 cells for the dispersions of neat and modified rGO-Am (**a**) as well as PCL-based composite films with rGO-Am and rGO-Am-oligo(Glu) as fillers (**b**).

The high biocompatibility for polymers of glutamic acid with cells was earlier demonstrated in several papers [38,39,51]. Taking into account the high biocompatibility of PCL and low content of the filler in composites, the absence of cytotoxicity for composite films could be expected. Indeed, PCL-based films filled with rGO-Am and rGO-Am-oligo(Glu) demonstrated a similar and comparable level of cell adhesion to the nonfilled PCL films for all tested composite specimens. Therefore, the addition of rGO-Am into PCL, as well as the modification of rGO-Am, did not affect the biocompatibility of PCL with cells.

4. Conclusions

In present work, an approach to the modification of aminated graphene by the "grafting from" technique based on ring-opening polymerization of N-carboxyanhydride of glutamic acid γ-benzyl ester was proposed. Grafting of glutamic acid oligomers was testified by both FTIR spectroscopy and XPS. According to XPS analysis, the approximate chain length of grafted oligomers was in the range of 4–8 monomer units. Modification of aminated graphene with oligo(Glu) contributed to a decrease in the average size of rGO-Am particles in the aqueous dispersion. A less pronounced aggregation of modified aminated graphene was also observed after its distribution in the PCL matrix. TGA and DTG analysis of modified aminated graphene revealed its slightly lower thermal stability in comparison with that of neat rGO-Am. However, the mechanical properties of PCL-based composites obtained with neat aminated graphene considerably decreased along with the filler content increase. At the same time, the better distribution of aminated graphene grafted with glutamic acid oligomers improved or preserved the mechanical properties of the composite film. The biocompatibility of PCL/rGO-Am-oligo(Glu) composites with human osteoblast-like cells seems to be very promising for their further use as functional scaffolds for bone tissue regeneration.

Supplementary Materials: The following are available online at https://www.mdpi.com/article/10.3390/polym13162628/s1, Figure S1: The analysis of the size distribution of rGO-Am. (a) The size distribution histogram derived from the laser diffraction measurements. SEM images of (b) an array and (c) a single flake of rGO-Am; S1. Calculations of the molar ratio and number of functional groups in rGO-Am; S2. Calculations of the oligo(Glu) content after the grafting; Figure S2. SEM images of nonfilled PCL and its composite films (×100).

Author Contributions: Conceptualization, E.K.-V.; methodology, M.S., M.R. and E.K.-V.; formal analysis, M.S., O.S., M.R., I.G. and E.K.-V.; investigation, M.S., O.S., I.A., I.G., Y.N., G.A., A.S., B.B. and A.N.; data curation, M.S., M.R. and E.K.-V.; writing—original draft preparation, M.R. and E.K.-V.;

writing—review and editing, M.R. and E.K.-V.; visualization, M.R. and E.K.-V.; supervision, M.R. and E.K.-V. All authors have read and agreed to the published version of the manuscript.

Funding: The work by M.R., G.A., A.S. and B.B. on the XPS studies was supported by the Russian Foundation for Basic Research (grant no. 20-04-60458). The work by M.S., O.S., I.A., I.G., Y.N. and E.K.-V. was carried out as part of a State Assignment.

Institutional Review Board Statement: Not applicable.

Informed Consent Statement: Not applicable.

Data Availability Statement: The data are contained within the article and supplementary materials.

Acknowledgments: The XPS and SEM studies of aminated graphene and its derivatives were carried out in the Joint Research Center "Materials science and characterization in advanced technology" (Ioffe Institute, St. Petersburg, Russia).

Conflicts of Interest: The authors declare no conflict of interest.

References

1. Yin, Y.; Zhao, L.; Jiang, X.; Wang, H.; Gao, W. Poly(lactic acid)-based biocomposites reinforced with modified cellulose nanocrystals. *Cellulose* **2017**, *24*, 4773–4784. [CrossRef]
2. Gredes, T.; Schönitz, S.; Gedrange, T.; Stepien, L.; Kozak, K.; Kunert-Keil, C. In vivo analysis of covering materials composed of biodegradable polymers enriched with flax fibers. *Biomater. Res.* **2017**, *21*, 1–12. [CrossRef] [PubMed]
3. Korzhikov, V.; Averianov, I.; Litvinchuk, E.; Tennikova, T.B. Polyester-based microparticles of different hydrophobicity: The patterns of lipophilic drug entrapment and release. *J. Microencapsul.* **2016**, *33*, 199–208. [CrossRef] [PubMed]
4. Manavitehrani, I.; Fathi, A.; Badr, H.; Daly, S.; Shirazi, A.N.; Dehghani, F. Biomedical Applications of Biodegradable Polyesters. *Polymers* **2016**, *8*, 20. [CrossRef] [PubMed]
5. Lou, C.W.; Yao, C.H.; Chen, Y.S.; Hsieh, T.C.; Lin, J.H.; Hsing, W.H. Manufacturing and Properties of PLA Absorbable Surgical Suture. *Text. Res. J.* **2008**, *78*, 958–965. [CrossRef]
6. Narayanan, G.; Vernekar, V.N.; Kuyinu, E.L.; Laurencin, C.T. Poly (lactic acid)-based biomaterials for orthopaedic regenerative engineering. *Adv. Drug Deliv. Rev.* **2016**, *107*, 247–276. [CrossRef] [PubMed]
7. Jiang, C.P.; Chen, Y.Y.; Hsieh, M.F. Biofabrication and in vitro study of hydroxyapatite/mPEG–PCL–mPEG scaffolds for bone tissue engineering using air pressure-aided deposition technology. *Mater. Sci. Eng. C* **2013**, *33*, 680–690. [CrossRef]
8. Seyednejad, H.; Gawlitta, D.; Dhert, W.J.A.; van Nostrum, C.F.; Vermonden, T.; Hennink, W.E. Preparation and characterization of a three-dimensional printed scaffold based on a functionalized polyester for bone tissue engineering applications. *Acta Biomater.* **2011**, *7*, 1999–2006. [CrossRef] [PubMed]
9. Tallawi, M.; Rosellini, E.; Barbani, N.; Cascone, M.G.; Rai, R.; Saint-Pierre, G.; Boccaccini, A.R. Strategies for the chemical and biological functionalization of scaffolds for cardiac tissue engineering: A review. *J. R. Soc. Interface* **2015**, *12*, 20150254. [CrossRef]
10. Murariu, M.; Dubois, P. PLA composites: From production to properties. *Adv. Drug Deliv. Rev.* **2016**, *107*, 17–46. [CrossRef]
11. Liu, X.; Wang, T.; Chow, L.C.; Yang, M.; Mitchell, J.W. Effects of inorganic fillers on the thermal and mechanical properties of poly(lactic acid). *Int. J. Polym. Sci.* **2014**, *2014*, 827028. [CrossRef]
12. Stepanova, M.; Averianov, I.; Serdobintsev, M.; Gofman, I.; Blum, N.; Semenova, N.; Nashchekina, Y.; Vinogradova, T.; Korzhikov-Vlakh, V.; Karttunen, M.; et al. PGlu-Modified Nanocrystalline Cellulose Improves Mechanical Properties, Biocompatibility, and Mineralization of Polyester-Based Composites. *Materials* **2019**, *12*, 3435. [CrossRef]
13. Murizan, N.I.S.; Mustafa, N.S.; Ngadiman, N.H.A.; Yusof, N.M.; Idris, A. Review on nanocrystalline cellulose in bone tissue engineering applications. *Polymers* **2020**, *12*, 2818. [CrossRef]
14. Stepanova, M.; Averianov, I.; Gofman, I.; Solomakha, O.; Nashchekina, Y.; Korzhikov-Vlakh, V.; Korzhikova-Vlakh, E. Poly(ε-caprolactone)-based biocomposites reinforced with nanocrystalline cellulose grafted with poly(L-lactic acid). *IOP Conf. Ser. Mater. Sci. Eng.* **2019**, *500*, 012021. [CrossRef]
15. Korzhikov-Vlakh, V.; Averianov, I.; Sinitsyna, E.; Nashchekina, Y.; Polyakov, D.; Guryanov, I.; Lavrentieva, A.; Raddatz, L.; Korzhikova-Vlakh, E.; Scheper, T.; et al. Novel Pathway for Efficient Covalent Modification of Polyester Materials of Different Design to Prepare Biomimetic Surfaces. *Polymers* **2018**, *10*, 1299. [CrossRef] [PubMed]
16. Karaman, O.; Kumar, A.; Moeinzadeh, S.; He, X.; Cui, T.; Jabbari, E. Effect of surface modification of nanofibres with glutamic acid peptide on calcium phosphate nucleation and osteogenic differentiation of marrow stromal cells. *J. Tissue Eng. Regen. Med.* **2016**, E132–E146. [CrossRef]
17. Shi, X.; Wu, H.; Yan, H.; Wang, Y.; Wang, Z.; Zhang, P. Electroactive Nanocomposite Porous Scaffolds of PAPn/op-HA/PLGA Enhance Osteogenesis in Vivo. *ACS Appl. Bio Mater.* **2019**, *2*, 1464–1476. [CrossRef]
18. Wang, Z.; Xu, Y.; Wang, Y.; Ito, Y.; Zhang, P.; Chen, X. Enhanced in Vitro Mineralization and in Vivo Osteogenesis of Composite Scaffolds through Controlled Surface Grafting of L-Lactic Acid Oligomer on Nanohydroxyapatite. *Biomacromolecules* **2016**, *17*, 818–829. [CrossRef] [PubMed]

19. Lv, Q.; Deng, M.; Ulery, B.; Nair, L.; Laurencin, C. Nano-ceramic composite scaffolds for bioreactor-based bone engineering. *Clin. Orthop. Relat. Res.* **2013**, *471*, 2422–2433. [CrossRef] [PubMed]
20. Valapa, R.B.; Pugazhenthi, G.; Katiyar, V. Effect of graphene content on the properties of poly(lactic acid) nanocomposites. *RSC Adv.* **2015**, *5*, 28410–28423. [CrossRef]
21. Camargo, J.C.; Machado, Á.R.; Almeida, E.C.; Silva, E.F.M.S. Mechanical properties of PLA-graphene filament for FDM 3D printing. *Int. J. Adv. Manuf. Technol.* **2019**, *103*, 2423–2443. [CrossRef]
22. Prasadh, S.; Suresh, S.; Wong, R. Osteogenic Potential of Graphene in Bone Tissue Engineering Scaffolds. *Materials* **2018**, *11*, 1430. [CrossRef]
23. Daneshmandi, L.; Barajaa, M.; Tahmasbi Rad, A.; Sydlik, S.A.; Laurencin, C.T. Graphene-Based Biomaterials for Bone Regenerative Engineering: A Comprehensive Review of the Field and Considerations Regarding Biocompatibility and Biodegradation. *Adv. Healthc. Mater.* **2021**, *10*, 2001414. [CrossRef]
24. Pang, W.; Ni, Z.; Chen, G.; Huang, G.; Huang, H.; Zhao, Y. Mechanical and thermal properties of graphene oxide/ultrahigh molecular weight polyethylene nanocomposites. *RSC Adv.* **2015**, *5*, 63063–63072. [CrossRef]
25. Gong, M.; Zhao, Q.; Dai, L.; Li, Y.; Jiang, T. Fabrication of polylactic acid/hydroxyapatite/graphene oxide composite and their thermal stability, hydrophobic and mechanical properties. *J. Asian Ceram. Soc.* **2017**, *5*, 160–168. [CrossRef]
26. Krystyjan, M.; Khachatryan, G.; Grabacka, M.; Krzan, M.; Witczak, M. Starch/Chitosan Polymer Composites Modified by Graphene Oxide, Designed as New Bionanomaterials. *Polymers* **2021**, *13*, 2327. [CrossRef]
27. Mohamad, S.N.K.; Ramli, I.; Abdullah, L.C.; Mohamed, N.H.; Islam, M.S.; Ibrahim, N.A.; Ishak, N.S. Evaluation on structural properties and performances of graphene oxide incorporated into chitosan/poly-lactic acid composites: Cs/pla versus cs/pla-go. *Polymers* **2021**, *13*, 1839. [CrossRef] [PubMed]
28. Luo, Y.; Shen, H.; Fang, Y.; Cao, Y.; Huang, J.; Zhang, M.; Dai, J.; Shi, X.; Zhang, Z. Enhanced Proliferation and Osteogenic Differentiation of Mesenchymal Stem Cells on Graphene Oxide-Incorporated Electrospun Poly(lactic-co-glycolic acid) Nanofibrous Mats. *ACS Appl. Mater. Interfaces* **2015**, *7*, 6331–6339. [CrossRef] [PubMed]
29. Podsiadlo, P.; Kaushik, A.; Arruda, E.; Waas, A.; Shim, B.; Xu, J.; Nandivada, H.; Pumplin, B.; Lahann, J.; Ramamoorthy, A.; et al. Ultrastrong and stiff layered polymer nanocomposites. *Science* **2007**, *318*, 80–83. [CrossRef] [PubMed]
30. Bonderer, L.J.; Studart, A.R.; Gauckler, L.J. Bioinspired Design and Assembly of Platelet Reinforced Polymer Films. *Science* **2008**, *319*, 1069–1073. [CrossRef]
31. Wei, J.; Liu, A.; Zhang, P.; Chen, L.; Chen, X.; Jing, X. The surface modification of hydroxyapatite nanoparticles by the ring opening polymerization of gamma-benzyl-l-glutamate N-carboxyanhydride. *Macromol. Biosci.* **2009**, *9*, 631–638. [CrossRef] [PubMed]
32. Averianov, I.V.; Stepanova, M.A.; Gofman, I.V.; Nikolaeva, A.l.; Korzhikov-Vlakh, V.A.; Karttunen, M.; Korzhikova-Vlakh, E.G. Chemical modification of nanocrystalline cellulose for enchanced interfacial compatibility with poly(lactic acid). *Mendeleev Commun.* **2019**, *29*, 220–222. [CrossRef]
33. Qu, P.; Zhou, Y.; Zhang, X.; Yao, S.; Zhang, L. Surface modification of cellulose nanofibrils for poly(lactic acid) composite application. *J. Appl. Polym. Sci.* **2012**, *125*, 3084–3091. [CrossRef]
34. Stepanova, M.; Averianov, I.; Solomakha, O.; Zabolotnykh, N.; Gofman, I.; Serdobintsev, M.; Vinogradova, T.; Korzhikov-Vlakh, V.; Korzhikova-Vlakh, E. Composite biomaterials based on poly(L-lactic acid) and functionalized cellulose nanocrystals. *J. Renew. Mater.* **2020**, *8*, 383–395. [CrossRef]
35. Gu, A.; Wu, J.; Shen, L.; Zhang, X.; Bao, N. High-Strength GO / PA66 Nanocomposite Fibers via In Situ. *Polymers* **2021**, *13*, 1688. [CrossRef]
36. Wang, L.-N.; Wang, P.-Y.G.; Wei, J.-C. Graphene Oxide-Graft-Poly(l-lactide)/Poly(l-lactide) Nanocomposites: Mechanical and Thermal Properties. *Polymers* **2017**, *9*, 429. [CrossRef] [PubMed]
37. Rabchinskii, M.K.; Ryzhkov, S.A.; Kirilenko, D.A.; Ulin, N.V.; Baidakova, M.V.; Shnitov, V.V.; Pavlov, S.I.; Chumakov, R.G.; Stolyarova, D.Y.; Besedina, N.A.; et al. From graphene oxide towards aminated graphene: Facile synthesis, its structure and electronic properties. *Sci. Rep.* **2020**, *10*, 6902. [CrossRef] [PubMed]
38. Vlakh, E.; Ananyan, A.; Zashikhina, N.; Hubina, A.; Pogodaev, A.; Volokitina, M.; Sharoyko, V.; Tennikova, T. Preparation, characterization, and biological evaluation of poly(glutamic acid)-b-polyphenylalanine polymersomes. *Polymers* **2016**, *8*, 212. [CrossRef]
39. Zashikhina, N.; Sharoyko, V.; Antipchik, M.; Tarasenko, I.; Anufrikov, Y.; Lavrentieva, A.; Tennikova, T.; Korzhikova-Vlakh, E. Novel Formulations of C-Peptide with Long-Acting Therapeutic Potential for Treatment of Diabetic Complications. *Pharmaceutics* **2019**, *11*, 27. [CrossRef]
40. Schultz, B.J.; Dennis, R.V.; Aldinger, J.P.; Jaye, C.; Wang, X.; Fischer, D.A.; Cartwright, A.N.; Banerjee, S. X-ray absorption spectroscopy studies of electronic structure recovery and nitrogen local structure upon thermal reduction of graphene oxide in an ammonia environment. *RSC Adv.* **2014**, *4*, 634–644. [CrossRef]
41. Lazar, P.; Mach, R.; Otyepka, M. Spectroscopic Fingerprints of Graphitic, Pyrrolic, Pyridinic, and Chemisorbed Nitrogen in N-Doped Graphene. *J. Phys. Chem. C* **2019**, *123*, 10695–10702. [CrossRef]
42. Ganguly, A.; Sharma, S.; Papakonstantinou, P.; Hamilton, J. Probing the thermal deoxygenation of graphene oxide using high-resolution in situ X-ray-based spectroscopies. *J. Phys. Chem. C* **2011**, *115*, 17009–17019. [CrossRef]

43. Rabchinskii, M.K.; Varezhnikov, A.S.; Sysoev, V.V.; Solomatin, M.A.; Ryzhkov, S.A.; Baidakova, M.V.; Stolyarova, D.Y.; Shnitov, V.V.; Pavlov, S.S.; Kirilenko, D.A.; et al. Hole-matrixed carbonylated graphene: Synthesis, properties, and highly-selective ammonia gas sensing. *Carbon* **2021**, *172*, 236–247. [CrossRef]
44. Rabchinskii, M.K.; Saveliev, S.D.; Stolyarova, D.Y.; Brzhezinskaya, M.; Kirilenko, D.A.; Baidakova, M.V.; Ryzhkov, S.A.; Shnitov, V.V.; Sysoev, V.V.; Brunkov, P.N. Modulating nitrogen species via N-doping and post annealing of graphene derivatives: XPS and XAS examination. *Carbon* **2021**, *182*, 593–604. [CrossRef]
45. Aguilar-Bolados, H.; Vargas-Astudillo, D.; Yazdani-Pedram, M.; Acosta-Villavicencio, G.; Fuentealba, P.; Contreras-Cid, A.; Verdejo, R.; López-Manchado, M.A. Facile and Scalable One-Step Method for Amination of Graphene Using Leuckart Reaction. *Chem. Mater.* **2017**, *29*, 6698–6705. [CrossRef]
46. Cheng, J.; Deming, T.J. Synthesis of Polypeptides by Ring-Opening Polymerization of α-Amino Acid N-Carboxyanhydrides. *Top. Curr. Chem.* **2012**, *310*, 1–26.
47. Yu, M.; Zhang, S.; Chen, Y.; Jin, H.; Zhang, Y.; Lu, L.; Shu, Z.; Hou, S.; Xie, B.; Cui, H. A green method to reduce graphene oxide with carbonyl groups residual for enhanced electrochemical performance. *Carbon* **2018**, *133*, 101–108. [CrossRef]
48. Kolosov, A.E.; Sivetskii, V.I.; Kolosova, E.P.; Vanin, V.V.; Gondlyakh, A.V.; Sidorov, D.E.; Ivitskiy, I.I.; Symoniuk, V.P. Use of physicochemical modification methods for producing traditional and nanomodified polymeric composites with improved operational properties. *Int. J. Polym. Sci.* **2020**, *2019*, 1258727. [CrossRef]
49. Daukiya, L.; Seibel, J.; Feyter, S. De Chemical modification of 2D materials using molecules and assemblies of molecules. *Adv. Phys. X* **2019**, *4*, 1625723.
50. Rive, C.; Reina, G.; Wagle, P.; Treossi, E.; Palermo, V.; Bianco, A.; Delogu, L.G.; Rieckher, M.; Schumacher, B. Improved Biocompatibility of Amino-Functionalized Graphene Oxide in Caenorhabditis elegans. *Small* **2019**, *15*, 1902699. [CrossRef]
51. Iudin, D.; Zashikhina, N.; Demyanova, E.; Korzhikov-Vlakh, V.; Shcherbakova, E.; Boroznjak, R.; Tarasenko, I.; Zakharova, N.; Lavrentieva, A.; Skorik, Y.; et al. Polypeptide self-assembled nanoparticles as delivery systems for polymyxins B and E. *Pharmaceutics* **2020**, *12*, 868. [CrossRef] [PubMed]

polymers

MDPI

Article

Evaluation of the Antimicrobial, Thermal, Mechanical, and Barrier Properties of Corn Starch–Chitosan Biodegradable Films Reinforced with Cellulose Nanocrystals

Claudio Alonso Díaz-Cruz [1], Carolina Caicedo [2], Enrique Javier Jiménez-Regalado [3], Ramón Díaz de León [3], Ricardo López-González [3] and Rocio Yaneli Aguirre-Loredo [3,4,*]

1 Departamento de Ingeniería Química, Facultad de Ciencias Químicas, Universidad Autónoma de Coahuila, Blvd. Venustiano Carranza SN, Saltillo 25280, Coahuila, Mexico; alonsotv12@hotmail.com
2 Grupo de Investigación en Química y Biotecnología (QUIBIO), Facultad de Ciencias Básicas, Universidad Santiago de Cali, Pampalinda, Santiago de Cali 760035, Colombia; carolina.caicedo03@usc.edu.co
3 Departamento de Procesos de Polimerización, Centro de Investigación en Química Aplicada (CIQA), Blvd. Enrique Reyna Hermosillo 140, Saltillo 25294, Coahuila, Mexico; enrique.jimenez@ciqa.edu.mx (E.J.J.-R.); ramon.diazdeleon@ciqa.edu.mx (R.D.d.L.); ricardo.lopez@ciqa.edu.mx (R.L.-G.)
4 Investigadora por México CONACYT-Centro de Investigación en Química Aplicada Blvd. Enrique Reyna Hermosillo 140, Saltillo 25294, Coahuila, Mexico
* Correspondence: yaneli.aguirre@ciqa.edu.mx

Abstract: Packaging materials play an essential role in the preservation and marketing of food and other products. To improve their conservation capacity, antimicrobial agents that inhibit bacterial growth are used. Biopolymers such as starch and chitosan are a sustainable alternative for the generation of films for packaging that can also serve as a support for preservatives and antimicrobial agents. These substances can replace packaging of synthetic origin and maintain good functional properties to ensure the quality of food products. Films based on a mixture of corn starch and chitosan were developed by the casting method and the effect of incorporating cellulose nanocrystals (CNC) at different concentrations (0 to 10% w/w) was studied. The effect of the incorporation of CNC on the rheological, mechanical, thermal and barrier properties, as well as the antimicrobial activity of nanocomposite films, was evaluated. A significant modification of the functional and antimicrobial properties of the starch–chitosan films was observed with an increase in the concentration of nanomaterials. The films with CNC in a range of 0.5 to 5% presented the best performance. In line with the physicochemical characteristics which are desired in antimicrobial materials, this study can serve as a guide for the development this type of packaging for food use.

Keywords: biodegradable film; corn starch; chitosan; cellulose nanocrystals; antimicrobial film; nanocomposite

Citation: Díaz-Cruz, C.A.; Caicedo, C.; Jiménez-Regalado, E.J.; Díaz de León, R.; López-González, R.; Aguirre-Loredo, R.Y. Evaluation of the Antimicrobial, Thermal, Mechanical, and Barrier Properties of Corn Starch–Chitosan Biodegradable Films Reinforced with Cellulose Nanocrystals. *Polymers* **2022**, *14*, 2166. https://doi.org/10.3390/polym14112166

Academic Editor: Evgenia G. Korzhikova-Vlakh

Received: 27 April 2022
Accepted: 18 May 2022
Published: 26 May 2022

1. Introduction

Plastic materials are widely used globally due to their low production cost and excellent barrier, mechanical and thermal properties. Such synthetic plastics are used in various areas and products, such as food packaging, supermarket bags, toys, electronic devices, kitchen appliances, automotive parts and medical devices, among many others. The extensive use of plastic materials derived from petroleum, coupled with their inadequate disposal, has negatively impacted the environment. It is estimated that the global consumption of plastic exceeds 700 million tons per year [1]. To reduce and avoid this environmental problem, the use of polymers obtained from renewable sources to produce plastic materials is being more widely considered, mainly due to their rapid biodegradation and viability of being compostable. Polymers that can produce biodegradable packages/films can be polysaccharides such as starch, chitosan, cellulose and their derivatives, alginate and pectin;

proteins such as gelatin, zein, and collagen; or polyesters such as polyhydroxybutyrate, polylactic acid, polycaprolactone [2–6].

The use of starch as a packaging material has been widely investigated in recent years due to its high availability in nature, multiple sources of production, and economic viability [7–9]. Meanwhile, chitosan, i.e., a biopolymer obtained from waste from the fishing industry, is a product with significant added value which can be obtained in large quantities [10–12]. It has been reported that the protonated amino groups of chitosan promote the formation of intermolecular bonds with the structural matrix of starch films, giving rise to a material with better mechanical performance, good thermal stability, and better water resistance [13,14]. This biopolymer has demonstrated antimicrobial, antifungal, anti-inflammatory, and antioxidant properties, making it an excellent alternative for the production of active materials for various industries such as food, packaging, and medicine [12,15]. Chitosan is a widely investigated polymer for the preparation of composite materials due to its excellent compatibility with compounds of both natural and synthetic origin, promoted by amino, hydroxyl, and carboxyl groups [13,15]. With the mixture of both biopolymers, new, low-cost packaging materials could be generated that could also improve the quality of packaged products.

Recent studies have evaluated the use of nanometric compounds for the reinforcement of materials made from natural polymers. It has been found that reducing the particle size of the reinforcing materials promotes more homogeneous dispersion and increases the specific surface of the reinforcements, such as nanocellulose and nanoclays [16–19]. Nanocellulose is produced from cellulose, the most abundant polymer in nature, which can be obtained from the residues of agricultural byproducts. Cellulose is subjected to processes that can be enzymatic, mechanical, or chemical, in which its amorphous regions are eliminated, giving rise to a more crystalline and ordered nanometric structure [20–22]. Nanocellulose is a renewable material which is friendly to the environment, and due to its size, it can provide exciting advantages for packaging materials. The use of nanocomposites improves the thermal, mechanical, and barrier properties of packaging materials by using lower reinforcement loads (1 to 5% by volume) compared to when micrometric or larger size reinforcements are used [23]. In addition to improving the physicochemical properties of the materials, some nanocomposites have exhibited significant antimicrobial and antioxidant capacities, which would benefit the food packaging industry [5,17,22,24,25]. The food industry is currently trying to transition to cleaner products by reducing the number of additives and preservatives of synthetic origin, promoting research using various types of preservatives and antibacterials. The use of nanoparticles has been widely investigated as a substitute for various antibacterials, mainly to reduce and combat the global problem caused by the extensive use of antibiotics, which generate resistance in microorganisms [26,27] The antibacterial activities of nanomaterials depend on the properties of the nanoparticles and the bacteria of interest. Therefore, more research should be carried out, since there are many types of nanoparticles with different effects on microorganisms, depending on their nature, morphology, size, and composition.

Various studies have been carried out on the development of packaging materials using natural polymers in pure form or as mixtures, as well as nanometric materials for the improvement of physicochemical properties. The polymer–nanomaterial relationship plays a fundamental role in the development, functional properties, and application of such materials. Previously, it was possible to develop starch–chitosan composite materials with good handling and barrier characteristics; however, when seeking to incorporate a nano-sized material such as cellulose crystals, the information reported in the literature was inconsistent (0.5 to 50% *w*/*w*) [20,23,28,29]. This is due to the wide ratio when nanomaterials are incorporated into packaging materials, since their effect depends on the type of polymer, its composition, and its interactions with the matrix. As such, each nanomaterial must be evaluated to find the appropriate concentration for the material to be used.

The objective of this study was to evaluate the antimicrobial, thermal, water vapor barrier, and mechanical performance of corn starch–chitosan-based films when a reinforcing material such as cellulose nanocrystals (CNC) was incorporated at different concentrations (0, 0.5, 2.5, 5, 7.5, and 10% *w/w* biopolymers). Nanocrystals could substantially improve the properties of these biodegradable films and, in the future, be a viable alternative to synthetic food packaging.

2. Materials and Methods

2.1. Materials

Corn starch (Grain Processing Corporation, Muscatine, IA, USA) and chitosan (90% deacetylation degree, Alfadelta Materias Primas, Naucalpan, Mexico) were used without modification. Glycerol (J.T. Baker, Ciudad de Mexico, Mexico) as a plasticizing agent. Glacial acetic acid was acquired at Productos Quimicos Monterrey (Monterrey, Mexico). Cellulose nanocrystal (CNC) powder was provided by CelluForce Inc. (spray-dried powder, needle-shaped, nano-sized 7.5 nm × 150 nm, named CelluForce NCC®. Montreal, QC, Canada). To simulate environments with different relative humidities (11, 22, 32, 43, 57, 75, 84, and 90%), supersaturated salt solutions of LiCl, CH_3COOK, $MgCl_2$, K_2CO_3, NaBr, NaCl, KCl, and $BaCl_2$ were used, which were purchased from Jalmek Cientifica (Monterrey, Mexico). Trypticase soy agar, casein peptone, and trypticase soy broth manufactured by MCD LAB S.A. of C.V. (Oaxaca, Mexico) were used for the cultivation of microorganisms in antimicrobial tests.

2.2. Film Forming Solutions Preparation

For the preparation of the polymeric matrix, solutions of acetylated corn starch (5% *w/v* dispersed in water) and chitosan at 1% (*w/v*) dissolved in acetic acid (1% *v/v*) were used. A 50:50 ratio of polymeric solutions and 30% glycerol was used; these concentrations were determined according to the results obtained in a previous study [13,26]. First, the starch dispersion was prepared, and the plasticizer was added. Then, the temperature was increased. When the temperature reached 50 °C, the chitosan solution (prepared by dissolving the polymer at room temperature and stirring for 24 h) was added. Cellulose nanocrystals (CNC) were added at concentrations of 0, 0.5, 2.5, 5, 7.5, and 10% (*w/w*) with regard to the total biopolymer content. The composite film-forming solutions were constantly stirred and maintained at a temperature of 90 °C for 10 min to achieved adequate gelatinization of the corn starch. They were then left to stand for 30 min to eliminate air bubbles that had formed during the process. The polymer nanocomposite solutions (0.75 mL/cm^2) were poured into acrylic molds and dried in an oven (Thermolyne, Blue M, Blue Island, IL, USA) at 65 °C for 5 h. Composite films with 0% CNC were used as the control.

2.3. Rheology

The apparent viscosity of the filmogenic solutions was determined as a function of the nanocrystal content. A controlled stress rheometer (MCR 501 model, Anton Paar Physics, Graz, Austria) with a stainless steel cone-plate geometry of 50 mm diameter and angle of 2° was used, at a shear rate from 0.01 to 1000 s^{-1} [27]. The analysis was performed at 25 °C.

2.4. Antimicrobial Activity

Pathogenic bacteria *Staphylococcus aureus* and *Listeria monocytogenes*, which are of significant concern in the food industry, were used to determine the antimicrobial capacity of the composite films with various contents of CNC. The bacterial strains were inoculated following the methodology proposed by Gómez-Aldapa, Díaz-Cruz [28]. For the evaluation of the films, the disc diffusion technique was used, using film samples of 5 mm in diameter placed directly on the inoculated trypticase soy agar plates. The plates were incubated at 35.5 °C for 24 h. The evaluations were carried out in triplicate, and the results are expressed as mean and standard deviation.

2.5. Morphology by SEM

The morphology of the top-surface of the films was observed to determine if there was any change in the texture with an increase in the content of nanocellulose. This was performed with a SM-510 scanning electron microscope (TOPCON, Tokyo, Japan) at 5 kV at a working distance of 18 mm. Composite films were gold-palladium coated for 90 s on a Denton Vacuum model DESK II sputter (Moorestown, NJ, USA).

2.6. Thermal Properties

The effect of incorporating the CNC on the thermal resistance of the starch–chitosan films was determined using a DSC 2500 Discovery differential scanning calorimeter (TA Instruments, New Castle, DE, USA) in a temperature range of 0 to 250 °C, in a nitrogen atmosphere (50 cm^3 min^{-1}).

2.7. X-ray Diffraction

To determine if the presence and increase of the nanomaterial content modified the structural arrangement of the polymeric matrix, the latter was determined using a Siemens D500 powder diffractometer (Munich, Germany), according to the methodology suggested by Fonseca-García, Jiménez-Regalado [29].

2.8. Water Vapor Adsorption Isotherms

The water vapor absorption capacity of the materials was evaluated at room temperature (≈25 °C) according to the methodology proposed by Jiménez-Regalado, Caicedo [13], in a relative humidity (RH) range of 11 to 90% (a_w = 0.1 to 0.9). The saline solutions were prepared according to the methodology of Aguirre-Loredo, Rodriguez-Hernandez [30]. The experimental data were fitted to the Guggenheim–Anderson–de Boer (GAB) mathematical model (Equation (1)). In Equation (1), X is the moisture content absorbed by the material, Xm is the content of water that adheres directly to the material, forming the first layer of molecules, also called monolayer; C is a constant related to the sorption in the monolayer; and K is a constant associated with the sorption of water molecules in the layers after the monolayer, which forms multilayers [30].

$$X = \frac{X_m C K a_w}{(1 - K a_w)(1 - K a_w + C K a_w)} \quad (1)$$

2.9. Mechanical Properties

The mechanical properties of tensile strength and elongation at break of the materials were determined in a texture analyzer (TA.XT Express Enhanced, Stable Micro Systems, Godalming, UK) equipped with clamps. The films were cut into strips 1 cm wide by 6 cm long, with a gap of 3 cm between clamps. The equipment operated at a speed of 1 $mm \cdot s^{-1}$ [13,31]. Results are reported as the mean and standard deviation of at least ten replicates.

2.10. Water Vapor Permeability

Water vapor permeability (WVP) was evaluated according to the methodology proposed by Jiménez-Regalado, Caicedo [13] and following the desiccant method [32]. A glass permeability test dish containing silica gel was used, on which the biodegradable films were mounted. The test dish was placed in a desiccator with NaBr, which generated a pressure gradient of 2854.23 Pa. The test dishes were weighed every 40 min for 8 h. WVP are reported as the mean and standard deviation of three replicates.

2.11. Statistical Analysis

Experimental data were analyzed for statistical significance by analysis of variance (ANOVA) and Tukey's test with a $p < 0.05$ significance level with the help of statistical software OriginPro 8.5.0 SR1 (OriginLab Corporation, Northampton, MA, USA).

3. Results and Discussion

3.1. Rheology Behavior of Film Forming Solutions

A rheological study of the biopolymeric film forming solutions was carried out to examine the behavior of the nanocrystals and their effect on viscosity. Figure 1 shows the apparent viscosity behavior as a function of the shear rate of the filmogenic solutions. The results showed that the apparent viscosity of the filmogenic solutions of starch–chitosan mixtures strongly depended on the shear rate they were subjected to. The decrease was due to the disentangling of polymeric chains. Similar behavior was observed in other formulations based on chitosan, both alone and in mix with other biopolymers [33]. The addition of the cellulose nanocrystals increased the apparent viscosity of the filmogenic solutions when these did not exceed a concentration of 5%. This was probably due to crosslinks generated by the CNC forming networks that caused resistance to flow. At higher percentages of nanomaterial, the viscosity no longer increased; it remained in a range similar to that of low concentrations of CNC, but always higher than that of the control film.

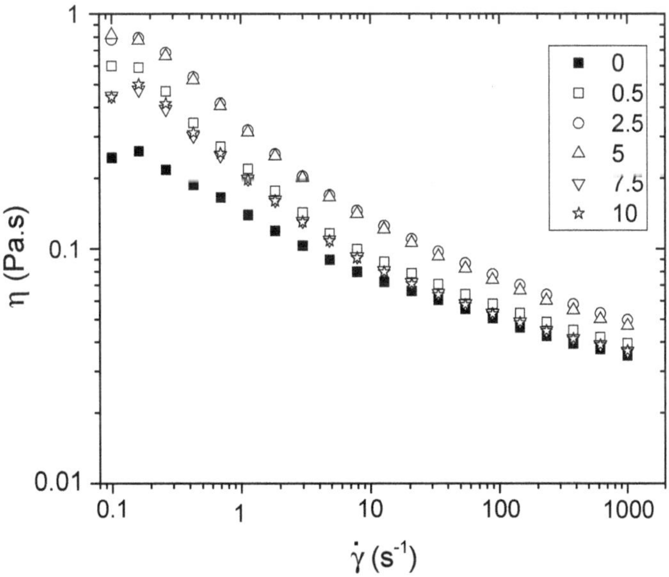

Figure 1. Log-log viscosity versus shear rate of film forming solutions of corn-starch and several ratios of CNC (*w*/*w*).

3.2. Antimicrobial Activity of the Nanocomposite Films

A package or film that is going to be in contact with food, in addition to protecting it from mechanical damage, can also prevent or delay the growth of spoilage microorganisms and some pathogenic microorganisms. In this study, the antimicrobial capacity of CNCs was evaluated when they were incorporated into corn starch–chitosan composite films. Two pathogenic microorganisms, *Listeria monocytogenes* and *Staphylococcus aureus*, were used due to their relevance to the food industry. The size of the inhibition halos (clear areas around the disks) generated by the nanocomposite film samples are presented in Table 1. An evident inhibition of the growth of both microorganisms was observed when the percentage of nanocomposites was low (0 to 5% *w*/*w*). The inhibitory effect of chitosan in the corn starch–chitosan composite film was confirmed in both *L. monocytogenes* and *S. aureus*; this test was consistent with results presented in other reports [34,35]. Chitosan is a biopolymer that has been shown to have some antibacterial capacity against both gram-positive and gram-negative bacteria [36,37]. This good antimicrobial effect has improved biodegradable materials made from natural polymers such as starch from several sources [12,38].

Table 1. Antimicrobial effect of cellulose nanocrystals (CNC) incorporated in corn starch–chitosan biodegradable films.

CNC Content (*w*/*w*) in Starch–Chitosan Films	Inhibition Halo Diameter (mm)	
	Listeria monocytogenes	*Staphylococcus aureus*
0% (control)	11.02 ± 0.59 [b]	11.00 ± 0.71 [b]
0.5%	12.52 ± 0.01 [c]	8.79 ± 0.49 [a]
2.5%	11.35 ± 0.49 [b]	11.32 ± 0.49 [b]
5.0%	9.39 ± 0.12 [a]	(--)
7.5%	(--)	(--)
10.0%	(--)	(--)

Mean ± standard deviation of three replicas. Values with different letters ([a, b, c]) in the same column denote significant differences (Tukey test; $p < 0.05$). (--): No inhibition halo was observed.

The maximum inhibitory effect was observed on *L. monocytogenes*, which showed the most extensive diameter inhibition halo (12.52 ± 0.01 mm.) when a concentration of 0.5% CNC was used. In this case, the antimicrobial effect of the content of CNC was demonstrated in contrast to the control. In *S. aureus*, on the other hand, although the evaluations showed an inhibitory effect of the starch–chitosan films with CNC concentrations lower than 5.0%, the impact was not more significant than that shown in the control films, revealing a difference in the sensitivity of both gram-positive microorganisms.

The nanomaterials (NM) or nano reinforcement composition, shape, and size play an essential role in their antibacterial capacity. The smaller the NMs, the greater their specific surface area, with the latter increasing the probability of interacting with and crossing the bacterial cell membrane [23,26,39,40].

The mechanism by which nanoparticles generate toxicity is unknown; however, efforts are still being made to elucidate it. One possible mechanism is adhesion by the electrostatic interaction of nanoparticles to the bacterial cell membrane, affecting its structural integrity [25,38]. Some NMs can also cause oxidative stress through the formation of free radicals, which alters the permeability of the cell membrane, damaging it and causing its death [24,25]. The use of nanometric-sized compounds incorporated in packaging materials has been investigated in recent years due to their interesting functional properties.

The use of nanomaterials with antimicrobial effects offers several advantages and has broad applications. Benefits include prolonged antimicrobial activity, very low environmental or health toxicity and the absence of resistance of microbes to antibiotics and other drugs [38]. These antimicrobial biopolymeric formulations can be used to develop materials for the food packaging industry, as well as in the development of various medical devices.

3.3. Surface Morphology

Visually, there was no change in the appearance of the starch–chitosan films with incorporated nanomaterials, nor with the increase in their concentration (Figure 2a, only one photography of the biodegradable materials is presented because those obtained with the other formulations were visually the same). However, when observed with a scanning electron microscope (SEM), a change in their morphology was observed with the addition of CNC. The top surface morphology of the films is shown in Figure 2.

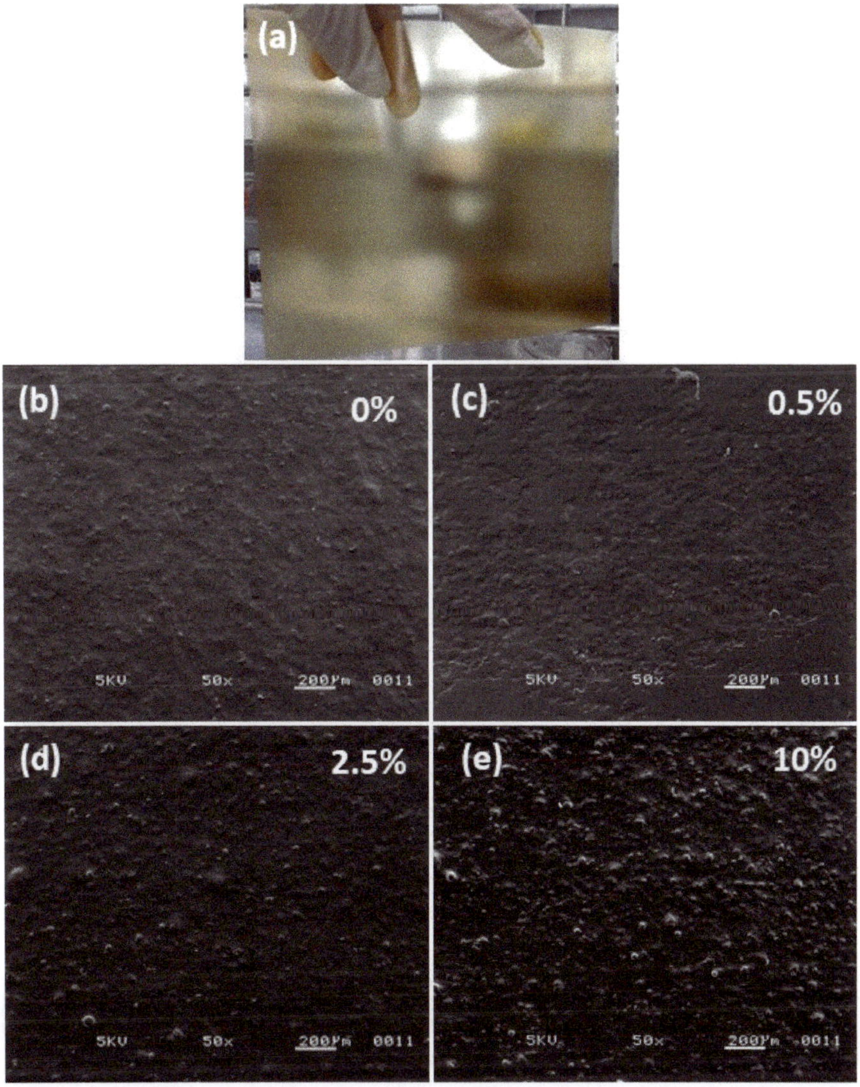

Figure 2. (**a**) Photograph and top surface morphology of corn starch–chitosan films with cellulose nanocrystals (CNC) at (**b**) 0%, (**c**) 0.5%, (**d**) 2.5%, and (**e**) 10% (*w*/*w*).

The control film (Figure 2b), that is, the film containing no nanomaterials, presented a more homogeneous appearance. An increase in surface roughness or heterogeneity was observed in the presence of CNC. The surface appeared to be better structured and less rough with low concentrations of CNC (0.5 and 2.5% *w*/*w*), as seen in Figure 2c,d. In contrast, with higher CNC content, the films had a more irregular surface (Figure 2e), which may be a consequence of the reduction in the dispersibility and an aggregation of nanomaterials [6,19,20,41]. This change in the morphology of the composite films with the addition of CNC was similar to that reported by Chen, Shi [20] for cassava starch films with cellulose crystals in micro- and nano-sizes.

3.4. Thermal Properties

Figure 3a,b show thermogravimetric analysis curves and their respective derivatives that detail the behavior of biopolymeric samples with varying contents, i.e., between 0 and 10%, of CNC. In general, the biopolymers had similar degradation temperatures (~300 °C); however, the plasticizer content and the addition of particles have been shown to significantly influence the thermal behavior at processing temperatures between 120 °C and 200 °C.

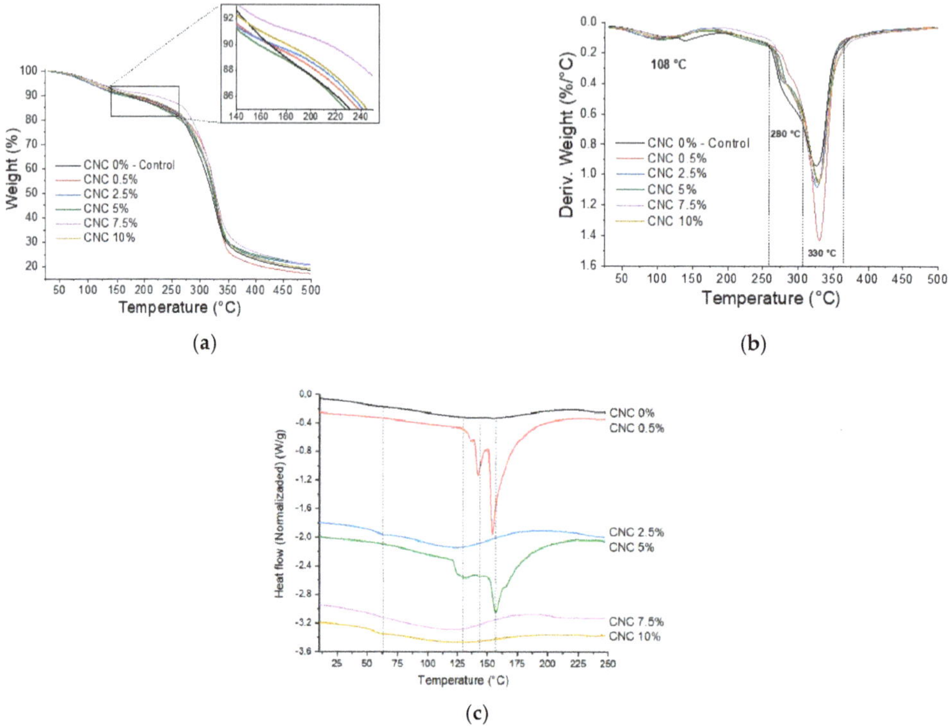

Figure 3. (**a**) TGA, (**b**) DTG curves, (**c**) DSC thermogram of biopolymer (corn starch–chitosan) and nanocomposite films incorporated with different concentrations of CNC (0 to 10% *w*/*w*).

In this study, the plasticizer content was constant. As such, the addition of nanoparticles maintained stability and, in some cases, improved it. The first zone of mass loss corresponded to the evaporation of water. Later, in the extended zone between 150 °C and 250 °C, a more pronounced negative slope (abrupt fall) of the control sample was observed. This second area represents the outlet of the plasticizer. This behavior could be explained by the interaction between the CNC and the plasticizer [13,20], or by the barrier effect of CNC that caused a delay in the release of the plasticizer [16]. Likewise, the degradation temperatures in the range of 255 °C and 365 °C (Figure 3b) presented bifurcation due to the degradation of each polymer. Similar degradation processes and degradation temperature ranges have been reported for other biomaterials based on starch and chitosan with cellulose nanoparticles [20,39]. The effect on the CNC sample was 0.5% (containing a low percentage of CNC). This result is consistent with other studies that described an adequate interaction through the formation of hydrogen bonds between the hydroxyl groups of cellulose and starch [40] and between the hydroxyl groups and the amino groups of chitosan [34].

The DSC thermogram (Figure 3c) shows curves with an amplified band between ~50 °C and ~160 °C, characteristic of partially gelatinized and plasticized starch materials. The transitions found in pure polymers are reported in a previous work showing Tg for starch and chitosan of around 57 °C and 112 °C, respectively [13]. In the same way, a chitosan-starch blend in the development of films by casting, which implies the prior mixing of film-forming solutions, revealed optimal interaction, resulting in a homogeneous morphology. Thus, a calorimetric analysis showed a decrease in the enthalpy of gelatinization (Table 2) when the destructuring of the starch granules was promoted with acid solutions (acetic acid was used in the preparation of the chitosan film-forming solution) [41]. The values of the glass transition temperature (T_g) are of low intensity (Table 2); however, the samples increased with the increase in the concentration of CNC due to the restriction in the mobility in the amorphous regions of the starch that generated intermolecular interactions between the nanocrystals and the matrix. On the other hand, the incorporation of CNC showed an irregular trend in the values of the melting temperatures due to possible variations in the crystalline domains that each concentration of CNC induced on the polymeric matrix. In the case of 0.5% and 5% (*w*/*w*) CNC, the fusion of the starch crystallites was observed around 155 °C; some authors have reported higher values (at 160 °C) [42]. The shift of this thermal transition slightly towards lower temperatures was probably due to the plasticizing effect of the CNCs on the matrix, which means that lower temperatures are needed to melt these types of nanocrystals.

Table 2. Temperature data (expressed in °C) related to thermal analysis of nanocomposite films incorporated with CNC (0 to 10% *w*/*w*).

Starch–Chitosan Film Sample	TGA		DTG	DSC	
	T_{10}	T_{30}	T_d	T_g	T_m
CNC 0%	166.9	290.7	326.1	54.7	144.9
CNC 0.5%	167.6	303.1	330.3	--	154.3
CNC 2.5%	170.5	296.4	326.9	63.4	122.5
CNC 5.0%	156.8	295.1	328.8	--	156.9
CNC 7.5%	214.6	304.9	327.8	--	120.3
CNC 10.0%	181.6	298.2	329.8	61.9	127.6

(--) It was not possible to determine.

3.5. X-ray Diffraction

Figure 4 shows the X-ray diffraction patterns of biodegradable films based on starch–chitosan with several proportions of CNC. In the starch–chitosan biodegradable films, the presence of a peak at 22.42° was an indication of the crystalline nature of the CNC nanoparticle [6,20,42], being more visible when the concentration of nanocrystals was 2.5% (*w*/*w*) and higher. These nanocrystals correspond mainly to type I cellulose, which has a high crystallinity index, as observed through a diffraction peak at 20.5° [43], which was visible in the films developed in this study with CNC contents of 5, 7 and 10% (*w*/*w*). In the control films, a behavior corresponding to a more amorphous material was observed, which became more crystalline, that is, more ordered with the incorporation of the CNC. The crystalline nature of nanocellulose helps to generate packaging materials with a more ordered structure; this is reflected in the physicochemical characteristics of the material, i.e., mainly in the thermal, mechanical, and water absorption properties, which are described in Sections 3.4, 3.6 and 3.7.

3.6. Water Vapor Adsorption Isotherms

The adsorption isotherm graphically represents the relationship between water activity (or the equilibrium relative humidity of air in the environment) and material moisture content under conditions of equilibrium and constant temperature. It also allows us to know how water binds to the material [13,44]. Various mathematical models provide information on parameters that could effectively monitor a material or food during its storage [30]. The films composed of starch and chitosan presented a water absorption capacity that changed depending on the concentration of CNC.

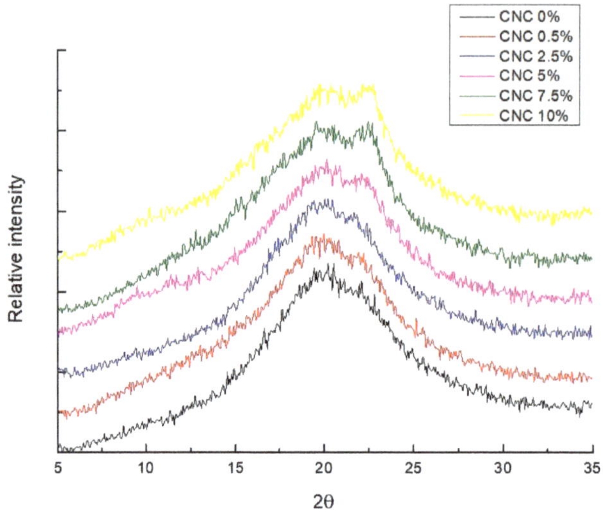

Figure 4. X-ray diffractograms of starch–chitosan films reinforced with CNC (0 to 10% w/w).

By incorporating CNC into biobased starch–chitosan materials, their water absorption capacity decreased, except when 7.5% CNC was added (Figure 5), as illustrated in the relevant X-ray diffraction pattern. The higher the crystallinity of the material, the lower the water absorption capacity, because water molecules first interact with the amorphous part of the material. However, the shape or type of isotherm was not modified with an increase in the content of nanomaterials, with each film maintaining the same behavior or slope in the total range of relative humidity used in the analysis of the water adsorption isotherms. It has been reported that when the proportion of chitosan increases in starch-based formulations, the shape of the isotherm and the moisture content that these materials can absorb increase significantly when subjected to RH conditions higher than 60% or a_w 0.6 [13]. According to Brunauer [45], the isotherms of the films evaluated in this study corresponded to those of type 2, where the adsorbate (water molecules) covers the film (adsorbent) until a monolayer is formed and the adsorption process continues in the form of a multilayer. This is a common profile in physical adsorption processes in which interactions are not very specific. Therefore, to produce this type of behavior, the affinity of the adsorbate for the adsorbent must be higher than the affinity of the adsorbate for itself [45,46]. In this type of isotherm, two identifiable regions (i.e., two different slopes on the curve) can be observed: one around a_w 0.1 to 0.4 and the other around a_w 0.6 to 0.9. This represents typical behavior of a regular food product [46]. All materials showed an increase in the amount of water they could absorb when subjected to high relative humidity, i.e., mainly in regions with a_w from 0.8 to 0.99. This range of a_w is a critical point for the preservation of food products due to the presence of free water, which is no longer directly linked to the material and is more available so that pathogenic and deteriorating bacteria can perform their vital functions and reproduce [46,47]. Foods within this range of a_w, such as eggs, meats, vegetables, and fresh fruits, are therefore potentially more susceptible to harboring pathogenic bacteria.

In this study, a reduction in the content of water that the material could adsorb was observed with an increase of CNC content in the structural matrix, with the sample containing 7.5% CNC being able to absorb even more moisture than the control film. This indicated that in this formulation, the polymeric chains were free, interacting little with each other, making them more available to bond with water molecules. This was reflected in the parameters calculated by the GAB mathematical model, as shown in Table 3.

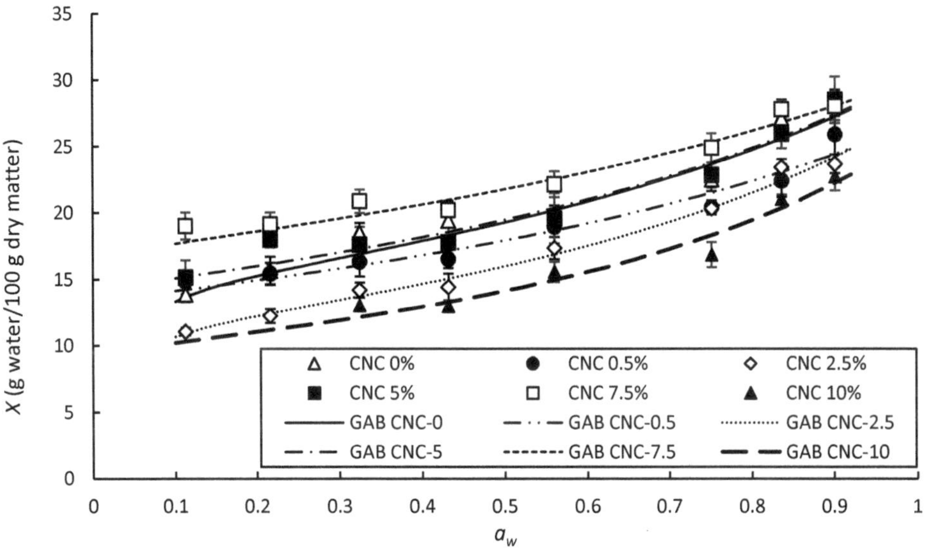

Figure 5. Water adsorption isotherms of starch–chitosan films reinforced with CNC (0, 0.5, 2.5, 5, 7.5, and 10% *w*/*w*) at 25 °C. Experimental data (symbols); fitted to the GAB mathematical model (solid and dotted lines).

Table 3. GAB mathematical model parameters and regression coefficient, R^2, calculated for composite starch–chitosan films with CNC.

CNC Content (*w/w*) in Starch–Chitosan Films	*Xm*	*C*	*K*	R^2
0%	14.67	151.59	0.53	0.984
0.5%	13.46	6283.33	0.50	0.993
2.5%	11.57	106.25	0.58	0.997
5.0%	14.32	7692.23	0.53	0.987
7.5%	17.04	2863.44	0.43	0.991
10.0%	9.73	958.54	0.62	0.987

The value of *Xm* for the sample with 7.5% CNC presented the highest value, with a water content of 17% forming the monolayer of water that interacted directly with the material. In the monolayer, it was assumed that each hydrophilic group on the material had a bound water molecule [46]. This water content, present in the monolayer phase, was much higher than that of the control film, as well as that absorbed by the films containing lower percentages of CNC, although its binding energy was lower, as expressed by the *K* parameter.

3.7. Mechanical and Gas Barrier Properties

In addition to their antimicrobial capacity, nanocomposites have been reported to improve the barrier properties against gases (water vapor and oxygen) and the mechanical properties of natural polymer-based packaging materials [16,48]. Table 4 shows the mechanical properties (tensile strength and elongation at break) and water vapor permeability of corn starch–chitosan films reinforced with CNC.

Table 4. Physicochemical properties of starch–chitosan biodegradable films reinforced with cellulose nanocrystals (CNC).

CNC Content (*w*/*w*) in Starch–Chitosan Films	Tensile Strength (MPa)	Elongation at Break (%)	WVP × 10^{-10} ($g{\cdot}m^{-1}s^{-1}Pa^{-1}$)
0%	3.49 ± 0.32 [b]	74.67 ± 5.48 [b]	3.01 ± 0.37 [a]
0.5%	2.89 ± 0.23 [a]	140.40 ± 10.62 [d]	4.84 ± 0.55 [b]
2.5%	6.16 ± 0.39 [c]	88.96 ± 8.99 [c]	2.46 ± 0.21 [a]
5.0%	6.72 ± 0.41 [d]	145.13 ± 12.74 [d]	4.33 ± 0.34 [b]
7.5%	13.61 ± 0.55 [e]	5.32 ± 0.53 [a]	4.35 ± 0.26 [b]
10.0%	6.00 ± 0.50 [c]	90.57 ± 5.08 [c]	4.30 ± 0.27 [b]

Values with different letters ([a, b, c, d, e]) in the same column denote significant differences (Tukey test; $p < 0.05$). Values are given as mean ± standard deviation (n = 8 for mechanical properties, and n = 4 for WVP).

A significant improvement in the mechanical properties of tension and elongation at fracture was observed with the addition of cellulose nanocrystals. Without presenting any trend or pattern of behavior as CNC concentration increased, in the present study, something similar to what was observed in other characterizations of these materials was observed. The addition of CNC at concentrations of 2.5% to 10% significantly increased the strength of the material, with the sample with 7.5% showing the highest value, i.e., 13.61 MPa, while the control (0% CNC) was 3.49 MPa. Regarding the elongation of the material, the film without CNC presented an elongation of 74.67%. When 0.5 and 5.0% of CNC were added to the formulation, the material increased its extension up to 140 and 145%, respectively. However, it was found that when nanocrystals were added in a proportion of 7.5%, the material was significantly affected, reducing its elongation to a value of 5.32%. In cassava starch films obtained by the casting method, it was found that incorporating only 1.5% of CNC (modified with stearic acid) improved the tensile strength of the materials by more than 300% [20]. This result was attributed to the fact that cellulose in micro- and nano-sizes has a high resistance to tension. When nanomaterials are compatible with the polymer matrix, a strong matrix-filler interaction is generated due to Van der Waals forces that can transfer stress, thus improving the resistance of the materials [49]. Nonetheless, the mechanical performance of the starch–chitosan-CNC films still cannot be compared to those of some films of synthetic origin, such as polyethylene [50].

As observed in the SEM microphotographs in Figure 2b–e, it is likely that the nanomaterials agglomerated when added in higher proportions [51]. These agglomerates may have influenced the mechanical properties of the materials. An excess of nanocomposites such as CNC negatively modified the mechanical properties of packaging materials [20], so it would be necessary to determine the optimal concentration according to the results or properties sought for a specific material.

According to the results obtained by Maradini, Oliveira [52], the tensile strength and Young's modulus of polyester can be improved with the addition of 1% (*w*/*w*) of CNC; however, when the concentration is increased to 2%, the mechanical performance is negatively affected. Although the addition of 4.5% CNC in PLA-gelatin multilayer films significantly reduced the tensile strength and elongation, this nanocomponent was not a viable alternative to reinforce this polymeric matrix [22]. When the nature of the reinforcement material and the polymeric matrix are not the same, this low compatibility can be seen as a weakness in terms of mechanical properties, i.e., the same effect as that generated when the nanomaterials have not been uniformly dispersed in the structural matrix material [53].

Water vapor permeability (WVP) is directly influenced by various material parameters, such as the hydrophilic or hydrophobic nature, the presence and type of additives, as well as the morphology of the resulting materials. Table 4 shows the WVP of the films made in this study based on their NCN content. Films containing 2.5% CNC (*w*/*w*) showed a lower WVP (2.46 × 10^{-10} $g{\cdot}m^{-1}s^{-1}Pa^{-1}$) than the control film and those with other nanocrystal concentrations. This behavior can be attributed to the hydrophobic and highly crystalline nature of the cellulose nanocrystals, in contrast to those of starch and chitosan. The addition of nanometric-sized additives can improve the structural matrix of biopolymers, generating a tortuous path and making it difficult for water molecules to pass. Other authors have reported similar results for starch-based biodegradable films reinforced with different nanometric particles [16,54]. However, a large amount of CNC could be considered excess, encouraging these nanomaterials to agglomerate and giving rise to a less homogeneous film, causing gases to flow more freely through the polymer chains and increasing their permeability [26,39]. Azeredo, Mattoso [55] reported that the presence and increase in the concentration of cellulose nanofibers in chitosan-glycerol films caused a significant reduction in WVP. A similar result was reported by Chou, Shi [56], who noted that the WVP of PVOH films decreased with the addition of CNC at 2.5% (*w*/*w*). Meanwhile, for biodegradable films made from a mixture of cassava starch, chitosan and gallic acid, the incorporation of 7.5% (*w*/*w*) cellulose nanofibers generated the material with the best water vapor barrier capacity [39].

4. Conclusions

This study made described a method for the production of biobased films from a mixture of corn starch and chitosan, which served as carriers of reinforcing nanocomposites, such as cellulose nanocrystals (CNC). It was observed that the cellulose nanomaterials were arranged differently with the polymeric chains of starch and chitosan, as reflected in their mechanical, rheological, thermal, and gas barrier performance. The small size of the CNCs helped them mix easily with the polymeric matrix; however, this interaction was limited to a certain extent. It was observed that when more nanomaterials were added, they could not disperse uniformly, which negatively affected mechanical performance, as manifest in a decrease in tensile stress. According to the results of the calorimetric analysis, a relationship was observed in the 0.5% and 5% CNC samples that presented greater elongation due to the plasticity effect and structural arrangements that allowed the structure recover under tension. The starch–chitosan films with CNC concentrations of between 0.5 and 5% presented the best functional properties, perhaps due to a more uniform distribution of the nanomaterials. The maximum inhibitory effect of the CNC nanocomposite films was observed with *Listeria monocytogenes* bacteria when 0.5% CNC was added. In comparison, the inhibitory impact on *Staphylococcus aureus* was very similar to that observed with the control films, which themselves showed a significant antimicrobial effect.

Among the critical factors in the selection of food packaging materials are the permeability and the speed of transmission of gases (water, light, oxygen, ethylene, aromas) and the mechanical performance that determines if the material will be able to maintain the integrity of the product. The films developed in this study were found to be permeable to water vapor to different degrees. The concentration of nanoparticles that can generate useful materials based on starch and chitosan can be determined according to the data presented herein. This study may be the starting point for the production of new nanocomposite materials; such antimicrobial films are currently being tested to preserve fresh fruits which are of economic importance but which are susceptible to the action of bacteria.

Author Contributions: Conceptualization, C.A.D.-C., E.J.J.-R. and R.Y.A.-L.; methodology, C.A.D.-C., E.J.J.-R. and R.Y.A.-L.; formal analysis, C.C., C.A.D.-C., E.J.J.-R. and R.Y.A.-L.; investigation, E.J.J.-R. and R.Y.A.-L.; writing—original draft preparation, C.C., C.A.D.-C., R.L.-G., R.D.d.L., E.J.J.-R. and R.Y.A.-L.; writing—review and editing, C.C., C.A.D.-C. and R.Y.A.-L.; visualization, E.J.J.-R. and R.Y.A.-L.; supervision, E.J.J.-R. and R.Y.A.-L.; project administration, R.Y.A.-L.; funding acquisition, C.C. and R.Y.A.-L. All authors have read and agreed to the published version of the manuscript.

Funding: This research was funded by Centro de Investigación en Química Aplicada (CIQA) under internal project 6610 (call 2021) and by Dirección General de Investigaciones (DGI) of Universidad Santiago de Cali under call no. 01-2022.

Institutional Review Board Statement: Not applicable.

Informed Consent Statement: Not applicable.

Data Availability Statement: The data presented in this study are available on request from the corresponding author.

Acknowledgments: Thanks to the technicians M.C. Blanca Huerta Martinez, M.C. Maria Guadalupe Mendez-Padilla, M.C. Myrna Salinas Hernández and Jesus Francisco Lara Sánchez for DRX, thermal, and SEM characterizations, respectively. Rocio Yaneli Aguirre-Loredo (R.Y.A.-L.) give thanks to CONACYT for their nomination as Investigadora por Mexico assigned to CIQA. Carolina Caicedo (C.C.) acknowledges financial support from DGI of Universidad Santiago de Cali under project No. 939-621120-2148.

Conflicts of Interest: The authors declare no conflict of interest.

References

1. ONU. *PLÁSTICOS DE UN SOLO USO: Una Hoja de Ruta Para la Sostenibilidad*; Programa de las Naciones Unidas para el Medio Ambiente: Nairobi, Kenia, 2018.
2. Cazón, P.; Velazquez, G.; Ramírez, J.A.; Vázquez, M. Polysaccharide-based films and coatings for food packaging: A review. *Food Hydrocoll.* **2017**, *68*, 136–148. [CrossRef]
3. Luo, Q.; Hossen, M.A.; Zeng, Y.; Dai, J.; Li, S.; Qin, W.; Liu, Y. Gelatin-based composite films and their application in food packaging: A review. *J. Food Eng.* **2022**, *313*, 110762. [CrossRef]
4. Chaudhary, B.U.; Lingayat, S.; Banerjee, A.N.; Kale, R.D. Development of multifunctional food packaging films based on waste Garlic peel extract and Chitosan. *Int. J. Biol. Macromol.* **2021**, *192*, 479–490. [CrossRef]
5. Wang, J.; Euring, M.; Ostendorf, K.; Zhang, K. Biobased materials for food packaging. *J. Bioresour. Bioprod.* **2022**, *7*, 1–13. [CrossRef]
6. Oyeoka, H.C.; Ewulonu, C.M.; Nwuzor, I.C.; Obele, C.M.; Nwabanne, J.T. Packaging and degradability properties of polyvinyl alcohol/gelatin nanocomposite films filled water hyacinth cellulose nanocrystals. *J. Bioresour. Bioprod.* **2021**, *6*, 168–185. [CrossRef]
7. Combrzyński, M.; Oniszczuk, T.; Kupryaniuk, K.; Wójtowicz, A.; Mitrus, M.; Milanowski, M.; Soja, J.; Budziak-Wieczorek, I.; Karcz, D.; Kamiński, D.; et al. Physical Properties, Spectroscopic, Microscopic, X-ray, and Chemometric Analysis of Starch Films Enriched with Selected Functional Additives. *Materials* **2021**, *14*, 2673. [CrossRef]
8. Tiozon, R.J.N.; Bonto, A.P.; Sreenivasulu, N. Enhancing the functional properties of rice starch through biopolymer blending for industrial applications: A review. *Int. J. Biol. Macromol.* **2021**, *192*, 100–117. [CrossRef] [PubMed]
9. Gómez-Aldapa, C.A.; Velazquez, G.; Gutierrez, M.C.; Castro-Rosas, J.; Jiménez-Regalado, E.J.; Aguirre-Loredo, R.Y. Characterization of Functional Properties of Biodegradable Films Based on Starches from Different Botanical Sources. *Starch-Stärke* **2020**, *72*, 1900282. [CrossRef]
10. Mohan, K.; Ganesan, A.R.; Ezhilarasi, P.N.; Kondamareddy, K.K.; Rajan, D.K.; Sathishkumar, P.; Rajarajeswaran, J.; Conterno, L. Green and eco-friendly approaches for the extraction of chitin and chitosan: A review. *Carbohydr. Polym.* **2022**, *287*, 119349. [CrossRef]
11. Pakizeh, M.; Moradi, A.; Ghassemi, T. Chemical extraction and modification of chitin and chitosan from shrimp shells. *Eur. Polym. J.* **2021**, *159*, 110709. [CrossRef]
12. Molina-Ramírez, C.; Mazo, P.; Zuluaga, R.; Gañán, P.; Álvarez-Caballero, J. Characterization of Chitosan Extracted from Fish Scales of the Colombian Endemic Species Prochilodus magdalenae as a Novel Source for Antibacterial Starch-Based Films. *Polymers* **2021**, *13*, 2079. [CrossRef] [PubMed]
13. Jiménez-Regalado, E.J.; Caicedo, C.; Fonseca-García, A.; Rivera-Vallejo, C.C.; Aguirre-Loredo, R.Y. Preparation and Physicochemical Properties of Modified Corn Starch-Chitosan Biodegradable Films. *Polymers* **2021**, *13*, 4431. [CrossRef] [PubMed]
14. Hasan, M.; Gopakumar, D.A.; Olaiya, N.G.; Zarlaida, F.; Alfian, A.; Aprinasari, C.; Alfatah, T.; Rizal, S.; Khalil, H.P.S.A. Evaluation of the thermomechanical properties and biodegradation of brown rice starch-based chitosan biodegradable composite films. *Int. J. Biol. Macromol.* **2020**. [CrossRef] [PubMed]
15. Madni, A.; Kousar, R.; Naeem, N.; Wahid, F. Recent advancements in applications of chitosan-based biomaterials for skin tissue engineering. *J. Bioresour. Bioprod.* **2021**, *6*, 11–25. [CrossRef]
16. Calambas, H.L.; Fonseca, A.; Adames, D.; Aguirre-Loredo, Y.; Caicedo, C. Physical-Mechanical Behavior and Water-Barrier Properties of Biopolymers-Clay Nanocomposites. *Molecules* **2021**, *26*, 6734. [CrossRef]
17. Reshmy, R.; Madhavan, A.; Philip, E.; Paul, S.A.; Sindhu, R.; Binod, P.; Pugazhendhi, A.; Sirohi, R.; Pandey, A. Sugarcane bagasse derived nanocellulose reinforced with frankincense (*Boswellia serrata*): Physicochemical properties, biodegradability and antimicrobial effect for controlling microbial growth for food packaging application. *Environ. Technol. Innov.* **2021**, *21*, 101335. [CrossRef]
18. Gao, C.; Wang, S.; Liu, B.; Yao, S.; Dai, Y.; Zhou, L.; Qin, C.; Fatehi, P. Sustainable Chitosan-Dialdehyde Cellulose Nanocrystal Film. *Materials* **2021**, *14*, 5851. [CrossRef]
19. Chen, Y.; Hanshe, M.; Sun, Z.; Zhou, Y.; Mei, C.; Duan, G.; Zheng, J.; E, S.; Jiang, S. Lightweight and anisotropic cellulose nanofibril/rectorite composite sponges for efficient dye adsorption and selective separation. *Int. J. Biol. Macromol.* **2022**, *207*, 130–139. [CrossRef]

20. Chen, Q.; Shi, Y.; Chen, G.; Cai, M. Enhanced mechanical and hydrophobic properties of composite cassava starch films with stearic acid modified MCC (microcrystalline cellulose)/NCC (nanocellulose) as strength agent. *Int. J. Biol. Macromol.* **2020**, *142*, 846–854. [CrossRef]
21. Shaikh, H.M.; Anis, A.; Poulose, A.M.; Al-Zahrani, S.M.; Madhar, N.A.; Alhamidi, A.; Alam, M.A. Isolation and Characterization of Alpha and Nanocrystalline Cellulose from Date Palm (*Phoenix dactylifera* L.) Trunk Mesh. *Polymers* **2021**, *13*, 1893.
22. Valdés, A.; Martínez, C.; Garrigos, M.C.; Jimenez, A. Multilayer Films Based on Poly(lactic acid)/Gelatin Supplemented with Cellulose Nanocrystals and Antioxidant Extract from Almond Shell By-Product and Its Application on Hass Avocado Preservation. *Polymers* **2021**, *13*, 3615. [CrossRef]
23. Butron, A.; Llorente, O.; Fernandez, J.; Meaurio, E.; Sarasua, J.-R. Morphology and mechanical properties of poly(ethylene brassylate)/cellulose nanocrystal composites. *Carbohydr. Polym.* **2019**, *221*, 137–145. [CrossRef] [PubMed]
24. Wang, L.; Hu, C.; Shao, L. The antimicrobial activity of nanoparticles: Present situation and prospects for the future. *Int. J. Nanomed.* **2017**, *12*, 1227–1249. [CrossRef]
25. Hajipour, M.J.; Fromm, K.M.; Akbar Ashkarran, A.; Jimenez de Aberasturi, D.; Larramendi, I.R.d.; Rojo, T.; Serpooshan, V.; Parak, W.J.; Mahmoudi, M. Antibacterial properties of nanoparticles. *Trends Biotechnol.* **2012**, *30*, 499–511. [CrossRef] [PubMed]
26. Dehnad, D.; Emam-Djomeh, Z.; Mirzaei, H.; Jafari, S.-M.; Dadashi, S. Optimization of physical and mechanical properties for chitosan–nanocellulose biocomposites. *Carbohydr. Polym.* **2014**, *105*, 222–228. [CrossRef] [PubMed]
27. Ascencio Carvente, P.; Maldonado Textle, H.; Soriano Moro, G.; Rivera Vallejo, C.C.; González Coronel, V.; Jiménez Regalado, E.J.; St Thomas, C. Synthesis of linear and branched hydrophobically associating multiblock copolymers via a one-pot process. *J. Polym. Res.* **2020**, *27*, 200. [CrossRef]
28. Gómez-Aldapa, C.A.; Díaz-Cruz, C.A.; Castro-Rosas, J.; Jiménez-Regalado, E.J.; Velazquez, G.; Gutierrez, M.C.; Aguirre-Loredo, R.Y. Development of Antimicrobial Biodegradable Films Based on Corn Starch with Aqueous Extract of *Hibiscus sabdariffa* L. *Starch-Stärke* **2021**, *73*, 2000096. [CrossRef]
29. Fonseca-García, A.; Jiménez-Regalado, E.J.; Aguirre-Loredo, R.Y. Preparation of a novel biodegradable packaging film based on corn starch-chitosan and poloxamers. *Carbohydr. Polym.* **2021**, *251*, 117009. [CrossRef]
30. Aguirre-Loredo, R.Y.; Rodriguez-Hernandez, A.I.; Velazquez, G. Modelling the effect of temperature on the water sorption isotherms of chitosan films. *Food Sci. Technol.* **2017**, *37*, 112–118. [CrossRef]
31. ASTM. *D882*; Standard Test Method for Tensile Properties of Thin Plastic Sheeting. ASTM International: West Conshohocken, PA, USA, 2012. Available online: http://www.randb.co.kr/wp-content/uploads/2017/12/ASTM-D882-Tensile-Properties-of-Thin-Plastic-Sheeting-%ED%95%9C%EA%B8%80.pdf?ckattempt=1 (accessed on 25 April 2022).
32. ASTM. *E96*; Standard Test Methods for Water Vapor Transmission of Materials. ASTM International: West Conshohocken, PA, USA, 2002. Available online: https://www.astm.org/standards/e96 (accessed on 25 April 2022).
33. Silva-Weiss, A.; Bifani, V.; Ihl, M.; Sobral, P.J.A.; Gómez-Guillén, M.C. Structural properties of films and rheology of film-forming solutions based on chitosan and chitosan-starch blend enriched with murta leaf extract. *Food Hydrocoll.* **2013**, *31*, 458–466. [CrossRef]
34. Costa, S.M.; Ferreira, D.P.; Teixeira, P.; Ballesteros, L.F.; Teixeira, J.A.; Fangueiro, R. Active natural-based films for food packaging applications: The combined effect of chitosan and nanocellulose. *Int. J. Biol. Macromol.* **2021**, *177*, 241–251. [CrossRef] [PubMed]
35. Serio, A.; Chaves-López, C.; Sacchetti, G.; Rossi, C.; Paparella, A. Chitosan Coating Inhibits the Growth of Listeria monocytogenes and Extends the Shelf Life of Vacuum-Packed Pork Loins at 4 °C. *Foods* **2018**, *7*, 155. [CrossRef] [PubMed]
36. Al-Tayyar, N.A.; Youssef, A.M.; Al-Hindi, R.R. Antimicrobial packaging efficiency of ZnO-SiO_2 nanocomposites infused into PVA/CS film for enhancing the shelf life of food products. *Food Packag. Shelf Life* **2020**, *25*, 100523. [CrossRef]
37. Ke, C.-L.; Deng, F.-S.; Chuang, C.-Y.; Lin, C.-H. Antimicrobial Actions and Applications of Chitosan. *Polymers* **2021**, *13*, 904. [CrossRef] [PubMed]
38. Beyth, N.; Houri-Haddad, Y.; Domb, A.; Khan, W.; Hazan, R. Alternative Antimicrobial Approach: Nano-Antimicrobial Materials. *Evid.-Based Complementary Altern. Med.* **2015**, *2015*, 246012. [CrossRef] [PubMed]
39. Zhao, Y.; Huerta, R.R.; Saldaña, M.D.A. Use of subcritical water technology to develop cassava starch/chitosan/gallic acid bioactive films reinforced with cellulose nanofibers from canola straw. *J. Supercrit. Fluids* **2019**, *148*, 55–65. [CrossRef]
40. Bangar, S.P.; Whiteside, W.S. Nano-cellulose reinforced starch bio composite films- A review on green composites. *Int. J. Biol. Macromol.* **2021**, *185*, 849–860. [CrossRef]
41. Caicedo, C.; Aguirre Loredo, R.Y.; Fonseca García, A.; Ossa, O.H.; Vázquez Arce, A.; Calambás Pulgarin, H.L.; Ávila Torres, Y. Rheological, Thermal, Superficial, and Morphological Properties of Thermoplastic Achira Starch Modified with Lactic Acid and Oleic Acid. *Molecules* **2019**, *24*, 4433. [CrossRef]
42. Nordmark, T.S.; Ziegler, G.R. Spherulitic crystallization of gelatinized maize starch and its fractions. *Carbohydr. Polym.* **2002**, *49*, 439–448. [CrossRef]
43. Díaz de León, R.; Guzmán, E.; López González, R.; Díaz Elizondo, A.; Magaña, I.; Neira, G.; Castañeda Facio, A.; Valencia, L. Surface Modification of Cellulose Nanocrystals with Lactone Monomers via Plasma-Induced Polymerization and Their Application in ABS Nanocomposites. *Polymers* **2021**, *13*, 2699. [CrossRef]
44. Aguirre-Loredo, R.Y.; Rodríguez-Hernández, A.I.; Morales-Sánchez, E.; Gómez-Aldapa, C.A.; Velazquez, G. Effect of equilibrium moisture content on barrier, mechanical and thermal properties of chitosan films. *Food Chem.* **2016**, *196*, 560–566. [CrossRef] [PubMed]

45. Brunauer, S. *The Adsorption of Gases and Vapors*; Physical adsorption; Princeton University Press: Princeton, NJ, USA, 1945; Volume 1.
46. Labuza, T.P.; Altunakar, B. Water Activity Prediction and Moisture Sorption Isotherms. *Water Act. Foods* **2020**, 161–205. [CrossRef]
47. Lin, S.-Y.; Wang, S.-L.; Wei, Y.-S.; Li, M.-J. Temperature effect on water desorption from methylcellulose films studied by thermal FT-IR microspectroscopy. *Surf. Sci.* **2007**, *601*, 781–785. [CrossRef]
48. Trache, D.; Hussin, M.H.; Haafiz, M.K.M.; Thakur, V.K. Recent progress in cellulose nanocrystals: Sources and production. *Nanoscale* **2017**, *9*, 1763–1786. [CrossRef]
49. Chen, J.; Long, Z.; Wang, J.; Wu, M.; Wang, F.; Wang, B.; Lv, W. Preparation and properties of microcrystalline cellulose/hydroxypropyl starch composite films. *Cellulose* **2017**, *24*, 4449–4459. [CrossRef]
50. Karel, M.; Lund, D.B. *Physical Principles of Food Preservation: Revised and Expanded*; CRC Press: Boca Raton, FL, USA, 2003; Volume 129.
51. Shih, Y.-T.; Zhao, Y. Development, characterization and validation of starch based biocomposite films reinforced by cellulose nanofiber as edible muffin liner. *Food Packag. Shelf Life* **2021**, *28*, 100655. [CrossRef]
52. Maradini, G.d.S.; Oliveira, M.P.; Carreira, L.G.; Guimarães, D.; Profeti, D.; Dias Júnior, A.F.; Boschetti, W.T.N.; Oliveira, B.F.d.; Pereira, A.C.; Monteiro, S.N. Impact and Tensile Properties of Polyester Nanocomposites Reinforced with Conifer Fiber Cellulose Nanocrystal: A Previous Study Extension. *Polymers* **2021**, *13*, 1878. [CrossRef]
53. Nascimento, N.R.d.; Pinheiro, I.F.; Alves, G.F.; Mei, L.H.I.; Macedo Neto, J.C.d.; Morales, A.R. Role of cellulose nanocrystals in epoxy-based nanocomposites: Mechanical properties, morphology and thermal behavior. *Polímeros Ciência E Tecnologia* **2022**, *31*. [CrossRef]
54. López, O.V.; Castillo, L.A.; García, M.A.; Villar, M.A.; Barbosa, S.E. Food packaging bags based on thermoplastic corn starch reinforced with talc nanoparticles. *Food Hydrocoll.* **2015**, *43*, 18–24. [CrossRef]
55. Azeredo, H.M.C.; Mattoso, L.H.C.; Avena-Bustillos, R.J.; Filho, G.C.; Munford, M.L.; Wood, D.; McHugh, T.H. Nanocellulose Reinforced Chitosan Composite Films as Affected by Nanofiller Loading and Plasticizer Content. *J. Food Sci.* **2010**, *75*, N1–N7. [CrossRef]
56. Chou, C.-T.; Shi, S.-C.; Chen, C.-K. Sandwich-Structured, Hydrophobic, Nanocellulose-Reinforced Polyvinyl Alcohol as an Alternative Straw Material. *Polymers* **2021**, *13*, 4447. [CrossRef] [PubMed]

polymers MDPI

Article

Biomimetic Hierarchical Structuring of PLA by Ultra-Short Laser Pulses for Processing of Tissue Engineered Matrices: Study of Cellular and Antibacterial Behavior

Albena Daskalova [1,*], Liliya Angelova [1], Emil Filipov [1], Dante Aceti [1], Rosica Mincheva [2], Xavier Carrete [2], Halima Kerdjoudj [3,4], Marie Dubus [3,4], Julie Chevrier [3,4], Anton Trifonov [5] and Ivan Buchvarov [5]

1 Laboratory of Micro and Nano-Photonics, Institute of Electronics, Bulgarian Academy of Sciences, 1784 Sofia, Bulgaria; lily1986@abv.bg (L.A.); emil.filipov95@gmail.com (E.F.); acetidm@gmail.com (D.A.)
2 Laboratory of Polymeric and Composite Materials (LPCM), Center of Innovation and Research in Materials and Polymers (CIRMAP), University of Mons, 7000 Mons, Belgium; Rosica.MINCHEVA@umons.ac.be (R.M.); Xavier.CARETTE@umons.ac.be (X.C.)
3 Bomatériaux et Inflammation en Site Osseux BIOS, Université de Reims Champagne Ardenne, EA 4691, 51100 Reims, France; halima.kerdjoudj@univ-reims.fr (H.K.); marie.dubus@hotmail.com (M.D.); julie.e.chevrier@orange.fr (J.C.)
4 UFR d'odontologie, Université de Reims Champagne Ardenne, 51100 Reims, France
5 Faculty of Physics, St. Kliment Ohridski University of Sofia, 1164 Sofia, Bulgaria; a.trifonov@phys.uni-sofia.bg (A.T.); ivan.buchvarov@phys.uni-sofia.bg (I.B.)
* Correspondence: albdaskalova@gmail.com

Citation: Daskalova, A.; Angelova, L.; Filipov, E.; Aceti, D.; Mincheva, R.; Carrete, X.; Kerdjoudj, H.; Dubus, M.; Chevrier, J.; Trifonov, A.; et al. Biomimetic Hierarchical Structuring of PLA by Ultra-Short Laser Pulses for Processing of Tissue Engineered Matrices: Study of Cellular and Antibacterial Behavior. *Polymers* **2021**, *13*, 2577. https://doi.org/10.3390/polym13152577

Academic Editor: Evgenia G. Korzhikova-Vlakh

Received: 9 July 2021
Accepted: 1 August 2021
Published: 3 August 2021

Abstract: The influence of ultra-short laser modification on the surface morphology and possible chemical alteration of poly-lactic acid (PLA) matrix in respect to the optimization of cellular and antibacterial behavior were investigated in this study. Scanning electron microscopy (SEM) morphological examination of the processed PLA surface showed the formation of diverse hierarchical surface microstructures, generated by irradiation with a range of laser fluences (F) and scanning velocities (V) values. By controlling the laser parameters, diverse surface roughness can be achieved, thus influencing cellular dynamics. This surface feedback can be applied to finely tune and control diverse biomaterial surface properties like wettability, reflectivity, and biomimetics. The triggering of thermal effects, leading to the ejection of material with subsequent solidification and formation of raised rims and 3D-like hollow structures along the processed zones, demonstrated a direct correlation to the wettability of the PLA. A transition from superhydrophobic ($\theta > 150°$) to super hydrophilic ($\theta < 20°$) surfaces can be achieved by the creation of grooves with V = 0.6 mm/s, F = 1.7 J/cm^2. The achieved hierarchical architecture affected morphology and thickness of the processed samples which were linked to the nature of ultra-short laser-material interaction effects, namely the precipitation of temperature distribution during material processing can be strongly minimized with ultrashort pulses leading to non-thermal and spatially localized effects that can facilitate volume ablation without collateral thermal damage The obtained modification zones were analyzed employing Fourier transform infrared (FTIR), X-ray photoelectron spectroscopy (XPS), Energy dispersive X-ray analysis (EDX), and optical profilometer. The modification of the PLA surface resulted in an increased roughness value for treatment with lower velocities (V = 0.6 mm/s). Thus, the substrate gains a 3D-like architecture and forms a natural matrix by microprocessing with V = 0.6 mm/s, F = 1.7 J/cm^2, and V = 3.8 mm/s, F = 0.8 J/cm^2. The tests performed with Mesenchymal stem cells (MSCs) demonstrated that the ultra-short laser surface modification altered the cell orientation and promoted cell growth. The topographical design was tested also for the effectiveness of bacterial attachment concerning chosen parameters for the creation of an array with defined geometrical patterns.

Keywords: cell matrices; Fs bioscaffolds structuring; ultra-short functionalization of cell matrices; tissue engineering; temporal scaffolds; PLA texturing

1. Introduction

According to data from World Population Prospects: the 2019 Revision, (reference date: 01.07 2020), 24.2% of all the people in the world are over the age of 50 [1]. This fact determines the recrudescence of diseases such as bone fragilities, osteoarthritis, cartilage, and cardiovascular pathologies as one of the major problems of modern societies. Tissue engineering, as an innovative field in biotechnology, deals with the creation, regeneration, and improvement of the function of injured or missing tissues. The design of smart porous biocompatible and biodegradable scaffolds, in association with specific tissue cells and growth factors, plays a crucial role in the regeneration of tissues and recovery of their functionality [2]. The main purpose of the engineered scaffold is to create a biomimetic environment that stimulates cell adhesion, proliferation, and differentiation; as a result, the ingrowing cells are reorganized into new three-dimensional tissue [3]. During this gradual biodegradation of the scaffold, a new tissue with the desired properties is naturally formed. The selection of appropriate biocompatible source materials is crucial for successful scaffold fabrication [4,5]. Polylactic acid (PLA) has excellent mechanical properties (stability, toughness, elasticity, and strength), biocompatibility, biodegradability, ease of processing, and thermal stability [6–8]. The up-listed properties are the main features for the successful application of PLA-engineered constructs in the regeneration of tissues, undergoing mechanical stress under normal physiological conditions [3,9]. PLA is an aliphatic polyester with backbone formula $(C_3H_4O_2)_n$ or $[–C(CH_3)HC(=O)O–]_n$, approved by the U.S. Food and Drug Administration (FDA) for different biomedical applications, [10] for which the synthetic biopolymer is in direct contact with biological fluids [6,11]. An important advantage of this polymer is that it is naturally resorbed as the physiological repair process takes over the main product of PLA metabolism in mammals is lactic acid [12]. It is incorporated in the carboxylic acid cycle and completely excreted by natural degradation pathways [13]. The biocompatibility of the scaffold generally refers to inflammatory response, as well as to the intrinsic micro and nanoscale structure of the biosurfaces. On the other hand, PLA is a hydrophobic polymer, which lowers its bioactivity, regarding the interaction of the scaffolds with cells, tissue, and fluids [14–22]. Additional surface functionalization is needed for the creation of effective bio-interfaces between the cells seeded and the PLA-based scaffold [23]. This could be achieved by different chemical methods, most of which, however, use organic compounds in order to control surface properties of the tissue scaffolds (by modifying surface chemistry), leaving residual toxic components in the cells environment and thus may lead to risks of cell toxicity and carcinogenicity [24]. As an alternative, the properties of various cell matrices could also be controlled by the means of different physical methods (by controlled removal, addition, or deformation of the biomaterial surface in order to increase its roughness) [25,26]. The most innovative and bio-friendly approach are the laser-based techniques due to their non-contact nature of interaction with the material and thus the absence of chemicals used for additional treatment of the created cell/tissue matrices. In the literature, there is scarce information concerning femtosecond laser modification of PLA. Irradiation with ultra-short laser pulses has been extensively studied in the last years, especially for the functionalization of tissue-engineered scaffolds, based on various biomaterials [27,28]. Femtosecond (Fs) laser-based processing is an alternative approach, characterized by extremely high processing accuracy and the absence of thermal deformations on the processed material due to its extremely high-peak power and pulse duration which is shorter than material thermal relaxation time. Ultra-short pulses lead to non-thermal and spatially localized effects, a result of which volume ablation occurs without collateral thermal damage to the surrounding zones. The effects of temperature distribution during the interaction with extremely short pulses is strongly minimized. The creation of micro and nanostructures on the surface of the biomaterial can strongly affect cell behavior, as diverse surface roughness (in macro, micro, and nanoscale) can be achieved by controlling the applied laser parameters [29]. Such surface feedback can be used to finely tune and control diverse properties of the processed material like wettability and biomimetics. Wettability is an important factor in cell adhesion, and the fs laser surface modification has the ability to aid in tuning it in respect to the effect of interest—namely achieving cell adhesive/antiadhesive scaffold surfaces, needed for diverse tissue regenerative applications [30–33] or even the creation of enhanced antibacterial cell surface environment [34]. Yada and Terakawa even achieved the generation of laser-induced periodic surface structures (LIPSS) on a highly biocompatible poly-L-lactic acid scaffold by femtosecond laser processing [35]. These structures are examined for cell behavior control. In the current study, PLA polymer surfaces were modified by Ti: sapphire femtosecond laser at diverse laser energies and scanning velocities were applied. Different surface treatments were used to obtain structures with hierarchical geometries for future preparation and design of porous PLA-based cellularized scaffolds and bio-interfaces between the tissues of the recipient and the foreign implant. Laser microstructured PLA scaffolds were investigated by SEM, EDX, XPS, optical profilometer studies, wettability, and FTIR. The thickness of the samples, before and after ultra-fast laser modification, was compared. In vitro degradation tests and preliminary cellular studies were accomplished. Mesenchymal stem cells from Wharton Jelly (MSCs) were cultured on PLA scaffolds for evaluating the effect of the patterned polymer surface on cell proliferation and cell morphology from day 1 to 14, compared to control PLA matrix. The adhesion of *Staphylococcus aureus* was also evaluated for the potential antibacterial effect which laser treatment could achieve. Thus, the aim of this examination will be directed towards achievement of conditions, associated with regime above modification threshold, and which gives a feedback concerning to which limit the PLA material could be processed without triggering chemical state alterations. This would potentially lead to the optimization of laser patterning conditions in order to achieve desired improved morphologies for cellular adhesion and bacterial rejection. The results obtained demonstrate that the precise control of the laser parameters applied to the PLA scaffolds could generate bioactive surfaces by means of the nondestructive Fs laser modification technique.

2. Materials and Methods

2.1. Preparation of PLA Samples

Polylactic acid (PLA) in the form of an amorphous compound was purchased from Nature Works, Nebraska, USA (PLA 4060D). Compression molding was used for the PLA samples preparation—the thermos compression of the PLA pellets was performed by Carver 4122 12-12H Manual Heated Press (Carver Inc., Wabash, IN, USA). First, the pellets were vacuum dried overnight at 60 °C (Tg PLA = 60 °C), then inserted in the mold (square of 0.5 mm thick). The procedure used for the compression consists of the following steps: 3 min of contact at 180 °C, several cycles of degassing, and 2 min at 12 bars. The as prepared PLA was cut into squares of 2 × 2 cm.

2.2. Fs Laser Experimental SETUP

The synthesized PLA samples were laser processed by Quantronix- Integra-C Ti: sapphire (Hamden, CT, USA) Fs laser system (τ = 150 fs) at a central wavelength of λ = 800 nm and 0.5 kHz repetition rate Figure 1. For samples positioning perpendicular to the laser beam, a high-precision XYZ translation stage was used. All the experiments were performed in the air by scanning the laser beam over the material surface at precisely defined separation intervals (in the X and Z axis) in order to optimize the PLA surface texturing. All modifications of the PLA plates, performed by the Integra-C laser system were precisely controlled by LabView software. On the basis of previously performed large scale preliminary experiments on PLA, part of which are already published [36], the following laser parameters were chosen for surface structuring of the material: fluence F = 1.7 and 0.8 J/cm^2 and scanning velocity V = 16, 3.8, 1.7, and 0.6 mm/s for both F values used. Each laser processed PLA scaffold was analyzed with respect to the control surface.

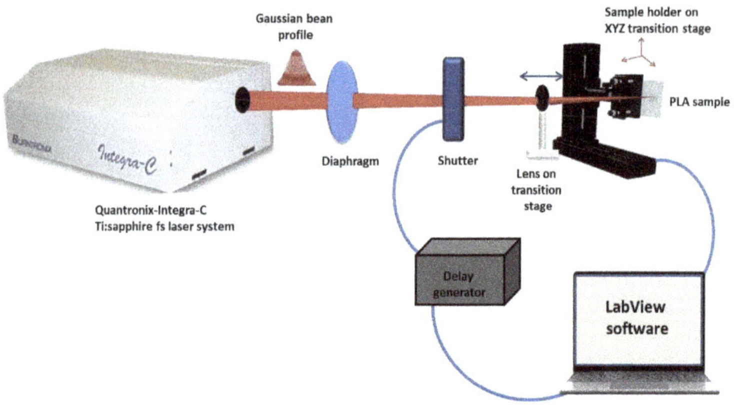

Figure 1. A scheme of the optical setup for Fs laser patterning of PLA sample positioned perpendicular to the laser beam on XYZ translation stage.

2.3. *Laser-Modified Samples Qualitative Characterization Methods*

2.3.1. Thickness Measurement

For determination of the thickness of the PLA samples before and after laser treatment, a coating thickness gauge VA 8042 coating meter (Shenzhen, China) was used, and every value presented (in μm) is averaged over 10 separate measurements.

2.3.2. Roughness Measurement

For an evaluation of the surface topography of the PLA samples, before and after fs laser treatment, 2D and 3D roughness analyes were performed via 3D Optical profiler, Zeta-20 (Zeta Instruments, KLA, Milpitas, CA, USA). It is a fully integrated microscope (magnifications from 5× to 100×) that provides 3D imaging as it scans a sample over specified vertical range. The Zeta Optics Module of the profilometer records the XY location and the precise Z height at each Z position scanned (vertical (Z) resolution < 1 nm). As a result, true color 3D and 2D composite images are created and roughness parameters R_a (arithmetical mean height of a line—the mean value of the deviations of the surface height from the median line, according to DIN4776 standards) and S_a (the extension of R_a to a surface) are obtained. Every value for R_a and S_a is averaged over 5 roughness measurements of every Fs-processed and control PLA sample surface area. For better visualization and data processing ProfilmOnline software (https://www.profilmonline.com, accessed on 2 June 2021) was additionally used.

2.3.3. WCA Evaluation

The water contact angle measurement measurements were performed in air with distilled water drop with volume of 1 μL on control and laser modified surfaces of the PLA samples for the time period of 0.5–7 s—Figure 2. The samples were positioned on XYZ horizontal translation stage. The measurements were performed according to a contact angle goniometry method—the droplet was deposited by a fixed above the sample micropipette (Dlab, Beijing, China ISO9001/13485 certified – 0.1–10 μL). The process was captured by a high-resolution camera (Huawei P40 pro camera in super macro mode with additional macro lens—Huawei Investment & Holding Co., Ltd., Shenzhen, China). For WCA evaluation, the images obtained were analyzed by ImageJ software equipped with contact angle measurement plug-in. Measurements were performed independently in two directions—along and perpendicular to the Fs grooves in respect to control PLA surface for a period of 7 s. Every WCA value presented is averaged over 10 separate measurements.

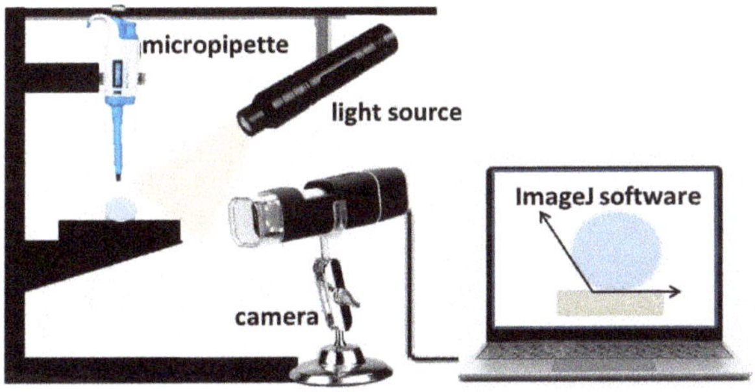

Figure 2. Experimental setup for measuring the water contact angle (WCA) of PLA samples.

2.3.4. SEM-EDX Analysis

The surface morphology and elemental composition, before and after Fs laser treatment were analyzed by Scanning Electron Microscopy (SEM) Hitachi SU8020 (Hitachi High-Technologies Corporation, Tokyo, Japan), with field emission gun with landing energy at 5 kV and SE (UL) detector. The samples were sputter- coated with a nm layer of gold (Au) and several images for every modification were taken at magnification 100×, 500×, 5000×. Energy Dispersive X-ray analysis (EDX) was also performed—the corresponding elemental composition [wt.%] was defined.

2.3.5. FTIR Analysis

Fourier-Transform Infrared (FTIR) spectrophotometer (IR Affinity-1, Shimadzu, Kyoto, Japan), with a resolution of 4 cm^{-1} and working range 4500–500 cm^{-1}, was used for obtaining the IR transmittance spectra [%] of the Fs treated PLA plates for evaluation of possible chemical bonds alterations after laser treatment. Every measurement was related to a control PLA spectrum [37].

2.3.6. XPS Analysis

AXIS Supra electron spectrometer (Kratos Analytical Ltd., Manchester, UK) was used for X-ray photoelectron spectroscopy (XPS) measurements of control and laser processed PLA samples. High-resolution detailed spectra of C_{1s} and O_{1s} (%) were taken for identifying the elemental composition and chemical state of the PLA samples of interest (laser treated with F = 1.7 and 0.8 J/cm^2, V = 0.6 and 3.8 mm/s, respectively), according to control surface.

2.3.7. In Vitro Degradation Test

A selected group of fs laser modified PLA samples, as well as control samples were subjected to in vitro degradation test—the samples were stored in an incubator at 37 °C for 8 weeks in PBS buffer saline (pH 7.2, Sigma-Aldrich, St. Louis, MO, USA). Each sample was put in 3 mL PBS, the pH of the solution was checked every week with a pH meter (VAT1011, V & A Instrument Co., Ltd., Shanghai, China) with an external sensor, and the PBS solution was replaced with a fresh buffer.

2.4. Cell Behavior on Laser-Textured PLA Matrices

Two groups of Fs treated samples were chosen for preliminary cellular experiments—Group 1 (G1)—PLA processed with F = 0.8 J/cm^2, V = 3.8 mm/s and Group 2 (G2)—PLA processed with F = 1.7 J/cm^2, V = 0.6 mm/s. Each group consisted of 30 samples; in order to confirm the obtained results the performed cell seeding experiments were repeated twice, as parallel cell seeding was performed on control (laser untreated PLA samples—Group 3 (G3)). PLA samples from the three groups were cleaned with 70% ethanol for 15 min in a 12-well plate Figure 3.

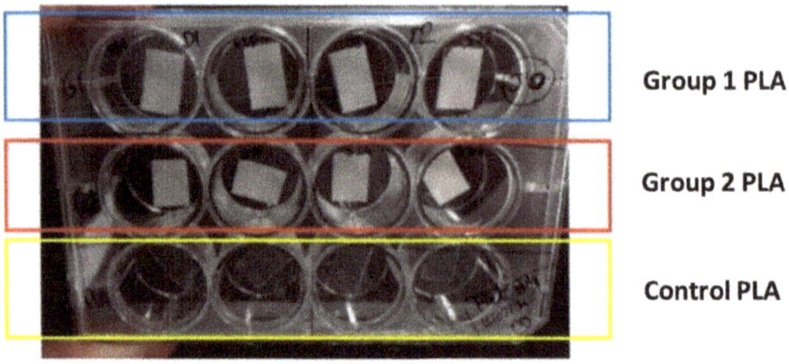

Figure 3. Preparation procedure of cell culture studies.

The plates were placed under a laminar flow hood to help the remaining ethanol to evaporate and then sterilized under a UV lamp, for 20 min, before any other manipulation. Mesenchymal stem cells from Wharton Jelly (MSCs) were seeded at a density of 30,000 cells per sample and were cultured for 14 days. WST-1 cell proliferation assay (Roche Diagnostics, Risch-Rotkreuz, Switzerland) was performed on day 7, day 10, and day 14 on MSCs in accordance with the manufacturer protocol—Figure 4. Absorbance was measured at 440 nm using a FLUOstar Omega microplate reader (BMG Labtech, Aylesbury, England) against a background control as blank. A wavelength of 750 nm was used as the correction wavelength.

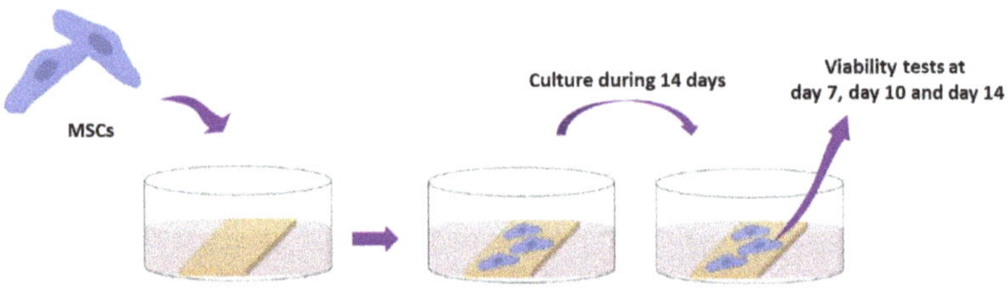

Figure 4. Cell viability test performed at day 7, day 10 and day 14.

After 14 days of culture, MSCs were fixed with 4% (*w*/*v*) paraformaldehyde (Sigma-Aldrich, Missouri, United States) at 37 °C for 10 min and permeabilized with 0.5% (*v*/*v*) Triton ×100 for 5 min. Alexa® Fluor-488 conjugated-Phalloidin® (Abcam, Cambridge, UK)—1/100 dilution in 0.1% Triton ×100 was used to stain F-actin cytoskeleton for 45 min at room temperature. Nuclei were counter-stained with 4,6-diamidino-2-phenylindole (DAPI, 100 ng/mL, 1/10,000 dilution) for 5 min.-Stained cells were mounted and imaged by epifluorescence microscopy (Zeiss microscope, ×20, Carl Zeiss AG, Jena, Germany).

2.5. Microbiology Studies

Duplicate PLA samples from the second group (**G2**)—processed with F = 1.7 J/cm^2, V = 0.6 mm/s were chosen for performing antibacterial studies. Parallel seeding was performed on control, laser untreated PLA samples—Group 3 (**G3**). Both groups consisted of 12 samples; experiments were performed twice in order to confirm the obtained results. All samples were incubated with 70% ethanol for 15 min in a 12-well plate. The remaining ethanol was thrown out, and plates were placed under a laminar flow hood to help ethanol evaporate. Plates were then sterilized under a UV lamp for 20 min. The adhesion of *Staphylococcus aureus* (SH1000 strain), commonly found in nosocomial infections, was evaluated by incubating the PLA scaffolds for 24 h with a *S. aureus* suspension. Then, *S. aureus* adhered on **G2** and **G3** PLA samples were enumerated to obtain the number of adhered bacteria on each sample. Three independent experiments were carried out, on samples from each group, by the protocol presented on Figure 5.

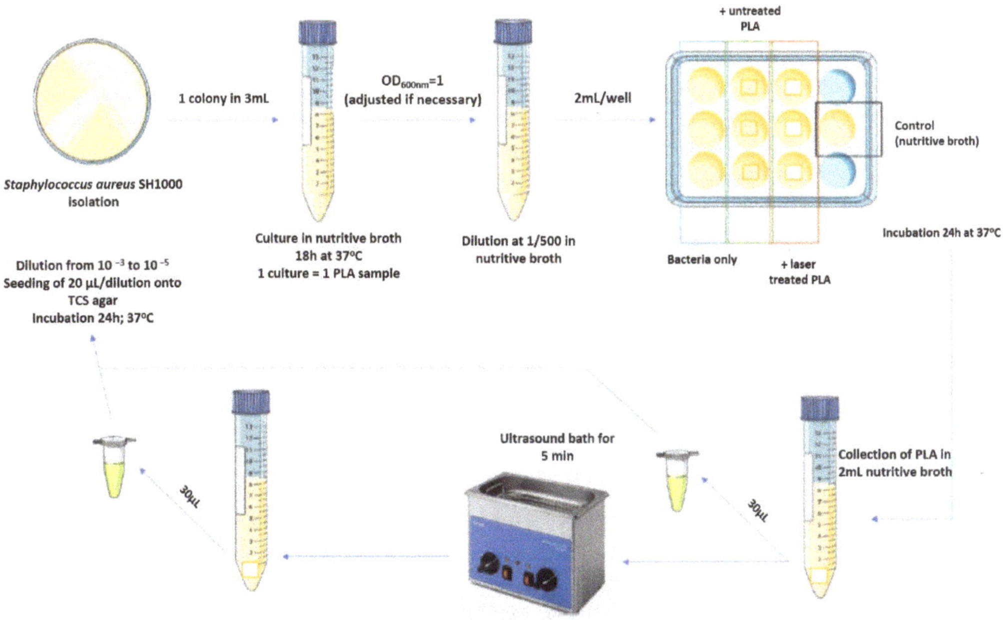

Figure 5. Scheme of the used microbiology protocol.

3. Results and Discussion

3.1. Morphological, Structural and Chemical Characterization of Fs Modified PLA Specimens

After synthesis and Fs modification of PLA samples, SEM analysis was performed for evaluating the surface morphological change, according to applied laser parameters, chosen for surface structuring of the samples: fluence F = 0.831 J/cm^2 and scanning velocity V = 16, 3.8, 1.72 and 0.551 mm/s (Figure 6a–d, respectively) and F = 1.663 J/cm^2 and V = 16, 3.8, 1.72 and 0.551 mm/s (Figure 6e–h, respectively). Each Fs modification of the PLA scaffolds is observed at 100×, 500× and 5000× magnification. The corresponding values of C [wt.%] and O [wt.%] of the performed EDX elemental composition analysis are given in Table 1.

Table 1. EDX elemental composition values [wt.%] of the control and Fs laser modified PLA samples.

Sample (PLA)	C [wt.%]	O [wt.%]
Control	74.27	25.73
V = 16 mm/s, F = 0.8 J/cm^2	68.05	31.95
V = 3.8 mm/s, F = 0.8 J/cm^2	61.51	38.49
V = 1.7 mm/s, F = 0.8 J/cm^2	60.52	39.48
V = 0.6 mm/s, F = 0.8 J/cm^2	58.82	41.18
V = 16 mm/s, F = 1.7 J/cm^2	62.48	37.52
V = 3.8 mm/s, F = 1.7 J/cm^2	60.42	39.58
V = 1.7 mm/s, F = 1.7 J/cm^2	59.09	40.91
V = 0.6 mm/s, F = 1.7 J/cm^2	58.32	41.68

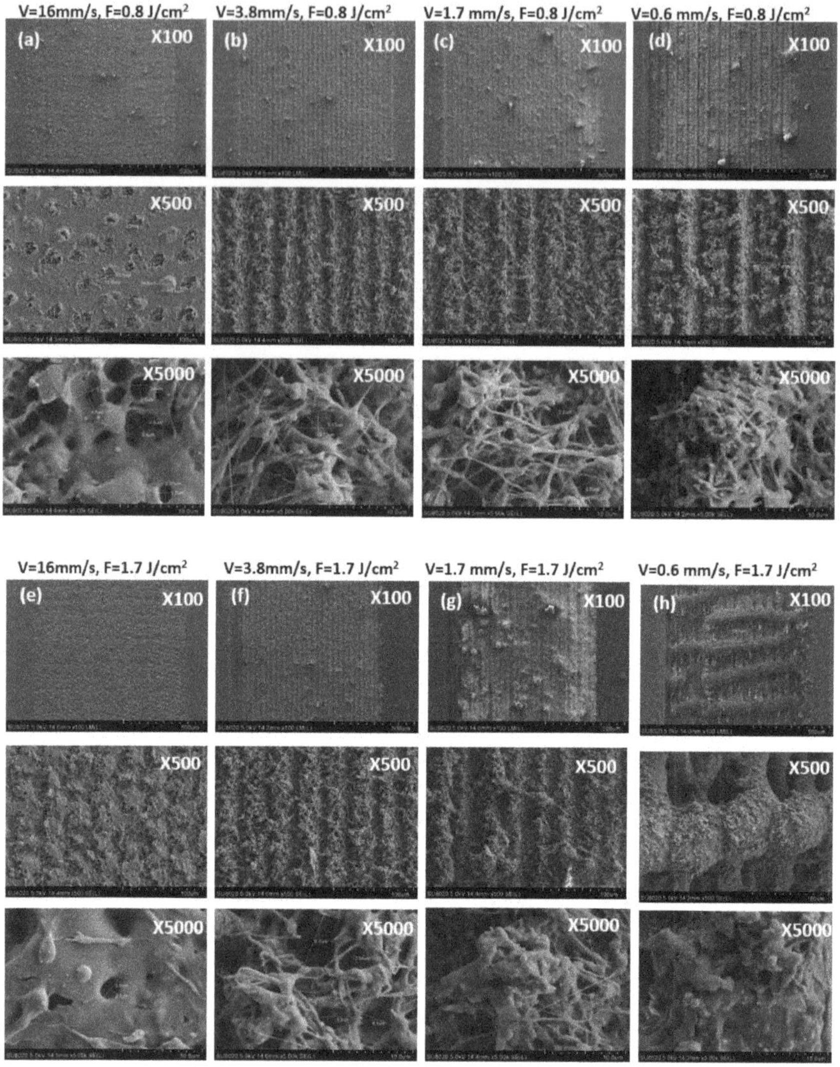

Figure 6. SEM images of PLA matrices: (**a**–**d**) patterned with F = 0.8 J/cm^2 and V = 16, 3.8, 1.7, 0.6 mm/s, respectively (**e**–**h**) and F = 1.7 J/cm^2 and V = 16, 3.8, 1.7, 0.6 mm/s respectively.

The surface of the control PLA matrix has a value of S_a = 0.41 µm, compared to the laser treated samples where change of the S_a values in the range of 0.81 ÷ 11.84 µm was monitored; see Table 2. By varying the laser parameters, very precise control over surface and even volume structuring could be achieved—from very gentle surface round formations at V = 16 mm/s, F = 0.8 J/cm^2, through holes formation and ejection of material around the zones of interaction and formation of grooves with different depths and widths (Table 2) for V = 3.8, 1.7, 0.6 mm/s, F = 0.8 J/cm^2 and V = 16, 3.8, 1.7 mm/s, F = 1.7 J/cm^2. We observed the creation of self-formed complex frameworks with two types of periodicities—parallel and perpendicular to the direction of the laser processing is achieved at V = 0.6 mm/s, F = 1.7 J/cm^2—Figure 6h. The interaction with the highest values of V (V = 16 mm/s, for both values of F) with the PLA surface leads to formation of single craters saturated with porous structures Figure 6a,e. By further increase of applied laser fluence and lowering the scanning velocity transformation into the groove-like structure is observed with increment in depth and width. Thus, a 3D foam-like porous structure is formed on the treated PLA surface. These hierarchical forms represent a very fine network of filaments, which is clearly seen in the images with the highest magnification (V = 16, 3.8 mm/s at both F—Figure 6b,c,f,g. In the case of V = 0.6 mm/s, F = 1.7 J/cm^2, Figure 6h, two types of perpendicular to each other periodical structures are observed. Such laser-induced surface structures are formed due to the elevation of the material and are known to be very effective for controlling cell behavior. In addition to that, several groups state that these structures could lead to an actual reduction in bacterial biofilm formation and thus contribute to the creation of scaffolds with enhanced antibacterial properties [34].

Table 2. Thickness [μm] and Roughness (Mean average roughness (R_a) and Areal surface roughness (S_a) [μm]) of the laser processed PLA samples, compared to control surface.

Sample (PLA)	Thickness [μm]	R_a [μm]
Control	250	0.024
V = 16 mm/s, F = 0.8 J/cm^2	228	0.53
V = 3.8 mm/s, F = 0.8 J/cm^2	215	1.92
V = 1.7 mm/s, F = 0.8 J/cm^2	219	2.4
V = 0.6 mm/s, F = 0.8 J/cm^2	236	4.24
V = 16 mm/s, F = 1.7 J/cm^2	227	3.35
V = 3.8 mm/s, F = 1.7 J/cm^2	223	5.8
V = 1.7 mm/s, F = 1.7 J/cm^2	218	4
V = 0.6 mm/s, F = 1.7 J/cm^2	240	5.19

EDX elemental composition [wt.%] values obtained from the laser processed and from control surfaces of PLA matrices are presented in Table 1. According to the EDX results, the presence of uncommon elements in the composition of all PLA samples was not detected. The breakage of side O-C=O chemical bonds could explain the slight variation in the percent concentration [wt.%] of oxygen [O] and carbon [C] observed with increasing values of laser fluence (F) and lowering the scanning velocity (V).

In parallel with these measurements thickness and roughness evaluation of the Fs processed PLA samples were performed. The results are summarized in Table 2. Each test was done in relation to the control surface.

The results presented in Table 2 demonstrate a tendency of thinning the samples with increasing laser fluence (F) and lowering the scanning velocity (V) applied to each sample's surface. Laser treatment with V = 0.6 mm/s at F = 0.8 J/cm^2 and F = 1.7 J/cm^2 represents a turning point since it leads to an elevation of the material around the laser treated zones, creating a local 3D structuring on the surface of the PLA plate. These findings could be explained by bumps formation from elevated material in the high energy regime of processing, which could be clearly seen on the SEM images Figure 6d,h. This effect could be attributed to the formation of melting and severe ablation of the solid PLA matrix leading to the formation of a structure with a hierarchy of repeating depths and heights. The monitored change could be explained by rapid increase in surface temperature, followed by sudden partial melting of the polymer matrix and confinement of phase transformation. This is affecting the thickness of the processed region in the case of the lowest values of scanning velocity (V) at F = 0.8 J/cm^2 and F = 1.7 J/cm^2. The high roughness values (S_a = 11.84 μm and 24.07 μm) achieved at the above mentioned laser processing conditions, could not have a positive effect only on cells attachment and guidance, but also on the orientation of the cellular ingrowth towards zones with a greater porosity in volume as provided by the two types of periodicities achieved at V = 0.6 mm/s, F = 1.7 J/cm^2. On the other hand, the surface roughness (S_a = 2.10 μm) achieved at F = 0.8 J/cm^2 at V = 3.8 mm/s, in respect to depth (6.30 μm) and width (1.19 μm) of the grooves, as well as to the observed morphology at the SEM images—Figure 6b and the R_a = 1.92 μm in this specific case seems appropriate for mesenchymal stromal cells (MSCs) seeding, according to Szmukler-Moncler et al., who reported R_a = 0.9–1.53 μm as optimal for MSCs [38]. The optimum roughness for stable bonding of the implant surface to recipient tissue, according to literature, is min. R_a = 1 μm [39,40].

For a better visualization of the detailed results, presented in Table 2, representative 3D reconstructed images of the control and processed PLA matrices are shown in Figure 7.

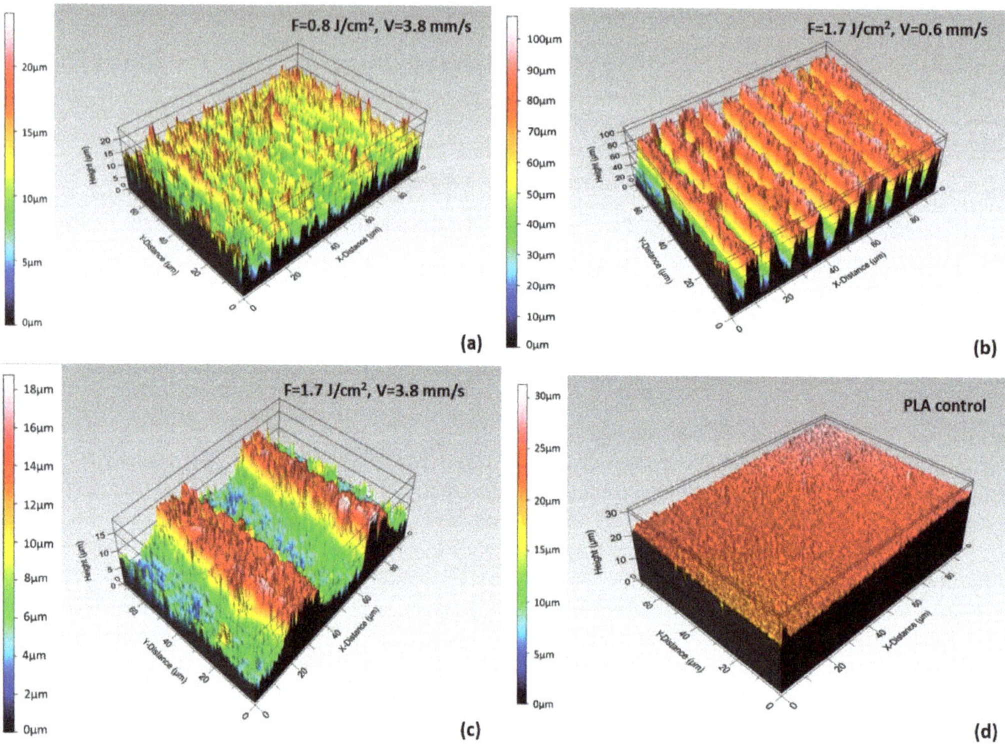

Figure 7. Representative 3D cross section images of PLA matrices: (**a**) F = 0.8 J/cm^2, V = 3.8 mm/s, (**b**) F = 1.7 J/cm^2, V = 0.6 mm/s, (**c**) F = 1.7 J/cm^2, V = 3.8 mm/s; (**d**) control surface.

The WCA measured on structured surfaces depends on the direction of water drop application in respect to the grooves' direction—Figure 8. Every WCA value presented on Figure 9 graphs is averaged over 10 separate measurements.

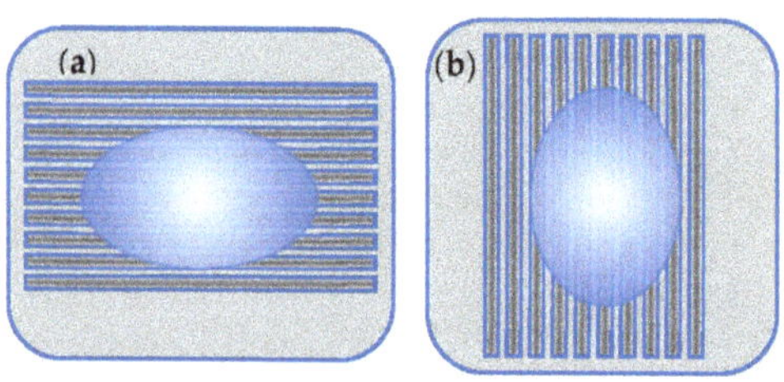

Figure 8. Scheme of water droplet positioning of Fs modified PLA samples: (**a**) perpendicular to the laser created grooves and (**b**) along the laser created grooves.

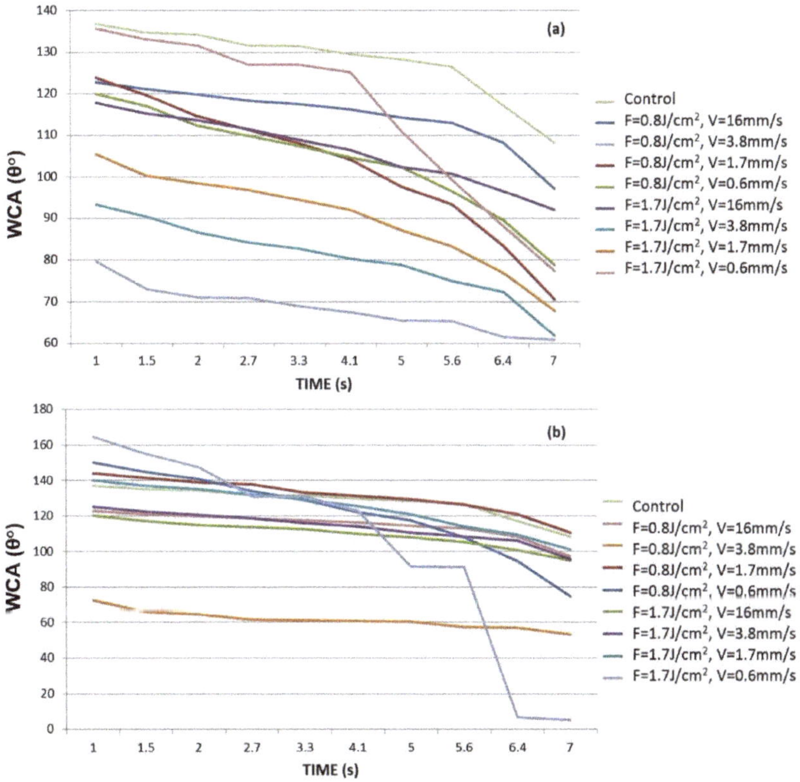

Figure 9. WCA evaluation of Fs modified PLA samples for a period of 7 s—(**a**) the dH_2O drop applied along the direction of the grooves; (**b**) the dH_2O drop applied perpendicular to the direction of created grooves.

As can be clearly seen from the graphically presented results in Figure 9, in both types of droplet application, a general trend of lowering the water contact angle values is observed by increasing the contact time of the water drop with the surface at higher values of applied laser fluence and lower scanning velocity. The monitored change of WCA (θ) values from 137° for control surface to 60°–20° for processed PLA transfers the surface wettability properties to more hydrophilic ones, thus making it "cell adhesion and proliferation friendly", which could improve the tissue integration of PLA implants [41]. The groups of Dekker et al. and Wan et al. for example, report that maximal cells attachment strength occurs at WCA (θ) of 20–55° for endothelial cells and fibroblasts [42,43].

However, some interesting cases stand out from the others. In the case of water application along with the Fs stripes, the control PLA surface possesses most hydrophobic features, which is not the case when perpendicular dropping is performed—at the first seconds of application, part of the Fs treated PLA surfaces exhibit greater hydrophobicity than the non-treated PLA. However, at the 7th second of both types of water drop application, each laser modified surface changes its wettability properties to more hydrophilic. This could be explained by the Wenzel model (1936), which describes the homogeneous wetting on the rough surfaces at the stable equilibrium state—the minimum free energy state for the system, which the water drop demands [44]. The WCA gradually decreases with time except for the PLA sample treated with V = 0.6 mm/s, F = 1.7 J/cm^2, where interesting bends (along and perpendicular the direction of the grooves—Figure 9a,b, respectively) in both WCA graphs are observed. These twists are more profoundly expressed when "perpendicular dropping" is performed. This could be explained by the model established by Cassie and Baxter (1944), according to which a heterogeneous wetting of the surface is observed [45]. As can be seen from SEM images (Figure 6h) and 3D cross section images (Figure 7b) of V = 0.6 mm/s, F = 1.7 J/cm^2 treated PLA samples, the formation of the porous periodical microstructured architecture of the PLA scaffold has a hierarchical origin and thus a preservation of air pockets in the volume of the material is possible, which prevent the initial uniform entry of the water droplet inside the laser created microstructure and could lead to the sudden jump-like decrease of the WCA values which forms the graph bends, observed in Figure 9. It was found out that superhydrophobic ($\theta > 150°$) or superhydrophilic ($\theta < 20°$) surfaces can be achieved by creation of grooves with V = 0.6 mm/s, F = 1.7 J/cm^2 PLA, where $\theta = 164.8°$ at the 1st second drops noticeably to $\theta = 5°$ at the 7th second of application. This drastic jump-like change in WCA could be explained with the Cassie-Baxter model [45]. The WCA evaluation shows that by fine-tuning laser parameters, the surface wettability of PLA matrices can be precisely controlled in order to achieve cell-adhesive surfaces.

The chemical composition after laser treatment of the PLA cell matrices was also investigated. The results of the performed FTIR analysis are presented in Figure 10.

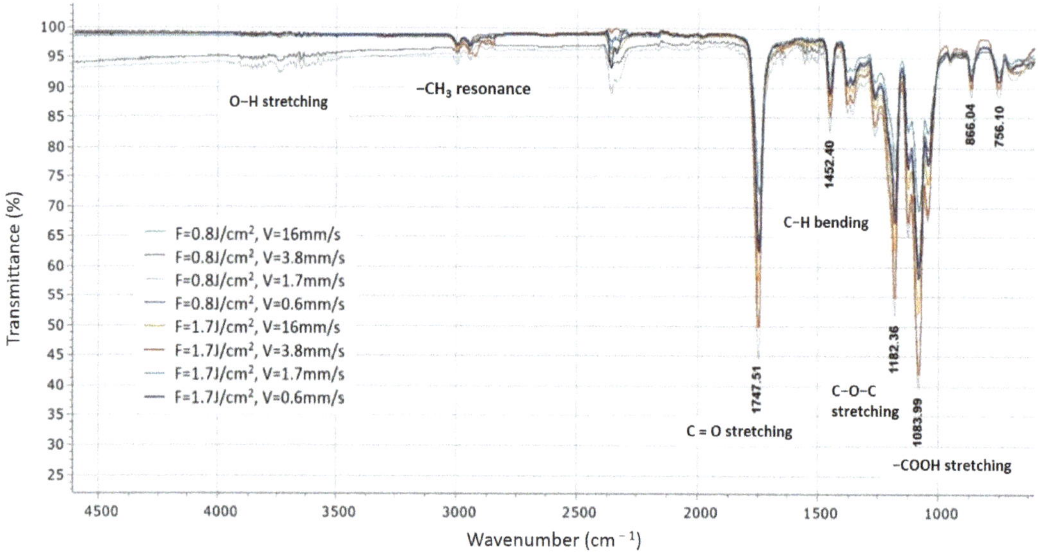

Figure 10. FTIR spectrum obtained for laser structured PLA surfaces treated by laser irradiation at F = 0.8 and 1.7 J/cm^2, V = 16, 3.8, 1.7 and 0.6 mm/s.

As can be seen from the results obtained for different treatment conditions of the performed FTIR spectroscopy, in respect to control PLA [37], slight deviations in the intensity of the acquired peaks are monitored, after Fs-laser treatment is observed. All characteristic chemical groups for the PLA spectra are detected, deviations are observed at the intensity of the occurring peaks. The peak detected at 3750 cm^{-1}, corresponds to O–H bond stretching and is characteristic for the PLA. The resonance peak of the CH_3 group is detected at 2925 cm^{-1}. The maximums at 1747.51 cm^{-1} and 1182 cm^{-1} correspond to the C=O stretching and the C–O–C stretching of PLA molecules. The carbonyl peak at 1452.4 cm^{-1} is well defined. The Carboxyl group was also detected at 1083.99 cm^{-1} [46].

The obtained data via FTIR were additionally characterized by means of Principal Component Analysis (PCA), which is used as a chemometric methodology for the analysis of data with high-dimensional sets. The basic aim of PCA is to gain a reduced set of principal components (PC), which provides explanation to the variability of obtained data. For visualizing clustering of multivariate data, the *Clustvis web tool* (Nucleic Acids Research,

43(W1):W566–W570, 2015) was employed—each treatment condition of the obtained FTIR spectra was assigned as a variable in PCA, which permits the discovery of changes of the samples characteristics in respect to physico-chemical properties.

The obtained variances are 77% and 20.8% for PC-1 and PC-2, respectively. The results for all sets of data for the first two principal components PC1 and PC2, jointly were calculated to be 97.8% of the data matrix variance. The scores scatter plot PC1 vs. PC2 (Figure 11a) showed peaks distribution and scattering along the PC1 axis. The V1, V3, V4, and V5 are positioned on the positive side of PC2. In Figure 11 the tendency for the assembling of scores in a pattern going to negative values of PC2 is monitored. In our specific case the data refers to the same materials used for processing, thus in this concrete case we want to highlight the similarities between different processing regimes and applied laser parameters. Our analysis is directed towards an estimation of possible contribution of separate laser treatment conditions to the deviation in chemical composition of the material, which under investigation of FTIR spectra remains almost equal, after different laser treatment conditions were applied. This is clearly visible in the provided data on (Figure 11a). To summarize these findings, the analyzed data suggest that each treatment condition was well separated, and scores showed the most significant variation lying along the PC2 axis. As a general tendency is observed, that the scores which appear in the negative part of the graph are related to the gentle regime of laser processing, associated with lower laser energy values applied, while the distribution of the components in the upper (positive part of the graph are related to laser treatment with increased laser fluence. The model prediction is shown on Figure 11b Where are plotted cumulative values for R2 (percent of variation of the training set—X with PCA—explained by the model) and Q2 (percent of variation of the training set—X with PCA—predicted by the model according to cross-validation). When R2 values are close to 1 then we can predict that our conditions are very close to a good model. Concerning Q2, it indicates how well the model predicts new data. A large value of Q2 (Q2 > 0.5) indicates good predictivity. In our specific case, we have for both values numbers approximately to 1, which defines a good approximation to the PCA model (Figure 11b).

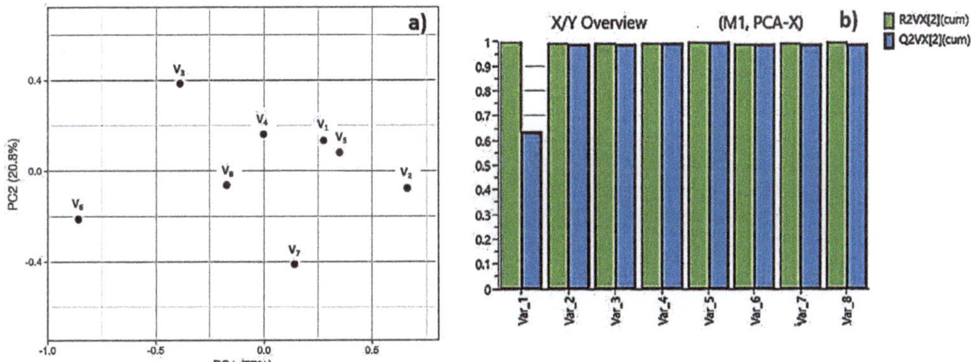

Figure 11. PCA scores scatter plots of FTIR spectra in the region from 600—4600 cm^{-1}; (**a**) V1—F = 1.663 J/cm^2, V = 1.7 mm/s; V2—F = 1.663 J/cm^2, V = 3.8 mm/s; V3—F = 1.663 J/cm^2, V = 16 mm/s; V4—F = 0.831 J/cm^2, V = 6.0 mm/s; V5—F = 0.831 J/cm^2, V = 1.7 mm/s; V6—F = 0.831 J/cm^2, V = 16 mm/s; V7—F = 0.831 J/cm^2, V = 3.8 mm/s; V8—F = 1.663 J/cm^2, V = 0.6 mm/s; (**b**) The Summary of plot fit of the PCA model for component R2 and Q2.

Additionally, XPS analysis was performed at the two most interesting cases (adapted to achieve laser modification of PLA surface, optimized to achieve enhanced cell adhesion conditions) of laser parameters applied to PLA samples- F = 0.8 J/cm^2, V = 3.8 mm/s, and F = 1.7 J/cm^2, V = 0.6 mm/s—Figure 12.

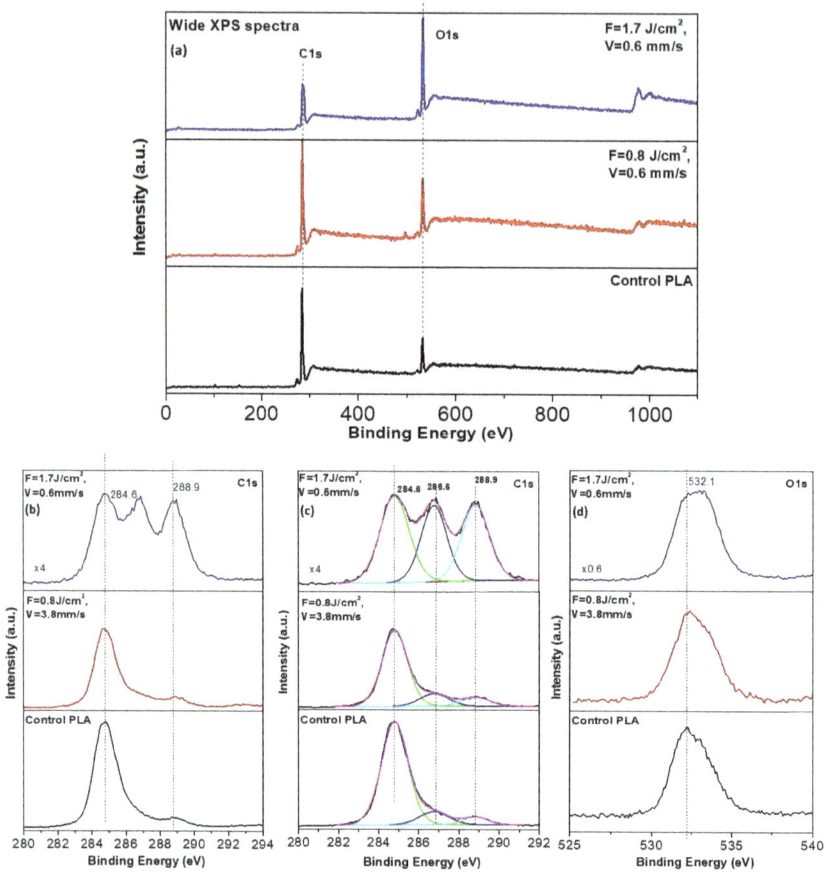

Figure 12. Comparison of XPS spectra of PLA surfaces treated by laser irradiation at F = 0.8 J/cm^2, V = 3.8 mm/s and F = 1.7 J/cm^2, V = 0.6 mm/s; (**a**) Wide XPS Spectra; (**b**) C_{1s} spectra; (**c**) C_{1s} decomposed spectra; (**d**) O_{1s} spectra.

Based on the PLA atomic composition, the C_{1s}, O_{1s} regions were explored with XPS analysis, as the biocompatible polymer is mainly composed of carbon and oxygen [47]. The wide XPS spectra show no deviation from the positions of the C_{1s}, O_{1s} peaks obtained from the laser-treated zones of PLA, compared to control surfaces—Figure 12a. C_{1s} spectra (b) and C_{1s} decomposed spectra (c). On the other hand, obtained spectra from treatments at F = 1.7 J/cm^2, V = 0.6 mm/s significantly differ from control spectrum and that from F = 0.8 J/cm^2, V = 3.8 mm/s—Figure 12b,c. In the case of F = 1.7 J/cm^2, V = 0.6 mm/s fs treated matrices, three components at the C_{1s} XPS spectra are observed—at 284.6, 286.6, and 288.9 eV. They can be assigned to C atoms in the (C–C)/(C–H), (C–O) and (O-C=O) bonds, respectively. In the case of the spectra detected from the control and the spectra obtained for irradiation conditions F = 0.8 J/cm^2, V = 3.8 mm/s, the second peak (286.6 eV) assigned to (C=O) bonds is almost not observed Figure 11b, while in the case of F = 1.7 J/cm^2, V = 0.6 mm/s this maximum is with significantly higher intensity, which could be related to an increase in the concentration of C–O and O-C=O bonds and corresponding incorporation of peroxyl, hydroxyl, ether or carbonyl groups on the PLA scaffold surface [47,48]. The significant increase in the third peak (288.9 eV) of C_{1s} spectra suggests an increase in the molar concentration of C=O bonds and accompanying incorporation of carbonyl groups on PLA samples surface treated with F = 1.7 J/cm^2, V = 0.6 mm/s. As can be seen from

Figure 12d, only a slight deviation in O_{1s} XPS spectra is observed after laser treatment—the peak at 532.1 eV is attributed to oxygen atoms in (C–O) and (O-C=O) bonds.

In order to see if the degradation rate of the PLA scaffolds matches the speed of the new tissue regeneration at the side of implantation, a preliminary in vitro degradation study in PBS, 37 °C of the PLA matrices chosen for cell seeding experiments, was carried out—Group1 (G1)—PLA F = 0.8 J/cm^2, V = 3.8 mm/s; Group2 (G2)—PLA F = 1.7 J/cm^2, V = 0.551 mm/s and Group3 (G3)—control PLA. The results of pH measurements obtained through a period of 8 weeks are summarized in Table 3 and Figure 13.

Table 3. pH change of PBS of Group1 (G1)—PLA F = 0.8 J/cm^2, V = 3.8 mm/s; Group2 (G2)—PLA F = 1.7 J/cm^2, V = 0.6 mm/s and Group3 (G3)—control PLA during the 8 week period of in vitro degradation test.

Week	Group 1 PBS pH	Group 2 PBS pH
0	7.2	7.2
1	7.2	7.2
2	7.2	7.2
3	7.2	7.2
4	7.2	7.2
5	7.2	7.2
6	7.1	6.9
7	6.9	6.7
8	6.5	6.3

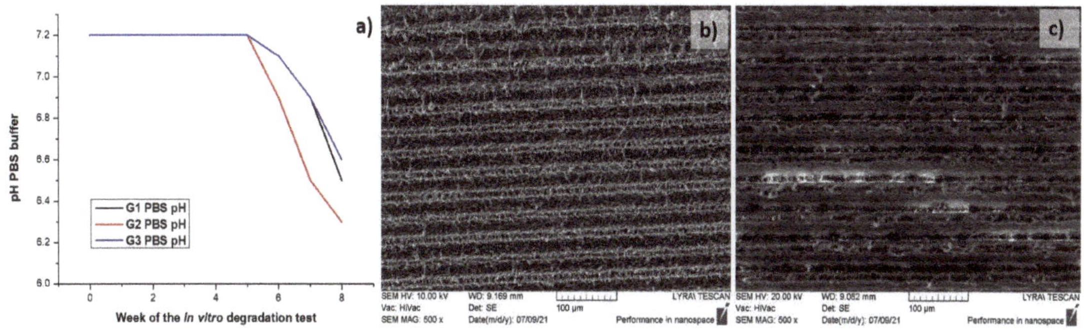

Figure 13. (**a**) Graphical presentation of pH change of PBS of Group1, 2 and 3 over time; (**b**) SEM images of G1, (**c**) G2 samples after the in vitro degradation test.

The graph presented on Figure 13a shows that the PBS pH values of the three groups remained unchanged until the end of the fifth week (~7.2). At the sixth week a decrease in the pH value of the buffer of the **G1** (pH = 6.9), **G2** (pH = 6.7) and **G3** (pH = 6.9) was observed, as it is more noticeable for **G2** samples. This tendency was kept until the end of the in vitro degradation test—the pH dropped with ~0.2 units respectively for each of the groups. The slight decrease of the pH values of the PBS saline could be attributed to PLA degradation process, which byproducts released (like lactic acid), make the medium more acidic. The SEM images taken for G1 and G2 after the period of 8 weeks buffer incubation showed the slight deviation (Figure 13b,c) in the surface morphology of PLA as compared to Figure 6. The results show that Fs laser processing with higher fluence and lower scanning velocity could speed up/accelerate the in vitro degradation rate of the polymer matrix, and thus assist in achieving an appropriate speed, matching the specific tissue regeneration rate [49]. The group of Guo et al., for example, demonstrated that the synthesized PLA scaffold were degraded completely into harmless products in simulated body fluid (SBF) at a slow degradation rate and that the scaffolds weight loss could reach

80% after 8 months in vitro degradation, which matches the speed of new bone tissue regeneration in the human body [50].

3.2. Microbiology Studies

In order to see if the created laser induced microchannel complex frameworks, obtained by fs laser structuring type G2 on PLA samples surface have potential antibacterial properties, microbiology studies with *S. aureus* were performed. The results of the conducted experiments with *S. aureus*, show that the adhesion of the bacteria is very high onto PLA samples treated with laser patterning type (G2) in comparison to the control (G3) scaffolds. The results from Fs laser surface structuring showed to promote bacterial adhesion, as a 5-fold increase in the number of adhered bacteria onto PLA-G2 is observed in comparison to control PLA G3 surface—Figure 14. The laser treatment creates "3D-cavities" into PLA G2 samples, which favor bacteria adhesion.

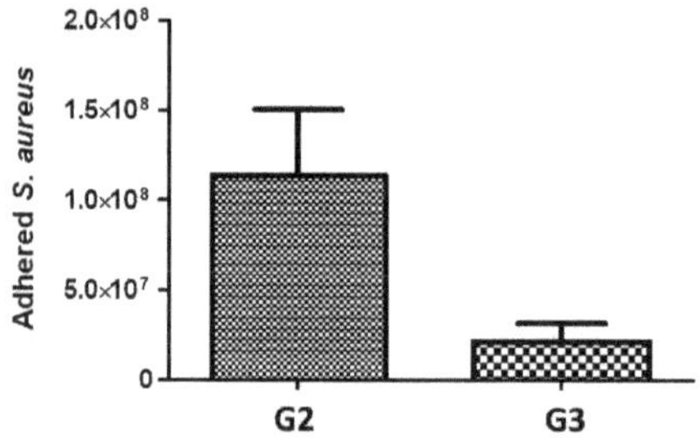

Figure 14. Number of adhered *S. aureus* bacteria on the grooved PLA samples surface (G2—PLA F = 1.7 J/cm^2, V = 0.6 mm/s; G3—control PLA samples).

The group of Alves et al. for example showed experimentally that Fs laser created grooves, separated by distance of 8–12 μm and with channel width in the range 0.54 μm to 1.33 μm favor *S. aureus* adhesion. A further increase in surface roughness did not favor the bacteria adhesion and biofilm formation [51]. It is reported by other research groups that an intact bacteria biofilm occurred much faster when the surface roughness was greater than 0.8 μm as the contact area between the bacteria cell and the surface is increased [52,53]. On the other hand, it is reported that such bacteria attach easier on hydrophilic surfaces (as G2) stronger, compared to hydrophobic, where the cells are easier to remove by body fluid circulation [54–56].

3.3. Mesenchymal Stem Cell Behavior

After full characterization of the laser processed PLA matrices, two groups of Fs treated samples were chosen for preliminary cell experiments—Group1 (G1)—PLA F = 0.8 J/cm^2, V = 3.8 mm/s (in respect to the obtained R_a and WCA values, which fall within the required conditions for MSCs adhesion and proliferation, according to the mentioned above literature survey [38–43]) and Group2 (G2)—PLA F = 1.7 J/cm^2, V = 0.6 mm/s, which exhibits interesting wetting behavior and superhydrophilic nature due to its double hierarchical porosity. Each group consisted of 30 identical samples. As already mentioned, the MSCs proliferation experiments performed for a period of 14 days were repeated twice in order to confirm the results obtained. Parallel cell seeding was performed on control PLA samples Group3 (G3). The results presented in Figure 15a demonstrate that MSCs proliferation rate on G1 is close to that of G3, while a slower MSCs proliferation rate was observed on G2 matrices in comparison to the G3. Moreover, after the 14 days of culture on the three groups of PLA samples, MSCs were fixed, and fluorescently labeled with Phalloidin® (colors cytoskeleton in green) and DAPI (colors nuclei in blue). The observation of labeled cells on samples under fluorescence microscope confirmed the presence of MSCs on the samples after the 14 days of culture Figure 15b. From the figure, it is observed that the Fs laser treatment influences MSCs morphology and organization —the cells cytoskeleton obtained an elongated shape simultaneously aligning along the grooves while on the PLA **G3** group cells keep a random organization. These preliminary results suggest that MSCs can adhere and orientate on both **G1** and **G2** types of patterning conditions and do not show differences between them in the context of cell viability.

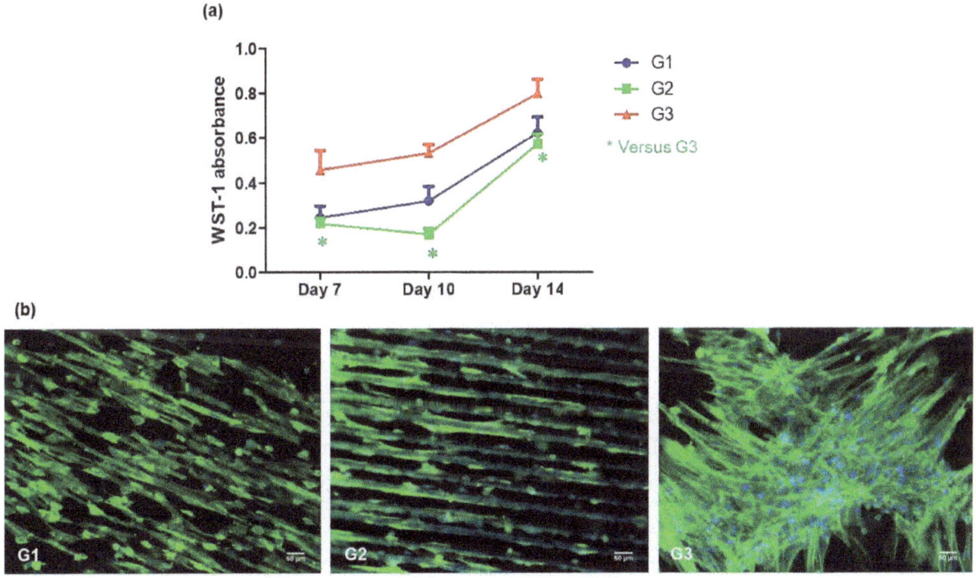

Figure 15. MSCs proliferation rate (**a**) and fluorescence images of morphology cellular cytoskeleton (**b**) on **G1**, **G2** and **G3** matrices. cellular cytoskeleton (in green) and nuclei (in blue). Scale bars = 50 μm.

Femtosecond laser processing of polymer-based cell scaffolds has been extensively studied in last decades as an alternative contactless and chemically clean method for matrix surface structuring and functionalization, thus enhancing the biocompatibility and biomimetic properties of diverse polymer scaffolds [57–62]. The ultrashort time of interaction between the femtosecond laser pulse and the transparent biopolymers significantly reduces the size and depth of the heat affected zone, thus fs structuring of polymers like PLA, that have low glass-transition and melting temperatures, ($T_{g\ PLA}$ = 60 °C) is possible to be achieved without triggering severe thermal side effects [35]. Not only surface micro and nano structuring, but also precise three-dimensional porous structures can be fabricated on top and in the interior of the processed biodegradable polymer [27]. Femtosecond laser systems enable the processing of transparent materials via nonlinear optical interaction in visible to near-IR light [63,64]. Ortiz et al. reported that the laser ablation of poly-L-lactic acid (PLLA) is governed by a multi-photon process in the three cases of interaction: with the fundamental (1064 nm), the second (532 nm) and the third harmonic generation wave (355 nm) of Nd: YVO_4 laser [65]. The group of Daskalova et al., for example, achieved 3D porous structures composed of μm-sized porous foam on the surface of collagen, elastin and chitosan thin films by femtosecond laser structuring at λ = 800 nm in which cells adhere, elongate, and orientate [66–68].

It has been shown that optimal cellular adhesion conditions in respect to protein absorption are achieved in the intermediate values of wettability—θ = 20–55 ° for fibroblasts and endothelial cells [42,43]. Femtosecond laser processing of polymers leads to rupture of weak hydrogen bonds (C-H). The unsaturated C- bond on the main carbon chain reacts with oxygen and forms new carboxyl, ester, hydroxyl, carbonyl, etc., bonds, making the polymer-based cell scaffolds hydrophilic [30]. Enhanced surface roughness is very important for MSCs adhesion and further engineering of hard tissues like bone and teeth. For example, the group of Cordero et al. showed a better MC3T3-E1 pre-osteoblast cell alignment on laser micro patterns formed with line separation distances below 50 μm [69]. Larger distances between structures prevent cells from developing focal adhesion plaques and bridge over the groove created on the polymer surface—the cells remain oriented along the length of the channels, but do not proliferate and do not express differentiation markers [70,71]. The roughness of 0.9–1.53 μm (R_a) is reported as optimal for orientation of MSCs [38]. If the surface does not possess high roughness values, this would lead to an insufficient formation of healthy tissue. Efficient implant to tissue bonding contacts could not be established, due to poor cell adhesion and tight fixation of the implant to the host tissue could fail [70]. On the other hand, roughness (R_a) values higher than 1 μm are experimentally proven to be the minimum required for efficient tight bonding between the implant/tissue interfaces [39].

One of the most common reasons for implant rejection is bacterial infection [72]. The main efforts to prevent it are aimed at rupturing the bacteria biofilm, not allowing bacteria proliferation [72]. This could be achieved by inducing suitable surface irregularities, which could lead to the stretching and disfiguration of bacteria, cell wall rupture, and death [73,74]. The optimal goal to obtain biocompatible and anti-bacterial cell scaffold surfaces could be achieved by means of Fs-laser treatment. In its elaborate review, N. Sirdeshmukh and G. Dongre outline the biocompatibility and antibacterial analysis of laser surface textured patterns made by different research groups [75]. The scientists are pointing out that the right combination of micro and nano roughness could lead to the desired antibacterial effect, by means of LIPSS generation in combination with different nanostructures-nanopillars [76], spike nanotextured patterns [77], nanoripples [78], conic and spherical nanostructures [79], etc.—all mimicking the topography of natural structured surfaces exhibiting antibacterial proper-ties [80]. On the other hand, as already mentioned, the generation of laser-induced periodic structures, on the polymer surface such as PLA could be a difficult task to achieve, which could be attributed to their low melting (T_m) and glass transition ($T_{g\ PLA}$ = 60 °C) temperature [35]. The importance of the nanostructures could be explained by the attachment point theory according to which, if the surface available is smaller or near the size of the bacteria, not enough attachment strength can be achieved, and bacteria can not adhere effectively (*S. aureus*—d = 0.5 ÷ 1.5 μm) [81,82]. Guenther et al. report that laser-induced pillar-like patterns with 0.5 μm spatial period significantly decrease *S. aureus* attachment on structured polymer surfaces [83].

We demonstrated experimentally that by simply tuning the Fs laser parameters applied—fluence and scanning velocity, precise control of PLA scaffold surface properties, like roughness, porosity, and wettability, without changing its elemental composition and chemical structure, could be achieved. The results presented clearly show no cell cytotoxicity and a tendency of cell adhesion and elongation along the grooves, which could orientate them in the desired direction of the tissue regeneration. The results from microbiology studies with *S. aureus* demonstrated that such surface modifications could find application as "bacteria traps" in clinical practice, "luring" microorganisms as an optimal environment, leaving in that way the desired surface (biosensors, biochips, drug carriers, prostheses for medical implants, etc.) clean of them. The preferential alignment of *S. aureus* to the periodic structures, created by the Fs laser on the PLA surface, demonstrates the microorganism's attempt to improve their surface retention by maximizing their attachment points with the laser structured PLA scaffold [84–86]. Additional nanostructuring of the achieved laser-induced micro-channel complex frameworks is therefore expected to reduce bacteria attachment by limiting the number of contact adhesion points with the PLA implant surface. Further optimization of Fs laser parameters used for surface structuring in the context of anti-bacterial effect and investigation of the results obtained should be made in order for bio-compatible and antibacterial PLA cell scaffolds and bio-interfaces to be achieved.

4. Conclusions

The experimental results obtained in the current study demonstrate that Fs micropatterning could improve biological properties of PLA cell scaffolds in the context of directional orientation, and biocompatibility. By controlling the laser parameters (laser fluence and scanning velocity), PLA bioactive surfaces can be achieved via a contactless and non-destructive laser modification process, which leads to precise control of PLA cell-matrix surface properties like roughness, wettability, and porosity.

PLA sheets were patterned in a form of rows using fs-laser emitting laser radiation with 150-fs pulse duration. In order to achieve specific patterning, the optimal laser fluence and scanning velocity were determined as F = 0.8 J/cm^2, V = 3.8 mm/s, and PLA F = 1.7 J/cm^2, V = 0.6 mm/s, respectively. At these applied laser parameters, the PLA polymer matrix formed a hierarchical structure. The examination of degradation properties before and after laser patterning, showed degradation of the processed and control PLA samples after the 6th week of the in vitro degradation test, with emphasis on the samples modified with increased laser fluence. These findings are explained by the decrease in crystallinity related to the occurrence of surface melting, in a high fluence regime. The obtained results of the wettability test at different laser processing parameters demonstrate that some laser-patterned designs have the potential to create stable hydrophilic properties, while in some specific cases, obtaining hierarchical structures, a sudden drop of the contact angle values was observed due to the formation of air pockets. Principal component analysis was employed to distinguish the main origin of variance in the FTIR spectra of PLA samples obtained under diverse treatment conditions. PCA permits the samples to be analyzed in terms of their similarities, in accordance with applied Fs laser processing, with respect to the acquired FTIR spectra.

The Fs laser treatment influences MSCs morphology and orientation along the grooves formed by the laser, compared to the chaotic spreading of cells, cultured on control Fs non-treated PLA scaffolds, without affecting cell viability. The generation of hierarchical structure on the PLA surface was monitored, and the creation of an ideal cytocompatible structure was realized.

Author Contributions: Conceptualization, A.D.; methodology, A.D.; writing, A.D.; original draft preparation, A.D.; project administration, A.D.; writing, L.A.; original draft preparation, L.A.; investigation L.A.; investigation, E.F.; investigation, D.A.; investigation, R.M.; investigation, X.C.; methodology, H.K.; visualization, H.K.; methodology, M.D.; visualization, M.D.; investigation, J.C.; visualization, J.C.; software, A.T.; supervision, I.B. All authors have read and agreed to the published version of the manuscript.

Funding: This work was supported by: EUROPEAN UNION'S H2020 research and innovation program under the Marie Sklodowska-Curie Grant Agreement AIMed No. 861138; BULGARIAN NATIONAL SCIENCE FUND (NSF) under grant number No. KP-06-H48/6 (2020–2023), "Development of hybrid functional micro/nanoporous biomaterial scaffolds by ultra-fast laser modification"; and H2020 FET Open METAFAST Grant Agreement No. 899673.

Institutional Review Board Statement: Not applicable.

Data Availability Statement: Not applicable.

Conflicts of Interest: The authors declare no conflict of interest.

References

1. World Population Prospects: The 2019 Revision. United Nations DESA/Population Division. 2019. Available online: https://population.un.org/wpp/ (accessed on 26 March 2021).
2. Langer, R. Tissue engineering. *Mol. Ther.* **2000**, *1*, 12–15. [CrossRef] [PubMed]
3. Scheinpflug, J.; Pfeiffenberger, M.; Damerau, A.; Schwarz, F.; Textor, M.; Lang, A.; Schulze, F. Journey into Bone Models: A Review. *Genes* **2018**, *9*, 247. [CrossRef]
4. Bose, S.; Roy, M.; Bandyopadhyay, A. Recent advances in bone tissue engineering scaffolds. *Trends Biotechnol.* **2012**, *30*, 546–554. [CrossRef] [PubMed]

5. Hutmacher, D.W.; Schantz, J.T.; Lam, C.X.F.; Tan, K.C.; Lim, T.C. State of the art and future directions of scaffold-based bone engineering from a biomaterials perspective. *J. Tissue Eng. Regener. Med.* **2007**, *1*, 245–260. [CrossRef]
6. Rasal, R.M.; Hirt, D.E. Poly (lactic acid) toughening with a better balance of properties. *Macromol. Mater. Eng.* **2010**, *295*, 204–209. [CrossRef]
7. Ratner, B.D.; Hoffman, A.S.; Schoen, F.J.; Lemons, J.E. *Biomaterials Science: An Introduction to Materials in Medicine*; Elsevier: San Diego, CA, USA, 2004; ISBN 9780080470368.
8. Santoro, M.; Shah, S.R.; Walker, J.L.; Mikos, A.G. Poly (lactic acid) nanofibrous scaffolds for tissue engineering. *Adv. Drug Deliv. Rev.* **2016**, *107*, 206–212. [CrossRef]
9. Senatov, F.S.; Niaza, K.V.; Zadorozhnyy, M.Y.; Maksimkin, A.V.; Kaloshkin, S.D.; Estrin, Y.Z. Mechanical properties and shape memory effect of 3D-printed PLA-based porous scaffolds. *J. Mech. Behav. Biomed. Mater.* **2016**, *57*, 139–148. [CrossRef]
10. Tyler, B.; Gullotti, D.; Mangraviti, A.; Utsuki, T.; Brem, H. Polylactic acid (PLA) controlled delivery carriers for biomedical applications. *Adv. Drug Deliv. Rev.* **2016**, *107*, 163–175. [CrossRef]
11. Carrasco, F.; Page's, P.; Gamez-Perez, J.; Santana, O.O.; Maspoch, M.L. Processing of poly (lactic acid): Characterization of chemical structure, thermal stability and mechanical properties. *Polym. Degrad. Stab.* **2010**, *95*, 116–125. [CrossRef]
12. Guntillake, P.A.; Adhikari, R. Biodegradable synthetic polymers for tissue engineering. *Eur. Cell Mater.* **2003**, *5*, 1–16. [CrossRef]
13. Barbieri, D.; de Bruijn, J.D.; Luo, X.; Fare, S.; Grijpma, D.W.; Yuan, H. Controlling dynamic mechanical properties and degradation of composites for bone regeneration by means of filler content. *J. Mech. Behav. Biomed. Mater.* **2013**, *20*, 162–172. [CrossRef] [PubMed]
14. Yang, S.; Leong, K.F.; Du, Z.; Chua, C.K. The design of scaffolds for use in tissue engineering. Part II. Rapid prototyping techniques. *Tissue Eng.* **2002**, *8*, 1–11. [CrossRef]
15. Schneider, G.B.; Zaharias, R.; Seabold, D.; Keller, J.; Stanford, C. Differentiation of preosteoblasts is affected by implant surface microtopographies. *J. Biomed. Mater. Res. Part. A.* **2004**, *69*, 462–468. [CrossRef] [PubMed]
16. Lim, J.Y.; Donahue, H.J. Cell sensing and response to micro- and nanostructured surfaces produced by chemical and topographic patterning. *Tissue Eng.* **2007**, *13*, 1879–1891. [CrossRef]
17. Yu, Y.; Zhang, Q.; Chang, C.C.; Liu, Y.; Yang, Z.; Guo, Y.; Wang, Y.; Galanakis, D.K.; Levon, K.; Rafailovich, M. Potentiometric sensors based on surface molecular imprinting: Detection of cancer biomarkers and viruses. *Chemical* **2016**, *141*, 5607–5617.
18. Guo, Y.; Chang, C.C.; Cuiffo, M.A.; Xue, Y.; Zuo, X.; Pack, S.; Zhang, L.; He, S.; Weil, E.; Rafailovich, M.H. Engineering flame retardant biodegradable polymer nanocomposites and their application in 3D printing. *Polym. Degrad. Stab.* **2017**, *137*, 205–215. [CrossRef]
19. Kim, D.; Kim, J.; Hyun, H.; Kim, K.; Roh, S. A nanoscale ridge/groove pattern arrayed surface enhances adipogenic differentiation of human supernumerary tooth-derived dental pulp stem cells in vitro. *Arch. Oral Biol.* **2014**, *59*, 765–774. [CrossRef] [PubMed]
20. Kolind, K.; Kraft, D.; Bøggild, T.; Duch, M.; Lovmand, J.; Pedersen, F.S.; Bindslev, D.A.; Bünger, C.; Fossand, M.; Besenbacher, F. Control of proliferation and osteogenic differentiation of human dental-pulp-derived stem cells by distinct surface structures. *Acta. Biomater.* **2014**, *10*, 641–650. [CrossRef]
21. Ching, Y.C.; Rosiyah, Y.Y.; Li, G. Preparation and characterization of nano particle reinforced polyactides composite. *J. Nano Res.* **2013**, *25*, 128–136.
22. Conde, C.M.; Demarco, F.F.; Casagrande, L.; Alcazar, J.C.; Nör, J.E.; Tarquinio, S.B.C. Influence of poly-l-lactic acid scaffold's pore size on the proliferation and differentiation of dental pulp stem cells. *Braz. Dent. J.* **2015**, *26*, 93–98. [CrossRef]
23. Serra, T.; Mateos-Timoneda, M.A.; Planell, J.A.; Navarro, M. 3D printed PLA-based scaffolds. *Organogenesis* **2013**, *9*, 239–244. [CrossRef] [PubMed]
24. Zhang, D.; Wu, X.; Chen, J.; Lin, K. The development of collagen based composite scaffolds for bone regeneration. *Bioact. Mater.* **2018**, *3*, 129–138. [CrossRef] [PubMed]
25. Ratner, B.D. Surface modification of polymers: Chemical, biological and surface analytical challenges. *Biosens. Bioelectron.* **1995**, *10*, 797–804. [CrossRef]
26. Wakelin, E.A.; Fathi, A.; Kracica, M.; Yeo, G.C.; Wise, S.G.; Weis, A.S.; McCulloch, D.G.; Dehghani, F.; Mckenzie, D.R.; Bilek, M.M. Mechanical properties of plasma immersion ion implanted PEEK for bioactivation of medical devices. *ACS Appl. Mater. Interfaces* **2015**, *7*, 23029–23040. [CrossRef]
27. Terakawa, M. Femtosecond laser processing of biodegradable polymers. *Appl. Sci.* **2018**, *8*, 1123. [CrossRef]
28. Riveiro, A.; Maçon, A.L.B.; del Val, J.; Comesaña, R.; Pou, J. Laser Surface Texturing of Polymers for Biomedical Applications. *Front. Phys.* **2018**, *6*, 16. [CrossRef]
29. Lippert, T. Laser application of polymers. *Adv. Polym. Sci.* **2004**, *168*, 51–246.
30. Bacáková, L.; Filová, E.; Rypácek, F.; Svorcík, V.; Starý, V. Cell adhesion on artificial materials for tissue engineering. *Physiol. Res.* **2004**, *53* (Suppl. 1), S35–S45.
31. Khalili, A.A.; Ahmad, M.R. A review of cell adhesion studies for biomedical and biological applications. *Int. J. Mol. Sci.* **2015**, *16*, 18149–18184. [CrossRef]
32. Li, H.; Wen, F.; Wong, Y.S.; Boey, F.Y.; Subbu, V.S.; Leong, D.T.; Ng, K.W.; Ng, G.K.; Tan, L.P. Direct laser machining-induced topographic pattern promotes up-regulation of myogenic markers in human mesenchymal stem cells. *Acta Biomater.* **2012**, *8*, 531–539. [CrossRef]

33. Mofokeng, J.P.; Luyt, A.S.; Ta'bi, T.; Kova'cs, J. Comparison of injection moulded, natural fibre-reinforced composites with PP and PLA as matrices. *J. Thermoplast. Compos. Mater.* **2012**, *25*, 927–948. [CrossRef]
34. Lee, B.; Jeon, H.; Wang, A.; Yan, Z.; Yu, J.; Grigoropoulos, C.; Li, S. Femtosecond laser ablation enhances cell infiltration into three-dimensional electrospun scaffolds. *Acta Biomater.* **2012**, *8*, 2648–2658. [CrossRef]
35. Yada, S.; Terakawa, M. Femtosecond laser induced periodic surface structure on poly-L-lactic acid. *Opt. Express.* **2015**, *23*, 5694–5703. [CrossRef]
36. Daskalova, A.; Angelova, L.; Carette, X.; Mincheva, R.; Raquez, J.M.; Trifonov, A.; Buchvarov, I. Influence of femtosecond laser processing parameters on surface morphology and wettability properties of polylactic acid (PLA). *IOP Mater. Sci. Eng.* **2021**, *1056*, 012001.
37. Lutey, A.H.A.; Gemin, L.; Romoli, L.; Lazzini, G.; Fuso, F.; Faucon, M.; Kling, R. Towards Laser-Textured Antibacterial Surfaces. *Sci. Rep.* **2018**, *8*, 10112. [CrossRef] [PubMed]
38. Szmukler-Moncler, S.; Perrin, D.; Ahossi, V.; Magnin, G.; Bernard, J.P. Biological properties of acid etched titanium implants: Effect of sandblasting on bone anchorage. *J. Biomed. Mater. Res. B. Appl. Biomater.* **2004**, *68*, 149–159. [CrossRef]
39. Ponsonnet, L.; Reybier, K.; Jaffrezic, N.; Comte, V.; Lagneau, C.; Lissac, M.; Martelet, C. Relationship between surface properties (roughness, wettability) of titanium and titanium alloys and cell behavior. *Mater. Sci. Eng. C* **2003**, *23*, 551–560. [CrossRef]
40. Riveiro, A.; Soto, R.; Comesaña, R.; Boutinguiza, M.; Del Val, J.; Quintero, F.; Lusquiños, F.; Pou, J. Laser surface modification of PEEK. *Appl. Surf. Sci.* **2012**, *258*, 9437–9442. [CrossRef]
41. Gittens, R.A.; Olivares-Navarrete, R.; Schwartz, Z.; Boyan, B.D. Implant osseointegration and the role of microroughness and nanostructures: Lessons for spine implants. *Acta Biomater* **2014**, *10*, 3363–3371. [CrossRef] [PubMed]
42. Dekker, A.; Reitsma, K.; Beugeling, T.; Bantjes, A.; Feijen, J.; Van Aken, W.G. Adhesion of endothelial cells adsorption of serum proteins on gas plasma-treated polytetrafluoroethylene. *Biomaterials* **1991**, *12*, 130–138. [CrossRef]
43. Wan, Y.; Yang, J.; Yang, J.; Bei, J.; Wang, S. Cell adhesion on gaseous plasma modified poly-(L-lactide) surface under shear stress field. *Biomaterials* **2003**, *24*, 3757–3764. [CrossRef]
44. Wenzel, R.N. Resistance of solid surfaces to wetting by water. *Ind. Eng. Chem.* **1936**, *28*, 988. [CrossRef]
45. Cassie, A.B.D.; Baxter, S. Wettability of porous surfaces. *Trans. Faraday Soc.* **1944**, *40*, 546. [CrossRef]
46. Lasprilla, A.J.R.; Martinez, G.A.R.; Lunelli, B.H.; Jardini, A.L.; Filho, R.M. Poly-lactic acid synthesis for application in biomedical devices—A review. *Biotechnol. Adv.* **2012**, *30*, 321–328. [CrossRef] [PubMed]
47. Shah, A.; Shah, S.; Mani, G.; Wenke, J.; Agrawal, M. Endothelial cell behavior on gas-plasma-treated PLA surfaces: The roles of surface chemistry and roughness. *J. Tissue Eng. Regen Med.* **2011**, *5*, 301–312. [CrossRef]
48. Wan, Y.; Qu, X.; Zhu, C.; Wan, L.; Yang, J.; Bei, J.; Wang, S. Characterization of surface property of poly (lactide-co-glycolide) after oxygen plasma treatment. *Biomaterials* **2004**, *25*, 4777–4783. [CrossRef] [PubMed]
49. Rodrigues, N.; Benning, M.; Ferreira, A.M.; Dixon, L.; Dalgarno, K. Manufacture and Characterisation of Porous PLA Scaffolds. *Procedia. CIRP* **2016**, *49*, 33–38. [CrossRef]
50. Guo, Z.; Yang, C.; Zhou, Z.; Chen, S.; Li, F. Characterization of biodegradable poly (lactic acid) porous scaffolds prepared using selective enzymatic degradation for tissue engineering. *RSC Adv.* **2017**, *7*, 34063. [CrossRef]
51. Estevam-Alves, R.; Ferreira, P.H.D.; Coatrini, A.C.; Oliveira, O.N.; Fontana, C.R.; Mendonca, C.R. Femtosecond Laser Patterning of the Biopolymer Chitosan for Biofilm Formation. *Int. J. Mol. Sci.* **2016**, *17*, 1243. [CrossRef] [PubMed]
52. Whitehead, K.A.; Verran, J. The Effect of Surface Topography on the Retention of Microorganisms. *Food Bioprod. Process.* **2006**, *84*, 253–259. [CrossRef]
53. García, S.; Trueba, A.; Vega, L.M.; Madariaga, E. Impact of the surface roughness of AISI 316L stainless steel on biofilm adhesion in a seawater-cooled tubular heat exchanger-condenser. *Biofouling* **2016**, *32*, 1185–1193. [CrossRef]
54. Dou, X.Q.; Zhang, D.; Feng, C.; Jiang, L. Bioinspired Hierarchical Surface Structures with Tunable Wettability for Regulating Bacteria Adhesion. *ACS Nano* **2015**, *9*, 10664–10672. [CrossRef]
55. Fadeeva, E.; Truong, V.K.; Stiesch, M.; Chichkov, B.N.; Crawford, R.J.; Wang, J.; Ivanova, E.P. Bacterial Retention on Superhydrophobic Titanium Surfaces Fabricated by Femtosecond Laser Ablation. *Langmuir* **2011**, *27*, 3012–3019. [CrossRef] [PubMed]
56. Privett, B.J.; Youn, J.; Hong, S.A.; Lee, J.; Han, J.; Shin, J.H.; Schoenfisch, M.H. Antibacterial Fluorinated Silica Colloid Superhydrophobic Surfaces. *Langmuir* **2011**, *27*, 9597–9601. [CrossRef]
57. Ortiz, R.; Moreno-Flores, S.; Quintana, I.; Vivanco, M.; Sarasua, J.R.; Toca-Herrera, J.L. Ultra-fast laser microprocessing of medical polymers for cell engineering applications. *Mater. Sci. Eng. C Mater. Biol. Appl.* **2014**, *1*, 241–250. [CrossRef] [PubMed]
58. Castillejo, M.; Rebollar, E.; Oujja, M.; Sanz, M.; Selimis, A.; Sigletou, M.; Psycharakis, S.; Ranella, A.; Fotakis, C. Fabrication of porous biopolymer substrates for cell growth by UV laser: The role of pulse duration. *Appl. Surf. Sci.* **2012**, *258*, 8919–8927. [CrossRef]
59. Li, H.; Wong, Y.S.; Wen, F.; Ng, K.W.; Ng, G.K.; Venkatraman, S.S.; Boey, F.Y.; Tan, L.P. Human mesenchymal stem-cell behavior on direct laser micropatterned electrospun scaffolds with hierarchical structures. *Macromol. Biosci.* **2013**, *13*, 299–310. [CrossRef] [PubMed]
60. Gaspard, S.; Oujja, M.; de Nalda, R.; Castillejo, M.; Bañares, L.; Lazare, S.; Bonneau, R. Nanofoaming dynamics in biopolymers by femtosecond laser irradiation. *Appl. Phys. A.* **2008**, *93*, 209–213. [CrossRef]
61. Gaspard, S.; Forster, M.; Huber, C.; Zafiu, C.; Trettenhahn, G.; Kautek, W.; Castillejo, M. Femtosecond laser processing of biopolymers at high repetition rate. *Phys. Chem. Chem. Phys.* **2008**, *10*, 6174–6181. [CrossRef]

62. Gaspard, S.; Oujja, M.; de Nalda, R.; Abrusci, C.; Catalina, F.; Bañares, L.; Lazare, S.; Castillejo, M. Nanofoaming in the surface of biopolymers by femtosecond pulsed laser irradiation. *Appl. Surf. Sci.* **2007**, *254*, 117–1184. [CrossRef]
63. Krüger, J.; Kautek, W. Ultrashort Pulse Laser Interaction with Dielectrics and Polymers. In *Polymers and Light*; Lippert, T., Ed.; Advances in Polymer Science; Springer: Berlin/Heidelberg, Germany, 2004; p. 168.
64. Sima, F.; Sugioka, K.; Vazquez, R.M.; Osellame, R.; Kelemen, L.; Ormos, P. Three-dimensional femtosecond laser processing for lab-on-a-chip applications. *Nanophotonics* **2018**, *7*, 613–647. [CrossRef]
65. Ortiz, R.; Quintana, I.; Etxarri, J.; Lejardi, A.; Sarasua, J.R. Picosecond laser ablation of poly-L-lactide: Effect of crystallinity on the material response. *J. Appl. Phys.* **2011**, *110*, 094902. [CrossRef]
66. Daskalova, A.; Nathala, C.S.R.; Bliznakova, I.; Stoyanova, E.; Zhelyazkova, A.; Ganz, T.; Lueftenegger, S.; Husinsky, W. Controlling the porosity of collagen, gelatin and elastin biomaterials by ultrashort laser pulses. *Appl. Surf. Sci.* **2014**, *292*, 367–377. [CrossRef]
67. Daskalova, A.; Bliznakova, I.; Angelova, L.; Trifonov, A.; Declercq, H.; Buchvarov, I. Femtosecond Laser Fabrication of Engineered Functional Surfaces Based on Biodegradable Polymer and Biopolymer/Ceramic Composite thin films. *Polymers* **2019**, *11*, 378. [CrossRef] [PubMed]
68. Daskalova, A.; Trifonov, A.; Bliznakova, I.; Nathala, C.; Ajami, A.; Husinsky, W.; Declercq, H.; Buchvarov, I. Selective cell response on natural polymer bio-interfaces textured by femtosecond laser. *Appl. Phys. A* **2018**, *124*, 207. [CrossRef]
69. Cordero, D.; López-Álvarez, M.; Rodríguez-Valencia, C.; Serra, J.; Chiussi, S.; González, P. In vitro response of pre-osteoblastic cells to laser microgrooved PEEK. *Biomed Mater.* **2013**, *8*, 055006. [CrossRef]
70. Bačáková, L.; Starý, V.; Kofroňová, O.; Lisá, V. Polishing and coating carbon fiber-reinforced carbon composites with a carbon-titanium layer enhances adhesion and growth of osteoblast-like MG63 cells and vascular smooth muscle cells in vitro. *J. Biomed. Mater. Res.* **2001**, *54*, 567–578. [CrossRef]
71. Bačáková, L.; Stary, V.; Hornik, J.; Glogar, P.; Lisa, V.; Kofronova, O. Osteoblast-like MG63 cells in cultures on carbon fibre-reinforced carbon composites. *Eng. Biomater.* **2001**, *4*, 11–12.
72. Veerachamy, S.; Yarlagadda, T.; Manivasagam, G.; Yarlagadda, P.K.D.V. Bacterial adherence and biofilm formation on medical implants: A review. *Proc. Inst. Mech. Eng. Pt. H J. Eng. Med.* **2014**, *228*, 1083–1099. [CrossRef] [PubMed]
73. Pogodin, S.; Hasan, J.; Baulin, V.A.; Webb, H.K.; Truong, V.K.; Nguyen, T.H.; Boshkovikj, V.; Fluke, C.J.; Watson, G.S.; Watson, J.A.; et al. Biophysical model of bacterial cell interactions with nanopatterned cicada wing surfaces. *Biophys. J.* **2013**, *104*, 835–840. [CrossRef]
74. Hasan, J.; Webb, H.K.; Truong, V.K.; Pogodin, S.; Baulin, V.A.; Watson, G.S.; Watson, J.A.; Crawford, R.J.; Ivanova, E.P. Selective bactericidal activity of nanopatterned superhydrophobic cicada Psaltoda claripennis wing surfaces. *Appl. Microbiol. Biotechnol.* **2013**, *97*, 9257–9262. [CrossRef]
75. Sirdeshmukh, N.; Dongre, G. Laser micro & nano surface texturing for enhancing osseointegration and antimicrobial effect of biomaterials: A review. *Mater. Today Proc.* **2021**, *44*, 2348–2355.
76. Cunha, A.; Elie, A.M.; Plawinski, L.; Serro, A.P.; do Rego, A.M.B.; Almeida, A.; Urdaci, M.C.; Durrieu, M.-C.; Vilar, R. Femtosecond laser surface texturing of titanium as a method to reduce the adhesion of Staphylococcus aureus and biofilm formation. *Appl. Surf. Sci.* **2016**, *360*, 485–493. [CrossRef]
77. Martínez-Calderon, M.; Manso-Silván, M.; Rodríguez, A.; Gómez-Aranzadi, M.; García-Ruiz, J.P.; Olaizola, S.M.; Martín-Palma, R.J. Surface micro- and nano-texturing of stainless steel by femtosecond laser for the control of cell migration. *Sci. Rep.* **2016**, *6*, 36296. [CrossRef]
78. Lazzini, G.; Romoli, L.; Lutey, A.H.A.; Fuso, F. Modelling the interaction between bacterial cells and laser-textured surfaces. *Surf. Coat. Technol.* **2019**, *375*, 8–14. [CrossRef]
79. Jalil, S.A.; Akram, M.; Bhat, J.A.; Hayes, J.J.; Singh, S.C.; Kabbash, M.E.; Guo, C. Creating superhydrophobic and antibacterial surfaces on gold by femtosecond laser pulses. *Appl. Surf. Sci.* **2020**, *506*, 144952. [CrossRef] [PubMed]
80. Jaggessar, A.; Shahali, H.; Mathew, A.; Yarlagadda, P.K.D.V. Bio-mimicking nano and micro-structured surface fabrication for antibacterial properties in medical implants. *J. Nanobiotechnol.* **2017**, *15*, 64. [CrossRef] [PubMed]
81. Scardino, A.J.; Guenther, J.; de Nys, R. Attachment point theory revisited: The fouling response to a microtextured matrix. *Biofouling* **2008**, *24*, 45–53. [CrossRef]
82. Lorenzetti, M.; Dogša, I.; Stošicki, T.; Stopar, D.; Kalin, M.; Kobe, S.; Novak, S. The influence of surface modification on bacterial adhesion to titanium-based substrates. *ACS Appl. Mater. Interfaces* **2015**, *7*, 1644–1651. [CrossRef] [PubMed]
83. Guenther, D.; Valle, J.; Burgui, S.; Gil, C.; Solano, C.; Arana, A.T.; Helbig, R.; Werner, C.; Lasa, I.; Lasagni, A.F. Direct laser interference patterning for decreased bacterial attachment. *Laser-Based Micro Nanoprocess. X* **2016**, *9736*, 11.
84. Hochbaum, A.; Aizenberg, J. Bacteria pattern spontaneously on periodic nanostructure arrays. *Nano Lett.* **2010**, *10*, 3717–3721. [CrossRef] [PubMed]
85. Vasudevan, R.; Kennedy, A.J.; Merritt, M.; Crocker, F.H.; Baney, R.H. Microscale patterned surfaces reduce bacterial fouling microscopic and theoretical analysis. *Colloids Surf. B Biointerfaces* **2014**, *117*, 225–232. [CrossRef]
86. Perera-Costa, D.; Bruque, J.M.; González-Martín, M.L.; Gómez-García, A.C.; Vadillo-Rodriguez, V. Studying the influence of surface topography on bacterial adhesion using spatially organized microtopographic surface patterns. *Langmuir* **2014**, *30*, 4633–4641. [CrossRef] [PubMed]

polymers

MDPI

Article

Effective Aging Inhibition of the Thermoplastic Corn Starch Films through the Use of Green Hybrid Filler

Di Sheng Lai [1,2], Azlin Fazlina Osman [1,2,*], Sinar Arzuria Adnan [1,2], Ismail Ibrahim [1,2], Midhat Nabil Ahmad Salimi [1,2] and Awad A. Alrashdi [3]

1 Faculty of Chemical Engineering Technology, Universiti Malaysia Perlis (UniMAP), Arau 02600, Perlis, Malaysia; disheng.lds@gmail.com (D.S.L.); sinar@unimap.edu.my (S.A.A.); ismailibrahim@unimap.edu.my (I.I.); nabil@unimap.edu.my (M.N.A.S.)
2 Biomedical and Nanotechnology Research Group, Center of Excellent Geopolymer and Green Technology (CEGeoTech), Universiti Malaysia Perlis (UniMAP), Arau 02600, Perlis, Malaysia
3 Chemistry Department, Umm Al-Qura University, Al-Qunfudah University College, Al-Qunfudah Center for Scientific Research (QCSR), Al Qunfudah 21962, Saudi Arabia; aarashdi@uqu.edu.sa
* Correspondence: azlin@unimap.edu.my

Abstract: Recently, hybrid fillers have been widely used to improve the properties of biopolymers. The synergistic effects of the hybrid fillers can have a positive impact on biopolymers, including thermoplastic corn starch film (TPCS). In this communication, we highlight the effectiveness of hybrid fillers in inhibiting the aging process of TPCS. The TPCS, thermoplastic corn starch composite films (TPCS-C), and hybrid thermoplastic corn starch composite film (TPCS-HC) were stored for 3 months to study the effect of hybrid filler on the starch retrogradation. TPCS-C and TPCS-HC were prepared by casting method with 5 wt% of fillers: nanocellulose (NC) and bentonite (BT). The alteration of the mechanical properties, aging behavior, and crystalline structure of the films were analyzed through the tensile test, Fourier transform infrared (FTIR), X-ray diffraction (XRD), differential scanning calorimetry (DSC), and water absorption analysis. The obtained data were correlated to each other to analyze the retrogradation of the TPCS, which is the main factor that contributes to the aging process of the biopolymer. Results signify that incorporating the hybrid filler (NC + BT) in the TPCS/4BT1NC films has effectively prevented retrogradation of the starch molecules after being stored for 3 months. On the contrary, the virgin TPCS film showed the highest degree of retrogradation resulting in a significant decrement in the film's flexibility. These findings proved the capability of the green hybrid filler in inhibiting the aging of the TPCS.

Keywords: thermoplastic starch; nanocellulose; bentonite; biocomposite; aging; retrogradation

Citation: Lai, D.S.; Osman, A.F.; Adnan, S.A.; Ibrahim, I.; Ahmad Salimi, M.N.; Alrashdi, A.A. Effective Aging Inhibition of the Thermoplastic Corn Starch Films through the Use of Green Hybrid Filler. *Polymers* **2022**, *14*, 2567. https://doi.org/10.3390/polym14132567

Academic Editor: Evgenia G. Korzhikova-Vlakh

Received: 28 May 2022
Accepted: 22 June 2022
Published: 24 June 2022

1. Introduction

The world is emphasizing the consumption of green energy and material for sustainability, thus biopolymers such as starch, polylactic acid, chitosan, and others are intensively developed to become the next potential material to replace the conventional plastics [1]. Starch-based biopolymers are highly attractive due to their high abundance on the earth, being environmentally friendly, and having the same process-structure property as fossil-fuel plastic, which makes them viable as packaging materials [2]. Starch is semi-crystalline and made up of two different macromolecules: amylose and amylopectin. The ratio of amylose to amylopectin varies according to the botanical origin of starch. The properties of the starch and starch-based plastic are strongly dependent on the ratio of these two macromolecules [3]. Typically, except for waxy starch and high amylose starch, starch will contain a higher amount of amylopectin (70–80%) than amylose (20–30%) [4]. Amylopectin is highly branched, while amylose possesses a linear molecular structure. The crystalline structure of starch is mainly contributed by double helix branched amylopectin. The neighboring branching of amylopectin chains with a degree of polymerization from

10 to 20 forms a double helix structure, causing the formation of high-order allomorphs. The interactions between double helix amylopectin and long chain amylose give rise to the semi-crystalline structure in the starch biopolymer. The alternate arrangement of the double helix crystalline lamella and the amylose amorphous lamella is known as a "growth ring", which becomes the basic structure of the starch granules.

It is clearly understood that the pristine form of starch has a poor processing ability due to an intense and strong hydrogen bonding network structure. Commonly, starch will degrade before reaching its melting temperature; therefore, plasticizers such as water, glycerol, polyol, and urea are required to weaken and break down the hydrogen bonding by forming the new hydrogen bonding with the plasticizer. Under shearing force, high temperature, and incorporation of plasticizers, granule starch can be processed into an amorphous and homogenous material known as thermoplastic starch (TPS). Glycerol and water are the most common plasticizers as they are compatible with the starch structure and were proven in many studies [5,6]. Glycerol and water can effectively reduce the intermolecular force of granule starch and loosen up the dense packing structure. Full disruption of the starch structure can be achieved through the presence of plasticizers and a sufficient supply of thermomechanical energy [7].

Virgin TPS films are rarely utilized due to their low mechanical properties and high moisture sensitivity. The high moisture absorption of the virgin TPS is due to the existence of a large number of hydroxyl groups in the starch structure [8]. Besides that, the metastability of the TPS structure may lead to recrystallization or retrogradation process during storage over a period. Retrogradation can deteriorate the characteristics of the TPS films as it will alter the mechanical properties and affect the quality of the final product. Consequently, this will limit the usage of TPS in many applications.

Many factors can lead to the retrogradation of the TPS films, such as humidity, storage time, and glass-transition temperature, Tg [7]. Retrogradation is a process of reconstruction of the crystalline structure of starch chains. Despite the controversy about the accurate mechanisms of retrogradation that happen in TPS films, retrogradation was identified as the main cause of the aging and brittleness of this bioplastic. Generally, the retrogradation process can be divided into three stages, (i) nucleation, (ii) propagation (growth of crystalline structure), and (iii) maturation (perfection of crystals) [8]. Two retrogradation processes occur during the aging of the TPS films: linear amylose structure recrystallizes into a single helix structure (fast retrogradation rate) and branched amylopectin recrystallizes into a double helix structure (slow retrogradation rate). The slow retrogradation of amylopectin compared to amylose is due to its limited chain dimensional and lower crystalline structure stability [8]. The retrogradation rate was highly dependent on the mobility freedom of starch molecules. The restoration of crystallinity structure, loss of water holding capacity, and increased stiffness of the films are the most commonly used methods to determine the retrogradation of the TPS films [9]. A large amount of research was conducted to prevent starch's retrogradation by incorporating different types of plasticizers such as glycerol, polyol, urea citric acid, formamide, and so on to reduce the interaction between the starch chains [9,10]. A plasticizer can be defined as a substance (typically a solvent) added to synthetic or natural polymers (biopolymers) to reduce brittleness by promoting plasticity and flexibility. Several papers have discussed the role of mono-plasticizer and co-plasticizer in controlling the retrogradation of the TPS films. The co-plasticization method is preferred to produce the TPS films as it can compensate for the disadvantages of each plasticizer to make better properties films compared to mono-plasticizer films. For instance, Esmaeili et al. employed glycerol and sorbitol as co-plasticizers to form the TPS films. The low mechanical strength of glycerol/TPS films was successfully enhanced by incorporating sorbitol. However, the film's ductility was greatly reduced by 80% compared to the glycerol/TPS films [11]. In another work, Khan et al. have proposed the use of boric acid and glycerol as co-plasticizer to reduce the starch retrogradation. They showed that the boric acid can provide a better anti-retrogradation effect compared to the glycerol, whereas the TPS films that contain co-plasticizers have higher moisture sensitivity compared to glycerol/TPS films [12]. Glycerol was also proposed to be a co-plasticizer with other components containing amide groups such as urea, formaldehyde, and formamide in order to reduce the retrogradation in the TPS by forming a stronger hydrogen bonding [13–15]. Nonetheless, this kind of plasticizer is rather toxic to human health; therefore it is not suitable to be applied in food packaging material. Several researchers have proposed the incorporation of a co-plasticizer that can form crosslinking with the TPS chains such as citric acid, malic acid, and choline salts–DES to reduce the retrogradation of the TPS [16,17]. However, the ductility of the co-plasticizer films was significantly reduced, making them brittle and unsuitable for film packaging applications. Therefore, researchers have investigated the incorporation of additives into the TPS structure, such as filler or nanofiller to restrict the mobility of the TPS chains. Nanofillers are those fillers that have one of their dimensions in the nano-size range (less than 100 nm). Nanocelluse and nanoclays such as bentonite and montmorillonite are examples of the nanofiller that can be used to reinforce and improve the properties of the TPS films. The solid and compact interfacial bonding between the filler/nanofiller and the TPS chains may form a highly stable thermodynamic 3D networking structure that can prevent retrogradation [18].

The crystalline structure of starch acts like a cement structure embedded in the amorphous regions, increasing the films' stiffness, and reducing the elongation at the break of the films. Starch's retrogradation happens during storage by forming hydrogen bonding with its adjacent starch chains. Therefore, incorporating nanocellulose and bentonite is expected to interfere with the alignment of the starch chains, thus preventing them from "self-interactions". Lendvoi et al. have successfully reduced the retrogradation of TPS starch through the incorporation of bentonite. Intercalation of bentonite in between the TPS molecular chains has successfully suppressed the retrogradation of starch, however, at the cost of the flexibility of the films [19]. Balakrishnan et al. studied the effect of nanocellulose on the properties of the TPS films. They have concluded that the nanocellulose confined the chain segment of the TPS chains, reducing the mobility of the starch macromolecules and suppressing the retrogradation of starch during aging [20]. Several review papers have included more detailed information on the effect of nanofillers on the aging of the TPS films [6,21,22]. Moreover, the size and number of the hydroxyl group of the filler were proposed to play a significant part in preventing the retrogradation of starch. Meanwhile, some researchers found out that the hydroxyl group's reactivity and bond-forming ability are more important in forming a stable hydrogen bond with a starch chain to prevent retrogradation. Nevertheless, most of the research concluded that the prevention of retrogradation depends on the hydrogen strength forming between the starch, fillers, and plasticizers. The stronger the hydrogen formed between the starch chain, fillers, and plasticizer, the slower the retrogradation rate [22].

Recently, hybrid fillers have been studied as potential alternative components to inhibit the retrogradation in TPS films. TPS films produced by hybrid fillers may combine the good properties of both fillers and optimize the best mechanical performance for the films. The hydrophilic type of fillers, such as nanocellulose and nano-bentonite, are compatible with the TPS, which is also hydrophilic. Hybrid fillers were studied extensively to enhance the mechanical properties of TPS films. It has the vast potential to develop into packaging material. Many research papers studied the effect of hybrid filler on the TPS films but mainly focused on the mechanical strength, structure, and humidity properties [1,23,24]. There is still a lack of studies and reports on the relationship between hybrid fillers on the retrogradation rate of TPS films due to time-consuming data collection. The effect of storage time on the properties of the TPS hybrid biocomposite is somewhat limited. Our previous study found that a hybrid of nanocellulose (NC) and nano-bentonite (BT) has a synergy effect in enhancing the toughness of the TPS films [25]. Thus, to close the research gap in this field, this current work aims to analyze the effect of green hybrid filler addition (NC + BT) on the TPS film's retrogradation during aging. In this study, corn starch powder was utilized as raw material. Corn starch is a type of starch that is derived from maize (corn) grain, which can be obtained from the kernel and endosperm part of the plant. In this article, thermoplastic starch derived from corn is referred to as thermoplastic corn starch (TPCS). The impact of hybrid filler on the TPCS film's mechanical property was studied throughout the storing period. A tensile test was applied to observe the mechanical stability throughout the 3-month storage. DSC, XRD, and FTIR analyses were employed to study the crystalline structural change, enthalpy crystalline melting, and crystalline melting temperature of the TPS films. These three tests are highly sensitive and have been proven in many studies as valuable and reliable tools to quantify the retrogradation that happens in TPS films [15,16]. All the data were compiled, analyzed, and correlated with the retrogradation process of the starch chains, to investigate the effectiveness of the hybrid filler in inhibiting the aging of the TPCS matrix.

2. Experimental

2.1. Materials

Corn starch (72% amylopectin, 28% amylose) was purchased from Sigma Aldrich (St. Louise, MO, USA). It was employed as a matrix material after being plasticized into thermoplastic form. The plasticized corn starch is referred to as thermoplastic corn starch (TPCS). The nanocellulose and bentonite were combined to form a hybrid filler (NC + BT) of the TPCS biocomposite film. The nanocellulose was extracted from the oil palm empty fruit bunch (OPEFB) fiber purchased from United Oil Palm Industries Sdn Bhd (Nibong Tebal, Malaysia). Natural nano-bentonite clay was obtained from Sigma-Aldrich (St. Louise, MO, USA). Details on the hybrid filler preparation can be found in our previous paper [14]. Glycerol was purchased from HmbG Chemicals (Hamburg, Germany). It was used as a plasticizer with deionized water.

2.2. Preparation of TPCS, TPCS-C, and TPCS-HC Films

The virgin TPCS films, TPCS biocomposite films (TPCS-C), and hybrid TPS biocomposite films (TPCS-HC) were prepared through the casting method. A total of 5 g of corn starch was dispersed in 100 mL distilled water and a 2 g glycerol mixture to form thermoplastic corn starch films. The mixture was kept stirring at 300 rpm using a heated magnetic stirrer for 30 min at 80 °C to obtain homogenous TPS gel. For TPCS-C and TPCS-HC, 5 wt% of single filler or hybrid filler were prepared and then subjected to an ultrasonication process before being incorporated into the TPCS matrix. There were two ratios of hybrid fillers selected: (i) 4BT:1NC and (ii) 2BT:3NC for producing the TPCS biocomposite films. These two ratios were selected based on our previous study where 4BT:1NC was found to be the optimum ratio to produce the toughest films, while 2BT:3NC was observed to result in the lowest toughness value to the TPCS-HC film.

In the preparation of the biocomposite films, the mixture of TPCS/BT/NC was continuously stirred for 10 min to achieve homogenous dispersion of hybrid filler in the matrix. The homogenous TPCS suspension was poured into an 8-inch round Teflon casting plate and dried in the oven for 24 h at 45 °C. The composition and the abbreviation of the samples are presented in Table 1.

Table 1. The formulation of TPCS, TPCS-C, and TPCS-HC films.

	Acronym	Starch (wt%)	Bentonite (BT) (wt%)	Nanocellulose (NC) (wt%)
Virgin Thermoplastic Corn Starch (TPCS)	TPCS	100	0	0
Thermoplastic Corn Starch Composite (TPCS-C)	TPCS/5BT	95	5	0
	TPCS/5NC	95	0	5
Hybrid Thermoplastic Starch (TPCS-HC)	TPCS/4BT1NC	95	4	1
	TPCS/2BT3NC	95	2	3

Storage Procedures for Aging Analysis

The TPCS, TPCS-C, and TPCS-HC films were kept in a chamber with a constant humidity of 53% at 25 °C and stored for 15, 30, 45, 60, and 90 days. The film samples were taken out for testing and analysis after the specified storage period. Figure 1 shows the films' appearance after being stored for 15, 45, and 90 days.

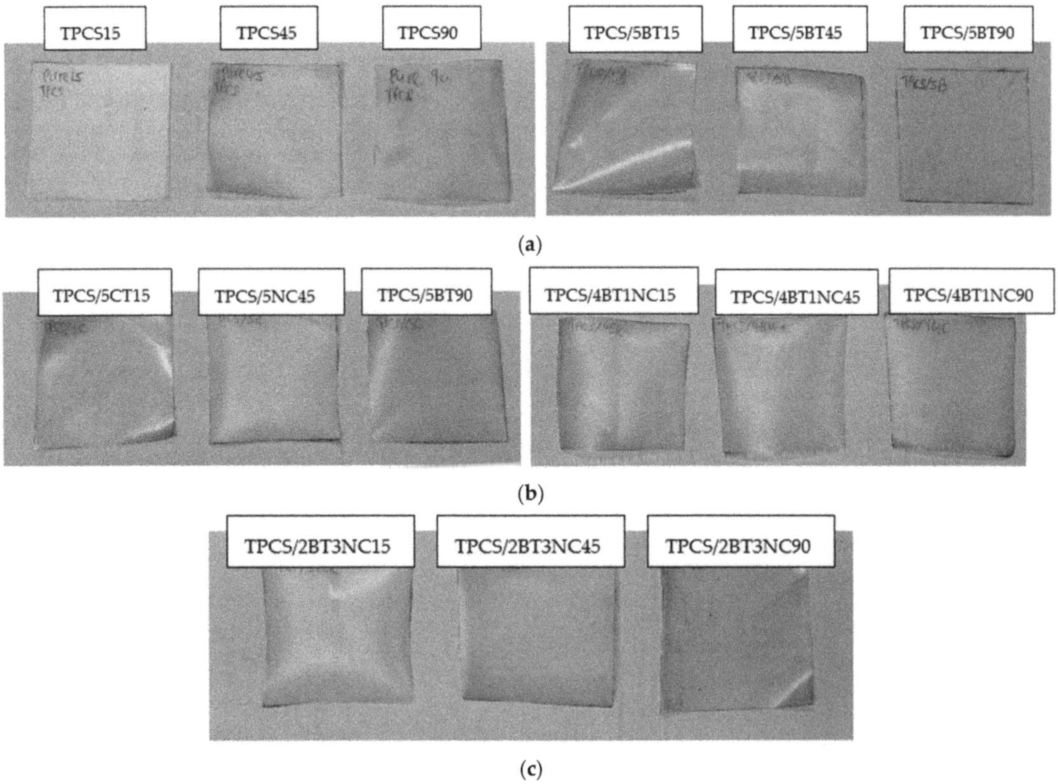

Figure 1. The physical appearance of virgin TPCS, TPCS-C, and TPCS-HC films aged for (**a**) 15, (**b**) 45, and (**c**) 90 days.

2.3. *Testing and Characterization of Films*

2.3.1. Tensile Test

A tensile test was carried out according to ASTM-638 Type V to determine the tensile properties of films with different storage duration using the Universal Instron Machine model-5582 (Norwood, MA, USA). The tensile test was performed with a 5 kN load sensor and 10 mm/min crosshead speed. The testing was carried out at room temperature (23 °C) with 53% humidity. Seven replicates were tested and the average values of tensile properties (tensile strength, Young's modulus, and elongation at break) were acquired from the Instron Merlin software Version 5.41.00 (Instron®, Norwood, MA, USA) and then recorded.

2.3.2. Differential Scanning Calorimetry (DSC)

Differential scanning calorimetry (DSC) analyses were carried out by TA instrument Q-10 (Lukens Drive, New Castle, DE, USA) to determine the rate of retrogradation of the TPCS-C and TPCS-HC. A total of 30 mg of samples were tested and analyzed between 25 °C to 180 °C with a heating rate of 10 °C/min in a nitrogen atmosphere to determine the enthalpy of crystallization of each sample. The reference pans used were 40 µL aluminum pans.

According to ASTM D3418-03, the enthalpy of the crystalline structure was determined by measuring the area of the main endothermic peak, while the melt temperature was determined as the midpoint of onset and end temperature of the endothermic peak. The enthalpy change in temperature was calculated and identified by using TA universal analysis 2000 software version 4.5A (TA-instruments-Water L.C.C, Lukens Drive, New Castle, DE, USA). All the films were stored under the same humidity (53%) for 15, 45, and 90 days.

2.3.3. X-ray Diffraction (XRD)

The crystalline structure of all the films was determined using the Bruker D2 Phaser X-ray diffractometer (Billerica, MA, USA) using Cu Kα X-rays. The samples were tested by using a scan rate of 0.1 s per step from $2\theta = 10 - 40^{\circ}$. The analysis and identification of the peak was analyzed by using X'Pert HighScore plus under PANalytical B.V. Almelo, The Netherlands. Meanwhile, the calculation for the degree of crystallinity was performed using Originpro 2019 (b) under OriginLab Corporation Northampton, Massachusetts, USA. The testing was carried out at room temperature, 25 °C with 53% humidity. The crystallinity of the films was determined by calculating the subtracting the amorphous halo from the XRD scan shown in Formulation (1)

$$\text{Degree of crystallinity } (\%) = \frac{I_o - I_a}{I_o} \times 100\% \quad (1)$$

where I_o = Total surface area of the peak from the XRD-based line, I_a = Amorphous halo from the XRD scan.

2.3.4. Fourier Transform Infrared (FTIR)

TPCS, TPCS-C, and TPCS-HC films were scanned by using a Perkin Elmer spectrum 65 (Waltham, MA, USA) FTIR spectrometer with a wavelength range of 650–4500 cm^{-1}, 16 scans and resolution of 4 cm^{-1}. All the films were stored at the humidity of 53%, 25 °C, and stored for 15, 45, and 90 days before conducting the FTIR analysis. The infrared spectra range of 900–1200 cm^{-1} was focused to calculate the intensity ratio of the spectra at a specific wavenumber (1045 cm^{-1}:1022 cm^{-1}). The purpose was to study the short recrystallize structure of the films. The infrared spectrum range 900–1200 cm^{-1} was deconvoluted, whereas the intensity ratio for 1045 cm^{-1}:1022 cm^{-1} was calculated by using the Origin 2019 (B) software (OriginLab Corporation Northampton, MA, USA).

2.3.5. Moisture Absorption

The moisture absorption test was determined by using the samples of 10 × 10 × 0.2 mm dimension to quantify the amount of moisture absorption during the 3-month storage duration. All the produced films were kept in the sealed humidity chamber with a control relative humidity (RH) of 53% at 25 °C. The film samples were immediately weighed after being taken out from the humidity chamber, and weighed in a specific period. The moisture content of the films was calculated by using Equation (2):

$$\text{Moisture content } (\%) = \frac{W - W_o}{W_o} \times 100\% \quad (2)$$

where W = the film's weights after taking out from the humidity chamber and W_o are the original mass of the films right after the films form. The test was repeated 3 times to obtain the average value of the moisture content [15].

3. Results and Discussion

3.1. Mechanical Analysis of TPCS, TPCS-C, and TPCS-HC Films Aged for 15, 30, 45, 60, and 90 Days

3.1.1. Tensile Strength

Measuring the change of mechanical properties across a period is one of the effective ways to postulate the aging properties. Figure 2 summarizes the mechanical properties of the TPCS, TPCS-C, and TPCS-HC biocomposite films for 15, 30, 45, 60, and 90 days of storage. After being stored for 3 months, the tensile strength of the virgin TPCS increased from 3.01 MPa to 5.3 MPa, almost 76% during the storage. The gradual increase of the tensile strength indicates the retrogradation of the TPCS occurred due to enhancement in crystallinity, as detected through XRD and DSC analyses. More crystalline structure in the starch chains requires greater force to break the molecular bonds; therefore the tensile strength value increases [26]. Meanwhile, the tensile strength of the TPCS/4BT1NC sample has been slightly increased (from 7.05 MPa (15 days) to 8.31 MPa (90 days)) throughout the storage period.

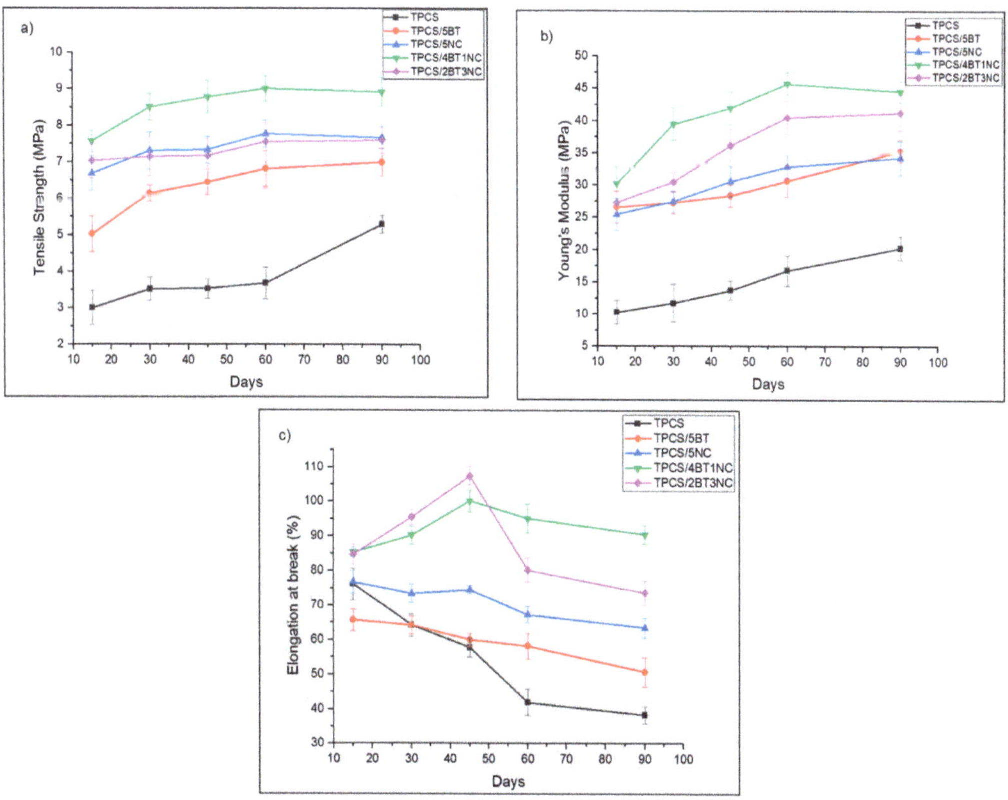

Figure 2. (**a**) Tensile strength, (**b**) Young's modulus, and (**c**) Elongation at break of TPS, TPSC, and HTPS films aged for 15, 30, 45, 60, and 90 days.

Generally, the trend shown in these tensile test data can be attributed to the structural evolution of the TPS, which is also influenced by the plasticizer and hybrid filler. Tensile strength for all the films increases in the first 15 days to 30 days, indicating migration of plasticizers and rearrangement of starch microstructure that occur within the first 30 days. This observation was in accordance with previous studies, and those results were proved by the XRD analysis which indicated the increase in the crystalline structure of the host

TPS in the first month of storage. An increase in the crystallinity was attributed to the migration of plasticizers, resulting in an increase in tensile strength [27–29]. However, the tensile strength of the TPCS/4BT1NC was relatively stable after the first 30 days compared to other films. This shows that the interaction between TPCS, plasticizer, and hybrid filler could effectively minimize the rearrangement of the starch molecular chain. TPCS/5BT and TPCS/5NC showed a gradual and constant increase in strength with increased storage time. Meanwhile, the TPCS films show a noticeable increase in tensile strength after 60 days. The rise of the TPCS films' tensile strength was contributed by the increased crystallization of the amylopectin structure in the TPCS matrix, as shown in DSC and XRD analyses. The increase of the recrystallized amylopectin region indicated that the 3D starch network formed was less stable compared to other films under the normal storage condition. The absence of filler/hybrid filler caused the migration of the plasticizer to the film's surface. The void left behind by the plasticizer causes amylopectin to form bonding with the adjacent molecular chains and form another crystalline structure.

It is worth mentioning that the retrogradation rate of the TPCS films is highly dependent on the TPCS's chain mobility. Strong interfacial bonding within the TPCS chains can reduce the starch chain mobility and thus, effectively hinder the retrogradation. The stronger the interfacial bonding, the slower the retrogradation rate. Therefore, incorporating a hybrid filler is postulated to reduce the mobility and retrogradation process of the starch chains, but at the same time increase the glass-transition temperature, Tg, of the TPCS. Meanwhile, for the virgin TPCS films, the high moisture absorption capability caused the decrease of Tg across the storage time. As the Tg decreases, the degree of plasticizer migration increases with the storage time [30]. Therefore, we can observe a sudden increase in tensile strength from 3.69 MPa (60 days) to 5.30 MPa (90 days). By comparing the tensile strength of the TPCS, TPCS-C, and TPCS-HC films for 3 months, the TPCS-HC was seen to show the best retention in the tensile strength value, suggesting that the hybrid fillers can effectively slow down or limit the retrogradation through solid interaction between the plasticizer and TPCS chains.

3.1.2. Young's Modulus

The Young's modulus trend is almost identical to tensile strength, where the TPCS films show the highest increase in Young's modulus after 90 days of storage, even though the value is still the lowest among all the samples. The virgin TPCS film exhibits a 100% increase in Young's modulus from 10.1 MPa to 20.3 MPa after 3 months of storage. Meanwhile, the TPCS/4BT1NC demonstrates the lowest increase of Young's modulus (20% increase of Young's modulus from 37.2 MPa to 44.6 MPa). By comparing the increment percentage, it can be postulated that the TPCS/4BT1NC has relatively higher structural stability than other films. The progressive increase of the TPCS film's Young's modulus can be attributed to the retrogradation. During the storage, the amylose and amylopectin recrystallize into a single helix and double helix crystal structure through the expulsion of plasticizers located within their adjacent chains. This enhances the crystallinity of the TPCS and leads to the stiffening and brittleness of the TPCS film [31,32]. Obviously, the modulus of the TPCS/4BT1NC was rather stable throughout the storing period, indicating there was no drastic alteration of the crystalline structure causing the fluctuation in Young's modulus value.

The Young's modulus of TPCS/2BT3NC has shown a different trend than its tensile strength. The tensile strength of the TPCS/2BT3NC showed a slight decrease from days 45 to 90. However, the Young's modulus value showed a slightly increased from 32.2 MPa to 38.3 MPa. The slightly increased Young's modulus indicating the less chain mobility of TPCS/2BT3NC can be associated with the increased crystalline structure as proved in XRD analysis. However, DSC showed that the crystalline structure in TPCS/2BT3NC is highly unstable, and low energy is required to break the crystalline structure. Therefore, the increase in crystalline structure is not reflected in the tensile strength.

TPCS/5BT and TPCS/5NC biocomposites demonstrate a gradual increase of Young's modulus through the storage time. The increase of the modulus can be due to the increase in the crystalline structure of the TPCS-C chain or can be related to the improved reinforcing effect of filler with increased storage time. As the migration of plasticizers happens during storage, it will trigger the recrystallization in TPCS-C and increase the film's stiffness. The migration of plasticizers will promote a "closer" interaction between the filler and starch chains, which can improve direct interfacial interaction and form the glassy structure in the TPCS matrix [15].

3.1.3. Elongation at Break

The elongation at break of the TPCS films has significantly decreased from 76.1% to 38.4% and became highly brittle as the storage time increased. The embrittlement of the TPCS films was one of the undesired characteristics caused by the retrogradation. It has been well documented that the amorphous structure of amylose and amylopectin can recrystallize into the different crystalline structures (single-crystalline or double helix crystalline structure) during the storage [3,16]. The different crystalline structure formed in TPCS films will cause heterogeneous phases occurs as amylose and amylopectin recrystallize through intra- and intermolecular bonding. The different extent of chain recrystallization rate and crystalline structure induces the hard and soft segment order in the TPCS matrix. The disparity hardness between these segment orders could increase the internal stress of TPCS films at the junction of crystalline structure when the film is stretched. The internal stress evolves into the microvoids, which results in premature failure and reduction in the elongation at break [16].

The elongation at the break of the TPCS/4BT1NC was gradually increased from 85.5% (15 days) to 112.3 % (30 days) but then decreased slightly to 92.6% (90 days). Interestingly, the TPCS/4BT1NC does not show a significant reduction in the elongation at break value compared to other films, indicating the hybrid fillers could stimulate stress relaxation in the hard segment order and enhance the soft segment order which reduces the internal stress build-up in the films as mentioned in our previous study [24]. Thus, the TPCS/4BT1NC is physically stable across the storage and aging.

Apparently, the results suggest that the stability toward the aging of the TPCS/2BT3NC film was less prominent as compared to the TPCS/4BT1NC film. The TPCS/2BT3NC exhibits a stable microstructure for the first 60 days; however, as the storage increased to 90 days, the microstructure stability was reduced, and retrogradation was detected after reaching 60 days. This shows that the ratio of NC/BT filler in the hybrid system would affect the long-term retrogradation of the TPCS films. It can be said that the high composition of the NC compared to BT in the biocomposite films can attract the migration of plasticizers toward the NC. The long storage period has caused an accumulation of plasticizers in the interface of NC and the amylopectin. This encouraged the crystallization of the amylopectin chains, forming the unstable transcrystalline region in the starch structure [33,34]. At the same time, the number of the plasticizer molecules in the amylopectin-rich region decreased, leading to reduced chain mobility and decreased elongation at break. As a result, elongation at break for the TPCS/5BT and TPCS/5NC films decreased along with the increase in storage time. However, the elongation at break for both films is still better than the virgin TPCS films, indicating that the single filler (BT or NC) can prevent retrogradation of the TPCS matrix to some extent, although not as good as the hybrid NC/BT filler.

3.2. *Thermal Analysis for TPCS, TPCS-C, and TPCS-HC Films Aged for 15, 45, and 90 Days*

Figure 3 presents the DSC thermal analysis of the virgin TPCS, TPCS-C, and TPCS-HC biocomposite films, which were stored for 15, 45, and 90 days. This thermal analysis was performed to study the thermal transition throughout the 3-month storage time. The thermal transition was detected by measuring the heat flow differences between the TPCS films and the reference pans, which can usually be related to the melting of the crystalline structure in the films.

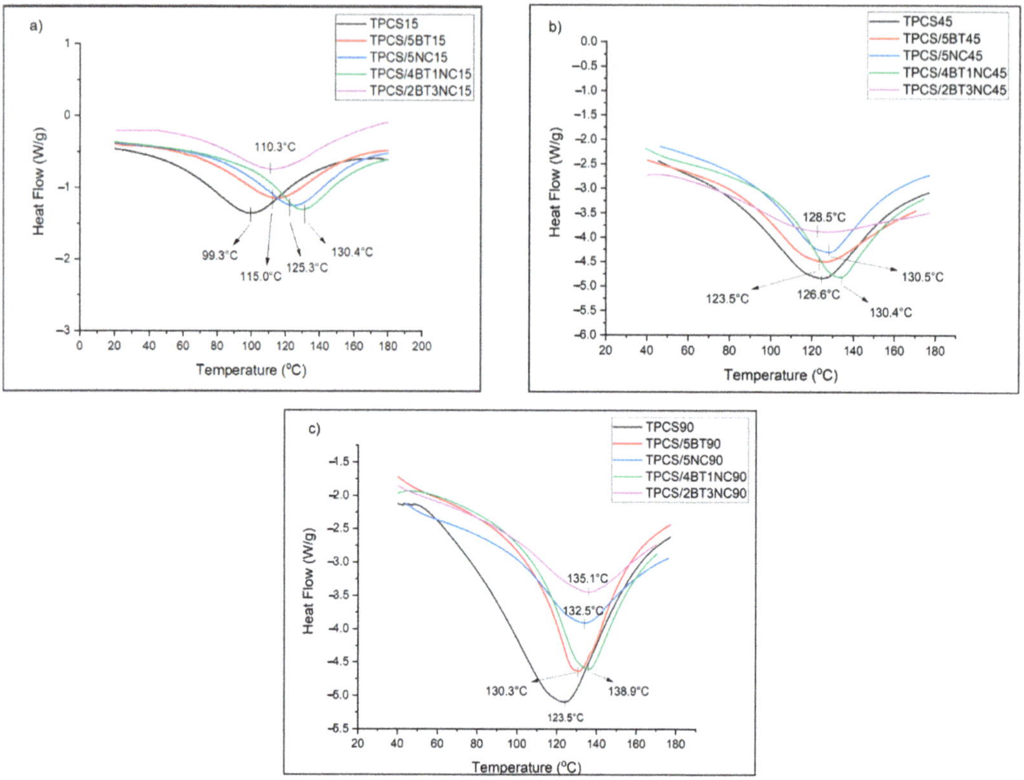

Figure 3. DSC heating curves of virgin TPCS, TPCS-C, and TPCS-HC films aged for (**a**) 15, (**b**) 45, and (**c**) 90 days.

The thermal transitions of the virgin TPCS and associate biocomposites were analyzed by observing the changes in the enthalpy of fusion of the melting endotherms (ΔHf) and melting temperature (T_m). The recrystallized structure of the retrograded starch films can be realized through the observed enthalpy changes. The increase of enthalpy melting can be related to the extent of recrystallization happening in the TPCS structure. The homogeneity and the quality of the crystallized structure formed by starch retrogradation also can be studied by the broadness and the intensity of the melting peak. A more complex, broad, and weak melting peak can be associated with a high number of the different crystalline structures formed in the matrix with varying morphologies. By comparing the DSC heating curves between the TPCS, TPCS-C, and TPCS-HC films, we can reveal the effect of single filler and hybrid filler on the recrystallization structure of TPCS. The melting temperatures (T_m) and enthalpy of fusion (ΔH_f) for TPCS, TPCS-C, and TPCS-HC films are summarized in Table 2, while the DSC heating curves are shown in Figure 3.

During the first 15 days, all the samples exhibited a broad and weak endothermic melting peak. TPCS15 films showed the highest enthalpy of melting (194.3 J/g) followed by TPCS/2BT3CN15 (186.6 J/g), TPCS/4BT1NC15 (178.5 J/g), TPCS/5BT15 (142.6 J/g) and TPCS/5NC15 (135.4 J/g). Even though the TPCS films have the highest melting enthalpy, it possesses the lowest T_m (99.3 °C) and a broad endothermic peak, suggesting that the crystalline structure in the starch is non-uniform and less stable compared to other films. During the retrogradation process, the TPCS chains may form different structures and morphologies and result in distinct retrograded crystallites [35]. The differences of distinct retrograded crystalline structures may cause the different extent of the crystalline structure formed in the TPCS matrix and result in the large and broad endothermic peak being detected. Meanwhile, incorporating single or hybrid fillers into the TPCS has increased the T_m of the crystalline structure. Even though incorporating the BT and NC can hinder the recrystallization in the TPS films, studies also showed that the BT and NC may provide the "thermal shielding effect" to the matrix phase, which causes a higher melting temperature needed to break the structure [24,36]. Moreover, the difference in morphology between the BT and NC may contribute to different ways of matrix-filler interactions, which causes a difference in T_m between the TPCS/5BT15 and TPCS/5NC15.

Table 2. Melting temperatures (Tm) and enthalpy of fusion (ΔHf) for TPCS, TPCS-C, and TPCS-HC films.

Sample Films	Days					
	15		45		90	
	T_m (°C)	ΔH_f (J/g)	T_m (°C)	ΔH_f (J/g)	T_m (°C)	ΔH_f (J/g)
TPCS	99.3	194.3	126.6	435.6	123.5	500.3
TPCS/5BT	115.0	142.6	123.5	329.6	130.3	400.5
TPCS/5NC	125.3	135.4	130.5	325.4	132.5	348.5
TPCS/4BT1NC	130.4	178.5	134.4	412.4	138.9	423.6
TPCS/2BT3NC	110.3	186.6	128.5	203.4	135.1	212.5

As the storage time increased to 45 days, the T_m of the TPCS, TPCS-C, and TPCS-HC films increased, and the endothermic peak became sharper and more intense. The T_m of TPCS/4BT1NC45 only shows a small increase (+4 °C), but the melting enthalpy showed a tremendous increase from 178.5 to 412.4 J/g. The significant growth of enthalpy could be due to the enhancement effect of hybrid fillers toward the crystalline structure which required higher thermal energy to melt the crystalline structure [24]. The melting temperature and enthalpy of the TPCS films also significantly increased by as much as 24.3 °C and 241.3 J/g from day 15 to 45. The increase of both parameters indicated the crystalline structure continued to develop into a more well-defined and stable crystal structure. Interestingly, the T_m of the TPCS/2BT3NC45 was only increased by 18.2 °C. However, it shows only a small increase of melting enthalpy (128.5 J/g) from day 15 to 45 days. The small increment and broad melting enthalpy peak of TPCS/2BT3NC45 indicate that the crystalline structure formed in TPCS/2BT3NC45 is highly unstable and non-uniform in size.

As the storage time increased to 90 days, the T_m and melting enthalpy of the films increased and become more well defined. The melting peak of the TPCS90 becomes broader and more intense compared to other films. The high-intensity TPCS90 melting peak can be related to a high retrogradation rate due to the high growth rate of the crystal structure. However, the broader peak observed could be due to the high number of imperfect and non-uniform crystals [23,37]. The higher Tm and sharper melting enthalpy of the TPCS/5NC90 and TPCS/5BT90 compared to the TPCS90 indicate that a more uniform and stable crystalline structure can be induced by adding NC or BT. Even though NC and BT can induce crystalline structure forming in the TPCS matrix, the overall increase in the melting enthalpy of TPCS/5NC and TPCS/5BT is rather smaller than the TPCS films from days 45 to 90. This shows that BT and NC can reduce the retrogradation of starch for a prolonged storage period. The DSC heating curve of the TPCS/2BT3NC90 revealed the lowest melting enthalpy of this sample compared to other samples, after 3 months of storage. The low melting enthalpy indicates that there is less thermal energy required to destroy the crystalline structure of the TPCS/2BT3NC90. Furthermore, the fluctuation in the melting temperature of the film suggests that the retrogradation process occurred to some extent. Meanwhile, the T_m and melting enthalpy of the TPCS/4BT1NC show relatively stable across the 3-month storage period. The sharp and low intensity of melting enthalpy from the beginning of storage indicates the hybrid filler with a 4:1 ratio is highly favored to form a stable crystalline structure at the early stage of film-forming and restrict the mobility of the amylopectin, preventing the retrogradation phenomenon.

3.3. *Crystalline Structure of TPCS, TPCS-C, and TPCS-HC Films Aged for 15, 30, 45, 60, and 90 Days by Using XRD Analysis*

Overall, the TPCS, TPCS-C, and TPCS-HC films' crystallinity increased with the storage time. The longer the storage time, the higher the crystalline percentage. However, incorporating a single or hybrid filler slowed down or inhibited the retrogradation rate. Table 3 and Figure 4 shows the changes in the crystalline percentage of the TPCS, TPCS-C, and TPCS-HC films after 15, 45, and 90 days of aging. The virgin TPCS film demonstrates the highest rate of the crystals growth as the crystalline percentage reaches 28.7% in the 3-month storage. Meanwhile, TPCS/5BT and TPCS/5NC exhibit a lower crystallization percentage compared to the TPCS films. For TPCS-HC films, TPCS/2BT3NC and TPCS/4BT1NC proved to effectively prevent the retrogradation in starch for the first 45 days as the crystalline percentage only showed a minor increase of 3.7% and 2.2%, respectively. The crystalline percentage of TPCS/4BT1NC experienced a slight increase in crystallinity structure (2.9%) even though the storage period increased to 3 months. However, for TPCS/2BT3NC, the efficacy to inhibit retrogradation was only seen to be dominant in the first 45 days. This is because the crystallinity of the TPCS/2BT3NC increased drastically to 20% after 3 months of storage, showing that the retrogradation process has taken place.

Table 3. The changes in the crystalline percentage of TPCS, TPCS-C, and TPCS-HC films after 15, 45, and 90 days of storing.

Sample	Crystallinity Percentage after Aging (%)		
	15 Days	45 Days	90 Days
TPCS	16.7	20.3	29.7
TPCS/5BT	13.5	19.4	24.4
TPCS/5NC	14.3	20.6	23.6
TPCS/4BT1NC	10.4	12.6	15.1
TPCS/2BT3NC	10.1	13.8	20.0

The XRD diffraction pattern of the TPCS films, TPCS-C films, and TPCS-HC films that were stored for 15, 45, and 90 days was recorded. Upon the film drying process, the amylose chain will undergo fast retrogradation to form a Va-type or Vh-type crystalline structure with the inclusion of plasticizers, which is typically detected in $2\theta = 7.8°, 15.0°, 20.8°$, and $2\theta = 17.0°, 18.3°, 22.1°$ respectively. However, the Va-type crystals are relatively unstable compared to the Vh-type crystalline structure; they would absorb water molecules and transform to the Vh-type crystalline during storage in ambient humidity [38]. Therefore, a small amount of Va-type crystalline residual structure was detected in TPCS films' XRD diffraction pattern. The complete transformation of crystalline usually happens in the first 20 days [11,39,40]. The crystalline structure change by absorbing water molecules agreed with the water absorption result of the pure TPS films. As the storage time increased to 45 days, the residual for Va-type crystalline disappeared and completely transformed into Vh-type crystalline. Meanwhile, as the TPS films were scanned on day 45, the intensity of the Vh seemed to reduce. This phenomenon was explained by Schmitt et al. where the Vh-type crystalline structure was superimposition with B-type crystalline during the slow retrogradation process, causing the intensity of the Vh to be reduced in the XRD diffraction diagram [16].

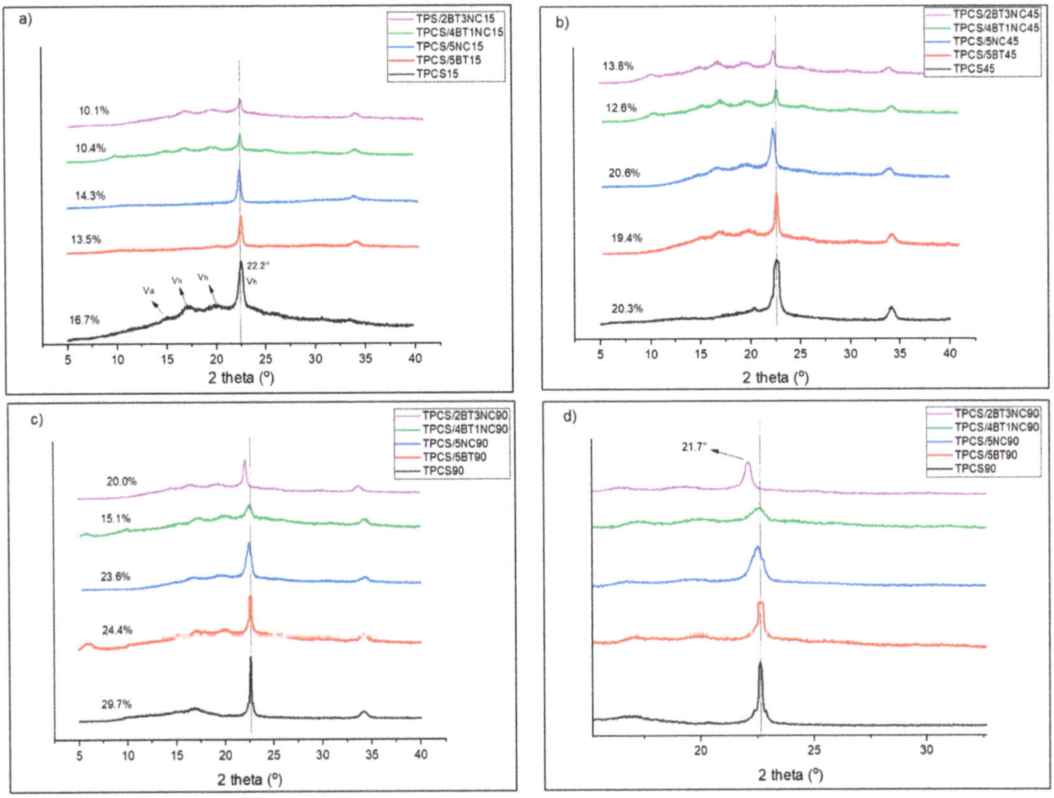

Figure 4. XRD scan for virgin TPCS, TPCS-C, and TPCS-HC films aged for (**a**) 15, (**b**) 45, and (**c**) 90 days from 2θ= 5.0–40.0° and (**d**) 90 days from 2θ = 10.0–30.2°.

Based on the XRD curves, the TPCS15 and TPCS/5NC15 and TPCS/5BT15 exhibited a firm Vh-type retrogradation peak compared to TPCS/4BT1NC15 and TPCS/2TB3NC15 for the first 15 days. The high intensity of the Vh structure strongly suggests the plasticizer was not effectively dispersed in the TPCS matrix, resulting in the formation of the Vh-plasticizer crystalline structure. Meanwhile, only a small Vh peak was detected in both TPCS-HC films, indicating that the hybrid filler more effectively interacts with the plasticizer and restricts the mobility of plasticizer molecules to prevent the formation of Vh-plasticizer crystalline structure.

In the case of TPCS/5NC and TPCS/5BT, the increased intensity of the peak at 2θ = 22.2° over time can be due to the nucleating effect of the fillers. Generally, slow retrogradation consists of three major steps (nucleation, propagation, and maturation) in recrystallization [8]. The nucleation of the crystalline structure started at the early stage of film-forming. It was propagated throughout the storing time. The crystalline structure was finally matured after three months of storage and increased the intensity of the peaks. The biocomposite incorporating the single filler, regardless of BT or NT, exhibited a lower crystalline percentage than the TPCS films after 3 months, indicating that a single filler can suppress the retrogradation in starch. However, Lendvai et al. showed a contrary result where the addition of filler increases the retrogradation rate in starch. The increase in crystalline percentage may be attributed to poor filler dispersion in the TPS matrix. The agglomerated filler served as the nucleating site to facilitate the recrystallization of the starch chain [30].

The crystalline structure of the TPCS/4BT1NC only experienced slight changes after 3 months of storage, evidently proved by the slight alteration of the diffraction peaks. This result was also in agreement with the stable mechanical performance of the films. All the films exhibited a more defined peak as the storage time increased. A pronounced increase in peak intensity at 2θ = 22.2° was observed in TPCS, TPCS/5NC, and TPCS/5BT after 45 days of storage. However, the increased intensity of this peak was not observed in the TPCS-HC films, suggesting that the hybrid fillers can effectively form stable TPCS structures and reduce the retrogradation. However, as the NC ratio in the hybrid filler increased to 3, the peak intensity at 2θ = 22.2° for TPS/2BT3NC has a noticeable increase after 3 months of storage. This increase in peak intensity indicates that the semi-crystalline structure of the TPCS/2BT3NC is not stable with a long storage time. The inhibition of the retrogradation process is only prominent during the early storage period. TPCS/2BT3NC is only effective in preventing the fast retrogradation happening in starch, and it was less effective in suppressing the slow retrogradation happening in the TPCS films compared to the TPCS/4BT1NC.

The effectiveness in inhibiting the retrogradation in TPCS/4BT1NC and TPCS/2BT3NC may be attributed to the different ratios of the hybrid filler component. The increase in NC ratio in the TPCS-HC films resulted in the low TPS film's stability in long-term storage. Previous studies showed that the NC has a higher plasticizer affinity than the starch molecules. Drakopoulos et al. have studied the interfacial effect of TPS, microcrystalline cellulose, and plasticizer by using broadband dielectric spectroscopy (BDS). They showed that even though the microcrystalline cellulose and starch chains have high similarities in chemical structure, they possess different dimensional structures that affect their polarization toward plasticizers. The difference in dimensional structure makes the microcrystalline cellulose more hydrophilic, thus water molecules can more easily enter and decrease the polymer chain's mobility [41]. As the storage time increases, the plasticizer in the TPCS/2BT3NC will slowly get away from the amylopectin-rich phases. The voids left behind by the plasticizer causes more water to be diffused into the amylopectin structure and facilitates the recrystallization. This result agrees with the water absorption data of the TPCS/2BT3NC as the water absorption of TPCS/2BT3NC is slightly decreased compared to the TPCS/4BT1NC after 3 months of storage. This explained the sudden increase in crystalline percentage and intensity peak of the TPCS/2BT3NC at 2θ = 22.2° after 45 days of storage. The unexpected crystalline alteration was also seen in the study of A. Ujcic et al. However, they showed the opposite trend where the crystallinity of the TPS composite films decreased in the middle of the storage period by incorporating metal oxide. They proposed that the metal oxide induced the early crystalline structure detected in the TPS films. However, the metal oxide caused an unstable crystalline structure due to low compatibility with the starch chains, and the plasticizer can quickly destroy it during the recrystallization process [41]. This result also showed that the migration of plasticizers toward the fillers highly affected the TPS's retrogradation process. Moreover, it also showed that the retrogradation process highly influences the relationship of plasticizers with the filler in the TPS structure.

The agglomeration of plasticizer at the interfacial interaction between the NC in TPCS/2BT3NC and amylopectin will induce the amylopectin chain to form a trans crystallization structure on the surface of the NC as reported in many studies [33,34,42,43]. According to their studies, the existence of trans crystallization can be detected through the appearance of a peak at 2θ = 21.5°. The peak magnitude of TPCS/2BT3NC at 2θ = 22.2° was observed to be shifted away from 2θ = 22.5° to 21.7° from day 45 to 90 in Figure 4d. This shifting of diffraction peak was only observed for TPCS/2BT3NC90 and TPCS/5NC90. However, the TPCS/5NC90 only shows a slight shifting of peak compared to TPCS/2BT3NC90. The shifting of the peak is only seen in the films that contain NC in the structure, therefore this shifting of the peak is most probably related to the existence of transcrystalline structure in TPCS/2BT3NC90. The transcrystalline structure formed as an oriented crystalline layer of amylopectin on the NC with the inclusion of a plasticizer at the interfacial surface [44].

The transcrystalline structure in TPCS/2BT3NC may contribute to low melting enthalpy, as detected through the DSC analysis.

3.4. Recrystallization of Short-Chain Amylopectin in the TPCS, TPCS-C, and TPCS-HC Films Aged for 15, 45, and 90 Days as Detected through FTIR Analysis

The FTIR spectra of TPCS, TPCS-C, and TPCS-HC in the range of 900–1200 cm^{-1} are shown in Figure 5. Many studies proved that the alteration of short-range molecular order could be obtained from the bands at 1022 cm^{-1} and 1047 cm^{-1} [45–47]. The number of amorphous regions and crystalline regions was represented by the bands at 1022 cm^{-1} and 1047 cm^{-1}, respectively. The intensity ratio between the 1047 cm^{-1}:1022 cm^{-1} (C values) can be used to index the starch crystalline structure region in the amorphous area. The higher the C values, the more organized the TPCS matrix structure. The infrared spectra from 900 to 1200 cm^{-1} were deconvoluted, and the 1047 cm^{-1}:1022 cm^{-1} values were calculated and presented in Table 4.

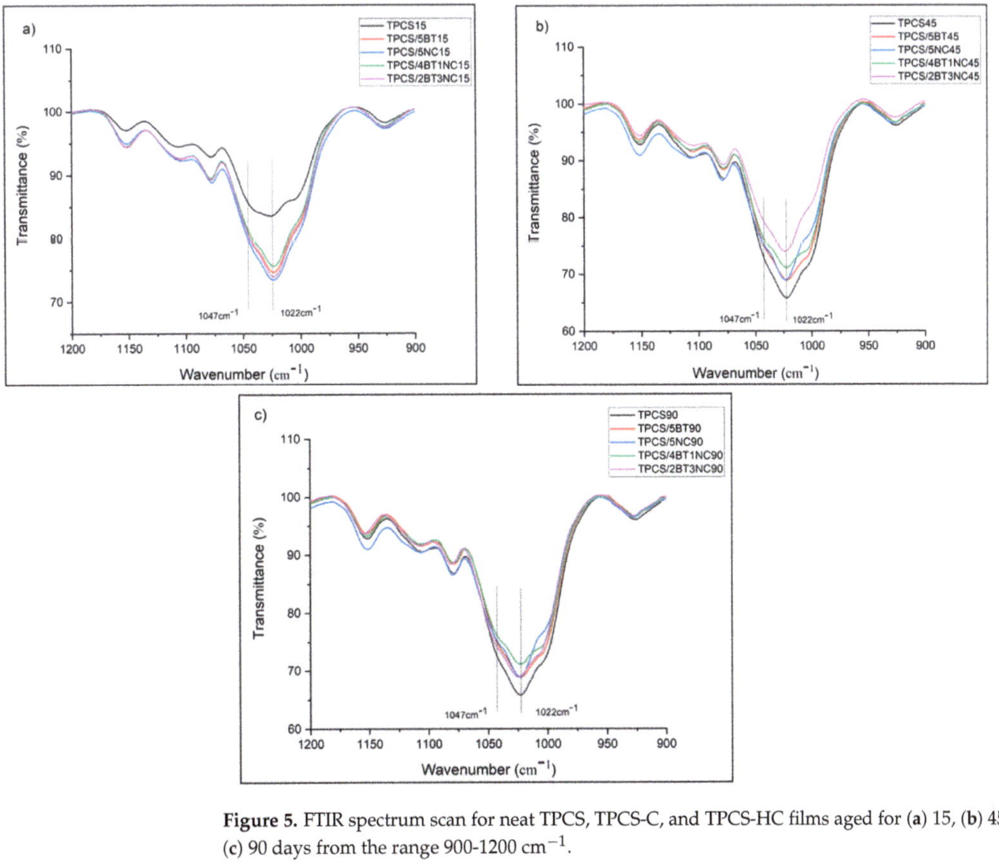

Figure 5. FTIR spectrum scan for neat TPCS, TPCS-C, and TPCS-HC films aged for (**a**) 15, (**b**) 45, and (**c**) 90 days from the range 900-1200 cm^{-1}.

After reaching 90 days of storage, the virgin TPCS film exhibits the highest increase in the C values compared to other films, indicating that the recrystallization of the short-chain amylopectin occurred more significantly. However, by adding a single filler or hybrid fillers, the recrystallization of short-chain amylopectin could be slowed down. The C values of the TPCS/4BT1NC, TPCS/5BT, and TPCS/5NC films have been consistently increased with the increase of storage time, which is in line with the increase in crystallinity percentage measured through the XRD analysis. However, the increase of the C value for TPCS/2BT3NC from 15 to 90 days showed little difference compared to the results obtained through the XRD analysis. The crystalline percentage calculated from XRD suggested that the TPCS/2BT3NC experienced a slow rate of crystal growth from 15 to 45 days, and the growth of crystalline structure was only detected after 45 days. However, the steady increase in the C-value as detected from the FTIR analysis indicates that the TPCS/2BT3NC possessed a consistent growth of crystalline structure from 15 to 90 days. The contradiction between the FTIR and XRD results could be due to the formation of a transcrystalline structure in the TPCS/2BT3NC sample, where the recrystallization of short-range molecular was more sensitive to be detected by FTIR. The transcrystalline structure in TPCS/2BT3NC was only detected in XRD analysis by shifting the peak at 2θ. The transcrystalline does not reflect the intensity of the XRD peak because XRD is more sensitive to detecting long-range molecular crystallized structures [48].

Table 4. The ratio of 1047 cm^{-1}:1022 cm^{-1} (C value) of TPCS, TPCS-C, and TPCS-HC films after 15, 45, and 90 days of aging.

Days	Films	1047 cm^{-1}:1022 cm^{-1} (C Value)
15 days	TPCS15	0.827
	TPCS/5BT15	0.837
	TPCS/5NC15	0.815
	TPCS/4BT1NC15	0.803
	TPCS/2BT3NC15	0.797
45 days	TPCS45	0.854
	TPCS/5BT45	0.849
	TPCS/5NC45	0.823
	TPCS/4BT1NC45	0.815
	TPCS/2BT3NC45	0.825
90 days	TPCS90	0.870
	TPCS/5BT90	0.853
	TPCS/5NC90	0.846
	TPCS/4BT1NC90	0.823
	TPCS/2BT3NC90	0.857

3.5. Moisture Absorption of TPCS, TPCS-C, and TPCS-HC Films Aged for 90 Days

The accelerated retrogradation can happen due to the high amount of water absorption during storage and will cause the alteration of films' crystalline structure. The changing of crystalline structure will affect the mechanical and biodegradation properties of the samples. Water can absorb into the TPCS films and accelerate the retrogradation rate. Absorbed water molecules can increase the plasticizer content in the films and may facilitate the retrogradation rate. This phenomenon has been also described by other researchers [26,27]. Figure 6 shows the moisture absorption of the TPCS, TPCS-C, and TPCS-HC films throughout the 3-month storage period. As shown in Figure 6, the moisture absorption rate of TPCS-C and TPCS-HC is lower compared to the virgin TPCS. This shows that filler could create a more tortuous path for water molecules into the films, which can reduce the TPCS films' hydrophilic properties. All the films experienced different amounts of water absorption from the beginning of storage and reached their equilibrium moisture absorption after 10 days. The virgin TPCS film demonstrates the highest water absorption rate for the first 10 days and reached its maximum water absorption at 18.86%.

The high water absorption may be attributed to the presence of a large amount of hydroxyl groups in the starch structure, which causes a high affinity of the biopolymer to absorb water into its structure. However, after being stored for 3 months, the water absorption of the TPCS films seems to experience a slight decrease to 16.87%. This may be due to the retrogradation in the TPCS film structure, as shown in the XRD analysis, causing the expulsion of water molecules, and reducing the water content in the films. This decreasing water absorption effect was somewhat challenging to observe in the other films. TPCS/5NC has the second-highest water content compared to TPCS/5BT, TPCS/4BT1NC and TPCS/2TB3NC. This may be attributed to the higher affinity of NC toward water molecules compared to BT. However, the high crystallinity structure of nanocellulose can reduce the water sensitivity and reduce the water permeation into the TPCS films, as reported in our previous study and other studies [49,50]. Apparently, the TPCS/4BT1NC film demonstrates the lowest water absorption rate compared to other films. The low water absorption of the TPCS/4BT1NC film indicates that BT and NC, when added in the ratio of 4:1, can most efficiently repel the water inclusion as the well-dispersed particles can create a more tortuous path for the entrance of the water.

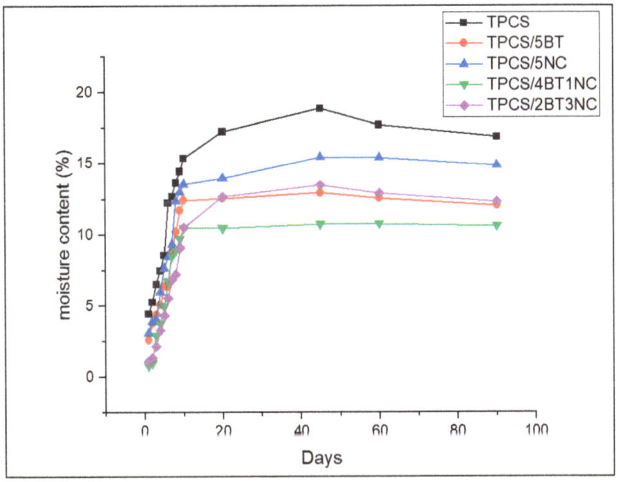

Figure 6. Moisture absorption percentage for virgin TPCS, TPCS-C, and TPCS-HC films aged for 90 days.

The significantly reduced water permeability of the TPCS/4BT1NC film, when benchmarked with the virgin TPCS, could also be due to its greater aging resistance and low retrogradation rate. Fewer chain breakage, voids, and surface embrittlement in this TPCS hybrid biocomposite film make it more resistant to water absorption as it can reduce the permeation and penetration of water molecules into the films.

4. Conclusions

The virgin thermoplastic corn starch (TPCS), thermoplastic corn starch composite (TPCS-C), and thermoplastic corn starch hybrid biocomposite (TPCS-HC) films were stored for 3 months to study the effect of hybrid filler on the starch retrogradation through the tensile test, DSC, XRD, FTIR, and water absorption analyses. Results indicate that the presence of single filler or hybrid fillers (nanocellulose + bentonite) can minimize the degree of starch retrogradation, thus inhibiting the aging of the TPS film. The interactions between the TPCS, plasticizer, and hybrid filler have reduced the mobility of the starch molecular chains, minimizing the rearrangement of the starch molecular chain into the crystal structure, and subsequently retarded the retrogradation process of the biopolymer. On the contrary, the absence of the hybrid filler caused the migration of the plasticizer to the film's surface, allowing more free movement and arrangement of molecules for the formation of a more dominant amylopectin crystalline structure. This mechanism led to the retrogradation of this starch-based film.

The TPCS-C films showed moderate efficacy in reducing the starch retrogradation compared to the TPCS-HC. The TPCS-HC film is proven to possess the most negligible starch retrogradation and can maintain its mechanical stability throughout the 3-month aging period. Apparently, the ratio of nanocellulose (NC):bentonite (BT) hybrid filler in the TPCS-HC system affects the retrogradation and aging of the film during a long storing period (3 months). The high composition of the NC compared to BT in the biocomposite films can attract the migration of plasticizer molecules that accumulate in the interface of the NC and the amylopectin. This encourages the formation of the oriented crystalline layer of the amylopectin chains in the interface of the NC/amylopectin (known as a transcrystalline structure). This unstable transcrystalline region in the starch structure is responsible for the retrogradation phenomenon.

Overall, results indicate that the TPCS film incorporating the BT:NC hybrid filler in the ratio of 4:1 is the best hybrid biocomposite sample as it has demonstrated the most stable microstructure and mechanical properties upon the 3-month storing period. This suggests that the use of a green hybrid filler of nanocellulose and bentonite with the appropriate ratio/composition can be an efficient and environmentally friendly approach to avoid early biodegradation or aging of the starch-based film in the development of a sustainable packaging material.

Author Contributions: Conceptualization, A.F.O. and S.A.A.; data curation, D.S.L. and A.A.A.; formal analysis, D.S.L. and A.A.A.; investigation, A.F.O. and D.S.L.; methodology, A.F.O., D.S.L. and I.I.; project administration, A.F.O., S.A.A., I.I. and M.N.A.S.; software, M.N.A.S.; supervision, A.F.O. and S.A.A.; writing—original draft, D.S.L.; writing—review and editing, A.F.O. and A.A.A. All authors have read and agreed to the published version of the manuscript.

Funding: The authors gratefully acknowledge the funding and the financial support from the Fundamental Research Grant Scheme (FRGS)—(FRGS/1/2019/TK10/UNIMAP/03/2) funded by the Ministry of Higher Education Malaysia.

Institutional Review Board Statement: Not applicable.

Informed Consent Statement: Not applicable for studies not involving humans.

Data Availability Statement: The data presented in this study are available on request from the corresponding author.

Conflicts of Interest: The authors declare no conflict of interest.

References

1. Sheng, L.; Adnan, S.; Osman, A.F.; Nabil, M.; Salimi, A.; Ibrahim, I.; Haq, N.; Salimi, N. Thermoplastic Starch Biocomposites with Cellulose and Bentonite Fillers. *AIP Conf. Proc.* **2021**, *2339*, 20045. [CrossRef]
2. Cheng, H.; Chen, L.; McClements, D.J.; Yang, T.; Zhang, Z.; Ren, F.; Miao, M.; Tian, Y.; Jin, Z. Starch-based biodegradable packaging materials: A review of their preparation, characterization, and diverse applications in the food industry. *Trends Food Sci. Technol.* **2021**, *114*, 70–82. [CrossRef]
3. Thakur, R.; Pristijono, P.; Scarlett, C.J.; Bowyer, M.; Singh, S.; Vuong, Q.V. Starch-based films: Major factors affecting their properties. *Int. J. Biol. Macromol.* **2019**, *132*, 1079–1089. [CrossRef] [PubMed]
4. Jiang, T.; Duan, Q.; Zhu, J.; Liu, H.; Yu, L. Starch-based biodegradable materials: Challenges and opportunities. *Adv. Ind. Eng. Polym. Res.* **2020**, *3*, 8–18. [CrossRef]
5. Csiszar, E.; Kun, D.; Fekete, E. The Role of Structure and Interactions in Thermoplastic Starch-Nanocellulose Composites. *Polymers* **2021**, *13*, 3186. [CrossRef] [PubMed]
6. Liu, Y.-X.; Liang, Z.-S.; Liang, J.-N.; Liao, L.-Y. Effect of different surface properties of nanosilica on retrogradation behavior and structures of thermoplastic cassava starch. *J. Polym. Res.* **2021**, *28*, 147. [CrossRef]
7. Rivadeneira-Velasco, K.E.; Utreras-Silva, C.A.; Díaz-Barrios, A.; Sommer-Márquez, A.E.; Tafur, J.P.; Michell, R.M. Green nanocomposites based on thermoplastic starch: A review. *Polymers* **2021**, *13*, 3227. [CrossRef]
8. Diyana, Z.; Jumaidin, R.; Selamat, M.Z.; Ghazali, I.; Julmohammad, N.; Huda, N.; Ilyas, R. Physical properties of thermoplastic starch derived from natural resources and its blends: A review. *Polymers* **2021**, *13*, 1396. [CrossRef] [PubMed]
9. Eerlingen, R.; Jacobs, H.; Delcour, J. Enzyme-resistant starch. 5. Effect of retrogradation of waxy maize starch on enzyme susceptibility. *Cereal Chem.* **1994**, *71*, 351–355.
10. Gudmundsson, M. Retrogradation of starch and the role of its components. *Thermochim. Acta* **1994**, *246*, 329–341. [CrossRef]
11. Esmaeili, M.; Pircheraghi, G.; Bagheri, R. Optimizing the mechanical and physical properties of thermoplastic starch via tuning the molecular microstructure through co-plasticization by sorbitol and glycerol. *Polym. Int.* **2017**, *66*, 809–819. [CrossRef]
12. Khan, B.; Niazi, M.B.K.; Jahan, Z.; Farooq, W.; Naqvi, S.R.; Ali, M.; Ahmed, I.; Hussain, A. Effect of ultra-violet cross-linking on the properties of boric acid and glycerol co-plasticized thermoplastic starch films. *Food Packag. Shelf Life* **2019**, *19*, 184–192. [CrossRef]
13. Ivanič, F.; Jochec-Mošková, D.; Janigová, I.; Chodák, I. Physical properties of starch plasticized by a mixture of plasticizers. *Eur. Polym. J.* **2017**, *93*, 843–849. [CrossRef]

14. Ivanič, F.; Kováčová, M.; Chodak, I. The effect of plasticizer selection on properties of blends poly (butylene adipate-co-terephthalate) with thermoplastic starch. *Eur. Polym. J.* **2019**, *116*, 99–105. [CrossRef]
15. Šmídová, N.; Šoltýs, A.; Hronský, V.; Olčák, D.; Popovič, L'.; Chodák, I. Aging-induced structural relaxation in cornstarch plasticized with urea and glycerol. *J. Appl. Polym. Sci.* **2021**, *138*, 50218. [CrossRef]
16. Montilla-Buitrago, C.E.; Gómez-López, R.A.; Solanilla-Duque, J.F.; Serna-Cock, L.; Villada-Castillo, H.S. Effect of Plasticizers on Properties, Retrogradation, and Processing of Extrusion-Obtained Thermoplastic Starch: A Review. *Starch-Stärke* **2021**, *73*, 2100060. [CrossRef]
17. Li, J.; He, H.; Zhang, H.; Xu, M.; Gu, Q.; Zhu, Z. Preparation of thermoplastic starch with comprehensive performance plasticized by citric acid. *J. Appl. Polym. Sci.* **2022**, *139*, e52401. [CrossRef]
18. De Freitas, A.d.S.M.; da Silva, A.P.B.; Montagna, L.S.; Nogueira, I.A.; Carvalho, N.K.; de Faria, V.S.; Dos Santos, N.B.; Lemes, A.P. Thermoplastic Starch Nanocomposites: Sources, Production and Applications-A review. *J. Biomater. Sci. Polym. Ed.* **2021**, *33*, 900–945. [CrossRef]
19. Balakrishnan, P.; Sreekala, M.; Kunaver, M.; Huskić, M.; Thomas, S. Morphology, transport characteristics and viscoelastic polymer chain confinement in nanocomposites based on thermoplastic potato starch and cellulose nanofibers from pineapple leaf. *Carbohydr. Polym.* **2017**, *169*, 176–188. [CrossRef]
20. Benito-González, I.; López-Rubio, A.; Martínez-Sanz, M.J.C.p. High-performance starch biocomposites with celullose from waste biomass: Film properties and retrogradation behaviour. *Prop. Retrogradation Behav.* **2019**, *216*, 180–188.
21. Bangar, S.P.; Whiteside, W.S. Nano-cellulose reinforced starch bio composite films-A review on green composites. *Int. J. Biol. Macromol.* **2021**, *185*, 849–860. [CrossRef] [PubMed]
22. Wu, Y.; Chen, Z.; Li, X.; Li, M. Effect of tea polyphenols on the retrogradation of rice starch. *Food Res. Int.* **2009**, *42*, 221–225. [CrossRef]
23. Vinod, A.; Gowda, Y.; Krishnasamy, S.; Sanjay, M.; Siengchin, S. Thermal Properties of Hybrid Natural Fiber-Reinforced Thermoplastic Composites. In *Natural Fiber-Reinforced Composites: Thermal Properties Applications*; Wiley: Hoboken, NJ, USA, 2022; pp. 17–30.
24. Osman, A.F.; Ashafee, A.M.T.L.; Adnan, S.A.; Alakrach, A. Influence of Hybrid Cellulose/Bentonite Fillers on Structure, Ambient, and Low Temperature Tensile Properties of Thermoplastic Starch Composites. *Polym. Eng. Sci.* **2020**, *60*, 810–822. [CrossRef]
25. Lai, D.S.; Osman, A.F.; Adnan, S.A.; Ibrahim, I.; Alrashdi, A.A.; Ahmad Salimi, M.N.; Ul-Hamid, A. On the Use of OPEFB-Derived Microcrystalline Cellulose and Nano-Bentonite for Development of Thermoplastic Starch Hybrid Bio-Composites with Improved Performance. *J. Polym.* **2021**, *13*, 897. [CrossRef] [PubMed]
26. Wang, S.; Li, C.; Copeland, L.; Niu, Q.; Wang, S. Starch Retrogradation: A Comprehensive Review. *Compr. Rev. Food Sci. Food Saf.* **2015**, *14*, 568–585. [CrossRef]
27. Liu, Y.; Fan, L.; Mo, X.; Yang, F.; Pang, J. Effects of nanosilica on retrogradation properties and structures of thermoplastic cassava starch. *J. Appl. Polym. Sci.* **2018**, *135*, 45687. [CrossRef]
28. Paiva, D.; Pereira, A.M.; Pires, A.L.; Martins, J.; Carvalho, L.H.; Magalhães, F.D.J.P. Reinforcement of thermoplastic corn starch with crosslinked starch/chitosan microparticles. *Polymers* **2018**, *10*, 985. [CrossRef]
29. Huang, S.; Chao, C.; Yu, J.; Copeland, L.; Wang, S. New insight into starch retrogradation: The effect of short-range molecular order in gelatinized starch. *Food Hydrocoll.* **2021**, *120*, 106921. [CrossRef]
30. Lendvai, L.; Sajó, I.; Karger-Kocsis, J. Effect of Storage Time on the Structure and Mechanical Properties of Starch/Bentonite Nanocomposites. *Starch-Stärke* **2019**, *71*, 1800123. [CrossRef]
31. Fekete, E.; Bella, E.; Csiszar, E.; Moczo, J. Improving physical properties and retrogradation of thermoplastic starch by incorporating agar. *Int. J. Biol. Macromol.* **2019**, *136*, 1026–1033. [CrossRef]
32. Zhang, K.; Su, T.; Cheng, F.; Lin, Y.; Zhou, M.; Zhu, P.; Li, R.; Wu, D. Effect of sodium citrate/polyethylene glycol on plasticization and retrogradation of maize starch. *Int. J. Biol. Macromol.* **2019**, *154*, 1471–1477. [CrossRef] [PubMed]
33. Schmitt, H.; Guidez, A.; Prashantha, K.; Soulestin, J.; Lacrampe, M.F.; Krawczak, P. Studies on the effect of storage time and plasticizers on the structural variations in thermoplastic starch. *Carbohydr. Polym.* **2015**, *115*, 364–372. [CrossRef]
34. Dufresne, A.; Castaño, J. Polysaccharide nanomaterial reinforced starch nanocomposites: A review. *Starch-Stärke* **2017**, *69*, 1500307. [CrossRef]
35. Teixeira, E.d.M.; Pasquini, D.; Curvelo, A.A.; Corradini, E.; Belgacem, M.N.; Dufresne, A. Cassava bagasse cellulose nanofibrils reinforced thermoplastic cassava starch. *Carbohydr. Polym.* **2009**, *78*, 422–431. [CrossRef]
36. Li, C.; Hu, Y. Antagonistic effects of amylopectin and amylose molecules on the starch inter-and intramolecular interactions during retrogradation. *LWT* **2021**, *148*, 111942. [CrossRef]
37. Ghanbari, A.; Tabarsa, T.; Ashori, A.; Shakeri, A.; Mashkour, M. Thermoplastic starch foamed composites reinforced with cellulose nanofibers: Thermal and mechanical properties. *Carbohydr. Polym.* **2018**, *197*, 305–311. [CrossRef] [PubMed]
38. Tian, Y.; Li, Y.; Xu, X.; Jin, Z. Starch retrogradation studied by thermogravimetric analysis (TGA). *Carbohydr. Polym.* **2011**, *84*, 1165–1168. [CrossRef]
39. Paluch, M.; Ostrowska, J.; Tyński, P.; Sadurski, W.; Konkol, M. Structural and Thermal Properties of Starch Plasticized with Glycerol/Urea Mixture. *J. Polym. Environ.* **2021**, *30*, 728–740. [CrossRef]
40. Drakopoulos, S.; Karger-Kocsis, J.; Kmetty, Á.; Lendvai, L.; Psarras, G. Thermoplastic starch modified with microfibrillated cellulose and natural rubber latex: A broadband dielectric spectroscopy study. *Carbohydr. Polym.* **2017**, *157*, 711–718. [CrossRef]

41. Ujcic, A.; Krejcikova, S.; Nevoralova, M.; Zhigunov, A.; Dybal, J.; Krulis, Z.; Fulin, P.; Nyc, O.; Slouf, M. Thermoplastic Starch Composites with Titanium Dioxide and Vancomycin Antibiotic: Preparation, Morphology, Thermomechanical Properties, and Antimicrobial Susceptibility Testing. *Front. Mater.* **2020**, *7*, 9. [CrossRef]
42. Dominic, C.D.M.; Dos Santos Rosa, D.; Camani, P.H.; Kumar, A.S.; Neenu, K.V.; Begum, P.M.S.; Dinakaran, D.; John, E.; Baby, D.; Thomas, M.M.; et al. Thermoplastic starch nanocomposites using cellulose rich Chrysopogon zizanioides nanofibers. *Int. J. Biol. Macromol.* **2021**, *191*, 572–583. [CrossRef]
43. Klemm, D.; Kramer, F.; Moritz, S.; Lindström, T.; Ankerfors, M.; Gray, D.; Dorris, A. Nanocelluloses: A new family of nature-based materials. *Angew. Chem. Int. Ed.* **2011**, *50*, 5438–5466. [CrossRef]
44. Teixeira, E.M.; Da Róz, A.L.; Carvalho, A.J.F.; Curvelo, A.A.S. The effect of glycerol/sugar/water and sugar/water mixtures on the plasticization of thermoplastic cassava starch. *Carbohydr. Polym.* **2007**, *69*, 619–624. [CrossRef]
45. Hu, Y.; He, C.; Zhang, M.; Zhang, L.; Xiong, H.; Zhao, Q. Inhibition from whey protein hydrolysate on the retrogradation of gelatinized rice starch. *Food Hydrocoll.* **2020**, *108*, 105840. [CrossRef]
46. Castillo, L.A.; Lopez, O.V.; Garcia, M.A.; Barbosa, S.E.; Villar, M.A. Crystalline morphology of thermoplastic starch/talc nanocomposites induced by thermal processing. *Heliyon* **2019**, *5*, e01877. [CrossRef] [PubMed]
47. Van Soest, J.J.; Tournois, H.; de Wit, D.; Vliegenthart, J.F. Short-range structure in (partially) crystalline potato starch determined with attenuated total reflectance Fourier-transform IR spectroscopy. *Carbohydr. Res.* **1995**, *279*, 201–214. [CrossRef]
48. Zhang, C.W.; Nair, S.S.; Chen, H.; Yan, N.; Farnood, R.; Li, F.Y. Thermally stable, enhanced water barrier, high strength starch bio-composite reinforced with lignin containing cellulose nanofibrils. *Carbohydr. Polym.* **2020**, *230*, 115626. [CrossRef] [PubMed]
49. Gray, N.; Hamzeh, Y.; Kaboorani, A.; Abdulkhani, A. Influence of cellulose nanocrystal on strength and properties of low-density polyethylene and thermoplastic starch composites. *Ind. Crops Prod.* **2018**, *115*, 298–305. [CrossRef]
50. Nazrin, A.; Sapuan, S.; Zuhri, M.; Ilyas, R.; Syafiq, R.; Sherwani, S.J.F.i.c. Nanocellulose reinforced thermoplastic starch (TPS), polylactic acid (PLA), and polybutylene succinate (PBS) for food packaging applications. *Front. Chem.* **2020**, *8*, 213. [CrossRef]

polymers MDPI

Article

Hydrothermal Ageing Effect on Reinforcement Efficiency of Nanofibrillated Cellulose/Biobased Poly(butylene succinate) Composites

Olesja Starkova [1,*], Oskars Platnieks [2], Alisa Sabalina [1] and Sergejs Gaidukovs [2]

[1] Institute for Mechanics of Materials, University of Latvia, Jelgavas 3, LV-1004 Riga, Latvia; alisasabalina@gmail.com
[2] Faculty of Materials Science and Applied Chemistry, Institute of Polymer Materials, Riga Technical University, P. Valdena 3/7, LV-1048 Riga, Latvia; oplatnieks@gmail.com (O.P.); sergejs.gaidukovs@rtu.lv (S.G.)
* Correspondence: olesja.starkova@lu.lv

Abstract: Nanofibrillated cellulose (NFC) is a sustainable functional nanomaterial known for its high strength, stiffness, and biocompatibility. It has become a key building block for the next-generation of lightweight, advanced materials for applications such as consumer products, biomedical, energy storage, coatings, construction, and automotive. Tunable and predictable durability under environmental impact is required for high performance applications. Bio-based poly (butylene succinate) (PBS) composites containing up to 50% NFC content were designed and aged in distilled water or at high relative humidity (RH98%). PBS/NFC composites are characterized by up to 10-fold increased water absorption capacity and diffusivity and the data are correlated with model calculations. Aged samples exhibited decreased crystallinity and melting temperature. Incorporation of NFC into PBS showed up to a 2.6-fold enhancement of the elastic modulus, although accompanied by a loss of strength by 40% and 8-fold reduction in the strain at failure of maximally loaded composites. Hydrothermal ageing had almost no influence on the tensile characteristics of PBS; however, there were considerable degradation effects in PBS/NFC composites. Altered reinforcement efficiency is manifested through a 3.7-fold decreased effective elastic moduli of NFC determined by applying the Halpin–Tsai model and a proportional reduction of the storage moduli of composites. The adhesion efficiency in composites was reduced by hydrothermal ageing, as measured Puckanszky's adhesion parameter for the strength, which decreased from 3 to 0.8. For the loss factor, Kubat's adhesion parameter was increased by an order. PBS filled with 20 wt.% NFC is identified as the most efficient composition, for which negative environmental degradation effects are counterbalanced with the positive reinforcement effect. The PBS matrix can be used to protect the NFC network from water.

Keywords: biodegradable polymer; biocomposite; water diffusion; modeling; environmental ageing; durability; mechanical properties; thermomechanical properties; adhesion parameter

Citation: Starkova, O.; Platnieks, O.; Sabalina, A.; Gaidukovs, S. Hydrothermal Ageing Effect on Reinforcement Efficiency of Nanofibrillated Cellulose/Biobased Poly(butylene succinate) Composites. *Polymers* **2022**, *14*, 221. https://doi.org/10.3390/polym14020221

Academic Editor: Evgenia G. Korzhikova-Vlakh

Received: 13 November 2021
Accepted: 4 January 2022
Published: 6 January 2022

Publisher's Note: MDPI stays neutral with regard to jurisdictional claims in published maps and institutional affiliations.

1. Introduction

Preserving our planet requires sustainable technologies with low environmental impact to produce high-value products. This has attracted research interest to abundant and renewable materials. Cellulose meets all the criteria, being by far the most abundant renewable resource in the biosphere, having exceptionally tunable morphology, and high mechanical properties [1,2]. Advanced material structure modeling and preparation requires the application of nanocellulose. Nanofibrillated cellulose (NFC) is a sustainable functional nanomaterial known for its high strength, stiffness, and biocompatibility. It has become a key building block for the next generation of lightweight, advanced materials for applications such as consumer products, biomedical, energy storage, functional and decorative coatings, construction, and automotive [3,4]. NFC can form rigid mechanical systems with unique structure that can accommodate various modifications. Combining NFC with polymer into the nanocomposite results in additional layer of protection and functionality from matrix [5,6]. The polymer selection highly benefits bio-based and biodegradable materials to expand upon sustainability and biomedical applications.

Poly (butylene succinate) (PBS) is a relatively new bio-based and biodegradable polyester with high dimensional stability and low water absorption during immersion, even at elevated temperatures [7,8]. This low water sensitivity is one of the key advantages for selecting PBS over other popular biopolyesters such as poly (lactic acid) and polyhydroxyalkanoates. PBS potential applications, such as biomedical, food packaging, agriculture, and automotive, have been evaluated in literature [9,10], showing promising results. PBS biodegradability has been assessed in various studies, and bio-based fillers such as cellulose slightly promote decomposition by promoting water absorption [11–13]. As a result, a composite with a higher loading of hydrophilic fillers has better biodegradation capabilities. It has also been demonstrated that PBS remains stable for everyday use items and only starts to degrade when buried in the soil [7]. The mechanisms of biodegradation and some methods for predicting the "end of life" of various bioplastics are discussed in [14].

Owing to excellent intrinsic properties and 2D geometry, NFC particles are considered as promising eco-friendly reinforcement fillers for polymer composites [1,15], aerogels and foams [16,17]. Incorporation of NFC into various biopolymers allows material designers to develop advanced sustainable yet durable bioplastics with tailored mechanical performance and biodegradability [5,7]. Depending on material applications and related requirements for specific properties, the amount of cellulose fillers in biopolymer-based composites can vary from a few percent up to 50–70% [18–21]. The reinforcement efficiency of NFC on composite stiffness normally increases with the higher content of the filler, although this improvement is often accompanied by material embrittlement and loss of adhesion efficiency at high loadings. This applies if NFC does not agglomerate. Coupling agents are used for addressing these shortcomings in order to improve stress transfer at the polymer/filler interface [8,21]. By achieving reasonable compatibility between the composite constituents and their effective dispersion, highly loaded composites with uniquely high stiffness can be developed for advanced applications, e.g., wood–plastic composites [21].

Various strategies have been applied in nanocomposite preparation, i.e., melt mixing, solvent casting, and more advanced approaches such as in situ polymerization [22,23]. Direct melt mixing has been shown to be ineffective in the case of NFC, resulting in poor dispersion, damage to the nanostructure of the filler, and even thermal oxidation [24,25]. Thus, it is generally recommended to use solvent processing techniques for the preparation of NFC composites [26]. Solvent casting is divided depending on polymer matrix solubility, i.e., water soluble or non-water soluble. In the case of PBS, solvent exchange for NFC can be applied to prepare the nanocomposite. A different approach is to use the masterbatch method, by preparing a highly loaded system that is then redispersed using melt processing methods. One masterbatch method involves using solvent exchange for NFC and another involves using a water-soluble polymer matrix such as poly (ethylene glycol) (PEG) or poly (vinyl alcohol) as a matrix followed by subsequent blending with PBS. The use of PEG/NFC dispersions in hydrophobic polymer matrix has been shown to be an inefficient strategy due to PEG acting like a plasticizer [27]. Further processing of the solvent cast nanocomposite using methods such as extrusion and injection molding can have a significant impact on properties by achieving fibril orientation [28]. While different processing methods can result in large differences in composite performance properties, in this study we focus on composites' ability to retain those properties in a PBS matrix when exposed to a hydrothermal ageing environment.

Advanced and functional applications impose exceptionally high requirements for materials' durability. NFC hybrid materials are highly moisture sensitive [6]. Water absorption from the environment (hydrothermal degradation) significantly reduce the durability and service lifetime of these materials. Insufficient long-term performance studies under environmental ageing restrict the usage of biopolymer hybrid systems. Some studies indicate a significant difference in resistance to environmental ageing between conventional non-biodegradable fossil polymers and biopolyesters [29,30]. The durability of biodegradable plastics, including PBS-based nanocomposite systems, is reported in [31,32]. Hydrothermal ageing of highly loaded cellulose-based biopolymer systems is studied in [8,33] demonstrating their particularly high susceptibility to environmental impact. Property degradation is normally related to water absorption capacity and barrier properties of bioplastics [34,35], thus making NFC less competitive in terms of environmental stability compared to hydrophobic materials, e.g., carbon, nanofillers. Development of sustainable biocomposites with tailored durability and biodegradability performance is a major task for the growing circular economy. Some modelling tools for predicting environmental ageing and degradation of polymers and polymer composites are reported in [14]. Despite continuously growing interest in the research of NFC-based biopolymer composites, most studies are focused on the characterization of their basic engineering properties and biodegradability, while environmental impacts on durability performance and the role of NFC in it is rarely reported.

The present study is aimed at quantitative characterization of the hydrothermal impact on the reinforcement and adhesion efficiency of PBS-based composites with NFC loading up to 50 wt.%. Water absorption and diffusivity were studied in composites under humid and warm water environments to model different degradation scenarios. The large disparity in NFC loading allows to identify critical structural transitions in the NFC network and their contribution to the environmental durability of composites. Structural changes and the role of NFC on composite hydrothermal degradation were revealed through calorimetric studies. Mechanical and thermomechanical properties were studied, while changes in the reinforcement and adhesion efficiency of NFC on the tensile stiffness and strength of composites were evaluated through model parameters. The results allow us to identify the most efficient composition, which is characterized by reasonable mechanical performance and hydrothermal stability to facilitate advanced hybrid systems.

2. Materials and Methods

2.1. Materials and Samples' Processing

PTT MCC Biochem Company Ltd. (Bangkok, Thailand) supplied biobased poly (butylene succinate) (PBS) in the form of pellets. BioPBS™ FZ71PB® grade was selected as it provides balanced properties for use in a wide range of applications. This grade is characterized by a 115 °C melting temperature and a 1.26 g/cm^3 density. Kraft pulp (hardwood birch) was purchased from Metsä Fibre. Pulp was mechanically treated and processed into a nanofibrillated cellulose (NFC) dispersion using a method described in our previous study [20]. The NFC used was from the same batch, as described in [20], with a diameter d in the range of 15–50 nm and a length l of 200–550 nm. The average aspect ratio for l/d is assumed to be equal to 10 in the present study. Chloroform and N,N-dimethylformamide (DMF) were purchased from Merck KGaA (Darmstadt, Germany).

Pure NFC films for water absorption studies were produced from NFC gel suspended in water and mixed with a high-shear mixer (2 min, 5500 rpm). The final concentration of suspension was adjusted to 1 wt% NFC. The NFC film was prepared by casting suspension in a polystyrene Petri dish at room temperature (20 °C). Films were dried in ambient conditions until the water evaporated, then in a thermostat at 50 °C for 48 h. The resulting thickness of the NFC films was 8 ± 1 µm.

2.2. Nanocomposite Preparation

PBS granules were dried in a vacuum furnace (J.P. Selecta, Barcelona, Spain) according to the manufacturer's instructions (60 °C, 8 h). PBS was dissolved in chloroform using magnetic stirring for 1 h. The NFC water suspension was diluted with DMF and solvent replacement was carried out using solvent-assisted centrifugation. The process was carried out three times, and a high-shear mixer L5M-A (Silverson Machines LTD, Waterside, UK) was used for homogenization (2 min, 5500 rpm). The NFC/DMF suspension was mixed with the PBS/chloroform solution using a high-shear mixer (2 min, 5500 rpm). Solution casting was used to obtain films. Dry films were further dried in a vacuum furnace (60 °C, 8 h). The obtained films were then compression molded (Carver CH 4386, Diamond, MO, USA) to thickness of a 0.2 mm (140 °C, preheating 2 min, 3 min compression under 3 MT). Films were cooled between steel plates (30 kg of thermal conductive mass).

Six compositions with different contents of NFC were produced (Table 1). The volume contents of NFC were calculated using the density $\rho_{cellulose}$ = 1.6 g/cm^3 [21]. Density of composite samples increased with NFC content up to 5% for maximally loaded samples. This increase correlates well with calculations according to the rule of mixture, indicating a reasonable structural homogeneity of the prepared composites.

Table 1. Prepared PBS/NFC composites and their densities.

Sample	NFC wt.%	NFC vol.%	ρ [1], g/cm^3
PBS	0	0	1.260 ± 0.0018
5NFC	5	4.8	1.266 ± 0.0025
10NFC	10	8.0	1.273 ± 0.0082
20NFC	20	16.4	1.296 ± 0.0039
30NFC	30	25.2	1.307 ± 0.0164
50NFC	50	44.1	1.326 ± 0.0072

[1] determined experimentally by hydrostatic weighing in isopropanol (ρ_{iso} = 0.786 g/cm^3).

Rectangular samples 10 mm × 55 mm were cut from the films for water absorption and tensile tests. Prior to testing, all samples were dried in a thermostat at 50 °C for 4 days until mass stabilization was achieved. Samples were divided into three groups depending on their storage conditions. As-produced samples stored in plastic zipped bags in a desiccator under silica gel at 22 °C were considered the reference samples (denoted as Ref). One group of samples was immersed in distilled water in a glass container and put into a thermostat at 50 °C: these samples are denoted as W50 in the text. The next group of samples were stored in a desiccator under a saturated salt solution of K_2SO_4 giving a relative humidity of RH = 98% at 22 °C (denoted as RH98).

2.3. Water Absorption Studies

An amount of absorbed water in samples was determined by gravimetric measurements. The samples were weighed on a regular basis using analytical scales XS205 (±0.01 mg Mettler Toledo, Columbus, OH, USA), and the relative weight changes [%] were calculated as weight gain per weight unit:

$$w = \frac{m_t - m_0}{m_0} \times 100 \tag{1}$$

where m_t is the weight of the immersed sample at time t, and m_0 is the weight of the reference sample. Replicate samples of each composition were taken out of the water by groups, wiped with blotting paper to remove water from the surface, and weighed following the same sequence. Measuring time of each of the group of samples did not exceed five minutes. The mean values were calculated from data of five replicate samples.

2.4. Tensile Tests

Uniaxial tensile tests were performed on a Zwick 2.5 kN universal testing machine at a constant crosshead speed of 2 mm/min according to ASTM D882. Tabs made from a paper tape were applied to samples to ensure smooth loading and prevent sample slippage during the test. Five replicate samples for each composition in the reference (Ref) and aged state, i.e., after conditioning in a humid atmosphere for 40 days (RH98) and warm water for 18 days (W50) were tested. Aged samples were tested 5–10 min after their removal from the environment. The secant elastic modulus was determined in the strain range of 0.2–0.5%.

2.5. Differential Scanning Calorimetry

Differential scanning calorimetry (DSC) was performed using the DSC-1 device (Mettler Toledo, Columbus, OH, USA). The measurements were made for samples with an average weight of 10–12 mg under a nitrogen atmosphere, from 25 °C to 150 °C, at a scan rate of ±10 K/min. Melting temperature and crystallinity were evaluated at the first heating scan. The crystalline fraction X_c [%] of PBS/NFC composites was calculated according to the equation:

$$X_c = \frac{\Delta H}{\Delta H_{PBS}\left(1 - \varphi_f\right)} \times 100 \quad (2)$$

where φ_f is the weight fraction of NFC filler in the composite, ΔH_{PBS} is the heat of fusion of the perfect crystal of PBS, and ΔH is the enthalpy of fusion of the studied samples, respectively. The value of ΔH_{PBS} is 200 J/g [25]. For water saturated samples (RH98 and W50), ΔH in Equation (2) was replaced by $\Delta H/(1 - w)$ in order to exclude the weight of absorbed water.

2.6. Dynamic Mechanical Thermal Analysis

Dynamic mechanical thermal analysis (DMTA) was performed in a tension mode on a Mettler Toledo SDTA861e device (Mettler Toledo, Columbus, OH, USA). Tests were carried out on the reference and RH98 samples with dimensions 4 mm × 8.5 mm. Frequency 1 Hz, heating rate of 3 K/min, displacement amplitude of 10 μm, force amplitude 10 N, and a temperature range from −70 to 70 °C measurement parameters were used.

3. Results

3.1. Water Absorption

Water absorption curves for RH98 and W50 samples are shown in Figure 1. PBS, as previously reported [8,35,36], has a low water absorption capacity with a total weight gain of less than 1%. Neat polymer and composites possess Fickian-type diffusion behaviour: linear weight gain versus square root of time dependence at the start of water uptake, whereupon saturation is approached. The data are fitted by the Fick's model assuming 1D through-thickness diffusion [37,38]

$$w(t) = w_\infty\left[1 - \frac{2}{\pi^2}\sum_{m=1}^{\infty}\frac{\left[1-(-1)^m\right]^2}{m^2}\exp\left(-\left(\frac{\pi m}{a}\right)^2 Dt\right)\right] \quad (3)$$

where w_∞ is the equilibrium water content, D is the diffusion coefficient, and a is the thickness of the sample. The initial water content at $t = 0$ is assumed to be zero. As seen from Figure 1, the model approximations are in good agreement with the experimental data.

The equilibrium water content of samples increases with the growing content of NFC in the polymer (Figure 2a) following the rule of mixture given by

$$w_\infty^c = w_\infty^m\left(1 - \varphi_f\right) + w_\infty^f\varphi_f \quad (4)$$

where w_∞^c, w_∞^m, and w_∞^f are the equilibrium water contents of PBS/NFC composite, PBS matrix and NFC filler, respectively.

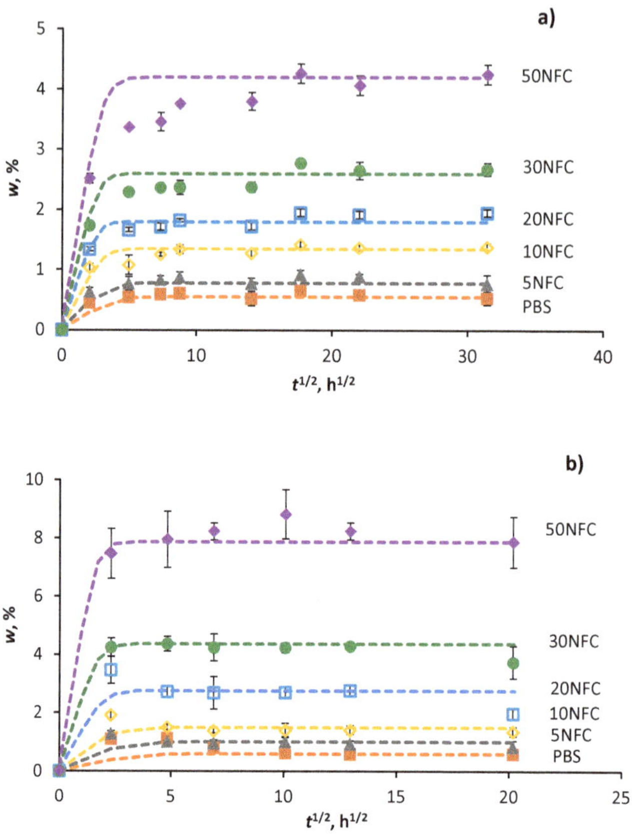

Figure 1. Weight gain of PBS/NFC composites conditioned under RH 98%, 22 °C (**a**) and in water at 50 °C (**b**). Lines are calculated by Equation (3).

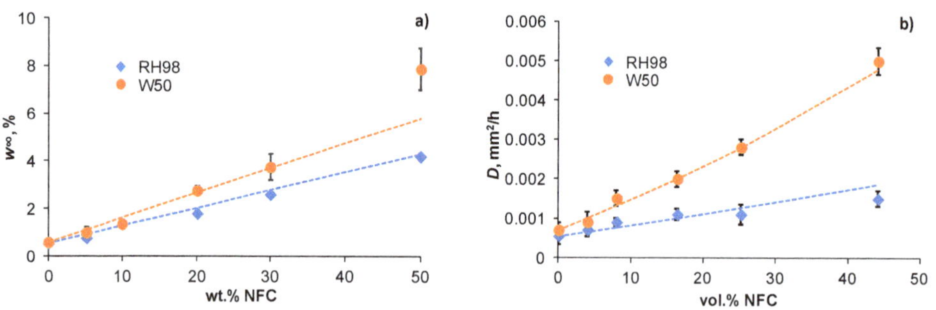

Figure 2. Equilibrium water content of PBS/NFC composites as a function of NFC weight content. Lines are approximations by Equation (4) (**a**); Diffusivity of PBS/NFC composites as a function of NFC volume content. Lines are approximations by Equation (5) (**b**).

Water sorption capacity increases with NFC content and more extensively for W50 samples compared to their RH98 counterparts (up to an 8- and 10-fold increase, respectively). Equation (4) gives reasonable approximations with w_{∞}^{f}(RH98) = 8% and w_{∞}^{f}(W50) = 11%.

The former value correlates well with the experimental value for pure NFC film (Figure 3). A noticeable deviation (overshoot) from the predicted line can be noticed for 50NFC-W50 samples. This fact could be related to structural inhomogeneities, e.g., voids due to imperfect adhesion, formed during processing or initiated by hydrothermal degradation of the material and facilitating additional water ingress [39].

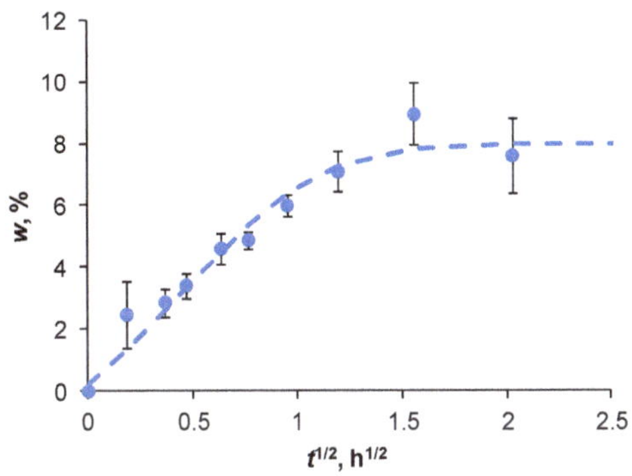

Figure 3. Weight gain of pure NFC film under RH 98%, 22 °C. Line is approximation by the Fick's model, Equation (3).

The diffusion coefficients of neat PBS and composites were determined by fitting the weight gain curves (Figure 1) by Equation (3). The obtained D values versus volume content of NFC are shown in Figure 2b. The diffusivity of PBS greatly increased with the addition of hydrophilic NFC, leading to up to a 3- and 10-fold gain of D for maximally loaded RH98 and W50 samples, respectively. The rate of diffusion is greatly accelerated at elevated temperatures and the greater with higher NFC content in PBS. D of neat PBS is 5.5 (±0.2) × 10^{-4} mm^2/h and 7.0 (±0.2) × 10^{-4} mm^2/h for RH98 and W50 samples, respectively. These values agree well with data reported elsewhere [35].

The Halpin–Tsai model, which is among the most widely used models to predict the elastic modulus, heat conductivity, and water diffusivity in heterogeneous multicomponent polymer systems [40–43], was applied to fit the data in Figure 2b. The Halpin–Tsai model for the diffusivity is given by Equation (5) [44]:

$$D_c = D_m\left(\frac{1+\xi\eta V_f}{1-\eta V_f}\right);\ \eta = \frac{D_f/D_m - 1}{D_f/D_m + \xi} \quad (5)$$

where V_f is the volume content of the filler, D_m and D_f are the diffusion coefficients of the matrix and filler, i.e., PBS and NFC, respectively. $\xi = 2l/d$ is the structural parameter related to the aspect ratio of the filler. For NFC under the study, ξ = 20 (Section 2.1). Figure 2b shows a reasonable correlation between the experimentally determined and calculated by Equation (5) diffusivities of PBS/NFC composites. The best fitting results are obtained for D_f equal to 0.004 (±0.0005) and 0.015 (±0.001) mm^2/h for RH98 and W50 samples, respectively. The diffusivity of 50NFC-RH98 samples (V_f = 44.1%) is slightly lower than that predicted by the Halpin–Tsai model and could be related to the high tortuosity effect in a highly loaded polymer. However, this effect is counterbalanced by additional water ingress into voids in W50 samples.

An attempt has also been made to estimate the water absorption characteristics of pure NFC. Weight gain data for NFC films conditioned under RH 98% at 22 °C are shown in Figure 3. As seen from the graph, the experimental data agrees well with the approximation of Fick's model given by Equation (3). NFC films reached saturation at w = 8% and this value is identical to w_∞^f determined by Equation (4) for PBS/NFC composites. The diffusivity of pure NFC determined from the Fick's approximation is very low:

1.0 ($\pm$0.5) 10^{-5} mm^2/h, which is two orders of magnitude lower than D_f determined by Equation (5). NFC in the deposited thin films forms a highly oriented structure, acting as an efficient barrier against the diffusion of water molecules and increasing the tortuosity factor contribution [2,35,43]. This effect is greatly diminished in NFC-filled polymer due to the counterbalancing contribution from the hydrophilic nature of cellulose and structural defects such as NFC agglomerates and microcracks at the interface between the filler and the matrix. These imperfections, which originated during the processing of the material or due to environmental impact, contribute to additional water ingress by the capillary mechanism [39]. The weaker the interfacial adhesion, the greater its contribution.

The diffusion characteristics of PBS/NFC composites correlate well with the results reported in literature for similar cellulose fiber/hydrophobic polymer systems [36,39], comparison of the absolute values is not actually reasonable due to their high sensitivity to the properties of constituents and processing conditions of composites. For example, saturation of polypropylene/cellulose fiber (30 wt.%) composites immersed in water at 23 °C was reached at w = 6%, while the diffusivity was estimated as D = 0.0032 mm^2/h (8.84 10^{13} m^2/s) [39]. Both characteristics increased considerably with temperature and the content of the fibers. In [45], the equilibrium water content of pure NFC was determined at RH90%, room temperature, and it was 16.6%.

3.2. Tensile Properties

Representative stress–strain diagrams for the reference and aged samples of neat PBS and 20NFC samples are shown in Figure 4. Incorporation of NFC into the polymer resulted in enhancement of its elastic properties, although accompanied by embrittlement of the composites (Figure 4a). The tensile properties of PBS were almost unaffected by hygro- and hydrothermal ageing (Figure 4b), whereas noticeable degradation effects are observed for PBS/NFC composites, where NFC content contributed to higher values. Ageing in warm water resulted in more deleterious property degradation compared to conditioning in a humid environment (Figure 4c).

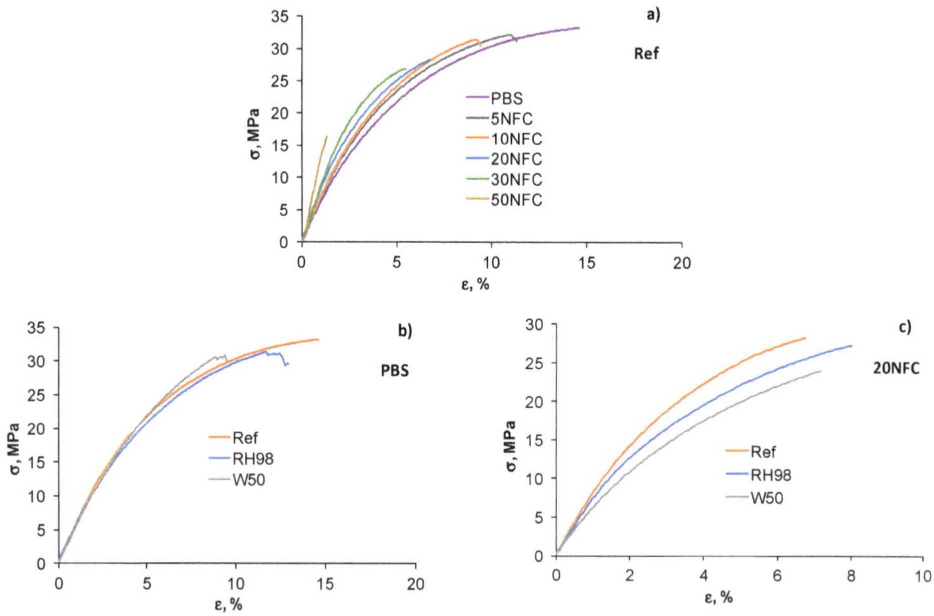

Figure 4. Stress–strain diagrams for reference PBS/NFC composites (**a**) and aged PBS (**b**) and 20NFC (**c**) samples.

The ultimate properties of PBS/NFC composites are shown in Figure 5. The strength (σ^*) of the reference samples decreases by 40% with the higher content of NFC: from 30.1 ± 3.2 MPa for neat PBS to 18.6 ± 4.5 MPa for 50NFC samples. However, the $\sigma^*$of PBS with moderate NFC loadings up to 10 wt.% was not influenced by the presence of the filler. The strain at failure (ε^*) decreased from 12.5 ± 2.4% to 1.6 ± 0.4% when comparing neat PBS and 50NFC samples, respectively. The observed embrittlement effect could be explained by inefficient stress transfer at the filler-matrix interface due to low compatibility of the constituents, inefficient dispersion, and agglomerates formed during processing of samples [46–49].

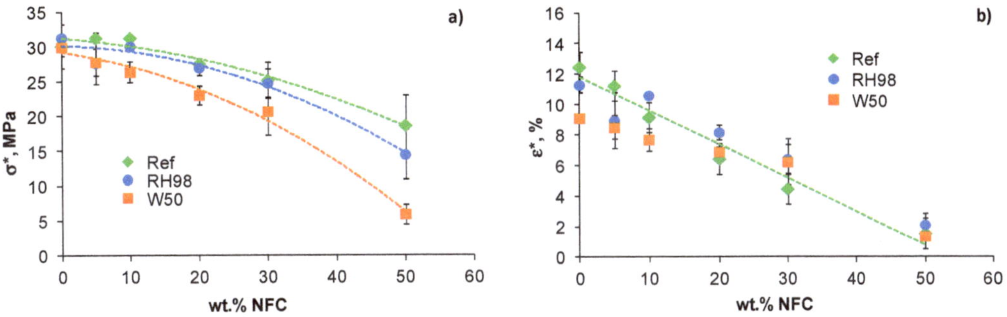

Figure 5. Strength (**a**) and strain at failure (**b**) of PBS/NFC composites as functions of NFC weight content. Lines are approximations.

Hydrothermal ageing resulted in even more progressive strength loss of PBS/NFC composites (Figure 5a). By comparing the σ^* values of reference and W50 samples, the drop is estimated to be 16% and 68% for 20NFC and 50NFC composites, respectively. The extent of property degradation increases with NFC content, which is partly explained by higher amounts of absorbed water in highly loaded samples (Figure 2a) and their higher structural defectiveness. Interestingly, the strain at failure is only slightly affected by hydrothermal ageing, and the data for the reference, RH98, and W50 samples are within the data scatter range (Figure 5b).

The elastic moduli (E) of composites as a function of NFC volume content in the reference and aged states are compared in Figure 6. NFC demonstrated rather high reinforcement efficiency with up to a 2.6-fold increase in the elastic modulus for 50NFC samples. Ageing caused significant stiffness degradation in composites, as well as a loss of the NFC reinforcing effect.

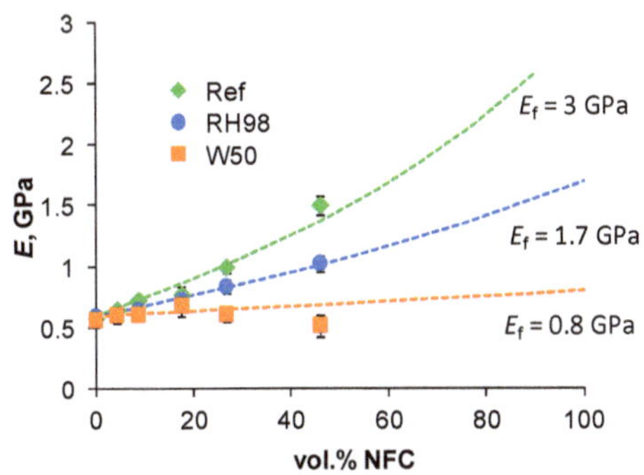

Figure 6. Elastic modulus as a function of NFC volume content for reference and aged samples. Lines are approximations by Equation (6) with different E_f.

The Halpin–Tsai model for randomly oriented short cylindrical fibers, was used to model the elastic moduli of PBS/NFC composites [15,34,45]:

$$\begin{aligned} E_c &= E_m\left[\frac{3}{8}\left(\frac{1+\xi\eta_L V_f}{1-\eta_L V_f}\right)+\frac{5}{8}\left(\frac{1+2\eta_T V_f}{1-\eta_T V_f}\right)\right] \\ \eta_L &= \frac{E_f/E_m-1}{E_f/E_m+\xi};\ \eta_T=\frac{E_f/E_m-1}{E_f/E_m+2} \end{aligned} \quad (6)$$

where E_c, E_m, and E_f are the elastic moduli of the composite, matrix (PBS), and filler (NFC); l and d are the length and diameter of the NFC, respectively. V_f and ξ are the same as in Equation (5).

Calculations of E_c by Equation (6) are shown in Figure 6. According to the experimental data, $E_m = 0.57 \pm 0.02$ GPa. E_f is determined by fitting the experimental data. Calculations with E_f = 3.0, 1.7, and 0.8 GPa give the best approximation results for the reference and RH98 and W50 samples, respectively. This method estimates the plasticization effect of absorbed water on the elastic modulus of NFC indirectly, however with certain simplified assumptions on perfect interfacial adhesion. E_f, and thus, the reinforcement efficiency decreases significantly for water-saturated material by 43% and 73% for RH98 and W50 samples, respectively. This reduction correlates well with higher w_∞^{NFC} values for the latter samples (Equation (4), Figure 2a).

Literature data on the elastic properties of NFC are very varied according to the source of cellulose and fiber aspect ratio. E_f is found in the range of 15–140 GPa [15,45,50]. Thus, the determined E_f values of NFC under study are highly underestimated compared to literature data. This could be related to the low properties of the produced NFC itself or the limited compatibility of PBS and NFC, resulting in dispersion and adhesion shortcomings. Because both factors are difficult to distinguish, reasonable data comparison within the same compositions is possible.

For composites, the strength versus the elastic modulus can be plotted to evaluate the correlation between reinforcement and adhesion efficiency. Figure 7 shows the data for the reference and aged PBS/NFC samples. Composites with low NFC content (up to 20 wt.% NFC) exhibit higher stiffness with moderate strength loss compared to neat PBS. The same trend is maintained after hydrothermal ageing. At higher loadings, the positive NFC reinforcing effect is counterbalanced by negative environmental degradation effects on both NFC and stress transfer at the NFC/PBS interface. The dramatic loss of both the mechanical characteristics observed for highly loaded samples makes them practically inapplicable in harsh environments. By correlating two opposite contributions, optimal NFC content could be determined for specific applications and service requirements.

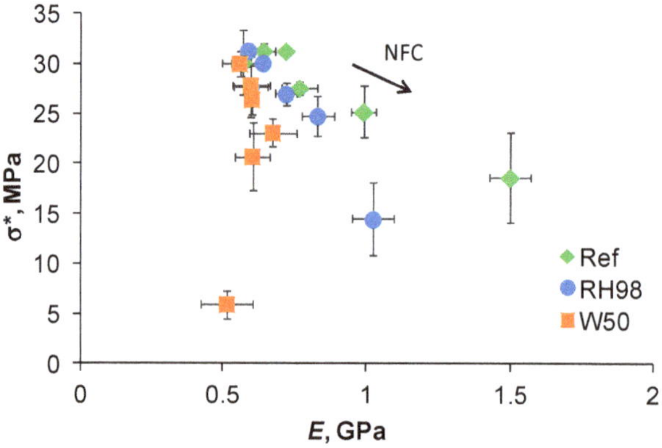

Figure 7. Strength versus elastic modulus of PBS/NFC reference and aged samples.

A decrease in the strength of particulate composites is related to the degree of interfacial adhesion between the polymer matrix and fillers. The adhesion efficiency could be quantitatively assessed by using Pukanszky's adhesion parameter [41,48,51]. The Pukanszky's model for strength is given by [51]

$$\sigma_c = \left[\frac{1-V_f}{1+2.5V_f}\sigma_m\right]\exp\left(BV_f\right) \tag{7}$$

where σ_c and σ_m are the strengths of the composite and polymer matrix, respectively. B is an empirical constant, also known as Pukanszky's adhesion factor. Parameter B is defined as a measure of the load carried by particles; it approaches zero for poorly bonded particles and increases for efficient polymer-particle interface adhesion interactions [48,51]. After some rearranging and applying a natural logarithm, Equation (7) takes the form

$$\ln\sigma_{red} = \ln\left[\frac{1+2.5V_f}{1-V_f}\sigma_c\right] = \ln\sigma_m + BV_f \tag{8}$$

where σ_{red} is the reduced strength. As follows from Equation (8), the plot $\ln\sigma_{red}$ versus V_f is a straight line characterized with a slope B.

Figure 8a shows the reduced strength of PBS/NFC as a function of NFC volume content. For comparison, data for other similar composites are also presented. Polyvinylalcohol (PVA) filled with NFC [50] and PBS filled with wine lees (WL) [40]. The adhesion factors B, according to Equation (8), are indicated by the slope of the linear dependences. PBS/NFC is characterized by reasonable adhesion efficiency with $B = 3$, which is higher than B of PBS/WL ($B = 1.6$) and lower than that of PVA/NFC ($B = 7.6$). The high B value for the latter material is most likely related to the good compatibility of hydrophilic PVA with hydrophilic NFC, in contrast to hydrophobic PBS. Similar B values in the range of 0.8–5.3 were reported in other studies [8,41,48,51].

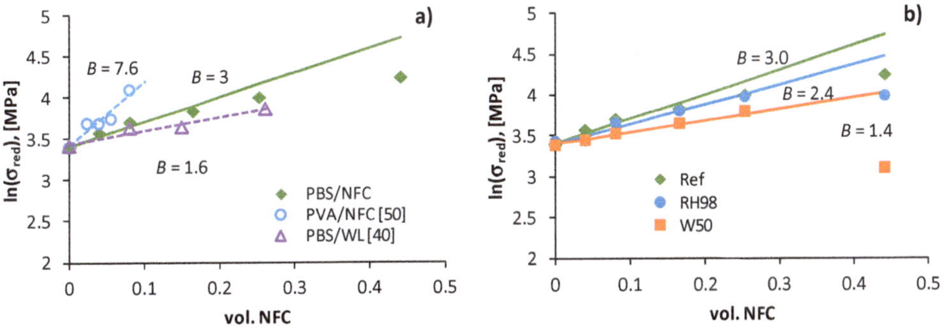

Figure 8. Reduced strength of composites as volume content of the filler for reference PBS/NFC, PVA/NFC [50] and PBS/WL [40] (**a**); and aged PBS/NFC (**b**). Lines are linear approximations with slope B.

Hydrothermal ageing results in deterioration of the polymer matrix/filler interface, reducing the strength and adhesion efficiency of composites. The reduced strength as a function of volume content of NFC for the reference, RH98 and W50 samples of PBS/NFC composites is shown in Figure 8b. Aged composites are characterized by $\ln\sigma_{red} vs. V_f$ dependence on a lower slope: B decreases to 2.4 and 1.4 for RH98 and W50 samples, respectively. A similar 1.7-fold decrease in B due to hydrothermal ageing of PBS/microcrystalline cellulose composites was reported in [8]. For other systems, epoxy filled with inorganic fillers, water absorption resulted in a 1.4-fold drop of B [52].

Noticeable deviations from the linear dependences $\ln \sigma_{red} - V_f$ are observed for high NFC loadings $V_f \geq 25.2\%$ (30 wt.%) (Figure 8b). The experimental data show highly underestimated values, indicating a poor adhesion efficiency of 30NFC and particularly 50NFC composites. The deviations are the most extensive for W50 aged samples (Figure 8b), which explain the dramatic loss of interfacial adhesion between PBS and NFC.

3.3. *Calorimetric Properties*

Representative DSC first heating scans for PBS/NFC reference and aged samples are shown in Figure 9. The calorimetric characteristics are listed in Table 2. The DSC curves of PBS often exhibit two endothermic peaks and a small exothermic peak [53]. This has been explained by the recrystallization process, where two competing processes occur at the same time-melting and recrystallization [20,27,54]. The first melting occurs for crystallites that have a defective (incomplete) structure, followed by a recrystallization process into a more perfect structure [55]. Thus, the observed process depends on how the polymer achieves crystallization. In our case, sample films were obtained with the rapid cooling method. Thus, all melting peaks are relatively broad, and perfect crystalline phase was not achieved. Exposure to water contributed to the appearance of an exothermic crystallization peak in all RH98 and W50 samples.

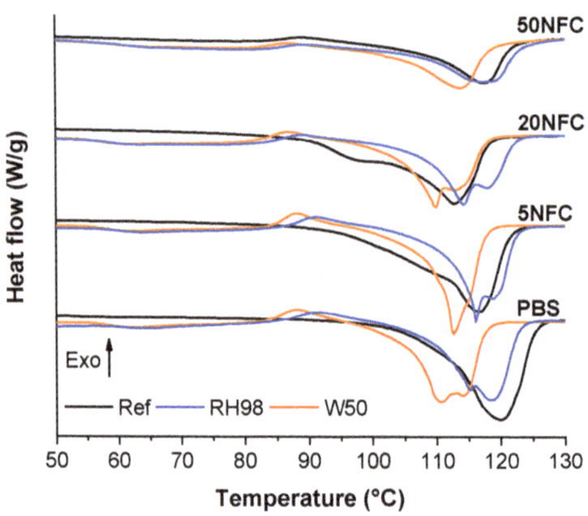

Figure 9. DSC first heating curves for reference and aged PBS/NFC samples.

Table 2. Calorimetric characteristics of PBS/NFC for reference and aged samples (1st heating scan).

Sample	χ (%)			T_m (°C)		
	Ref	**RH98**	**W50**	**Ref**	**RH98**	**W50**
PBS	54.6	38.0	37.6	118	118	111
5NFC	54.7	36.8	37.5	116	116	113
10NFC	49.6	36.7	34.4	115	114	111
20NFC	50.5	33.9	35.8	113	114	110
30NFC	53.6	34.3	36.3	114	115	110
50NFC	43.1	32.4	39.7	117	117	114

NFC nucleates the formation of spherulites, which contributes to additional disruption of the PBS crystalline structure [56]. This effect is visible for 5NFC and 20NFC reference samples, where a pronounced shoulder from the low endothermic peak appears. However, in the case of 50NFC, the high NFC loading results in PBS acting more as an interphase modifier than a continuous polymer matrix. Absorbed water interfered with the recrystallization process, thus the melting peak split in two endothermic peaks. Nanocellulose has been shown to form dense percolation network after reaching critical concentration, which is around few percent loading depending on aspect ratio of filler [57,58]. Thus, the initial formation of spherulites is promoted by the NFC filler, but high loading restricts polymer chain arrangement. The high quality of NFC filler contributes to a relatively low impact on the crystallinity of PBS even at high loadings.

Reduced recrystallization decreased melting enthalpy values. This is visible as a significant drop in crystallinity for aged compositions (Table 2). The highest χ reduction was for RH98 samples with a decrease around 17% for all compositions except 50 NFC. 5NFC samples showed almost identical changes to PBS. Similar observations on the reduced crystallinity caused by hydrothermal ageing of PBS-based wood plastics are reported in [36]. Furthermore, all W50 samples showed a significant melting peak shift to lower temperatures by 3–7 °C. However, no specific trend related to NFC and water content in composites are noticed. As discussed in Section 3.1, W50 samples showed about a 2-fold higher water uptake than RH98. In this case, water significantly penetrated the structure of the nanocomposite. When small water molecules get inside a polymer structure, they reduce intermolecular bonding between macromolecule chains. However, because NFC promoted this process, an even greater amount of water was absorbed. This results in swelling, which induces internal stresses in the structure and promotes the breakdown of intermolecular bonds. Reduced intermolecular bonding then promotes the breakdown of the crystalline structure during melting. Absorbed water reduced intermolecular bonding and interfered with the recrystallization process, shifting PBS/NFC composite melting process to a lower temperature.

3.4. Thermomechanical Properties

The addition of NFC to PBS greatly improved its thermomechanical characteristics. Similar results on DMTA analysis for unaged PBS/NFC prepared by solution casting and melt processing methods are presented in a previous study [20]. Temperature dependences of the storage modulus (E') and loss factor (tan δ) of RH98 samples are shown in Figure 10. Aged composites demonstrated principally similar thermomechanical behavior to the reference samples. Absorbed water caused small changes in the glass transition temperature (T_g), which were in the 2–5 °C range with the maximally loaded sample having the greatest impact. T_g of the reference samples determined by tan δ peaks are −16.2 °C for neat PBS and −17.8, −17.0, −16.6, −17.3, and −20.2 °C for 5NFC, 10NFC, 20NFC, and 50NFC samples, respectively.

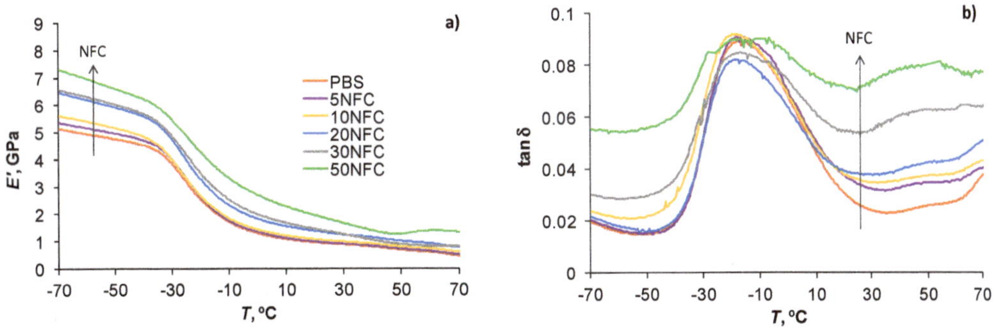

Figure 10. Storage modulus (**a**) and tan δ curves (**b**) for PBS/NFC samples aged under RH98%.

The storage moduli of PBS, in both the glassy and rubbery regions, increased with the higher content of NFC. The reinforcement efficiency of NFC is compared in Figure 11 determined at different temperatures: −70 °C (glassy state), 22 °C (room or service temperature that corresponds to the glass-to-rubbery transition region of PBS), and 70 °C (rubbery state). The slope of the linear dependences E'_c/E'_m (c is composite, m is matrix) is related to the reinforcement efficiency factor: the higher the slope, greater the filler contribution to the stiffness improvement of the polymer matrix [20,38]. Increased crystallinity of composites could also contribute to the E' increase [36,41]. Although this is not the case for this study, since the crystallinity of PBS is only slightly and negatively affected by NFC (Table 2). The rubbery moduli are, to a great extent, correlated with the degree of interaction between the polymer and NFC, while the reinforcing effect of the filler is eliminated. Thus, an increasing E'_c/E'_m dependence for 70 °C (up to 60% improvement for 50NFC), although with a much lower slope than that for −70 °C, indicates a reasonably high interaction degree at the PBS-NFC interface. Almost a 3-fold increase in the rubbery E' (70 °C) is obtained with the addition of NFC. Identical improvements of E' are also observed at 22 °C, i.e., in the transition region.

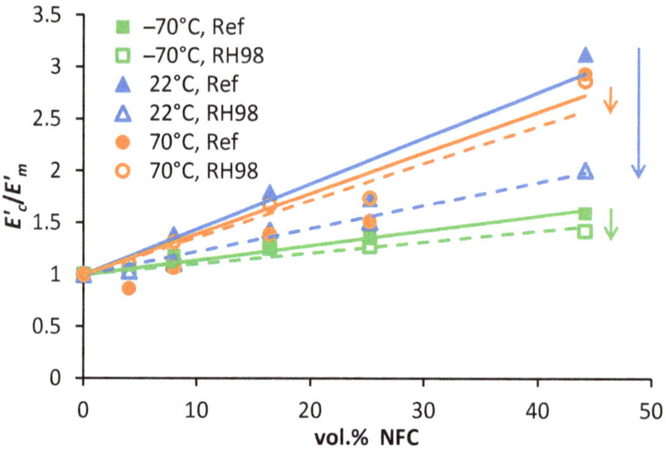

Figure 11. Relative storage moduli as a function of volume content of NFC of the reference and RH98 samples determined at −70, 22, and 70 °C. Solid and dotted lines are approximations.

The glassy and rubbery storage moduli of PBS/NFC were only slightly affected by ageing in a humid environment, resulting in minor decrease in the slopes of the dependences E'_c/E'_m versus NFC content for both characteristic temperatures (Figure 11, shown by errors). At the same time, a great decrease in E' of aged samples is observed at 22 °C. This could be explained by the higher plasticization effect of absorbed moisture in this temperature range. When the temperature is too low (−70 °C) water molecules are inactive and do not contribute to molecular relaxations of the polymer, while under elevated temperatures (70 °C) the plasticizing water effect is negated by extensive structural rearrangements of PBS. E'_c/E'_m decrease at 22 °C is finely correlated with tensile test data (Figure 6). It should also be noted that the service temperatures of these types of composites are more likely to be in the range of 20 ± 10 °C, so testing material properties at conditions that are similar to those found in real-world applications is critical.

The addition of NFC resulted in an enhancement of the damping properties of PBS that appeared in the decreased tan δ peaks [20]. Similar improvements are reported elsewhere for different types of composites [25,59]. Absorbed moisture, acting as an efficient plasticizer, further contributes to increased energy dissipation and the lowering and widening of the loss factor [38,60]. Assessment of the tan δ of composites also provides valuable data on interfacial adhesion between the polymer matrix and filler. The Kubat's adhesion parameter is among these quantitative measures of the adhesion efficiency [47,59].

For rigid inclusions, the Kubat's adhesion parameter A is given as follows [61]:

$$A = \left[\frac{1}{1 - V_f} \times \frac{\tan \delta_c}{\tan \delta_m} \right] - 1 \quad (9)$$

where $\tan \delta_c$ and $\tan \delta_m$ are the loss factors of the composite and polymer matrix, respectively. The lower A values are characteristic of stronger interfacial adhesion between the polymer and filler since the ideal interface does not contribute to dampening. The increase of A results from the enhanced energy dissipation and damping ability of the composite related to the polymer's viscoelastic nature and structural defects, e.g., voids that appear due to poor interfacial interactions between the matrix and filler [47].

The adhesion factors of PBS/NFC increased with the higher content of NFC and the more significantly for aged samples. A values are in the range of 0.006–0.1 for reference and 0.06–0.8 for aged PBS/NFC samples. Adhesion efficiency is strongly correlated with the mechanical performance of composites. The strength and elastic modulus of PBS/NFC as functions of the adhesion factor A are shown in Figure 12. The strength decreases with increasing A, while an opposite, almost mirroring, trend is observed for the elastic modulus.

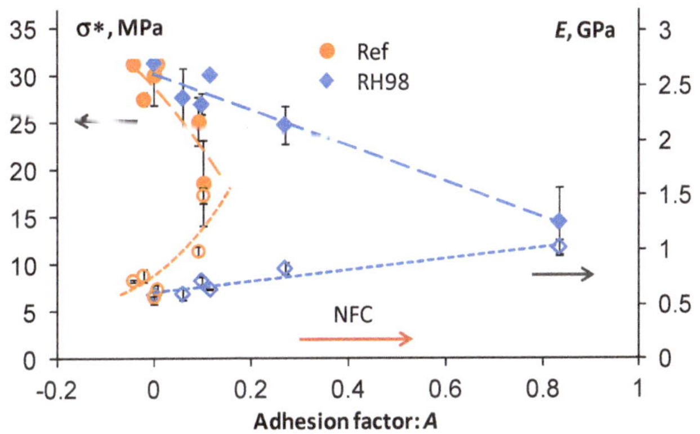

Figure 12. Strength (filled symbols) and elastic modulus (open symbols) as functions of the adhesion factor A for reference and RH98 PBS/NFC samples. Lines are approximations.

The greatest differences between the A values for reference and aged samples are observed for 30NFC and 50NFC compositions. In other words, the inherently low adhesion efficiency of highly loaded composites makes them greatly susceptible to environmental impacts. Similar observations are made by comparing the Pukanszky's adhesion parameters B (Figure 8b, Section 3.2). The stiffness of composites is mainly related to the reinforcement efficiency of the inclusions, rather than interfacial adhesion between the constituents. NFC single particles and their agglomerates form a rigid network in PBS, resulting in an E increase but, at the same time, promoting crack propagation and lower strength [47].

4. Conclusions

Fully biobased PBS/NFC composites have been designed with high NFC content (up to 50 wt.%) and reasonable reinforcement and adhesion efficiency are retained after hydrothermal ageing. Ageing in a humid RH98% environment at 22 °C and in water at 50 °C were studied to mimic different degradation scenarios.

Water absorption behaviour of all composites follows the Fick's model, while water absorption capacity and diffusivity increase considerably with NFC loading. Up to a 10-fold increase in the equilibrium water content and diffusion coefficient is obtained for maximally loaded PBS/NFC samples immersed in warm water. The barrier properties of the composite constituents were estimated by applying the rule of mixture and the Halpin–

Tsai model. A reasonable correlation between the calculated and experimental values of NFC water saturation content is obtained, while the diffusion coefficients determined by the micromechanical model fitting for composites are considerably higher compared to the experimental values determined for pure NFC films.

The structure of pristine PBS was only slightly affected by NFC addition, with no extensive differences in its crystallinity (except for 50NFC composition) and melting temperature. Both thermal characteristics decreased due to absorbed water interference, although less extensively in the case of 50NFC.

Incorporation of NFC into PBS resulted in up to 2.6-fold enhancement of the elastic modulus, while accompanied by a loss of strength by 40% and an 8-fold reduction in the strain at failure of maximally loaded composites. Tensile properties of PBS were almost not affected by hydrothermal ageing, while noticeable degradation effects were observed for PBS/NFC composites. A greater decrease in the mechanical characteristics was observed for compositions with higher NFC content. This fact is related to the higher amount of absorbed water in composites. The reinforcement efficiency of NFC on the stiffness of PBS decreases significantly after hydrothermal ageing. The effective elastic modulus of NFC in composites was reduced from 3 GPa for reference samples to 1.7 and 0.8 GPa for hygro- and hydro-thermally aged samples, respectively. Calculated from the Halpin–Tsai model. The adhesion efficiency in PBS/NFC was evaluated through the Puckanszky adhesion parameter. $B = 3$ was found for the reference PBS/NFC, while water ageing resulted in its 2-fold reduction.

Absorbed water resulted in moderate plasticization effects in composites, which appeared in a shift of T_g for 2–5 °C and a decrease of the glassy storage moduli by 10%. Ageing effects on the reinforcement efficiency on the storage moduli of PBS/NFC composites were evaluated at three characteristic temperatures corresponding to the glassy, rubbery, and glass-to-rubbery transition. The greatest changes in the composite storage moduli related to those of the matrix are revealed at an ambient temperature corresponding to the transition region and service temperature range. Three-fold moduli improvements are obtained for the reference samples, while the reinforcement efficiency decreased by a factor of 2 after the conditioning of samples in a humid environment. The damping properties of PBS were altered by NFC addition and environmental ageing. The Kubat's parameter, used as a quantitative measure of the adhesion efficiency of composites, was in the range of 0.06–0.1 for the reference samples and increased by almost an order due to degradation. These findings are consistent with the decrease in tensile strength and the decrease in the Puckanszky adhesion parameter.

The results will contribute to the development of advanced sustainable biocomposites with excellent biodegradability and durability performance, extending their applications and thus benefiting the growth of the circular economy.

Author Contributions: Conceptualization, O.S. and S.G.; methodology, O.S., O.P., and S.G.; formal analysis, O.S., O.P., A.S. and S.G.; investigation, O.S., O.P. and A.S.; writing—original draft preparation, O.S. and O.P.; writing—review and editing, O.S., O.P., A.S. and S.G.; project administration, O.S.; funding acquisition, O.S. and S.G. All authors have read and agreed to the published version of the manuscript.

Funding: This research was funded by Latvian Research Council, grant No. lzp-2020/2-0207 "Durability of sustainable bio-based polymer nanocomposites under the combined effects of mechanical loading and environmental ageing".

Institutional Review Board Statement: Not applicable.

Informed Consent Statement: Not applicable.

Data Availability Statement: The data presented in this study are openly available in Mendeley Data at DOI:10.17632/8pnvjtcbwc.1, 2021.

Acknowledgments: This research was funded by Latvian Research Council, grant No. lzp-2020/2-0207.

Conflicts of Interest: The authors declare no conflict of interest.

References

1. Omran, A.A.B.; Mohammed, A.A.B.A.; Sapuan, S.M.; Ilyas, R.A.; Asyraf, M.R.M.; Koloor, S.S.R.; Petru, M. Micro- and nanocellulose in polymer composite materials: A review. *Polymers* **2021**, *13*, 231. [CrossRef]
2. Helanto, L.; Matikainen, L.; Talja, R.; Rojas, O.J. Bio-based polymers for sustainable packaging and bio-barriers: A critical review. *Bioresources* **2019**, *14*, 4902–4951. [CrossRef]
3. Kalia, S.; Boufi, S.; Celli, A. Nanofibrillated cellulose: Surface modification and potential applications. *Colloid Polym. Sci.* **2014**, *292*, 5–31. [CrossRef]
4. Tan, Y.; Du, X.; Du, X.; Wang, H.; Cheng, X. Form-stable phase change composites based on nanofibrillated cellulose/polydopamine hybrid aerogels with extremely high energy storage density and improved photothermal conversion efficiency. *RSC Adv.* **2021**, *11*, 5712–5721. [CrossRef]
5. Lee, K.Y.; Tammelin, T.; Schulfter, K.; Kiiskinen, H.; Samela, J.; Bismarck, A. High performance cellulose nanocomposites: Comparing the reinforcing ability of bacterial cellulose and nanofibrillated cellulose. *ACS Appl. Mater. Interfaces* **2012**, *4*, 4078–4086. [CrossRef]
6. Naderi, A. Nanofibrillated cellulose: Properties reinvestigated. *Cellulose* **2017**, *24*, 1933–1945. [CrossRef]
7. Platnieks, O.; Gaidukovs, S.; Thakur, V.K.; Barkane, A.; Beluns, S. Bio-based poly (butylene succinate): Recent progress, challenges and future opportunities. *Eur. Polym. J.* **2021**, *161*, 110855. [CrossRef]
8. Gaidukovs, S.; Platnieks, O.; Gaidukova, G.; Starkova, O.; Barkane, A.; Beluns, S.; Thakur. Understanding the impact of microcrystalline cellulose modification on durability and biodegradation of highly loaded biocomposites for woody like materials applications. *J. Polym. Environ.* **2021**, 1–6. [CrossRef]
9. Gigli, M.; Fabbri, M.; Lotti, N.; Gamberini, R.; Rimini, B.; Munari, A. Poly (butylene succinate)-based polyesters for biomedical applications: A review. *Eur. Polym. J.* **2016**, *75*, 431–460. [CrossRef]
10. Rafiqah, S.A.; Khalina, A.; Harmaen, A.S.; Tawakkal, I.A.; Zaman, K.; Asim, M.; Nurrazi, M.N.; Lee, C.H. A review on properties and application of bio-based poly (butylene succinate). *Polymers* **2021**, *13*, 1436. [CrossRef]
11. Terzopoulou, Z.N.; Papageorgiou, G.Z.; Papadopoulou, E.; Athanassiadou, E.; Reinders, M.; Bikiaris, D.N. Development and study of fully biodegradable composite materials based on poly (butylene succinate) and hemp fibers or hemp shives. *Polym. Compos.* **2016**, *37*, 407–421. [CrossRef]
12. Huang, Z.; Qian, L.; Yin, Q.; Yu, N.; Liu, T.; Tian, D. Biodegradability studies of poly (butylene succinate) composites filled with sugarcane rind fiber. *Polym. Test.* **2018**, *66*, 319–326. [CrossRef]
13. Cho, S.Y.; Lee, M.E.; Kwak, H.W.; Jin, H.J. Surface-modified cellulose nanocrystal-incorporated poly (butylene succinate) nanocomposites. *Fibers Polym.* **2018**, *19*, 1395–1402. [CrossRef]
14. Krauklis, A.E.; Karl, C.W.; Rocha, I.B.C.M.; Burlakovs, J.; Ozola-Davidane, R.; Gagani, A.I.; Starkova, O. Modelling of environmental ageing of polymers and polymer composites—modular and multiscale methods. *Polymers* **2022**, *14*, 216. [CrossRef]
15. Jithin, J.; Cintil, J.; Srirama, B.V.; Lovely, M.P.; Sabu, T.; Srikanth, P. Preparation and characterization of poly (butylene succinate) bionanocomposites reinforced with cellulose nanofiber extracted from helicteres isora plant. *J. Renew. Mater.* **2016**, *4*, 351–364. [CrossRef]
16. Andersons, J.; Kirpluks, M.; Cabulis, U. Reinforcement efficiency of cellulose microfibers for the tensile stiffness and strength of rigid low-density polyurethane foams. *Materials* **2020**, *13*, 2725. [CrossRef]
17. Beluns, S.; Gaidukovs, S.; Platnieks, O.; Gaidukova, G.; Mierina, I.; Grase, L.; Starkova, O.; Brazdausks, P.; Thakur, V.K. From wood and hemp biomass wastes to sustainable nanocellulose foams. *Ind. Crop. Prod.* **2021**, *170*, 113780. [CrossRef]
18. Nazrin, A.; Sapuan, S.M.; Zuhri, M.Y.M.; Ilyas, R.A.; Syafiq, R.; Sherwani, S.F.K. Nanocellulose reinforced thermoplastic starch (TPS), polylactic acid (PLA), and polybutylene succinate (PBS) for food packaging applications. *Front. Chem.* **2020**, *8*, 213. [CrossRef]
19. Zinge, C.; Kandasubramanian, B. Nanocellulose based biodegradable polymers. *Eur. Polym. J.* **2020**, *133*, 109758. [CrossRef]
20. Platnieks, O.; Sereda, A.; Gaidukovs, S. Adding value to poly (butylene succinate) and nanofibrillated cellulose-based sustainable nanocomposites by applying masterbatch process. *Ind. Crop. Prod.* **2021**, *169*, 113669. [CrossRef]
21. Platnieks, O. Highly Loaded cellulose/poly (butylene succinate) sustainable composites for woody-like advanced materials application. *Molecules* **2020**, *25*, 121. [CrossRef]
22. Kim, T.; Jeon, H.; Jegal, J.; Kim, J.H.; Yang, H.; Park, J.; Oh, D.X.; Hwang, S.Y. Trans crystallization behavior and strong reinforcement effect of cellulose nanocrystals on reinforced poly (butylene succinate) nanocomposites. *RSC Adv.* **2018**, *8*, 15389–15398. [CrossRef]
23. Spoljaric, S.; Salminen, A.; Luong, N.D.; Lahtinen, P.; Vartiainen, J.; Tammelin, T.; Seppälä, J. Nanofibrillated cellulose, poly (vinyl alcohol), montmorillonite clay hybrid nanocomposites with superior barrier and thermomechanical properties. *Polym. Compos.* **2014**, *35*, 1117–1131. [CrossRef]
24. Alloin, F.; D'Aprea, A.; Dufresne, A.; El Kissi, N.; Bossard, F. Poly(oxyethylene) and ramie whiskers based nanocomposites: Influence of processing: Extrusion and casting/evaporation. *Cellulose* **2011**, *18*, 957–973. [CrossRef]

25. Platnieks, O.; Gaidukovs, S.; Barkane, A.; Sereda, A.; Gaidukova, G.; Grase, L.; Thakur, V.K.; Filipova, I.; Fridrihsone, V.; Skute, M.; et al. Bio-based poly (butylene succinate)/microcrystalline cellulose/nanofibrillated cellulose-based sustainable polymer composites: Thermo-mechanical and biodegradation studies. *Polymers* **2020**, *12*, 1472. [CrossRef] [PubMed]
26. Kargarzadeh, H.; Mariano, M.; Huang, J.; Lin, N.; Ahmad, I.; Dufresne, A.; Thomas, S. Recent developments on nanocellulose reinforced polymer nanocomposites: A review. *Polymer* **2017**, *132*, 368–393. [CrossRef]
27. Ludueña, L.N.; Fortunati, E.; Morán, J.I.; Alvarez, V.A.; Cyras, V.P.; Puglia, D.; Manfredi, L.B.; Pracella, M. Preparation and characterization of polybutylene-succinate/poly(ethylene-glycol)/cellulose nanocrystals ternary composites. *J. Appl. Polym. Sci.* **2016**, *133*, 43302. [CrossRef]
28. Peng, N.; Huang, D.; Gong, C.; Wang, J.; Zhou, J.; Chang, C. Controlled arrangement of nanocellulose in polymeric matrix: From reinforcement to functionality. *ACS Nano* **2020**, *14*, 16169–16179. [CrossRef] [PubMed]
29. Lajarrige, A.; Gontard, N.; Gaucel, S.; Peyron, S. Evaluation of the food contact suitability of aged bio-nanocomposite materials dedicated to food packaging applications. *Appl. Sci.* **2020**, *10*, 877. [CrossRef]
30. Bazan, P.; Mierzwiński, D.; Bogucki, R.; Kuciel, S. Bio-based polyethylene composites with natural fiber: Mechanical, thermal, and ageing properties. *Materials* **2020**, *13*, 2595. [CrossRef] [PubMed]
31. Glaskova-Kuzmina, T.; Starkova, O.; Gaidukovs, S.; Platnieks, O.; Gaidukova, G. Durability of biodegradable polymer nanocomposites. *Polymers* **2021**, *13*, 3375. [CrossRef]
32. Nam, T.H.; Ogihara, S.; Nakatani, H.; Kobayashi, S.; Song, J.I. Mechanical and thermal properties and water absorption of jute fiber reinforced poly (butylene succinate) biodegradable composites. *Adv. Compos. Mater.* **2012**, *21*, 241–258. [CrossRef]
33. Kamau-Devers, K.; Kortum, Z.; Miller, S.A. Hydrothermal aging of bio-based poly(lactic acid) (PLA) wood polymer composites: Studies on sorption behavior, morphology, and heat conductance. *Constr. Build. Mater.* **2019**, *214*, 290–302. [CrossRef]
34. Mazur, K.; Kuciel, S. Mechanical and hydrothermal aging behaviour of polyhydroxybutyrate-co-valerate (PHBV) composites reinforced by natural fibres. *Molecules* **2019**, *24*, 3538. [CrossRef] [PubMed]
35. Cosquer, R.; Pruvost, S.; Gouanvé, F. Improvement of barrier properties of biodegradable polybutylene succinate/graphene nanoplatelets nanocomposites prepared by melt process. *Membranes* **2021**, *11*, 151. [CrossRef] [PubMed]
36. Petchwattana, N.; Sanetuntikul, J.; Sriromreun, P.; Narupai, B. Wood plastic composites prepared from biodegradable poly(butylenes succinate) and burma padauk sawdust (pterocarpus macrocarpus): Water absorption kinetics and sunlight exposure investigations. *J. Bionic Eng.* **2017**, *14*, 781–790. [CrossRef]
37. Crank, J. *The Mathematics of Diffusion. Oxford;* Clarendon Press: Oxford, UK, 1956.
38. Starkova, O.; Gaidukovs, S.; Platnieks, O.; Barkane, A.; Garkusina, K.; Palitis, E.; Grase, L. Water absorption and hydrothermal ageing of epoxy adhesives reinforced with amino-functionalized graphene oxide nanoparticles. *Polym. Degrad. Stab.* **2021**, *191*, 109670. [CrossRef]
39. Espert, A.; Vilaplana, F.; Karlsson, S. Comparison of water absorption in natural cellulosic fibres from wood and one-year crops in polypropylene composites and its influence on their mechanical properties. *Compos. Part A Appl. Sci. Manuf.* **2004**, *35*, 1267–1276. [CrossRef]
40. Nanni, A.; Messori, M. Thermo-mechanical properties and creep modelling of wine lees filled polyamide 11 (PA11) and polybutylene succinate (PBS) bio-composites. *Compos. Sci. Technol.* **2020**, *188*, 107974. [CrossRef]
41. Battegazzore, D.; Noori, A.; Frache, A. Natural wastes as particle filler for poly (lactic acid)-based composites. *J. Compos. Mater.* **2019**, *53*, 783–797. [CrossRef]
42. Jiang, Y.P.; Wu, J.; Zhu, Y. A mesoscopic model for particle-filled composites with network microstructures. *Mech. Compos. Mater.* **2021**, *57*, 415–424. [CrossRef]
43. Nair, S.S.; Zhu, J.Y.; Deng, Y.; Ragauskas, A.J. High performance green barriers based on nanocellulose. *Sustain. Chem. Process.* **2014**, *2*, 1–7. [CrossRef]
44. Lei, Y.; Luo, L.; Kang, Z.; Zhang, J.; Zhang, B. Modified Halpin–Tsai equation for predicting interfacial effect in water diffusion process. *Sci. Eng. Compos. Mater.* **2021**, *28*, 180–189. [CrossRef]
45. Ansari, F.; Galland, S.; Johansson, M.; Plummer, C.J.G.; Berglund, L.A. Cellulose nanofiber network for moisture stable, strong and ductile biocomposites and increased epoxy curing rate. *Compos. Part A Appl. Sci. Manuf.* **2014**, *63*, 35–44. [CrossRef]
46. Nakajima, H.; Dijkstra, P.; Loos, K. The recent developments in biobased polymers toward general and engineering applications: Polymers that are upgraded from biodegradable polymers, analogous to petroleum-derived polymers, and newly developed. *Polymers* **2017**, *9*, 523. [CrossRef] [PubMed]
47. Mysiukiewicz, O.; Kosmela, P.; Barczewski, M.; Hejna, A. Mechanical, thermal and rheological properties of polyethylene-based composites filled with micrometric aluminum powder. *Materials* **2020**, *13*, 1242. [CrossRef] [PubMed]
48. Bartos, A.; Kócs, J.; Anggono, J.; Móczó, J.; Pukánszky, B. Effect of fiber attrition, particle characteristics and interfacial adhesion on the properties of PP/sugarcane bagasse fiber composites. *Polym. Test.* **2021**, *98*, 107189. [CrossRef]
49. Bhasney, S.M.; Kumar, A.; Katiyar, V. Microcrystalline cellulose, polylactic acid and polypropylene biocomposites and its morphological, mechanical, thermal and rheological properties. *Compos. Part B-Eng.* **2020**, *184*, 107717. [CrossRef]
50. Andersons, J.; Filipova, I.; Kirpluks, M.; Cabulis, U. Evaluation of the apparent interfacial shear strength of nanocellulose/PVA composites. *Key Eng. Mater.* **2018**, *774*, 54–59. [CrossRef]
51. Danyadi, L.; Janecska, T.; Szaboc, Z.; Nagy, G.; Moczo, J.; Pukanszky, B. Wood flour filled PP composites: Compatibilization and adhesion. *Compos. Sci. Technol.* **2007**, *67*, 2838–2846. [CrossRef]

52. Sugiman, S.; Salman, S.; Maryudi, M. Effects of volume fraction on water uptake and tensile properties of epoxy filled with inorganic fillers having different reactivity to water. *Mater. Today Commun.* **2020**, *24*, 101360. [CrossRef]
53. Yasuniwa, M.; Satou, T. Multiple melting behavior of poly (butylene succinate). I. Thermal analysis of melt-crystallized samples. *J. Polym. Sci. Pol. Phys.* **2002**, *40*, 2411–2420. [CrossRef]
54. Papageorgiou, D.G.; Zhuravlev, E.; Papageorgiou, G.Z.; Bikiaris, D.; Chrissafis, K.; Schick, C. Kinetics of nucleation and crystallization in poly(butylene succinate) nanocomposites. *Polymer* **2014**, *55*, 6725–6734. [CrossRef]
55. Yao, S.F.; Chen, X.T.; Ye, H.M. Investigation of structure and crystallization behavior of poly(butylene succinate) by fourier transform infrared spectroscopy. *J. Phys. Chem. B* **2017**, *121*, 9476–9485. [CrossRef] [PubMed]
56. Auad, M.L.; Contos, V.S.; Nutt, S.; Aranguren, M.I.; Marcovich, N.E. Characterization of nanocellulose- reinforced shape memory polyurethanes. *Polym. Int.* **2008**, *57*, 651–659. [CrossRef]
57. Camarero-Espinosa, S.; Boday, D.J.; Weder, C.; Foster, E.J. Cellulose nanocrystal driven crystallization of poly(d,l-lactide) and improvement of the thermomechanical properties. *J. Appl. Polym. Sci.* **2015**, *132*, 41607. [CrossRef]
58. Rastogi, V.K.; Samyn, P. Novel processing of polyhydroxybutyrate with micro- to nanofibrillated cellulose and effect of fiber morphology on crystallization behaviour of composites. *eXPRESS Polym. Lett.* **2020**, *14*, 115–133. [CrossRef]
59. Jyoti, J.; Singh, P.B.; Arya, A.K.; Dhakate, S.R. Dynamic mechanical properties of multiwall carbon nanotube reinforced ABS composites and their correlation with entanglement density, adhesion, reinforcement and C factor. *RSC Adv.* **2016**, *6*, 3997–4006. [CrossRef]
60. Perotto, G.; Simonutti, R.; Ceseracciu, L.; Mauri, M.; Besghini, D.; Athanassiou, A. Water-induced plasticization in vegetable-based bioplastic films: A structural and thermo-mechanical study. *Polymer* **2020**, *200*, 122598. [CrossRef]
61. Kubat, J.; Rigdhal, M.; Welander, M. Characterization of interfacial interaction in high density polyethylene filled with glass spheres using dynamic mechanical analysis. *J. Appl. Polym. Sci.* **1990**, *39*, 1527–1539. [CrossRef]

polymers

MDPI

Article

Electrospun Polymer-Fungicide Nanocomposites for Grapevine Protection

Nasko Nachev [1], Mariya Spasova [1,*], Petya Tsekova [1], Nevena Manolova [1], Iliya Rashkov [1] and Mladen Naydenov [2]

[1] Laboratory of Bioactive Polymers, Institute of Polymers, Bulgarian Academy of Sciences, Acad. G. Bonchev St., Bl. 103A, BG-1113 Sofia, Bulgaria; nachev_n@polymer.bas.bg (N.N.); cekovapetya@polymer.bas.bg (P.T.); manolova@polymer.bas.bg (N.M.); rashkov@polymer.bas.bg (I.R.)

[2] Department of Microbiology, Agricultural University, BG-4000 Plovdiv, Bulgaria; mladen@au-plovdiv.bg

* Correspondence: mspasova@polymer.bas.bg; Fax: +359-(0)2-8700309

Abstract: Nowadays, diseases in plants are a worldwide problem. Fungi represent the largest number of plant pathogens and are responsible for a range of serious plant diseases. Esca is a grapevine disease caused mainly by fungal pathogens *Phaeomoniella chlamydospora* (*P. chlamydospora*) and *Phaeoacremonium aleophilum* (*P. aleophilum*). The currently proposed methods to fight esca are not curative. In this study, polymer composites based on biodegradable polymer containing chemical fungicides with antifungal activity were successfully prepared by electrospinning. The obtained materials were hydrophobic with good mechanical properties. In vitro studies demonstrated that the fungicide release was higher from PLLA/K5N8Q fibrous mats (ca. 72% for 50 h) compared to the released drug amount from PLLA/5-Cl8Q materials (ca. 52% for 50 h), which is due to the better water-solubility of the salt. The antifungal activity of the fibrous materials against *P. chlamydospora* and *P. aleophilum* was studied as well. The incorporation of the fungicide in the biodegradable fibers resulted in the inhibition of fungal growth. The obtained materials are perspective candidates for the protection of vines from the penetration and growth of fungal pathogens.

Keywords: biodegradable polymer; electrospinning; fungicide; *Phaeomoniella chlamydospora*; *Phaeoacremonium aleophilum*; esca

Citation: Nachev, N.; Spasova, M.; Tsekova, P.; Manolova, N.; Rashkov, I.; Naydenov, M. Electrospun Polymer-Fungicide Nanocomposites for Grapevine Protection. *Polymers* **2021**, *13*, 3673. https://doi.org/10.3390/polym13213673

Academic Editor: Evgenia G. Korzhikova-Vlakh

Received: 30 September 2021
Accepted: 22 October 2021
Published: 25 October 2021

1. Introduction

Grapevine trunk diseases reduce the lifespan of vineyards and increase the costs of producing wine grapes [1]. They are caused mainly by fungal pathogens, with the major pathogens including *Phaeomoniella chlamydosporum*, *Phaeoacremonium aleophilum*, *Botryosphaeria* spp., *Cylindrocarpon* spp., *Eutypa lata*, and *Phomopsis viticolaand* [2].

Esca is a destructive grapevine trunk disease first described over 100 years ago, which occurs worldwide and induces heavy economic losses [3]. The disease could be developed by intensive pruning, frost, and other mechanical injuries. The first symptoms of esca appear as dark red or yellow stripes on leaves, which eventually dry and become necrotic. The disease can then progress, potentially causing the entire plant to die [4]. Esca was first successfully controlled in 1903, when sodium arsenite was used as an insecticide on grapes. However, sodium arsenite was noted as being highly toxic and carcinogenic, and since 2003 has been banned in Europe. Nowadays, in practice, there is no curative approach for fighting esca. This fact challenges researchers to find a solution to effectively fight with this complex disease.

Electrospinning is a facile and efficient technique for fabricating functional nanofibrous materials possessing a large specific surface area and a fine porous structure [5,6]. These fibers have received much attention for use in many applications: Biomedical applications such as drug delivery [7], tissue engineering [8], and wound dressing, as well as cosmetics and functional materials and devices such as composite reinforcement, filters, protective clothing, and smart textiles, and even energy and electronics such as batteries/cells, capacitors, sensors, and catalysts.

Recently, novel varieties of the electrospinning technique were developed in order to generate more complex nanostructures with desired features such as coaxial electrospinning [9], side-by-side [10] and tri-axial [11] electrospinning, and other multiple fluid processes. Such methods of fabrication can lead to composite structures such as core–shell, Janus, tri-layer core–shell, and other complex structures. Another recently used strategy is to combine electrospinning with other traditional methods to fabricate novel nanofibers [12]. Regardless of the direction, the final objective is a suitable application of the resultant nanofibers [13]. The present study highlights a new potential application of electrospun nanofibers for grapevine protection.

Phytopathogenic fungal infections have become a serious problem in agricultural production, reducing food yield and quality. Therefore novel antifungal agents with high efficiency and low toxicity are needed. Modern plant protection products should be designed to achieve the desired biological effect without harmful impact or side effects. The use of nanomaterials in agriculture and, in particular, for the protection of vineyards is an emerging field of interest. Sett et al. created rayon membranes on which nanofibers of soy protein/polyvinyl alcohol and soy protein/polycaprolactone are electrospun. This material aims to physically block the penetration of fungal spores [14]. However, the authors commented that the blocking was insufficient and an antifungal component should be incorporated. Furthermore, electrospun materials from two copolymers loaded with polyhexamethylene guanidine have been fabricated to serve as bandages for vineyards against the penetration of esca-causing fungi [15]. The authors reported, however, that more effective polymers and antifungal agents should be used.

Some authors have also proposed an original approach and have successfully obtained fibrous materials from poly(3-hydroxybutyrate), nanosized TiO_2-anatase, and chitosan oligomers (COS) with antifungal activity for plant protection via a combination of electrospinning and electrospraying [16]. The obtained eco-friendly materials possess high roughness, hydrophobicity, and antifungal activity against *P. chlamydospora*.

It is known that compounds containing 8-hydroxyquinoline exhibit anticancer [17,18], antimicrobial [19,20], antiviral [21], and antifungal activities [22,23]. Due to their antifungal properties, these biologically active compounds have found application in agriculture. We created stable solutions on the basis of a water-soluble polymer and a fungicide with antifungal activity suitable for applications in agriculture [24].

The incorporation of low molecular weight derivatives of 8-hydroxyquinoline into fibrous polymer materials obtained by electrospinning is of interest because it allows combining the valuable biological properties of 8-hydroxyquinoline derivatives with the advantages of electrospun materials. In our previous study, we obtained fibrous membranes on the basis on cellulose acetate loaded with an antifungal agent for active protection against spore penetration and plant infection in vineyards [25]. However, the incorporation of low molecular weight derivatives significantly lower the physicochemical properties and therefore they need to be improved.

In the last few decades, considerable research interest has been dedicated to the use of biodegradable polymer materials for various applications, such as in medicine, as well as in industry to replace conventional petrochemical-based polymers. Due to its merits, thermoplastic aliphatic polyesters are the most commonly explored synthetic biodegradable polymers. PLA has an extensive mechanical property profile and is a highly biocompatible and biodegradable polymer [26].

Therefore, the present study aimed to prepare electrospun composite materials from PLA and 8-hydroxyquinoline derivatives with antifungal activities. The effect of the incorporated biologically active compound on the morphology, wetting, crystallinity, thermal, and physicochemical properties was studied. Microbiological tests against *P. chlamydospora* and *P. aleophilum* were performed.

2. Materials and Methods

2.1. Materials

Poly(L-lactide) (PLA; Ingeo™ Biopolymer 4032D, NatureWorks LLC—USA, Minnetonka, MN, USA; M_W = 259,000 g/mol; M_W/M_n = 1.94; as determined by size-exclusion chromatography using polystyrene standards), 5-nitro-8-hydroxyquinoline (Pharmachim, Sofia, Bulgaria), and 5-chloro-8-hydroxyquinoline (Sigma-Aldrich, St. Louis, MO, USA) were used. Dichloromethane (DCM; Merck, Darmstadt, Germany) and ethanol (abs. EtOH; Merck, Darmstadt, Germany) were of an analytical grade of purity.

Potato dextrose agar medium was obtained from Merck, Darmstadt, Germany. The disposable consumables were purchased from Orange Scientific, Braine-l'Alleud, Belgium.

2.2. Procedures

The potassium 5-nitro-8-hydroxyquinoline was prepared as described by Ermakov and coworkers [27].

2.3. Preparation of Electrospun Fibrous Materials

Spinning solutions in DCM/EtOH (DCM/EtOH = 90/10) were prepared for PLA, PLA/5-Cl8Q, and PLA/K5N8Q. The total polyester concentration was 10 wt% 5-Cl8Q and 10 wt% K5N8Q.

A Brookfield LVT viscometer equipped with an adaptor for small samples, a spindle, and a camera SC 4-18/13 R at 20 ± 0.1 °C was used to measure the solution viscosities. The spinning solutions were measured in triplicate and the mean values with their standard deviations were used.

Electrospinning was performed using a high-voltage power supply (up to 30 kV), a grounded metal drum collector, an infusion pump (NE-300 Just InfusionTM Syringe Pump, New Era Pump Systems Inc., Farmingdale, NY, USA) for delivering the spinning solution at a constant rate, and a syringe equipped with a metal needle (gauge: 20GX1$\frac{1}{2}$"). The applied voltage was 25 kV, the distance to the collector was 15 cm, the collector rotating speed was 1000 rpm, the humidity was 50%, and the temperature was 20 °C.

2.4. Characterization of the Electrospun Materials

The morphology of the fibers was examined by a scanning electron microscope (SEM). The samples were vacuum-coated with gold and observed by a Jeol JSM-5510 SEM (Jeol Ltd., Tokyo, Japan) at acceleration voltage of 10 kV with 1000×, 2500×, and 5000× magnification. The fiber morphology was evaluated using the criteria for complex evaluation of electrospun materials as described elsewhere [28] via ImageJ software by measuring the diameters of at least 20 fibers from each SEM micrograph [29].

Attenuated total reflection Fourier transform infrared (ATR-FTIR) spectra were recorded using an IRAffinity-1 spectrophotometer (Shimadzu Co., Kyoto, Japan) equipped with a MIRacle™ATR (diamond crystal with a depth of penetration of the IR beam into the sample of approximately 2 μm) accessory (PIKE Technologies, Fitchburg, WI, USA) in the range of 600–4000 cm^{-1} and a resolution of 4 cm^{-1}. All spectra were corrected for H_2O and CO_2 using an IRsolution software program.

The absence/presence of a crystalline phase in the electrospun materials was assessed by X-ray diffraction analysis (XRD). XRD spectra were recorded at r.t. using a computer-controlled D8 Bruker Advance powder diffractometer (Bruker, Billerica, MA, USA) with a filtered CuKα radiation source and a luminescent detector. The analyses were performed in the 2θ range from 5° to 50° with a step of 0.02° and a counting time of 1 s/step.

Static contact angle measurements of the membranes were performed using an Easy Drop DSA20E Krűss GmbH drop shape analysis system (Hamburg, Germany) at 20 ± 0.2 °C. A sessile drop of deionized water with a volume of 10 μL controlled by a computer dosing system was deposited onto the electrospun fibrous materials. The contact angles were calculated by computer analysis of the acquired images of the droplet. The data are an average from 10 measurements for each sample.

Mechanical properties were evaluated by tensile measurements performed on the fibrous materials using a single-column system for mechanical testing INSTRON 3344, equipped with a loading cell 50 N and Bluehill universal software (Instron Bluehill Universal V4.05 (2017) software, Norwood, MA, USA). The stretching rate was 10 mm/min, the initial length between the clamps was 40 mm, and the room temperature was 21 °C. All samples were cut into dimensions of 20 × 60 mm^2 with a thickness of ca. 200 µm. For the sake of statistical significance, 10 specimens of each sample were tested, after which the average values of Young's modulus, the ultimate stress, and the maximum deformation at break were determined.

5-Cl8Q and K5N8Q release was studied in vitro at 37 °C in acetate buffer (CH_3COONa/CH_3COOH) containing lactic acid (acetate buffer/lactic acid = 96/4 v/v) at pH 3 and an ionic strength of 0.1. Fibrous materials loaded with 5-Cl8Q or K5N8Q (4 mg) were immersed in 100 mL of buffer solution under stirring in a water bath (Julabo, Seelbach, Germany). The release kinetics were determined by withdrawing aliquots (2 mL) from the solution at determined time intervals, then adding back the same amount of fresh buffer and recording the absorbance of the aliquots by a DU 800 UV–vis spectrophotometer (Beckman Coulter, Brea, USA) at wavelengths of 255 nm and 364 nm. The amount of released 5-Cl8Q or K5N8Q was calculated using calibration curves (correlation coefficient R = 0.999) for the membranes in acetate buffer/lactic acid = 96/4 v/v with a pH of 3 and an ionic strength of 0.1. The data are the average values from three measurements.

2.5. In Vitro Antifungal Assay

The antifungal activity of the fibrous materials was monitored against the fungi *P. chlamydospora* CBS 239.74 and *P. aleophilum* CBS 631.94. *P. chlamydospora* CBS 239.74 and *P. aleophilum* CBS 631.94 were purchased from Westerdijk Fungal Biodiversity Institute, Utrecht, the Netherlands.

P. chlamydospora and *P. aleophilum* grow normally on potato dextrose agar [30] and malt extract agar [31].

In the present study, in vitro studies were performed using potato dextrose agar medium (PDA; Merck, Germany). The surface of the solid agar was inoculated with a suspension of fungi culture with a fungi concentration of 1 × 10^5 cells/mL, and on the surface of the agar in each Petri dish, one electrospun material (17 mm in diameter) was placed. The Petri dishes were incubated for 96 h for *P. chlamydospora* and *P. aleophilum* at 28 °C, and subsequently, the zones of inhibition around the disks were measured. The average diameters of the zones of inhibition were determined using ImageJ software based on 15 measurements in 15 different directions for each zone.

For the preparation of the conidia suspension test, microorganisms were grown on potato dextrose agar (PDA) medium for 14 days. Conidia were obtained by pouring 5 mL of sterile water onto the plate and washing it off with a sterile loop. Conidia suspensions were filtered through two layers of sterile round cloth to remove mycelial fragments. The final concentration of conidia was adjusted to 10^7 conidia/mL with sterile water. The fibrous materials were cut into disks with a diameter of 45 mm and a thickness of ~1 µm. Digital Thickness Gauge FD 50 (Käfer GmbH, Villingen-Schwenningen, Germany) was used to determine the thickness of the fibrous materials. All fibrous materials were sterilized for 30 min under UV light in a laminar box before being used for further experiments. Then, the fibrous material in disk form was placed between the two parts of the filtration device supplied with a pump. The two parts of the device were pinched with a clip. After this, 20 mL of the spore conidia suspension was passed through each type of fibrous material. Then, every used disk was taken with pincers and placed on a surface of solid PDA medium in a Petri dish. The Petri dishes were maintained at 28 °C for 96 h. Then, the fungal growth was assessed. The concentration of conidia passed through the materials was determined using a hemocytometer.

2.6. Statistical Analysis

The data are displayed as means ± standard deviations (SDs). To determine the statistical significance of the data, one-way analysis of variance (ANOVA) followed by Bonferroni's post-hoc test were performed. Values of * $p < 0.05$, ** $p < 0.01$, and *** $p < 0.001$ were considered significant.

3. Results and Discussion

3.1. Morphological Analysis

In our previous studies, we showed the successful fabrication of fibrous materials obtained from polyesters by electrospinning [20,32,33]. In concrete, we found that for PLA with M_W = 259,000 g/mol and M_W/M_n = 1.94, the optimal total polymer concentration for conducting electrospinning resulting in preparation of defect-free cylindrical fibers was 10 wt% in a DCM/EtOH solvent system [20]. However, it is necessary to study the effect of incorporation of a biologically active substance(s) on the morphologies of fibers, their physical–chemical properties, and their ability to inhibit the growth and penetration of pathogenic fungi.

SEM pictures of the obtained PLLA, PLLA/5-Cl8Q, and PLLA/K5N8Q fibers are presented in Figure 1. The electrospinning of PLLA resulted, reproducibly, in the fabrication of fibers with average fiber diameters of 1045 ± 320 nm (Figure 1a). The obtained diameters of the PLA fibers are in fairly good agreement with the literature data [34]. As can be easily seen using the selected conditions (concentration, solvent system, applied voltage, feeding rate, collector rotating speed, etc.), fibers with a cylindrical shape without defects and pores were obtained.

The addition of chemical fungicides (5-Cl8Q or K5N8Q) at a concentration of 10 wt% resulted in the preparation of stable solutions that did not alter the process of electrospinning and resulted in the fabrication of composite fibers with a cylindrical shape with a mean diameter close to that of the neat PLLA (Figure 1b,c). The average diameter of the fibers of the fibrous materials based on PLLA/5-Cl8Q and PLLA/K5N8Q was 1125 ± 300 nm and 1065 ± 250 nm, respectively. This is an indication that the addition of low molecular fungicides (derivatives of 8-hydroxyquinoline) did not lead to a significant change in the fiber morphology or diameters and distribution. These findings were confirmed by the measured values of the dynamic viscosities of the prepared solutions as well. The dynamic viscosity for PLLA, PLLA/5-Cl8Q, and PLLA/K5N8Q were relatively close and were 1350 cP, 1500 cP, and 1420 cP, respectively.

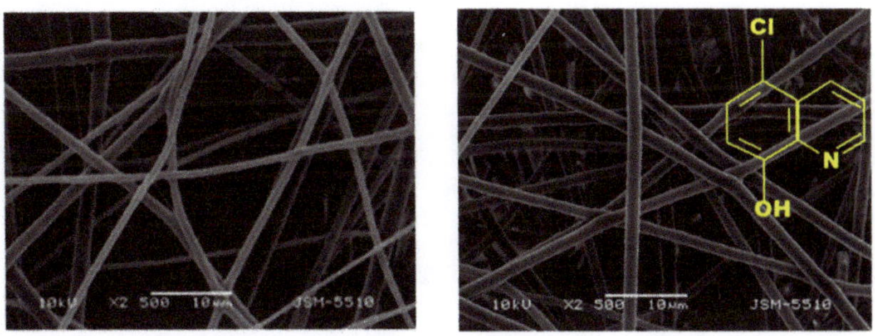

(a) (b)

Figure 1. *Cont.*

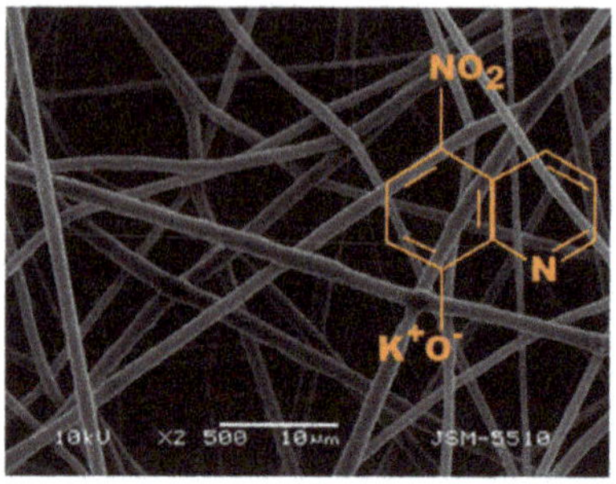

(c)

Figure 1. Representative SEM images of the fibers of electrospun fibrous materials of PLLA (**a**), PLLA/5-Cl8Q (**b**), and PLLA/K5N8Q (**c**); magnification ×2500.

3.2. Contact Angle Measurements

It is well known that bacterial and fungal adhesion is influenced by the surface characteristics and the hydrophilic/hydrophobic balance of the host surface. For this reason, it is important to determine the values of the contact angle of the prepared electrospun materials that will contact the fungal species. The values of the water contact angles for all obtained samples were determined using distilled water droplets, and representative images of the droplets are shown in Figure 2. The PLLA fibrous material was hydrophobic, with a water contact angle of 117° ± 2.5° (Figure 2a). The measured value for the pure PLLA was close to the values found in the literature [35]. The measured contact angle values of the PLLA/5-Cl8Q and PLLA/K5N8Q composite fibrous materials were 120° ± 3° and 118.0° ± 2°, respectively (Figure 2b,c). The measured water contact angle values were close to those measured for the PLLA electrospun material. All of the obtained and studied electrospun fibers had water contact angle values ca. 120° and were hydrophobic.

(a) (b) (c)

Figure 2. Images of water droplets deposited on the surface of fibrous materials: (**a**) PLLA, (**b**) PLLA/5-Cl8Q, and (**c**) PLLA/K5N8Q.

3.3. FTIR Spectroscopic Analysis

FTIR spectroscopy was performed to characterize the prepared PLLA, PLLA/5-Cl8Q, and PLLA/K5N8Q fibrous materials, and the recorded spectra are shown in Figure 3. Characteristic bands for PLLA appeared at 1751 cm^{-1} for the C=O groups and at 1182 cm^{-1} for the C–O–C groups.

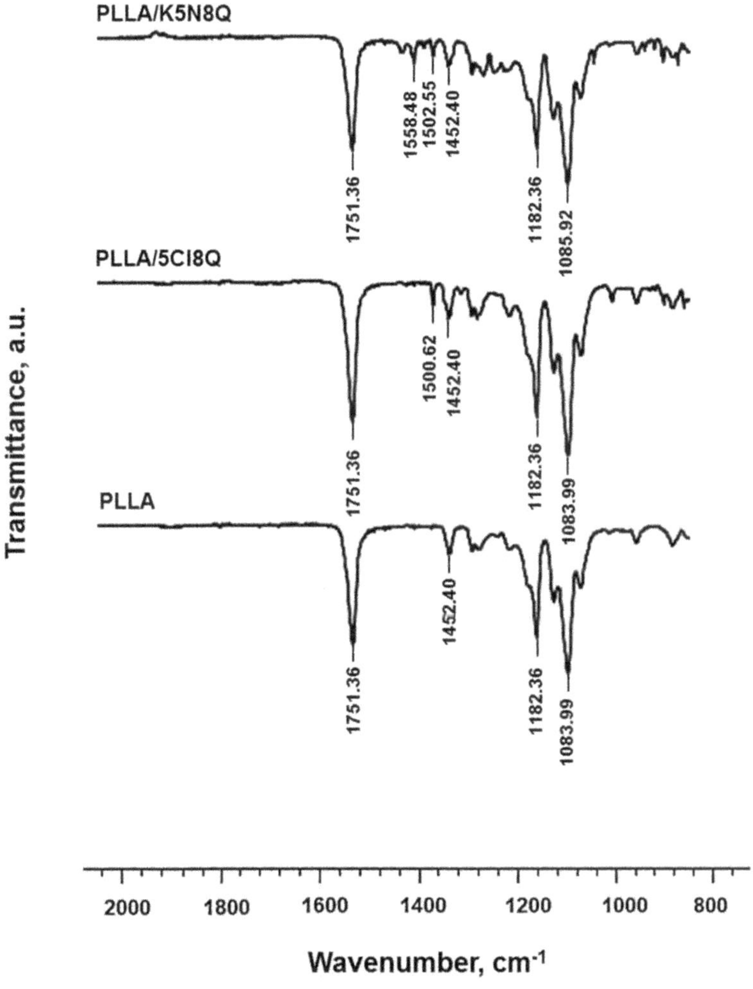

Figure 3. FTIR spectra of electrospun fibrous materials of PLLA, PLLA/5-Cl8Q, and PLLA/K5N8Q.

The characteristic stretching frequencies for C–O at 1080 cm^{-1} and the bending frequencies for $-CH_3$ asymmetric and $-CH_3$ symmetric at 1452 cm^{-1} and 1361 cm^{-1}, respectively, were identified, also in accordance with the literature [36].

A new band appeared at 1500 cm^{-1}, characteristic for the aromatic ring of the chemical fungicide in the PLLA/5-Cl8Q and PLLA/K5N8Q fibrous materials (Figure 3), in addition to the characteristic bands of PLLA [37], proving its presence in the electrospun composite materials.

Clearly, no molecular interaction between PLLA and the used fungicides was detected on the FT-IR spectra of the PLLA/5-Cl8Q and PLLA/K5N8Q composite fibrous materials.

3.4. XRD Analysis

Delivery systems based on nano- and microcarriers have been proven to be promising candidates for the delivery of poorly water-soluble or non-water-soluble compounds (drugs), wherein amorphization during thier encapsulation by the electrospinning process improves the dissolution of these compounds [12]. Therefore, it was of interest to study the changes in the crystallinity of chemical fungicides after their incorporation in composite electrospun fibrous materials. The crystallinity of the obtained fibers was determined by X-ray diffraction (XRD) analysis (Figure 4). The XRD pattern of PLLA and PLLA/5-Cl8Q materials and 5-Cl8Q powder, as well as of PLLA/K5N8Q fibrous material and K5N8Q powder, are presented in Figure 4a,b respectively. XRD patterns of 5-Cl8Q and K5N8Q (powder) with characteristic sharp diffraction peaks of the compounds were observed. These peaks showed that the fungicides (powders) were highly crystalline. The XRD spectra of the PLLA fibers showed a strong amorphous halo, proving that these materials have a typical amorphous structure. Moreover, in the spectra of the PLLA/5-Cl8Q and PLLA/K5N8Q composite materials, an amorphous halo was detected as well. This result indicates that each component in the composite fibrous materials prepared by electrospinning was in an amorphous state. This observation could be explained by the rapid drying of the jet during electrospinning, which impeded molecular motion. The obtained results are in accordance with the literature concerning the amorphization of poorly water-soluble drugs by electrospinning [38].

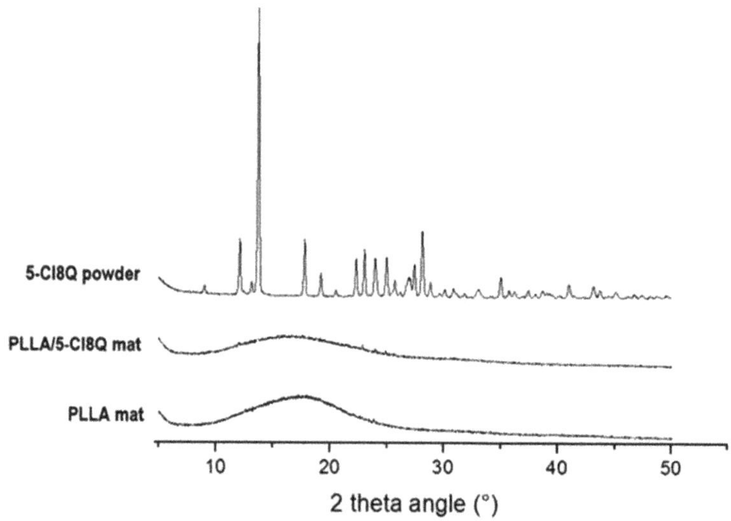

(a)

Figure 4. *Cont.*

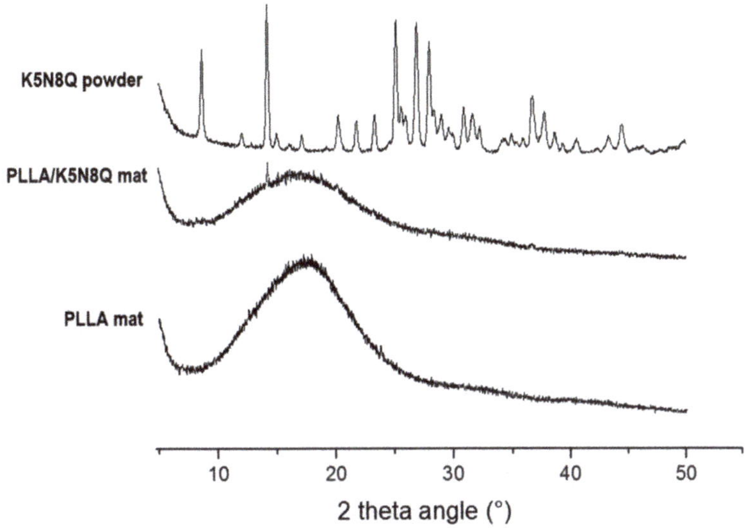

(b)

Figure 4. X-ray diffraction pattern of (**a**) 5-Cl8Q powder and PLLA and PLLA/5-Cl8Q materials, and (**b**) K5N8Q powder and PLLA and PLLA/K5N8Q materials.

3.5. Mechanical Properties

Stress–strain curves of the PLLA, PLLA/5-Cl8Q, and PLLA/K5N8Q electrospun materials are presented in Figure 5. The PLLA material showed the highest tensile strength values. The obtained values are in good agreement with the literature data [20]. The determination of the mechanical characteristics of the PLLA/5-Cl8Q and PLLA/K5N8Q composite fibers showed that these materials possess similar mechanical properties, albeit a bit lower than those of the PLLA fibrous materials. This result indicates that the incorporation of 5-Cl8Q in fibrous membranes does not considerably change the mechanical characteristics of these membranes. The tensile strength of the PLLA/5-Cl8Q and PLLA/K5N8Q composite fibrous materials was ca. 2.5 MPa, while the tensile strength of the PLLA fibrous materials reached 3.4 MPa. The slight decrease in mechanical characteristics might be due to the incorporation of low molecular weight chemical fungicides in the PLA matrix, which might have generated weak spots when the tensile test was carried out.

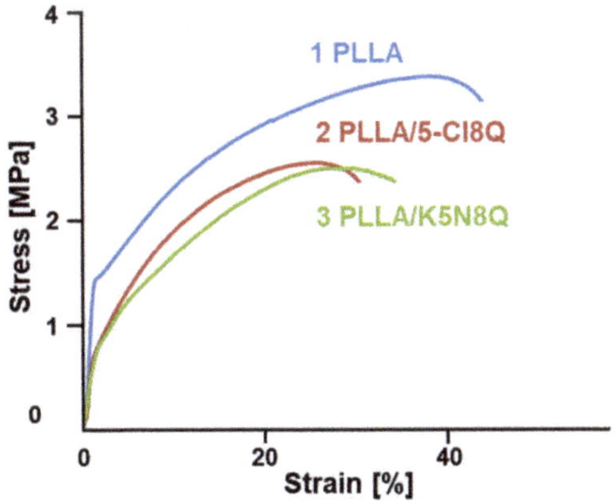

Figure 5. Stress–strain curves of electrospun materials: (1) PLLA, (2) PLLA/5-Cl8Q, and (3) PLLA/K5N8Q.

3.6. Cumulative Drug Release Analysis

The electrospinning method is often used for encapsulation of drugs for delivery. There are many data in the literature concerning the release rates of incorporated drugs in different polymer matrixes showing diverse behavior: Some show an initial burst release of the drug, while others show a more controlled release of the drug over a longer duration. This is due to the fact that many parameters could influence drug release, e.g., the molecular characteristics of the polymer, the polymer crystallinity and hydrophobicity, the nature of the drug and its crystallinity, the compatibility of the drug with the polymers matrix, the fiber morphology and diameters, the presence of defects along the fibers. Therefore, the same drug loaded in different polymer matrixes or different drugs incorporated in same polymer matrix may exhibit different release profiles.

The release of 5-Cl8Q and K5N8Q from PLLA fibrous matrixes was studied, and their release profiles are shown in Figure 6. Initially, both drugs showed a relatively burst release from the PLLA fibrous matrix. However, K5N8Q was released in a higher amount compared to 5-Cl8Q for the same duration. For instance, the released K5N8Q was 10.5% and 21.4% for 30 and 60 min. For the same time durations, the released 5-Cl8Q was 6.7% and 9.3%, respectively. This difference in the release profiles could be due to the different natures of the drugs and their water solubility. K5N8Q is a partially water-soluble drug favoring a more rapid release. On the contrary, 5-Cl8Q is water-insoluble, which hampers its release. After 50 h, the amounts of the released 5-Cl8Q and K5N8Q were 52.8% and 72.5%, respectively.

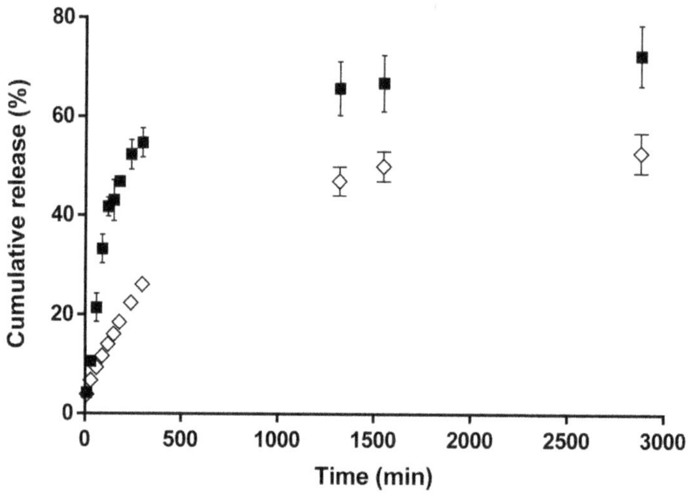

Figure 6. Release profiles of 5-Cl8Q and K5N8Q from PLLA fibers: PLLA/5-Cl8Q (◊) and PLLA/K5N8Q (■). The results are presented as the average values from three separate measurements with the respective standard deviation; acetate buffer/lactic acid (96/4 *v*/*v*), pH 3, 37 °C, ionic strength 0.1.

3.7. Antifungal Activity of the Fibrous Membranes

The two main fungi causing esca disease are *P. chlamydospora* and *P. aleophilum*. The wounds on vines caused during pruning are the main point of entry for the penetration of fungal spores in grapevines. Because there is no direct way to fight esca, there is a rising demand for the development of novel plant protective agents and materials that are non-toxic but are efficient against esca.

Although there are some data with respect to the antifungal activity of 8-hydroxyquilonine derivatives against *Candida* species, there are no data concerning their effects on *P. chlamydospora* and *P. aleophilum*, which are the main causative agent of esca disease. In our previous study, we determined the minimum inhibitory concentration (MIC) of 5-Cl8Q against *P. chlamydospora* and *P. aleophilum*, and it was found to be 0.75 μg/mL for both strains [24]. The MIC determined by us for K5N8Q was 12.5 μg/mL and 25 μg/mL for *P. chlamydospora* and *P. aleophilum*, respectively.

The antifungal activity of the electrospun fibrous materials (diameter 17 mm) was determined by carrying out antifungal tests against *P. chlamydospora* and *P. aleophilum*.

Figure 7 presents the observed zones of inhibition after contact of the fibrous materials with the fungal cells. The loading of 5-Cl8Q in the composite fibrous materials that were laid in contact with *P. chlamydospora* resulted in complete inhibition of fungal growth. Moreover, there was a wide inhibition zone around the PLLA/5-Cl8Q disc put in contact with *P. aleophilum* (4.7 cm). Additionally, the incorporation of K5N8Q resulted in wide zones of inhibition as well. The diameters of the inhibition zones around the PLLA/K5N8Q discs were 6.2 cm and 4.0 cm against *P. chlamydospora* and *P. aleophilum*, respectively. From the obtained results, it is easily seen that *P. chlamydospora* is more vulnerable to treatment with the used 8-hydroxyquinoline derivatives.

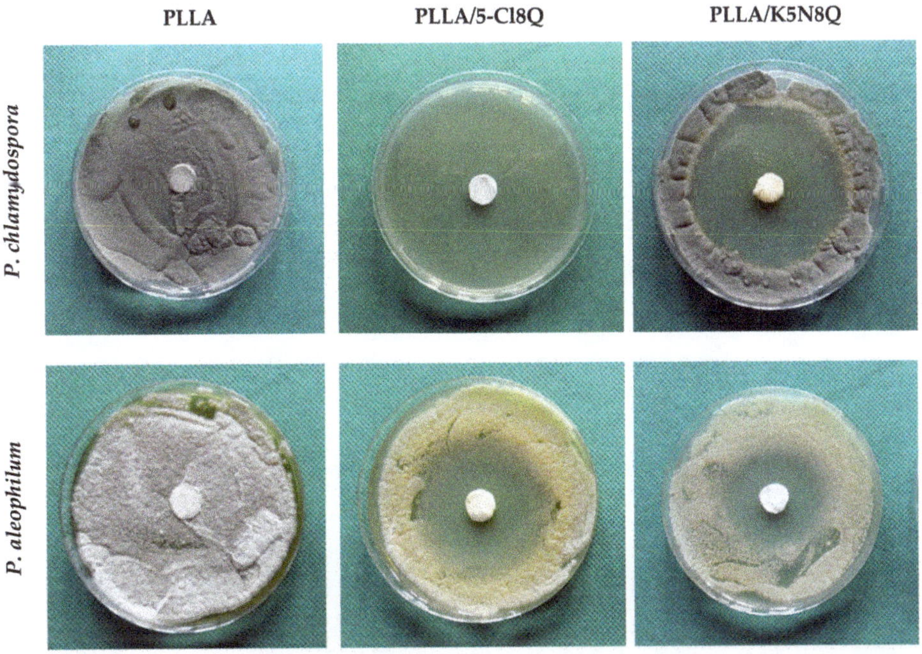

Figure 7. Digital pictures of the zones of inhibition against *P. chlamydospora* and *P. aleophilum* after contact of the fibrous materials with fungi cells. The material type is indicated at the top of each column. The cell type is marked in the left of each row.

The results obtained in the present study demonstrate that composite fibrous materials containing hydroxyquinoline derivatives have strong antifungal activity. In contrast, neat PLLA fibrous materials do not change the fungal growth or exhibit any antifungal activity.

In the present study, the barrier efficacy of PLLA, PLLA/5-Cl8Q, and PLLA/K5N8Q electrospun fibrous materials were studied as well. For this purpose, 20 mL of conidia suspension was passed through each fibrous material (diameter 45 mm) using a filtration device. Initially, we determined the size of the *P. chlamydospora* and *P. aleophilum* conidia in the fungal suspension using SEM analysis. Figure 8 presents the used conidia.

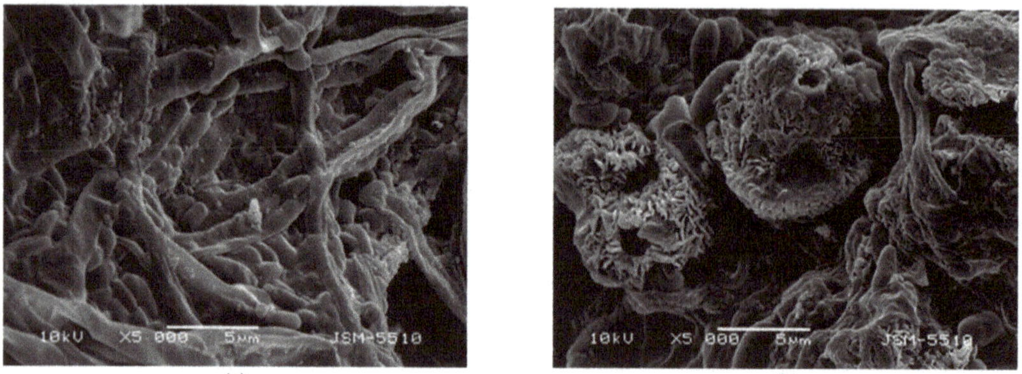

(a) (b)

Figure 8. SEM micrographs of the (**a**) *P. chlamydospora* and (**b**) *P. aleophilum* conidia.

The diameter and length of the *P. chlamydospora* conidia were ~0.75–1.2 μm and 1.8–2.3 μm, respectively. The measured diameter and length of the *P. aleophilum* conidia were ~1.1–1.5 μm and 2.5–3.5 μm, respectively. The initial concentration in the filtration experiments was 1×10^7 conidia/mL for both strains. After passing through the electrospun discs, the determined spore concentration was 1.6×10^3, 1.3×10^3, and 1.4×10^3 for the PLLA, PLLA/5-Cl8Q, and PLLA/K5N8Q materials, respectively. This result reveals that the final conidia concentration decreased significantly. However, some conidia passed through all of the fibrous materials in this study.

It was of interest to determine not only the ability of the materials to impede the penetration of fungal spores, but also to study if the fibrous materials impede the growth of pathogenic fungi remaining in the material after filtration. Therefore, after filtration, we placed the used disks on a surface of solid agar in a Petri dish in order to determine the growth of the remaining fungi in the fibrous discs. The Petri dishes were incubated for 96 h at 28 °C, and then the fungal growth was determined. Figure 9 presents the growth of *P. chlamydospora* on the fibrous materials' surface. It was found that the PLLA fibrous material used in the filtration experiments developed colonies of *P. chlamydospora* (Figure 9a). The developed colonies showed that this material did not possess antifungal activity. PLLA/5-Cl8Q and PLLA/K5N8Q, which were placed in suitable conditions for the development of remaining spores in the materials, impeded the fungal growth, resulting in compete fungal inhibition (Figure 9b,c).This result indicates that the fungi remaining in the PLLA/5-Cl8Q and PLLA/K5N8Q materials after filtration could not grow due to the antifungal activity of the obtained composite materials.

(a) (b) (c))

Figure 9. Digital images of the growth *P. chlamydospora* on the fibrous materials after spore filtration: (**a**) PLLA, (**b**) PLLA/5-Cl8Q, and (**c**) PLLA/K5N8Q.

4. Conclusions

Novel micro- and nanofibrous materials of PLLA, PLLA/5-Cl8Q, and PLLA/K5N8Q were successfully electrospun. The obtained composite materials were hydrophobic with good mechanical properties. The incorporation of 5-Cl8Q or K5N8Q into the PLLA fibers imparted to them a considerable antifungal activity against *P. chlamydospora* and *P. aleophilum*. These features demonstrate that the obtained composite fibrous materials could be a potential candidate for application in agriculture for grapevine protection against esca-associated fungi.

Author Contributions: M.S., I.R., and N.M. conceived the original concept. N.N. and M.S. conducted the experiments and characterized the electrospun fibrous materials. P.T. performed the drug release analysis. M.N. performed the microbiological assessments of the obtained materials. M.S., N.M., and I.R. wrote the manuscript. M.S., N.M., I.R., and M.N. revised the manuscript. All authors have read and agreed to the published version of the manuscript.

Funding: This research was funded by the Bulgarian National Science Fund, grant number KP-06-OPR03/2.

Institutional Review Board Statement: Not applicable.

Informed Consent Statement: Not applicable.

Data Availability Statement: The data presented in this study are available on request from the corresponding author.

Acknowledgments: The financial support from the Bulgarian National Science Fund (grant KP-06-OPR03/2) is gratefully acknowledged.

Conflicts of Interest: The authors declare no conflict of interest.

References

1. Hofstetter, V.; Buyck, B.; Croll, D.; Viret, O.; Couloux, A.; Gindro, K. What if esca disease of grapevine were not a fungal disease? *Fungal Divers.* **2012**, *54*, 51–67. [CrossRef]
2. Díaz, G.; Auger, J.; Besoain, X.; Bordeu Schwarze, E.; Latorre, B. Prevalence and pathogenicity of fungi associated with grapevine trunk diseases in Chilean vineyards. *Cien. Inv. Agric.* **2013**, *40*, 327–339. [CrossRef]
3. Fuente, M.; Fontaine, F.; Gramaje, D.; Armengol, J.; Smart, R.; Nagy, Z.; Borgo, M.; Rego, C.; Corio-Costet, M. *Grapevine Trunk Diseases. A Review*, 1st ed.; OIV Publications: Paris, France, 2016.
4. Gramaje, D.; Úrbez-Torres, J.; Sosnowski, M. Managing Grapevine Trunk Diseases With Respect to Etiology and Epidemiology: Current Strategies and Future Prospects. *Plant Dis.* **2018**, *102*, 12–39. [CrossRef] [PubMed]
5. Wang, S.; Yap, C.; He, J.; Chen, C.; Wong, S.; Li, X. Electrospinning: A facile technique for fabricating functional nanofibers for environmental applications. *Nanotechnol. Rev.* **2016**, *5*, 51–73. [CrossRef]
6. Paneva, D.; Spasova, M.; Stoyanova, N.; Manolova, N.; Rashkov, I. Electrospun fibers from polylactide-based stereocomplex: Why? *Int. J. Polym. Mater.* **2021**, *70*, 270–286. [CrossRef]
7. Stoyanova, N.; Spasova, M.; Manolova, N.; Rashkov, I.; Georgieva, A.; Toshkova, R. Antioxidant and antitumor activities of novel quercetin-loaded electrospun cellulose acetate/polyethylene glycol fibrous materials. *Antioxidants* **2020**, *9*, 232. [CrossRef]
8. Rahmati, M.; Mills, D.; Urbanska, A.; Saeb, M.; Venugopal, J.; Ramakrishna, S.; Mozafari, M. Electrospinning for tissue engineering applications. *Prog. Mater. Sci.* **2021**, *117*, 100721. [CrossRef]
9. Han, D.; Steckl, A. Coaxial electrospinning formation of complex polymer fibers and their applications. *ChemPlusChem* **2019**, *84*, 1453–1497. [CrossRef]
10. Wang, M.; Li, D.; Li, J.; Li, S.; Chen, Z.; Yu, D.; Liu, Z.; Guo, J. Electrospun Janus zein–PVP nanofibers provide a two-stage controlled release of poorly water-soluble drugs. *Mater. Des.* **2020**, *196*, 109075. [CrossRef]
11. Ghosal, K.; Augustine, R.; Zaszczynska, A.; Barman, M.; Jain, A.; Hasan, A.; Kalarikkal, N.; Sajkiewicz, P.; Thomas, S. Novel drug delivery systems based on triaxial electrospinning based nanofibers. *React. Funct. Polym.* **2021**, *163*, 10489. [CrossRef]
12. Kamath, S.; Sridhar, K.; Jaison, D.; Gopinath, V.; Ibrahim, B.; Gupta, N.; Sundaram, A.; Sivaperumal, P.; Padmapriya, S.; Patil, S. Fabrication of tri-layered electrospun polycaprolactone mats with improved sustained drug release profile. *Sci. Rep.* **2020**, *10*, 18179. [CrossRef] [PubMed]
13. Kowalczyk, T. Functional micro- and nanofibers obtained by nonwoven post-modification. *Polymers* **2020**, *12*, 1087. [CrossRef] [PubMed]
14. Sett, S.; Lee, M.; Weith, M.; Pourdeyhim, B.; Yarin, A. Biodegradable and biocompatible soy protein/polymer/adhesive sticky nano-textured interfacial membranes for prevention of esca fungi invasion, into pruning cuts and wounds of vines. *J. Mater. Chem. B.* **2015**, *3*, 2147–2162. [CrossRef]

15. Buchholz, V.; Molnar, M.; Wang, H.; Reich, S.; Agarwal, S.; Fischer, M.; Greiner, A. Protection of vine plants against Esca disease by breathable electrospun antifungal nonwovens. *Macromol. Biosci.* **2016**, *16*, 1391–1397. [CrossRef]
16. Spasova, M.; Stoilova, O.; Manolova, N.; Rashkov, I.; Naydenov, M. Electrospun Eco-Friendly Materials Based on Poly(3-hydroxybutyrate) (PHB) and TiO_2 with Antifungal Activity Prospective for Esca Treatment. *Polymers* **2020**, *12*, 1384. [CrossRef] [PubMed]
17. Chan-On, W.; Huyen, N.; Songtawee, N.; Suwanjang, W.; Prachayasittikul, S.; Prachayasittikul, V. Quinoline-based clioquinol and nitroxoline exhibit anticancer activity inducing FoxM1 inhibition in cholangiocarcinoma cells. *Drug Des. Devel. Ther.* **2015**, *9*, 2033. [PubMed]
18. Ignatova, M.; Stoyanova, N.; Manolova, N.; Rashkov, I.; Kukeva, R.; Stoyanova, R.; Toshkova, R.; Georgieva, A. Electrospun materials from polylactide and Schiff base derivative of Jeffamine ED® and 8-hydroxyquinoline-2-carboxaldehyde and its complex with Cu^{2+}: Preparation, antioxidant and antitumor activities. *Mater. Sci. Eng. C.* **2020**, *116*, 111185. [CrossRef]
19. Spasova, M.; Manolova, N.; Paneva, D.; Rashkov, I. Preparation of chitosan-containing nanofibres by electrospinning of chitosan/poly(ethylene oxide) blend solutions. *e-Polymers* **2004**, *4*, 56. [CrossRef]
20. Stoyanova, N.; Paneva, D.; Mincheva, R.; Toncheva, A.; Manolova, N.; Dubois, P.; Rashkov, I. Poly(L-lactide) and poly(butylene succinate) immiscible blends: From electrospinning to biologically active materials. *Mater. Sci. Eng. C* **2014**, *41*, 119–126. [CrossRef] [PubMed]
21. Guardia, C.; Stephens, D.; Dang, H.; Quijada, M.; Larionov, O.; Lleonart, R. Antiviral activity of novel quinoline derivatives against dengue virus serotype 2. *Molecules* **2018**, *23*, 672. [CrossRef] [PubMed]
22. Saadeh, H.; Sweidan, K.; Mubarak, M. Recent advances in the synthesis and biological activity of 8-hydroxyquinolines. *Molecules* **2020**, *25*, 4321. [CrossRef] [PubMed]
23. Pippi, B.; Lopes, W.; Reginatto, P.; Silva, F.; Joaquim, A.; Alves, R.; Silveira, G.; Vainstein, M.; Andrade, S.; Fuentefri, A. New insights into the mechanism of antifungal actionof 8-hydroxyquinolines. *Saudi Pharm. J.* **2019**, *27*, 41–48. [CrossRef] [PubMed]
24. Spasova, M.; Manolova, N.; Rashkov, I. Composition of Plant Protection Product. Office of Republic of. Bulgaria Patent No. 4353, 8 July 2019.
25. Spasova, M.; Manolova, N.; Rashkov, I.; Naydenov, M. Electrospun 5-chloro-8-hydroxyquinoline-loaded cellulose acetate/polyethylene glycol antifungal membranes against esca. *Polymers* **2019**, *11*, 1617. [CrossRef]
26. Singhvi, M.; Zinjarde, S.; Gokhale, D. Polylactic acid: Synthesis and biomedical applications. *J. Appl. Microbiol.* **2019**, *127*, 1612–1626. [CrossRef]
27. Ermakov, A.; Voronin, V.; Nelyubin, V.; Oblapenko, E.; Sorokin, A.; Epshtein, N.; Muravskaya, I. Study of certain structural features and properties of 8-hydroxyquinoline derivatives. *Pharm. Chem. J.* **1985**, *19*, 129–134. [CrossRef]
28. Spasova, M.; Mincheva, R.; Paneva, D.; Manolova, N.; Rashkov, I. Perspectives on: Criteria for complex evaluation of the morphology and alignment of electrospun polymer nanofibers. *J. Bioact. Compat. Polym.* **2006**, *21*, 465–479. [CrossRef]
29. Rasband, W.S.; ImageJ, U.S. National Institutes of Health, Bethesda, Maryland, USA. Available online: https://imagej.nih.gov/ij/ (accessed on 3 January 2016).
30. Haleem, R.; Abdullah, S.; Jubraeel, J. Morphological and molecular identification of *Phaeoacremonium aleophilum* associated with grapevines decline phenomenon in Duhok governorate. *J. Basrah Res. Sci.* **2011**, *37*, 1–8.
31. Armengol, J.; Vicent, A.; Torné, L.; García-Figueres, F.; García-Jiménez, J. Fungi Associated with Esca and Grapevine Declines in Spain: A Three-Year Survey. *J. Phytopathol. Mediterr.* **2001**, *40*, 325–329.
32. Spasova, M.; Stoilova, O.; Manolova, N.; Rashkov, I.; Altankov, G. Preparation of PLLA/PEG nanofibers by electrospinning and potential applications. *J. Bioact. Compat. Polym.* **2007**, *22*, 62–76. [CrossRef]
33. Spasova, M.; Paneva, D.; Manolova, N.; Radenkov, P.; Rashkov, I. Electrospun chitosan-coated fibers of poly(L-lactide) and poly(L-lactide)/poly(ethylene glycol): Preparation and characterization. *Macromol. Biosci.* **2008**, *8*, 153–162. [CrossRef]
34. Fan, T.; Daniels, R. Preparation and characterization of electrospun polylactic acid (PLA) fiber loaded with birch bark triterpene extract for wound dressing. *AAPS PharmSciTech* **2021**, *22*, 205. [CrossRef] [PubMed]
35. Sun, X.; Yang, S.; Xue, B.; Huo, K.; Li, X.; Tian, Y.; Liao, X.; Xie, L.; Qin, S.; Xu, K.; et al. Super-hydrophobic poly (lactic acid) by controlling the hierarchical structure and polymorphictransformation. *Chem. Eng. J.* **2020**, *397*, 125297. [CrossRef]
36. Leonés, A.; Peponi, L.; Lieblich, M.; Benavente, R.; Fiori, S. In-vitro degradation of plasticized PLA electrospunFiber mats: Morphological, thermal and crystalline evolution. *Polymers* **2020**, *12*, 2975. [CrossRef] [PubMed]
37. Patel, H.; Oza, K. Synthesis, characterization and antimicrobial activity of metal chelates of 5-[1(H)-benzotriazole methylene]-8-quinolinol. *E-J. Chem.* **2009**, *6*, 371–376. [CrossRef]
38. Mohac, L.; Keating, A.; Pina, M.; Raimi-Abraham, B. Engineering of nanofibrous amorphous and crystalline solid dispersions for oral drug delivery. *Pharmaceutics* **2019**, *11*, 7. [CrossRef] [PubMed]

polymers

Article

Improvement of the UV Barrier and Antibacterial Properties of Crosslinked Pectin/Zinc Oxide Bionanocomposite Films

Karina Dyasti Hari [1], Coralia V. Garcia [1], Gye-Hwa Shin [2] and Jun-Tae Kim [1,*]

[1] Department of Food Science and Technology, Keimyung University, Daegu 42601, Korea; karina.dyastihari@gmail.com (K.D.H.); cvalentinag@yandex.com (C.V.G.)
[2] Department of Food and Nutrition, Kunsan National University, Gunsan 54150, Korea; winnie19@kunsan.ac.kr
* Correspondence: jtkim92@kmu.ac.kr

Abstract: Pectin-based antibacterial bionanocomposite films were prepared by crosslinking with calcium chloride ($CaCl_2$) and mixing with zinc oxide nanoparticles (ZnO-NPs) at various concentrations (0.5%, 1%, and 1.5% *w/w*, based on pectin). Crosslinking with 1% $CaCl_2$ significantly ($p < 0.05$) improved the tensile strength of the pectin films, although their elongation at break was decreased. The UV-light barrier property of the pectin/ZnO bionanocomposite films was significantly ($p < 0.05$) improved with increasing ZnO-NP concentrations. In addition, the bionanocomposite films incorporating 1.5% ZnO-NPs showed excellent antibacterial effects against both *Escherichia coli* and *Staphylococcus aureus*, inhibiting over 99% of the bacteria. Therefore, the developed crosslinked pectin/ZnO bionanocomposite films show great potential as active packaging materials with excellent UV-blocking and antibacterial properties.

Keywords: antibacterial; bionanocomposite; pectin; zinc oxide; crosslinking

Citation: Hari, K.D.; Garcia, C.V.; Shin, G.-H.; Kim, J.-T. Improvement of the UV Barrier and Antibacterial Properties of Crosslinked Pectin/Zinc Oxide Bionanocomposite Films. *Polymers* **2021**, *13*, 2403. https://doi.org/10.3390/polym13152403

Academic Editor: Evgenia G. Korzhikova-Vlakh

Received: 2 July 2021
Accepted: 19 July 2021
Published: 22 July 2021

1. Introduction

In recent years, antimicrobial packaging has attracted a great deal of attention from the food industry because of the increase in the consumer demand for chemical preservative-free products [1]. Moreover, there is growing interest in biodegradable packaging as an eco-friendly replacement for petroleum-based polymers. As abundant, renewable, environmentally friendly, and sustainable materials, biopolymers have been used to fabricate biodegradable plastics and edible films [2,3]. Among biopolymers, polysaccharides are attractive materials because of their good film-forming properties, sustainability, and abundance [4,5].

Pectin is the major structural component of the plant cell wall and is one of the largest constituents in citrus by-products. This polysaccharide is generally recognized as safe (GRAS) by the United States Food and Drug Administration (FDA), and is used in food processing, mostly as a gelling, stabilizing, or thickening agent in products such as jams, yogurt, fruit drinks, and ice cream [6]. Pectin is mainly composed of galacturonic acid units. Depending on its degree of esterification (DE), pectin can be classified as high methoxyl pectin (DE > 50%) and low methoxyl pectin (DE < 50%), with both being able to be used for film formation [7]. Some of the advantages of pectin edible films are biocompatibility, biodegradability, and non-toxicity [8]. Nevertheless, neat pectin films do not show satisfactory functionality because of their high hydrophilicity, poor mechanical and barrier properties, and low water resistance [9,10]. Hence, various strategies have been tested for improving the properties of pectin-based films, including crosslinking, blending with other polymers, and incorporating nanoparticles and essential oil nanoemulsions [7,10]. Such films can also be engineered to exhibit functionalities. For instance, pectin films incorporating silica nanoparticles extended the shelf life of strawberries, and pectin/corn flour/beetroot powder films extended the shelf life of tomatoes [11,12]. Pectin films incorporating lime antioxidants hindered soybean oil oxidation [13]. Films made of a combination of pectin and hydrocolloids from chia seeds showed antioxidant properties, and electrospun pectin films were used in multilayer structures with aroma barrier properties [14,15]. Moreover, pectin films incorporating ZnO nanoparticles (ZnO-NPs) were reported to show enhanced strength and water barrier properties as well as antimicrobial effects [9].

Compared with neat pectin films, crosslinked pectin films exhibit increased tensile strength and water resistance [16]. The crosslinking mechanism of pectin with divalent metal ions such as Ca^{2+} and Zn^{2+} is explained by the "egg-box" model, in which the crosslinks are formed by divalent ions occupying electronegative cavities in the ribbon structure of carboxylic groups [17]. Hence, the functional and mechanical properties of pectin films could be improved by incorporating both calcium ions and ZnO-NPs.

ZnO-NPs exhibit advantages such as non-toxicity, availability, low cost, high ultraviolet absorption capacity, and strong antimicrobial activity [18,19]. Compared to other natural antibacterial materials such as nisin, essential oils, grapefruit seed extract (GFSE), etc., ZnO-NPs have high thermal stability and do not lose their antibacterial activity during the processing of most food packaging films such as polyethylene (PE), polypropylene (PP), polyethylene terephthalate (PET), etc. It is thought that the antimicrobial activity of ZnO-NPs is dependent on the release of Zn^{+2} ions and the formation of reactive oxygen species (ROS), which damage the integrity of the microbial cells [20]. ZnO is recognized as a GRAS substance by the FDA, although no differentiation is made between the bulk and nano form. Moreover, ZnO-NPs are approved for certain food contact applications in the European Union [21].

Therefore, here, we aimed to optimize the development of crosslinked pectin-based films incorporating ZnO-NPs. The effects of calcium chloride and ZnO-NP concentration on the mechanical, physical, optical, UV barrier, and antibacterial ability of the bionanocomposite films were evaluated. In particular, the goal of this study was to produce bionanocomposite films with enhanced UV-blocking and antibacterial properties.

2. Materials and Methods

2.1. Materials

Pectin powder from citrus was purchased from Daejung Chemicals & Metals Co., Ltd. (Siheung, Korea). Glycerol was obtained from Duksan Pure Chemicals Co., Ltd. (Ansan, Korea). Calcium chloride ($CaCl_2$) and ZnO nanopowder were purchased from Sigma Aldrich Chemicals (St. Louis, MO, USA). Nutrient broth (NB) was purchased from DB Difco^{TM} (Sparks, MD, USA). *Escherichia coli* ATCC 8739 and *Staphylococcus aureus* ATCC 6538 P were procured from the American Type Culture Collection (ATCC).

2.2. Film Preparation

2.2.1. Crosslinked Pectin Films

To prepare crosslinked pectin films (CPF), 4 g of pectin was dissolved in 170 mL of distilled water. As a plasticizer, glycerol was added at 20% (*w/w*, based on pectin) to prevent brittleness. Then, a $CaCl_2$ solution (30 mL; 0.5%, 1%, 2%, and 5% *w/v*) was added to the pectin solution and mixed using a magnetic stirrer at 70 °C for 20 min. The film-forming solution was cooled down at room temperature for 30 min before casting it onto a Teflon-coated glass plate (13.5 cm × 25 cm) and drying at 30 °C in a convection oven for 48 h. The fully dried pectin films were conditioned at 25 °C and 50% relative humidity (RH) for 48 h before testing their mechanical properties.

2.2.2. Crosslinked Pectin/ZnO Bionanocomposite Films

For the preparation of crosslinked pectin/ZnO bionanocomposite films (CPZBF), a ZnO-NP dispersion was prepared by adding ZnO nanopowder (0.5%, 1%, and 1.5%, *w/w*, based on pectin) to 170 mL of distilled water containing 20% (*w/w*) glycerol, followed by ultrasonication at 40% amplitude for 10 min. 4 g of pectin was dissolved in the homogeneous dispersion of ZnO-NPs and stirred at room temperature for 3 h. Then, 30 mL of a $CaCl_2$ solution (1% *w/w*, based on pectin) was added and mixed using a magnetic stirrer at 70 °C for 20 min. The film-forming solution was cooled down at room temperature for 30 min before casting it onto the Teflon-coated glass plate (13.5 cm × 25 cm) and dried as described in Section 2.2.1.

2.3. Characterization of Bionanocomposite Films

2.3.1. Optical Properties

The color of the CPZBF was determined using a Chroma Meter (CR-400, Minolta, Tokyo, Japan). A white color standard plate (L^* = 96.76, a^* = 0.05, b^* = 1.94) was used as a background. The total color difference (ΔE) was calculated using Equation (1)

$$\Delta E = \sqrt{(\Delta L)^2 + (\Delta a)^2 + (\Delta b)^2} \quad (1)$$

where ΔL, Δa, and Δb represent the differences in lightness (L), redness (a), and yellowness (b) values of the control (crosslinked pectin film) and CPZBF samples. At least five points were measured for the color parameters for each film, and the average values were used.

The UV absorbance of the CPZBF was measured using a UV–visible spectrophotometer (UV-2600, Shimadzu, Kyoto, Japan) at wavelengths of 200–550 nm.

2.3.2. Mechanical Properties

The bionanocomposite films were cut into 10 mm × 10 mm strips and conditioned at 25 °C and 50% RH for 48 h before their mechanical test. Film thickness was measured in five random locations in each sample using a micro-caliper (MDC-25MJ, Mitutoyo Co., Kanagawa, Japan), and the average was calculated. Tensile strength, elongation at break, and Young's modulus of the films were tested using a Universal Testing Machine (UTM; Zwick 2010TN, Zwick GmbH & Co. KG, Ulm, Germany) according to the standard method of ASTM D882-12 [22].

2.3.3. Moisture Content

To determine the water content, the crosslinked pectin films were cut into 2 cm × 2 cm squares and weighed (W_1). The water content of the films was determined by drying the film samples in a convection oven at 105 °C for 24 h, and weighing again (W_2) after cooling them to room temperature. The moisture content was calculated using Equation (2):

$$\text{Moisture content } (\%) = \frac{W_1 - W_2}{W_1} \times 100 \quad (2)$$

2.3.4. Water Solubility

Water solubility was determined using a previously published method with slight modifications [23]. Film samples (2 cm × 2 cm) were dried at 105 °C in a convection oven for 24 h and weighed (W_i). Then, the film samples were put into flasks (50 mL) containing 20 mL distilled water with constant stirring at 25 °C for 6 h. The liquid was filtered and the remaining film mass was dried at 105 °C for 24 h until it showed a constant weight (W_f). Film solubility was calculated using Equation (3)

$$\text{Film solubility } (\%) = \frac{W_i - W_f}{W_i} \times 100 \quad (3)$$

2.3.5. Water Vapor Permeability (WVP)

The water vapor permeability of the films was determined using a modified cup method [24]. Film samples were cut into 6.7 cm × 6.7 cm squares and covered the cups containing distilled water. Samples were stored in a chamber at 25 °C and 25% RH for

12 h. Changes in the weight of the film samples were recorded every 2 h. The water vapor permeability was calculated using Equation (4)

$$WVP = \frac{WVTR \times \delta}{P_1 - P_2} \times 100 \quad (4)$$

where WVTR is the water vapor transmission rate, δ is the film thickness, and p_1 and p_2 are the partial pressures of water vapor inside and outside the cup, respectively.

2.3.6. FTIR Analysis

The Fourier-transform infrared (FTIR) spectra of the films were recorded using attenuated total reflection Fourier transmission infrared (ATR-FTIR) equipment (Nicolet Is5, Thermo Fisher Scientific, Waltham, MA, USA). The film samples were cut into rectangular strips (4 cm × 4 cm) and were directly put onto the ATR cell. The spectra were recorded as 64 scans at a 4 cm^{-1} resolution ranging from 600 to 4000 cm^{-1} wavenumber.

2.3.7. Differential Scanning Calorimetry (DCS)

Thermal properties of the films were evaluated using a DSC 25 (TA Instruments, New Castle, DE, USA) as previously reported, with modifications [25]. Roughly 5–10 mg of each film sample was put in an aluminum pan and heated from 0 to 300 °C at a constant heating rate of 10 °C/min under a nitrogen flow of 50 mL/min. DSC thermograms were recorded.

2.3.8. Antibacterial Activity

The antibacterial activity of the bionanocomposite films was examined against *E. coli* (Gram-negative) and *S. aureus* (Gram-positive). The antimicrobial test was performed by the plate counting method with slight modifications [18]. *S. aureus* and *E. coli* were individually grown on NB at 37 °C. The sample films were cut into rectangular shapes (1 cm × 6 cm) and placed in test tubes. Diluted broth (10 mL, around 10^5 CFU/mL) was poured into a test tube containing a film sample, and incubated at 37 °C for 24 h. Diluted broth alone was used as the control. After incubation, the cell viability of each bacterium was calculated by counting colonies on the plates, and reporting the values as CFU/mL. The antibacterial rate was calculated using Equation (5):

$$\text{Antibacterial rate } (\%) = \frac{(Nc - Ns)}{Nc} \times 100 \quad (5)$$

where Nc is the number of viable bacteria on the control film (pure LDPE film) and Ns is the number of viable bacteria on the bionanocomposite films after 24 h of incubation. The antibacterial activity was calculated as the difference in logarithmic values of viable bacteria between the control film and bionanocomposite films as described in the JIS Z2801:2010 regulation [26].

2.4. Statistical Analysis

Experiments were duplicated at least three times, and data are expressed as mean ± standard deviation (SD). Statistical significance was determined by analysis of variance (ANOVA) and Duncan's multiple range test. The level of significance was set at $p < 0.05$ using SPSS version 16 (SPSS Inc., Chicago, IL, USA).

3. Results and Discussion

3.1. Characterization of Crosslinked Pectin Films

3.1.1. Mechanical Properties

Table 1 shows the effects of $CaCl_2$ concentration on the thickness, tensile strength (TS), Young's modulus (YM), elongation at break (*E*), moisture content, and water solubility of pure pectin (control) and crosslinked pectin films. Adding $CaCl_2$ increased the thickness of the films as compared to the control film, with the films crosslinked with 1% $CaCl_2$ having a thickness of 101.5 ± 11.9 μm and the control films of 85.1 ± 8.1 μm. The TS of pectin film crosslinked with 1% $CaCl_2$ (22.7 ± 1.4 MPa) was significantly ($p < 0.05$) higher than that of the control (14.3 ± 1.4 MPa), although further increases in $CaCl_2$ resulted in a slight but not significant ($p > 0.05$) decrease in TS. On the other hand, crosslinking caused a significant ($p < 0.05$) decrease in the *E* of the films, with the control and the film crosslinked with 1% $CaCl_2$ having *E* values of 8.3% and 6.1%, respectively. Nevertheless, *E* was not significantly ($p > 0.05$) changed with further increases in the $CaCl_2$ content.

Table 1. Effect of $CaCl_2$ concentration on the mechanical properties of the crosslinked pectin films (CPF).

Sample *	Thickness (μm)	Young's Modulus (MPa)	Tensile Strength (MPa)	Elongation at Break (%)	Moisture Content (%)	Water Solubility (%)
Control	85.1 ± 8.1 [a]	535.8 ± 54.8 [a]	14.3 ± 1.4 [a]	8.3 ± 1.1 [c]	19.6 ± 1.9 [b]	51.3 ± 3.4 [b]
CPF-0.5%	90.6 ± 10.3 [a]	721.8 ± 34.5 [b]	18.3 ± 1.4 [b]	7.4 ± 0.8 [b]	18.3 ± 0.7 [b]	47.8 ± 1.0 [b]
CPF-1.0%	101.5 ± 11.9 [b]	857.3 ± 56.5 [c]	22.7 ± 1.4 [c]	6.1 ± 1.0 [a]	14.2 ± 1.2 [a]	31.8 ± 2.8 [a]
CPF-2.0%	93.5 ± 17.9 [a]	833.2 ± 121.6 [c]	22.0 ± 2.5 [c]	5.7 ± 0.7 [a]	15.9 ± 0.7 [a]	33.4 ± 0.8 [a]
CPF-5.0%	88.1 ± 6.4 [a]	763.3 ± 95.5 [b]	21.9 ± 1.4 [c]	5.6 ± 1.0 [a]	16.0 ± 1.3 [a]	32.6 ± 2.6 [a]

* Concentrations represent % $CaCl_2$ solution (*w*/*v*). Data represent the mean ± standard deviation. Different letters in the same column indicate significant differences at $p < 0.05$ by Duncan's multiple range test.

The increase in TS by the addition of $CaCl_2$ is due to the crosslinking between the carboxyl acid (COO^-) of pectin and calcium ions (Ca^{2+}), described in the "egg-box" model [27,28]. The multivalent cations penetrate the film matrix and enhance the formation of bonds between the polymer chains, resulting in an increase in TS but also less flexibility and a more rigid polymer structure. Based on the results obtained, the film crosslinked with 1% $CaCl_2$ was used for further experiments.

3.1.2. Moisture Content and Water Solubility

Crosslinking with $CaCl_2$ decreased both the moisture content and water solubility of the pectin films (Table 1). The pure pectin film (control) had 19.6% ± 1.9% moisture, which significantly ($p < 0.05$) decreased to 14.2% ± 1.2% after the addition of 1% $CaCl_2$. Nevertheless, further increases in $CaCl_2$ did not significantly ($p > 0.05$) affect the moisture content. Water solubility showed the same trend, decreasing significantly ($p < 0.05$) from 51.3% ± 3.4% to 31.8% ± 2.8% after crosslinking with 1% $CaCl_2$. $CaCl_2$-crosslinked pectin resulted in decreased moisture content and water solubility, indicating that the Ca^{2+} ions promoted greater cohesion of the intermolecular bonds, leaving less available spaces for accommodating water molecules [29]. Crosslinking agents can thus help to overcome the poor moisture resistance of polysaccharide films by decreasing their hydrophilic properties through the formation of bonds between the polysaccharide molecules, reducing water absorption [30].

3.1.3. FTIR Analysis

The FTIR spectra of neat and crosslinked pectin films are shown in Figure 1. The main functional groups were identified as carbomethoxy (C=O, ~1740 cm^{-1}), methyl ($-CH_3$, ~2938 cm^{-1}), and carboxylate (COO– asymmetric and symmetric stretching vibration at ~1630 cm^{-1} and ~1437 cm^{-1}) [31]. The intensity of the C=O, $-CH_3$, and COO–became higher with calcium concentrations, indicating that Ca^{2+} interacted with the C=O group of pectin [32]. In addition, the increase in intensity may also be due to the electrostatic interactions between Ca^{2+} and COO– according to the "egg-box" model, and demonstrate that crosslinking occurred effectively [27,33].

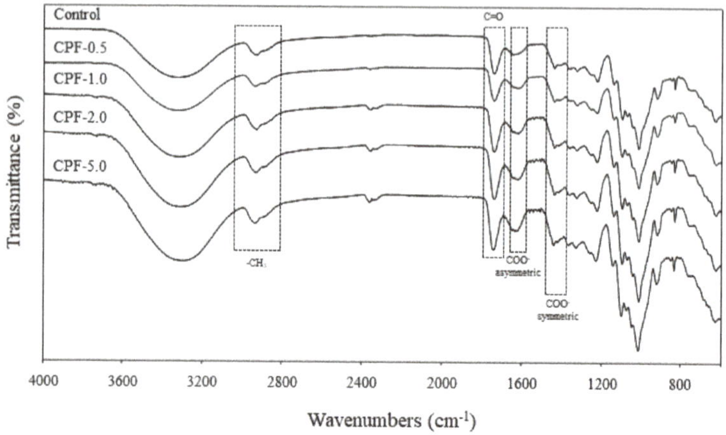

Figure 1. FTIR spectra of pure pectin film (control) and crosslinked pectin films (CPF) with 0.5, 1, 2, and 5% $CaCl_2$.

3.2. *Characterization of Crosslinked Pectin/ZnO Bionanocomposite Films*

3.2.1. Surface Color and Optical Properties

Figure 2 shows the photographs of the crosslinked pectin film (CPF) and the crosslinked pectin/ZnO bionanocomposite films (CPZBF) containing 0.5~1.5% ZnO-NPs. The appearance of the composite films changed to a more yellow color with increasing ZnO-NP concentrations. The chromaticity of the bionanocomposite films is shown in Table 2. The control film (crosslinked pectin films without ZnO-NPs) was more transparent than the bionanocomposite films combined with ZnO-NPs. As expected, the total color difference (ΔE) in the bionanocomposite films increased with the addition of ZnO-NPs. As the content of ZnO increased, the L and a value decreased and the b value increased significantly ($p < 0.05$), corresponding to a decrease in transparency and increase in yellowness. Control films exhibited L, a, and b values of 93.3 ± 0.6, -0.23 ± 0.04, and 6.31 ± 0.67, respectively. By contrast, the L, a, and b values of the bionanocomposite film with 1.5% ZnO-NPs were 91.4 ± 0.2, -0.34 ± 0.03, and 8.52 ± 0.22, respectively. These changes were reflected in the more intense dark green coloration of the bionanocomposite films as the ZnO-NP concentration increased. The increase in opacity resulting from the incorporation of ZnO-NPs has been reported by other researchers [9,34].

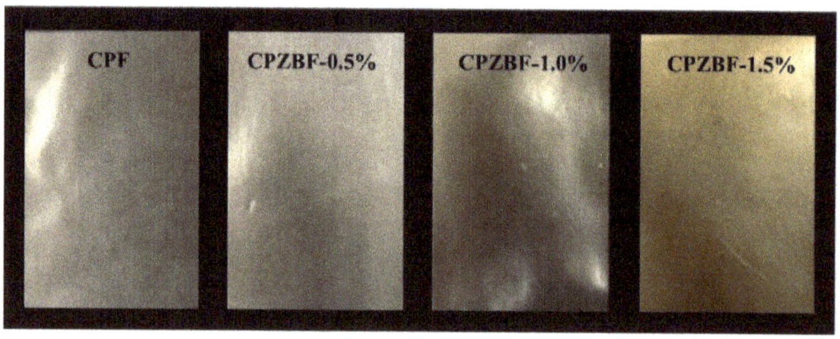

Figure 2. Photographs of the crosslinked pectin film (CPF) and the crosslinked pectin/ZnO bionanocomposite films (CPZBF) containing 0.5~1.5% ZnO-NPs.

Table 2. Effect of ZnO concentration on the surface color of the crosslinked pectin/ZnO bionanocomposite films (CPZBF).

Sample *	L	a	b	ΔE
CPF	93.3 ± 0.6 [c]	-0.23 ± 0.04 [c]	6.31 ± 0.67 [a]	-
CPZBF-0.5%	92.1 ± 0.6 [b]	-0.29 ± 0.03 [b]	8.07 ± 0.54 [b]	2.18 ± 0.74 [a]
CPZBF-1.0%	91.9 ± 0.5 [b]	-0.36 ± 0.03 [a]	8.16 ± 0.54 [bc]	2.35 ± 0.65 [a]
CPZBF-1.5%	91.4 ± 0.2 [a]	-0.34 ± 0.03 [a]	8.52 ± 0.22 [c]	2.96 ± 0.25 [b]

* Concentrations represent % ZnO (*w/w*). Data represent the mean ± standard deviation. Different letters in the same column indicate significant differences at $p < 0.05$ by Duncan's multiple range test.

The UV barrier properties of the bionanocomposite films were evaluated by measuring their absorbance in the wavelength range of 200–550 nm (Figure 3). The absorbance of the bionanocomposite films increased after the incorporation of ZnO-NPs. Furthermore, two absorbance peaks were visible, including a sharp peak in the 280–300 nm region, which is characteristic of pectin [35], and a broad peak at approximately 360 nm, corresponding to ZnO-NPs [12]. The increase in UV absorbance as a result of ZnO incorporation agrees with previous reports [18,34,36,37]. The results obtained indicate that incorporating ZnO-NPs into pectin improves the UV barrier properties of the resulting bionanocomposite films. Therefore, the developed bionanocomposite films could be used as UV-screening food packaging materials.

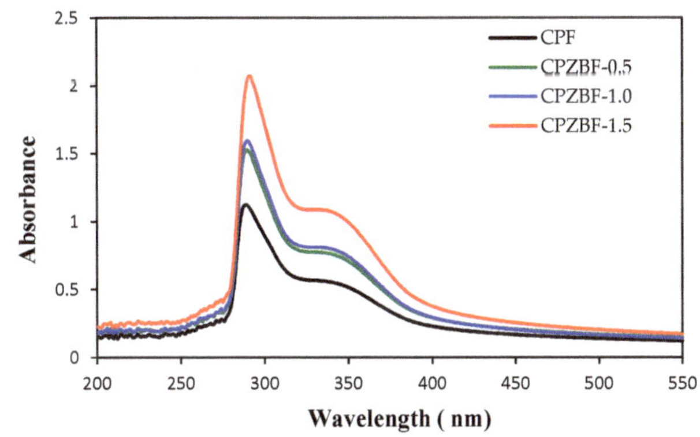

Figure 3. UV–visible spectra of crosslinked pectin film (CPF) and crosslinked pectin/ZnO bionanocomposite films (CPZBF) with 0.5, 1, and 1.5% ZnO nanoparticles.

3.2.2. Mechanical Properties

Table 3 shows the effects of the ZnO-NP concentration on the thickness, YM, TS, and *E* of the pectin/ZnO bionanocomposite films. Increasing the ZnO-NP concentration did not have significant ($p > 0.05$) effects on the bionanocomposite film thickness. However, the TS was significantly ($p < 0.05$) increased from 22.7 ± 1.4 MPa for the control to 24.6 ± 1.8 MPa for the bionanocomposite film with 1.5% ZnO-NPs. The Young's modulus also increased from 857.3 ± 56.5 MPa for the control to 899.4 ± 65.8 MPa for the nanocomposite with 1% ZnO-NPs. On the other hand, *E* tended to decrease as the ZnO-NP content increased, from 6.1% ± 1.0% for the control to 4.9% ± 0.5% for the sample with 1.5% ZnO-NPs. The bionanocomposite films developed here were stronger than pea starch/ZnO nanocomposites (TS of 10.8 MPa) [37], although weaker than pectin/alginate/ZnO nanocomposites (TS of 41.6 MPa) [34] and pectin/ZnO nanocomposites incorporating 5% ZnO-NPs (TS of 45.4 MPa) [9].

Table 3. Effect of ZnO concentration on the mechanical and water vapor barrier properties of the crosslinked pectin/ZnO bionanocomposite films (CPZBF).

Sample *	Thickness (μm)	Young's Modulus (MPa)	Tensile Strength (MPa)	Elongation at Break (%)	WVP (mm·g/m² kPa·h)
CPF	101.5 ± 11.9 ^{a}	857.3 ± 56.5 ^{ab}	22.7 ± 1.4 ^{a}	6.05 ± 1.01 ^{b}	3.71 ± 0.07 ^{a}
CPZBF-0.5%	97.9 ± 13.9 ^{a}	838.8 ± 70.4 ^{a}	24.1 ± 2.2 ^{ab}	5.92 ± 0.89 ^{b}	3.45 ± 0.09 ^{a}
CPZBF-1.0%	95.4 ± 15.6 ^{a}	884.0 ± 48.4 ^{b}	24.4 ± 2.1 ^{b}	4.94 ± 0.65 ^{a}	3.56 ± 0.08 ^{ab}
CPZBF-1.5%	106.0 ± 13.2 ^{a}	899.4 ± 65.8 ^{b}	24.6 ± 1.8 ^{b}	4.87 ± 0.49 ^{a}	3.63 ± 0.14 ^{ab}

* Concentrations represent % ZnO (*w/w*). Data represent the mean ± standard deviation. Different letters in the same column indicate significant differences at $p < 0.05$ by Duncan's multiple range test.

The mechanical properties of bionanocomposite films can be influenced by several factors, including the dispersion of the nanofillers and nanofiller–matrix interactions [37]. Some studies agree with the observed increase in TS as a result of the incorporation of ZnO-NPs [33,36]; however, the opposite effect was reported for nanocomposites based on agar, carrageenan, carboxymethyl cellulose, and gelatin [18,36].

The increase in TS observed here suggests that Zn^{2+} binds to the hydroxyl (OH^-) and carboxylate (COO^-) groups of pectin chains in a similar way as that described in the "egg-box" model, although Ca^{2+} is reported to interact only with carboxylate groups [38]. The effectiveness of the reinforcement material and its dispersion in the polymer matrix are crucial for effective stress transfer at the matrix–nanofiller interface, and ultimately increase the TS of polymeric biocomposite materials, although the trade-off is a decrease in elasticity [39,40]. The TS of the bionanocomposite films developed is comparable to that of polypropylene and ethylene-polypropylene (26 MPa), although their E is substantially lower than that of most other polymers used in packaging except polystyrene (1.6%) [41]. Thus, further improvements in the mechanical properties of the bionanocomposite films are necessary.

3.2.3. Water Vapor Permeability (WVP)

The WVP of the bionanocomposite films is shown in Table 3. No significant ($p > 0.05$) changes in the WVP of the films were observed, suggesting that the incorporation of ZnO-NPs did not affect the water affinity of the pectin-based bionanocomposite films. This result may be due to the high hydrophilicity of pectin, the low concentration of ZnO-NPs incorporated into the films, and the high concentration of glycerol used to prepare them, as it has been reported that glycerol may increase the spacing between polymer chains, enabling the passage of water vapor [29]. The WVP obtained here is higher than that of synthetic plastics as well as pure hydroxypropyl methylcellulose films [24], which are known for their hydrophilicity, and thus indicates that decreasing the WVP of the pectin/ZnO bionanocomposite films remains a challenge.

3.2.4. DSC Analysis

The DSC thermograms of the crosslinked pectin film and pectin/ZnO bionanocomposite films are shown in Figure 4. Only small changes were observed in the thermograms, suggesting that ZnO incorporation only slightly improved the thermal resistance of the films. This may be explained by the low concentration of ZnO incorporated [42]. The thermograms of the crosslinked pectin film (CPF) showed two endotherms, one at around 110 °C and the other at around 180 °C. These temperatures increased to 115 and 190 °C, respectively, for the bionanocomposite film with 1.5% ZnO (CPZBF-1.5). The first endotherm represents the glass transition temperature (T_g) of the film and is associated with water loss, whereas the second one represents the melting temperature (T_m) of the film [43,44]. In addition, an exothermic peak was observed at around 245 °C for the bionanocomposite film with 1.5% ZnO (240 °C for the crosslinked pectin film). This peak represents the thermal depolymerization of pectin [45] and, as the results show, the addition of ZnO hindered the degradation of the polymer, albeit slightly.

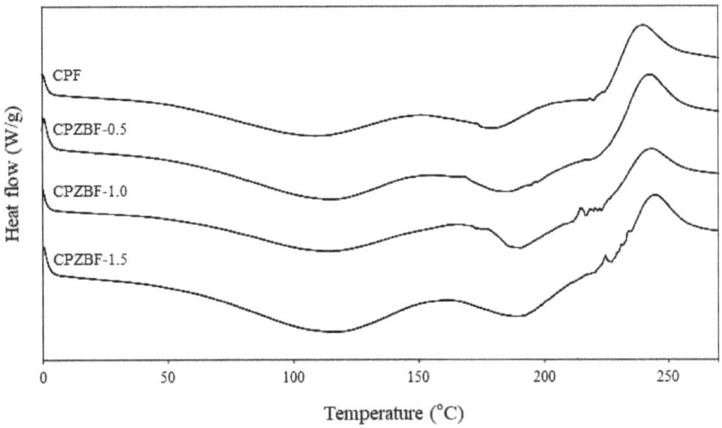

Figure 4. DSC thermograms of crosslinked pectin film (CPF) and crosslinked pectin/ZnO bionanocomposite films (CPZBF).

3.2.5. Antibacterial Activity

The antibacterial activity of the bionanocomposite films against *E. coli* and *S. aureus* is shown in Table 4 and Figure 5. The bionanocomposite films showed antibacterial activity against both *E. coli* and *S. aureus*, although their effectiveness was slightly greater against the latter. Other studies have also reported the stronger antimicrobial effects of ZnO-NPs against Gram-positive bacteria [18,36]. Moreover, antifungal effects of ZnO-NPs were also reported [9]. The antibacterial activity of the bionanocomposite films increased significantly ($p < 0.05$) with an increase in the concentration of ZnO-NPs, with the bionanocomposite film containing 1.5% ZnO showing the greatest antibacterial rate (>99%) against both strains. At this concentration, the antibacterial activity was 2.04 ± 0.04 and 2.11 ± 0.05 for *E. coli* and *S. aureus*, respectively, indicating that the films were effective according to the Z2801:2010 regulation, which requires a minimum antibacterial activity of 2.0 [26].

Table 4. Antibacterial rate and activity of the crosslinked pectin/ZnO bionanocomposite films (CPZBF) against *E. coli* and *S. aureus*.

Sample *	*E. coli*		*S. aureus*	
	Antibacterial Rate (%)	Antibacterial Activity	Antibacterial Rate (%)	Antibacterial Activity
CPF	1.9 ± 0.2 [a]	0.01 ± 0.01 [a]	0.9 ± 0.5 [a]	0.00 ± 0.01 [a]
CPZBF-0.5	94.1 ± 0.4 [b]	1.23 ± 0.03 [b]	96.1 ± 0.3 [b]	1.41 ± 0.03 [b]
CPZBF-1.0	98.2 ± 0.2 [c]	1.74 ± 0.05 [c]	99.0 ± 0.1 [c]	1.98 ± 0.05 [c]
CPZBF-1.5	99.1 ± 0.1 [c]	2.04 ± 0.04 [d]	99.2 ± 0.1 [c]	2.11 ± 0.05 [d]

* Concentrations represent % ZnO (*w/w*). Data represent the mean ± standard deviation. Different letters in the same column indicate significant differences at $p < 0.05$ by Duncan's multiple range test.

The antibacterial properties of ZnO-NPs have been reported by other researchers [18,36,42,46]. The antibacterial activity of ZnO-NPs appears to be due to the release of Zn^{2+} ions, which penetrate the cell wall of bacteria, killing them. Because Gram-positive bacteria have a thick cell wall to which ZnO-NPs can bind, they are more susceptible to these antibacterial effects. ZnO-NPs can also generate reactive oxygen species (ROS) such as hydrogen peroxide (H_2O_2), which damage the cell membrane of bacteria [47]. It is assumed that both the production of ROS and the deposition of Zn^{2+} ions within the cytoplasm lead to either the inhibition or killing of bacterial cells [20,48]. Thus, the films developed in this study could be applied as wrapping to hinder bacteria and extend the shelf life of fresh produce, as reported in the literature for similar films [7]. Considering the lower mechanical and water barrier properties of pectin films compared to polymers commonly used in packaging, another strategy to be considered would be to incorporate these films into multilayer structures, aiming to provide specific functions, namely antimicrobial effects and UV-barrier properties, to the packaging [15,20].

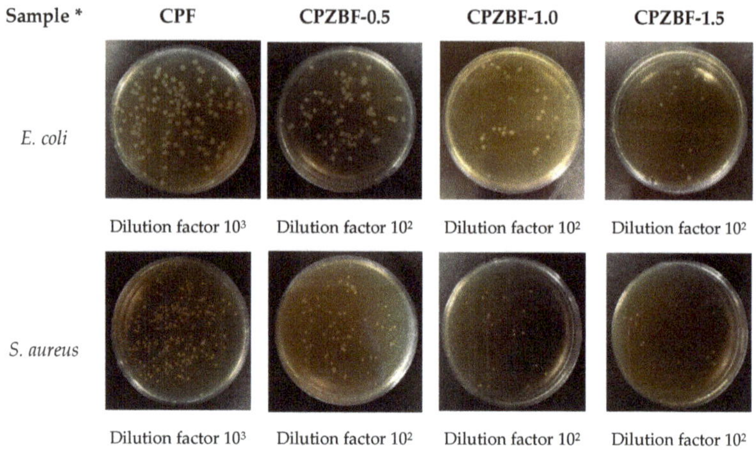

Figure 5. Colony forming units (CFU) of *E. coli* and *S. aureus* after exposure to crosslinked pectin film (CPF) and crosslinked pectin/ZnO bionanocomposite films (CPZBF). * Concentrations represent % ZnO (*w/w*).

4. Conclusions

Crosslinked pectin films showed greater tensile strength than did pure pectin films, although their elongation at break was lower. FTIR analysis revealed the interaction between pectin and $CaCl_2$ and confirmed the formation of crosslinking. Adding ZnO-NPs to the crosslinked pectin films significantly improved their UV barrier and mechanical properties, resulting in stronger bionanocomposite films able to block UV irradiation, although the trade-off of this outcome was darkening and loss of flexibility. In addition, the crosslinked pectin/ZnO bionanocomposite films exhibited strong antibacterial activity against food-borne pathogenic bacteria such as *E. coli* and *S. aureus*. Therefore, the developed bionanocomposite films show potential to be used as active food packaging with antibacterial and UV-light barrier properties.

Author Contributions: K.D.H.: conceptualization, methodology, software, validation, formal analysis, investigation, writing—original draft; C.V.G.: writing—original draft, writing—review and editing; G.-H.S.: writing—review and editing; J.-T.K.: conceptualization and supervision, writing—review and editing. All authors have read and agree to the published version of the manuscript.

Funding: This research received no external funding.

Data Availability Statement: All the data will be available to the readers.

Acknowledgments: This research was supported by the Keimyung University Research Grant of 2020.

Conflicts of Interest: The authors declare no conflict of interest.

References

1. Jo, Y.; Garcia, C.V.; Ko, S.; Lee, W.; Shin, G.H.; Choi, J.C.; Park, S.J.; Kim, J.T. Characterization and antibacterial properties of nanosilver-applied polyethylene and polypropylene composite films for food packaging applications. *Food Biosci.* **2018**, *23*, 83–90. [CrossRef]
2. Neira, L.M.; Agustinelli, S.P.; Ruseckaite, R.A.; Martucci, J.F. Shelf life extension of refrigerated breaded hake medallions packed into active edible fish gelatin films. *Packag. Technol. Sci.* **2019**, *32*, 1–10. [CrossRef]
3. Duncan, T.V. Application of nanotechnology in food packaging and food safety: Barrier materials, antimicrobials and sensors. *J. Colloid Interface Sci.* **2011**, *363*, 1–24. [CrossRef] [PubMed]
4. Rhim, J.W.; Park, H.M.; Ha, C.S. Bio-nanocomposites for food packaging applications. *Prog. Polym. Sci.* **2013**, *38*, 1629–1652. [CrossRef]
5. Kim, J.T.; Netravali, A.N. Physical properties of biodegradable films of soy protein concentrate/gelling agent blends. *Macromol. Mater. Eng.* **2012**, *297*, 176–183. [CrossRef]
6. Laurent, M.A.; Boulenguer, P. Stabilization mechanism of acid dairy drinks (ADD) induced by pectin. *Food Hydrocoll.* **2003**, *17*, 445–454. [CrossRef]
7. Mellinas, C.; Ramos, M.; Jiménez, A.; Garrigós, M.C. Recent trends in the use of pectin from agro-waste residues as a natural-based biopolymer for food packaging applications. *Materials* **2020**, *13*, 673. [CrossRef]
8. Aldana, D.S.; Andrade-Ochoa, S.; Aguilar, C.N.; Contreras-Esquivel, J.C.; Nevárez-Moorillón, G.V. Antibacterial activity of pectic-based edible films incorporated with Mexican lime essential oil. *Food Control* **2015**, *50*, 907–912. [CrossRef]
9. Suyatma, N.E.; Ishikawa, Y.; Kitazawa, H. Nanoreinforcement of pectin film to enhance its functional packaging properties by incorporating ZnO nanoparticles. *Adv. Mat. Res.* **2014**, *845*, 451–456. [CrossRef]
10. Chaichi, M.; Hashemi, M.; Badii, F.; Mohammadi, A. Preparation and characterization of a novel bionanocomposite edible film based on pectin and crystalline nanocellulose. *Carbohydr. Polym.* **2017**, *157*, 167–175. [CrossRef] [PubMed]
11. Al-Asmar, A.; Giosafatto, C.V.L.; Sabbah, M.; Sanchez, A.; Santana, R.V.; Mariniello, L. Effect of mesoporous silica nanoparticles on the physicochemical properties of pectin packaging material for strawberry wrapping. *Nanomaterials* **2020**, *10*, 52. [CrossRef] [PubMed]
12. Sucheta; Chaturvedi, K.; Sharma, N.; Yadav, S.K. Composite edible coatings from commercial pectin, corn flour and beetroot powder minimize post-harvest decay, reduces ripening and improves sensory liking of tomatoes. *Int. J. Biol. Macromol.* **2019**, *133*, 284–293. [CrossRef]
13. Rodsamran, P.; Sothornvit, R. Lime peel pectin integrated with coconut water and lime peel extract as a new bioactive film sachet to retard soybean oil oxidation. *Food Hydrocoll.* **2019**, *97*, 105173. [CrossRef]
14. da Silva, I.S.V.; de Sousa, R.M.F.; de Oliveira, A.; de Oliveira, W.J.; Motta, L.A.C.; Pasquini, D.; Otaguro, H. Polymeric blends of hydrocolloid from chia seeds/apple pectin with potential antioxidant for food packaging applications. *Carbohydr. Polym.* **2018**, *202*, 203–210. [CrossRef] [PubMed]
15. Balik, B.A.; Argin, S.; Lagaron, J.M.; Torres-Giner, S. Preparation and characterization of electrospun pectin-based films and their application in sustainable aroma barrier multilayer packaging. *Appl. Sci.* **2019**, *9*, 5136. [CrossRef]
16. Azeredo, H.M.C.; Waldron, K.W. Crosslinking in polysaccharide and protein films and coatings for food contact-A review. *Trends Food Sci. Technol.* **2016**, *52*, 109–122. [CrossRef]
17. Fraeye, I.; Duvetter, T.; Doungla, E.; Loey, A.V.; Hendrickx, M. Fine-tuning the properties of pectin-calcium gels by control of pectin fine structure, gel composition and environmental conditions. *Trends Food Sci. Technol.* **2010**, *21*, 219–228. [CrossRef]
18. Shankar, S.; Teng, X.; Li, G.; Rhim, J.W. Preparation, characterization, and antimicrobial activity of gelatin/ZnO nanocomposite films. *Food Hydrocoll.* **2015**, *45*, 264–271. [CrossRef]
19. Marcous, A.; Rasouli, S.; Ardestani, F. Low-density polyethylene films loaded by titanium dioxide and zinc oxide nanoparticles as a new active packaging system against *Escherichia coli* O157:H7 in fresh calf minced meat. *Packag. Technol. Sci.* **2017**, *30*, 693–701. [CrossRef]
20. Espitia, P.J.P.; Soares, N.dF.F.; Coimbra, J.S.dR.; de Andrade, N.J.; Cruz, R.S.; Medeiros, E.A.A. Zinc oxide nanoparticles: Synthesis, antimicrobial activity and food packaging applications. *Food Bioprocess Technol.* **2012**, *5*, 1447–1464. [CrossRef]
21. Garcia, C.V.; Shin, G.H.; Kim, J.T. Metal oxide-based nanocomposites in food packaging: Applications, migration, and regulations. *Trends Food Sci. Technol.* **2018**, *82*, 21–31. [CrossRef]
22. American Society for Testing and Material. *Standard Test Method for Tensile Properties of Thin Plastic Sheeting*; American Society for Testing and Material: West Conshohocken, PA, USA, 2012.
23. Hosseini, M.H.; Razavi, S.H.; Mousavi, M.A. Antimicrobial, physical and mechanical properties of chitosan-based films incorporated with thyme, clove and cinnamon essential oils. *J. Food Process Preserv.* **2009**, *33*, 727–743. [CrossRef]
24. Lee, J.Y.; Garcia, C.V.; Shin, G.H.; Kim, J.T. Antibacterial and antioxidant properties of hydroxypropyl methylcellulose-based active composite films incorporating oregano essential oil nanoemulsions. *LWT* **2019**, *106*, 164–171. [CrossRef]
25. Nisar, T.; Wang, Z.C.; Yang, X.; Tian, Y.; Iqbal, M.; Guo, Y. Characterization of citrus pectin films integrated with clove bud essential oil: Physical, thermal, barrier, antioxidant and antibacterial properties. *Int. J. Biol. Macromol.* **2018**, *106*, 670–680. [CrossRef]
26. Japanese Industrial Standard. *Antimicrobial Products: Test for Antimicrobial Activity and Efficacy*; Japanese Industrial Standard: Tokyo, Japan, 2010.

27. Braccini, I.; Pérez, S. Molecular basis of Ca^{2+}- induced gelation in alginates and pectins: The egg box model revisited. *Biomacromolecules* **2001**, *2*, 1089–1096. [CrossRef]
28. Pavlath, A.E.; Voisin, A.; Robertson, G.H. Pectin-based biodegradable water insoluble films. *Macromol. Symp.* **1999**, *140*, 107–113. [CrossRef]
29. Seixas, F.L.; Turbiani, F.R.B.; Salomão, P.G.; Souza, R.P.; Gimenes, M.L. Biofilms composed of alginate and pectin: Effect of concentration of crosslinker and plasticizer agents. *Chem. Eng. Trans.* **2013**, *32*, 1693–1698.
30. Garavand, F.; Rouhi, M.; Razavi, S.H.; Cacciotti, I.; Mohammadi, R. Improving the integrity of natural biopolymer films used in food packaging by crosslinking approach: A review. *Int. J. Biol. Macromol.* **2017**, *104*, 687–707. [CrossRef]
31. Guo, X.; Duan, H.; Wang, C.; Huang, X. Characteristics of two calcium pectinates prepared from citrus pectin using either calcium chloride or calcium hydroxide. *J. Agric. Food Chem.* **2014**, *62*, 6354–6361. [CrossRef]
32. Nurjaya, S.; Wong, T.W. Effects of microwave on drug release properties of matrices of pectin. *Carbohydr. Polym.* **2005**, *62*, 245–257. [CrossRef]
33. Thibault, J.F.; Rinaudo, M. Chain association of pectic molecules during calcium-induced gelation. *Biopolymers* **1986**, *25*, 455–468. [CrossRef]
34. Ngo, T.M.P.; Dang, T.M.Q.; Tran, T.X.; Rachtanapun, P. Effects of zinc oxide nanoparticles on the properties of pectin/alginate edible films. *Int. J. Polym. Sci.* **2018**, *2018*, 5645797. [CrossRef]
35. Shankar, S.; Tanomrod, N.; Rawdkuen, S.; Rhim, J.W. Preparation of pectin/silver nanoparticles composite films with UV-light barrier and properties. *Int. J. Biol. Macromol.* **2016**, *92*, 842–849. [CrossRef] [PubMed]
36. Kanmani, P.; Rhim, J.W. Properties and characterization of bionanocomposite films prepared with various biopolymers and ZnO nanoparticles. *Carbohydr. Polym.* **2014**, *106*, 190–199. [CrossRef] [PubMed]
37. Ma, X.; Chang, P.R.; Yang, J.; Yu, J. Preparation and properties of glycerol plasticized-pea starch/zinc oxide-starch bionanocomposites. *Carbohydr. Polym.* **2009**, *75*, 472–478. [CrossRef]
38. Assifaoui, A.; Lerbret, A.; Uyen, H.T.D.; Neiers, F.; Chambin, O.; Loupiac, C.; Cousin, F. Structural behaviour differences in low methoxy pectin solutions in the presence of divalent cations (Ca^{2+} and Zn^{2+}): A process driven by the binding mechanism of the cation with the galacturonate unit. *Soft Matter* **2015**, *11*, 551–560. [CrossRef] [PubMed]
39. Díez-Pascual, A.M.; Díez-Vicente, A.L. ZnO-reinforced poly(3-hydroxybutyrate-co-3-hydroxyvalerate) bionanocomposites with antimicrobial function for food packaging. *ACS Appl. Mater. Interfaces* **2014**, *6*, 9822–9834. [CrossRef] [PubMed]
40. Ahmed, J.; Arfat, Y.A.; Castro-Aguirre, E.; Auras, R. Mechanical, structural and thermal properties of Ag-Cu and ZnO reinforced polylactide nanocomposite films. *Int. J. Biol. Macromol.* **2016**, *86*, 885–892. [CrossRef]
41. Crompton, T.R. *Physical Testing of Plastics*; Smithers Rapra Technology: Ravenna, OH, USA, 2012; pp. 1–134.
42. Shankar, S.; Wang, L.F.; Rhim, J.W. Incorporation of zinc oxide nanoparticles improved the mechanical, water vapor barrier, UV-light barrier, and antibacterial properties of PLA-based nanocomposite films. *Mat. Sci. Eng. C* **2018**, *93*, 289–298. [CrossRef]
43. Mishra, R.K.; Sutar, P.B.; Singhal, J.P.; Banthia, A.K. Graft copolymerization of pectin with polyacrylamide. *Polym. Plast. Technol. Eng.* **2007**, *46*, 1079–1085. [CrossRef]
44. Einhorn-Stoll, U.; Kunzek, H.; Dongowski, G. Thermal analysis of chemically and mechanically modified pectins. *Food Hydrocoll.* **2007**, *21*, 1101–1112. [CrossRef]
45. Shi, L.; Gunasekaran, S. Preparation of pectin-ZnO nanocomposite. *Nanoscale Res. Lett.* **2008**, *3*, 491–495. [CrossRef]
46. Li, J.H.; Hong, R.Y.; Li, M.Y.; Li, H.Z.; Zheng, Y.; Ding, J. Effects of ZnO nanoparticles on the mechanical and antibacterial properties of polyurethane coatings. *Prog. Org. Coat.* **2009**, *64*, 504–509. [CrossRef]
47. Tayel, A.A.; El-Tras, W.F.; Moussa, S.; El-Baz, A.F.; Mahrous, H.; Salem, M.F.; Brimer, L. Antibacterial action of zinc oxide nanoparticles against foodborne pathogens. *J. Food Saf.* **2011**, *31*, 211–218. [CrossRef]
48. Zhang, L.; Jiang, Y.; Ding, Y.; Daskalakis, N.; Jeuken, L.; Povey, M.; O'Neill, A.J.; York, D.W. Mechanistic investigation into antibacterial behavior of suspensions of ZnO nanoparticles against *E. coli*. *J. Nanopart. Res.* **2010**, *12*, 1625–1636. [CrossRef]

polymers

MDPI

Article

Biodegradation of Polylactic Acid-Based Bio Composites Reinforced with Chitosan and Essential Oils as Anti-Microbial Material for Food Packaging

Teuku Rihayat [1,*], Agung Efriyo Hadi [2], Nurhanifa Aidy [3], Aida Safitri [4], Januar Parlaungan Siregar [5], Tezara Cionita [6], Agustinus Purna Irawan [7], Mohammad Hazim Mohamad Hamdan [8] and Deni Fajar Fitriyana [9]

1 Department of Chemical Engineering, Politeknik Negeri Lhokseumawe, Lhokseumawe 24301, Indonesia
2 Mechanical Engineering Department, Faculty of Engineering, Universitas Malahayati, Bandar Lampung 35153, Indonesia; efriyo@malahayati.ac.id
3 Department of Renewable Energy Engineering, Universitas Malikussaleh, Muara Batu 24355, Indonesia; nurhanifaaidy@gmail.com
4 Department of Chemical Engineering, Faculty of Engineering, Universitas Sumatera Utara, Kota Medan 20222, Indonesia; aidasafitri853@gmail.com
5 College of Engineering, Universiti Malaysia Pahang, Gambang 26300, Malaysia; januar@ump.edu.my
6 Department of Mechanical Engineering, Faculty of Engineering and Quantity Surveying, INTI International University, Seremban 71800, Malaysia; tezara.cionita@newinti.edu.my
7 Faculty of Engineering, Universitas Tarumanagara, Jakarta Barat 11440, Indonesia; agustinus@untar.ac.id
8 Faculty of Engineering and Computing, First City University College, Petaling Jaya 47600, Malaysia; hazim.hamdan@firstcity.edu.my
9 Department of Mechanical Engineering, Universitas Negeri Semarang, Semarang 50229, Indonesia; deniifa89@mail.unnes.ac.id
* Correspondence: teukurihayat@yahoo.com

Citation: Rihayat, T.; Hadi, A.E.; Aidy, N.; Safitri, A.; Siregar, J.P.; Cionita, T.; Irawan, A.P.; Hamdan, M.H.M.; Fitriyana, D.F. Biodegradation of Polylactic Acid-Based Bio Composites Reinforced with Chitosan and Essential Oils as Anti-Microbial Material for Food Packaging. *Polymers* **2021**, *13*, 4019. https://doi.org/10.3390/polym13224019

Academic Editor: Evgenia G. Korzhikova-Vlakh

Received: 14 October 2021
Accepted: 9 November 2021
Published: 20 November 2021

Abstract: This study aims to produce and investigate the potential of biodegradable Polylactic Acid (PLA)-based composites mixed with chitosan and Turmeric Essential Oil (TEO) as an anti-microbial biomaterial. PLA has good barrier properties for moisture, so it is suitable for use as a raw material for making packaging and is included in the GRAS (Generally Recognized As Safe). Chitosan is a non-toxic and antibacterial cationic polysaccharide that needs to be improved in its ability to fight microbes. TEO must be added to increase antibacterial properties due to a large number of hydroxyl (-OH) and carbonyl functional groups. The samples were prepared in three different variations: 2 g of chitosan, 0 mL TEO and 0 mL glycerol (Biofilm 1), 3 g of chitosan, 0.3 mL TEO and 0.5 mL of glycerol (Biofilm 2), and 4 g of chitosan, 0.3 of TEO and 0.5 mL of glycerol (Biofilm 3). The final product was characterized by its functional group through Fourier transform infrared (FTIR); the functional groups contained by the addition of TEO are C-H, C=O, O-H, and N-H with the extraction method, and as indicated by the emergence of a wide band at 3503 cm^{-1}, turmeric essential oil interacts with the polymer matrix by creating intermolecular hydrogen bonds between their terminal hydroxyl group and the carbonyl groups of the ester moieties of both PLA and Chitosan. Thermogravimetric analysis (TGA) of PLA as biofilms, the maximum temperature of a biofilm was observed at 315.74 °C in the variation of 4 g chitosan, 0.3 mL TEO, and 0.5 mL glycerol (Biofilm 3). Morphological conditions analyzed under scanning electron microscopy (SEM) showed that the addition of TEO inside the chitosan interlayer bound chitosan molecules to produce solid particles. Chitosan and TEO showed increased anti-bacterial activity in the anti-microbial test. Furthermore, after 12 days of exposure to open areas, the biofilms generated were able to resist *S. aureus* and *E. coli* bacteria.

Keywords: polylactic acid; turmeric essential oil; chitosan; TGA; FT-IR; antimicrobial and antioxidant properties; *S. aureus*; *E. coli*

1. Introduction

Plastic manufacturers are facing a new challenge in the food industry as a result of the growing demand for high-quality food that is free of preservatives. This is because there is a greater emphasis on developing conservation products and anti-microbials for renewable and environmentally friendly products. Furthermore, manufacturers' use of natural packaging materials (biofilms) as an alternative extends the life of a product [1,2]. Food packaging protects food products from external factors such as microorganisms, moisture, and ultraviolet (UV) light, thereby extending their shelf life. Recently, biopolymer-based food packaging has attracted a lot of attention in terms of environmental issues due to the fact that plastics constitute a considerable portion of buried garbage in the natural ecosystem and are regarded as emerging contaminants with significant environmental effects due to their high concentration, extensive dispersion, and non-biodegradability. This is considerable evidence of aquatic plastic pollution, including plastic islands and microplastics [3]. Conventional plastics are a vital commodity owing to their particular lightweight, thermostable, crystalline, and easy-to-mold properties, resulting in a wide range of items that combine comfort and quality in our lives. Increased usage of plastic materials in the home and industrial sectors has outpaced worldwide production by up to 400 Mt/year, raising severe issues about disposal, environmental pollution, toxicity to the ecosystem, and human health [4].

The selection of anti-microbial agents used as packaging materials must adhere to established regulations, particularly in terms of toxicological effects [5]. Chitosan, a biodegradable polymer, has been widely used in the production of edible films because it is non-toxic and environmentally friendly. Chitosan also has excellent film-forming ability, massive antimicrobial activity, selective permeability to gases (CO_2 and O_2), and it is compatible with other substances, such as vitamins and minerals [6]. Furthermore, the use of antimicrobial agents in biodegradable food packaging systems to prevent microbial growth on food surfaces has been the focus of recent research. Several previous researchers [7,8] have investigated the use of chitosan as an additional material in packaging manufacturing. The use of chitosan in food packaging represents a renewal based on a biodegradable food packaging concept in which the spread of packaging can reduce microorganisms in foods and prevent their development [9]. Chitosan-derived natural food packaging (biofilm) can improve melons' quality and extend their shelf life. It can also inhibit the growth of microbes, bacteria, and fungi.

Chitosan is an antimicrobial agent that contains polysaccharides and amino groups. To enhance chitosan's antimicrobial activity, several other chemicals were added, including a biopolymer, polysaccharides, lipids, or a mixture of these, as well as fatty acids and essential oils [10,11]. According to previous research, the inclusion of turmeric (essential oils) in film materials may protect L. inocula from UV light contamination. This is due to the fact that UV light can be rendered inactive. When compared to the three-hour exposure period for pure chitosan, adding turmeric extract to chitosan can significantly boost anti-microbial action and reduce the number of *Staphylococcus aureus* and *Salmonella bacteria* [12]. Formalized paraphrase Curcumin is also one of the naturally occurring hydrophobic polyphenols and is an ayurvedic treatment in the Indian traditional medicine system as well as in many Asian countries. Curcumin can be found in turmeric and it is well known for its antimicrobial, antioxidant, anticancer, and anti-inflammatory properties [13].

Renewable aliphatic polyesters of homopolymers and copolymer-type polyhydroxylic acid composed of PLA, Poly Glycolic Acid (PGA), and Poly e-Caprolactone (PCL) are the most promising materials for biofilm applications. PLA has received special attention as a replacement for traditional petroleum-based plastics. Based on a study [14], biodegradable films with antioxidant activity for active food packaging were developed using PLA film with a weight of 5% Poly-Caprolaktone (PCL) employing thymol, carvacrol fillers, and their blend with supercritical CO_2 fluids at 400 °C and 10 MPa for five hours. The PLA film is 5% by weight. The findings showed that due to the plasticizing impact of the compound induced by the supercritical CO_2 fluid, the resulting film's traction strength dropped. This supercritical CO_2 fluid forms more porous microstructures and modifies its mechanical characteristics due to structural discontinuities which result in reduced flexibility and cracking resistance [15,16].

The antioxidant activity of the samples submerged in this coating type was greater. The improvement in physical-mechanical properties with the addition of filler must meet the following requirements: the filler must be miscible with PLA in order to form a homogeneous mixture; the filler must not be too volatile because it will cause evaporation when the temperature is raised (elevated temperature) during the process and the filler must not be easily migratory [17]. Natural extracts and other compounds, as previously stated, can be added to edible films, and the films can gain active properties, extending the shelf life of food through the migration of bioactive compounds or intelligent properties. This also enables the detection of food contamination by changing the color of the packaging [18].

Until now, there has been a lot of research into using PLA in food packaging as a biodegradable polymer to improve the mechanical properties of the film. The interaction of PLA and chitosan as antimicrobial agents against various target organisms such as algae, bacteria, yeast, and fungi in experiments involving in vivo and in vitro interactions with chitosan in various forms (solutions, films, and composites). TEO, because of their non-toxicity, biocompatibility, and high antibacterial and antioxidant capabilities, natural essential oils are also ideal candidates to replace frequently used food preservatives. The use of essential oils in active packaging applications is a cost-effective solution that can help to decrease food safety hazards. Furthermore, the use of natural essential oils in active packaging materials can help to preserve packed food from oxidation and protect against contamination and spoilage by microbes that cause food to perish and harmful free radicals. Antimicrobials have an effect on the properties of blocking materials from being exposed to bacteria in food packaging, thereby increasing shelf life and product quality. Therefore, the novelty of this study is in determining and examining the effect of the addition of turmeric essential oil mixed with chitosan on making food packagings using their "compatible" characteristics as a vitamin and a mineral. However, on the basis of literature investigations, no study has yet been carried out to change the physical and mechanical characteristics of environmentally friendly polymers in PLA and chitosan–TEO "green fillers." The objective of the study was to investigate whether concentrations of turmeric extract were effective against *E. coli* and *Staphylococcus aureus*. *Staphylococcus aureus* is the most common cause of food poisoning globally. These bacteria can infect some foods, such as minimally processed ready-to-eat vegetables and processed meat products, and create a variety of enterotoxins. From the standpoint of public health, *Escherichia coli* as an intestinal pathogen is becoming increasingly significant, particularly the psychrotropic strain of *E. coli O157:H7* that can thrive on minimally processed vegetables and processed meat products.

2. Materials and Methods

2.1. Materials

Glycerol, curcumin, and gelatin were supplied by Sigma-Aldrich (Jakarta Timur, Indonesia), while turmeric, shrimp skin, 1% acetic acid, and commercial cassava flour were supplied by PT, Fugha Pratama Mandiri (Lhokseumawe, Indonesia). Phosphate-buffered saline (PBS), tryptic soybean broth (TSB), and tryptic soy agar (TSA) were purchased from Fisher Scientific (Jakarta, Indonesia). Medium molecular weight chitosan (molecular weight 190–310 kDa, 75–85% deacetylation), grafted polyethylene maleic anhydride (PEgMA) with 3.5% maleic anhydride content, and glycerol were supplied from Sigma Aldrich (Jakarta Timur, Indonesia). Furthermore, the crystalline PLA was supplied from NatureWorks Co. (Tangerang, Indonesia), while NaOH reagent was supplied from Sigma Aldrich (Jakarta Timur, Indonesia) without dilution.

2.2. *Methods*

2.2.1. Chitosan Synthesis

An amount of 100 g of shrimp skin was cleaned with running water. Next, the shrimp skin was dried in an oven for 2 h at 160 °C, which was then smoothed. The demineralization process of shrimp skin powder was carried out using HCl concentrations of 0.25 M–2 M (ratio of 1:10 (b/v)) by heating and stirring for 1–2 h. After that, it was filtered and dried for 24 h. The deproteinization process uses 0.5 M–2 M NaOH solution with an immersion time of 10–400 min at temperatures of 20–100 °C. After the two procedures, the shrimp shell was filtered, washed with distilled water, and dried again to produce chitin powder. Then, the decolorization process was carried out using an acetone 1:10 (b/v) ratio and then immersed for 10 min. Chitin was dried in an oven for 2 h at 28 °C. The chitin powder obtained was sequentially bleached with 0.315% NaOCl for 5 min. Next, in the last deacetylation process, chitin powder was washed using 50% NaOH with a ratio of 1:20 (b/v). Then, it was heated for 3–5 h at 80–100 °C and washed with distilled water and 80% alcohol. Finally, chitin powder was then filtered. The chitin powder produced was analyzed using FTIR to determine the chitosan functional group [19].

2.2.2. Extraction of Turmeric

A total of 20 g of turmeric was mashed, which was then placed into Soxhlet. The process of isolating turmeric oil was carried out by adding 200 mL ethanol for 8 h until no condensate drops again, and the process temperature was maintained at 78 °C. Thus, turmeric insulation is maintained with the distillation process until the oil was obtained.

2.2.3. Biofilms Manufacture

The biofilm manufacturing process in [20] was a success.Three samples were formed from the composition of 2 g chitosan, 0 mL TEO, and 0 mL glycerol (Biofilm 1); 3 g chitosan, 0.3 mL TEO, and 0.5 mL glycerol (Biofilm 2); and 4 g chitosan, 0.3 mL TEO and 0.5 mL glycerol (Biofilm 3). The chitosan powder obtained was then dissolved in 30 mL of 1% acetic acid and 70 mL of distilled water, which was then mixed with 20 g PLA and glycerol according to the composition. The solution was homogenized and stirred at 70 °C for 60 min until the film solution was entirely homogeneous. In the final step, the film solution was added to TEO. Then, the film solution was homogenized for 30 min with a magnetic stirrer. The film solution was poured into a mold cleaned with 96% ethanol. It was then dried in the oven at 35 °C for 45 min. The remaining dried film was removed from the mold, which was ready for analysis.

2.2.4. Surface Morphological Analysis

Scanning electron microscopy is a tool that can form shadows on the surface of broken microscopic specimens. The surface structure of the fibers was observed using a JEOL-T20 microscope at State Polytechnic of Lhokseumawe laboratory (Lhokseumawe, Indonesia). To form a conductive sample, it is necessary to coat the sample with a thin layer of gold. Scanning electron analysis was carried out at 5–20 kV. Tensile specimens tested were placed on a glass preparation part which was then placed on an optical lamp enlarged up to 1000 times. It was then photographed on each surface of the broken specimen or its fracture. The analysis of the treatment effect on the surface structure of the fiber was carried out using a microscope.

2.2.5. Fourier Transform Infrared (FTIR)

The sample's infrared spectroscopy was obtained on a KBr pallet (measurement method) using a Shimadzu FTIR Spectrophotometer from Chemical Engineering's Laboratory at State Polytechnic of Lhokseumawe (Lhokseumawe, Indonesia). The spectrum was seen in the range of 500–4000 cm^{-1} with a resolution of 2 cm^{-1} with an empty KBr melting background.

2.2.6. Thermogravimetric Analysis (TGA)

Thermogravimetric analysis (TGA) was used to determine the decomposition temperature of the sample, which is c using a Perkin–Elmer instrument from Chemical Engineering's Laboratory at State Polytechnic of Lhokseumawe (Lhokseumawe, Indonesia) at a frequency of 1 Hz from −98 °C to 140 °C and a heating rate of 5 °C/min. Samples were then cut into small sizes. Finally, the position of the maximum tan value was used to determine the decomposition temperature of the sample.

2.2.7. Anti-Microbial Test

A method was developed for determination of the antibiotic susceptibility of anaerobic bacteria by use of a single-disc diffusion technique and incorporation of the inoculum in pour plates from Chemical Engineering's Laboratory at State Polytechnic of Lhokseumawe (Lhokseumawe, Indonesia). The method was standardized by the correlation of zone diameters with minimal inhibitory concentrations determined in the samples [21]. The antibacterial activity of biofilms was investigated using the agar media method in Petri dishes. Bacterial cultures grown in the mid-logarithmic phase were placed on agar media. *Escherichia coli* and *Staphylococcus aureus* were injected into agar media. After solidification of the agar coating, perfume solutions (15 mm in diameter) with different concentrations (5% and 10% by weight) were placed on the surface of the agar. The layers were incubated at 25 °C for 0 days to 9 days.

2.2.8. Tensile Strength

The tensile strength of PLA–Chitosan–TEO was tested using a tensile strength tool according to the ASTM D 638–99 procedure from Chemical Engineering's Laboratory at State Polytechnic of Lhokseumawe (Lhokseumawe, Indonesia).

2.2.9. Statistical Analysis Using Statistical Statistical Package for Social Science (SPSS)

To find out the results of the analysis of whether the characteristics of the biofilms have a significant effect or not, further analysis was carried out using SPSS (Statistical Package for Social Science) version 22.0 using the one-way ANOVA method. If the data are normally distributed and homogeneous, a one-way ANOVA test was carried out with a 95% confidence level.

All data regarding the mechanical test of the sample are listed in Tables 1–4. Table 1 presents tensile strength data for all samples using ANOVA method, Table 2 presents tensile strength data for all samples using ANOVA method for difference samples), Table 3 presents data % elongation for all samples using ANOVA method and Table 4 presents the data for % elongation for all samples using ANOVA method for difference samples.

Table 1. Tensile strength for all samples using ANOVA method.

	Sum of Square	df	Mean Square	F	Sig.
Between Groups	879.069	4	219.76725	76.156	0
Within Groups	23.953	7	3.421857143	–	–
Total	903.022	11	–	–	–

Table 2. Tensile strength for all samples using ANOVA method for difference samples.

Chitosan Concentration (gr)	Glycerol Concentration (mL)	TEO Concentration (mL)	Mean Difference	Std. Error	Sig.	95% Confidence Interval	
						Lower Bound	Upper Bound
2	0	0	−20.37551	1.40253	0	−23.9106	−17.2414
	0.3	0.5	−21.17437	1.40253	0	−24.7604	−17.1339
3	0	0	−16.75093	1.40253	0	−19.9763	−13.9132
	0.3	0.5	−20.83454	1.40253	0	−17.4488	−23.0916
4	0	0	−20.03342	1.40253	0.54	4.0089	−2.7487
	0.3	0.5	−19.29869	1.40253	0.033	−6.7871	−7.2353

Table 3. % elongation for all samples using ANOVA method.

	Sum of Square	df	Mean Square	F	Sig.
Between Groups	0.029	4	0.01	51.736	0
Within Groups	0.001	7	0	–	–
Total	0.03	11	–	–	–

Table 4. % elongation for all samples using ANOVA method for difference samples.

Chitosan Concentration (gr)	Glycerol Concentration (mL)	TEO Concentration (mL)	Mean Difference	Std. Error	Sig.	95% Confidence Interval	
						Lower Bound	Upper Bound
2	0	0	−0.09319	0.01308	0	−0.2197	−0.0721
	0.3	0.5	−0.11412	0.01308	0	−0.1164	0.0958
3	0	0	−0.09319	0.01308	0	−0.5295	−0.0923
	0.3	0.5	−0.15497	0.01308	0.08	−0.3756	−0.1211
4	0	0	−0.09998	0.01308	0.07	−0.5297	−0.0496
	0.3	0.5	−0.15497	0.01308	0.08	−0.4292	−0.0217

Scheme 1 depicts the procedures required to produce chitin powder from shrimp shell waste using a variety of ways of methods.

Scheme 2 describes the steps that need to be taken to produce chitosan, the process of converting the acetyl group ($NHCOCH_3$) in chitin to an amine group (NH_2) in chitosan with the addition of NaOH.

Scheme 3 is a soxhletation method carried out by installing a soxhletation device tools. A number of raw materials are put into the cladding and 96% ethanol is added as a solvent.

Scheme 4 shows the manufacture of the biofilms is done by applying the principle of solution thermodynamics where the initial state of the solution is stable and then undergoes instability (addition of filler) in the phase change process (demixing), solidification (solidification) and phase transition so that at the stage it becomes solid after the addition of a high concentration polymer.

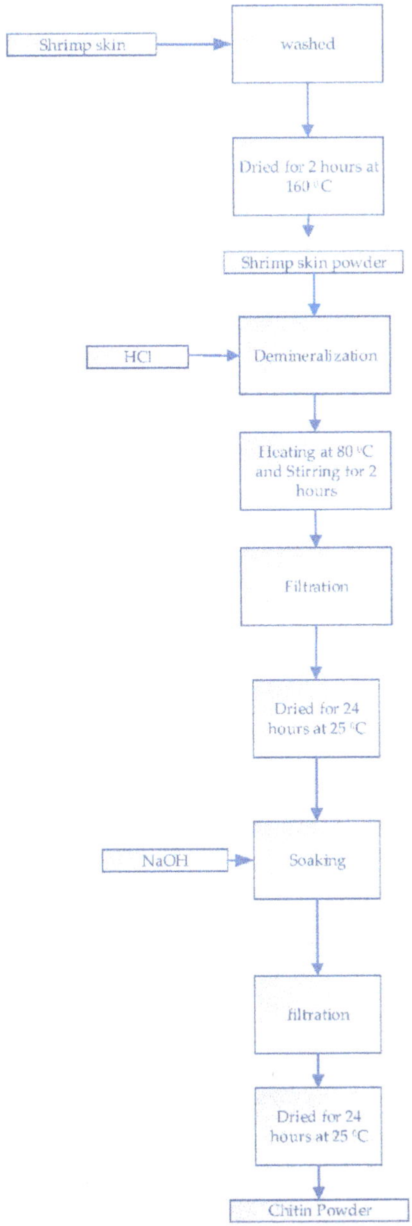

Scheme 1. Chitin production scheme [22].

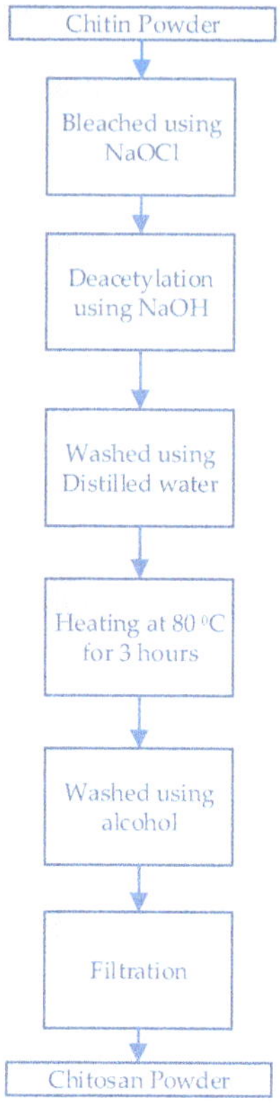

Scheme 2. Chitosan production [23].

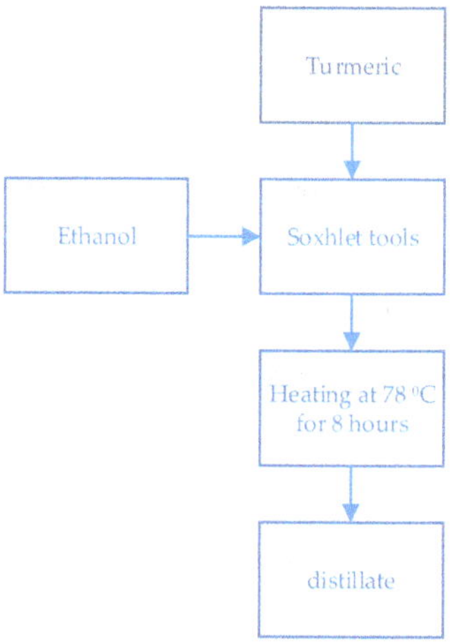

Scheme 3. Turmeric essential oil extraction [24].

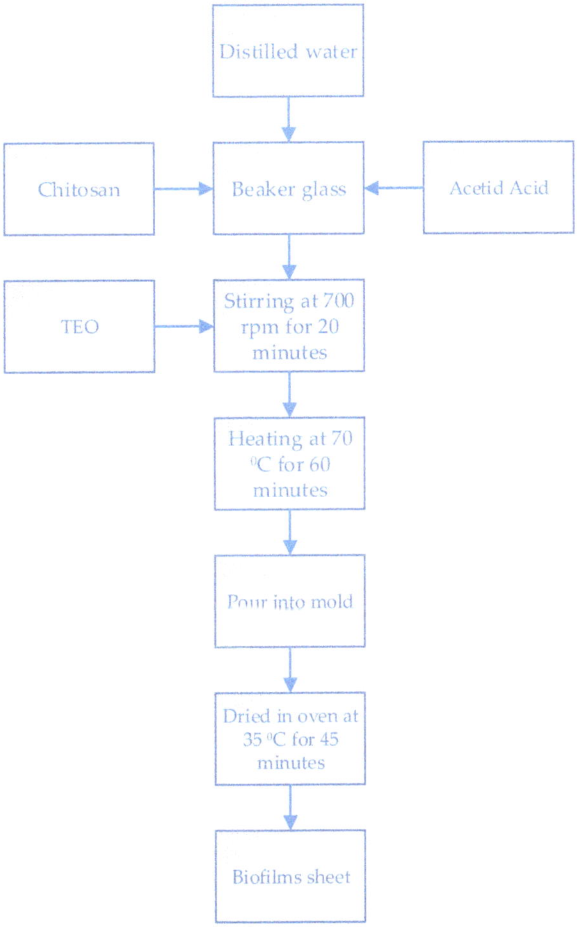

Scheme 4. Biofilm manufacture [25].

3. Results and Discussion

3.1. Scanning Electron Microscopic Analysis (SEM)

The scanning process of electron microscopy is based on the principle that an electron pistol produces an anode electron beam. In the direction of the sample, the magnet lens focuses on electrons. The whole sample is scanned by focused electron shafts through the scanner coil. The sample emits new electrons that are then received and sent back to the monitor where the particle is sampled.

Based on Figure 1, there were no chitosan particles in the analyses of Biofilm 1. The morphology of biofilms on the surface had space as no addition of glycerol as a binder had been found. Chitosan spreads evenly with Biofilm 2 particles and has a tight structure by adding TEO and glycerol. The difference between images (a) and (b) is shown in the image taken at 50× magnification. From the picture, it can be seen that the surface of the sample is hollow and cracked. The (a) sample surface is quite dominant with white spots and cracks, and in Figure 1b, the white spots are more clearly visible than in the previous image. These white spots are chitosan that has not been spread evenly and the mixing has not been uniform. A larger hollow in biofilm 3's surface appears. This proves that the mixture of materials in this sample has a fairly good level of homogeneity. In the morphology of this second sample, there are also just small numbers of white spots, which are chitosan and

TEO that have been evenly distributed [26,27]. Biofilm 3's particles are evenly distributed on the surface but have a smaller film layer because the glycerol is lower than in biofilm 2. According to [28], insufficient compositions in chitosan could spread evenly on the PLA polymer matrix with plasticizers and oregano oil. They were able to form homogeneous bonds that affected the properties. According to the findings of this study, TEO has been shown to bind biopolymer elements in chitosan into good bonds and produce complex biofilm surfaces when added to biofilms. The interaction of the two biopolymer elements demonstrates an attraction capable of replacing inorganic filmmakers [29,30]. The pore structure is a site where cell growth or proliferation occurs. Among the three sample outcomes, the third sample (biofilm 3) with a total composition variation of 4 g chitosan, 0.3 mL TEO, and 0.5 mL glycerol is the best sample based on morphology with the greatest structure and pores with greater porosity.

3.2. Fourier Infrared Spectroscopy Analysis (FTIR)

The results of the analysis of the characteristics of turmeric essential oil from the extraction method by spectroscopy to obtain the value of the absorption results are used in Table 5. Qualitative analysis of different organic compounds can be confirmed from the characteristics of the vibration band appearing in the infrared spectrum region at a certain frequency influenced by certain functional groups. The transmittance percentage corresponding to the wave number is summed up in the total attenuated in the reflected IR spectrum, as shown in the figure above. There is an intense broad peak in the range of 2500–3200 cm^{-1}, specifically at 2414.88 cm^{-1}, corresponding to the polymer hydroxyl (OH) group. Another intense and branching peak in the range of 2256.71–2086.96 cm^{-1} corresponds to methyl OH stretching and C-H bending, i.e., most of the aromatic compound and alcohol components. Other studies have shown that there is a functional group of turmeric essential oil that will be produced will depend on the species of curcuma used [31].

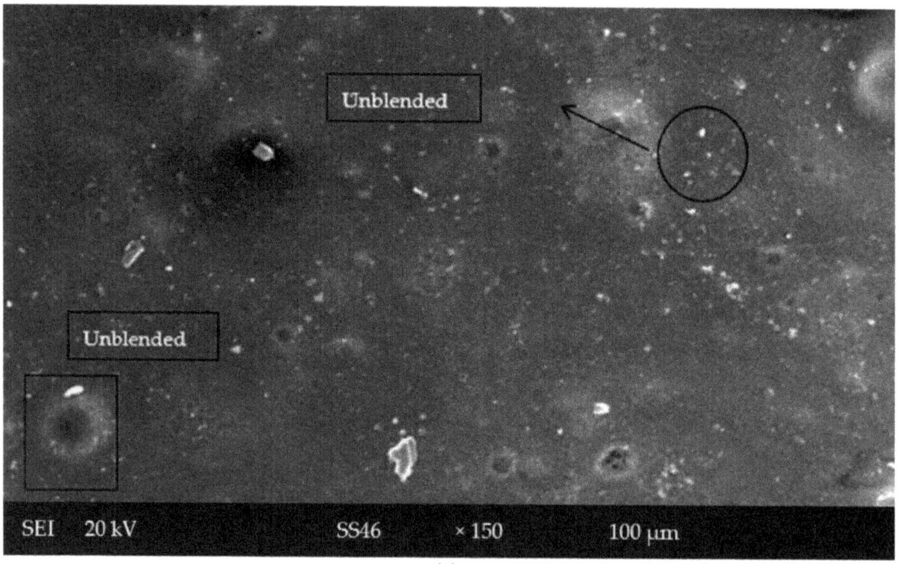

(a)

Figure 1. *Cont.*

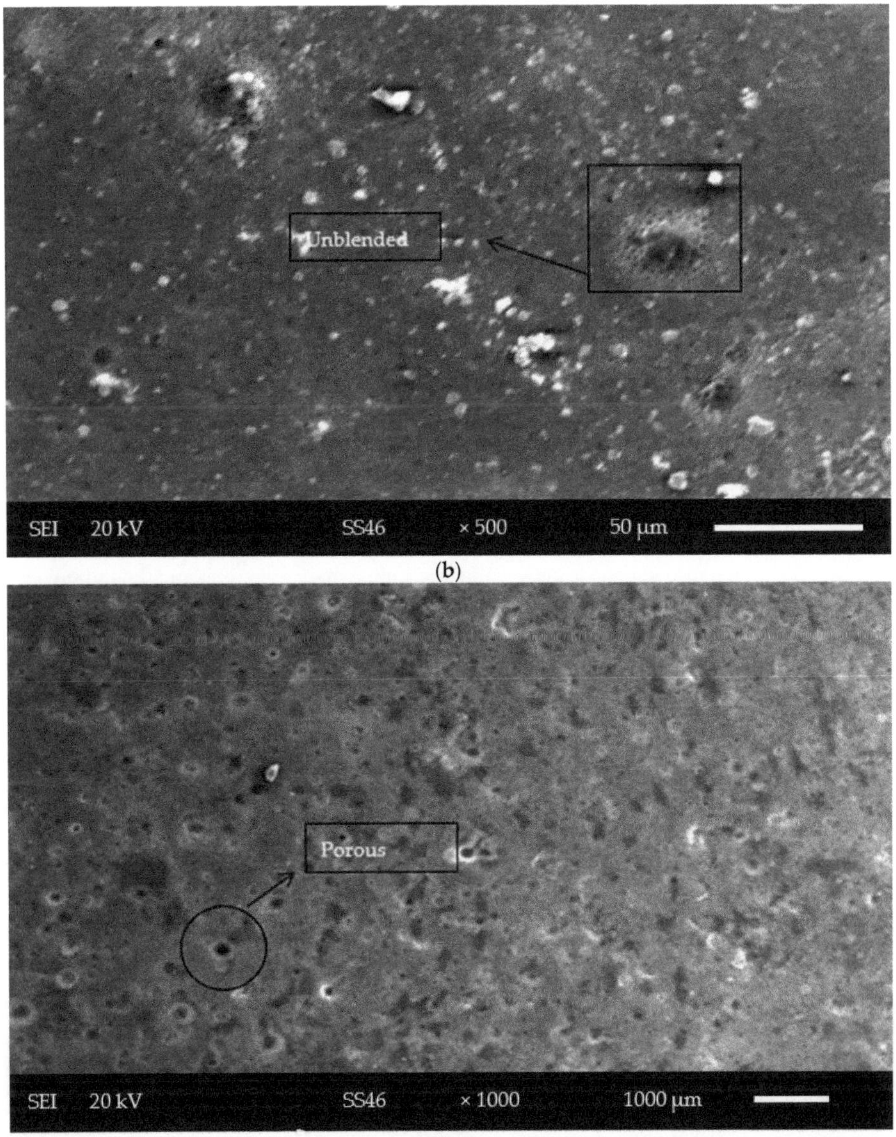

(**b**)

(**c**)

Figure 1. Analysis result on the PLA variations of (**a**) biofilm 1, (**b**) 3 biofilm 2, and (**c**) biofilm 3 to observe the blendable properties between matrix (PLA) and filler (chitosan–TEO).

The figure below shows a graph of the analysis of the PLA functional group. The graph shows the results of the FTIR test that from the optimum sample tensile strength in PLA, there exists N-H, C-H, O-H and C=O mentioned in Table 6. Where the NH group is present at the wave number 3336.85 cm^{-1} with a wavelength range of 3300–3400 cm^{-1}, the OH group is found at the number 3630.03 cm^{-1} in the wave range of 3584–3700 cm^{-1} and the CH group at the number 3630.03 cm^{-1} in the wave range 3584–3700 cm^{-1}. The C=O group is discovered at 1793.80 cm^{-1} with a wave range of 1540–1870 cm^{-1}, while the following group is located at 3294.42 cm^{-1} with a wave range of 3267–3333 cm^{-1}. This indicates that no new functional groupings, but only PLA pure characteristics, have been discovered in line with research by [32].

Table 5. Wave number for functional groups of turmeric essential oil.

Functional Groups	Wavelength cm^{-1}
N-H	3209.85
C-H (aromatic)	2912.51
N=C=O (TDI)	2276
C=O	1712.9
O-H	1492.9
C=N	636.51

Table 6. Wave number for functional groups of pure PLA.

Functional Groups	Wavelength cm^{-1}
N-H	3336.85
O-H	3630.03
C-H	3294.42
C=O	1793.80

Turmeric essential oil interacts with the polymer matrix by forming intermolecular hydrogen bonds between their terminal hydroxyl group and the carbonyl groups of the ester moieties of both PLA and chitosan as mentioned in Table 7., in line with research [33] C-O stretching of alcohols and carboxylic acids, and N-H wagging of primary and secondary amines. These functional groups were predicted because chitosan is made up mostly of them. As evidenced by the appearance of a broad band at 3503 cm^{-1} corresponding to phenolic OH stretching vibrations and an increased intensity and shift of bands attributed to C–O bending vibrations as show in Figure 2. The bands ascribed to isoprenoids' out-of-plane C-H wagging vibrations emerged at 2918 cm^{-1} and 2922 cm^{-1} when turmeric oil was added and mixed at a high speed to bind the matrix and filler.

Table 7. Wave number for each functional group of chitosan.

Functional Groups	Wavelength cm^{-1}
C-H	1766.62
C-O	2409.03
C-O-C	2202.40

3.3. *Thermogravimetric Analysis (TGA)*

We examined TGA as a tool to evaluate differences in mass loss patterns in order to quantify differences in biofilm development. To the best of our knowledge, this is the first time TGA has been used to characterize a biofilm in Indonesia. Biofilm thickness is generally measured using confocal microscopy, which is a very efficient approach but necessitates expensive equipment and extensive training to produce high-quality pictures. TGA is frequently used for material characterisation since it needs little training and sample preparation and produces findings quickly. This is the temperature range in which bacterial organic matter is most likely to breakdown and burn; therefore, the difference is ascribed to a bacterial biofilm [34]. The TGA results are consistent with our findings. An active biofilm in direct contact with the carbon foam surface should generate a current by creating a potential difference between the anode and the cathode, as seen.

The researchers discovered a mass drop in humidity from 50 °C to 150 °C and a mass decrease from 250 °C to 400 °C owing to thermal deterioration of the two components. This is the temperature range in which bacterial organic matter is most likely to break down and burn; therefore, the difference is ascribed to a bacterial biofilm. As a result, the temperature of the biofilm breakdown was about 285 °C. Figure 3 depicts biofilms degraded at 285.55 °C with a mass removal of 1.313 mg at 2 g chitosan, 0 mL TEO, and 0 mL glycerol. Biofilm 2, on the other hand, disintegrated at 274.02 °C with a mass removal of 1.54 mg. Furthermore, at 315.74 °C, samples with a weight removal of 1.74 mg were degraded using biofilm 3.

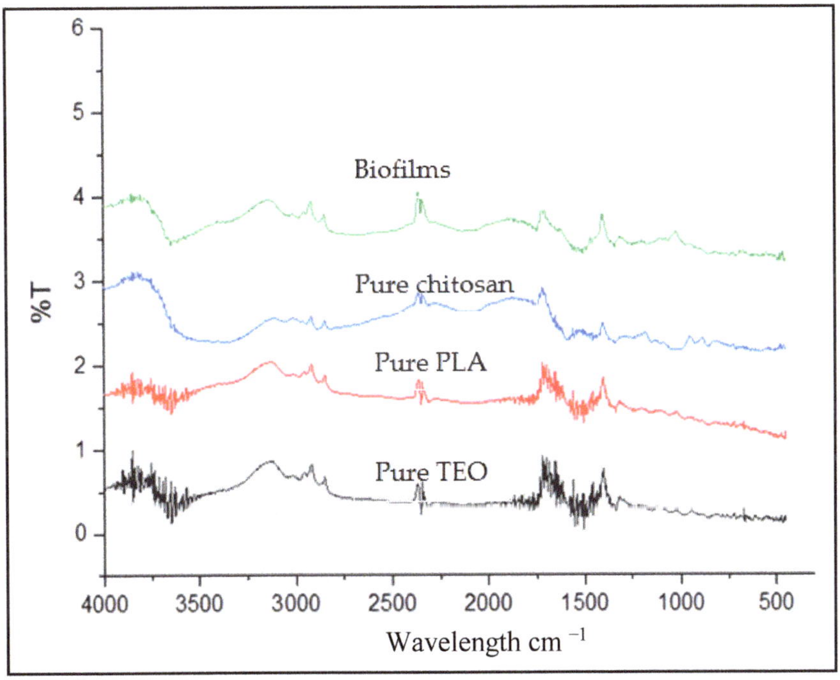

Figure 2. Spectrum of FT-IR for each sample.

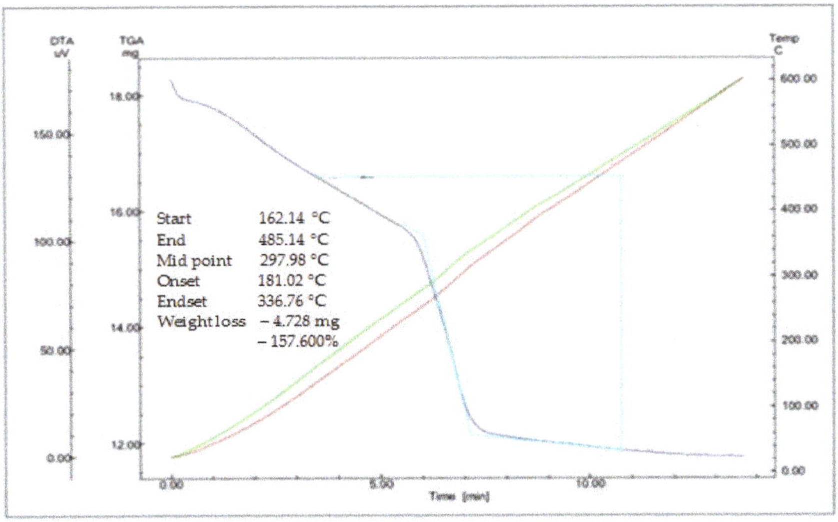

Figure 3. TGA analysis on biofilms to investigate sample resistance at a high temperature range before being applied as a food packaging.

This research is in line with [35], in which weight loss and DTG (a derivative of TGA curves) of chitosan and Mandarin Essensial Oil–chitosan nanoparticles are presented. Chitosan went through two distinct stages. The first stage with green line (50–150 °C) was associated with free water loss. The second step used temperatures ranging from 250 °C to 400 °C with pink line for chitosan dehydration and breakdown. Material structural modifications, such as the addition of Mandarin Essensial Oil, Tween20, and chitosan cross-linked with sodium tripolyphosphate show by black line, resulted in a new stage blue line (150–300 °C) in the chitosan nanoparticles and Mandarin Essensial Oil–chitosan nanoparticles.

The comparison between samples (a) and (b) based on Figure 4 shows that the physical change in the biofilm tested may have been the most degraded in the biofilm. The initial golden color became black after being burned. The sample's weight was also lost throughout the breakdown process. Each sample was examined, and it is known that the sample was broken down, owing to the combustion process in the TGA test equipment, resulting in the loss of carbon, water, or volatile components throughout the analysis method. Other combinations, such as chitosan, TEO, and glycerol, also had an effect on the sample's heat resistance.

(a) (b)

Figure 4. The result of biofilm samples (**a**) before the bioplastic thermal resistance test, (**b**) after the bioplastic thermal resistance test.

3.4. Anti-Microbial Test

TEO additives to chitosan biofilms have been studied to enhance the temperature of decomposition as one of the criteria for biological degradation. Furthermore, because of its antiseptic properties for food, biofilms with antimicrobial compounds have great food preservation abilities. As a result, in addition to preserving food, it also ensures the safety, freshness, and longer shelf life of these goods. This is due to the presence of antimicrobial action in natural substances and extracts utilized as additions in chitosan products. As a result, the usage of chitosan as a packaging material has several applications and is aimed at food safety.

It is critical to determine the antimicrobial activity, composition, structure, and functional groups of such extracts. Clove oil, thyme, cinnamon, rose salad, sage, and vanillin are the most active substances in the battle against germs. Several investigations have shown that they have antimycotic, non-toxic, and anti-parasitic effects. They can also be linked to the function of those chemicals in plants, as well as the antibacterial capabilities of essential oils and their constituents [16,36].

Figure 5 presents microbial development through the 12-day testing process that occurred in B1 (biofilm 1), B2 (biofilm 2), and B3 (biofilm 3). The results showed that the lowest colony growth rate of 61 colonies/g *S. aureus* was found in B1 with the composition of 2 g of chitosan, 0 mL of TEO, and 0 mL of glycerol, followed by B2 with 3 g of chitosan, 0.3 mL of TEO, and 0.5 mL of the glycerol growth rate of 77 colonies/g *E. coli*. B3 biofilm showed the best results, wherein 4 g of chitosan, 0.3 mL of TEO and 0.5 mL of glycerol have been mixed. The test results showed a microbial reduction with the addition of chitosan compounds, and the addition of essential oil from turmeric increased the anti-bacterial activity. *S. aureus* is a Gram-positive bacteria that can attach to glass, metal, and plastic as abiotic surfaces and host tissues as biotic surfaces. The attachment of *S. aureus* to surfaces depends on the surface components of the bacterial microbes recognizing the adhesive matrix molecules for host proteins. To prevent the attachment of *S. aureus* to a surface through the matrix, the surface must be coated with anti-adhesion agents such as arylrhodamins, calcium chelators, silver nanoparticles, and chitosan [37–39].

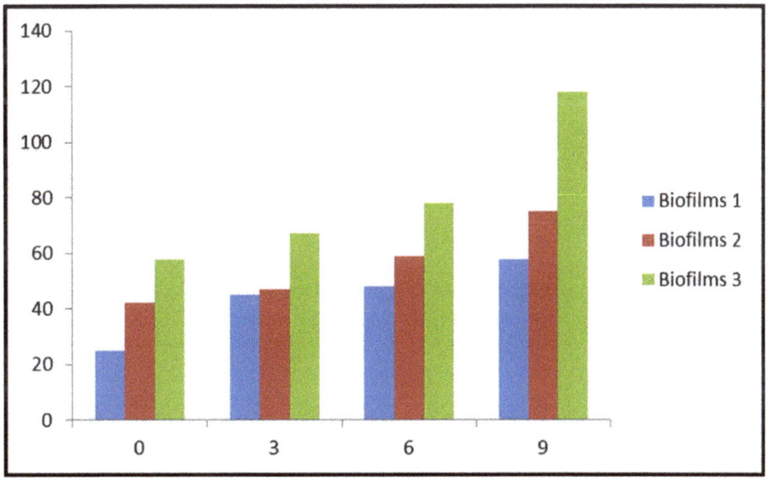

Figure 5. Anti-Microbial test result on PLA the variated of B1 (Biofilm 1), B2 (Biofilm 2) and (Biofilm 3) B3.

Bacteria adhere to surfaces in the form of a biofilm; therefore, one of the procedures that can prevent *S. aureus* pathogenesis is to prevent it from adhering to biotic and abiotic surfaces. Antibiofilm vaccinations are increasingly being used to inhibit *S. aureus* biofilm development [40]. Extracellular polysaccharides or cell wall-associated proteins identified in the biofilm matrix have the ability to stimulate an immune response that protects against *S. aureus* infection. Resistance to *S. aureus* can be acquired by active or passive vaccination with surface polysaccharides [41]. In contrast to *S. aureus*, *E. coli* bacteria live in the digestive system as part of the microbiota and may be a dangerous adversary. The presence of this bacterium can cause intestinal and extraintestinal infections in people and animals [42]. Based on the information gathered, *E. coli* has been added to the list of microorganisms of worldwide concern that cause the most prevalent illnesses in a variety of contexts, including communities, hospitals, and foodborne diseases [43]. These studies stated that combining chitosan and oregano oil as an essential oil could decrease water vapor permeability, puncture, and tensile strength. Still, they increased anti-bacterial and microbial properties due to the curcumin content and its essential oils, which inhibited the growth of causative bacteria, such as Bacillus sp, Shigella dysmetria, *S. aureus*, and

E. coli [44,45]. Compared to the previously explained study, it can be concluded that the addition of TEO had shown an increased anti-bacterial activity, which was to inhibit microbial growth in the sample. That was because the composite material was directly contaminated with the air, containing various microbe types that can affect both physical and chemical ingredients.

3.5. Effect of Variations in PLA–Chitosan–TEO Concentration on Mechanical Characteristics of Food Packaging

Tensile strength testing of food packaging is an important factor that must be studied further to determine its application in the food industry, particularly for applications to vegetables and fruits. Mobility mechanical properties are critical because they are required to support in vitro culture and implantation processes [46]. Furthermore, testing the mechanical properties of packaging materials is critical for determining the homogeneity of a polymer mixture and the mixed materials used in the manufacturing process. In this study, the mechanical characteristics of the PLA–chitosan–TEO material were analyzed by testing the tensile strength (Tensile Strength) of the sample using the UTM (Universal Testing Machine) in the Tensile Test Laboratory of the Chemical Engineering Department, Lhokseumawe State Polytechnic. The dimensions of the specimens used follow the dimensions of the specimens in ASTM D 638–99. In this graph, the highest tensile strength value was obtained in sample biofilm 3. The composition of each biofilm 1 material is 13.03 Mpa. Meanwhile, biofilm 2 (3 g of chitosan, 0.3 mL TEO and 0.5 mL of glycerol) and 3 (4 g of chitosan, 0.3 of TEO and 0.5 mL of glycerol) have tensile strength values of 15.15 MPa and 17.67 Mpa, respectively.

Figure 6 based on the three samples, the third sample has a large tensile strength value, which is because the high concentration of chitosan causes the tensile strength of the analysis to increase. The increase in tensile strength was due to the reduced water content in the bioplastic because it had been absorbed by chitosan as an organic polysaccharide. Therefore, the molecular structure of bioplastics becomes denser and more homogeneous, which causes the tensile strength to increase [47,48]. This is also in line with conducted research which states that the addition of chitosan in PLA causes the tensile strength to increase based on research that has been carried out by [49] regarding the effect of the homogenization method and the content of carvacrol on the microstructure and physical properties of chitosan-based films. Film-forming emulsions were made with chitosan (1.5%), Tween 80 (0.5%), and carvacrol (0.25%, 0.5%, and 1.0%). The homogenization method used is the rotor-stator with high- and low-pressure homogenization. The results showed that the use of carvacrol had a significant impact on the mechanical properties of the emulsified films. The tensile strength produced by the film was significantly reduced. The results of adding chitosan to bioplastics showed that there was an interaction in the mixed film (starch–chitosan bioplastic). Chitosan has a good filming property and has extensive water absorption in the acidic medium due to the chain relaxation effect generated in the macromolecular networks by protonated amino groups. The addition of chitosan can increase the tensile strength of bioplastics [50]. However, if more chitosan is added, the more the tensile strength value of the bioplastics will decrease. In other words, the resulting bioplastic will have brittle properties [51]. The table below is in line with research conducted by [52]: the strength of the prepared samples was not affected by the type of natural extract, but only by the used basic matrix/polymer. The explanation of different textural properties can be affected by the different pH since the preparation of films included PLA.

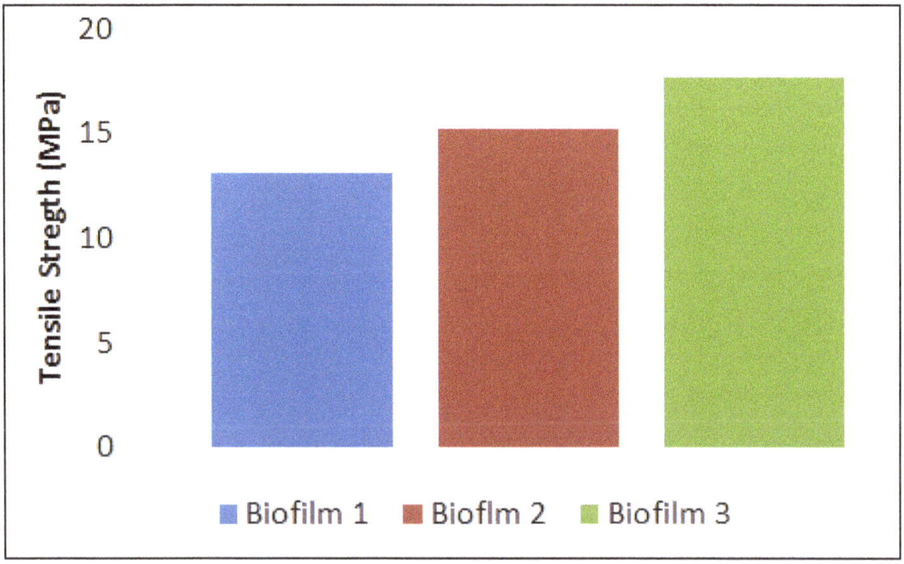

Figure 6. Comparison of the mechanical properties of each biofilm produced based on variations in chitosan.

The concentration of chitosan also affected the % elongation in PLA–chitosan–TEO biofilms samples. Elongation is a value that states the ability of a sample to be able to extend from its original shape or size as mentioned in Table 8. Figure 7 shows the elongation (%) of the PLA–chitosan–TEO biofilms obtained ranged from 18% to 25%. At a sample concentration of 2 gr chitosan, the lowest % extension value is obtained at a concentration of 2 g is only 18%. The chitosan concentration of 3 g, namely 22.7%, was the middle elongation obtained. The highest elongation came from biofilm 3 with 25% elongation. This is in line with the study of [53], where the minimum amount of chitosan used was set at 2 (by wt). It should be noted that starch could not be formed into films by glycerol solution. However, chitosan has an exceptional capacity to create film. As a result, it is also mentioned that biodegradable films were created using starch, and chitosan will provide a much better film because the films exhibited high strength and optimal elongation values. The results show that adding extracts to TEO and chitosan matrices improves the elongation and tensile strength of the biofilms formed. To decrease the resultant error value, each sample test was performed three times.

Table 8. Mean values (± standard deviation) of elongation (%) obtained from biofilm formations of: (B1) 2 g chitosan; (B2) 3 g chitosan; (B3) 4 g chitosan after the elongation test each biofilms.

Sample	Mean ± Standard Deviations	Biofilms Composition
B1	18.66 ± 0.05	2 g chitosan, 0 mL TEO, and 0 mL glycerol
B2	22.07 ± 0.05	3 g chitosan, 0.3 mL TEO, and 0.5 mL glycerol
B3	25.04 ± 0.05	4 g chitosan, 0.3 mL TEO, and 0.5 mL glycerol

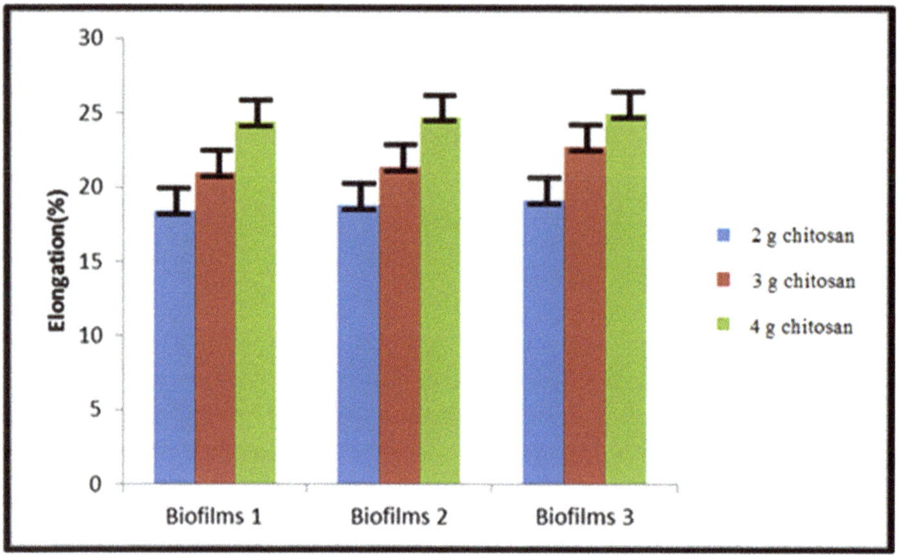

Figure 7. Mean values (±standard deviation) of elongation (%) obtained from biofilm formations of: 2 gr chitosan; 3 gr chitosan; 4 gr chitosan after the elongation test each biofilms. A statistical difference from SPSS is indicated ($p < 0.05$).

3.6. Techno Economic Challenges of the Developed and Future Research Directions

Biofilms are complex microbial communities composed of one or more species submerged in an extracellular matrix with varying compositions based on the kind of food-producing environment and the invading species. Bacteria and fungi are examples of microorganisms that may form these biofilms. The presence of many bacterial species in a biofilm has significant ecological benefits since it facilitates the biofilm's adhesion to a surface. This can also happen in the absence of specific fimbriae in some species. Disinfectants such as quaternary ammonium compounds and other biocides are more resistant to mixed biofilms. In the food business, biofilms may build swiftly. The first two phases condition the material's surface and the reversible cell binding to that surface. The binding then becomes permanent, and the formation of microcolonies begins. Finally, the tridimensional structure of the biofilm is established, giving rise to a sophisticated ecosystem suitable for dispersal. Some biofilm-forming bacteria in food factory settings cause human diseases, which is very important in the food business. These pathogens may form biofilm structures on a variety of artificial substrates used in the food business, including stainless steel, polyethylene, wood, glass, and polypropylene.

Prior to selection, the in situ relevance of the biofilm model must also be reviewed. For example, before employing a specific model, it should be evaluated if the inoculum, flow regime (if any), and surfaces/interfaces fit the environment in the system being represented. To achieve in situ relevance, the essential drivers and factors in the system and environment must be recognized. Tools for evaluating how the model and results may be translated to practice should be established, and practitioners (including doctors and engineers) should be included in model review to guarantee in situ relevance. With an ever-increasing number of biofilm-related articles being published and a wide range of various methodologies becoming accessible, it can be difficult for researchers new to the subject (or to a specific subdiscipline within the field) to select the best appropriate model system. In addition, some instruction on data interpretation may be required (e.g., how much biofilm reduction is needed in a particular model before considering it biologically meaningful). Although biofilms have an influence on many elements of life and biological sciences, they are considered a specific discipline.

3.7. Comparative Table of Materials Used for Food Packaging Applications

Table 9 is represent the difference process of producing food packaging with essential oil fillers as anti-bacterial and anti-oxidant chemicals that can extend the shelf life of components and protect product quality from free radicals.

Table 9. Comparative Table of Materials Used for Food Packaging Applications.

Lukic Dkk, 2021	This Research
This study aims to produce a biodegradable film with antioxidant activity for active food packaging. A poly lactic acid (PLA) film with 5% by weight poly-caprolactone (PCL), filled with thymol, carvacrol, and their mixture using supercritical CO_2 at 40 °C and 10 MPa for 5 h, with 27.53, 21.18, respectively, and 21.49% was used. The morphological, structural and mechanical properties of the obtained PLA-5 films were analyzed. The mixture of thymol and carvacrol incorporated into the PLA-5 film showed synergy and a low percentage of migration to distilled water, indicating great potential for application as a food packaging material with the aim of reducing the need for the amount of active ingredients. The addition of Carvacrol and a Thymol–Carvacrol mixture improved flexibility, extensibility, and ductility, but subsequently lower tensile strength of the PLA-5 film due to the plasticizing effect of loaded compounds promotedby $scCO_2$.	The samples were prepared in three different variations: 2 g of chitosan, 0 mL TEO, and 0 mL glycerol (biofilm 1), 3 g of chitosan, 0.3 mL TEO, and 0.5 mL of glycerol (biofilm 2), and 4 g of chitosan, 0.3 of TEO, and 0.5 mL of glycerol (biofilm 3). The concentration of chitosan affects the tensile strength of the PLA–chitosan–TEO sample; a higher concentration value of chitosan will make the bioscaffold material have an increased tensile strength value. The highest tensile strength value was in biofilm 3 with a composition of 4 g of chitosan concentration. Microbial development results showed that the lowest colony growth rate of 61 colonies/g *S. aureus* was found in B1 followed by B2 with 3 g of chitosan, 0.3 mL of TEO, and 0.5 mL of the glycerol growth rate of 77 colonies/g *E. coli*. B3 biofilm showed the best microbial reduction with the addition of chitosan compounds and essential oils from turmeric.

3.8. Comparison between Produced Materials and Commercials Product

The FTIR study of chitosan yielded the absorption regions of functional groups, as shown in Figure 8. The the red line shows that stretching OH group absorption emerged at a wave number of 3189.37 cm^{-1} with a deacetylation temperature treatment of 100 °C for two hours, and the hydroxyl group (-OH) appeared at a wave number of 2980.20 cm^{-1}. The N-H stretching group may be found at wave number 3004.61 cm^{-1}. The vibration of the C-H range on aliphatic CH_2 is shown by the absorption at a wave number of 2779.37 cm^{-1}. The appearance of bending vibration absorption CH_2 at a wave number of 1365.75 cm^{-1} supports this. The absorption at 1567.09 cm^{-1} indicates the C=O group (amide peak) that remains because the chitosan generated has not been entirely deacetylated. In the isolated chitosan IR spectra, N-H bending vibration of NH_2 is represented by the 1376 cm^{-1}. The bending absorption of -CH_3 is modest and visible at the wave number 1389.45 cm^{-1}. At the wave number 1306.09 cm^{-1}, C-N stretching vibrations with low intensity were detected. The C-O bond range was discovered at wave numbers 1155.36 cm^{-1} and 1114.86 cm^{-1}.

Chitosan at a deacetylation temperature of 100 °C from commercial chitosan for two hours (black line) showed absorption of the stretching OH group at a wave number of 2992.11 cm^{-1} and a hydroxyl group (-OH) at a wave number of 3180.91 cm^{-1}. The N-H stretching group emerges at the wave number 3003.43 cm^{-1}. Absorption at wavelength numbers 2945.87 cm^{-1} and 2790. cm^{-1} demonstrates the C-H stretching vibration on aliphatic CH_2, which is reinforced by the presence of bending vibration absorption CH2 at wave number 1398.44 cm^{-1}. The absorption at the wave number 1590.34 cm^{-1} indicates the C=O group (amide peak) that remains because the chitosan generated has not been entirely deacetylated.

The area for pure substances is greater than for commercial products. The Figure 9 shows that the yield of the resultant extract increases, as well as the number of extraction steps, with the extraction stage, yielding more extract than the 1-stage extraction. Extraction with fewer solvents will be more effective than extraction with all the solvents at once.

This is due to the fact that at each stage, there will be contact with a new solvent, which provides a driving force in the form of variations in concentration and solubility at each stage, ensuring that a solute is always transferred from solid to solvent. The extract yield will eventually fall like Figure 10. This is due to the fact that in crosscurrent, multistage extraction, the solids employed at each stage are the same solids; therefore, the longer the extract is extracted, the more saturated it becomes until the extract gain no longer grows and decreases.

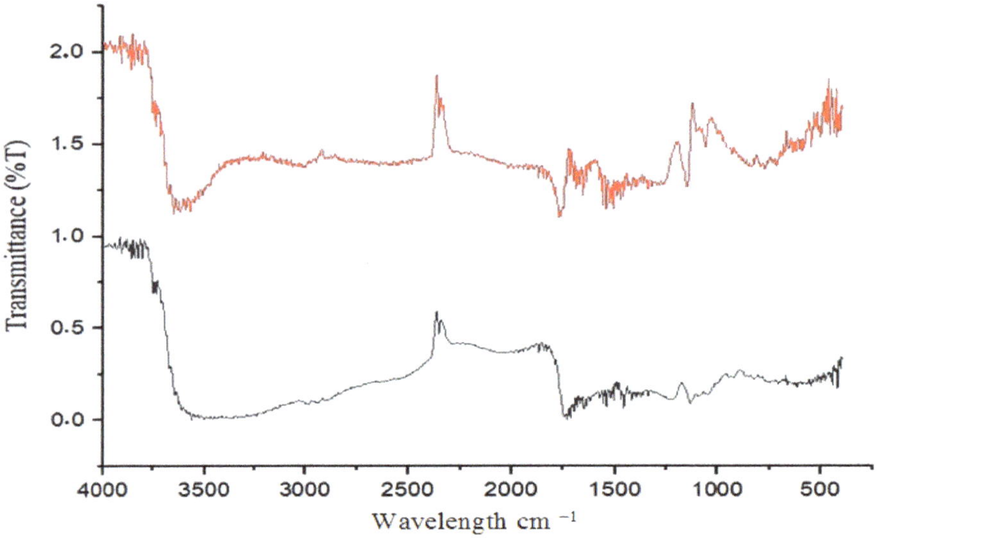

Figure 8. Comparison between extracted chitosan and commercial chitosan using FT-IR tools.

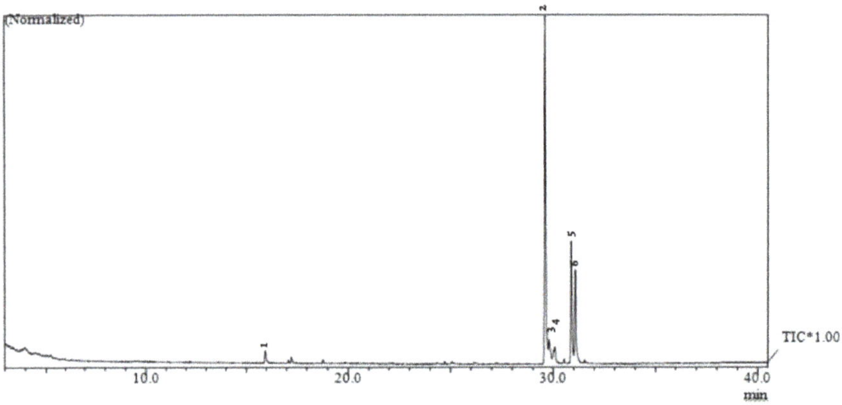

Peak Report TIC

Peak#	R.Time	Area	Area%	Name
1	15.945	220494	2.30	Beta turmerone
2	29.611	5159635	53.73	ar-curcumene
3	29.812	615964	6.41	curlone
4	30.092	281173	2.93	Beta bisabolen
5	30.891	1802015	18.77	Turmerone
6	31.080	1522774	15.86	Alpha humulene
		9602055	100.00	

Figure 9. Purity of extracted essential oil tested using gas chromatography-mass spectroscopy.

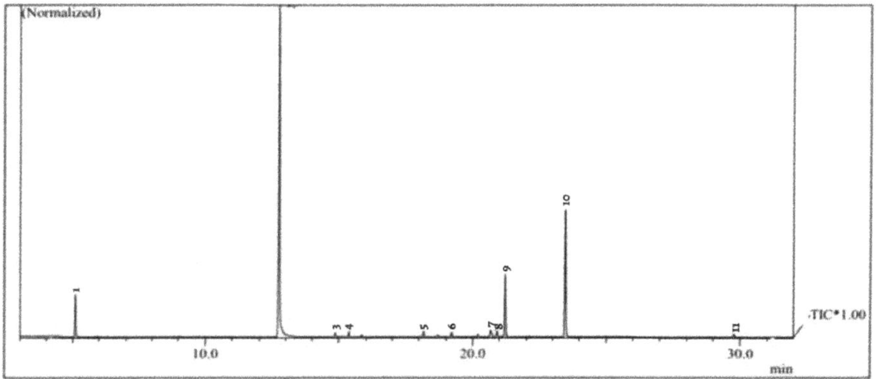

Peak#	R.Time	Area	Area%	Name
1	14.190	6850123	61.01	(-)-.alpha, Naphthalene,
2	19.722	248293	1.98	Alpha humulene
3	25.341	359802	16.76	Ar-curcumene, Azulenemethanol,
4	26.098	281173	12.32	(-)Naphthalenemethanol
5	30.120	231891	2.68	Alpha turmerone, 1,2,3,4,4a,5,6,8a-octahydro
6	31.672	119380	5.25	2-Naphthalenemethanol
		9193412	100.00	

Figure 10. Purity of comerrcially essential oil tested using gas chromatography-mass spectroscopy.

4. Conclusions

Based on the research results on PLA as a biofilm, it was concluded that the maximum decomposition temperature obtained was at the temperature of 315.74 °C with variations of 4 g chitosan, 0.3 mL TEO, and 0.5 mL glycerol (biofilm 3). Molecules produce solid particles and are evenly distributed around the surface, affecting anti-microbes' properties. Based on the anti-microbial test, chitosan and TEO showed an increase in anti-bacterial activity. Moreover, the biofilms produced were able to fight *S. aureus* and *E. coli* bacteria in 9 days of exposure to open spaces. The concentration of chitosan affects the tensile strength of the PLA–chitosan–TEO sample: a higher concentration value of chitosan will make the bioscaffold material have an increased tensile strength value. The highest tensile strength value was in biofilm 3, with a composition of 4 g of chitosan concentration. However, if the addition of chitosan is continuous, it will increase the brittleness through higher water absorption so that further studies on the maximum amounts of chitosan as a filler in the manufacture of food packaging need to be considered. Biofilms in the food business are a severe economic and health concern. On the one hand, the presence of biofilms of food processing surfaces might result in financial losses due to corrosion on metal surfaces caused by certain bacteria, necessitating the replacement of these parts. Furthermore, certain bacterial species, such as Pseudomonas spp. and Bacillus spp., express a variety of proteolytic and lipolytic enzymes that can produce unpleasant smells and tastes (rancid, bitter). The impacted production batches must be removed and destroyed in these cases. On the other hand, and more importantly, biofilm development in food manufacturing is a critical public health issue. These biofilms may comprise bacterial (and occasionally fungal) species known to be harmful in healthy people, or they may solely target the immunocompromised (such as organ transplant recipients, oncology or HIV patients, etc.). These bacteria can cause food poisoning *(S. aureus)* as well as gastroenteritis *(E. coli)* and systemic illnesses in some situations *(E. coli* O157:H7).

Author Contributions: This study was carried out with the help of numerous persons who had the following responsibilities: Conceptualization, T.R.; methodology, J.P.S.; software, T.C.; validation, N.A.; formal analysis, M.H.M.H.; investigation, D.F.F.; resources, A.S.; data curation, A.E.H.;

writing—original draft preparation, A.P.I. All authors have read and agreed to the published version of the manuscript.

Funding: This research was funded by The Ministry of Education, Culture, Research, and Technology of the Republic Indonesia, grant number: 852/E4.1/AK.04.PT/PNL/2021.

Institutional Review Board Statement: Not appicable.

Informed Consent Statement: Not appicable.

Data Availability Statement: The data presented in this study are available on request from the corresponding author.

Acknowledgments: We would like to thank A.E.H., N.A., A.S., J.P.S., T.C., A.P.I., M.H.M.H. and D.F.F. for providing support of experimental instruments.

Conflicts of Interest: The authors declare no conflict of interest.

References

1. Begum, S.; Yuhana, N.Y.; Saleh, N.M.; Kamarudin, N.N.; Sulong, A.B. Review of chitosan composite as a heavy metal adsorbent: Material preparation and properties. *Carbohydr. Polym.* **2021**, *259*, 117613. [CrossRef] [PubMed]
2. Chisenga, S.M.; Tolesa, G.N.; Workneh, T.S. Biodegradable food packaging materials and prospects of the fourth industrial revolution for tomato fruit and product handling. *Int. J. Food Sci.* **2020**, *2020*, 8879101. [CrossRef] [PubMed]
3. Sohn, Y.J.; Kim, H.T.; Baritugo, K.A.; Jo, S.Y.; Song, H.M.; Park, S.Y.; Park, S.K.; Pyo, J.; Cha, H.G.; Kim, H.; et al. Recent advances in sustainable plastic upcycling and biopolymers. *Biotechnol. J.* **2020**, *15*, 1900489. [CrossRef] [PubMed]
4. Saratale, R.G.; Cho, S.-K.; Saratale, G.D.; Kadam, A.A.; Ghodake, G.S.; Kumar, M.; Bharagava, R.N.; Kumar, G.; Kim, D.S.; Mulla, S.I.; et al. A comprehensive overview and recent advances on polyhydroxyalkanoates (PHA) production using various organic waste streams. *Bioresour. Technol.* **2021**, *325*, 124685. [CrossRef]
5. Moyo, M.; Kanny, K.; Mohan, T.P. Thermo-mechanical response of kenaf/PLA biocomposites to clay nanoparticles infusion. *Mater. Today Proc.* **2021**, *38*, 609–613. [CrossRef]
6. Altan, A.; Aytac, Z.; Uyar, T. Carvacrol loaded electrospun fibrous films from zein and poly (lactic acid) for active food packaging. *Food Hydrocoll.* **2018**, *81*, 48–59. [CrossRef]
7. Anggarini, F. Aplikasi plasticizer gliserol pada pembuatan plastik biodegradable dari biji nangka. *Bioeksperimen* **2018**, *28*, 4–5.
8. Abdul Khalil, A.K.; Saurabh, C.; Adnan, A.S.; Fazita, M.N.; Syakir, M.; Davoudpour, Y.; Rafatullah, M.; Abdullah, C.; Haafiz, M.M.; Dungani, R. A review on chitosan-cellulose blends and nanocellulose reinforced chitosan biocomposites: Properties and their applications. *Carbohydr. Polym.* **2016**, *150*, 216–226. [CrossRef]
9. Borkar, T.; Goenka, V.; Jaiswal, A.K. Application of poly-ε-caprolactone in extrusion-based bioprinting. *Bioprinting* **2021**, *21*, e00111. [CrossRef]
10. Boudries, H.; Nabet, N.; Chougui, N.; Souagui, S.; Loupassaki, S.; Madani, K.; Dimitrov, K. Optimization of ultrasound-assisted extraction of antioxidant phenolics from Capparis spinosa flower buds and LC–MS analysis. *J. Food Meas. Charact.* **2019**, *13*, 2241–2252. [CrossRef]
11. Chen, P.; Liang, X.; Xu, Y.; Zhou, Y.; Nie, W. Enhanced thermal and mechanical properties of PLA/MoS2 nanocomposites synthesized via the in-situ ring-opening polymerization. *Appl. Surf. Sci.* **2018**, *440*, 1143–1149. [CrossRef]
12. Chen, W.; Qi, C.; Li, Y.; Tao, H. The degradation investigation of biodegradable PLA/PBAT blend: Thermal stability, mechanical properties and PALS analysis. *Radiat. Phys. Chem.* **2021**, *180*, 109239. [CrossRef]
13. Karthikeyan, C.; Varaprasad, K.; Akbari-Fakhrabadi, A.; Hameed, A.S.H.; Sadiku, R. Biomolecule chitosan, curcumin and ZnO-based antibacterial nanomaterial, via a one-pot process. *Carbohydr. Polym.* **2020**, *249*, 116825. [CrossRef] [PubMed]
14. Lukic, I.; Vulic, J.; Ivanovic, J. Antioxidant activity of PLA/PCL films loaded with thymol and/or carvacrol using $scCO_2$ for active food packaging. *Food Packag. Shelf Life* **2020**, *26*, 100578. [CrossRef]
15. Costa, S.M.; Ferreira, D.P.; Teixeira, P.; Ballesteros, L.F.; Teixeira, J.A.; Fangueiro, R. Active natural-based films for food packaging applications: The combined effect of chitosan and nanocellulose. *Int. J. Biol. Macromol.* **2021**, *177*, 241–251. [CrossRef] [PubMed]
16. Flores, Z.; San-Martin, D.; Beldarraín-Iznaga, T.; Leiva-Vega, J.; Villalobos-Carvajal, R. Effect of Homogenization Method and Carvacrol Content on Microstructural and Physical Properties of Chitosan-Based Films. *Foods* **2021**, *10*, 141. [CrossRef]
17. Gere, D.; Czigany, T. Future trends of plastic bottle recycling: Compatibilization of PET and PLA. *Polym. Test.* **2019**, *81*, 106160. [CrossRef]
18. Gerometta, M.; Rocca-Smith, J.R.; Domenek, S.; Karbowiak, T. Physical and Chemical Stability of PLA in Food Packaging. *Ref. Modul. Food Sci.* **2019**, *12*, 231. [CrossRef]
19. Guarino, V.; Gentile, G.; Sorrentino, L.; Ambrosio, L. Polycaprolactone: Synthesis, Properties, and Applications. *Encycl. Polym. Sci. Technol.* **2002**, *7*, 36. [CrossRef]
20. Elsabee, M.Z.; Morsi, R.E.; Fathy, M. *Chitosan-Oregano Essential Oil Blends Use as Antimicrobial Packaging Material*; Elsevier Inc.: Amsterdam, The Netherlands, 2016; ISBN 9780128007235.

21. Balouiri, M.; Sadiki, M.; Ibnsouda, S.K. Methods for in vitro evaluating antimicrobial activity: A review. *J. Pharm. Anal.* **2016**, *6*, 71–79. [CrossRef]
22. Radwan, M.A.; Farrag, S.A.; Abu-Elamayem, M.M.; Ahmed, N.S. Extraction, characterization, and nematicidal activity of chitin and chitosan derived from shrimp shell wastes. *Biol. Fertil. Soils* **2012**, *48*, 463–468. [CrossRef]
23. De Queiroz Antonino, R.S.C.M.; Lia Fook, B.R.P.; de Oliveira Lima, V.A.; de Farias Rached, R.I.; Lima, E.P.N.; da Silva Lima, R.J.; Lia Fook, M.V. Preparation and characterization of chitosan obtained from shells of shrimp (*Litopenaeus vannamei* Boone). *Mar. Drugs* **2017**, *15*, 141. [CrossRef] [PubMed]
24. Mwaurah, P.W.; Kumar, S.; Kumar, N.; Attkan, A.K.; Panghal, A.; Singh, V.K.; Garg, M.K. Novel oil extraction technologies: Process conditions, quality parameters, and optimization. *Compr. Rev. Food Sci. Food Saf.* **2020**, *19*, 3–20. [CrossRef]
25. Gonçalves, C.; Gonçalves, I.C.; Magalhães, F.D.; Pinto, A.M. Poly (lactic acid) composites containing carbon-based nanomaterials: A review. *Polymers* **2017**, *9*, 269. [CrossRef]
26. Arfat, Y.A.; Ahmed, J.; Ejaz, M.; Mulla, M.F. Polylactide/graphene oxide nanosheets/clove essential oil composite films for potential food packaging applications. *Int. J. Biol. Macromol.* **2018**, *107*, 194–203. [CrossRef]
27. Dubreuil, J.D. Fruit extracts to control pathogenic *Escherichia coli*: A sweet solution. *Heliyon* **2020**, *6*, e03410. [CrossRef]
28. Aradmehr, A.; Javanbakht, V. A novel biofilm based on lignocellulosic compounds and chitosan modified with silver nanoparticles with multifunctional properties: Synthesis and characterization. *Colloids Surf. A Physicochem. Eng. Asp.* **2020**, *600*, 124952. [CrossRef]
29. Narayanan, K.B.; Park, G.T.; Han, S.S. Biocompatible, antibacterial, polymeric hydrogels active against multidrug-resistant *Staphylococcus aureus* strains for food packaging applications. *Food Control* **2020**, *123*, 107695. [CrossRef]
30. Song, X.; Wang, L.; Liu, T.; Liu, Y.; Wu, X.; Liu, L. Mandarin (*Citrus reticulata* L.) essential oil incorporated into chitosan nanoparticles: Characterization, anti-biofilm properties and application in pork preservation. *Int. J. Biol. Macromol.* **2021**, *185*, 620–628. [CrossRef]
31. Dosoky, N.S.; Setzer, W.N. Chemical composition and biological activities of essential oils of Curcuma species. *Nutrients* **2018**, *10*, 1196. [CrossRef] [PubMed]
32. Yuniarto, K.; Purwanto, Y.A.; Purwanto, S.; Welt, B.A.; Purwadaria, H.K.; Sunarti, T.C. Infrared and Raman studies on polylactide acid and polyethylene glycol-400 blend. In *AIP Conference Proceedings*; AIP Publishing LLC: New York, NY, USA, April 2016; Volume 1725, p. 020101.
33. Ayodele, O.; Okoronkwo, A.E.; Oluwasina, O.O.; Abe, T.O. Utilization of blue crab shells for the synthesis of chitosan nanoparticles and their characterization. *Songklanakarin J. Sci. Technol.* **2018**, *15*, 220–223.
34. Rihayat, T. Morphology Properties of Polyurethane/clay Nanocomposites Base on Palm Oil Polyol Paint. In *Advanced Material Research*; Trans Tech Publications Ltd.: Bäch, Switzerland, 2013; Volume 647, pp. 701–704.
35. Tosati, J.V.; de Oliveira, E.F.; Oliveira, J.V.; Nitin, N.; Monteiro, A.R. Light-activated antimicrobial activity of turmeric residue edible coatings against cross-contamination of Listeria innocua on sausages. *Food Control* **2018**, *84*, 177–185. [CrossRef]
36. Haghighi, H.; Licciardello, F.; Fava, P.; Siesler, H.W.; Pulvirenti, A. Recent advances on chitosan-based films for sustainable food packaging applications. *Food Packag. Shelf Life* **2020**, *26*, 100551. [CrossRef]
37. Ou, C.; Shang, D.; Yang, J.; Chen, B.; Chang, J.; Jin, F.; Shi, C. Prevalence of multidrug-resistant *Staphylococcus aureus* isolates with strong biofilm formation ability among animal-based food in Shanghai. *Food Control* **2020**, *112*, 107106. [CrossRef]
38. Shahid, A.H.; Nazir, K.N.H.; El Zowalaty, M.E.; Kabir, A.; Sarker, S.A.; Siddique, M.P.; Ashour, H.M. Molecular detection of vancomycin and methicillin resistance in *Staphylococcus aureus* isolated from food processing environments. *One Health* **2021**, *13*, 100276. [CrossRef]
39. Chen, Q.; Xie, S.; Lou, X.; Cheng, S.; Liu, X.; Zheng, W.; Zheng, Z.; Wang, H. Biofilm formation and prevalence of adhesion genes among *Staphylococcus aureus* isolates from different food sources. *MicrobiologyOpen* **2019**, *9*, e00946. [CrossRef]
40. Fahmy, H.M.; Eldin, R.E.S.; Abu Serea, E.S.; Gomaa, N.M.; Aboelmagd, G.M.; Salem, S.A.; Elsayed, Z.A.; Edrees, A.; Shams-Eldin, E.; Shalan, A.E. Advances in nanotechnology and antibacterial properties of biodegradable food packaging materials. *RSC Adv.* **2020**, *10*, 20467–20484. [CrossRef]
41. Salim, S.; Rihayat, T.; Riskina, S.; Safitri, A. Physical and mechanical properties of bamboo/flax fibre reinforced epoxy composite water absorption behaviour and high-temperature conditions. *Plast. Rubber Compos.* **2021**, *50*, 415–424. [CrossRef]
42. Kumari, S.; Annamareddy, S.H.K.; Abanti, S.; Rath, P.K. Physicochemical properties and characterization of chitosan synthesized from fish scales, crab and shrimp shells. *Int. J. Biol. Macromol.* **2017**, *104*, 1697–1705. [CrossRef]
43. Łopusiewicz, L.; Kwiatkowski, P.; Drozłowska, E.; Trocer, P.; Kostek, M.; Śliwiński, M.; Polak-Śliwińska, M.; Kowalczyk, E.; Sienkiewicz, M. Preparation and Characterization of Carboxymethyl Cellulose-Based Bioactive Composite Films Modified with Fungal Melanin and Carvacrol. *Polymers* **2021**, *13*, 499. [CrossRef]
44. Filipini, G.D.S.; Romani, V.P.; Martins, V.G. Biodegradable and active-intelligent films based on methylcellulose and jambolão (Syzygium cumini) skins extract for food packaging. *Food Hydrocoll.* **2020**, *109*, 106139. [CrossRef]
45. Singh, J.; Kaur, T.; Singh, N.; Pandey, P.M. Biological and mechanical characterization of biodegradable carbonyl iron powder/polycaprolactone composite material fabricated using three-dimensional printing for cardiovascular stent application. *Proc. Inst. Mech. Eng. Part H J. Eng. Med.* **2020**, *234*, 975–987. [CrossRef]

46. Schugens, C.; Maquet, V.; Grandfils, C.; Jérôme, R.; Teyssie, P. Biodegradable and macroporous polylactide implants for cell transplantation: 1. Preparation of macroporous polylactide supports by solid-liquid phase separation. *Polymer* **1996**, *37*, 1027–1038. [CrossRef]
47. Jancikova, S.; Dordevic, D.; Tesikova, K.; Antonic, B.; Tremlova, B. Active Edible Films Fortified with Natural Extracts: Case Study with Fresh-Cut Apple Pieces. *Membranes* **2021**, *11*, 684. [CrossRef] [PubMed]
48. Gisan, K.A.; Chan, M.Y.; Koay, S.C. Solvent-cast Biofilm from Poly (lactic) Acid and Durian Husk Fiber: Tensile, Water Absorption, and Biodegradation Behaviors. *J. Nat. Fibers* **2020**, *121*, 1857894. [CrossRef]
49. Wilkins, T.D.; Holdeman, L.V.; Abramson, I.J.; Moore, W.E.C. Standardized Single-Disc Method for Antibiotic Susceptibility Testing of Anaerobic Bacteria. *Antimicrob. Agents Chemother.* **1972**, *1*, 451–459. [CrossRef]
50. Sohn, Y.J.; Kim, H.T.; Jo, S.Y.; Song, H.M.; Baritugo, K.A.; Pyo, J.; Park, S.J. Recent advances in systems metabolic engineering strategies for the production of biopolymers. *Biotechnol. Bioprocess Eng.* **2020**, *64*, 848–861. [CrossRef]
51. de Donno, M.L.; Moreno, S.; Rene, E.R. Polyhydroxyalkanoate (PHA) production via resource recovery from industrial waste streams: A review of techniques and perspectives. *Bioresour. Technol.* **2021**, *80*, 124985. [CrossRef]
52. Gasti, T.; Dixit, S.; Sataraddi, S.P.; Hiremani, V.D.; Masti, S.P.; Chougale, R.B.; Malabadi, R.B. Physicochemical and biological evaluation of different extracts of edible Solanum nigrum L. leaves incorporated chitosan/poly (vinyl alcohol) composite films. *J. Polym. Environ.* **2020**, *28*, 2918–2930. [CrossRef]
53. Cavallaro, G.; Micciulla, S.; Chiappisi, L.; Lazzara, G. Chitosan-based smart hybrid materials: A physico-chemical perspective. *J. Mater. Chem. B* **2021**, *9*, 594–611. [CrossRef]

polymers MDPI

Article

Chitosan/Poly Vinyl Alcohol/Graphene Oxide Based pH-Responsive Composite Hydrogel Films: Drug Release, Anti-Microbial and Cell Viability Studies

Muhammad Umar Aslam Khan [1,2,3,*], Zahida Yaqoob [4], Mohamed Nainar Mohamed Ansari [5,*], Saiful Izwan Abd Razak [1,6], Mohsin Ali Raza [4], Amna Sajjad [7], Sajjad Haider [8] and Fauzi Mh Busra [9]

1 BioInspired Device and Tissue Engineering Research Group, School of Biomedical Engineering and Health Sciences, Faculty of Engineering, Universiti Teknologi Malaysia, Skudai 81300, Malaysia; saifulizwan@utm.my
2 Institute for Personalized Medicine, School of Biomedical Engineering, Med-X Research Institute, Shanghai Jiao Tong University, Shanghai 200030, China
3 Nanoscience and Technology Department (NS & TD), National Center for Physics, Islamabad 44000, Pakistan
4 Institute of Metallurgy and Materials Engineering, Faculty of Chemical and Materials Engineering, University of the Punjab, Lahore 54590, Pakistan; Zahidayaqoob20@gmail.com (Z.Y.); mohsin.imme@pu.edu.pk (M.A.R.)
5 Institute of Power Engineering, Universiti Tenaga Nasional, Kajang 43000, Malaysia
6 Centre for Advanced Composite Materials, Universiti Teknologi Malaysia, Skudai 81300, Malaysia
7 Department of Zoology, Government College University Faisalabad, Faisalabad 38000, Pakistan; amnasajjad7@yahoo.com
8 Department of Chemical Engineering, College of Engineering, King Saud University, P.O. Box 800, Riyadh 11421, Saudi Arabia; shaider@ksu.edu.sa
9 Tissue Engineering Centre, UKM Medical Centre, Jalan Yaacob Latiff, Bandar Tun Razak, Cheras, Kuala Lumpur 56000, Malaysia; fauzibusra@ukm.edu.my
* Correspondence: umar007khan@gmail.com (M.U.A.K.); ansari@uniten.edu.my (M.N.M.A.)

Citation: Khan, M.U.A.; Yaqoob, Z.; Nainar, M.A.M.; Razak, S.I.A.; Raza, M.A.; Sajjad, A.; Haider, S.; Busra, F.M. Chitosan/Poly Vinyl Alcohol/Graphene Oxide Based pH-Responsive Composite Hydrogel Films: Drug Release, Anti-Microbial and Cell Viability Studies. *Polymers* **2021**, *13*, 3124. https://doi.org/10.3390/polym13183124

Academic Editor: Evgenia G. Korzhikova-Vlakh

Received: 27 June 2021
Accepted: 13 August 2021
Published: 16 September 2021

Abstract: The composite hydrogels were produced using the solution casting method due to the non-toxic and biocompatible nature of chitosan (CS)/polyvinyl alcohol (PVA). The best composition was chosen and crosslinked with tetraethyl orthosilicate (TEOS), after which different amounts of graphene oxide (GO) were added to develop composite hydrogels. Fourier transform infrared spectroscopy (FTIR), scanning electron microscopy (SEM), atomic force microscopy (AFM) and contact angle was used to analyze the hydrogels. The samples were also evaluated for swelling abilities in various mediums. The drug release profile was studied in phosphate-buffered saline (PBS) at a pH of 7.4. To predict the mechanism of drug release, the data were fitted into kinetic models. Finally, antibacterial activity and cell viability data were obtained. FTIR studies revealed the successful synthesis of CS/PVA hydrogels and GO/CS/PVA in hydrogel composite. SEM showed no phase separation of the polymers, whereas AFM showed a decrease in surface roughness with an increase in GO content. 100 μL of crosslinker was the critical concentration at which the sample displayed excellent swelling and preserved its structure. Both the crosslinked and composite hydrogel showed good swelling. The most acceptable mechanism of drug release is diffusion-controlled, and it obeys Fick's law of diffusion for drug released. The best fitting of the zero-order, Hixson-Crowell and Higuchi models supported our assumption. The GO/CS/PVA hydrogel composite showed better antibacterial and cell viability behaviors. They can be better biomaterials in biomedical applications.

Keywords: antibacterial activity; biocompatible; chitosan; composite hydrogel; drug release; drug delivery

1. Introduction

With increased environmental awareness and an emphasis on eco-friendly products, the goal of research efforts is in the production of biocompatible, biodegradable and low-cost film-forming materials for biomedical applications [1–3]. Due to their biocompatibility, nontoxicity, biodegradability, ease of availability and low cost, natural polymers have played an important role in producing biomaterials. CS, a natural biopolymer, is regarded as an outstanding candidate for film formation [4]. Alkaline de-acetylation is used to obtain a naturally occurring biopolymer (CS) with a poly-cationic structure from the chitin shells of shrimp and other crustaceans. It's a polysaccharide made up of *N*-acetyl D-glucosamine and D-glucosamine units linked by a (1–4) glycosidic bond that's abundant in nature [5]. CS has been widely used in biomedical and drug delivery applications. It possesses biocompatibility, low toxicity, mucoadhesive properties, antibacterial activities and permeation-enhancing characteristics [5]. However, on the other hand, CS has low mechanical properties that limiting its applications. Several methods have recently been developed to improve the mechanical properties of CS by using nanofillers, for instance, graphene and their derivatives [6]. GO is commonly used in biomedical applications and other carbon-based nanomaterials such as carbon nanotubes (CNTs). It has unique chemical properties, including such as high biocompatibility, low toxicity, a large surface area for effective drug binding, oxygen-containing functionalities and improved conductivity [7]. GO is graphite-derived two-dimensional (2D) carbon material. It has a long history of use as a precursor to chemically converted graphene. Many hydrophilic oxygenated functional groups in GO, such as hydroxyl (–OH), epoxy (–C–O–C–) and carboxyl (–COOH), enable it to disperse in water solution and be easily exfoliated into monolayer sheets [8]. Due to these functional classes, GO is amphiphilic with hydrophilic edges and hydrophobic basal planes. GO contains oxygen-enriched functional groups, which react with various polymers to form polymer-based graphene oxide nanocomposite [9,10]. The large surface area of GO is used to carry and distribute drugs through contact between GO functionalities and drug groups. The importance of GO-based hydrogels in drug release is well known [11], and GO-based drug delivery systems have been extensively researched due to their pH-sensitive drug release behavior. By adjusting the pH of the GO-based hydrogel, controlled drug release was achieved [12]. Polyvinyl alcohol (PVA) is known as green material due to its nontoxicity and biocompatibility. PVA is biodegradable, chemically stable, biocompatible, eco-friendly, as well as water-soluble. It has been extensively researched in various fields and has several available hydroxyl groups (OH) on its polymeric chain [13]. As a result, PVA-based hydrogels find use in the pharmaceutical industry for managed drug delivery. Kim and Kim et al. [14] studied the relationship between the percentage of PVA and drug release time. They developed double-layered composite PVA beads for managed drug delivery. They found that as the amount of PVA increased, the release of therapeutic agents was delayed due to increased cross-linking. Crosslinking is used to enhance the resistance of hydrogel to disintegration in any solution [15]. TEOS is well-known cross-linker and has been widely used in biomedical applications. Several cross-linkers (TEOS, formaldehyde, acetaldehyde and glutaraldehyde, etc.) have been used to cross-link different polymeric chains to develop biomaterials. However, TEOS is an ideal cross-linker since it provides covalent bonding between inorganic and polymer chains [16].

Combining different polymers as a blend is a valuable strategy to create new materials with low cost and synergy in properties. For instance, CS/GO-based hydrogel has been developed without the addition of cross-linker by non-covalent interactions for the application of rapidly self-healing performance. The GO contains several oxygen-based functional groups that formed hydrogen bonds with an amino group of CS to improve the mechanical properties of hydrogel [17]. PVA/CS hydrogel is reported in the literature for its biomedical applications since it has appropriate biocompatibility and nontoxicity. Flexible and robust PVA/GO composite films were prepared with a layered structure by vacuum filtration [18].

Similarly, CS/PVA blended films cross-linked via glutaraldehyde for skin tissue repair are also reported in the literature. These hydrogels have potential applications in artificial muscle and controlled drug release [19]. Various research studies have demonstrated the excellent pH-sensitive ability of PVA and GO. These CS/GO composites films have other widespread applications for bone tissue engineering, drug delivery and water treatment [20]. CS/GO becomes a stable and biocompatible composite with excellent thermal and mechanical properties. Due to the strong hydrogen bonds and electrostatic attraction between negatively charged GO sheets and positively charged polysaccharides groups in CS. Under proper pH conditions, uniformly dispersed CS/ GO films could be fabricated. It is known that the amino group of CS interacts with oxygen functionalities of GO through hydrogen bonding, which results in improved mechanical properties [21].

In this work, a series of hydrogels were synthesized by blending the PVA/CS. The performing samples were selected for the suspension of GO. The prepared hydrogels were characterized by Fourier transform infrared spectroscopy (FTIR), Scanning electron microscope (SEM), Atomic Force Microscope (AFM) and contact angle. The hydrogels were also tested for their swelling capabilities in different media. The drug release profile was investigated in phosphate buffer saline (PBS) at pH = 7.4. Finally, antibacterial activity and cell viability data were obtained.

2. Materials and Methods

2.1. Materials

Chitosan (CS), Tetraethoxysilane (TEOS) and Polyvinyl Alcohol (PVA) were purchased from Merck, Darmstadt, Germany. Graphite powder, NaCl, Na_2HPO_4, KH_2PO_4, KCl, NaOH, HCl, Acetic acid, Phosphoric acid, Sulphuric acid, Potassium Permanganate ($KMnO_4$), Hydrogen peroxide (H_2O_2) and ethanol were purchased from Sigma-Aldrich, Burlington, Massachusetts, USA. Deionized water was used for the preparation of solutions and suspensions. Paracetamol was obtained from a local pharmaceutical company. Mouse pre-osteoblast (*MC3T3-E1*) cell lines were purchased from the American Type Culture Collection (ATCC-USA), and Hyclone Laboratories Inc, Logan, Utah, USA, supplied Alpha-MEM (α-MEM). The fetal bovine serum (FBS) and L-glutamine penicillin/streptomycin were obtained from ThermoFisher Scientific, Gloucester, United Kingdom.

2.2. Experimental Details

2.2.1. Synthesis of Graphene Oxide

GO was prepared by a well-known Hummer method [22]. Graphite powder was used as a precursor for the synthesis of GO. Briefly, 3 g of graphite powder was added to the mixture of H_3PO_4/H_2SO_4 (1:9). Then, 18 g of $KMnO_4$ was added slowly to the mixture under continuous stirring for 12 h using a hotplate. When the reaction is completed, deionized water (400 mL) was added to the mixture, and the mixture was titrated with 3 mL (30% H_2O_2). Finally, the solution was filtered using filter paper, and the filtrated residues were washed with 30% HCl, ethanol, and deionized water repeatedly until neutral pH 7. The resultant residues were dried at 60 °C for 12 h in a vacuum oven and stored in sample containers [23].

2.2.2. Synthesis of Hydrogels

A series of CS/PVA blended films were fabricated via solution casting technique. CS (0.6 g) was dissolved in 2% acetic acid, and PVA (0.4 g) was dissolved in 50 mL deionized water. CS and PVA solutions were mixed under continuous stirring at 55 °C for 2 h. TEOS (100 µL) dissolved in10 mL ethanol was added dropwise into the stirring homogeneous solution at 55 °C. The stirring was continued at 55 °C for another 3 h. After that, the solution was poured carefully into well-cleaned and dried Petri dishes. The Petri dishes were kept in the oven at 50 °C for 3 days. The films with 100 and 150 µL TEOS were prepared via repeating the same procedure. The 100 and 150 µL were coded as CP-1, CP-2. The controlled film sample was synthesized using the same procedure but without the addition of TEOS. The controlled sample was coded as CP. The formulation is tabulated in Table 1.

Table 1. Formulations used for the preparation of hydrogels.

Samples	Formulation	CS (%)	PVA (%)	GO (%)	TEOS (μL)
CP 1	CS/PVA	80	20	0.00	100
CP 2	CS/PVA	60	40	0.00	100
CP 3	CS/PVA	20	80	0.00	100
CP 4	CS/PVA	40	60	0.00	100
CP 5	CS/PVA	60	40	0.00	0
CP 6	CS/PVA	60	40	0.00	150
CP 7	CS/PVA/GO-0.1%	60	40	0.1	100
CP 8	CS/PVA/GO-0.5%	60	40	0.5	100

Formulation of CS/PVA/GO Hydrogels

0.1 % and 0.5% of the prepared GO was dispersed via ultrasonication in deionized water (DI) until it was completely dispersed. The homogeneously dispersed solution of GO had light brown and dark brown colors (Figure 2b,c). Composite films with a varying weight ratio of GO (0, 0.1% and 0.5%) were fabricated via the solution casting technique. The proposed chemical mechanism has been presented in the schematic diagram. The samples were coded as CP-1, CPGO-0.1% and CPGO-0.5%. The proposed mechanism of the fabrication of the hydrogels and GO/ hydrogel composite is depicted in Figure 1, and the digital photographs of composite films are shown in Figure 2g–i. The formulation is tabulated in Table 1.

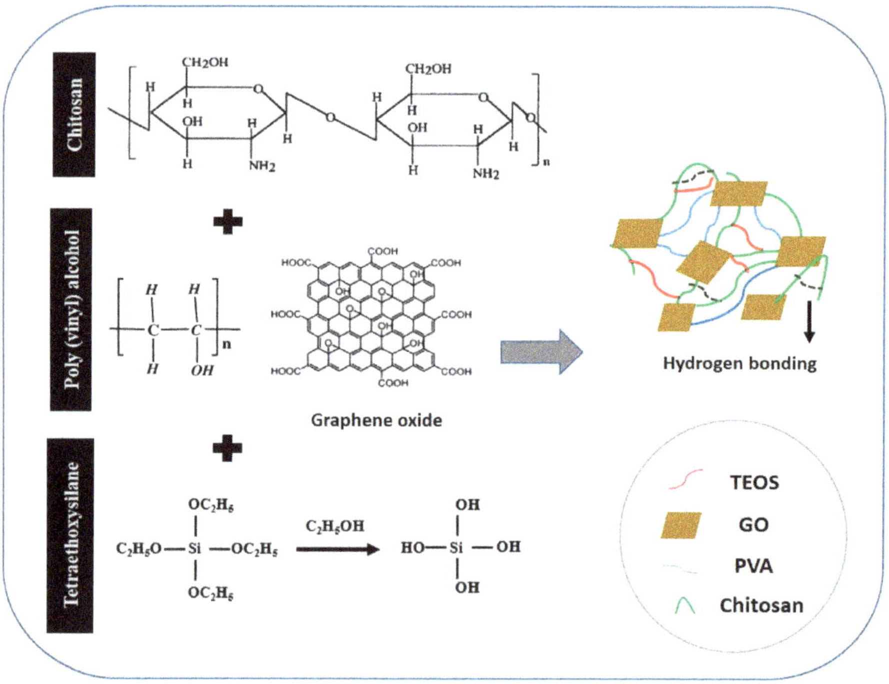

Figure 1. Proposed chemical mechanism of chitosan/polyvinyl alcohol/Graphene oxide composite films.

Figure 2. Photographs of CS/PVA/GO based hydrogel (**a**) graphite oxide in glass vial before dispersion, (**b**) 0.1% dispersed GO in aqueous solution, (**c**) 0.5% dispersed GO, (**d**) CP-1 solution, (**e**) CPGO-0.1%, (**f**) CPGO-0.5%, (**g**) CP-1 composite film, (**h**) CPGO-0.1% composite film and (**i**) CPGO-0.5 % composite film.

2.2.3. Drug-Loading into CS/PVA/GO Hydrogels

CS, PVA and GO solutions were prepared and then blended as mentioned in Section 2.2.2. After 1 h of blending, 20 mg of paracetamol dissolved in 10 mL deionized water was added dropwise to the blended solution by a dropper. The solution was further stirred for 1 h until all the drugs become dissolved. 100 µL of TEOS was dissolved in 10 mL of ethanol and was added dropwise into the drug-loaded blend. The final blend was further stirred for 3 h and poured carefully into the well cleaned and dried Petri dish. The casted films were kept in a vacuum drying oven at 55 °C for two or three days.

3. Characterization

3.1. Fourier Transform Infrared Spectroscopy (FTIR)

The functional group and interactions among all of the components of the prepared hydrogels were investigated using FTIR spectroscopy in ATR mode (Agilent Cary 630). The FTIR study was carried out at a scan rate of 60 scans and wavenumber ranging from 4000 to 400 cm^{-1}, with 4 cm^{-1} resolutions.

3.2. SEM Morphology

Scanning electron microscopy (SEM, JEOL-JSM-6480, Akishima, Japan) was used to examine the morphologies of the well-dried hydrogels, and hydrogel films were sliced and gold-sputtered. These were placed on a stub before being placed in the vacuum chamber to observe surface morphology.

3.3. Swelling and Degradation Analysis

The swelling experiments of the hydrogel (CS/PVA) and the composite (CS/PVA/G0-0.1% and CS/PVA/GO-0.5%) were conducted in DI water, buffer and salt solutions. The blended dried films were first to cut into small pieces, dried and weighed. Then, the samples were placed in DI water in separate containers and allowed to swell. The samples were removed from the containers at pre-decided intervals, adequately cleaned with filter paper to remove the surface water, and weighed again. The swelling experiments were conducted in DI, Buffer solution (at pH 2, 4, 7 and 10) and electrolyte solution (NaCl and

$CaCl_2$ (0.1 M, 0.3 M, 0.7 M and 0.9 M). Swelling (DS) was calculated using Equation (1) [24]. The swelling analysis was performed in triplicated, and average values were taken to calculate the swelling analysis.

$$\text{Swelling index} = \frac{Ws - Wd}{Wd} \quad (1)$$

where Ws = swollen weight of films, and Wd = the dried weight of the films.

The well-dried hydrogels films were cut into square form and weighted (45 mg) carefully. The degradation analysis was performed in PBS solution with pH = 7.4, and these samples were incubated at 37 °C for different periods (1, 2, 3, 5 and 7 days). The degradation of hydrogels was calculated using Equation (2). [24]. The degradation analysis was performed in triplicates, and average values were taken to calculate degradation analysis.

$$\text{Weight loss } (\%) = \frac{W_0 - W_t}{W_0} \times 100 \quad (2)$$

where W_0 = initial hydrogel weight, and W_t = hydrogel weight at the time "t".

3.4. Atomic Force Microscopy (AFM)

AFM (Nano-Solver, NT-MDT) equipped with a silicon nitride tip was used to investigate the surface roughness of the hydrogel samples at ambient conditions. AFM was performed using Nova-Px software over an area of 5 μm × 5 μm in the semi-contact mode. Dried samples (thin films) were used and stuck to the sample holder for AFM analysis.

3.5. In-Vitro Drug Release Procedure

The dried drug-loaded hydrogel was immersed into the beaker containing 100 mL of simulated PBS (pH = 7.4), composed of ions with concentrations close to those of ions present in human blood plasma. The beaker was placed on a magnetic stirrer to maintain the temperature of the system at 37 °C. The samples were withdrawn at various time intervals such as 10, 20, 30, 40, 50 and 60 minutes by syringe, and an equal volume of fresh dissolution medium of PBS pH 7.4 was replaced after each sampling. The sampling process was continued until 180 minutes to evaluate drug release behavior. At the above-mentioned periodic intervals, the drug content in the sample was analyzed at 243 nm in the UV- visible spectrophotometer (Spectro-115U, Reference 3000, Gamry Instruments, Warminster, PA, USA) against reference standard (PBS pH 7.4 as a blank). The amount of the drug was calculated using a standard working curve.

3.6. Contact Angle Measurement

The contact angle measurement was performed via a contact angle analysis system (JY-82, Dingsheng, Chengde, China). Water contact angle (θ) was measured for each sample (1 μL with a flow rate of 0.1 μL/s), and contact angle values were recorded within 0–5 s after placing the drop. The average contact angle values were obtained at least 10 drops in various areas to test the surface, and three test surfaces for each condition were used. All measurements were conducted in air and at ambient temperature.

3.7. Antibacterial Activities

The antimicrobial activities of composite hydrogels were observed against the Gram +ive *Staphylococcus aureus* (S. A.) and Gram -ive Escherichia coli (E. C.) by a well-reported method [22]. A disc diffusion method was employed by placing 80 μL on agar plates. Zone inhibition was measured in mm after incubation for 24 h. After 24 h, the circular inhibition zones were formed and photographed with no bacterial growth.

3.8. *In-Vitro Studies*

3.8.1. Cell Culture and Morphology

The pre-osteoblast (*MC3T3-E1*) cell lines were maintained into *α-MEM* without ascorbic acid containing 10% FBS, 1% (2 mM) L-glutamine, 1% penicillin/streptomycin. The cell density was assumed to be 5000 cells/cm^2 in a 100-mm culture plate. At the same time, gelatin (0.1% by conc.) was employed as a coating agent and controlled. These cell lines with hydrogels were incubated under standard in-vitro conditions (37 °C, 5% CO_2 and 90% humidity).

3.8.2. Cell Viability

MC3T3-E1 cell lines were cultured in varying concentrations of hydrogels (0.500, 1.000 and 2.000 mg/mL), 0.1% gelatin (+ive control) and 1% DMSO (–ive control) and incubated under standard in-vitro conditions for 72 h. After 72 h, these cultured cells were treated with a neutral red agent by a well-reported method by Repetto et al. [25]. These experiments were performed in triplicate, and these treated cells were incubated in a neutral red medium (40 µg/mL) for 2 h. The excess neutral red stain was removed by washing these cells with PBS solution and incubating for 2 h. Later, these cell lines were treated with destaining solution (50% distilled water, 49% absolute ethanol and 1% glacial acetic acid) at 37 °C for 10 min to destained cell lines. An absorbance microplate reader observed the optical density was observed at 570 nm by an absorbance microplate reader (Bio-Tek, ELx-800, Winooski, VT, USA). The cell viability percentage was calculated by Equation (3).

$$\text{Cell viability} = \frac{OD_S}{OD_C} \times 100 \tag{3}$$

whereas; OD_S = optical density of sample concentration and OD_C = the optical density of the positive control.

3.9. *Statistical Analysis*

The obtained experimental data has been presented in mean and standard errors (mean ± *S.E.*), and it is calculated by statistical software (IBM, SPSS Statistics 21). The mean, standard errors of means have been shown as Y-error bars in figures. ($p < 0.05$; n = 3).

4. Results and Discussion

4.1. *Morphology Study*

The surface SEM images of the CS/PVA and CS/PVA/GO hydrogels are shown in Figure 3. The SEM micrographs indicate that all of the CS/PVA hydrogels have homogenous surfaces; this indicates that both polymers mixed effectively. Furthermore, not a single hydrogel displayed phase separation when the CS to PVA ratio was changed, confirming that concentration changes did not affect the miscibility of the two polymers. Surfaces of CS/PVA/GO hydrogels indicated the presence of an exterior phase. This exterior phase comprises GO particles that have grown from the hydrogel surface [26].

4.2. *Swelling and Degradation Analysis*

Figure 4a–f illustrates the hydrogels' swelling behavior as a function of polymer ratio, cross-linker concentration, GO concentration and salt concentration and type. As shown in Figure 4a, as the ratio of the polymers in the hydrogel was changed, keeping the cross-linker constant, swelling is increased. The maximum swelling was obtained at CS/PVA (2:8). It was also found that the sample was unstable in the aqueous solution and was challenging to handle. It can be easily observed by the large error bars. The sample's instability is attributed to the fact that just one polymer, CS, is cross-linked, resulting in a loosely bounded semi-interpenetrating network (semi IPN). As the CS contraction in the hydrogel increased, the hydrogel became more stable. The swelling and stability of the hydrogels were optimized by keeping ratio 6 (CS): 4 (PVA), and the relationship between stability and swelling behavior of the hydrogel was investigated. The swelling behavior of a hydrogel with the same polymer ratio, with different cross-linker concentrations (100 and 150 L), was investigated further. Figure 4b clearly shows that the sample without cross-linking did not exhibit proper swelling behavior, but the hydrogel sample with a 100 µL cross-linker demonstrated optimal swelling. It was demonstrated that 100 µL of cross-linker was the critical concentration since the sample displayed excellent swelling and preserved its structure. The decrease in the swelling of hydrogels (with a cross-linking concentration of 150 µL) can be ascribed to the contraction of pore size and tremendous interconnectivity, which may explain why water was unable to enter the spaces, resulting in the decrease in swelling. This tendency is further verified and supported by reducing swelling as the cross-linker quantity has increased. Similar results were also obtained by Wang et al. [22]. The cross-linked hydrogel having CS/PVA 6:4 was selected for the fabrication of the composite hydrogels.

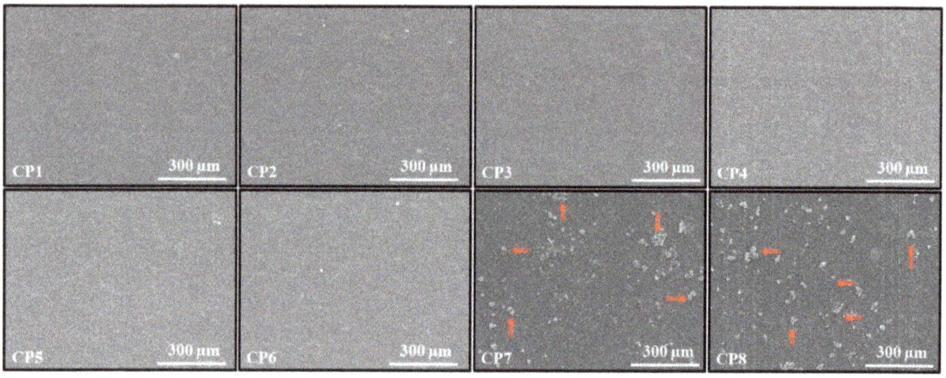

Figure 3. Surface SEM micrographs of the CS/PVA and CS/PVA/GO hydrogels film.

The effect of GO content on hydrogel swelling is depicted in Figure 4c. The swelling ratio of hydrogels with the maximum GO concentration tends to be lower than CP-1 and CPGO-0.1%. This result is attributed to the increasing cross-linking density of hydrogel due to the hydrogen bonds between GO, CS and PVA, which act as a physical cross-linker [27]. As shown in Figure 4d, the hydrogels were also exposed to varied pH (2, 4, 7 and 10) conditions, which resulted in different swelling behaviors. The pH of the swelling media affects the swelling ratio, which can be ascribed to the amino groups in the hydrogel network structure. In an acidic environment, CPG0–0.1% swells more than CP, CP-1 and CPG0–0.5%. The significant swelling in acidic conditions is due to the charged amino groups (NH^{3+}) in the polymeric structure, which causes electrostatic repulsion between the polymer chains, enabling maximal water absorption in the network structure [28]. Deprotonation of the amine functional groups occurs in a neutral or alkaline media, and the NH^{3+} groups return to NH_2. These hydrogels are pH-sensitive due to a carboxylic group of graphene oxide and amino groups of CS. Figure 4e,f shows two trends for CP, CP-1, CPGO-0.1% and CPGO-0.5% in varied molar concentrations of NaCl and CaCl2 solutions. The swelling ratio of hydrogels reduced as the concentrations of both electrolytes increased. All hydrogels displayed better swelling at 0.1 mol/L NaCl and $CaCl_2$ concentrations. The formation of ion pairs is responsible for reducing swelling of the hydrogels as the salt concentration increases. Ion pairs act as new cross-linking sites, which causes the polymer chain to shrink and decreasing swelling (24 h).

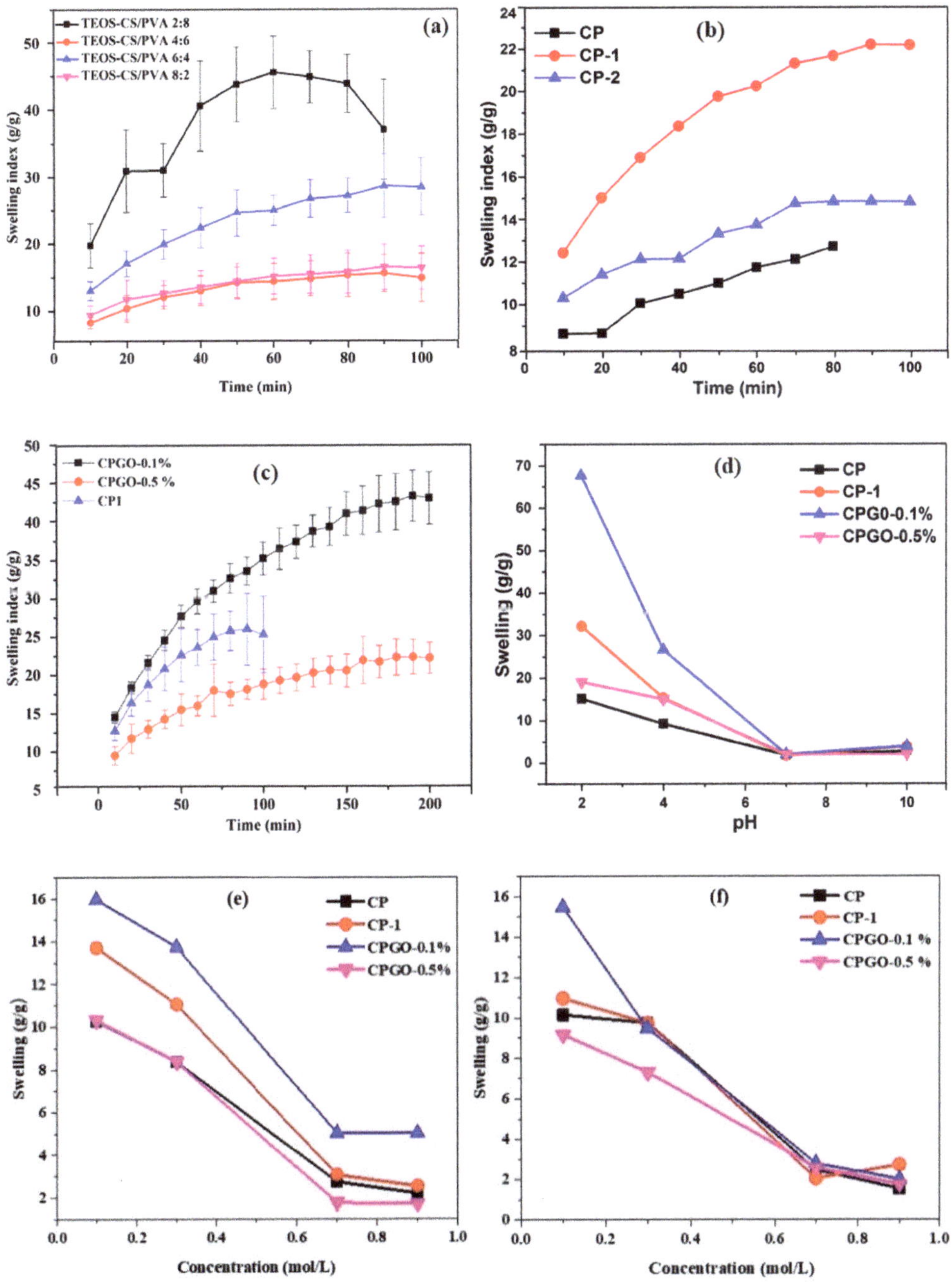
(a)
TEOS-CS/PVA 2:8
TEOS-CS/PVA 4:6
TEOS-CS/PVA 6:4
TEOS-CS/PVA 8:2
Swelling index (g/g)
Time (min)
(b)
CP
CP-1
CP-2
Swelling index (g/g)
Time (min)
(c)
CPGO-0.1%
CPGO-0.5 %
CP1
Swelling index (g/g)
Time (min)
(d)
CP
CP-1
CPG0-0.1%
CPGO-0.5%
Swelling (g/g)
pH
(e)
CP
CP-1
CPGO-0.1%
CPGO-0.5%
Swelling (g/g)
Concentration (mol/L)
(f)
CP
CP-1
CPGO-0.1 %
CPGO-0.5 %
Swelling (g/g)
Concentration (mol/L)

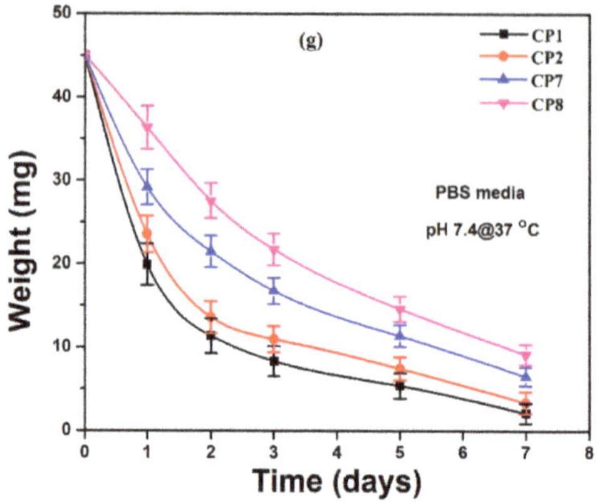

Figure 4. Swelling behavior of the hydrogels as a function of polymer ratio, cross-linker concentration, different GO amount, different pH, salt concentration and degradation (**a**) polymers ratio, (**b**) cross-linker concentration, (**c**) different GO amount, (**d**) different pH, (**e**,**f**) salt concentration and (**g**) degradation analysis of hydrogel films in PBS media (pH = 7.4 at 37 °C).

The degradation of hydrogel films was conducted in PBS media to determine the degradation behavior of well-dried hydrogel films. It was observed that hydrogel samples CP1 and CP2 degraded faster than the hydrogel samples CP7 and CP8. All these hydrogel samples have the same amount of TEOS (crosslinkers). However, hydrogel samples CP1 and CP2 are without GO. Even the degradation is different for the hydrogels sample CP7 and CP8. It is attributed to the different amounts of GO in these samples. Hydrogel sample CP7 with less amount of GO degraded more than CP8. Hence, different degradation behavior of the hydrogel sample confirms that they are successfully crosslinked and have developed differently types of interaction due to GO in hydrogels sample CP7 and CP8. Therefore we can say that an increasing the amount of GO may cause a delay in the degradation of the hydrogel samples.

4.3. FTIR Analysis

FTIR spectrum of different samples was investigated, as shown in Figure 5. The functional groups confirmed the oxidation process on the GO surface. The FTIR spectrum of CPG0–0.1% displays a broadband between 3212 to 3258 cm^{-1}, which showed –OH stretching of intermolecular and intramolecular hydrogen bonding. The characteristic peaks of alkyl groups (–CH stretching) were observed at 2920 cm^{-1}. The peak at 1246 cm^{-1} was attributed to –C–O–C stretching [29]. The peaks between 1400 cm^{-1} to 1500 cm^{-1} which attributed to –NH stretching. The FTIR spectrum of CS/PVA hydrogels revealed the characteristic peaks of siloxane at 1065 cm^{-1}. The FTIR investigations confirmed the presence of siloxane linkage among chitosan, polyvinyl alcohol and GO through the intermolecular hydrogen bond formation along with TEOS [30].

4.4. AFM Topography

AFM images of the prepared hydrogels (Figure 6) show a fascinating behavior in terms of surface roughness. The controlled film without cross-linker (CP) shows the highest roughness as compared to other samples. It is well understood as the non-crosslinked sample has a relatively open and porous structure. Adding GO in the range of 0.1 to 0.5 µg/mL to the cross-linked hydrogel (CP-1) causes the roughness of hydrogel samples to decrease gradually (Table 2). This reduction in roughness is due to the large surface area of GO which covers the pores, and a more compact structure of the hydrogel that offers a smooth surface morphology [31].

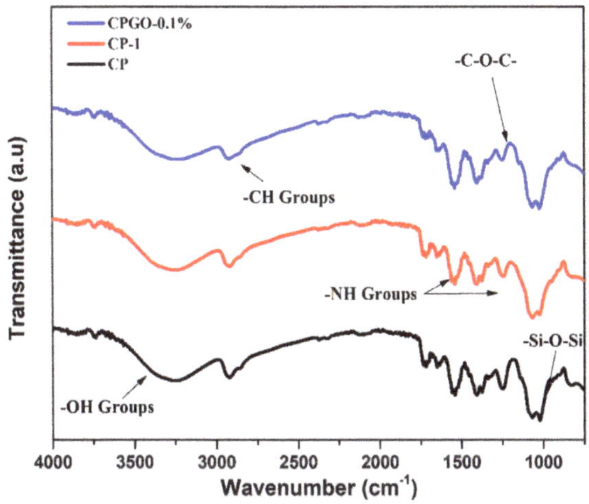

Figure 5. FTIR spectra of CS/PVA and CS/PVA/GO hydrogels.

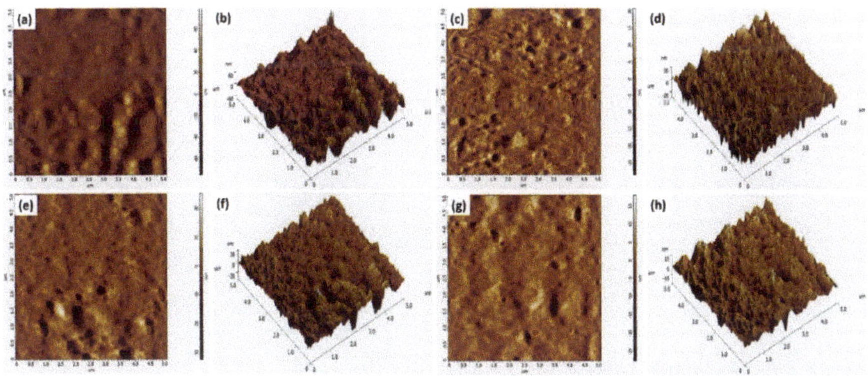

Figure 6. AFM images of (**a**,**b**) CP-1, (**c**,**d**) CP-2 (**e**,**f**) CP-7 and (**g**,**h**) CP-8.

Table 2. Avg. roughness of different hydrogel samples.

Sample	Avg. Roughness (nm)	Roughness (Centerline at Y-Axis) nm
CP1 (a)	2.401 ± 0.17	2.245 ± 0.19
CP2 (c)	3.861 ± 0.23	3.186 ± 0.27
CP7 (e)	4.644 ± 0.36	3.376 ± 0.43
CP8 (g)	7.532 ± 0.49	7.161 ± 0.57

4.5. In Vitro Drug Release Analysis

Hydrogels offer different release mechanisms to entrap drugs, such as swelling controlled, chemically controlled and diffusion-controlled mechanisms. The most acceptable one is the diffusion-controlled mechanism, and it obeys Fick's law of diffusion for drug released [32]. The diffusion coefficient of the hydrogels is related to the porosity of the hydrogels if the pore size of porous hydrogels is greater than the molecular dimensions of the drug molecule. Drug release only follows the swelling controlled mechanism if drug release time exceeds the time of swelling [21]. The release behavior of paracetamol from the hydrogel with and without GO is shown in Figure 7. The paracetamol releases slowly from

GO, and the release rate gradually declines after 140 min. Figure 7c,d presents the SEM analysis of CP7 to determine surface morphology before and after drug release analysis.

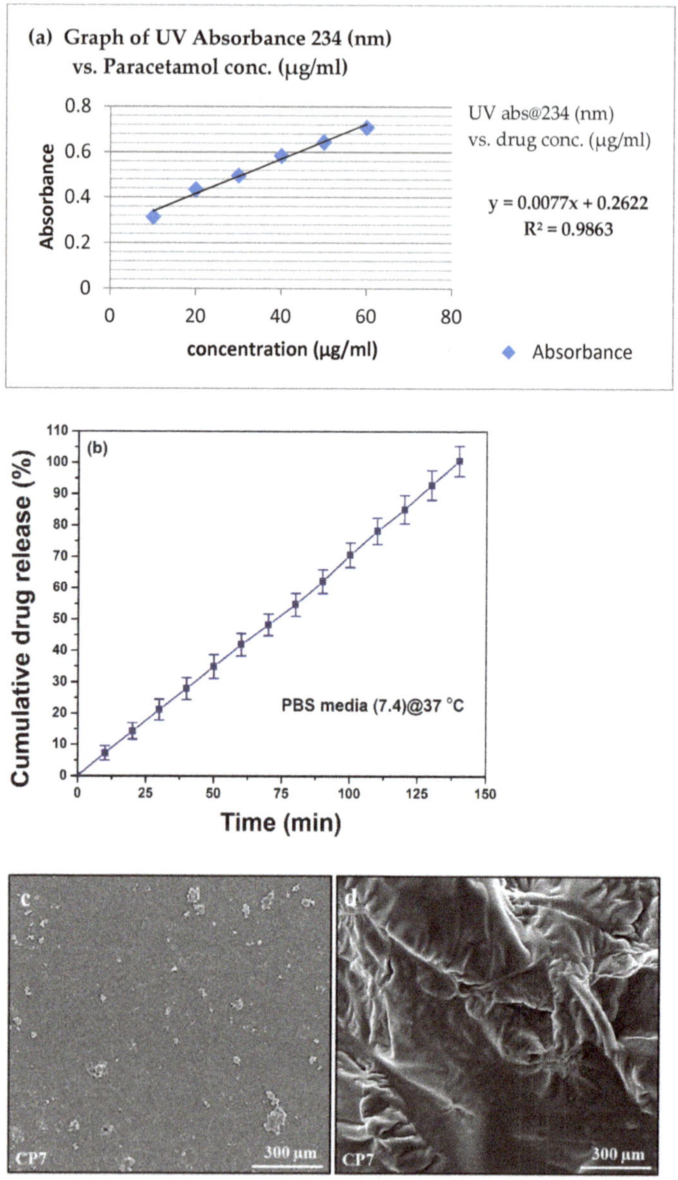

Figure 7. (**a**) Calibration curve of paracetamol in PBS pH 7.4 (**b**) Cumulative percentage of drug release with time (min) for CP7 hydrogel, (**c**) SEM analysis of C7 before drug release and (**d**) SEM analysis of CP7 after drug release.

On the other hand, CP1 released 90% drug within 120 min. The sustained release of drug in CPGO-0.1% attributed to the incorporation of GO in the polymer matrix. A six-point calibration graph was constructed in the concentration range of 10–60 µg/mL to accurately measure the released drug amount. Paracetamol drug release kinetic studies were investigated against various models (zero-order, first-order, Higuchi, Korsmeyer-Peppas and Hixson-Crowell, models) [21,32]. The mathematical model proves to be highly effective in forecasting drug release and predicting the mechanism of release.

$$\text{Zero-order } M_t = M_o + K_o t \tag{4}$$

$$\text{First order}: \ logC_o - \frac{kt}{2.303} \tag{5}$$

$$\text{Higuchi model}: \ ft \ = Q = K_H \ \times \ t^{1/2} \tag{6}$$

$$\text{Hixson Crowell model}: \ W^{1/3} - W^{1/3} = \ kt \tag{7}$$

$$\text{Korsmeyer-Peppas model}: \ ln\frac{M_t}{M_o} \ = n \ lnt \ + lnK \tag{8}$$

whereas M_t = drug release amount at time t, K_H, K and Ko are constants.

To explain the dissolution profile, model-dependent approaches use several mathematical functions. To investigate the drug release mechanism of paracetamol, model-dependent methods such as zero order, first order, Higuchi, Hixson-Crowell and Korsmeyer-Peppas models were used. Figure 8 clearly shows that best fitting was obtained with zero order, followed by Hixson-Crowell and Higuchi models. The zero-order model says that the dissolution of dosage forms does not disaggregate and slowly releases a constant drug. According to Hixson-Crowell, drug release is a process that is governed by diffusion, dissolution or both. Generally, diffusion-controlled drug release is predicted by the Higuchi square root of time model. Higuchi model is based on the hypotheses that (i) initial drug concentration in the matrix is much higher than drug solubility; (ii) drug diffusion takes place only in one dimension (edge effect must be negligible); (iii) matrix swelling and dissolution are negligible; (iv) drug diffusivity is constant; and (v) in the release environment, ideal sink conditions are always achieved. Since our results are obeying all three models (Figure 8) and since all of these models are very close when it comes to diffusion; hence it can be concluded that the release of paracetamol from the present system is a controlled process that is unaffected by drug concentration [21,33,34].

4.6. Water Contact Angle Measurement

The water contact angle is vital for determining hydrogel hydrophilicity or hydrophobicity, an essential factor in drug delivery, cell proliferation and adherence. Consequently, the wetting behavior of hydrogel samples (CP1, CP2, CP3, CP4, CP5, CP6, CP7 and CP8) was analyzed using water contact angles of hydrogels. The water contact angle is vital for determining hydrogel hydrophilicity or hydrophobicity, an essential factor in drug delivery, cell proliferation and adherence [33]. TEOS cross-linking played a crucial role in studying the water contact angle, as seen in Figure 9. The increasing amount of PVA shifts wetting behavior from hydrophilicity to hydrophobicity with an optimized quantity of cross-linker [34].

In comparison, contact angle values, different behavior of CP7 and CP8 were observed due to different amounts of GO sheets. F8 is more hydrophilic than CP7 due to the amount of GO sheet [35]. The hydrophilic behavior of hydrogels (CP7 and CP8) issue to various functional groups available over the surface that encourages Hydrogen bonding. The hydrogel samples showed improved water contact angle values due to the increasing amount of TEOS quantities. The increasing cross-linking changes the wetting behavior of hydrophilicity to hydrophobicity [36,37]. Since the close packing of polymeric structure and free functional are not available due to more covalent bonding, very few functional groups are available for hydrogen bonding and increasing cross-linking due to the increasing amount of TEOS, the hydrophilic character was shifted towards hydrophobic character [38]. Hence, hydrophilicity can be optimized using appropriate TEOS for required cross-linking for controlled drug delivery, swelling of hydrogel cell proliferation and adherence.

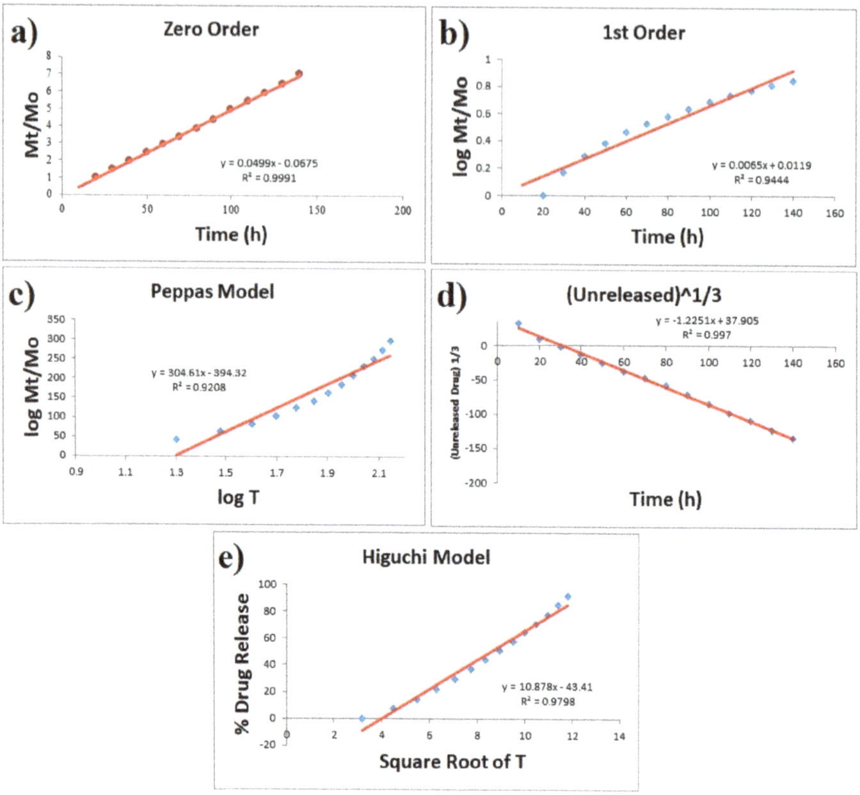

Figure 8. Paracetamol release drug studied various kinetic models ((**a**) zero-order, (**b**) first-order, (**c**) Higuchi, (**d**) Korsmeyer-Peppas and (**e**) Hixson-Crowell).

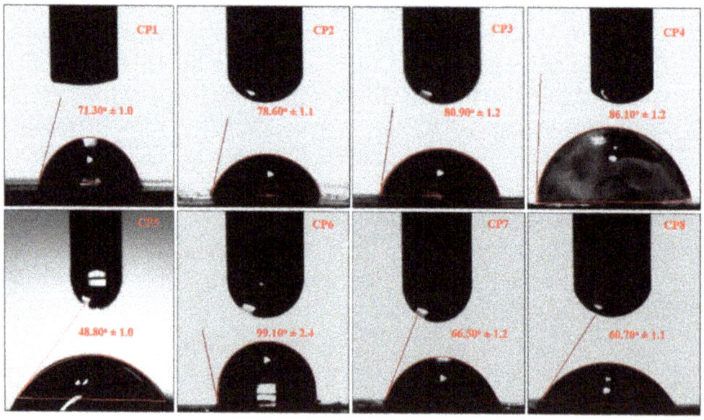

Figure 9. The wetting behavior of all samples of hydrogels.

4.7. *Antimicrobial activity*

The antimicrobial activities of all hydrogels are shown in Figure 10—the antimicrobial activities against Gram +ive and –ive bacterial *S. aureus* and *E. coli* were studied using the agar well diffusion method. The hydrogel samples with different compositions demonstrated different inhibition zones (mm) against the used pathogens. The hydrogel samples with GO (CP7 and CP8) showed maximum antibacterial activity, whereas those without

GO (CP1, CP2, CP3, CP4, CP5 and CP6) showed lower antibacterial activity. Both the increase in CS and GO in the hydrogels enhanced the antibacterial activities. The different zones of inhabitations against *E. coli* for samples were CP1 = 19.9 ± 1.0, CP2 = 17.3 ± 1.2, CP3 = 9.9 ± 1.1, CP4 = 11.2 ± 1.1, CP5= 13.3 ± 1.2, CP6 = 16.7 ± 1.3, CP7 = 29.9 ± 1.3 and CP 8 = 34.7 ± 1.5 and against *S. aureus* were CP1 = 24.1 ± 1.4, CP2 = 21.4 ± 1.2, CP3 = 12.2 ± 0.90, CP4 = 14.7 ± 1.0, CP5 = 17.6 ± 1.2, CP6 = 18.1 ± 1.1, CP7 = 28.6 ± 1.2 and CP8 = 32.4 ± 1.4. These results are attributed to the functionalities of Cs and GO. The functional groups interacted with the charged component of the bacterial surface membrane (i.e., phospholipids and lipopolysaccharides) [39], which interrupted the bacterial activity of both pathogens. CS has advantages over other polysaccharides due to the $-NH_2$ functionality that easily bonds with bacterial DNA. That is one of the reasons that an increasing amount of CS causes more antibacterial activities. By forming new bonds with bacterial DNA, hydrogel controls the whole bacteria and hinders its further growth. Due to their sharp edges, GO nanosheets rapture the bacterial membranes and their functional groups and control the bacterial activities. Hence, an increasing amount of GO caused even better antibacterial activities than CS [22].

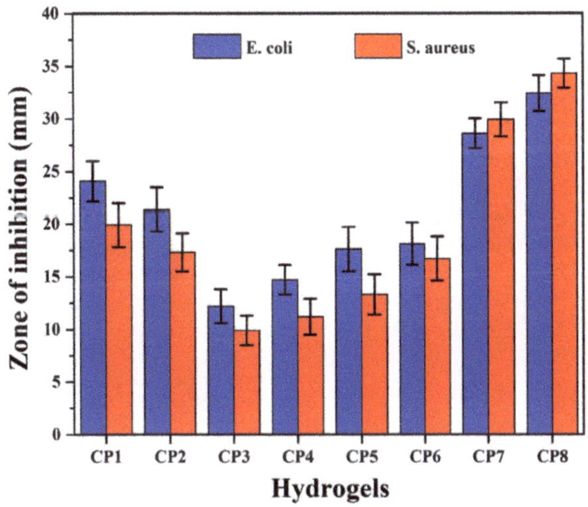

Figure 10. Antibacterial activities of all samples of hydrogels against severe disease-causing pathogens.

4.8. Cell Morphology

MC3T3-E1 cell lines were used to evaluate the in-vitro biocompatibility of hydrogels. The results are presented in Figure 11. Different functional groups enhance biocompatibility and cell differentiation on the substrate surfaces [28]. The increasing amount of CS causes more cell growth rather than an increasing amount of PVA (CP1 > CP2 > CP4 > CP3 > CP5 > CP6) due to the biocompatible affinity of CS with the MC3T3-E1 cell lines. On the other hand, an increasing GO facilitates H-bonding (CP8 > CP7) due to its various functional groups. The increasing amount of GO increases the functionalities and surface area that helps cell adherence and cell growth. It, in turn, confirms the enhancement of the micro-environment that expedites extracellular matrix (ECM) and biomaterials [40]. The matrix of the composite hydrogels has different chargeable functional groups ($-COOH$, $-NH_2$, $-H$ and $-OH$ groups), and hydrogels have several active sites for cell adhesion, growth and spreading. Protein adsorption modulation via integrin binding to negative modified surfaces could be used to regulate cell adhesion. The fibronectin adsorption and integrin-binding have been reported in the following order $-OH > -COOH > -NH_2 > -CH_3$ to enhance osteoblasts cells [41].

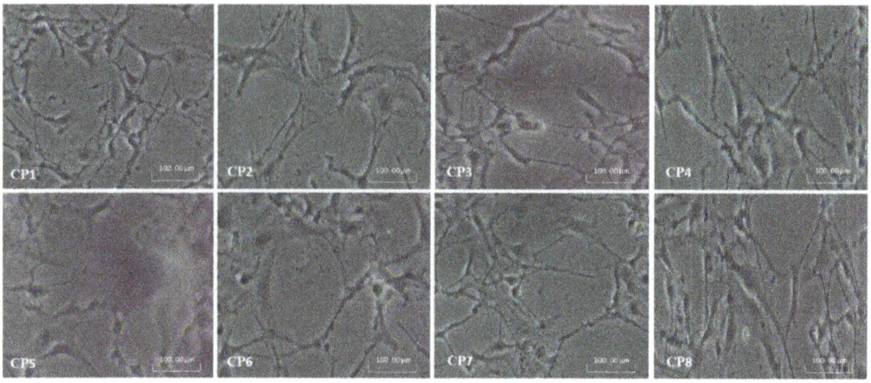

Figure 11. Shows the cell morphology of all hydrogels against pre-osteoblast (MC3T3-E1) cell lines.

4.9. Cell Viability and Optical Density

The cell viability and optical density assay of hydrogel samples were observed against *MC3T3-E1* cell lines with alongside concentrations (0.500, 1.000 and 2.000 µg/mL) after 72 h of incubation at 37 °C [42,43] as shown in Figure 12a,b. Among all hydrogels, CP1 and CP2, CP7 and CP8 exhibited maximum cell viability and optical density. Consequently, concentration 3 µg/mL was more appropriate for better results [2,43]. It was found that increasing the concentration of CS facilitates cell viability and optical density. Furthermore, a similar effect is observed with the increase in the amount of GO. As mentioned, these results were also attributed to the biocompatible affinity of CS with the *MC3T3-E1* cell lines and enhance the micro-environment (that expedite extracellular matrix (*ECM*)) provided by the functionalities of GO. Hence, all hydrogels had cell viability and nontoxicity towards pre-osteoblast cells with different values.

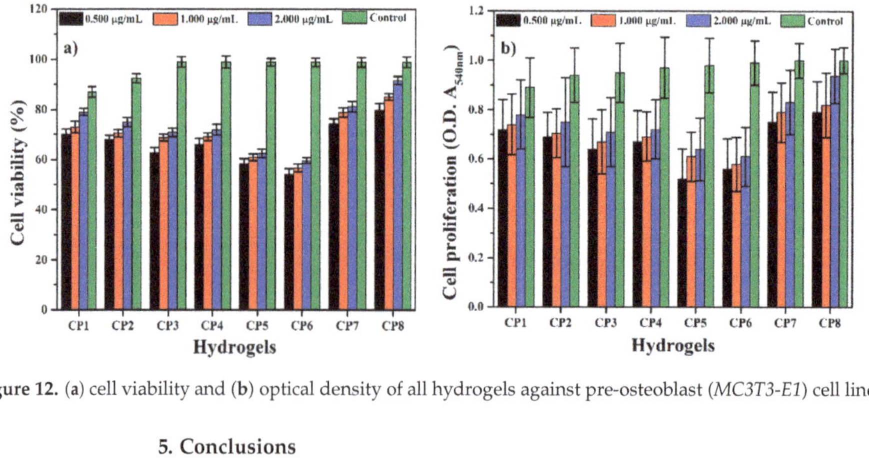

Figure 12. (**a**) cell viability and (**b**) optical density of all hydrogels against pre-osteoblast (*MC3T3-E1*) cell lines.

5. Conclusions

In the present work, novel hydrogels were successfully fabricated using the solution casting method. The best composition was selected and crosslinked with a critical concentration of tetraethyl orthosilicate (TEOS). For the fabrication of the GO/CS/PVA composite, this mixture was used. The samples were also tested for their ability to swell in a variety of media. The degradation analysis reveals that CP8 has a delayed degradation pattern, whereas CP1 has a rapid degradation behavior. The drug release profile was studied in phosphate-buffered saline (PBS) at a pH of 7.4. The data were fitted into kinetic models to predict the mechanism of drug release. Finally, information on antibacterial activity and cell viability was obtained. The successful synthesis of CS/PVA hydrogels and GO/CS/PVA composite hydrogel was revealed by FTIR studies. SEM revealed no phase separation between the polymers, whereas AFM revealed decreased surface roughness as GO content increased. The critical crosslinker concentration was 100 μL, at which the sample showed excellent swelling and structure preservation. The crosslinked and composite hydrogels both swelled, showing different swelling behavior. Diffusion-controlled drug release is the most acceptable mechanism, and it follows Fick's law of diffusion for drug release. Our hypothesis was supported by the best fitting zero-order, Hixson-Crowell and Higuchi models. The antibacterial and cell viability properties of the GO/CS/PVA hydrogels composite were better wound dressing biomaterial for wound healing and treatment applications.

Author Contributions: Conceptualization, M.U.A.K. and S.H.; methodology, M.U.A.K., S.H. and M.A.R.; software, M.U.A.K., Z.Y., S.H. and A.S.; validation, M.U.A.K., S.H., S.I.A.R. and M.A.R.; formal analysis, M.U.A.K., S.H. and Z.Y.; investigation, S.I.A.R. and M.A.R.; resources, M.U.A.K., S.I.A.R. and M.A.R.; data curation, M.U.A.K., Z.Y., S.H. and F.M.B.; writing—original draft preparation, M.U.A.K.; writing—review and editing, M.U.A.K., S.H. and S.I.A.R.; visualization, M.U.A.K., Z.Y. and S.H.; supervision, M.U.A.K., Saiful Izwan and M.A.R.; project administration, M.U.A.K., S.H. and M.A.R.; and funding acquisition, M.U.A.K., S.I.A.R. and M.N.M.A. All authors have read and agreed to the published version of the manuscript.

Funding: This research was funded by the Innovative Research and Management Centre (iRMC), Universiti Tenaga Nasional, Malaysia, research publication BOLD grant (J510050002), and Universiti Teknologi Malaysia for research grant number 02M44.

Institutional Review Board Statement: Not applicable.

Informed Consent Statement: Not applicable.

Data Availability Statement: The data presented in this study are available on request from the corresponding author.

Acknowledgments: The authors would like to thank the Innovative Research and Management Centre (iRMC), Universiti Tenaga Nasional, Malaysia for providing research publication BOLD grant (J510050002), the Universiti Teknologi Malaysia for research grant number 02M44 and Legasi Megajaya Sdn Bhd (M) for providing the facilities.

Conflicts of Interest: All authors declare no conflict.

References

1. Satyanarayana, K.G.; Arizaga, G.G.; Wypych, F. Biodegradable composites based on lignocellulosic fibers—An overview. *Prog. Polym. Sci.* **2009**, *34*, 982–1021. [CrossRef]
2. Khan, M.U.A.; Haider, S.; Shah, S.A.; Abd Razak, S.I.; Hassan, S.A.; Kadir, M.R.A.; Haider, A. Arabinoxylan-co-AA/HAp/TiO_2 nanocomposite scaffold a potential material for bone tissue engineering: An in vitro study. *Int. J. Biol. Macromol.* **2020**, *151*, 584–594. [CrossRef] [PubMed]
3. Khan, M.U.A.; Raza, M.A.; Mehboob, H.; Kadir, M.R.A.; Abd Razak, S.I.; Shah, S.A.; Iqbal, M.Z.; Amin, R. Development and in vitro evaluation of κ-carrageenan based polymeric hybrid nanocomposite scaffolds for bone tissue engineering. *RSC Adv.* **2020**, *10*, 40529–40542. [CrossRef]
4. Kondaveeti, S.; Damato, T.C.; Carmona-Ribeiro, A.M.; Sierakowski, M.R.; Petri, D.F.S. Sustainable hydroxypropyl methylcellulose/xyloglucan/gentamicin films with antimicrobial properties. *Carbohydr. Polym.* **2017**, *165*, 285–293. [CrossRef]
5. Moura, D.; Mano, J.F.; Paiva, M.C.; Alves, N.M. Chitosan nanocomposites based on distinct inorganic fillers for biomedical applications. *Sci. Technol. Adv. Ma TerialS* **2016**, *17*, 626–643. [CrossRef]
6. Li, S.; Wang, L.; Yu, X.; Wang, C.; Wang, Z. Synthesis and characterization of a novel double cross-linked hydrogel based on Diels-Alder click reaction and coordination bonding. *Mater. Sci. Eng. C* **2018**, *82*, 299–309. [CrossRef] [PubMed]
7. Liu, Z.; Robinson, J.T.; Tabakman, S.M.; Yang, K.; Dai, H. Carbon materials for drug delivery & cancer therapy. *Mater. Today* **2011**, *14*, 316–323.
8. Huang, Y.; Zeng, M.; Ren, J.; Wang, J.; Fan, L.; Xu, Q. Preparation and swelling properties of graphene oxide/poly (acrylic acid-co-acrylamide) super-absorbent hydrogel nanocomposites. *Colloids Surf. A Physicochem. Eng. Asp.* **2012**, *401*, 97–106. [CrossRef]
9. Chee, W.; Lim, H.; Huang, N.; Harrison, I. Nanocomposites of graphene/polymers: A review. *Rsc Adv.* **2015**, *5*, 68014–68051. [CrossRef]

10. Khan, M.U.A.; Abd Razak, S.I.; Mehboob, H.; Abdul Kadir, M.R.; Anand, T.J.S.; Inam, F.; Shah, S.A.; Abdel-Haliem, M.E.; Amin, R. Synthesis and characterization of silver-coated polymeric scaffolds for bone tissue engineering: Antibacterial and in vitro evaluation of cytotoxicity and biocompatibility. *ACS Omega* **2021**, *6*, 4335–4346. [CrossRef]
11. Khan, M.U.A.; Haider, S.; Haider, A.; Abd Razak, S.I.; Kadir, M.R.A.; Shah, S.A.; Javed, A.; Shakir, I.; Al-Zahrani, A.A. Development of porous, antibacterial and biocompatible GO/n-HAp/bacterial cellulose/β-glucan biocomposite scaffold for bone tissue engineering. *Arab. J. Chem.* **2021**, *14*, 102924. [CrossRef]
12. Rana, V.K.; Choi, M.C.; Kong, J.Y.; Kim, G.Y.; Kim, M.J.; Kim, S.H.; Mishra, S.; Singh, R.P.; Ha, C.S. Synthesis and drug-delivery behavior of chitosan-functionalized graphene oxide hybrid nanosheets. *Macromol. Mater. Eng.* **2011**, *296*, 131–140. [CrossRef]
13. Hassan, C.M.; Peppas, N.A. Structure and morphology of freeze/thawed PVA hydrogels. *Macromolecules* **2000**, *33*, 2472–2479. [CrossRef]
14. Kim, C.-J.; Lee, P.I. Composite poly (vinyl alcohol) beads for controlled drug delivery. *Pharm. Res.* **1992**, *9*, 10–16. [CrossRef] [PubMed]
15. Singh, A.; Narvi, S.; Dutta, P.; Pandey, N. External stimuli response on a novel chitosan hydrogel crosslinked with formaldehyde. *Bull. Mater. Sci.* **2006**, *29*, 233–238. [CrossRef]
16. Khan, M.U.A.; Raza, M.A.; Razak, S.I.A.; Abdul Kadir, M.R.; Haider, A.; Shah, S.A.; Mohd Yusof, A.H.; Haider, S.; Shakir, I.; Aftab, S. Novel functional antimicrobial and biocompatible arabinoxylan/guar gum hydrogel for skin wound dressing applications. *J. Tissue Eng. Regen. Med.* **2020**, *14*, 1488–1501. [CrossRef]
17. Hu, H.; Wang, X.; Wang, J.; Liu, F.; Zhang, M.; Xu, C. Microwave-assisted covalent modification of graphene nanosheets with chitosan and its electrorheological characteristics. *Appl. Surf. Sci.* **2011**, *257*, 2637–2642. [CrossRef]
18. Aslam Khan, M.U.; Haider, A.; Abd Razak, S.I.; Abdul Kadir, M.R.; Haider, S.; Shah, S.A.; Hasan, A.; Khan, R.; Khan, S.U.D.; Shakir, I. Arabinoxylan/graphene-oxide/nHAp-NPs/PVA bionano composite scaffolds for fractured bone healing. *J. Tissue Eng. Regen. Med.* **2021**, *15*, 322–335. [CrossRef]
19. Costa-Júnior, E.S.; Barbosa-Stancioli, E.F.; Mansur, A.A.; Vasconcelos, W.L.; Mansur, H.S. Preparation and characterization of chitosan/poly (vinyl alcohol) chemically crosslinked blends for biomedical applications. *Carbohydr. Polym.* **2009**, *76*, 472–481. [CrossRef]
20. Abolhassani, M.; Griggs, C.S.; Gurtowski, L.A.; Mattei-Sosa, J.A.; Nevins, M.; Medina, V.F.; Morgan, T.A.; Greenlee, L.F. Scalable chitosan-graphene oxide membranes: The effect of GO size on properties and cross-flow filtration performance. *ACS Omega* **2017**, *2*, 8751–8759. [CrossRef]
21. Nazir, S.; Khan, M.U.A.; Al-Arjan, W.S.; Abd Razak, S.I.; Javed, A.; Kadir, M.R.A. Nanocomposite hydrogels for melanoma skin cancer care and treatment: In-vitro drug delivery, drug release kinetics and anti-cancer activities. *Arab. J. Chem.* **2021**, *14*, 103120. [CrossRef]
22. Wang, X.; Du, Y.; Luo, J.; Yang, J.; Wang, W.; Kennedy, J.F. A novel biopolymer/rectorite nanocomposite with antimicrobial activity. *Carbohydr. Polym.* **2009**, *77*, 449–456. [CrossRef]
23. Raza, M.A.; Rehman, Z.U.; Ghauri, F.A. Corrosion study of silane-functionalized graphene oxide coatings on copper. *Thin Solid Film.* **2018**, *663*, 93–99. [CrossRef]
24. Haider, S.; Park, S.-Y.; Lee, S.-H. Preparation, swelling and electro-mechano-chemical behaviors of a gelatin–chitosan blend membrane. *Soft Matter* **2008**, *4*, 485–492. [CrossRef] [PubMed]
25. Repetto, G.; Del Peso, A.; Zurita, J.L. Neutral red uptake assay for the estimation of cell viability/cytotoxicity. *Nat. Protoc.* **2008**, *3*, 1125. [CrossRef]
26. Tayefeh, A.; Mousavi, S.; Wiesner, M.; Poursalehi, R. Synthesis and surface characterization of magnetite-titania nanoparticles/polyamide nanocomposite smart RO membrane. *Procedia Mater. Sci.* **2015**, *11*, 342–346. [CrossRef]
27. Jin, X.; Zhang, G.; Li, H. A pH sensitive graphene oxide/poly (N-methylolacrylamide-methyl acrylate) composite hydrogel. In *IOP Conference Series: Materials Science and Engineering*; IOP Publishing: Bristol, UK, 2017; Volume 274, p. 012114.
28. Bet, M.; Goissis, G.; Vargas, S.; Selistre-de-Araujo, H. Cell adhesion and cytotoxicity studies over polyanionic collagen surfaces with variable negative charge and wettability. *Biomaterials* **2003**, *24*, 131–137. [CrossRef]
29. Maqsood, M.F.; Raza, M.A.; Ghauri, F.A.; Rehman, Z.U.; Ilyas, M.T. Corrosion study of graphene oxide coatings on AZ31B magnesium alloy. *J. Coat. Technol. Res.* **2020**, *17*, 1321–1329. [CrossRef]
30. Wang, Z.; Hu, Y.; Gui, Z.; Zong, R. Halogen-free flame retardation and silane crosslinking of polyethylenes. *Polym. Test.* **2003**, *22*, 533–538. [CrossRef]
31. Zhu, Y.; Murali, S.; Cai, W.; Li, X.; Suk, J.W.; Potts, J.R.; Ruoff, R.S. Graphene and graphene oxide: Synthesis, properties, and applications. *Adv. Mater.* **2010**, *22*, 3906–3924. [CrossRef]
32. Lin, C.-C.; Metters, A.T. Hydrogels in controlled release formulations: Network design and mathematical modeling. *Adv. Drug Deliv. Rev.* **2006**, *58*, 1379–1408. [CrossRef]
33. Raucci, M.; Alvarez-Perez, M.; Demitri, C.; Giugliano, D.; De Benedictis, V.; Sannino, A.; Ambrosio, L. Effect of citric acid crosslinking cellulose-based hydrogels on osteogenic differentiation. *J. Biomed. Mater. Res. Part A* **2015**, *103*, 2045–2056. [CrossRef]
34. Koyano, T.; Koshizaki, N.; Umehara, H.; Nagura, M.; Minoura, N. Surface states of PVA/chitosan blended hydrogels. *Polymer* **2000**, *41*, 4461–4465. [CrossRef]
35. Wang, Z.; Yu, H.; Xia, J.; Zhang, F.; Li, F.; Xia, Y.; Li, Y. Novel GO-blended PVDF ultrafiltration membranes. *Desalination* **2012**, *299*, 50–54. [CrossRef]

36. Mamidi, N.; Zuníga, A.E.; Villela-Castrejón, J. Engineering and evaluation of forcespun functionalized carbon nano-onions reinforced poly (ε-caprolactone) composite nanofibers for pH-responsive drug release. *Mater. Sci. Eng. C* **2020**, *112*, 110928. [CrossRef]
37. Mamidi, N.; Velasco Delgadillo, R.M.; Gonzáles Ortiz, A.; Barrera, E.V. Carbon nano-onions reinforced multilayered thin film system for stimuli-responsive drug release. *Pharmaceutics* **2020**, *12*, 1208. [CrossRef]
38. Han, X.; Huo, P.; Ding, Z.; Kumar, P.; Liu, B. Preparation of Lutein-Loaded PVA/Sodium Alginate Nanofibers and Investigation of Its Release Behavior. *Pharmaceutics* **2019**, *11*, 449. [CrossRef]
39. Zykwinska, A.; Tripon-Le Berre, L.; Sinquin, C.; Ropartz, D.; Rogniaux, H.; Colliec-Jouault, S.; Delbarre-Ladrat, C. Enzymatic depolymerization of the GY785 exopolysaccharide produced by the deep-sea hydrothermal bacterium Alteromonas infernus: Structural study and enzyme activity assessment. *Carbohydr. Polym.* **2018**, *188*, 101–107. [CrossRef] [PubMed]
40. Dadsetan, M.; Pumberger, M.; Casper, M.E.; Shogren, K.; Giuliani, M.; Ruesink, T.; Hefferan, T.E.; Currier, B.L.; Yaszemski, M.J. The effects of fixed electrical charge on chondrocyte behavior. *Acta Biomater.* **2011**, *7*, 2080–2090. [CrossRef] [PubMed]
41. Tang, L.; Thevenot, P.; Hu, W. Surface chemistry influences implant biocompatibility. *Curr. Top. Med. Chem.* **2008**, *8*, 270–280. [CrossRef]
42. Aslam Khan, M.U.; Al-Arjan, W.S.; Binkadem, M.S.; Mehboob, H.; Haider, A.; Raza, M.A.; Abd Razak, S.I.; Hasan, A.; Amin, R. Development of Biopolymeric Hybrid Scaffold-Based on AAc/GO/nHAp/TiO2 Nanocomposite for Bone Tissue Engineering: In-Vitro Analysis. *Nanomaterials* **2021**, *11*, 1319. [CrossRef] [PubMed]
43. Aslam Khan, M.U.; Mehboob, H.; Abd Razak, S.I.; Yahya, M.Y.; Mohd Yusof, A.H.; Ramlee, M.H.; Sahaya Anand, T.J.; Hassan, R.; Aziz, A.; Amin, R. Development of polymeric nanocomposite (xyloglucan-co-methacrylic acid/hydroxyapatite/sio2) scaffold for bone tissue engineering applications—in-vitro antibacterial, cytotoxicity and cell culture evaluation. *Polymers* **2020**, *12*, 1238. [CrossRef] [PubMed]

polymers

MDPI

Article

Preparation of a Dmap-Catalysis Lignin Epoxide and the Study of Its High Mechanical-Strength Epoxy Resins with High-Biomass Content

Lingxia Song [1,2,3,4,5], Yeyun Meng [1,2], Peng Lv [1,2], Weiqu Liu [1,*] and Hao Pang [1,*]

1 Guangzhou Institute of Chemistry, Chinese Academy of Sciences, Guangzhou 510650, China; jnssllxx@163.com (L.S.); mengyeyun15@mails.ucas.ac.cn (Y.M.); lvpengtust@163.com (P.L.)
2 University of Chinese Academy of Sciences, Beijing 100049, China
3 Guangdong Provincial Key Laboratory of Organic Polymer Materials for Electronics, Guangzhou 510650, China
4 CAS Engineering Laboratory for Special Fine Chemicals, Guangzhou 510650, China
5 CASH GCC(Nanxiong) Research Institute of New Materials Co., Ltd., Nanxiong 512400, China
* Correspondence: liuwq@gic.ac.cn (W.L.); panghao@gic.ac.cn (H.P.)

Abstract: The depletion of limited petroleum resources used for the fabrication of epoxy resins calls for the development of biomass-based epoxides as promising alternatives to petroleum-derived epoxides. However, it is challenging to obtain an epoxy resin with both high lignin content and excellent mechanical performance. Herein, a 4-dimethylaminopyridine (DMAP)-lignin epoxide with a certain epoxy value and a small molecular weight is obtained by the catalysis of DMAP for the macromolecular lignin. It was discovered that compared to the prepared composite resin of benzyltriethylammonium chloride (BTEAC)-lignin epoxide, there is a better low-temperature storage modulus for the DMAP-lignin epoxide resin and its composite resins with high-biomass contents, and higher tensile strength for its composite resins. In particular, the DMAP-lignin epoxide/bisphenol A diglycidyl ether (BADGE) (DB) composite resin with DMAP-lignin epoxide replacement of 80 wt% BADGE, containing up to 58.0 wt% the lignin epoxide, exhibits the tensile strength of 76.3 ± 3.2 MPa. Its tensile strength is 110.2% of BTEAC-lignin epoxide/BADGE (BB) composite resins and is comparable to that of petroleum-based epoxy resins. There are good application prospects for the DB composite resin in the engineering plastics, functional composite, grouting, and other fields.

Keywords: DMAP-lignin epoxide; bio-based epoxy resin; composite resin; tensile strength

Citation: Song, L.; Meng, Y.; Lv, P.; Liu, W.; Pang, H. Preparation of a Dmap-Catalysis Lignin Epoxide and the Study of Its High Mechanical-Strength Epoxy Resins with High-Biomass Content. *Polymers* **2021**, *13*, 750. https://doi.org/10.3390/polym13050750

Academic Editor: Evgenia G. Korzhikova-Vlakh

Received: 3 February 2021
Accepted: 24 February 2021
Published: 28 February 2021

1. Introduction

As renewable biomass, lignin is abundant in nature and cheap to obtain. It is the second most abundant biopolymer after cellulose and the most abundant natural aromatic compound [1,2]. Nearly 50 million tons of lignin are produced each year, but only 2% are used for the production of value-added materials because of its complex construction and poor processability. With the efforts of researchers, lignin has been researched in many fields. The high carbon content in lignin can be used to prepare carbon-based supercapacitors and carbon fiber materials [3]. The phenolic hydroxyl group of lignin has a good reductivity for the synthesis of silver nanoparticles, and its macromolecular structure can be used as micro-carriers for the resulting silver nanoparticles [4]. Lignin has been transformed into polymers such as polyhydroxyalkanoates by means of microbial degradation [5,6]. Lignin as a macromolecular composed of benzene ring has been composited with petroleum-based compounds (such as resorcinol or phenol, urea, epoxy resins, and polyurethane), to improve the strength and toughness of composite resin and realize its application in the engineering field [7,8]. In particular, lignin, through chemical modification, can participate in the thermal curing reaction, which not only effectively improves the mechanical properties of other resins, but also realizes the partial and total replacement of the petroleum-based compounds, and solves the problems of the petroleum resource crisis and environmental pollution [9,10].

Lignin contains a large number of active hydroxyl groups and carboxyl groups, instead of bisphenol A to prepare the epoxy compound. The obtained lignin-based epoxy compound has thermosetting properties. However, compared with bisphenol A epoxy compound, the lower content of the epoxy group and the steric hindrance for the macromolecular lignin epoxides lead to poor mechanical properties for the resin, so it is very necessary to composite with bisphenol A epoxy compound. The lignin epoxide with the macromolecular structure is introduced into the bisphenol A epoxy resin matrix and participates in the chemical crosslinking of resin. As a result, it can effectively block the crack propagation fracture of bisphenol A epoxy resin and improve the strength and toughness of resin [1,11,12]. In addition to the above advantages, lignin epoxide itself will undergo certain changes of its chemical structure during continuous thermal processing, and condensation between the lignin fragments could occur, making the resin structure more packed to increase the strength [13–15]. Daniel Jason et al. synthesized the lignin-derived epoxy prepolymers via both the mild hydrogenolysis and the epoxidation of lignin and used it to replace 25–75% of the bisphenol A diglycidyl ether equivalent, discovering increases of up to 52% in the flexural modulus and up to 38% in the flexural strength [16]. Fatemeh Ferdosian et al. used the bio-based epoxy systems as polymer matrices for manufacturing fiber-reinforced plastics (FRPs) and coatings. The tensile and flexural strengths of the prepared FRPs using bio-based epoxy composites were superior or comparable to those of the FRP with the pure bisphenol A epoxy resin [17]. Claudio Gioia et al. investigated the relationship between thermomechanical properties and the chemical structure of the well-characterized lignin-based epoxy resins. It was discovered that compared to eucalyptus-based resins, the spruce-based lignin resin results in somewhat higher modulus and higher tensile strength because of the more condensed structures [18].

The above research work is mainly about the mechanical properties of the epoxy compound composite resin with the small molecule state lignin. Namely, The lignin is degraded or refined to a small molecular weight (M_W < 2000 Da). Then the lignin is epoxidized, and the composite resin is prepared. The epoxidation of lignin with higher molecular weight is rarely studied (M_W > 2000 Da). The main reason is that the intermolecular etherification crosslinking (crosslinking reaction between the epoxy group and hydroxyl group) easily occurs in the lignin epoxidation reaction, leading to an obvious increase in molecular weight and a severe decrease in compatibility with bisphenol A epoxide. Therefore, it is very important to control the molecular weight of lignin (M_W > 2000 Da) during the epoxidation and guarantee the epoxy group content.

The catalyst has a significant influence on the epoxy value and molecular weight of the reaction product [19,20]. For the catalysts, the selection is generally based on experimental design. There are two main methods: the one-step catalytic method of sodium hydroxide solution and the two-step catalytic method of ammonium salt-sodium hydroxide phase transfer [21,22]. The one-step method easily causes the etherification crosslinking reaction between molecules, resulting in the obvious increase of molecular weight. The two-step method, as a common method, can effectively promote the epoxidation of lignin and control the molecular weight effect. Herein, we found that 4-dimethylaminopyridine (DMAP) can be used to epoxidize lignin, with the catalytic effect comparable to that of quaternary ammonium salt catalyst, this can achieve the epoxidation of lignin and control the molecular weight of lignin epoxides.

In this paper, DMAP-lignin epoxide was synthesized by catalysis of DMAP, and its molecular weight and epoxy value were evaluated. The mechanical properties of amine-cured DMAP-lignin epoxy resin and its composite resin with bisphenol A diglycidyl ether (BADGE) were studied. In order to better illustrate DMAP's catalytic effect and its performance advantage as an alternative to the BADGE, the benzyltriethylammonium chloride (BTEAC)-lignin epoxide was compared with DMAP-lignin epoxide in terms of molecular features and resin properties.

2. Experimental

2.1. Materials and Chemicals

Lignin was purchased from Yanghai Environmental Protection material Co., Ltd. (Alibaba, Jinan, China). The lignin was from the straw plant. The parameters obtained for the purchased lignin are given in Table 1, and the typical structural units of lignin are described in Figure S1 [23].

Table 1. The parameters obtained for the as-purchased lignin.

Type	M_n	M_w	Đ	ph–OH Content (mmol/g)	–COOH Content (mmol/g)	Ratio of G to S (G/S)
Softwood lignin	2827	3153	1.12	3.05	1.71	>50%

Note: M_n is the number-average molecular weight, M_w is the weight-average molecular weight, Đ represents molecular weight distribution, ph–OH denotes the phenolic hydroxyl group, –COOH is the carboxyl group, G and S represent guaiacyl and syringyl, respectively.

ECH (99.5%), NaOH (99.5%), and dimethyl sulfoxide (DMSO, 99.5%) were supplied by Aladdin Biochemical Technology Co., Ltd (Shanghai, China). DMAP (99.5%), triethylene teramine (TETA, 70.0%), BTEAC (99.5%), DMF (99.0%), and BDDGE (95.0%) were obtained from Shanghai Macklin Biochemical Co., Ltd. (Shanghai, China). BADGE (85%) was purchased from Sigma-Aldrich Trading Co., Ltd. (Shanghai, China). All of the chemicals were of analytical grade and used as received unless stated otherwise.

2.2. Preparation of Lignin Epoxides

A given amount of lignin (4.80 g), DMAP (3.9 mmol, 0.48 g), and DMSO (50 mL) were placed in a round-bottom flask (250 mL) equipped with a thermometer, a stirrer, and a condenser. The mixture was heated to 117 °C under N_2 bubbling conditions and was stirred for 30 min. Afterward, ECH (100 mL, 1.27 mol) was added while holding the reaction at 117 °C for 180 min. Then, the reactants were cooled to 70 °C before a NaOH aqueous solution (5 M, 18 mL) was added at the speed of 0.1 mL/min over a 180 min duration using a peristaltic pump. Subsequently, the reaction mixture was cooled to room temperature and was followed by transferring it to a beaker (500 mL). For the removal of the water-soluble impurities, including the catalyst, the deionized water (350 mL) was poured into the beaker, and intensely stirred. The mixture was allowed to stand for a sufficient length of time until a brown oily mixture formed at the bottom of the aqueous solution, then the water phase of the separated liquids was decanted. This process was repeated thrice. Next, dichloromethane (50 mL) was poured into the beaker under adequate agitation. The product was separated out of the mixture containing a small amount of residual water, which was then allowed to stand until the emergence of the layered liquids. The water and insoluble solid layers were discarded, and the deep brown dichloromethane solution, with the product dissolved, was obtained. In order to facilitate the next purification of the product, 30 mL of dichloromethane was removed from the dichloromethane solution by rotary evaporation. The dichloromethane solution was added drop-by-drop to ethyl acetate (200 mL) under magnetic agitation, and a brown precipitate was separated out of ethyl acetate. By filtering and blow-drying the brown precipitate at room temperature, the DMAP-lignin epoxide sample was finally obtained. The prepared sample was stored in DMF, with a solid content of 55–65%, to avoid an intermolecular crosslinking reaction that would decrease its solubility.

The BTEAC-lignin epoxide was prepared according to a similar procedure as for the synthesis of the DMAP-lignin epoxide sample, using lignin (4.80 g), BTEAC (3.9 mmol, 0.89 g), DMSO (50 mL), ECH (100 mL, 1.27 mol), and NaOH (5 M, 18 mL) as the starting materials.

2.3. Preparation of Epoxy Resins

2.3.1. Preparation of the DMAP-Lignin Epoxide Resin, BTEAC-Lignin Epoxide Resin, and BADGE Resin

Into the DMF solution of the epoxide, a certain amount of the curing agent (i.e., TETA) was dropwise added while keeping agitation for homogeneous mixing at room temperature. The mixture was allowed to stand until the bubbles disappeared and was then cast into a polytetrafluoroethylene (PTFE)-lined mold to cure in an air-circulating oven. The following curing procedure was sequentially adopted: heating at 70 °C for 24 h, 120 °C for 4 h, and finally at 180 °C for 4 h. The epoxy resin sample was produced after demolding at room temperature. Different kinds of epoxy resin samples were prepared according to the recipe provided in Table 2.

Table 2. The specifications of different types of epoxy resins: the 4-dimethylaminopyridine (DMAP)-lignin epoxide resin, benzyltriethylammonium chloride (BTEAC)-lignin epoxide resin, and bisphenol A diglycidyl ether (BADGE) resin.

Sample	DMAP-Lignin Epoxide	BTEAC-Lignin Epoxide	BADGE	TETA
	(g)	(g)	(g)	(g)
DMAP-lignin epoxide resin	22.00	-	-	3.00
BTEAC-lignin epoxide resin	-	22.00	-	3.00
BADGE resin	-	-	22.00	3.00

2.3.2. Preparation of the Lignin Epoxide/BADGE Composite Epoxy Resin

The DMF solution of the lignin epoxide, BDDGE, and BADGE were successively dissolved in DMF solvent (3.5 mL); and the DMF solution with well-dissolved epoxides was obtained by adequate stirring. The amount of each added component is provided in Table 3. To promote the crosslinking and curing, we volatilized a certain amount of DMF, which was determined by the weight loss of the beaker loaded with the DMF solution during heating. After the reaction was continuously heated at 80 °C until 3.0 mL DMF was volatilized, the heating was stopped, and the reaction mixture was naturally cooled to room temperature. A mixture of epoxides with good fluidity was obtained. After that, the amine-based curing agent (TETA) was homogenized to the mixture by stirring; and the sequential curing procedures were adopted. Specifically, the mixture was first cured at room temperature until it was immobile, followed by heating at 70 °C for 12 h, 120 °C for 8 h, and, finally, 150 °C for 6 h. The lignin epoxide/BADGE composite epoxy resin was obtained after demolding at room temperature.

Table 3. The formulation adopted for the preparation of a series of the lignin epoxide/BADGE composite resin samples with different ratios of the lignin epoxide to BADGE.

Sample	DMAP-Lignin Epoxide	BTEAC-Lignin Epoxide	BADGE	BDDGE	TETA
	(g)	(g)	(g)	(g)	(g)
DB0/BB0	0.00	0.00	5.00	1.25	0.65
DB10	0.50	-	4.50	1.25	0.65
DB20	1.00	-	4.00	1.25	0.65
DB40	2.00	-	3.00	1.25	0.65
DB60	3.00	-	2.00	1.25	0.65
DB80	4.00	-	1.00	1.25	0.65
BB10	-	0.50	4.50	1.25	0.65
BB20	-	1.00	4.00	1.25	0.65

Table 3. *Cont.*

Sample	DMAP-Lignin Epoxide	BTEAC-Lignin Epoxide	BADGE	BDDGE	TETA
	(g)	(g)	(g)	(g)	(g)
BB40	-	2.00	3.00	1.25	0.65
BB60	-	3.00	2.00	1.25	0.65
BB80	-	4.00	1.00	1.25	0.65

Note: DB10 represents the prepared DMAP-lignin epoxide/BADGE resin with 10 wt% DMAP-lignin epoxide and 90 wt% BADGE, respectively, while the other samples are named accordingly based on the weight percentage of lignin epoxide.

2.4. Characterizations

Fourier transform infrared spectroscopy (FTIR) was adopted to analyze the functional groups of various samples using an infrared spectrometer (Bruker Tensor 27, Karlsruhe, Germany), with a scanning range of 400–4000 cm^{-1} and scanning resolution of 4 cm^{-1}. The molecular structure of the prepared samples was also investigated using a ^{1}H nuclear magnetic resonance (^{1}H NMR) spectrometer (Avance III 400, Bruker Co., Ltd., Karlsruhe, Germany). For the ^{1}H NMR test, the DMSO-d6 (0.6 mL) was adopted as the solvent, while *p*-nitrobenzaldehyde (*p*-NBD, 5.0 mg) was used as the internal standard. The mass of the specimen for the ^{1}H NMR test was 25.0 mg. The locations of the ^{1}H NMR peaks associated with the solvent were found at δ_H 2.46–2.54 ppm, while those of the internal standard were observed at δ_H 8.10–8.20 ppm, 8.30–8.50 ppm, and 10.10–10.20 ppm.

2.5. Analysis of the Lignin Epoxidation

Based on the ^{1}H NMR spectra, the epoxy values were calculated by a comparison between the peak area of hydrogen arising from the epoxy group and the peak area indexed to the internal standard (*p*-NBD, δ_H 10.10–10.20 ppm) [24]. The area in the chemical shift range of 2.60–2.90 ppm is the peak area of hydrogen arising from the epoxy group [25,26].

2.6. Molecular Weight Analysis

Gel permeation chromatography (GPC) was employed to measure the molecular weight of the prepared samples using a Waters chromatograph (Milford, MA, USA) equipped with a Waters 2414 refractive index detector. The polymethyl methacrylate (PMMA) was adopted as the standard specimen and DMF as eluent. A DMF solution of the sample (3 mg/mL) was prepared, and a portion (50 uL) was withdrawn and was passed through a single PLgel MIXED-C column after being filtered with a 0.22 μm permeable membrane. The samples were then eluted using DMF at an elution rate of 1 mL/min and a column temperature of 35 °C. The molecular weight was finally measured by a comparison between the eluted sample and the standard sample.

2.7. Test on the Mechanical, Thermomechanical, and Thermal Performance

Dynamic mechanical analysis (DMA) was performed using a DMA+300 analyzer (Metravib Co., Ltd., Limonest, France); and the storage modulus, loss modulus, and Tan δ were measured for the prepared samples. The single cantilever method was adopted for the DMA test on the sample with a size of 40.0 × 8.0 × 3.0 mm^3, with a temperature ramp of −80 to 200 °C at a heating rate and vibration frequency of 3 °C/min and 1 Hz.

The calculation of the crosslinking density was carried out according to the following method: the temperature at the maximum of the peak on the curve of Tan δ vs. temperature was assigned to the glass transition temperature (T_g), and according to the classical rubber elasticity theory, the crosslinking density of the thermosetting epoxy resin was calculated by Equation (1) [1,27].

$$\rho = E'/3RT \quad (1)$$

where E' represents the storage modulus of the sample at the temperature of $T_g + 30$ °C, and R and T denote gas constant and absolute temperature of the sample at the temperature of $T_g + 30$ °C, respectively.

The tensile test was conducted using the 10 kN universal testing machine (Reger Co., Ltd., Shenzhen, China) according to the ASTM D638-2003 standard. The specimens were prepared in a dumbbell shape (with the width, thickness, and fixed length of 4.0, 2.0, and 25.0 mm, respectively), and the tension loading rate was set to 5 mm/min. Each sample was tested in a parallel fashion four times, and the average value was reported in this paper.

The fractured cross-sectional morphologies of the samples after finishing the tensile test were observed using an S-4800 scanning electron microscope (SEM, Hitachi Co., Ltd., Tokyo, Japan). Before the SEM observation, a thin layer of gold was splattered onto the sample surface.

The thermal stability of the sample was evaluated using a TGA Q500 thermogravimetric analyzer (TA Co., Ltd., New Castle, DE, USA), with a temperature ramp of 30 to 800 °C at the heating rate of 10 °C/min. Thermal weight losses were measured under both N_2 and air atmospheres at a gas flow rate of 50 mL/min.

3. Results and Discussion

3.1. Synthesis and Characterization of the Lignin Epoxides

Figure 1 shows the structure of the pristine lignin and the lignin epoxides examined by the FTIR spectrum. There are similar characteristic absorption bands before and after the epoxidation of lignin. The assignments for the peaks of lignin are listed in Table 4 [28]. The absorption band at 3450 cm^{-1} for the lignin, which can be assigned to the –OH stretching vibrations, becomes weakened in the DMAP-lignin epoxide, and the absorption peaks of C=O stretching at 1700 cm^{-1} and C–O stretching at 1260 cm^{-1} for lignin disappear. While the C=O stretching peak at 1710 cm^{-1} is attributed to the ester group and the C–O stretching peak at 1248 cm^{-1} representing the phenoxy group appear, and the absorption peak at 910 cm^{-1} appears corresponding to the epoxy group. The changes of active groups and appearance of the epoxy group are consistent with the characterization of Chizuru Sasaki et al. for the lignin epoxide [29]. These results confirm the introduction of the epoxy groups into the molecular structure of lignin as the result of an effective catalytic epoxidation reaction. A similar result can also be observed in the FTIR spectrum of the BTEAC-lignin epoxide sample, revealing that the catalysis of the lignin epoxidation can also be achieved with the BTEAC catalyst.

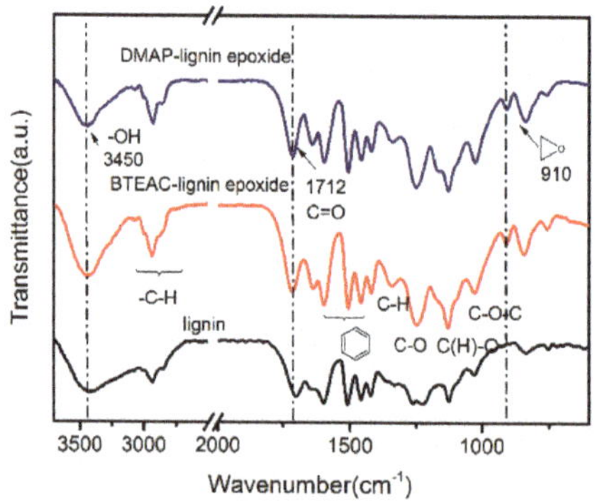

Figure 1. FTIR spectrum of the lignin, DMAP-lignin epoxide, and BTEAC-lignin epoxide.

Table 4. Vibrational modes for FTIR absorbance peaks of lignin.

FTIR Band Position (cm^{-1})	Assignment
3450 cm^{-1}	stretching of –OH in hydroxyl group and carboxyl group
3000, 2927, and 2846 cm^{-1}	stretching of alkyl C–H
1700, and 1648 cm^{-1}	stretching of C=O in carboxyl group
1600, 1506, and 1455 cm^{-1}	stretching of C=C–C bond in benzene ring
1338 cm^{-1}	symmetric deformation of C–H bond
1260 cm^{-1}	stretching of C–O in phenol group
1222 cm^{-1}	stretching of C–O in phenoxy group
1130 cm^{-1}	stretching of C–O in alkyl alcohol
1030 cm^{-1}	stretching of C–O in ether group
846, and 748 cm^{-1}	deformation of C–H of benzene ring

In order to further confirm the formation of the epoxy groups, the ^{1}H NMR spectra of the DMAP-lignin epoxide and BTEAC-lignin epoxide samples are provided in Figure 2. The characteristic protons of the main groups of lignin are assigned in the regions of δ_H 3.50–4.00 ppm (CH_3–O–), δ_H 4.01–4.89 ppm (–CH_2–O–), δ_H 6.14–8.00 ppm (benzene ring), δ_H 8.61–9.80 ppm (phenolic hydroxyl group), and δ_H 12.04–13.00 ppm (–COOH). The signal peaks of carboxyl protons and phenolic protons disappear in the epoxidation lignin, and the signal peaks attributed to the epoxy group appear in the chemical shift range of δ_H 2.60–2.90 ppm and δ_H 3.20–3.50 ppm. These demonstrate the successful synthesis of the lignin epoxides. The earlier reports of Shou Zhao et al. support our results and show the same shift signal peaks of epoxy groups [25,26].

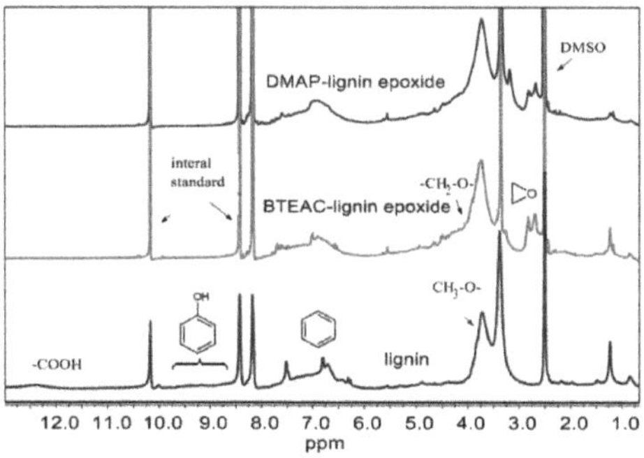

Figure 2. ^{1}H-NMR spectra of the lignin, DMAP-lignin epoxide, and BTEAC-lignin epoxide.*p*-NBD was adopted as the internal standard, with the characteristic ^{1}H-NMR peaks at 8.10–8.20, 8.30–8.50, and 10.10–10.20 ppm.

3.2. Evaluation of the Performance of the Prepared DMAP-Lignin Epoxide and Its Cured Resin

In addition to the epoxy value, the molecular weight and molecular weight distribution of the DMAP-lignin epoxide and BTEAC-lignin epoxide samples are also summarized in Table 5. Compared to the BTEAC-lignin epoxide, the DMAP-lignin epoxide possesses a lower molecular weight, narrower molecular weight distribution, and comparable epoxy value. Although DMAP and BTEAC have a similar epoxidized capability for lignin epoxidation, the DMAP catalyst can better control the molecular weight of the lignin epoxide as compared to BTEAC. The DMAP-lignin epoxide also exhibits better dispersibility and mobility during manufacturing processes that involve melting and curing.

Table 5. The epoxy value, molecular weight, and Đ of the lignin epoxides.

Samples	Epoxy Value (mmol/g)	Molecular Weight		
		M_n	M_W	Đ
DMAP-lignin epoxide	2.09	8244	17,085	2.07
BTEAC-lignin epoxide	2.16	9775	22,439	2.30

The crosslinking density, storage modulus, and glass transition temperature (T_g) of the three kinds of epoxy resins prepared by amine-based curing of the DMAP-lignin epoxide, BTEAC-lignin epoxide, and BADGE are summarized in Table 6. The order of the crosslinking density of these obtained resins is as follows: DMAP-lignin epoxide resin <BTEAC-lignin epoxide resin <BADGE resin. Such a sequence is consistent with the order of the epoxy value for the corresponding epoxides, namely the DMAP-lignin epoxide, BTEAC-lignin epoxide, and BADGE.

Table 6. Dynamic mechanical properties and crosslink density (ρ) of the DMAP-lignin epoxide resin, BTEAC-lignin epoxide resin, and BADGE resin.

Sample	Storage Modulus (MPa)			T_g (°C)	ρ (10^{-3} mol·cm^{-3})
	Glassy Region at −50 °C	Glassy Region at 25 °C	Rubbery Region at T_g + 30 °C		
BADGE resin	11,300	8190	146.3	75	15.52
DMAP-lignin epoxide resin	6460	1200	28.4	46	3.26
BTEAC-lignin epoxide resin	5280	4286	43.8	90	4.47

The variation of the storage modulus of the DMAP-lignin epoxide, BTEAC-lignin epoxide, and BADGE resin samples as a function of temperature is presented in Figure 3a. The storage modulus of the DMAP- and BTEAC-lignin epoxide resins are markedly lower than the BADGE resin at −50 °C, which is due to the much lower crosslinking density of the former (Table 6). When one compares the lignin-based resins, the DMAP-lignin epoxide resin shows a larger storage modulus as compared to the BTEAC-lignin epoxide resin, albeit with a lower crosslinking density. For the resin samples possessing similar crosslinking density, their packing structure can have a more significant impact on the storage modulus. The small molecular weight of DMAP-lignin epoxide resin can promote the packing density of resin (Figure S2), and further enhance the storage modulus in the glassy state. The influence of molecular weight on the dense packing and the storage modulus has been reported by Claudio Gioia et al., which is consistent with the conclusion in this paper [24].

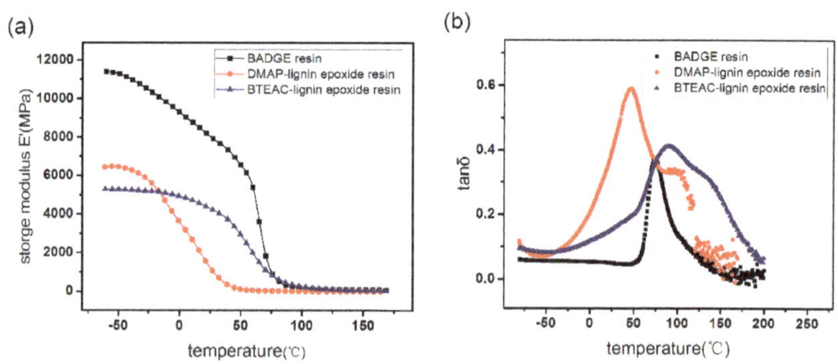

Figure 3. The storage modulus (**a**) and tanδ (**b**) measured for the DMAP-lignin epoxide, BTEAC-lignin epoxide, and BADGE resins as a function of temperature.

The order of the T_g of the three kinds of epoxy resins is as follows: BTEAC-lignin epoxide resin >BADGE resin >DMAP-lignin epoxide resin. All of T_g is larger than room temperature, making these kinds of resins applicable at room temperature. The T_g of DMAP-lignin epoxide resin is notably smaller than the BTEAC-lignin epoxide resin and BADGE resin due to its lower crosslink density [30]. Such a low T_g of the DMAP-lignin epoxide resin makes it promising for toughening the petroleum-derived epoxy resin (e.g., BADGE resin).

Although the BADGE resin has both high T_g and high storage modulus (Figure 3a,b, and Table 6), it is still expected to achieve the co-curing composite of lignin epoxides with BADGE. For one thing, the lignin epoxide as an active macromolecule can strengthen the skeleton structure of BADGE resin by chemical reaction, and further improve the strength and toughness. For another, it is beneficial to promote the T_g and the crosslink strength of lignin epoxide resins and achieve a partial substitute for the BADGE.

In order to co-cure DMAP-lignin epoxide and BADGE, the compatibility between them should be the foremost problem to be resolved. The lignin epoxide is a polar macromolecular epoxide bearing hydroxyl groups, while BADGE is a viscous and aprotic polar epoxide with a low molecular weight. This discrepancy causes poor compatibility between them. In this study, the BDDGE diluent and the DMF solvent were employed to dissolve these two kinds of epoxides and improve their compatibility (see Experimental). After the lignin epoxide and BADGE had been dissolved, the DMF solvent was removed to facilitate the reaction between the epoxides and amine-based curing agent (in this case, TEAE). With the volatilization of the DMF solvent, the lignin epoxide was readily separated out of the BDDGE/DMF mixture because of its larger molecular weight. It is also worth noting that the higher content of DGEBA caused the lignin epoxide to be more easily separated out of the BDDGE/DMF mixed solution. Therefore, the lignin epoxide, with a larger molecular weight, exhibited a lower solubility during the manufacturing process with the lower content, resulting in the formation of the epoxy resin with obvious phase segregation. Unsatisfactory mechanical properties were subsequently measured. To ensure the good dissolution of the lignin epoxide before being cured, we added the TEAE for administering the curing reaction at room temperature once the DMF solvent was volatilized to a certain extent. The initial crosslinking reaction among the three kinds of epoxides, namely the lignin epoxide, BADGE, and BDDGE, resulted in the formation of an epoxy resin featuring a gel structure and integrated with a certain amount of the solvent that remained after the volatilization. Afterward, the solvent was further removed by heating and was accompanied by ring-opening polymerization. Finally, the epoxy resin was completely cured. Due to the lower molecular weight of the DMAP-lignin epoxide (as compared to that of the BTEAC-lignin epoxide), it was more resistant to be separated out of the BDDGE/DMF mixed solution, thus indicating its better solubility and processability.

The BADGE and DMAP- or BTEAC-lignin epoxide at a series of weight ratios were mixed into the BDDGE/DMF solution, resulting in the DMAP-lignin epoxide/BADGE (DB) and BTEAC-lignin epoxide/BADGE (BB) resins with various compositions. As you can see from Figure 4a, with the increase of weight of lignin epoxides, the tensile strength of composite resin first increased and then decreased. The tensile strength of DB60 was the highest, reaching up to 88.5 ± 0.8 MPa. When the lignin epoxides substituted 80 wt% of BADGE, the tensile strength of DB80 resin still remained 76.3 ± 3.2 MPa. At this point, the lignin epoxide content in the resin was up to 58.0%. The results showed that the composite resins with a high-biomass content still possess high tensile strength. This is because the DMAP-lignin epoxide participates in the crosslinking reaction of composite resin, changes the resin structure of composite resin without lignin epoxides (DB0), disperses the tensile stress effectively, and blocks the linear diffusion of cracks (Figure 5). Moreover, Continuous high curing temperature not only promotes complete crosslinking among the components but also enhances the binding of lignin epoxide itself. The high heating temperature is conducive to the increase of condensation between lignin fragments in composite resin, which in turn increases the bonding strength of the resin and improves the tensile strength of the resin. Although there have been reports on the effect of heating on the mechanical strength of lignin-based resins [13–15], the research is not systematic enough, and more studies on the heating effect on the properties of lignin-based resins should be conducted in the future. DB resin has better tensile strength than BB resin in the same mass ratio, which is due to its smaller molecular weight and narrower distribution of DMAP-lignin epoxide, resulting in better compatibility with BADGE and better homogeneity with each component of composite resin. With the ongoing rise of the lignin epoxide content, the elongation at break of DB resin first increased and then decreased similar to the change of its tensile strength, but the DB resin has the highest elongation at break with the 20% DMAP-lignin epoxide content (DB20) (Figure 4b). This result elucidates the fact that DB resin presents a better enhancement of toughening at lower biomass content.

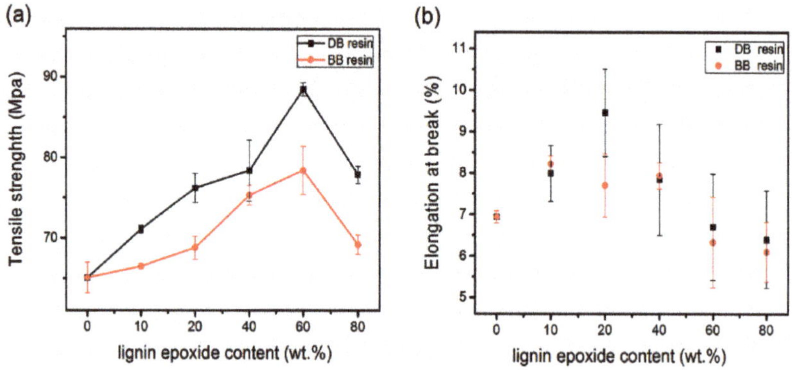

Figure 4. Tensile strength (**a**) and elongation at break (**b**) of the DB and BB resins with the different weight percentages of lignin epoxide. The lignin epoxide content is the weight percentages of DMAP-lignin epoxide replacement of BADGE.

After the tensile test, the fractured surface morphology of the DB resin was also studied by SEM observation; and the results are provided in Figure 5. No obvious phase interface can be observed in the composite resin, indicating that there is good homogeneity for each component of the composite resin. The DB0 without lignin epoxide is prone to form the brittle cracks of a radial pattern under the action of tensile stress (Figure 5a). When the lignin epoxide is added to the composite resin, the line crack propagation is blocked, forming a large number of regional areas with flabellate fracture surface morphology because of enhancement of the toughness (Figure 5b). When the content of the lignin epoxide compound is high, the crosslinking strength of the composite significantly decreases under the influence of the low epoxy value of lignin epoxide; and the tensile strength decreases accordingly, forming a fracture feature, like water-ripple morphology perpendicular to the tensile direction (Figure 5d). Only lignin epoxide and BADGE have a suitable mass ratio and uniform dispersion, and because the composite construct of their resin is stressed uniformly, the fracture surface morphology of the pretty turtle crack type will be formed. (Figure 5c). The BB resin presents a fractured surface morphology similar to the DB resin at the same weight percentage for lignin epoxide to BADGE (Figure S3).

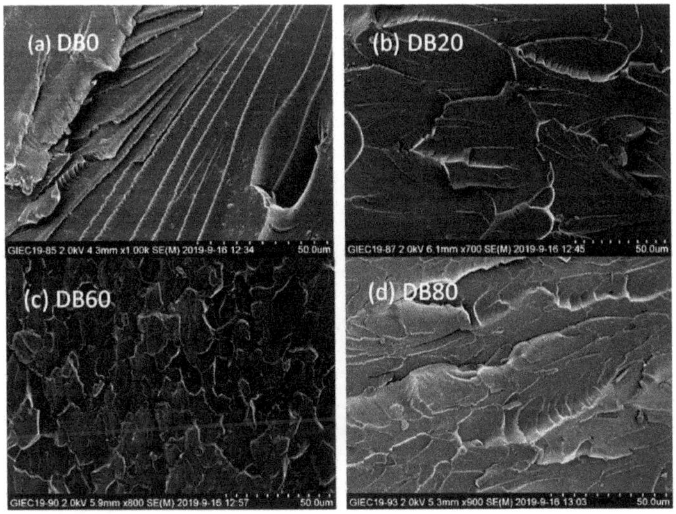

Figure 5. SEM images of the fractured surface of the DB0 (**a**), DB20 (**b**), DB60 (**c**) and DB80 (**d**) resin after tensile tests.

Since the DB and BB resins were prepared by co-curing of the lignin epoxide, BADGE, and BDDGE, their storage moduli are mainly affected by the factors in terms of crosslinking density, phase structure, and the molecular weight and structure of the epoxides [11,24,31–34]. The DB80 resin shows a higher storage modulus compared to the DB40 resin at a low temperature (i.e., −50 °C), as shown in Figure 6a and Table 7. The crosslinking density can be decreased with the rise of the DMAP-lignin epoxide added amount; so the storage modulus should decrease. However, the DMAP-lignin-epoxide chain segment as the main component of DB80 resin, possessing a lower molecular weight, formed the more efficient packing of a dense 3D network structure with a smaller free volume. Moreover, the condensation reaction of the DMAP-lignin-epoxide chain segment in the curing by heating can also increase the compactness of the construct. This is why there is a 1461 MPa-larger storage modulus of the DB80 resin as compared to that of DB40 at −50 °C (Table 7). The low-temperature (−50 °C) storage modulus of the BB80 resin is close to that of its BB40 counterpart, but with a lower 260 MPa difference (Table 7). This is because the poorer stacking structure of BB composite resin has sacrificed the storage modulus. Compared to the DMAP-lignin epoxide resin, the DB80 resin shows 1.5- and 2.7-fold increased storage moduli at −50 and 25 °C, respectively. These results demonstrate that the DB80 resin with a high-biomass content has a good storage modulus at room temperature by the co-curing of DMAP-lignin epoxide and BADGE.

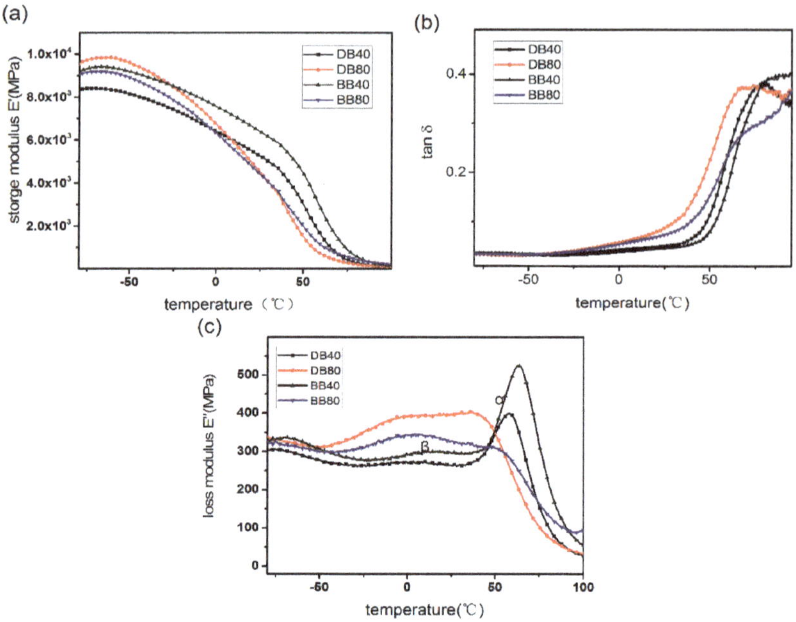

Figure 6. Plots of the storage modulus (**a**), tanδ (**b**), and the loss modulus (**c**) as a function of temperature for various DB and BB resin samples with different added amounts of lignin epoxide.

Table 7. Storage modulus at the glassy region and transition temperatures for the DB40, DB80, BB40, and BB80 resins.

Sample	E′ at −50 °C	1st Transition T_β from E″	E′ at 25 °C	2nd Transition T_α from E″
	(MPa)	(°C)	(MPa)	(°C)
DB40	8168	4.8	5239	59.1
DB80	9629	−4.6	4426	37.6
BB40	9200	16.4	6336	62.9
BB80	8940	4.2	4350	52.5

The DB resins exhibit poor thermomechanical properties at temperatures higher than 100 °C, and no intact peak can be discerned from the curve of the tan δ vs. temperature, thereby making it difficult to obtain the T_g (Figure 6b). Nevertheless, the T_g values of the DB80 and DB40 samples are over 46 °C, which indicates that the co-curing DMAP-lignin epoxide with BADGE can improve the T_g of the DMAP-lignin epoxide resin.

Since the multicomponent co-curing reaction was involved in the preparation of the DB resins, there are more than two characteristic peaks on the loss modulus curves. They mainly correspond to β and α peaks as generated by the motion of the lignin epoxide and BADGE chain segments, respectively (Figure 6c). The increase in the lignin epoxide content intensifies β peak and weakens α peak. Phase transition temperatures (namely T_α and T_β) are mainly affected by the crosslinking construct of the DB resin, and the increase in the lignin epoxide content makes the T_α and T_β shift to lower temperatures (Table 7). The lower T_α and T_β of the DB resin (vs. BB resin) likely results from the lower molecular weight of the DMAP-lignin epoxide as compared to that of the BTEAC-lignin epoxide.

The thermal stabilities of the DB and BB resins are further investigated, and the results are presented in Figure 7. Under an air atmosphere, the DB resin shows a variation tendency in the thermal degradation behavior similar to the BB resin (Figure 7a). There is a

5% weight loss at 250 °C for the DB resin; this is mainly due to the volatile components containing alkyl groups [35]. The thermal degradation occurring over the temperature range of 250–350 °C primarily arises from the alkyl ether chains. The thermal degradation over 250–350 °C begins to slow down, which can be associated with the alkyl groups connected to the phenyl oxygen occurring. With the ongoing temperature rise, the benzene ring structure starts to be thermally degraded, along with an increase in the thermal degradation rate. Finally, the DB resin is completely degraded at 650 °C [1,36]. Under the N_2 atmosphere, a large number of residues remain at 650 °C for the DB resin (e.g., 30% and 20% masses that remain for the DB80 and DB40 resins, receptively), as shown in Figure 7b. The increase of the DMAP-lignin epoxide content leads to an increased quantity of the residues, which indicates that the residual chain segments of the DMAP-lignin epoxide are more resistant to the thermal degradation in the N_2 atmosphere [37]. Under the conditions of the same lignin epoxide content, fewer residues can be found for the DB resin as compared to that for the BB resin. This result is likely due to the lower molecular weight of the DMAP-lignin epoxide composition, which probably causes the resulting DB resin to expose a large, thermally liable surface.

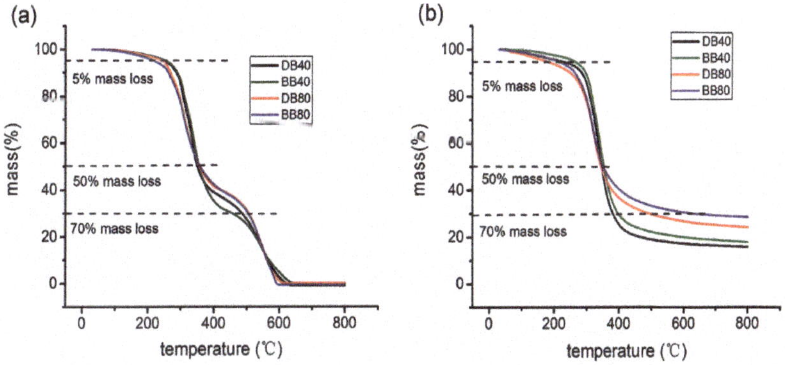

Figure 7. TG curves of the DB40, BB40, DB80, and BB80 resins under air (**a**) and N_2 (**b**) conditions.

4. DMAP-Lignin Epoxide Perspectives and Challenges for Practical Application

In the face of the petroleum resource crisis and the awareness of ecological protection, it is very necessary to study and utilize lignin as a renewable, degradable, and environmentally-friendly polymer resource. Although there are many methods to degrade lignin into a small molecular state, the engineering utilization of macromolecular lignin is of great significance in reducing the cost of raw materials and enhancing the mechanical property of materials. DMAP catalyzes the epoxidation of macromolecular lignin, effectively controls the molecular weight of the epoxidation lignin, and then performs better in the molding process of the resin, forming a composite resin with high-biomass and high mechanical properties. The DMAP-lignin epoxide can be used in engineering plastics, functional composite, grouting, and other fields for development and utilization. However, it is still a big challenge to achieve the complete substitution of macromolecular lignin for petroleum-based raw materials to prepare the resin with strong mechanical properties and environmental friendliness. Because of the molecular steric hindrance of lignin epoxides, the bulk density of resin is affected; and the common solvents and curing agents will cause some environmental hazards. The control of both molecular weight and molecular structure of lignin should be further explored by the catalysts such as DMAP. At the same time, a green solvent such as ionic liquid should be developed for dissolving and curing the lignin epoxides and solving the existing environmental pollution problems.

5. Conclusions

This study has demonstrated that the DMAP is an active and useful catalyst for the epoxidation of the lignin biomass with ECH. The resulting DMAP-lignin epoxide has an epoxy value of as high as 2.09 mmol/g. Compared to the lignin epoxide produced with the BTEAC catalyst (corresponding to the BTEAC-lignin epoxide), the DMAP-lignin epoxide presents advantages in terms of structure, manufacturability, and performance of its resin. It has a smaller molecular weight, narrower molecular weight distribution, and better flowability during the manufacturing of the DMAP-lignin epoxide. Better low-temperature storage modulus and a more compact structure are shown for the corresponding DMAP-lignin epoxide resin. The DB composite epoxy resins with different biomass content have different performance advantages. The DB20 resin has the best toughness; the DB60 resin has the highest tensile strength, reaching up to 88.5 ± 0.8 MPa; the tensile strength of DB80 resin with 58.0 wt% lignin epoxide has 76.3 ± 3.2 Mpa, and the storage modulus has 4426 Mpa at 25 °C. The different types of composite resins meet the various mechanical performance requirements for practical applications. Therefore, such a low-cost biomass-based DMAP-lignin epoxide holds substantial potential for the partial substitution of the petroleum-based epoxy resin.

Supplementary Materials: Supplementary data to this article can be found online at https://www.mdpi.com/2073-4360/13/5/750/s1. Figure S1: The typical structural units of lignin, and G, S, and H represent guaiacyl, syringyl, and *p*-hydroxypheny units, respectively. Figure S2: The structure morphology of both DMAP-lignin epoxide resin (a) and BTEAC-lignin epoxide resin (b). Figure S3: The SEM images of the fractured surface of the BB prepared with different addition amounts of BTEAC-lignin epoxide after tensile tests.

Author Contributions: Conceptualization, L.S., and H.P.; methodology, L.S.; investigation, L.S., Y.M., and P.L.; data curation, L.S.; validation, H.P., and L.S.; resources, L.S., Y.M., and P.L.; writ-ing-original draft preparation, L.S.; writing—review and editing, H.P., W.L., and Y.M.; project administration, H.P., and W.L.; funding acquisition, H.P. All authors have read and agreed to the published version of the manuscript.

Funding: This research was funded by the National Key Research and Development Program of China (2017YFD0601003), and GDAS' Project of Science and Technology Development (2020GDASYL-0102003).

Data Availability Statement: The data presented in this study are available on request from the corresponding author.

Conflicts of Interest: The authors declare no conflict of interest.

Abbreviations

DMAP	Dimethylaminopyridine
BTEAC	Benzyltriethylammonium chloride
BADGE	Bisphenol A diglycidyl ether
DB	DMAP-lignin epoxide/ BADGE
BB	BTEAC-lignin epoxide/BADGE
FRPs	Fiber-reinforced plastics
M_n	Number-average molecular weight
M_w	Weight-average molecular weight
Đ	Molecular weight distribution
ECH	Epichlorohydrin
DMSO	Dimethyl sulfoxide
TETA	Triethylene teramine
DMF	Dimethyl formamide
BDDGE	Butanediol diglycidyl ether
N_2	Nitrogen

PTFE	Polytetrafluoroethylene
FTIR	Fourier transform infrared spectroscopy
^{1}H NMR	^{1}H nuclear magnetic resonance
p-NBD	*p*-Nitrobenzaldehyde
GPC	Gel permeation chromatography
PMMA	Polymethyl methacrylate
DMA	Dynamic mechanical analysis
E′	Storage modulus
E″	Loss modulus
Tg	Glass transition temperature
ρ	Crosslink density
Tα	1st Phase transition temperature
Tβ	2nd Phase transition temperatures
G	Guaiacyl
S	Syringyl
DB80	DMAP-lignin epoxide/ BADGE (DB) with DMAP-lignin epoxide replacement of 80 wt% BADGE, while the other samples are named accordingly based on the weight percentage of lignin epoxide.
BB80	BTEAC -lignin epoxide/ BADGE (BB) with BTEAC -lignin epoxide replacement of 80 wt% BADGE, while the other samples are named accordingly based on the weight percentage of lignin epoxide.

References

1. Liu, W.; Zhou, R.; Goh, H.L.S.; Huang, S.; Lu, X. From waste to functional additive: Toughening epoxy resin with lignin. *ACS Appl. Mater. Interfaces* **2014**, *6*, 5810–5817. [CrossRef] [PubMed]
2. Wang, Z.; Ganewatta, M.S.; Tang, C. Sustainable polymers from biomass: Bridging chemistry with materials and processing. *Prog. Polym. Sci.* **2020**, *101*, 101197. [CrossRef]
3. Adam, A.A.; Ojur Dennis, J.; Al-Hadeethi, Y.; Mkawi, E.M.; Abubakar Abdulkadir, B.; Usman, F.; Mudassir Hassan, Y.; Wadi, I.A.; Sani, M. State of the art and new directions on electrospun lignin/cellulose nanofibers for supercapacitor application: A systematic literature review. *Polymers* **2020**, *12*, 2884. [CrossRef] [PubMed]
4. Saratale, R.G.; Saratale, G.D.; Ghodake, G.; Cho, S.-K.; Kadam, A.; Kumar, G.; Jeon, B.-H.; Pant, D.; Bhatnagar, A.; Shin, H.S. Wheat straw extracted lignin in silver nanoparticles synthesis: Expanding its prophecy towards antineoplastic potency and hydrogen peroxide sensing ability. *Int. J. Biol. Macromol.* **2019**, *128*, 391–400. [CrossRef] [PubMed]
5. Kumar, P.; Maharjan, A.; Jun, H.-B.; Kim, B.S. Bioconversion of lignin and its derivatives into polyhydroxyalkanoates: Challenges and opportunities. *Appl. Biochem. Biotechnol.* **2019**, *66*, 153–162. [CrossRef]
6. Ganesh Saratale, R.; Cho, S.-K.; Dattatraya Saratale, G.; Kadam, A.A.; Ghodake, G.S.; Kumar, M.; Naresh Bharagava, R.; Kumar, G.; Su Kim, D.; Mulla, S.I.; et al. A comprehensive overview and recent advances on polyhydroxyalkanoates (PHA) production using various organic waste streams. *Bioresour. Technol.* **2021**, *325*, 124685. [CrossRef]
7. Collins, M.N.; Nechifor, M.; Tanasa, F.; Zanoaga, M.; McLoughlin, A.; Strozyk, M.A.; Culebras, M.; Teaca, C.-A. Valorization of lignin in polymer and composite systems for advanced engineering applications—A review. *Int. J. Biol. Macromol.* **2019**, *131*, 828–849. [CrossRef] [PubMed]
8. Thakur, V.K.; Thakur, M.K.; Raghavan, P.; Kessler, M.R. Progress in green polymer composites from lignin for multifunctional applications: A review. *ACS Sustain. Chem. Eng.* **2014**, *2*, 1072–1092. [CrossRef]
9. Kalami, S.; Chen, N.; Borazjani, H.; Nejad, M. Comparative analysis of different lignins as phenol replacement in phenolic adhesive formulations. *Ind. Crops Prod.* **2018**, *125*, 520–528. [CrossRef]
10. Mahmood, N.; Yuan, Z.; Schmidt, J.; Xu, C. Preparation of bio-based rigid polyurethane foam using hydrolytically depolymerized Kraft lignin via direct replacement or oxypropylation. *Eur. Polym. J.* **2015**, *68*, 1–9. [CrossRef]
11. Sun, J.; Wang, C.; Stubbs, L.P.; He, C. Carboxylated Lignin as an Effective Cohardener for Enhancing Strength and Toughness of Epoxy. *Macromol. Mater. Eng.* **2017**, *302*, 1700341. [CrossRef]
12. Huang, F.; Liu, Y.Q.; Zhang, X.H.; Wei, G.S.; Gao, J.M.; Song, Z.H.; Zhang, M.L.; Qiao, J.L. Effect of elastomeric nanoparticles on toughness and heat resistance of epoxy resins. *Macromol. Rapid. Commun.* **2002**, *23*, 786–790. [CrossRef]
13. Kim, J.Y.; Hwang, H.; Oh, S.; Kim, Y.S.; Kim, U.J.; Choi, J.W. Investigation of structural modification and thermal characteristics of lignin after heat treatment. *Int. J. Biol. Macromol.* **2014**, *66*, 57–65. [CrossRef]
14. Cho, M.J.; Karaaslan, M.A.; Renneckar, S.; Ko, F. Enhancement of the mechanical properties of electrospun lignin-based nanofibers by heat treatment. *J. Mater. Sci.* **2017**, *52*, 9602–9614. [CrossRef]
15. Zhang, Y.; Wu, J.Q.; Li, H.; Yuan, T.Q.; Wang, Y.Y.; Sun, R.C. Heat treatment of industrial alkaline lignin and its potential application as an adhesive for green wood-lignin composites. *ACS Sustain. Chem. Eng.* **2017**, *5*, 7269–7277. [CrossRef]
16. Van de Pas, D.J.; Torr, K.M. Biobased epoxy resins from deconstructed native softwood lignin. *Biomacromolecules* **2017**, *18*, 2640–2648. [CrossRef] [PubMed]

17. Ferdosian, F.; Zhang, Y.; Yuan, Z.; Anderson, M.; Xu, C. Curing kinetics and mechanical properties of bio-based epoxy composites comprising lignin-bised epoxy resins. *Eur. Polym. J.* **2016**, *82*, 153–165. [CrossRef]
18. Gioia, C.; Colonna, M.; Tagami, A.; Medina, L.; Sevastyanova, O.; Berglund, L.A.; Lawoko, M. Lignin-based epoxy resins: Unravelling the relationship between structure and material properties. *Biomacromolecules* **2020**, *21*, 1920–1928. [CrossRef]
19. Ferdosian, F.; Yuan, Z.S.; Anderson, M.; Xu, C. Synthesis of lignin-based epoxy resins: Optimization of reaction parameters using response surface methodology. *RSC Adv.* **2014**, *4*, 31745–31753. [CrossRef]
20. Kuo, P.Y.; Barros, L.D.; Sain, M.; Tjong, J.S.Y.; Yan, N. Effects of reaction parameters on the glycidyl etherification of bark extractives during bioepoxy resin synthesis. *ACS Sustain. Chem. Eng.* **2016**, *4*, 1016–1024. [CrossRef]
21. Asada, C.; Basnet, S.; Otsuka, M.; Sasaki, C.; Nakamura, Y. Epoxy resin synthesis using low molecular weight lignin separated from various lignocellulosic materials. *Int. J. Biol. Macromol.* **2015**, *74*, 413–419. [CrossRef] [PubMed]
22. Jiang, Y.; Ding, D.C.; Zhao, S.; Zhu, H.Y.; Kenttamaa, H.I.; Abu-Omar, M.M. Renewable thermoset polymers based on lignin and carbohydrate derived monomers. *Green Chem.* **2018**, *20*, 1131–1138. [CrossRef]
23. Li, W.; Zhang, Y.; Das, L.; Wang, Y.; Li, M.; Wanninayake, N.; Pu, Y.; Kim, D.Y.; Cheng, Y.-T.; Ragauskas, A.J.; et al. Linking lignin source with structural and electrochemical properties of lignin-derived carbon materials. *RSC Adv.* **2018**, *8*, 38721–38732. [CrossRef]
24. Gioia, C.; Lo Re, G.; Lawoko, M.; Berglund, L. Tunable thermosetting epoxies based on fractionated and well-characterized lignins. *J. Am. Chem. Soc.* **2018**, *140*, 4054–4061. [CrossRef] [PubMed]
25. Zhao, S.; Abu-Omar, M.M. Synthesis of renewable thermoset polymers through successive lignin modification using lignin-derived phenols. *ACS Sustain. Chem. Eng.* **2017**, *5*, 5059–5066. [CrossRef]
26. Zhao, S.; Abu-Omar, M.M. Renewable epoxy networks derived from lignin-based monomers: Effect of cross-linking density. *ACS Sustain. Chem. Eng.* **2016**, *4*, 6082–6089. [CrossRef]
27. Flory, P.J. Molecular theory of rubber elasticity. *Polymer* **1979**, *20*, 1317–1320. [CrossRef]
28. Karumuri, S.; Hiziroglu, S.; Kalkan, A.K. Thermoset-cross-linked lignocellulose: A moldable plant biomass. *ACS Appl. Mater. Interfaces* **2015**, *7*, 6596–6604. [CrossRef] [PubMed]
29. Sasaki, C.; Wanaka, M.; Takagi, H.; Tamura, S.; Asada, C.; Nakamura, Y. Evaluation of epoxy resins synthesized from steam-exploded bamboo lignin. *Ind. Crops Prod.* **2013**, *43*, 757–761. [CrossRef]
30. Stutz, H.; Illers, K.H.; Mertes, J. A Generalized theory for the glass-transition temperature of cross-linked and uncrosslinked polymers. *J. Polym. Sci. Pol. Phys.* **1990**, *28*, 1483–1498. [CrossRef]
31. Nicastro, K.H.; Kloxin, C.J.; Epps, T.H., III. Potential lignin-derived alternatives to bisphenol a in diamine-hardened epoxy resins. *ACS Sustain. Chem. Eng.* **2018**, *6*, 14812–14819. [CrossRef]
32. Patil, P.N.; Rath, S.K.; Sharma, S.K.; Sudarshan, K.; Maheshwari, P.; Patri, M.; Praveen, S.; Khandelwal, P.; Pujari, P.K. Free volumes and structural relaxations in diglycidyl ether of bisphenol-A based epoxy-polyether amine networks. *Soft Matter* **2013**, *9*, 3589–3599. [CrossRef]
33. Tang, B.; Kong, M.; Yang, Q.; Huang, Y.; Li, G. Toward simultaneous toughening and reinforcing of trifunctional epoxies by low loading flexible reactive triblock copolymers. *RSC Adv.* **2018**, *8*, 17380–17388. [CrossRef]
34. Liu, T.; Han, B.; Zhang, L.; Wu, M.; Xing, A.; Miao, X.; Meng, Y.; Li, X. Environmentally friendly high performance homopolymerized epoxy using hyperbranched epoxy as a modifier. *RSC Adv.* **2016**, *6*, 14211–14221. [CrossRef]
35. Zhang, M.; Resende, F.L.P.; Moutsoglou, A.; Raynie, D.E. Pyrolysis of lignin extracted from prairie cordgrass, aspen, and Kraft lignin by Py-GC/MS and TGA/FTIR. *J. Anal. Appl. Pyrolysis* **2012**, *98*, 65–71. [CrossRef]
36. Liu, W.-J.; Jiang, H.; Yu, H.-Q. Thermochemical conversion of lignin to functional materials: A review and future directions. *Green Chem.* **2015**, *17*, 4888–4907. [CrossRef]
37. Morandim-Giannetti, A.A.; Agnelli, J.A.M.; Lancas, B.Z.; Magnabosco, R.; Casarin, S.A.; Bettini, S.H.P. Lignin as additive in polypropylene/coir composites: Thermal, mechanical and morphological properties. *Carbohydr. Polym.* **2012**, *87*, 2563–2568. [CrossRef]

polymers

Article

Improvement in Thermal Stability of Flexible Poly(L-lactide)-*b*-poly(ethylene glycol)-*b*-poly(L-lactide) Bioplastic by Blending with Native Cassava Starch

Yaowalak Srisuwan and Yodthong Baimark *

Biodegradable Polymers Research Unit, Department of Chemistry and Centre of Excellence for Innovation in Chemistry, Faculty of Science, Mahasarakham University, Mahasarakham 44150, Thailand
* Correspondence: yodthong.b@msu.ac.th

Abstract: High-molecular-weight poly(L-lactide)-*b*-poly(ethylene glycol)-*b*-poly(L-lactide) triblock copolymer (PLLA-PEG-PLLA) is a promising candidate for use as a biodegradable bioplastic because of its high flexibility. However, the applications of PLLA-PEG-PLLA have been limited due to its high cost and poor thermal stability compared to PLLA. In this work, native cassava starch was blended to reduce the production cost and to improve the thermal stability of PLLA-PEG-PLLA. The starch interacted with PEG middle blocks to increase the thermal stability of the PLLA-PEG-PLLA matrix and to enhance phase adhesion between the PLLA-PEG-PLLA matrix and dispersed starch particles. Tensile stress and strain at break of PLLA-PEG-PLLA films decreased and the hydrophilicity increased as the starch content increased. However, all the PLLA-PEG-PLLA/starch films remained more flexible than the pure PLLA film, representing a promising candidate in biomedical, packaging and agricultural applications.

Keywords: poly(lactic acid); block copolymer; cassava starch; biocomposites; thermal stability

Citation: Srisuwan, Y.; Baimark, Y. Improvement in Thermal Stability of Flexible Poly(L-lactide)-*b*-poly(ethylene glycol)-*b*-poly(L-lactide) Bioplastic by Blending with Native Cassava Starch. *Polymers* **2022**, *14*, 3186. https://doi.org/10.3390/polym14153186

Academic Editor: Evgenia G. Korzhikova-Vlakh

Received: 14 June 2022
Accepted: 2 August 2022
Published: 4 August 2022

Publisher's Note: MDPI stays neutral with regard to jurisdictional claims in published maps and institutional affiliations.

1. Introduction

Among bioplastics, poly(L-lactic acid) or poly(L-lactide) (PLLA) is a promising substitute for traditional petroleum-based plastics in many fields, such as biomedical, tissue engineering, drug delivery and packaging applications [1–4]. This is due to its excellent biodegradability, biocompatibility and compostability, as well as being environmentally friendly and having good processability and feasibility for an increased production scale [5,6]. However, the low flexibility of PLLA, due to its high glass-transition temperature (T_g, around 60 °C), limits its use in many applications [7–9]. High-molecular-weight PLLA-*b*-poly(ethylene glycol)-*b*-PLLA triblock copolymers (PLLA-PEG-PLLA) have exhibited more flexibility than PLLA because of the high flexibility of PEG middle blocks [10–12].

Unfortunately, these pure PLLA-PEG-PLLAs had high melt flow ability (low melt strength), which is not appropriate for many processing applications, such as injection molding, blow film molding and extrusion molding [11,12]. The melt flow properties of PLLA-PEG-PLLA can be controlled by reaction with a chain extender to form long-chain branching structures. However, our previous works reported that the thermal stability of PLLA-PEG-PLLA decreased after the chain extension reaction [11–13]. The branching structures of the chain-extended PLLA and PLLA-PEG-PLLA suppressed its thermal stability by preventing molecular interactions. This resulted in a narrower window of processing for the chain-extended PLLA-PEG-PLLA (the range between its melting and thermal decomposition temperatures was narrower). The 60/40 chain-extended PLLA-PEG-PLLA/thermoplastic starch (TPS) blends were co-continuous phase structures, indicating that they had good phase compatibility [14]. The PEG middle blocks enhanced the phase compatibility between PLLA end blocks and TPS phases. Moreover, the TPS blending improved the thermal stability of PLLA-PEG-PLLA. However, the effect of native starch on the thermal stability of PLLA-PEG-PLLA has not yet been explored in detail.

Native starches are low-cost and non-scarce biopolymers and have been blended with various biodegradable plastics, such as PLLA [15–17], poly(ε-caprolactone) [18] and polyhydroxybutyrate [19], for reducing their production costs and for maintaining their biodegradability. However, the highly brittle PLLA/native starch composites are limited in practical applications due to the poor mechanical properties of native starch and poor phase compatibility between PLLA and native starch particles [15,20]. Therefore, the aim of this work is to investigate the influence of native cassava starch (5, 10 and 20 wt%) on the thermal stability, phase morphology, mechanical properties and hydrophilicity of PLLA-PEG-PLLA compared to PLLA/native cassava starch blends.

2. Experimental Section

2.1. Materials

PLLA and PLLA-PEG-PLLA were synthesized through ring-opening polymerization of L-lactide monomer (96% L-enantiomer content) at 165 °C in bulk under a nitrogen atmosphere, as described in our previous works [11,12,14]. Stannous octoate (95%, Sigma, St. Louis, MO, USA) was used as a catalyst. Moreover, 1-dodecanol (98%, Fluka, Buchs, Switzerland) and poly(ethylene glycol) with molecular weight of 20,000 g/mol (Sigma, St. Louis, MO, USA) were used as initiators for the synthesis of PLLA and PLLA-PEG-PLLA, respectively. Number-average molecular weight (M_n) and dispersity index ($Đ$) of PLLA, analyzed by gel permeation chromatography (GPC, e2695 separations module, Waters, Milford, MA, USA), were 88,400 g/mol and 2.3, respectively, while the M_n and $Đ$ values of PLLA-PEG-PLLA were 89,900 g/mol and 2.8, respectively. Melt flow index (MFI) of obtained PLLA was 24 g/10 min at 190 °C under 2.16 kg load. However, the MFI of obtained PLLA-PEG-PLLA was too high. The melt flow property of PLLA-PEG-PLLA was then adjusted to be close to the value for PLLA obtained by chain extension reaction with 4 parts per hundred of resin by weight (phr) of Joncryl® ADR4368 chain extender (BASF, Bangkok, Thailand) [11,14]. The MFI of the obtained chain-extended PLLA-PEG-PLLA was 23 g/10 min. Native cassava starch (food-grade, 22% amylose content) was supplied by Kriangkrai Co., Ltd. (Nakornprathom, Thailand). A scanning electron microscope (SEM, JSM-6460LV, JEOL, Tokyo, Japan) image of native cassava starch is presented in Figure 1. The starch particles were in the range of 5–20 μm.

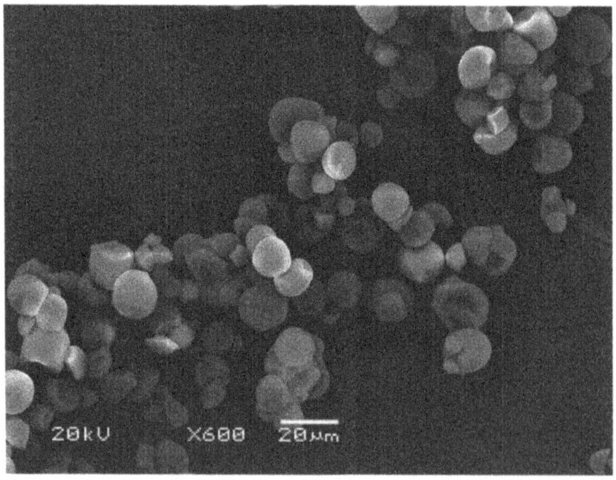

Figure 1. SEM image of cassava starch (bar scale = 20 μm).

2.2. Preparation of PLLA/Starch and PLLA-PEG-PLLA/Starch Composites

Chain-extended PLLA-PEG-PLLA and cassava starch were dried in a vacuum oven at 50 °C for 24 h before fabricating the composites. PLLA-PEG-PLLA/starch composites were prepared using a Rheomix batch mixer (HAAKE Polylab OS, Thermo Fisher Scientific, Waltham, MA, USA) at 180 °C with a rotor speed of 100 rpm for 5 min. The composites with PLLA-PEG-PLLA/starch ratios of 100/0, 95/5, 90/10 and 80/20 (*w*/*w*) were investigated. PLLA/starch composites were also prepared using the same process for comparison.

Before film forming, the composites were dried at 50 °C in a vacuum oven for 24 h. Composite films (0.2 mm in thickness) were prepared using a compression molding machine (Auto CH, Carver, Inc., Savannah, GA, USA). The composite pellets were filled into a mold and preheated at 180 °C for 3 min without compression force, followed by hot pressing at the same temperature under 5.0 MPa load for 1 min. Afterwards, the mold was rapidly cooled in a water-cooled press under 5.0 MPa load for 1 min. The obtained film samples were stored in a desiccator for 24 h before characterization.

2.3. Characterization of PLLA/Starch and PLLA-PEG-PLLA/Starch Composites

Thermal transition properties of the samples were investigated using a differential scanning calorimeter (DSC, Pyris Diamond, PerkinElmer, Waltham, MA, USA). First, samples were heated at 200 °C for 3 min to erase the thermal history, before fast quenching to 0 °C and heating from 0 °C to 200 °C at a rate of 10 °C/min under a nitrogen gas flow. The degree of crystallinity (X_c) was calculated according to the following equation:

$$X_c\ (\%) = [(\Delta H_m - \Delta H_{cc})/(93.6 \times W_{PLLA})] \times 100 \quad (1)$$

where ΔH_m and ΔH_{cc} are enthalpies of melting and cold crystallization, respectively. Moreover, 93.6 J/g is the theoretical ΔH_m for 100% crystalline PLLA [14]. W_{PLLA} is the PLLA weight fraction of the samples calculated from PLLA fractions (PLLA = 1.00 and PLLA-PEG-PLLA = 0.83 obtained from ^{1}H-NMR [11]) and the starch content.

Thermal decomposition behaviors of the samples were determined using a thermogravimetric analyzer (TGA, SDT Q600, TA Instruments, New Castle, DE, USA). Samples (5–10 mg) were heated from 50 °C to 800 °C at a rate of 20 °C/min under a nitrogen flow.

An SEM (JSM-6460LV, JEOL, Tokyo, Japan) operated at 20 kV was introduced to observe the phase morphology of the film samples. The films were cryogenically fractured after immersion in liquid nitrogen for 10 min and were sputter-coated with gold before SEM analysis. Film morphology of the starch-free composite films was also investigated after the cryogenically fractured films were immersed in 6 N HCl aqueous solution at room temperature for 3 h to remove the starch phase [21].

The tensile testing of film samples (100 mm × 10 mm) was carried out using a universal mechanical testing machine (LY-1066B, Dongguan Liyi Environmental Technology Co., Ltd., Dongguan, China) with a 100 kg load cell according to ASTM D882. A crosshead speed of 50 mm/min and a gauge length of 50 mm were used. The tensile properties were averaged from at least five measurements.

Water contact angle of each film surface was recorded using a contact angle goniometer (Ramé-Hart Instrument Co., Succasunna, NJ, USA) after 15 s. Each sample was averaged over five different determinations.

Moisture uptake of film samples (20 mm × 20 mm) was investigated as follows. The film samples were weighed after drying at room temperature in a vacuum oven for 48 h. They were kept in a desiccator with 90 ± 5% relative humidity, maintained with a saturated sodium chloride solution at 25 °C. The sample films were weighed again after being kept in the desiccator for an interval of time. The moisture uptake of film samples was calculated from Equation (2). The averaged moisture uptake was obtained from five different determinations [14].

$$\text{Moisture uptake } (\%) = [(M_f - M_i)/M_i] \times 100 \quad (2)$$

where M_i and M_f are the weights of the film before and after the test, respectively.

3. Results and Discussion

3.1. Thermal Transition Properties

The thermal transition properties of the composites were determined by DSC, and the DSC heating curves of PLLA/starch and PLLA-PEG-PLLA/starch composites are given in Figure 2. The DSC results are summarized in Table 1. The T_g, T_{cc} and T_m values of PLLA and PLLA-PEG-PLLA matrices did not change significantly when the starch was blended and the starch content was increased, indicating the immiscibility of these matrices and native starch [18]. The X_c values of the pure PLLA (X_c = 23.6%) and pure PLLA-PEG-PLLA (X_c = 15.3%) slightly increased to 27.1% and 20.3%, respectively, as the 5 wt% starch was blended. Further increase in starch content in the composites led to a decrease in X_c. This indicates that the presence of starch particles as a minority fraction enhanced the nucleation of PLLA-based matrices at the particle interface [18,22,23].

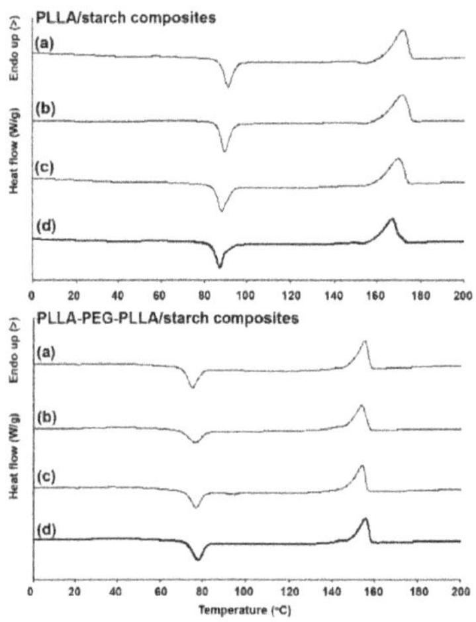

Figure 2. DSC heating curves of (**above**) PLLA/starch and (**below**) PLLA-PEG-PLLA/starch composites with blend ratios of (**a**) 100/0, (**b**) 95/5, (**c**) 90/10 and (**d**) 80/20 (*w*/*w*).

Table 1. Thermal transition properties of PLLA/starch and PLLA-PEG-PLLA/starch composites obtained from DSC heating curves in Figure 2.

Sample	T_g (°C) [a]	T_{cc} (°C) [b]	ΔH_{cc} (J/g) [c]	T_m (°C) [d]	ΔH_m (J/g) [e]	X_c (%) [f]
PLLA/starch (*w*/*w*)						
100/0	53	90	29.5	172	51.6	23.6
95/5	53	90	29.6	172	53.7	27.1
90/10	52	88	29.5	170	48.5	22.6
80/20	52	88	26.2	168	40.2	18.7
PLLA-PEG-PLLA/starch (*w*/*w*)						
100/0	32	76	17.9	155	29.8	15.3
95/5	32	78	14.7	154	29.7	20.3
90/10	32	77	15.0	154	28.6	19.4
80/20	32	78	16.6	155	26.6	16.1

[a] Glass transition temperature (T_g); [b] Cold crystallization temperature (T_{cc}); [c] Enthalpy of cold crystallization (ΔH_{cc}); [d] Melting temperature (T_m); [e] Enthalpy of melting (ΔH_m); [f] Degree of crystallinity (X_c) calculated from Equation (1).

However, starch particles with higher starch content (10 and 20 wt%) may aggregate, leading to reduced nucleation efficiency [18]. It should be noted that the X_c values of the PLLA-PEG-PLLA-based composites with 10 wt% (X_c = 19.4%) and 20 wt% (X_c = 16.1%) starch content were still larger than those of the pure PLLA-PEG-PLLA (X_c = 15.3%), whereas the X_c values of both the PLLA/starch composites with 10 wt% (X_c = 22.6%) and 20 wt% (X_c = 18.7%) starch content were lower than those of the pure PLLA (X_c = 23.6%). The results suggested that the phase compatibility between PLLA-PEG-PLLA matrices and starch particles was better in terms of the nucleation effect [24,25]. The dispersed component could promote the crystallization of the crystalline matrix due to the nucleation activity of the matrix/starch interface [26]. The dispersed starch particles were well compatible with the PLLA-PEG-PLLA matrices on matrix/starch interfaces and induced a nucleation barrier lower than that in pure PLLA-PEG-PLLA to accelerate the crystallization rate of PLLA-PEG-PLLA. In addition, blending of native starch particles in this work revealed a better nucleation effect for the PLLA-PEG-PLLA matrix compared to the blending of TPS in our previous work [14]. The X_c values of PLLA-PEG-PLLA/TPS blends steadily decreased as the TPS content increased.

3.2. Thermal Decomposition Behaviors

The thermal decomposition behaviors of the samples were determined from their TG thermograms, as shown in Figure 3. The pure PLLA showed a single step of thermal decomposition in the range of 250–450 °C. The TG thermogram of native cassava starch had two steps of thermal decomposition in the ranges 50–150 °C and 250–500 °C, as presented in Figure 4, due to the evaporation of residue moisture and pyrolysis of starch, respectively [24]. The decomposition temperature at 5% weight loss (5%-T_d) of pure PLLA was at 291 °C, as reported in Table 2. The 5%-Td values of PLLA/starch composites shifted to lower temperatures as the starch content increased due to the evaporation of moisture from the starch. The native cassava starch in this work had around 10 wt% residue ash at 800 °C. The TG thermograms of PLLA/starch composites in Figure 3 (above) also exhibit a single step of thermal decomposition in the range 250–450 °C.

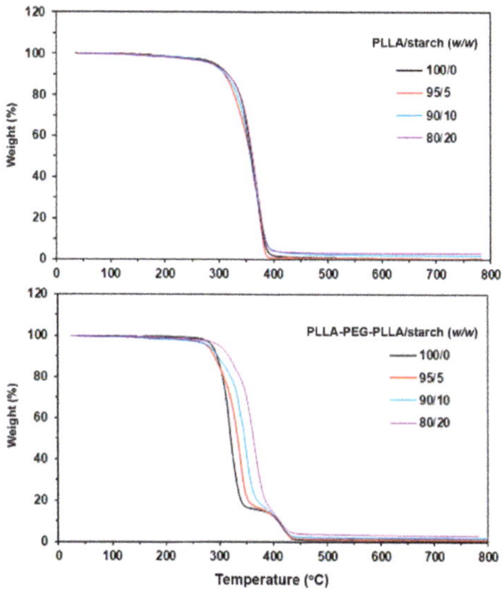

Figure 3. TG thermograms of (**above**) PLLA/starch and (**below**) PLLA-PEG-PLLA/starch composites with various blend ratios.

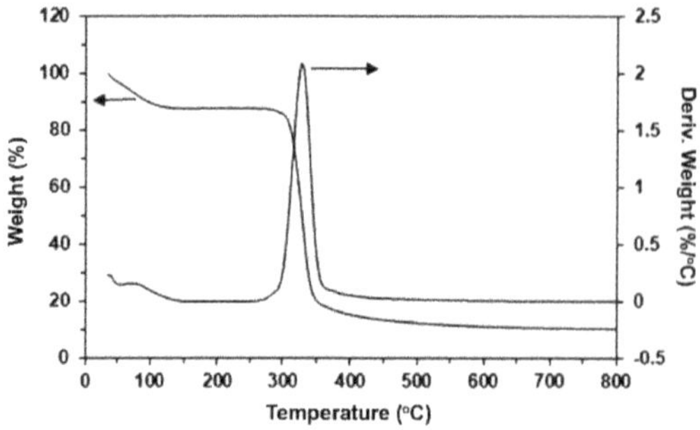

Figure 4. TG and DTG thermograms of native cassava starch.

Table 2. Residue ash and temperature of maximum decomposition rate ($T_{d,max}$) of the composites.

Sample	5%-T_d (°C) [a]	Residue ash (wt%) [b]	PLLA-$T_{d,max}$ (°C) [c]	PEG-$T_{d,max}$ (°C) [d]
PLLA/starch (*w*/*w*)				
100/0	291	0.06	365	-
95/5	291	0.59	369	-
90/10	287	1.64	366	-
80/20	285	2.61	369	-
PLLA-PEG-PLLA/starch (*w*/*w*)				
100/0	286	0.11	319	421
95/5	278	0.79	336	421
90/10	280	1.74	349	418
80/20	300	3.05	363	419

[a] Decomposition temperature at 5% weight loss obtained from TG thermograms in Figure 3; [b] Obtained from TG thermograms at 800 °C in Figure 3; [c] Temperature of maximum decomposition rate for PLLA blocks (PLLA-$T_{d,max}$) obtained from DTG thermograms in Figure 5; [d] Temperature of maximum decomposition rate for PEG blocks (PEG-$T_{d,max}$) obtained from DTG thermograms in Figure 5.

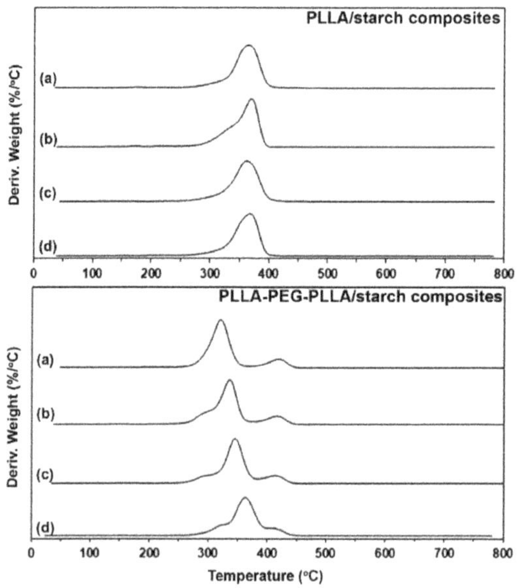

Figure 5. DTG thermograms of (**above**) PLLA/starch and (**below**) PLLA-PEG-PLLA/starch composites with blend ratios of (**a**) 100/0, (**b**) 95/5, (**c**) 90/10 and (**d**) 80/20 (*w*/*w*).

The pure PLLA-PEG-PLLA showed two steps of thermal decomposition in the ranges 250–350 °C and 350–450 °C due to thermal decomposition of PLLA and PEG blocks, respectively [11,12]. The 5%-T_d of PLLA-PEG-PLLA shifted from 286 °C to 278 °C when the 5% starch was blended. However, the 5%-T_d of PLLA-PEG-PLLA/starch composites shifted to higher temperatures when the starch content increased up to 10 and 20 wt%. It could be clearly seen that the thermal decomposition steps of PLLA end blocks dramatically shifted to higher temperatures as the starch content increased, as shown in Figure 3 (below). This indicates that the blending of native starch effectively improved the thermal stability of the PLLA end blocks of PLLA-PEG-PLLA, similar to the blending of TPS [14]. The residue ash in both the PLLA/starch and PLLA-PEG-PLLA/starch composites increased steadily with the starch content, as summarized in Table 2.

The derivative TG (DTG) thermograms in Figure 5 also enabled the determination of the thermal decomposition behaviors of composite samples. The DTG peak was assigned to the temperature of maximum decomposition rate ($T_{d,max}$) of the composites, as is also reported in Table 2. The pure PLLA and all the PLLA/starch composites had a single $T_{d,max}$ peak in the range 365–369 °C, attributed to the "unzipping" mechanism at the chain ends of PLLA [27]. The native cassava starch had a $T_{d,max}$ peak at 327 °C, as assigned in Figure 4. The small $T_{d,max}$ peaks of starch minor fractions may be overlapped with the large $T_{d,max}$ peaks of PLLA major fractions. Thus, the addition of starch did not significantly change the thermal decomposition behaviors of the PLLA matrix.

The pure PLLA-PEG-PLLA had two $T_{d,max}$ peaks at 319 °C and 422 °C, attributed to the thermal decomposition of PLLA (PLLA-$T_{d,max}$) and PEG (PEG-$T_{d,max}$) blocks, respectively [11,27]. It was found that the PLLA-$T_{d,max}$ peaks of all the PLLA-PEG-PLLA/starch composites (336–363 °C) were at higher temperatures than that of the pure PLLA-PEG-PLLA (PLLA-$T_{d,max}$ = 319 °C). The PLLA-$T_{d,max}$ peaks of the composites dramatically shifted to higher temperatures as the starch content increased. However, the PEG-$T_{d,max}$ peaks of the composites were in the range 418–422 °C and did not shift significantly with starch blending. This may be due to the starch fraction being completely decomposed before the thermal decomposition of PEG.

The TG and DTG results indicated that the interactions between PLLA-PEG-PLLA and starch were stronger than those between PLLA and starch and suggested that the starch acted as a low-cost and effective thermal stabilizer for PLLA-PEG-PLLA. This may be explained by the formation of hydrogen bonds between the oxygen of PEG middle blocks and hydroxyl groups of starch [14,25,28]. In addition, the thermal stability of compatible polymer/starch blends was improved by products from the thermal decomposition of starch [29,30].

3.3. Phase Morphology

SEM images of the composite films were used to observe their phase morphology, as shown in Figure 6. Both the pure PLLA and PLLA-PEG-PLLA films illustrated in Figure 6a,e, respectively, had no phase separation. For all the PLLA/starch and PLLA-PEG-PLLA/starch films in Figure 6b–d and f–h, respectively, starch particles were found to be dispersed in the film matrices in a typically heterogeneous structure. Micro-voids on the film matrices occurred from the detachment of starch particles during the cryo-fracture step, which suggested poor interfacial adhesion between the film matrix and starch particles [25,31]. It appears that the number of micro-voids (or starch particles) increased with the starch content. It could be seen that there were more adhered starch particles for PLLA-PEG-PLLA/starch films than on the PLLA/starch films for the same starch content; this supported the hypothesis of stronger interactions between PLLA-PEG-PLLA and starch, corresponding to the TG/DTG results, as described above [28].

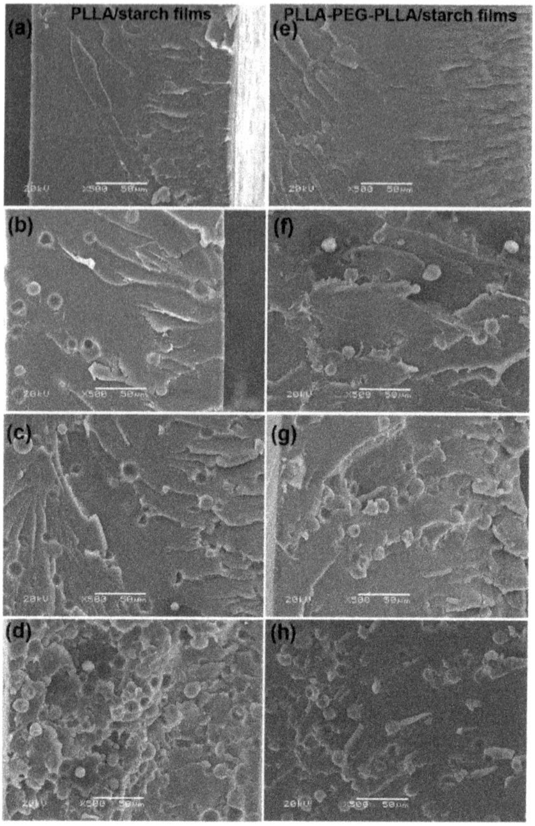

Figure 6. SEM images of cryogenically fractured surfaces of (**a**) pure PLLA film and PLLA/starch films with blend ratios of (**b**) 95/5, (**c**) 90/10 and (**d**) 80/20 (*w*/*w*), as well as (**e**) pure PLLA-PEG-PLLA film and PLLA-PEG-PLLA/starch films with blend ratios of (**f**) 95/5, (**g**) 90/10 and (**h**) 80/20 (*w*/*w*) (all bar scales = 50 μm).

Figure 7 presents SEM images of the cryogenically fractured 80/20 composite films after selective etching with HCl solution to remove the starch phases. Many micro-voids from the removal of the starch particles were obtained on the film matrices. This confirms that phase separation between film matrices and dispersed starch particles occurred for both the PLLA/starch and PLLA-PEG-PLLA/starch films. It should be noted that the film matrix of the 80/20 PLLA/starch composite was cracked, as indicated by the white arrows in Figure 7 (above), but the 80/20 PLLA-PEG-PLLA/starch film in Figure 7 (below) was not cracked. This implied that the 80/20 PLLA/starch film was brittle because the starch content was too high and also supported the supposition that interactions between PLLA and starch were poor.

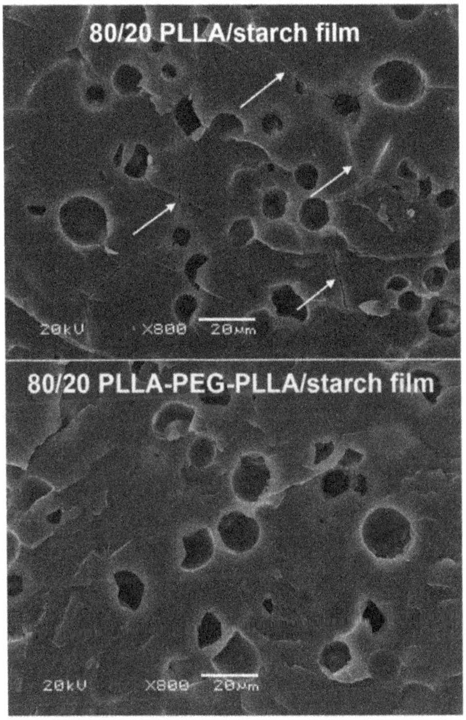

Figure 7. SEM images of cryogenically fractured surfaces for (**above**) 80/20 PLLA/starch and (**below**) 80/20 PLLA-PEG-PLLA/starch films after immersion in 6 N HCl solution for 3 h (some crack surfaces are indicated by white arrows; all bar scales = 20 μm).

3.4. Tensile Properties

Figure 8 shows tensile curves of the composite films, and the tensile results were averaged and are summarized in Table 3. Addition of starch decreased all determinations of tensile stress, strain at break and Young's modulus for the PLLA films. This was due to the poor mechanical properties of starch and poor phase compatibility between hydrophobic PLLA and hydrophilic starch [15,20]. The 80/20 PLLA/starch film in this work was very brittle. It could not be cut to the desired shape for tensile testing. It has been reported that dispersed thermoplastic starch (TPS) phases act as defect sites to decrease the tensile properties of the PLLA film matrices [32].

All the tensile curves of the PLLA-PEG-PLLA/starch films in Figure 8 (below) exhibited a yield point that suggested that they were flexible. The tensile stresses and Young's moduli of the films slightly decreased, and the strain at break largely decreased, as the starch content increased. It is well known that the tensile properties of starch materials are poor. From Table 3, it is seen that the tensile properties of all the PLLA/starch and PLLA-PEG-PLLA/starch films were lower than those of the pure PLLA and PLLA-PEG-PLLA films, respectively. Therefore, the decreasing tensile properties of both the PLLA/starch and PLLA-PEG-PLLA/starch film types were due to the poor tensile properties of starch and followed the "rule of mixtures" [32,33]. However, all the PLLA-PEG-PLLA/starch films still had higher strain at break (14.4–60.6%) than that of the pure PLLA film (3.4%). Thus, the more flexible PLLA-PEG-PLLA/starch composites have broader applications, such as flexible packaging, etc., than the pure PLLA and PLLA/starch composites.

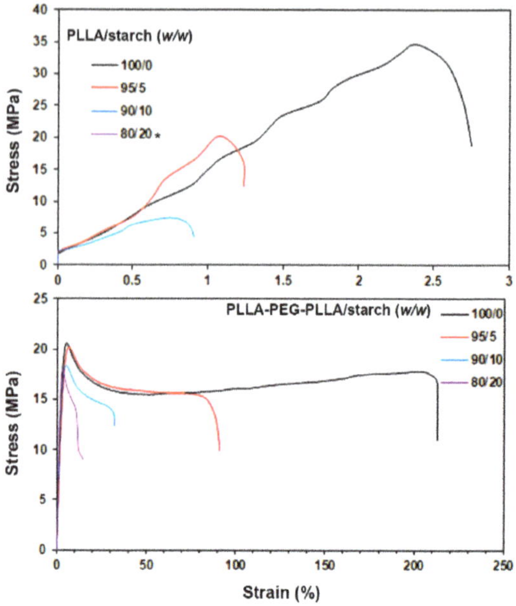

Figure 8. Selected tensile curves of (**above**) PLLA/starch and (**below**) PLLA-PEG-PLLA/starch films with various blend ratios (* 80/20 PLLA/starch film was not determined).

Table 3. Tensile properties and water contact angles of PLLA/starch and PLLA-PEG-PLLA/ starch films.

Film Samples	Ultimate Tensile Stress (MPa)	Strain at Break (%)	Young's Modulus (MPa)	Water Contact Angle (°)
PLLA/starch (*w*/*w*)				
100/0	38.4 ± 3.1	3.4 ± 0.7	957 ± 57	81.5 ± 3.3
95/5	24.6 ± 2.5	1.8 ±0.4	782 ±68	81.1 ± 4.1
90/10	6.2 ± 2.2	0.8 ± 0.5	229 ± 62	77.8 ± 3.8
80/20	- *	- *	- *	75.1 ± 2.7
PLLA-PEG-PLLA/starch (*w*/*w*)				
100/0	20.4 ± 1.8	239.6 ± 15.1	321 ± 31	68.9 ± 4.2
95/5	20.6 ± 1.5	90.6 ± 12.5	251 ± 16	67.5 ± 3.1
90/10	19.8 ± 1.9	31.2 ± 8.2	238 ± 25	64.4 ± 2.5
80/20	17.4 ± 0.8	14.4 ± 5.3	212 ± 12	63.9 ± 3.2

* was not determined because it was very brittle.

3.5. Water Contact Angle and Moisture Uptake

The hydrophilicity of film samples was determined from their water contact angles, as shown in Figure 9 and also summarized in Table 3. The water contact angles of pure PLLA and PLLA-PEG-PLLA films were 81.5° and 68.9°, respectively, implying that the hydrophilicity of PLLA-PEG-PLLA was higher than that of PLLA because of the hydrophilic PEG middle blocks. The water contact angles of the composite films decreased steadily (hydrophilicity increased) as the starch content increased. This was due to the high hydrophilicity of starch phases [15,20].

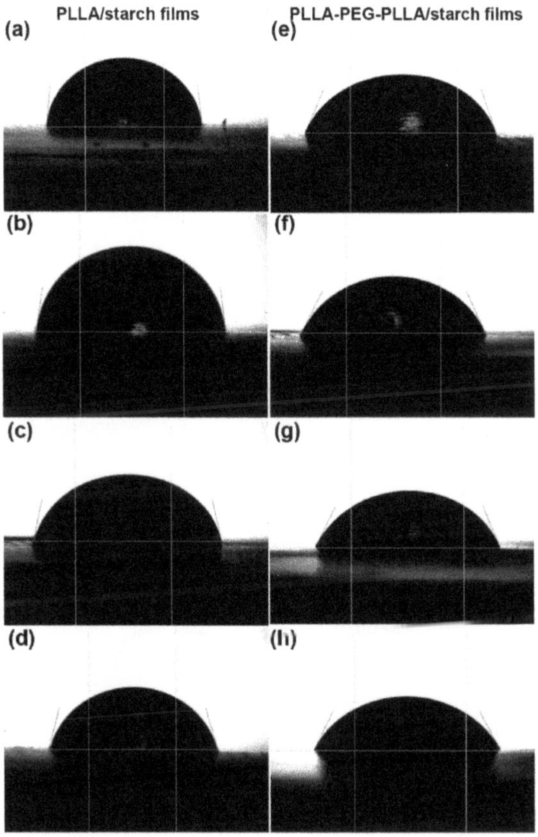

Figure 9. Images of water contact angles of (**a**) pure PLLA film and PLLA/starch films with blend ratios of (**b**) 95/5, (**c**) 90/10 and (**d**) 80/20 (*w*/*w*), as well as (**e**) pure PLLA-PEG-PLLA film and PLLA-PEG-PLLA/starch films with blend ratios of (**f**) 95/5, (**g**) 90/10 and (**h**) 80/20 (*w*/*w*).

The moisture uptake of the film samples was also determined for 48 h, as presented in Figure 10. The moisture uptake of pure PLLA and PLLA-PEG-PLLA films at 48 h was 7.7% and 0.6%, respectively, meaning that PLLA-PEG-PLLA was more hydrophilic than PLLA [14,34] corresponding to the results for the water contact angle. The moisture uptake of both the PLLA and PLLA-PEG-PLLA composite films increased as the starch content increased. The results of water contact angle and moisture uptake suggest that the PLLA-PEG-PLLA/starch composites had higher hydrophilicity than the PLLA/starch composites. In our previous work [14], PLLA/TPS and PLLA-PEG-PLLA/TPS composites were found to have higher hydrophilicity than the PLLA/native starch and PLLA-PEG-PLLA/native starch composites for the same blend ratio. This may be explained by the PLLA/TPS blends having smaller TPS particles in their film matrices than did native starch particles, whereas PLLA-PEG-PLLA/TPS blends were of the co-continuous phase type.

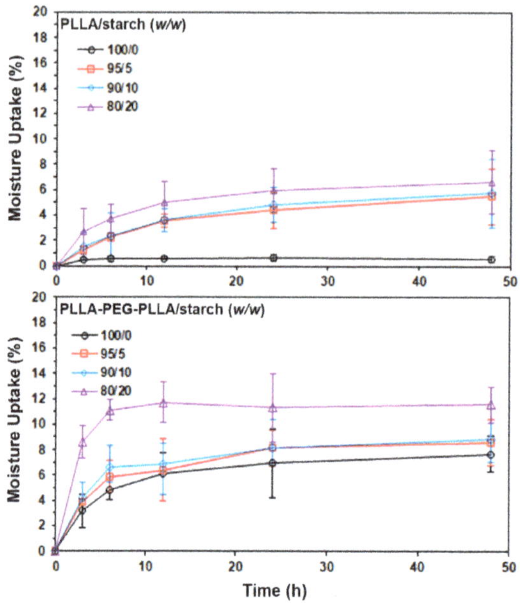

Figure 10. Moisture uptake of (**above**) PLLA/starch and (**below**) PLLA-PEG-PLLA/starch films with various blend ratios.

4. Conclusions

In this study, the effect of native cassava starch on the properties of PLLA-based and PLLA-PEG-PLLA-based composites was investigated. For the PLLA-PEG-PLLA/starch composites, improved thermal stability of PLLA-PEG-PLLA matrices and good compatibility between PLLA-PEG-PLLA and starch were observed compared to PLLA/starch composites. The PLLA-$T_{d,max}$ peaks of PLLA-PEG-PLLA at 319 °C dramatically shifted to higher temperatures by 17 °C, 30 °C and 44 °C when the starch content was 5, 10 and 20 wt%, respectively, but PLLA did not shift, as determined from TGA. The tensile properties of both the PLLA/starch and PLLA-PEG-PLLA/starch films decreased as the native starch content increased. However, all the PLLA-PEG-PLLA/starch films still exhibited a yield point and showed higher extensibility than the pure PLLA film. The hydrophilicity of both PLLA and PLLA-PEG-PLLA increased with the addition of native starch, as revealed by investigation of their water contact angles and moisture uptake.

The flexible PLLA-PEG-PLLA/starch composites, with balanced thermal stability, mechanical properties, hydrophilicity and cost-effectiveness, are very promising for the development of fully biodegradable biomedical, packaging and agricultural applications.

Author Contributions: Conceptualization, Y.B.; methodology, Y.S.; Investigation, Y.S.; Methodology, Y.S.; Resources, Y.S.; Visualization, Y.B.; Writing—original draft, Y.B.; Writing—review and editing, Y.B. All authors have read and agreed to the published version of the manuscript.

Funding: This research project was financially supported by Thailand Science Research and Innovation (TSRI) 2022. Y.B. is also grateful for the partial support provided by the Centre of Excellence for Innovation in Chemistry (PERCH-CIC), Office of the Higher Education Commission, Ministry of Education, Thailand.

Institutional Review Board Statement: Not applicable.

Informed Consent Statement: Not applicable.

Data Availability Statement: The data presented in this study are available on request from the corresponding author.

Conflicts of Interest: The authors declare no conflict of interest.

References

1. He, J.; Hu, X.L.; Cao, J.; Zhang, Y.; Xiao, J.; Peng, L.J.; Chen, D.; Xiong, C.; Zhang, L. Chitosan-coated hydroxyapatite and drug-loaded polytrimethylene carbonate/polylactic acid scaffold for enhancing bone regeneration. *Carbohydr. Polym.* **2021**, *253*, 117198. [CrossRef] [PubMed]
2. Daskalova, A.; Angelova, L.; Filipov, E.; Aceti, D.; Mincheva, R.; Carrete, X.; Kerdjoudj, H.; Dubus, M.; Chevrier, J.; Trifonov, A.; et al. Biomimetic Hierarchical Structuring of PLA by Ultra-Short Laser Pulses for Processing of Tissue Engineered Matrices: Study of Cellular and Antibacterial Behavior. *Polymers* **2021**, *13*, 2577. [CrossRef] [PubMed]
3. Ahmad Ruzaidi, D.A.; Mahat, M.M.; Shafiee, S.A.; Mohamed Sofian, Z.; Mohmad Sabere, A.S.; Ramli, R.; Osman, H.; Hamzah, H.H.; Zainal Ariffin, Z.; Sadasivuni, K.K. Advocating Electrically Conductive Scaffolds with Low Immunogenicity for Biomedical Applications: A Review. *Polymers* **2021**, *13*, 3395. [CrossRef] [PubMed]
4. Kamarudin, S.H.; Rayung, M.; Abu, F.; Ahmad, S.; Fadil, F.; Karim, A.A.; Norizan, M.N.; Sarifuddin, N.; Mat Desa, M.S.Z.; Mohd Basri, M.S.; et al. A Review on Antimicrobial Packaging from Biodegradable Polymer Composites. *Polymers* **2022**, *14*, 174. [CrossRef]
5. Silva, D.D.; Kaduri, M.; Poley, M.; Adir, O.; Krinsky, N.; Shainsky-Rotiman, J.; Schroeder, A. Biocompatibility, biodegradation and excretion of polylactic acid (PLA) in medical implants and theranostic systems. *Chem. Eng. J.* **2018**, *340*, 9–14. [CrossRef]
6. Rihayat, T.; Hadi, A.E.; Aidy, N.; Safitri, A.; Siregar, J.P.; Cionita, T.; Irawan, A.P.; Hamdan, M.H.M.; Fitriyana, D.F. Biodegradation of Polylactic Acid-Based Bio Composites Reinforced with Chitosan and Essential Oils as Anti-Microbial Material for Food Packaging. *Polymers* **2021**, *13*, 4019. [CrossRef]
7. Jin, F.L.; Hu, R.R.; Park, S.J. Improvement of thermal behaviors of biodegradable poly(lactic acid) polymer: A review. *Compos. B Eng.* **2019**, *164*, 287–296. [CrossRef]
8. Chen, Z.-J.; Tsou, C.-H.; Tsai, M.-L.; Guo, J.; De Guzman, M.R.; Yang, T.; Gao, C.; Lei, Y.; Gan, P.-W.; Chen, S.; et al. Barrier Properties and Hydrophobicity of Biodegradable Poly(lactic acid) Composites Reinforced with Recycled Chinese Spirits Distiller's Grains. *Polymers* **2021**, *13*, 2861. [CrossRef]
9. Li, M.-X.; Ren, Y.; Lee, D.; Choi, S.-W. Crystallization Behavior and Electrical Properties of Nanoparticle-Reinforced Poly(lactic Acid)-Based Films. *Polymers* **2022**, *14*, 177. [CrossRef]
10. Yun, X.; Li, X.; Jin, Y.; Sun, W.; Dong, T. Fast crystallization and toughening of poly(L-lactic acid) by incorporating with poly(ethylene glycol) as a middle block chain. *Polym. Sci. Ser. A* **2018**, *60*, 141–155. [CrossRef]
11. Baimark, Y.; Rungseesantivanon, W.; Prakymorama, N. Improvement in melt flow property and flexibility of poly(L-lactide)-*b*-poly(ethylene glycol)-*b*-poly(L-lactide) by chain extension reaction for potential use as flexible bioplastics. *Mater. Des.* **2018**, *154*, 73–80. [CrossRef]
12. Baimark, Y.; Srisuwan, Y. Thermal and mechanical properties of highly flexible poly(L-lactide)-*b*-poly(ethylene glycol)-*b*-poly(L-lactide) bioplastics: Effects of poly(ethylene glycol) block length and chain extender. *J. Elastomers Plast.* **2020**, *52*, 142–158. [CrossRef]
13. Baimark, Y.; Rungseesantivanon, W.; Prakymoramas, N. Improvement in crystallization and toughness of poly(*L*-lactide) by melt blending with poly(*L*-lactide)-*b*-polyethylene glycol-*b*-poly(L-lactide) in the presence of chain extender. *Polym. Sci. Ser. A* **2021**, *63*, S34–S45. [CrossRef]
14. Srisuwan, Y.; Baimark, Y. Thermal, morphological and mechanical properties of flexible poly(l-lactide)-*b*-polyethylene glycol-*b*-poly(l-lactide)/thermoplastic starch blends. *Carbohydr. Polym.* **2022**, *283*, 119155. [CrossRef]
15. Muller, J.; González-Martínez, C.; Chiralt, A. Combination of poly(lactic acid) and starch for biodegradable food packaging. *Materials* **2017**, *10*, 952. [CrossRef]
16. Rogovina, S.Z.; Aleksanyan, K.V.; Loginova, A.A.; Ivanushkina, N.E.; Vladimirov, L.V.; Prut, E.V.; Berlin, A.A. Influence of PEG on mechanical properties and biodegradability of composites based on PLA and starch. *Starch* **2018**, *70*, 1700268. [CrossRef]
17. Rogovina, S.Z.; Prut, E.V.; Aleksanyan, K.V.; Krasheninnikov, V.G.; Perepelitsina, E.; Shashkin, D.P.; Berlin, A.A. Composites based on starch and polylactide. *Polym. Sci. Ser. B* **2019**, *61*, 334–340. [CrossRef]
18. Nevoralová, M.; Koutný, M.; Ujčić, A.; Starý, Z.; Šerá, J.; Vlková, H.; Šlouf, M.; Fortelný, I.; Kruliš, Z. Structure characterization and biodegradation rate of poly(ε-caprolactone)/starch blends. *Front. Mater.* **2020**, *7*, 141. [CrossRef]
19. Adorna, J.A.; Aleman, C.K.A.; Gonzaga, I.L.E.; Pangasinan, J.N.; Sisican, K.M.D.; Dang, V.D.; Doong, R.-A.; Ventura, R.L.G.; Ventura, J.-R.S. Effect of lauric acid on the thermal and mechanical properties of polyhydroxybutyrate (PHB)/starch composite biofilms. *Int. J. Polym. Sci.* **2020**, *2020*, 7947019. [CrossRef]
20. Lago, R.C.; Oliveira, A.L.M.; Santos, A.A.; Zitha, E.Z.; Carvalho, E.E.N.; Tonoli, G.H.D.; Boas, E.V.B.V. Addition of wheat straw nanofibrils to improve the mechanical and barrier properties of cassava starch–based bionanocomposites. *Ind. Crops Prod.* **2021**, *170*, 113816. [CrossRef]
21. Noivoil, N.; Yoksan, R. Oligo(lactic acid)-grafted starch: A compatibilizer for poly(lactic acid)/thermoplastic starch blend. *Int. J. Biol. Macromol.* **2020**, *160*, 506–517. [CrossRef] [PubMed]
22. Sakai, F.; Nishikawa, K.; Inoue, Y.; Yazawa, K. Nucleation enhancement effect in poly(L-lactide) (PLLA)/poly(ε-caprolactone) (PCL) blend induced by locally activated chain mobility resulting from limited miscibility. *Macromolecules* **2009**, *42*, 8335–8342. [CrossRef]

23. Zhang, K.; Ran, X.; Wang, X.; Han, C.; Han, L.; Wen, X.; Zhuang, Y.; Dong, L. Improvement in toughness and crystallization of poly(l-lactic acid) by melt blending with poly(epichlorohydrin-co-ethylene oxide). *Polym. Eng. Sci.* **2011**, *51*, 2370–2380. [CrossRef]
24. Jariyasakoolroj, P.; Chirachanchai, S. Silane modified starch for compatible reactive blend with poly(lactic acid). *Carbohydr. Polym.* **2014**, *106*, 255–263. [CrossRef]
25. Ferrarezi, M.M.F.; de Oliveira Taipina, M.; Escobar da Silv, L.C.; Goncalves, M.C. Poly(ethylene glycol) as a compatibilizer for poly(lactic acid)/thermoplastic starch blends. *J. Polym. Environ.* **2013**, *21*, 151–159. [CrossRef]
26. Zhang, X.; Li, Y.; Han, L.; Han, C.; Xu, K.; Zhou, C.; Zhang, M.; Dong, L. Improvement in toughness and crystallization of poly(L-lactic acid) by melt blending with ethylene/methyl acrylate/glycidyl methacrylate terpolymer. *Polym. Eng. Sci.* **2013**, *53*, 2498–2508. [CrossRef]
27. Mothe, C.G.; Azevedo, A.D.; Drumond, W.S.; Wang, S.H. Thermal properties of amphiphilic biodegradable triblock copolymer of L,L-lactide and ethylene glycol. *J. Therm. Anal. Calorim.* **2010**, *101*, 229–233. [CrossRef]
28. Pereira, A.G.B.; Gollveia, R.F.; de Carvalho, G.M.; Rubira, A.F.; Muniz, E.C. Polymer blends based on PEO and starch: Miscibility and spherulite growth rate evaluated through DSC and optical microscopy. *Mater. Sci. Eng. C* **2009**, *29*, 499–504. [CrossRef]
29. Nguyen, D.M.; Vu, T.T.; Grillet, A.-C.; Ha Thuc, H.; Ha Thuc, C.N. Effect of organoclay on morphology and properties of linear low density polyethylene and Vietnamese cassava starch biobased blend. *Carbohydr. Polym.* **2016**, *136*, 163–170. [CrossRef]
30. Sessini, V.; Arrieta, M.P.; Raquez, J.-M.; Dubois, P.; Kenny, J.M.; Peponi, L. Thermal and composting degradation of EVA/thermoplastic starch blends and their nanocomposites. *Polym. Degrad. Stab.* **2019**, *159*, 184–198. [CrossRef]
31. Zaaba, N.F.; Ismail, H. A review on tensile and morphological properties of poly(lactic acid) (PLA)/thermoplastic starch (TPS) blends. *Polym. Plast. Technol. Mater.* **2019**, *58*, 1945–1964. [CrossRef]
32. Akrami, M.; Ghasemi, I.; Azizi, H.; Karrabi, M. A new approach in compatibilization of the poly(lactic acid)/thermoplastic starch (PLA/TPS) blends. *Carbohydr. Polym.* **2016**, *144*, 254–262. [CrossRef] [PubMed]
33. Esmaeili, M.; Pircheraghi, G.; Bagheri, R.; Altstädt, V. Poly(lactic acid)/coplasticized thermoplastic starch blend: Effect of plasticizer migration on rheological and mechanical properties. *Polym. Adv. Technol.* **2019**, *30*, 839–851. [CrossRef]
34. Xu, G.; Chen, S.; Yan, X.; Yang, C.; Chen, Z. Synthesis and hydrophilic performance of poly(lactic acid)-poly(ethylene glycol) block copolymers. *Am. J. Anal. Chem.* **2016**, *7*, 299–305. [CrossRef]

polymers MDPI

Article

Mechanical, Dynamic-Mechanical, Thermal and Decomposition Behavior of 3D-Printed PLA Reinforced with $CaCO_3$ Fillers from Natural Resources

Cristina Pavon [1], Miguel Aldas [2], María Dolores Samper [1], Dana Luca Motoc [3,*], Santiago Ferrandiz [1] and Juan López-Martínez [1]

1. Instituto de Tecnología de Materiales (ITM), Universitat Politècnica de València (UPV), 03801 Alcoy, Spain; crisppavonv@gmail.com (C.P.); masammad@upvnet.upv.es (M.D.S.); sferrand@mcm.upv.es (S.F.); jlopezm@mcm.upv.es (J.L.-M.)
2. Departamento de Ciencia de Alimentos y Biotecnología, Facultad de Ingeniería Química y Agroindustria, Escuela Politécnica Nacional, Quito 170517, Ecuador; miguel.aldas@epn.edu.ec
3. Faculty of Mechanical Engineering, Transilvania University of Brasov (UniTBv), 500036 Brasov, Romania

* Correspondence: danaluca@unitbv.ro

Abstract: This study evaluates the effect of $CaCO_3$ fillers extracted from waste eggshells on 3D-printed PLA performance. Samples of neat PLA and PLA reinforced with $CaCO_3$ fillers embedded with different wt.% were prepared using an FDM (fused deposition modeling) technology. The samples were examined using mechanical, dynamic mechanical, thermal, and thermal decomposition analyses. The results revealed increasing elastic moduli, tensile strength, and flexure as the filler content increased. The rheological results from the MFR tests showed that the filler content did not influence the PLA-based samples' processability. Further, the thermal degradation of neat and various $CaCO_3$-wt.%-reinforced PLA specimens revealed relatively small discrepancies in their exposure to the temperature increase, mainly concerning the eggshell organic components and volatile components, from their processability up to 300 °C. By contrast, the increased filler content positively shifted the peaks along the temperature scale at the maximum degradation rate. Additionally, the weight content of the natural reinforcement strongly influenced the surface wettability and appearance of the samples. Further, the SEM analysis featured both the presence of interlayer disturbances and the interfacial compatibility the PLA with the selected fillers.

Keywords: 3D-printing; PLA polymer; biological filler; mechanical; dynamic-mechanical; thermal; decomposition; wettability; surface appearance

Citation: Pavon, C.; Aldas, M.; Samper, M.D.; Motoc, D.L.; Ferrandiz, S.; López-Martínez, J. Mechanical, Dynamic-Mechanical, Thermal and Decomposition Behavior of 3D-Printed PLA Reinforced with $CaCO_3$ Fillers from Natural Resources. *Polymers* **2022**, *14*, 2646. https://doi.org/10.3390/polym14132646

Academic Editor: Evgenia G. Korzhikova-Vlakh

Received: 9 June 2022
Accepted: 27 June 2022
Published: 29 June 2022

1. Introduction

Additive manufacturing (AM), or 3-dimensional printing (3D printing), is a technology that translates a digitized solid model into physical models without cutting or casting machines [1]. The physical object is formed through the combination of 2-D cross-sections of a finite thickness in a layer-by-layer addition sequence [2,3]. In this way, AM allows the production of complex-shape objects or multiple-component objects in a shorter time and at low costs compared to the traditional manufacturing process [4]. Additionally, AM maximizes savings on raw materials during the process [5].

Additive manufacturing allows the processing of a wide range of materials, including metals, polymers, ceramics, and concrete [6]. In particular, polymers receive significant attention due to their ease of production, availability, and low cost [2,7]. AM processing uses polymers in diverse forms, such as reactive, liquid solutions, and thermoplastic melts [2]. Furthermore, they can be processed in a significant number of methods. The most frequently used techniques are stereolithography (SLA), fused deposition modelling (FDM), selective laser melting (SLM), and selective laser sintering (SLS). SLA is a technique that requires the raw material in a liquid form. By contrast, the other techniques use a solid material [3,5].

Stratasys. Inc. developed FDM in the USA in 1990 for processing traditional thermoplastics [8,9]. FDM is one of the most commonly used techniques throughout the world. FDM is a low-cost, easy technique [4,5,10], and can produce complex geometrical parts neatly and safely in an office-friendly environment [8]. The FDM process involves heating the polymer in a solid-state (e.g., filament) at the nozzle to reach a semi-liquid state and then extruding and depositing it on the platform or into the previous layers. When the nozzle deposits the whole layer, the platform moves down by the height of one layer and begins to print the next layer [3]. The nozzle temperature must be adapted to the material's melting point. Heating the deposition platform (build-plate) reduces thermal shrinkage after deposition [2].

The most common thermoplastics processed by FDM are poly(lactic acid) (PLA), polystyrene (PS), acrylonitrile butadiene styrene (ABS), polyamide (PA), polyethylene terephthalate (PET) [4,10], polycarbonate (PC), and PC-ABS blends [8]. One of the most favored thermoplastics for FDM 3-D printing is PLA [10] because of its relatively low melting point (150–160 °C), which implies energy savings [11]. Moreover, PLA is considered a user-friendly material that can be processed without producing toxic fumes [12,13]. PLA is a biodegradable and compostable polymer produced via fermentation or chemical synthesis from the bio-derived monomer, lactic acid (2-hydroxy propionic acid) [14].

PLA's stereochemical structure can easily be modified by controlling the ratio between the L- and D-isomers to yield high-molecular-weight amorphous or crystalline polymers [14]. The stereochemistry and thermal history of PLA directly influence its crystallinity and, therefore, its properties in general [14,15]. As a result, lactic-acid-based polymers present different grades with a wide variety of mechanical properties, ranging from soft and elastic plastics to stiff and high-strength materials [10,15]. Semi-crystalline PLA provides higher mechanical properties. It has an approximate tensile modulus of 3 GPa, a tensile strength of 50–70 MPa, a flexural modulus of 5 GPa, a flexural strength of 100 MPa, and an elongation at break of about 4% [16]. However, PLA is rigid and brittle at room temperature due to its T_g (~60 °C) [15], and it has some issues regarding its low thermal stability, high degradation rate during processing, and drawability [10,13]. Drawability and processability are essential in FDM technology because they influence feedstock filament production and layer deposition during printing [10]. To enhance its processability, thermal stability, physical properties, or/and appearance for specific applications and reducing its costs, PLA is combined with (1) inorganic fillers as calcium carbonate nanofillers [10], hydroxyapatite [17], (2) fibers as sheep's wool [18], or (3) polysaccharides as starch [19].

In this context, the present work aims to determine the material properties (e.g., mechanical, dynamic-mechanical, thermal, decomposition, etc.), surface wettability, and appearance of 3D-printed neat PLA thermoplastics and those reinforced with various wt.% (e.g., 3 and 5 wt.%) calcium carbonate ($CaCO_3$) fillers. The main aim is to improve the thermal stability, overcome problems related to the mechanical properties of these 3D-printed specimens, and support the enlargement of their application area, as other material properties seem appropriate. Additionally, this study brings forth a deployment of calcium carbonate fillers acquired directly from natural sources, namely eggshells, considered a residue in the industry.

2. Materials and Methods

2.1. Materials

Commercial poly(lactic acid) (PLA) pellets Ingeo Biopolymer 2003D were acquired from NatureWorks LLC. (Minnetonka, MN, USA) and deployed as the polymer matrix. Poly(lactic acid) was selected, accounting for its general purpose, high molecular weight, transparency, ease of processing, and wide range of potential applications. Table 1 lists the physical and mechanical properties of the pellets as provided by the supplier. Eggshell was the source for the calcium carbonate ($CaCO_3$) selected as filler. For this purpose, the eggshells were dried for 8 h at 90 °C, mechanically triturated, and sieved to 400 microns as mesh size [20]. No additional purification step of the eggshell as $CaCO_3$ primary source applied. The natural fillers were embedded with the neat PLA matrix in 3 and 5 wt.% and further referred to as PLA-0% $CaCO_3$, PLA-3% $CaCO_3$, and PLA-5% $CaCO_3$, respectively.

Table 1. The physical and mechanical properties of as-received PLA Ingeo 2003D pellets *.

Property	Value	Property	Value
Density	1.24 g/cm^3	Izod impact	0.16 J/cm
Tensile strength	53.0 MPa	Heat distortion temperature	55 °C
Flexural strength	82.7 MPa	Glass transition temperature	55–60 °C
Tensile modulus	3.61 GPa	Melting temperature	140–160 °C
Flexural modulus	3.83 GPa	Transmission (visible)	90%

* Ingeo™ biopolymer 2003D technical data sheet [21].

2.2. Filament Production and Temperature Printing Parameters

The PLA–$CaCO_3$ formulations were manually premixed in plastic containers and conditioned at 25 °C and 50 ± 5% relative humidity to produce the filament required for the 3D-printing process. The materials were processed in a co-rotating twin-screw extruder Dupra S.L (Castalla, Spain) at 50 rpm, using the following temperature profile from hopper to die: 160, 170, 180, and 190 °C, respectively. Next, the resulting material was milled into pellets. Finally, the pellets were processed in a filament extruder, model EX2, from Filabot (Barre, PA, USA) at 190 °C using a 2.95-millimeter nozzle diameter. Additionally, neat PLA pellets were identically processed to enable further property comparison. The resulting filaments' diameters varied between 2.80 and 3.00 mm.

2.3. Manufacturing of the Test Specimens

A 3D printer BCN3D (Barcelona, Spain) manufactured the test specimens. The nozzle temperature and build-plate temperature were selected based upon the results from differential scanning calorimetry (DSC) run on all PLA-based batches. The parameter used to print the test specimens were: nozzle diameter, 0.6 mm; infill density, 20%; raster angle, 45°; number of extruders, 1; layer height, 0.1 mm; printing speed, 20 mm/s; number of layers, 10; printing bed, glass. Figure 1 represents the geometry and dimensions of the printed test specimens associated with the *n* and *n* + 1 layer, respectively.

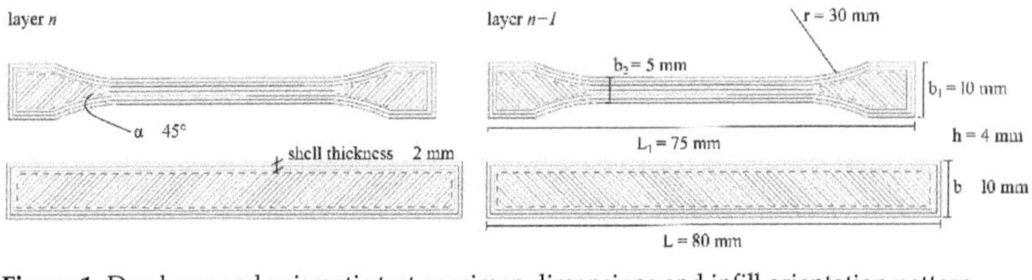

Figure 1. Dog-bone and prismatic test specimen dimensions and infill orientation pattern.

2.4. Tensile and Flexural Testing

Tensile and flexural tests were performed according to ISO 527-1:2012 and ISO 178:2019, respectively [22,23]. Testing devices LS100 and LR5K Plus from Lloyd Instruments (Bognor Regis, UK) running in tensile and 3-point bending modes, respectively, were operated to retrieve the mechanical properties of specimens. The tensile tests use a gauge length of 30 mm and 1 mm/min crosshead rate. At the same time, a 10-kilonewton load cell was operated to retrieve flexural properties of the specimens under a bending rate of 10 mm/min and length span of 64 mm. Five specimens of each formulation were selected and characterized for tensile and flexural strength, respectively, and this manuscript reports their mean and standard deviation. The significant differences were assessed at 95% confidence level according to Tukey's test using a one-way analysis of variance (ANOVA).

2.5. Differential Scanning Calorimetry (DSC)

Differential scanning calorimetry (DSC) tests determined the nozzle and glass-plate temperatures for 3D printing. The DSC analyses were performed on a DSC 821 calorimeter from Mettler-Toledo (Schwerzenbach, Switzerland) under a controlled nitrogen atmosphere (flow rate 30 mL/min). PLA-based samples (5–6 mg) were subjected to the following thermal cycle: dynamic heating from 25 °C to 150 °C, to remove thermal history from the printing process, followed by cooling to −50 °C, and then further dynamic heating up to 250 °C. The peak of the DSC curve of PLA-based specimens made it possible to identify the melting temperature (T_m).

2.6. Melt Flow Index (MFI)

The rheology characterization was performed through melt mass-flow-rate (MFR) measurement, following ISO 1133-1:2005 standard, using a Metrotec GmbH brand plastometer model Ars Faar, with 2.16 kg at 190 °C. The MFR value reported corresponds to the mean value of eight measurements.

2.7. Thermogravimetric Analysis (TGA)

Thermogravimetric analysis (TGA) was conducted on a TGA PT1000 device from Linseis (Selb, Germany), weighing 15 and 20 mg samples. Samples were heated under TGA dynamic mode at 10 °C/min, from 35 °C to 700 °C, under a controlled nitrogen atmosphere (flow rate 30 mL/min). The onset ($T_{5\%}$) and endset ($T_{95\%}$) degradation temperatures were retrieved for 5% and 95% sample mass loss, respectively. Further, the maximum degradation rate temperature (T_{max}) was identified from the minimum of the TGA curve first derivative (DTG).

2.8. Dynamic Mechanical Thermal Analysis (DMTA)

Dynamic-mechanical thermal analysis (DMTA) was performed on a TA Instrument AR G2 rheometer (New Castle, DE, USA). Samples were exposed to a temperature sweep from 35 °C to 140 °C with a heating rate of 2 °C/min, an oscillation frequency of 1 Hz, and 0.1% of maximum deformation. The samples were prismatic specimens with dimensions of 40 ×10 × 4 mm^3 (L × l × h).

2.9. Microstructural Characterization

Field emission scanning electron microscopy (FESEM) from the fracture surface of the impact specimens was carried out on a Carl Zeiss AG Ultra 55 microscope (Oberkochen, Germany) at 1 kV, displaying a resolution of 1.7 nm. Prior to investigations, each sample was sputter-coated with a thin layer of gold to avoid electrostatic charging under the electron beam with of a Sputter Mod Coater Emitech SC7620 from Quorum Technologies (East Sussex, UK).

2.10. Surface Characterization

Surface-color evaluation was performed on a Colorflex-Diff2 458/08 colorimeter from HunterLab (Reston, VA, USA) under the CIE L*a*b* color space. The L*, a*, and b* coordinates and the yellowness index (YI) were reported. Furthermore, the total color difference (ΔE) was calculated using Equation (1) [24]:

$$\Delta E = \sqrt{\Delta a^2 + \Delta b^2 + \Delta L^2} \quad (1)$$

Five individual measurements were performed on each specimen, and this manuscript reports their average and standard deviation.

An optical goniometer, EasyDrop-FM140 from Kruss Equipments (Hamburg, Germany), was used to assess samples' wettability and roughness. Images were processed using the Drop Shape Analysis software. Two values for water contact angles were obtained, accounting for two different surfaces, one in direct contact with the build plate and the other corresponding to the up-side surface. The measurements were carried out at room temperature using distilled water.

Further comparison with the color formulation and WCA of reference (PLA) was carried out by one-way ANOVA using Tukey's test, with 95% significant differences in the confidence level.

3. Results and Discussion

3.1. *Differential Scanning Calorimetry and Melt-Flow Index*

Figure 2 shows the differential scanning calorimetry curves of the PLA and PLA–$CaCO_3$ formulations. The $CaCO_3$ constitutive did not change the glass transition temperature values (T_g) irrespective of its wt.% (3% or 5%, respectively). This result is in accordance with the results reported by Nekhamanurak et al. (2012) [25]. Furthermore, the melting (T_m) and the endset (T_{end}) temperatures of the formulations were identical to those of the neat PLA sample.

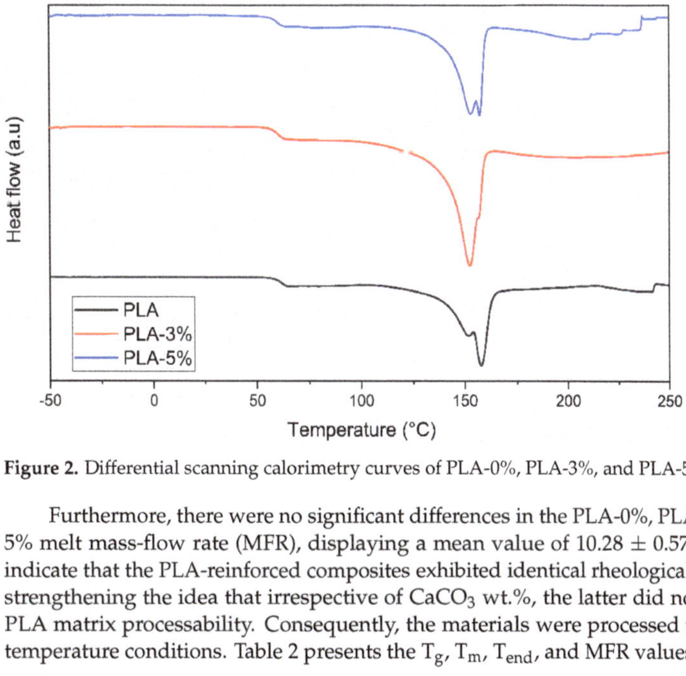

Figure 2. Differential scanning calorimetry curves of PLA-0%, PLA-3%, and PLA-5%$CaCO_3$.

Furthermore, there were no significant differences in the PLA-0%, PLA-3%, and PLA-5% melt mass-flow rate (MFR), displaying a mean value of 10.28 ± 0.57. These results indicate that the PLA-reinforced composites exhibited identical rheological characteristics, strengthening the idea that irrespective of $CaCO_3$ wt.%, the latter did not influence the PLA matrix processability. Consequently, the materials were processed under identical temperature conditions. Table 2 presents the T_g, T_m, T_{end}, and MFR values.

Table 2. Glass transition temperature, melting temperature, endset temperature, and melt mass-flow rate of neat PLA, PLA-3%, and PLA-5% $CaCO_3$.

Specimen	T_g (°C)	T_m (°C)	T_{end} (°C)	MFR (g/10 min at 190 °C)
PLA-0%	60	158	185	10.10 ± 0.60
PLA-3%	60	152	183	10.43 ± 0.40
PLA-5%	60	157	186	10.32 ± 0.72

All the formulations selected the printing temperature above the end melting temperature. The extrusion temperature of the 3D printing process is generally higher than that of filament extrusion (190 °C) [26] because of the short residence time of the material in the 3-D printer chamber and the reduced contribution of the shear stress generated by the loading gear [13,26]. Based on prior trials, different nozzle and build-in plate temperatures for the first and the other printed layers were imposed, as follows: first layer, 215 °C and 55 °C; bulk-printed layers, 210 °C and 40 °C, respectively. The temperatures selected for the first layer were slightly higher than those of the following bulk layers to allow PLA and PLA-based composites to melt better and adhere to the built-in plate. Figure 3 shows the printed samples. As can be seen, all the printed surfaces were uniform irrespective of the reinforcement wt.%, with minor differences on the 5% $CaCO_3$ specimen.

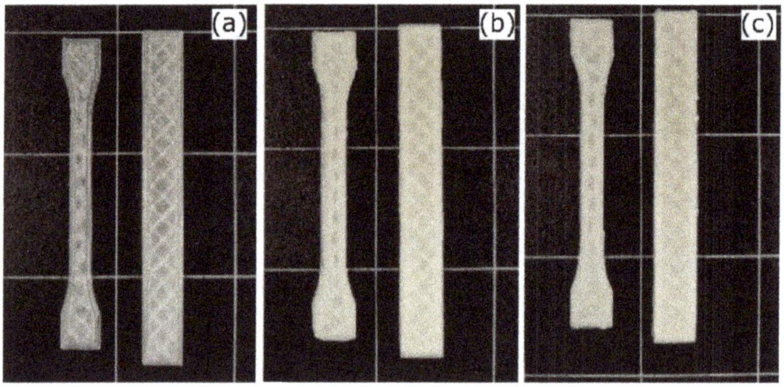

Figure 3. Test specimens of (**a**) PLA-0%, (**b**) PLA-3%, and (**c**) PLA-5%$CaCO_3$.

The thermo-plasticity of the polymer filament is an essential property of this method, which allows the filaments to fuse during printing and solidify at room temperature after printing. The layer thickness, width, and orientation of filaments and the air gap (in the same layer or between layers) are the main processing parameters that affect the mechanical properties of printed parts [6,8].

3.2. *Tensile, Flexural, and Impact Testing*

The results of the mechanical characterization of the neat PLA and PLA composites herein are listed in Table 3. The Young's moduli from both the tensile and the flexural test revealed an increasing tendency with the addition of $CaCO_3$ fillers, as expected, and followed the rule of mixture (RoM). Correspondingly the values retrieved from the static mechanical tests differed significantly between 28% to 42% in the tensile and about 1.50% to 2.25% in the flexure test. These discrepancies were due to the filament material composition and morphology, the manufacturing technology, and, less probably, to the testing procedure and equipment [14,17].

Concerning the filler distribution within the PLA matrix, the morphological studies from the literature revealed an increasing trend for the elongation at break based upon the improved interaction with the matrix [27,28]. The results reported by Nekhamanurak et al. (2012) on their $CaCO_3$-reinforced PLA composites were similar with herein findings [25]. Further, Nekhamanurak et al. found that a 5 wt.% fraction of embedded nano-fillers decreases the elongation at break and acts as a stress concentrator, since fillers may tend to agglomerate [25,29].

Table 3. Tensile, flexural, and impact-resistance properties and area of tensile stress–strain curve (Toughness) of 3-D printed neat PLA, PLA-3%, and PLA-5%$CaCO_3$.

	Tensile Properties			Flexural Properties		
Material	Young's Modulus (MPa)	Tensile Strength (MPa)	Elongation at Break (%)	Young's Modulus in Flexure (MPa)	Flexural Strength (MPa)	Charpy's Impact Energy (kJ/m^2)
PLA-0%	1201.2 ± 95.76	42.04 ± 1.98	9.79 ± 1.57	805.64 ± 10.113	45.53 ± 4.70	7.75 ± 0.13 ^{a}
PLA-3%	1534.6 ± 185.71	39.26 ± 2.98	4.26 ± 0.63	817.89 ±21.54	43.21 ± 5.65	6.33 ± 0.91 ^{b}
PLA-5%	1708.1 ± 43.22	31.22 ± 1.28	4.32 ± 0.45	823.43 ± 14.98	40.59 ± 6.67	6.19 ± 0.11 ^{b}

^{a,b} Different letters within the same property show statistically significant differences between formulations ($p < 0.05$).

Compared with the mechanical properties of raw pellets, the tensile and flexural properties of samples such as those reported above are strongly influenced by the manufacturing technology. In support of this argument, Silva et al. (2007) showed that the mechanical properties of polymers experience a decrease of 30% when processed by FDM technology compared to those obtained by deploying a compression-molding technique [30]. Further, the infill degree contributes to the degradation of mechanical properties, as shown by Silva et al. and Ahmed et al. [30–32]. According to Ahmed et al. (2019)'s studies on commercial PLA filaments, the Young's modulus of 3-D printed PLA at an infill level of 20% and a printing angle of 45° retrieved from tensile tests was about 889.33 ± 49.34 MPa [31]. These values closely match our values from the flexural tests. Nevertheless, the filaments described herein were produced under a laboratory-controlled environment, and their retrieved mechanical properties were influenced by the factors described above.

Yao et al. (2019) underlined that the flexural properties of 3D-printed parts depend mainly on the design parameters and printing settings [5]. Further, it is widely acknowledged by thermoplastic composite manufacturers that the addition of $CaCO_3$ fillers within the polymer resin contributes to inner-stress release during the manufacturing process [10]. According to the mechanical tests, the direct consequences of the factors mentioned above can be observed in the decreasing trend of the tensile and flexural strength. The tensile strength decreased from 6% to 25%, whereas the flexural strength was roughly 5% and 10% relative to the neat PLA values.

Further, the relative toughness of the neat and $CaCO_3$-wt.%-reinforced PLA-based 3D-printed samples experienced a decrease in line with the increase in the natural filler content, to about 18% and 20%, respectively. Consequently, the increase in the wt.% of this natural filler significantly affected the toughness of 3D-printed PLA-reinforced specimens.

3.3. *Thermograviemtric Analysis*

Figure 4 presents the weight loss of the PLA-0%, PLA-3%, and PLA-5% $CaCO_3$ as a function of temperature. The first step reveals the loss of moisture, greasy, and other volatile compounds within the fillers. We further referred to the temperature range up to 300 °C as the first weight-loss interval. Thus, 2% of the total weight was lost within the 30–120 °C temperature range, mainly due to volatile compounds from the manufacturing process of the samples [33]. The filler volume fraction seems to have influenced the weight loss as the temperature increased. Thus, within the 150–300 °C temperature range, the PLA-3% $CaCO_3$ exhibited an extra 1% weight loss, while the PLA-5% $CaCO_3$ was about 3%. This supplementary weight loss can be related to the decomposition of organic material [34] within the eggshell composition [35].

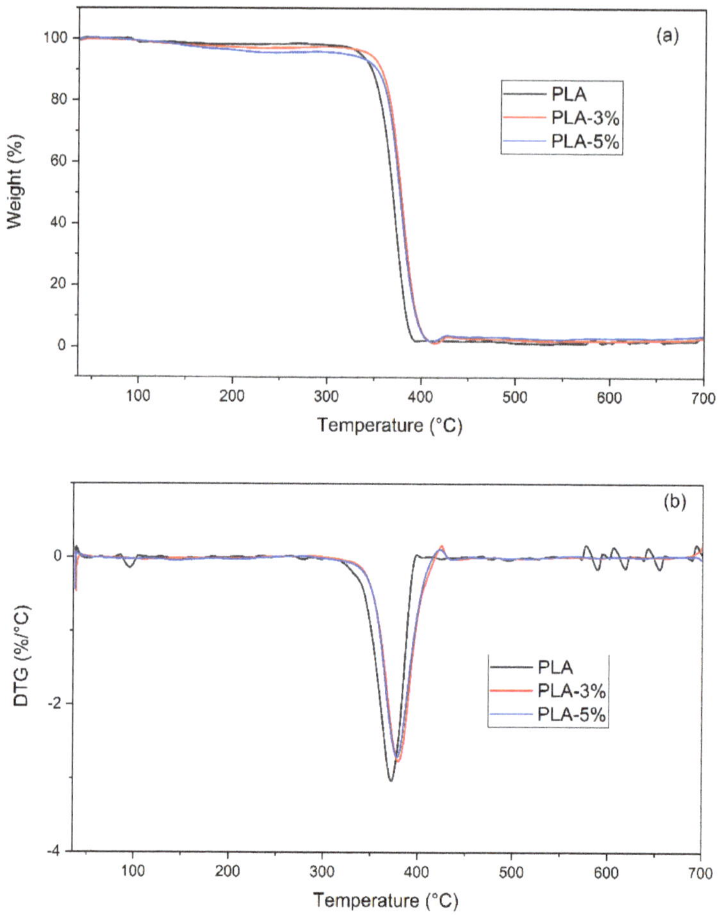

Figure 4. (**a**) TGA and (**b**) DTG of PLA, PLA-3%, and PLA-5% $CaCO_3$ fillers.

The second step in the neat and PLA-reinforced excerpt degradation revealed the association between the highest weight loss and the PLA chain scission, between 300 and 400 °C [36]. In the PLA with $CaCO_3$, this degradation step started at about 320 °C, while for the PLA–$CaCO_3$ formulations, it began around 340 °C. A shift towards higher temperatures was determined through the peak values for the degradation rates, accounting for the presence of the $CaCO_3$ fillers [33,37]. Finally, the last degradation step revealed a slightly higher endset degradation temperature ($T_{95\%}$) for the PLA-reinforced $CaCO_3$ formulations. Table 4 lists the main temperature values of the degradation process (e.g., T_{max} and $T_{95\%}$). The tendencies encountered in these variations and the retrieved values may enable us to present a statement regarding the filler's effect upon the thermal stability of the PLA-reinforced excerpts. Thus, the addition of $CaCO_3$ enabled improvements in thermal stability that contrasted with the results reported by Nekhamanurak et al. (2014) [36]. These discrepancies may have been related to the particle agglomeration which, in our case, did not influence the properties under discussion [36].

Table 4. Onset degradation temperature (T5%), temperature of maximum degradation rate (Tmax), and endset degradation temperature (T95%) for PLA and the formulations with $CaCO_3$.

Material	$T_{5\%}$ (°C)	T_{max} (°C)	$T_{95\%}$ (°C)
PLA-0%	335	372	387
PLA-3%	334	379	400
PLA-5%	321	380	401

3.4. Dynamic-Mechanical Analysis

Figure 5 shows the storage moduli (G′) and loss factor (Tan δ) of the neat and PLA-reinforced formulations. The DMA revealed two transitions and three different states (e.g., glassy, leathery, and rubbery) for the neat and PLA-reinforced composites. Table 5 presents the DMA parameters. The glassy region was relatively narrow, covering the 35–60 °C temperature range. The glass transition temperatures (T_g) of the neat PLA, 3% and 5-wt.%-$CaCO_3$-reinforced PLA, retrieved as the maximum of the Tan(δ) curve, were about 60.9 °C, 61.2 °C, and 60.8 °C, respectively. These values were consistent with those from the DSC analysis. Further, in the rubbery plateau, as the temperatures exceeded 120 °C, the storage moduli increased for all the specimens. This increase suggests an increase in the stiffness of the materials, which can be attributed to the cold crystallization of the PLA structure [38–40]. Moreover, adding $CaCO_3$ shifted the temperature at the beginning of the cold crystallization to lower values, suggesting a nucleating effect. Furthermore, in the temperature range beyond 100 °C, the storage moduli (G′) of the PLA–$CaCO_3$ formulations were lower than those of the neat PLA. These discrepancies in the storage module can be ascribed to the rearrangement into crystallites of the PLA polymer chains above the glass transition (T_g) [41,42]. Therefore, the addition of the $CaCO_3$ fillers interrupted the rearrangement, leading to an increase in the elastic behavior of the PLA-based composites.

Table 5. Storage modulus before glass transition and crystallization of PLA, PLA-3%, and PLA-5% $CaCO_3$ fillers, and temperature at Tan δ peak.

Material	G′ at 40 °C (MPa)	G′ at 80 °C (MPa)	G′ at 120 °C (MPa)	T at Tan δ Peak
PLA-0%	824	1.0	18	60.9
PLA-3%	721	1.1	12	61.2
PLA-5%	890	1.2	12	60.8

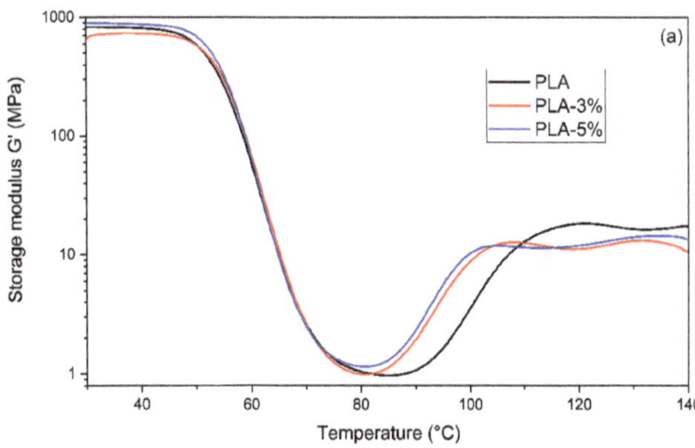

Figure 5. *Cont.*

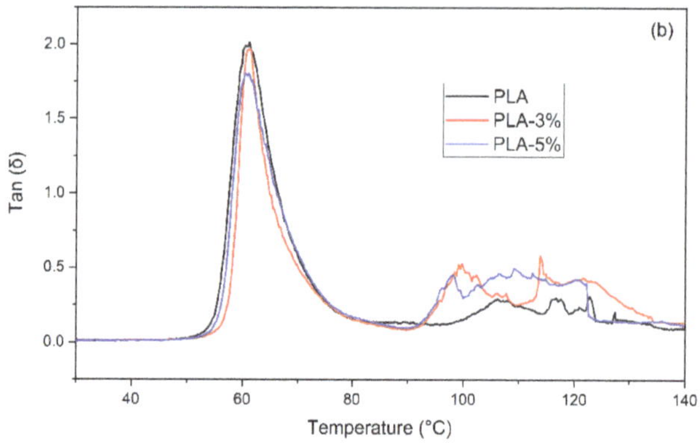

Figure 5. (**a**) Storage modulus and (**b**) tan(δ) variation as a function of temperature of PLA, PLA-3%, and PLA-5% $CaCO_3$ fillers.

3.5. Microstructural Characterization

Figure 6 presents the study samples' scanning electron microscopy (FESEM) results. This analysis assessed the final 3D-printed samples. In Figure 6a, on the left side, it is possible to note the fragile fracture section of the PLA. In Figure 6b,c, the 3 wt.% and 5 wt.% $CaCO_3$ materials, respectively, are shown. In both cases, the presence of the natural filler (e.g., eggshells) is evident, since the surface of the materials is not as smooth as the PLA sample. In both cases, the samples' layers are not pasted, contrary to the shape detected in the PLA sample. The lack of adherence is explained by the presence of the $CaCO_3$ fillers, which affect the printing process of these formulations. Moreover, in Figure 6c, it is possible to observe the presence of porosity due to the filler content. The defects in the structure suggest a non-optimal adhesion between the $CaCO_3$ and PLA matrix [43]. These defects in the microstructure of the formulations led to the decrease in the mechanical properties (e.g., tensile strength and elongation at break) discussed above.

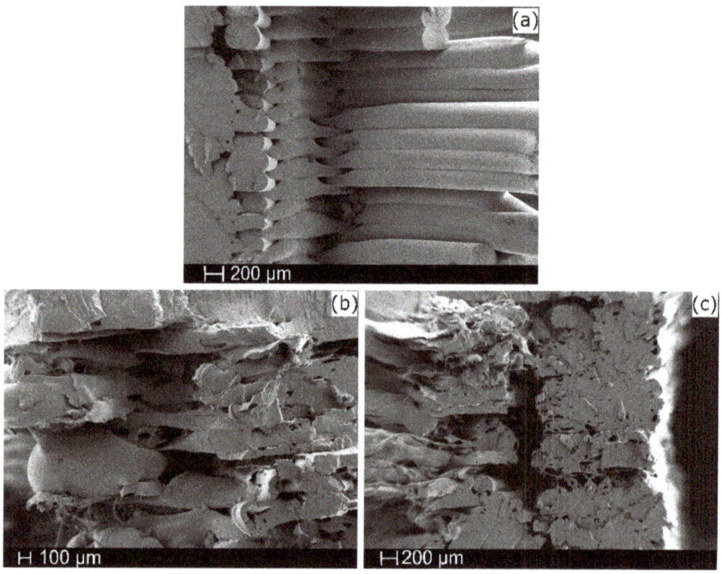

Figure 6. SEM microstructural characterization of (**a**) neat PLA, (**b**) PLA-3%, and (**c**) PLA-5%.

3.6. Surface Characterization

Table 6 presents the water contact angle (WCA) and the color coordinates in the CIEL*a*b* space of the neat PLA, PLA-3%, and PLA-5%. Regarding the water contact angle, adding the $CaCO_3$ fillers significantly reduced ($p < 0.05$) the WCA of the PLA-based composites on both surfaces, free and build-plate. The WCA reduced about 17% for the 3 wt.% $CaCO_3$ reinforced PLA on the build-plate surface and 24% for the 5 wt.% specimen. By contrast, on the free surface, the WCA experienced a 3% reduction in the PLA-3% and about 12% in the PLA-5%. The decrease in the WCA was attributed to the hydrophilic sites on the surface of the $CaCO_3$ fillers [44], and, to a greater extent, to an increase in the superficial roughness of the specimens [10]. Moreover, the differences between the WCA values reflect the roughness differences from one surface to the other. As already acknowledged, free surfaces exhibited higher roughness than the build-plate surfaces because the former does not have a mechanism to contains or shape its surface.

Table 6. Water contact angle (WCA) and color parameters for the CIEL*a*b* space for PLA-0%, PLA-3%, and PLA-5% $CaCO_3$ fillers.

Material	WCA Build-Plate Surface	WCA Free Surface	L*	a*	b*	YI	ΔE
PLA-0%	54.5 ± 2.3 [a]	81.1 ± 2.3 [a]	59.6 ± 1.5 [a]	-0.9 ± 0.1 [a]	3.2 ± 0.2 [a]	8.5 ± 0.7 [a]	0.7 ± 0.3 [a]
PLA-3%	44.9 ± 2.8 [b]	77.9 ± 2.5 [b]	72.1 ± 1.4 [b]	-0.6 ± 0.1 [b]	8.7 ± 0.4 [b]	20.6 ± 0.9 [b]	13.4 ± 1.0 [b]
PLA-5%	41.5 ± 1.8 [c]	71.2 ± 2.0 [c]	78.3 ± 0.5 [c]	-0.2 ± 0.1 [c]	12.6 ± 0.3 [c]	27.9 ± 0.6 [c]	21.6 ± 0.7 [c]

[a–c] Different letters within the same property show statistically significant differences between formulations ($p < 0.05$)

The color measurement results show that incorporating $CaCO_3$ fillers increases the brightness of PLA ($p < 0.05$) significantly. Adding 3 wt.% $CaCO_3$ increases the lightness by about 13 points, while adding 5 wt.% $CaCO_3$ increases it by about 19 points. This result means that the formulations obtained a whiter coloration due to the inherent coloration of the $CaCO_3$ fillers. The a* coefficient shows that the samples exhibited a green hue that decreased significantly ($p < 0.05$) with the increase in the $CaCO_3$ percentage [45]. As a* approaches zero, the samples do not reveal coloration [18]. The b* coordinates revealed the yellow hue exhibited by the samples. The retrieved values of the yellow index (YI) increased significantly ($p < 0.05$) with the $CaCO_3$ wt.% content [45]. Finally, the total difference in color (ΔE) shows statistical differences between the samples ($p < 0.05$), with values that differ by more than two units. Therefore, the color change is appreciable to the human eye [46,47].

4. Conclusions

This paper presents the results of tensile, flexural, and impact tests on 3D-printed neat PLA and PLA reinforced with $CaCO_3$ fillers from natural resources (e.g., eggshells) and their thermal and decomposition behavior. In addition, the samples' wettability and appearance were measured to aid in the characterization of the different PLA-based and wt.%-$CaCO_3$-reinforced composites.

The results revealed a decreasing tendency in the tensile (e.g., 6% and 25%) and flexural strength (e.g., 5% and 10%) concerning the neat PLA, performed under a controlled loading mode. These values diminish the application area's attractiveness; the material selection criteria may include specifications as to the direction of the strength variation. On the other hand, the elastic moduli, both tensile and flexural, increased with the increase in the wt.% of the fillers (e.g., tensile, 28% and 42%, respectively; flexural—1.5% and 2.5%) compared with the neat PLA and following the rule of mixture (RoM).

Depending on material properties, such as dynamic-mechanical, thermal, or decomposition, the results revealed less or no variations due to the changes in the wt.% content of the $CaCO_3$ fillers. Thus, the glass transition and melting temperatures of the neat and wt.%-$CaCO_3$-reinforced PLA composites retrieved from the DSC runs showed no differences. The decomposition of the PLA–$CaCO_3$ composites experienced identical weight loss tendencies, developed in three steps that were identified according to different temperature ranges. Up to 120 °C, there were no differences in the recorded weight loss for the neat and PLA-based composites related to the filler contents. Between 150 and 300 °C, the weight loss increased by about 1% and 3% of the total size. The increase in filler content accounts for the decomposition of the organic components with the eggshell. The maximum decomposition rate occurred around 370 °C, with a positive shift in the peaks sized on the temperature scale for the PLA-reinforced specimens. The dynamic-mechanical analysis provided results that were consistent with the above findings. The glass transition temperatures, as retrieved for the peaks of the loss factor, were in line with the results from the DSC tests, about 60 °C for all the PLA and PLA-based composites. Furthermore, the loss tangent peak decreased with the increase in the content of the $CaCO_3$ filler, without broadening its curve envelope over the temperature range to a greater extent. The SEM analysis revealed interlayer disturbances of the PLA matrix by the fillers, and an increasingly porous appearance of the deposited layers.

As a result of this research, 3D-printed PLA reinforced with calcium carbonate ($CaCO_3$) from eggshells, under different wt.%, can be considered a viable alternative to other natural reinforcements as its mechanical, dynamic-mechanical, thermal, and decomposition properties prove its competitiveness.

Author Contributions: Conceptualization, S.F. and D.L.M.; methodology, C.P.; validation, M.A. and M.D.S.; formal analysis, C.P.; investigation, C.P., M.A. and M.D.S.; resources, J.L.-M.; writing—original draft preparation, C.P.; writing—review and editing, M.A. and D.L.M.; visualization, M.D.S.; supervision, S.F. and D.L.M.; project administration, J.L.-M.; funding acquisition, J.L.-M. All authors have read and agreed to the published version of the manuscript.

Funding: This research is a part of the grant, PID2020-116496RB-C22, funded by MCIN/AEI/10.13039/501100011033.

Institutional Review Board Statement: Not applicable.

Informed Consent Statement: Not applicable.

Data Availability Statement: The data presented in this study are available upon request from the corresponding author.

Acknowledgments: C.P. acknowledges the Santiago Grisolía fellowship (GRISOLIAP/2019/113) from Generalitat Valenciana. The Microscopy Services at UPV are also acknowledged for their help in collecting and analyzing the images.

Conflicts of Interest: The authors declare no conflict of interest.

References

1. Parandoush, P.; Lin, D. A Review on Additive Manufacturing of Polymer-Fiber Composites. *Compos. Struct.* **2017**, *182*, 36–53. [CrossRef]
2. Hofmann, M. 3D Printing Gets a Boost and Opportunities with Polymer Materials. *ACS Macro Lett.* **2014**, *3*, 382–386. [CrossRef] [PubMed]
3. Gibson, I.; Rosen, D.W.; Stucker, B. *Additive Manufacturing Technologies: Rapid Prototyping to Direct Digital Manufacturing*; Springer: New York, NY, USA, 2010; ISBN 9781441911193.
4. Mohan, N.; Senthil, P.; Vinodh, S.; Jayanth, N. A Review on Composite Materials and Process Parameters Optimisation for the Fused Deposition Modelling Process. *Virtual Phys. Prototyp.* **2017**, *12*, 47–59. [CrossRef]
5. Yao, T.; Deng, Z.; Zhang, K.; Li, S. A Method to Predict the Ultimate Tensile Strength of 3D Printing Polylactic Acid (PLA) Materials with Different Printing Orientations. *Compos. Part B Eng.* **2019**, *163*, 393–402. [CrossRef]
6. Ngo, T.D.; Kashani, A.; Imbalzano, G.; Nguyen, K.T.Q.; Hui, D. Additive Manufacturing (3D Printing): A Review of Materials, Methods, Applications and Challenges. *Compos. Part B Eng.* **2018**, *143*, 172–196. [CrossRef]
7. de Leon, A.C.; Chen, Q.; Palaganas, N.B.; Palaganas, J.O.; Manapat, J.; Advincula, R.C. High Performance Polymer Nanocomposites for Additive Manufacturing Applications. *React. Funct. Polym.* **2016**, *103*, 141–155. [CrossRef]
8. Mohamed, O.A.; Masood, S.H.; Bhowmik, J.L. Optimization of Fused Deposition Modeling Process Parameters: A Review of Current Research and Future Prospects. *Adv. Manuf.* **2015**, *3*, 42–53. [CrossRef]

9. Stratasys Stratasys: 3D Printing & Additive Manufacturing | Stratasys. Available online: https://www.stratasys.com/ (accessed on 26 March 2020).
10. Coppola, B.; Cappetti, N.; Di Maio, L.; Scarfato, P.; Incarnato, L. 3D Printing of PLA/Clay Nanocomposites: Influence of Printing Temperature on Printed Samples Properties. *Materials* **2018**, *11*, 1947. [CrossRef]
11. Wittbrodt, B.; Pearce, J.M. The Effects of PLA Color on Material Properties of 3-D Printed Components. *Addit. Manuf.* **2015**, *8*, 110–116. [CrossRef]
12. Groenendyk, M.; Gallant, R. 3D Printing and Scanning at the Dalhousie University Libraries: A Pilot Project. *Libr. Hi Tech* **2013**, *31*, 34–41. [CrossRef]
13. Gkartzou, E.; Koumoulos, E.P.; Charitidis, C.A. Production and 3D Printing Processing of Bio-Based Thermoplastic Filament. *Manuf. Rev.* **2017**, *4*, 1. [CrossRef]
14. Garlotta, D. A Literature Review of Poly(Lactic Acid). *J. Polym. Environ.* **2001**, *9*, 63–84. [CrossRef]
15. Farah, S.; Anderson, D.G.; Langer, R. Physical and Mechanical Properties of PLA, and Their Functions in Widespread Applications—A Comprehensive Review. *Adv. Drug Deliv. Rev.* **2016**, *107*, 367–392. [CrossRef] [PubMed]
16. Södergård, A.; Stolt, M. Properties of Lactic Acid Based Polymers and Their Correlation with Composition. *Prog. Polym. Sci.* **2002**, *27*, 1123–1163. [CrossRef]
17. Liu, X.; Wang, T.; Chow, L.C.; Yang, M.; Mitchell, J.W. Effects of Inorganic Fillers on the Thermal and Mechanical Properties of Poly(Lactic Acid). *Int. J. Polym. Sci.* **2014**, *2014*, 827028. [CrossRef] [PubMed]
18. Pawlak, F.; Aldas, M.; López-Martínez, J.; Samper, M.D. Effect of Different Compatibilizers on Injection-Molded Green Fiber-Reinforced Polymers Based on Poly(Lactic Acid)-Maleinized Linseed Oil System and Sheep Wool. *Polymers* **2019**, *11*, 1514. [CrossRef] [PubMed]
19. Jang, W.Y.; Shin, B.Y.; Lee, T.J.; Narayan, R. Thermal Properties and Morphology of Biodegradable PLA/Starch Compatibilized Blends. *J. Ind. Eng. Chem.* **2007**, *13*, 457–464.
20. Toro, P.; Quijada, R.; Yazdani-Pedram, M.; Arias, J.L. Eggshell, a New Bio-Filler for Polypropylene Composites. *Mater. Lett.* **2007**, *61*, 4347–4350. [CrossRef]
21. NatureWorks. *IngeoTM Biopolymer 2003D Technical Data Sheet for Fresh Food Packaging and Food Serviceware*; Minnetonka: Minneapolis, MN, USA, 2013.
22. *ISO 527-1:2012*; Plastics—Determination of Tensile Properties—Part 1: General Principles. International Standards Organization: Geneva, Switzerland, 2012.
23. *ISO 178:2011*; Plastics—Determination of Flexural Properties. International Standards Organization ISO: Geneva, Switzerland, 2011; Volume 19.
24. Atodiresei, G.-V.; Sandu, I.G.; Tulbure, E.-A.; Vasilache, V.; Butnaru, R. Chromatic Characterization in Cielab System for Natural Dyed Materials, Prior Activation in Atmospheric Plasma Type DBD. *Rev. Chim.* **2013**, *64*, 165–169.
25. Nekhamanurak, B.; Patanathabutr, P.; Hongsriphan, N. Thermal-Mechanical Property and Fracture Behaviour of Plasticised PLA-$CaCO_3$ Nanocomposite. *Plast. Rubber Compos.* **2012**, *41*, 175–179. [CrossRef]
26. Melocchi, A.; Parietti, F.; Maroni, A.; Foppoli, A.; Gazzaniga, A.; Zema, L. Hot-Melt Extruded Filaments Based on Pharmaceutical Grade Polymers for 3D Printing by Fused Deposition Modeling. *Int. J. Pharm.* **2016**, *509*, 255–263. [CrossRef] [PubMed]
27. Zilles, J.U. Micas. In *Polymers and Polymeric Composites: A Reference Series Fillers for Polymer Applications*; Springer: Cham, Switzerland, 2016; pp. 177–194.
28. Móczó, J.; Pukánszky, B. Particulate Fillers in Thermoplastics. In *Polymers and Polymeric Composites: A Reference Series Fillers For Polymer Applications*; Springer: Cham, Switzerland, 2016; pp. 51–94.
29. Nekhamanurak, B.; Patanathabutr, P.; Hongsriphan, N. Mechanical Properties of Hydrophilicity Modified $CaCO_3$-Poly (Lactic Acid) Nanocomposite. *Int. J. Appl. Phys. Math.* **2012**, *2*, 98–103. [CrossRef]
30. Silva, A.F.; Carneiro, O.S.; Gomes, R. 3D Printing of Polypropylene Using the Fused Filament Fabrication Technique. *AIP Conf. Proc.* **2017**, *1896*, 1–5. [CrossRef]
31. Ahmed, A.A.; Susmel, L. Static Assessment of Plain/Notched Polylactide (PLA) 3D-Printed with Different Infill Levels: Equivalent Homogenised Material Concept and Theory of Critical Distances. *Fatigue Fract. Eng. Mater. Struct.* **2019**, *42*, 883–904. [CrossRef]
32. Fernandez-Vicente, M.; Calle, W.; Ferrandiz, S.; Conejero, A. Effect of Infill Parameters on Tensile Mechanical Behavior in Desktop 3D Printing. *3D Print. Addit. Manuf.* **2016**, *3*, 183–192. [CrossRef]
33. Essabir, H.; Bensalah, M.O.; Rodrigue, D.; Bouhfid, R.; el Kacem Qaiss, A. A Comparison between Bio- and Mineral Calcium Carbonate on the Properties of Polypropylene Composites. *Constr. Build. Mater.* **2017**, *134*, 549–555. [CrossRef]
34. Dasgupta Adak, M.; Purohit, K.M. Synthesis of Nano-Crystalline Hydroxyapatite from Dead Snail Shells for Biological Implantation. *Trends Biomater. Artif. Organs* **2011**, *25*, 101–106.
35. Hernández-Hernández, A.; Vidal, M.L.; Gómez-Morales, J.; Rodríguez-Navarro, A.B.; Labas, V.; Gautron, J.; Nys, Y.; García Ruiz, J.M. Influence of Eggshell Matrix Proteins on the Precipitation of Calcium Carbonate ($CaCO_3$). *J. Cryst. Growth* **2008**, *310*, 1754–1759. [CrossRef]
36. Nekhamanurak, B.; Patanathabutr, P.; Hongsriphan, N. The Influence of Micro-/Nano-$CaCO_3$ on Thermal Stability and Melt Rheology Behavior of Poly(Lactic Acid). *Energy Procedia* **2014**, *56*, 118–128. [CrossRef]
37. Gao, W.; Ma, X.; Liu, Y.; Wang, Z.; Zhu, Y. Effect of Calcium Carbonate on PET Physical Properties and Thermal Stability. *Powder Technol.* **2013**, *244*, 45–51. [CrossRef]

38. Pluta, M. Morphology and Properties of Polylactide Modified by Thermal Treatment, Filling with Layered Silicates and Plasticization. *Polymer* **2004**, *45*, 8239–8251. [CrossRef]
39. Pluta, M.; Murariu, M.; Alexandre, M.; Galeski, A.; Dubois, P. Polylactide Compositions. The Influence of Ageing on the Structure, Thermal and Viscoelastic Properties of PLA/Calcium Sulfate Composites. *Polym. Degrad. Stab.* **2008**, *93*, 925–931. [CrossRef]
40. Ren, Z.; Dong, L.; Yang, Y. Dynamic Mechanical and Thermal Properties of Plasticized Poly(Lactic Acid). *J. Appl. Polym. Sci.* **2006**, *101*, 1583–1590. [CrossRef]
41. Djellali, S.; Sadoun, T.; Haddaoui, N.; Bergeret, A. Viscosity and Viscoelasticity Measurements of Low Density Polyethylene/Poly(Lactic Acid) Blends. *Polym. Bull.* **2015**, *72*, 1177–1195. [CrossRef]
42. Menard, K.P. *Dynamic Mechanical Analysis: A Practical Introduction*, 2nd ed.; CRC Press: Boca Raton, FL, USA, 2008; ISBN 1420053132.
43. Fiore, V.; Botta, L.; Scaffaro, R.; Valenza, A.; Pirrotta, A. PLA Based Biocomposites Reinforced with Arundo Donax Fillers. *Compos. Sci. Technol.* **2014**, *105*, 110–117. [CrossRef]
44. Zuhaira, N.; Aziz, A.B.D.; Mohamed, R. Calcium Carbonate Composition Effect Upon Morphology, Water Absorption and Flexural. In Proceedings of the 63rd The IIER International Conference, Barcelona, Spain, 4 March 2016; pp. 20–24.
45. Prez-Magario, S.; Gonzalez-SanJose, M.L. Prediction of Red and Rosé Wine CIELab Parameters from Simple Absorbance Measurements. *J. Sci. Food Agric.* **2002**, *82*, 1319–1324. [CrossRef]
46. Pavon, C.; Aldas, M.; López-Martínez, J.; Ferrándiz, S. New Materials for 3D-Printing Based on Polycaprolactone with Gum Rosin and Beeswax as Additives. *Polymers* **2020**, *12*, 334. [CrossRef]
47. Arrieta, M.P.; Peltzer, M.A.; López, J.; del Carmen Garrigós, M.; Valente, A.J.M.; Jiménez, A. Functional Properties of Sodium and Calcium Caseinate Antimicrobial Active Films Containing Carvacrol. *J. Food Eng.* **2014**, *121*, 94–101. [CrossRef]

polymers

MDPI

Article

Physicochemical Properties and Atomic-Scale Interactions in Polyaniline (Emeraldine Base)/Starch Bio-Based Composites: Experimental and Computational Investigations

Soufiane Boudjelida [1,2], Souad Djellali [3,4,*], Hana Ferkous [5,6], Yacine Benguerba [7], Imane Chikouche [8] and Mauro Carraro [2,9,*]

1 Laboratory LMSE, University Mohamed El Bachir El Ibrahimi, Bordj Bou Arreridj 34030, Algeria; soufianeboudj@live.fr
2 Department of Chemical Sciences, University of Padova, Via Marzolo 1, 35131 Padova, Italy
3 Laboratoire de Physico-Chimie des Hauts Polymères, University Ferhat Abbas Setif-1, Setif 19000, Algeria
4 Department of Chemistry, Faculty of Sciences, University Ferhat Abbas Setif-1, Setif 19000, Algeria
5 Laboratoire de Génie Mécanique et Matériaux, Faculté de Technologie, Université de 20 août 1955 de Skikda, Skikda 21000, Algeria; hanaferkous@gmail.com
6 Département de Technologie, Université de 20 août 1955 de Skikda, Skikda 21000, Algeria
7 Department of Process Engineering, Faculty of Technology, University Ferhat Abbas Setif 1, Setif 19000, Algeria; yacinebenguerba@univ-setif.dz
8 Laboratoire Croissance et Caractérisation de Nouveaux Semi-Conducteurs, Faculté de Technologie, Université Sétif 1, Setif 19000, Algeria; chikouche_imene@yahoo.fr
9 ITM-CNR, UoS of Padova, Via Marzolo 1, 35131 Padova, Italy
* Correspondence: souad.djellali@univ-setif.dz (S.D.); mauro.carraro@unipd.it (M.C.)

Citation: Boudjelida, S.; Djellali, S.; Ferkous, H.; Benguerba, Y.; Chikouche, I.; Carraro, M. Physicochemical Properties and Atomic-Scale Interactions in Polyaniline (Emeraldine Base)/Starch Bio-Based Composites: Experimental and Computational Investigations. *Polymers* **2022**, *14*, 1505. https://doi.org/10.3390/polym14081505

Academic Editors: Evgenia G. Korzhikova-Vlakh and Pablo Marcelo Stefani

Received: 28 January 2022
Accepted: 4 April 2022
Published: 7 April 2022

Abstract: The processability of conductive polymers still represents a challenge. The use of potato starch as a steric stabilizer for the preparation of stable dispersions of polyaniline (emeraldine base, EB) is described in this paper. Biocomposites are obtained by oxidative polymerization of aniline in aqueous solutions containing different ratios of aniline and starch (% *w*/*w*). PANI-EB/Starch biocomposites are subjected to structural analysis (UV-Visible, RAMAN, ATR, XRD), thermal analysis (TGA, DSC), morphological analysis (SEM, Laser Granulometry), and electrochemical analysis using cyclic voltammetry. The samples were also tested for their solubility using various organic solvents. The results showed that, with respect to starch particles, PANI/starch biocomposites exhibit an overall decrease in particles size, which improves both their aqueous dispersion and solubility in organic solvents. Although X-ray diffraction and DSC analyses indicated a loss of crystallinity in biocomposites, the cyclic voltammetry tests revealed that all PANI-EB/Starch biocomposites possess improved redox exchange properties. Finally, the weak interactions at the atomic-level interactions between amylopectin–aniline and amylopectin–PANI were disclosed by the computational studies using DFT, COSMO-RS, and AIM methods.

Keywords: biocomposites; polyaniline colloids; starch; steric stabilizer; cyclic voltammetry; DFT; AIM study; COSMO-RS

1. Introduction

A variety of reports have emerged in recent decades documenting the use of sterically stabilized colloidal dispersions of conductive polymers, which are much more processable than traditional ones [1]. In the case of dispersion polymerization, and when the reaction medium contains a soluble stabilizer, the macroscopic aggregation and precipitation of the resulting conductive polymers can be avoided, leading to submicrometer well-dispersed particles [2]. In this context, several water-soluble polymers, such as cellulosic derivatives [3,4], poly(vinyl alcohol) [5], poly(ethylene oxide) [6], and poly(methyl vinyl ether) [7], were used as steric stabilizers to overcome the processing hardness of conductive polymers, which is related to their very low solubility in common solvents and/or infusibility, limiting their applications.

Polyaniline (PANI) is among the most studied and developed organic conductive polymers. Its large-scale applications stem from its good redox exchange properties, its simple and cheap preparation methods as well as its great environmental stability [8,9]. However, as with other conductive polymers, polyaniline is very difficult to be processed and the preparation of colloidal PANI-based dispersions seems to be a promising way to overcome such a problem [10]. Several workers have taken advantage of this approach to produce processable dispersions including PANI/poly(methyl methacrylate) particles stabilized in hydrocarbons [11], PANI stabilized with partially hydrolyzed poly(vinyl alcohol) [12] or with poly(N-vinylpyrrolidone) [13], and PANI/hydroxypropyl cellulose [14].

Numerous studies were also conducted to exploit biopolymers issued from renewable natural and agricultural resources, to develop novel bio-composites with conductive polymers, thus leading to entirely new directions in the research of advanced functional materials for a sustainable future [14,15]. In such a case, the water-soluble or water-dispersible biopolymers bind PANI chains and lead to stabilized colloids [16–18]. In this context, the "intelligent behavior" of PANI colloids prepared with chitosan or poly(N-isopropyl acrylamide) [19,20], cellulose derivatives [21], nanocellulose [22] and pectin [23,24], as steric stabilizers, was reported.

The resulting biocomposites could not only improve the processability of PANI, but also enhance its biocompatibility [25–27]. For example, PANI-based materials have been used for the detection and analysis of bio-molecules (enzymes, antibodies, DNA, proteins, etc.), making them suitable candidates for biomedical applications [25,27,28]. On the other hand, even if in vivo studies have shown that the emeraldine base form of PANI with electroactive properties [29] did not provoke inflammatory responses in a rodent model, suggesting good tolerance and bio-/histo-compatibility [30], risks may arise from the presence of polymerization by-products (benzidine, aniline dimers, and oligomers). Such by-products, indeed, can be cytotoxic, also with carcinogenic effects. All in all, although their presence can be avoided by suitable washing steps [31], the use of PANI-based materials still deserves many precautions.

In our work, we have used starch, which is the most abundant polysaccharide produced by higher plants, and it is extracted intracellularly in the form of granules of 2–100 μm in diameter [32]. In general, starches are semi-crystalline polymers made of a mixture of amylose and amylopectin chains, with a crystallinity of about 20–45% [33].

Starch was used by several authors to develop stabilized latexes [34–36]. In the work of Pradeep et al. [34], in particular, flexible semiconducting thin films from pure vulcanized natural rubber latex with enhanced electrical conductivity were produced. Other starch-stabilized colloids were also developed with bimetallic nanoparticles to improve the stability and reactivity of nanoparticles [37–39], PLGA (polylactic-co-glycolic acid) nanospheres [40], and PMMA (polymethylmethacrylate) lattices [41].

The usage of starch as a non-toxic and biodegradable biopolymer to elaborate PANI-based composites has led to the formation of advanced functional materials for various applications (water remediation, energy generation, and storage, electrochemical, medical and biomedical applications, etc.) [15,42–46]. In the same framework, we have attempted to elaborate water dispersible and processable polyaniline using starch as a steric stabilizer. The oxidative chemical polymerization of polyaniline was performed in an aqueous solution containing various ratios of aniline to starch (w/w). The obtained PANI/starch biocomposites were characterized by spectroscopic, microscopic, thermal, and electrochemical techniques, in order to assess the most promising PANI: starch ratio. Moreover, computational calculations based on DFT were carried out to find missing information about the interactions between starch–aniline and starch–polyaniline during the synthesis and in the final form of the biocomposites.

2. Materials and Methods

2.1. Synthesis of PANI/Starch Biocomposites

Polyaniline/starch materials (Table 1) were prepared by in situ polymerization of aniline in the presence of aqueous starch solutions under magnetic stirring. In a typical procedure [47], in a three-necked flask, aniline was added to the aqueous dispersions of starch at various ratios of aniline: starch (*w*/*w*) at room temperature (20 °C) and stirred for 4 h. The chemical polymerization of aniline was carried out at low temperatures in an ice bath (T~0–4 °C) using an APS (ammonium persulfate) oxidative solution. The mixture was maintained under magnetic stirring for 3 h, then left without stirring for 48 h to allow the majority of aniline to react, after which the dark powder obtained was filtered, washed with water and acetone to remove low-molecular-weight reaction by-products, i.e., aniline dimers and oligomers, then dried at 40 °C for 48 h.

Table 1. Codes of the synthesized biocomposites.

Codes	M_{11}	M_{21}	M_{31}
Aniline/Starch (*w*/*w*) ratio	1:1	2:1	3:1

2.2. Characterization

Structural characterization of samples was carried out by FTIR spectroscopy using a Nicolet 5700 spectrometer (Thermo Fisher, Waltham, MA, USA), equipped with a diamond ATR sampling accessory, on the powdered material, UV-Visible spectroscopy using a Cary 5000 UV−vis−NIR spectrometer (Varian, Palo Alto, CA, USA) on solutions prepared by dissolving 2 mg in 10 mL of DMF, and RAMAN spectroscopy using a DXR2 Raman spectrometer (Thermo Scientific, Madison, WI, USA), operating at 532 nm. The solubility tests were performed at different temperatures (20, 40, 50, 60, 80 °C) by dispersing the same weight of all materials in 10 mL of solvents, to obtain a dilute concentration (0.2% *w*/*v*). The solvents used are water, acetone, methanol, ethanol, glycerol, ethyl acetate, chloroform, toluene, hexane, cyclohexane, DMSO, and DMF. While colorless supernatants were obtained with insoluble mixtures, the presence of residual solids, in the case of partially soluble mixtures, was observed by a lens.

Homopolymers and biocomposites were subjected to laser granulometry analysis by using a CILAS 1190 analyzer (CPS Us, Inc., Madison, WI, USA). Thermal gravimetric analyses (with a TGA Q5000 instrument, TA Instruments, Hüllhorst, Germany) were performed until 700 °C (10 °C/min), while DSC (with a TA instrument Q20) was set from room temperature to 300 °C at 10 °C/min, under a nitrogen atmosphere; thermodynamic parameters were calculated by using the software TA Universal Analysis. X-ray diffraction analysis was performed using CuKα radiation (λ = 1.54 Å) (X'Pert3 Powder apparatus) from 2θ = 10° to 60°, at a scan rate of 0.02 s^{-1}; the peak crystalline areas were determined by using the X'pert HighScore software. SEM images of the composites were obtained after metallization with sputter quorum Q 150R E settings using Quanta 200 apparatus (FEI, Hillsboro, OR, USA). Cyclic voltammetry tests (with PGZ 301 potentiometer, Radiometer Analytical SAS, Lyon, France) were examined in a three-electrode system constituted of a reference electrode (Ag/AgCl), an auxiliary platinum electrode, and a carbon electrode as the working electrode. The electrolyte was a KCl solution (1 M) and the voltammograms were recorded in a voltage window between −0.8 and +0.8 mV at various scan rates (10, 50, and 100 mV s^{-1}).

3. Computational Study

Geometric optimization of the molecules was performed with the Dmol3 module [48] utilizing a modeling package created by Accelrys Incorporation (Cambridge, UK): Materials Studio 2017TM [49], B3LYP functional [50], and DNP basis set (4.4 basis file) [51].

The HOMO (Highest Occupied Molecular Orbital) and LUMO (Lowest Unoccupied Molecular Orbital) frontier molecular orbitals (FMOs) were investigated [52]. The high value of the energy ξ_{HOMO} justifies the inclination to transfer electrons to a suitable acceptor molecule. In contrast, the low value of the energy ξ_{LUMO} justifies the molecule's capacity to accept electrons from donor molecules [53].

The electronic chemical potential (μ) and global hardness (η) were determined using the ξ_{HOMO} and ξ_{LUMO} energies [54–56]:

$$\mu = (\xi_{HOMO} + \xi_{LUMO})/2 \quad (1)$$

$$\eta = (\xi_{LUMO} - \xi_{HOMO})/2 \quad (2)$$

The chemical potential μ is related to the electrophilicity index ω by the following relationship [57]:

$$\omega = (\mu^2)/(2\,\eta) \quad (3)$$

ω indicates an electrophile's capacity to gain an extra electrical charge [58].

The optimal fraction of transferred electrons (ΔN) is given by [59]:

$$\Delta N = -\mu/\eta \quad (4)$$

COSMO-RS (Conductor-like Screening Model for Real Solvents) is a thermodynamic-based quantum chemistry technique for determining chemical potentials in solutions [60]. For each species in the solution, this technique can estimate sigma charge densities and chemical potentials. The computation is divided into two parts: first, the molecule was geometrically optimized; then, using the COSMOtherm program [61] (p. 2), sigma profiles and sigma potentials were computed using the acquired cosmo-files.

Intermolecular interactions may be calculated using the Atoms in Molecule (AIM) method. The main goal of AIM is to examine the character and strength of the bonding interaction in molecular systems by using the electron density $\rho(r)$ of the molecules as a tool. Based on the second derivative, ∇^2 $\rho(r)$ sign [62,63], the kind of chemical bonds may be identified at the bond critical point (BCP) where it is minimum. As a result, if the ρ value is large and ∇^2 $\rho(r)$ is negative, the bond is covalent (polar). On the other hand, positive ∇^2 $\rho(r)$ indicates that the kinetic energy G(r) is greater than the potential energy V(r) [64]. The $|V|/G$ ratio indicates that the system's ability to group electrons, V(r), and its ability to dilute them through electronic mobility, G(r), compete within the system. As a result, the ratio $|V|/G > 1$ indicates that an excess of electric charges is utilized to generate the interaction, whereas $|V|/G < 1$ indicates the opposite. The total energy density, $H(r) = V(r) + G(r)$, in the first situation, has a negative value, whereas in the second case, it has a positive value. Thus, the categorization of interactions is based on the values of that indicator: (i) pure interactions with closed layers, $|V|/G < 1$; (ii) interactions with closed layers, $1 < |V|/G < 2$; and (iii) interactions with the shared layer, $|V|/G > 2$. The first two forms of interaction are ascribed to hydrogen bonding, whereas the third type is a covalent interaction. Furthermore, positive, ∇^2 $\rho(r)$, and H values indicate electrostatic contact, whereas negative values indicate a covalent bond. A partly covalent bond is defined as a positive ∇^2 $\rho(r)$ combined with a negative H value [65].

The Amsterdam Density Functional (ADF) program [66,67] was used to perform AIM calculations. For molecular structure optimization, the DFT method was employed. The AIM research [67,68] used the def-TZVP basis set to describe exchange correlation effects between these variables (Becke 3-Parameter, Lee, Yang, and Parr).

4. Results and Discussion

4.1. *Physico-Chemical Properties*

The scanning electron micrographs for polyaniline, starch, and polyaniline/starch biocomposites are shown in Figure 1a–d. The SEM micrograph of starch in Figure 1a reveals pseudo-spherical granules with smooth and well-defined surfaces. Figure 1b–d represent SEM images of starch/polyaniline composites with different ratios of aniline to starch. The figure shows that, in the case of a composite, when aniline is polymerized in the presence of starch, the polyaniline chains produced are grafted or adsorbed on this steric stabilizer. At the lowest concentration of aniline, the surface of the starch granules is rougher than that of pure starch, suggesting that aniline polymerization occurs on the surface of the granules of starch, forming an irregular coating over it [42,47]. However, as the concentration of aniline increases, more of the starch surface is covered and an overgrowth of polyaniline is observed over the starch particles [45].

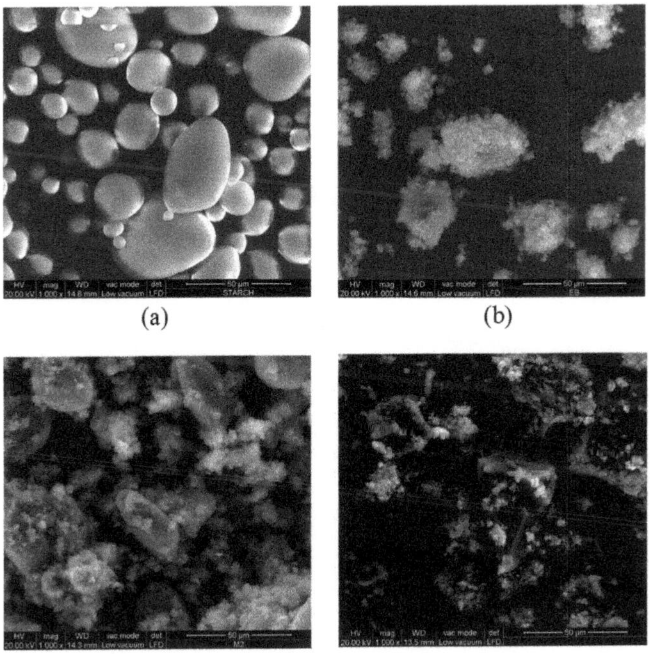

(a) (b) (c) (d)

Figure 1. SEM images of (**a**): starch, (**b**) PANI and PANI/starch biocomposites. (**c**): (1:1) and (**d**): (3:1).

From the analysis of particle size by laser granulometry, it can be seen that the particles of the obtained PANI/starch materials exhibit a wide distribution of particle sizes revealing a high polydispersity of particles (relative mean variance of the particle size distribution), which may indicate the occurrence of some aggregation of the biocomposites particles [69]. This statement is consistent with the findings of other authors [70,71]. Furthermore, as shown in Table 2, homopolymers and biocomposites particles exhibit average diameters ranging from 28.8 µm to 68.9 µm. It is also worth noting that all composites possess particle sizes lower than that of polyaniline, whereby the composite with the highest aniline content exhibits the lowest average diameter. Thus, the growing polyaniline chains may affect the intramolecular forces along the polymer chains of amylose and amylopectin, leading to smaller particles [15].

Table 2. Average particle diameters of PANI, STARCH, and PANI/starch biocomposites.

	PANI EB	Starch	M_{11}	M_{21}	M_{31}
Average diameter (µm)	68.9	38.9	40.2	31.5	28.8

The capacity of the various solvents (water, acetone, methanol, ethanol, glycerol, ethyl acetate, chloroform, toluene, hexane, cyclohexane, DMSO, and DMF) to dissolve the prepared materials was examined at different temperatures. The results obtained (Table 3) revealed that PANI-EB, starch, and PANI-EB/starch composites are insoluble in the majority of the solvents used even at high temperatures (80 °C). However, starch is soluble in DMSO but slightly soluble in water and DMF, whereas PANI-EB is slightly soluble in DMSO and soluble in DMF, and it is well known that it is not soluble in most common solvents [72]. The solubility behavior of PANI-EB/starch biocomposites differed from that of homopolymers where we noticed a good solubility in DMF and DMSO and an enhanced dispersion in water at 20–40 °C before more homogeneous mixtures were achieved at higher temperatures (60–80 °C). A further enhancement in the solubility was observed with glycerol and chloroform mainly at higher temperatures (>40 °C) with good dispersions at low temperatures. These findings indicate that the surfaces of the products have, at different temperatures, different affinities for each solvent. Nevertheless, the remaining solvents were not able to dissolve or even disperse PANI/starch biocomposites.

Table 3. Solubility test of PANI, STARCH, and PANI/starch biocomposites.

T °C	Materials	Acetone	Methanol	Ethanol	Glycerol	Ethyl Acetate	Chloro-Form	Toluene	Hexane	Cyclohexane	DMSO	Water	DMF
20	M_{11}	Δ	Δ	Δ	Δ	Δ	Δ	Δ	Δ	Δ	O	Δ	X
	M_{21}	Δ	Δ	Δ	Δ	Δ	Δ	Δ	Δ	Δ	O	Δ	X
	M_{31}	Δ	Δ	Δ	Δ	Δ	Δ	Δ	Δ	Δ	O	Δ	X
	PANI EB	Δ	Δ	Δ	Δ	Δ	Δ	Δ	Δ	Δ	O	Δ	X
	Starch	Δ	Δ	Δ	Δ	Δ	Δ	Δ	Δ	Δ	X	O	O
40	M_{11}	Δ	Δ	Δ	Δ	Δ	Δ	Δ	Δ	Δ	X	Δ	X
	M_{21}	Δ	Δ	Δ	Δ	Δ	Δ	Δ	Δ	Δ	X	Δ	X
	M_{31}	Δ	Δ	Δ	Δ	Δ	Δ	Δ	Δ	Δ	X	Δ	X
	PANI EB	Δ	Δ	Δ	Δ	Δ	Δ	Δ	Δ	Δ	X	Δ	X
	Starch	Δ	Δ	Δ	Δ	Δ	Δ	Δ	Δ	Δ	X	O	O
60	M_{11}	-	Δ	Δ	O	Δ	O	Δ	Δ	Δ	X	Δ	X
	M_{21}	-	Δ	Δ	O	Δ	O	Δ	Δ	Δ	X	Δ	X
	M_{31}	-	Δ	Δ	O	Δ	O	Δ	Δ	Δ	X	Δ	X
	PANI EB	-	Δ	Δ	Δ	Δ	Δ	Δ	Δ	Δ	X	Δ	X
	Starch	-	Δ	Δ	Δ	Δ	Δ	Δ	Δ	Δ	X	X	O
80	M_{11}	-	-	-	O	Δ	-	Δ	Δ	Δ	X	Δ	X
	M_{21}	-	-	-	O	Δ	-	Δ	Δ	Δ	X	Δ	X
	M_{31}	-	-	-	O	Δ	-	Δ	Δ	Δ	X	Δ	X
	PANI EB	-	-	-	Δ	Δ	-	Δ	Δ	Δ	X	Δ	X
	Starch	-	-	-	Δ	Δ	-	Δ	Δ	Δ	X	X	X

Δ: non-soluble, O: slightly (soluble/dispersion), X: soluble/good dispersion, -: evaporation of the solvents.

4.2. Structural Characterization

The UV-visible, FTIR, and RAMAN spectra of our products are shown in Figures 2–4, respectively. The UV-visible absorption spectra in Figure 2 display two broad absorption bands around 315–360 nm and 560–600 nm for all composites, which confirm the occurrence of the PANI-EB backbone structure. The first peak is assigned to π–π^* transition occurring from the highest occupied molecular orbital to the lowest unoccupied molecular orbital of the amine benzenoid rings, while the second peak is ascribed to benzenoid–quinoid rings' charge transfer [73]. The difference in the relative intensities of the peaks observed can be due to the variation in the quantities of benzenoid and quinoid units present in the materials as the ratio of aniline to starch varies. Moreover, the red shift for the 3:1 sample may be due to a higher amount of longer chains as well as to a growing contribution of a less oxidized form of PANI [74,75].

Figure 3 shows the FTIR-ATR spectra of polyaniline, starch, and their biocomposites. For the native starch, the broadband characteristic for the O-H groups of starch appears at 3252 cm^{-1} and the peak around 2924 cm^{-1} is attributed to an asymmetric stretching vibration of the C-H bond in pyranoid rings. Several other adsorption bands between 854 cm^{-1} and 1415 cm^{-1} are attributed to the contribution of various functional groups such as C-O and C-O-C [47]. Moreover, the peaks around 991 to 994 cm^{-1} are ascribed to the "C-O" bonds resulting from "C-O-H" and those of the anhydrous glucose rings in the starch backbone [76]. The characteristic peaks of pure PANI have been reported in previous works [42,47]. The bands appearing at 1539 cm^{-1} and 1456 cm^{-1} can be assigned to the vibration of quinoid rings ("N=Q=N") and benzenoid rings ("N-B-N"), respectively. The bands at 1282 cm^{-1} and 1230 cm^{-1} correspond to the "C-H" stretching vibration within the aromatic conjugation. The presence of PANI is confirmed in all composites' spectra in which PANI and starch are present at different concentrations. As PANI is more exposed on the surface of the particles, its signals dominate the spectra; moreover, the strong interaction between the chains may be responsible for the disappearance of some typical starch bands [77].

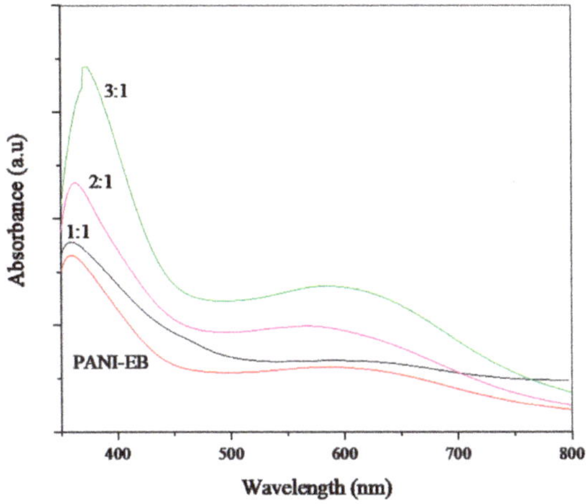

Figure 2. UV-vis spectra of PANI, starch, and PANI/starch biocomposites in DMF.

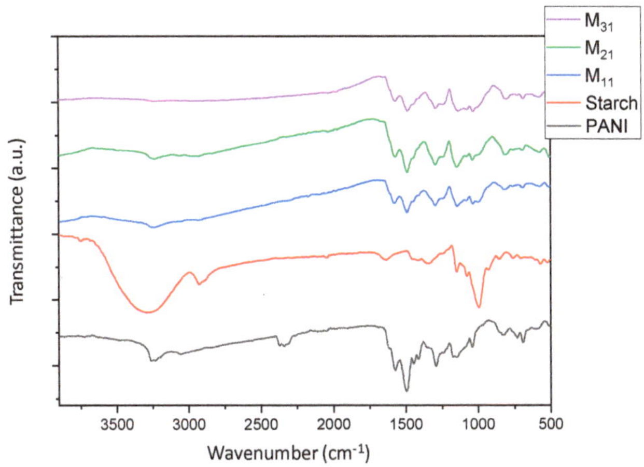

Figure 3. FTIR-ATR spectra of PANI, starch, and PANI/starch biocomposites. The residual CO_2 signal may be due to the adsorption properties of PANI.

The Raman spectra of PANI, starch, and their biocomposites, obtained using an exciting wavelength of 532 nm, are displayed in Figure 4. This method is very efficient for understanding the interaction among various components and provides information about the chemical structure of different domains in the composite systems.

From the spectrum of PANI, we can observe some characteristic bands rinsing from carbon–carbon, carbon–hydrogen, and carbon–nitrogen vibration modes. Benzenoid rings exhibited C-C stretching modes between 1520 cm^{-1} and 1650 cm^{-1} and C-H bending modes between 1100 cm^{-1} and 1210 cm^{-1}; however, quinoid rings give rise to C=N, and C-C ring stretching vibrations, which appeared at 1382 cm^{-1} and 1595 cm^{-1}, respectively. Further bands are observed in the region 1210–1520 cm^{-1} for C-N stretching modes (amines, imines) and in the region 1100–1140 cm^{-1} characteristic of the PANI backbone. A weak band at 1170 cm^{-1} has been assigned for in-plane deformation of the C-C bond of the quinoid ring of PANI [46,78].

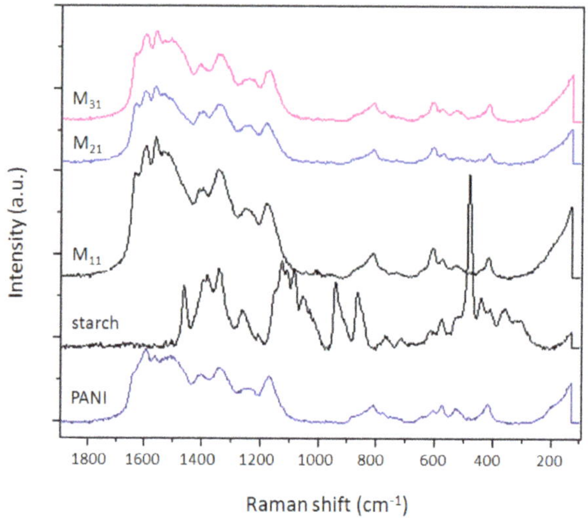

Figure 4. RAMAN spectra of PANI, starch, and PANI/starch biocomposites.

The observed bands in the Raman spectrum of starch, which are related to C-H and CH_2 vibrations (deformation, bending), appear in the region 1460–1340 cm^{-1} and at 862 cm^{-1}. The signals related to C-O-H bending vibrations arise at the wavenumbers 1335, 1123, and 1080 cm^{-1}. At 1257 cm^{-1}, the band related to vibrational modes of CH_2OH appears, while those associated with the skeletal mode involving α-(1–4) linkage appear at 942 cm^{-1} and between 770–360 cm^{-1} [79].

Similar to FTIR-ATR, the Raman curves of PANI/starch biocomposites display the typical peaks of PANI (Figure 4), with no meaningful shift. Moreover, knowing that Raman spectroscopy is a surface thickness-dependent tool [80,81], the response of PANI/starch biocomposites to laser excitation could be taken as a piece of further evidence that starch grains are covered by a thick PANI chain layer, which prevents the penetration of the laser beam to starch macromolecules and consequently impedes the appearance of starch characteristic peaks. These results concord with the SEM images in Figure 1.

4.3. X-ray Diffraction Pattern (XRD)

The X-ray diffraction patterns of starch, polyaniline, and PANI/starch biocomposites are presented in Figure 5. A semi-crystalline pattern can be observed for potato starch that exhibits strong reflection at $2\theta = 17.2°$, moderate reflections at $2\theta = 5.7°$, 15.04°, 22.22°, and 24.1°, in addition to weak reflections at $2\theta = 15.2°$ and 19.6°. Such a pattern is typical for B-type crystalline structure starches (Tuber starches) [82,83]. The XRD profile of PANI-EB representing the crystalline parts shows many broad reflection peaks, located at $2\theta = 6.4°$, 9.1°, 15.7°, 18.4°, 19.8°, and 25.4°. Larger peaks are believed to be formed by many reflections of wider clusters [84].

The interplanar distance 'd' is calculated for each peak using Bragg's equation:

$$n\lambda = 2d \sin\theta \quad (5)$$

where $\lambda = 1.542$ Å, n = 1 (first-order reflection), θ = the Bragg angle for the corresponding peak.

The diffraction spacing and Miller indices (hkl) for our materials are displayed in Table 4.

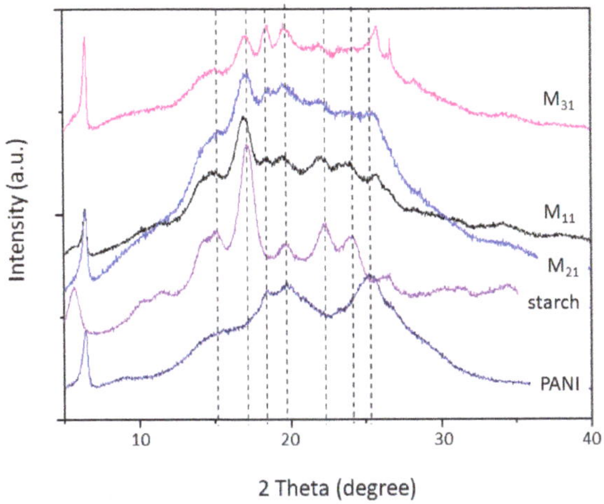

Figure 5. XRD diffractograms of starch, PANI, and PANI/starch biocomposites.

Table 4. Miller indices and diffraction spacing "d" of starch and PANI-EB.

Product	2θ (°)	hkl	d (Å)
PANI-EB	6.4	(001), (0–11)	13.81
	9.1	(021), (0–21), (003)	9.72
	15.7	(023), (030)	5.64
	19.8	(0–13), (−113)	4.48
	25.4	(1–24), (0–16)	3.51
Starch	5.7	(001)	15.51
	11.4	(111)	7.76
	15.04	(140)	5.89
	17.25	(131)	5.14
	19.59	(103)	4.53
	22.22	(113)	4.00
	24.1	(132)	3.69
	26.58	(142)	3.35

The values of the degree of crystallinity (Xc%) were obtained as a ratio between the peaks crystalline area and the total area of the XRD pattern under the curve. The following equation is employed [85]:

$$Xc\ (\%) = Hc/(Hc + Ha) \times 100 \tag{6}$$

where Hc and Ha are the intensities for the crystalline and amorphous profiles.

The crystallinity degree values of polyaniline and starch, as well as of biocomposites are summarized in Table 5. The native starch shows a high crystallinity (37%) when compared with polyaniline (19%). For PANI–starch biocomposites, a contribution due to different starch content can be confirmed. Changes in XRD patterns could be noticed, which may reveal alterations in the ultrastructure of the microdomains of PANI and starch mainly due to changes in the crystallinity. Such an observation is supported by the disappearance of starch peaks at 2θ = 22.22° and 26.58° and all PANI-EB peaks. Thus, it appears that the existence of aniline in the reaction medium and the PANI chains in propagation, partially destroy the original crystalline lattice of starch structure, and this is particularly evident for M_{21} and M_{31}. On the other hand, it seems that polyaniline chains can form an organized structure, which is also reflected in the diffractograms, although with a lack of peaks attributed to the PANI-EB crystalline structure.

Table 5. Degree of crystallinity of starch, PANI-EB, and their biocomposites.

Material	Degree of Crystallinity (%)
Starch	37
M_{11}	28
M_{21}	25
M_{31}	21
PANI EB	19

4.4. Thermal Characterization

The thermal behaviors of PANI, starch, and PANI/starch biocomposites were studied, under nitrogen, by TGA and DSC analyses. The corresponding thermogravimetry/differential thermogravimetry (TG/DTG) curves, depicted in Figure 6a–e, point out a weight loss within a broad range for polyaniline and as a sharp step for the degradation of starch. PANI-EB exhibits the first weight loss (Figure 6a) between 40 °C and 100 °C, which could be due to the evaporation of water and residual volatile molecules, followed by a small endothermic peak observed between 210 °C and 250 °C. A massive weight loss starts from 400 °C, owing to the degradation of the polyaniline chains [86]. The native starch undergoes 50% weight loss at 290 °C, while for polyaniline, a corresponding weight loss is noticed at 461 °C (Table 6). Furthermore, the DTG curves also show an overall decrease in the degradation rate for biocomposites with the rise in aniline amount, indicating a strong influence of polyaniline on the thermal stability of starch. As a matter of fact, the typical decomposition feature of PANI, with a strong DTG peak at 500 °C, is only present for M11, while M21 (not shown) and M31 show a quite regular weight loss till the limit of the explored range.

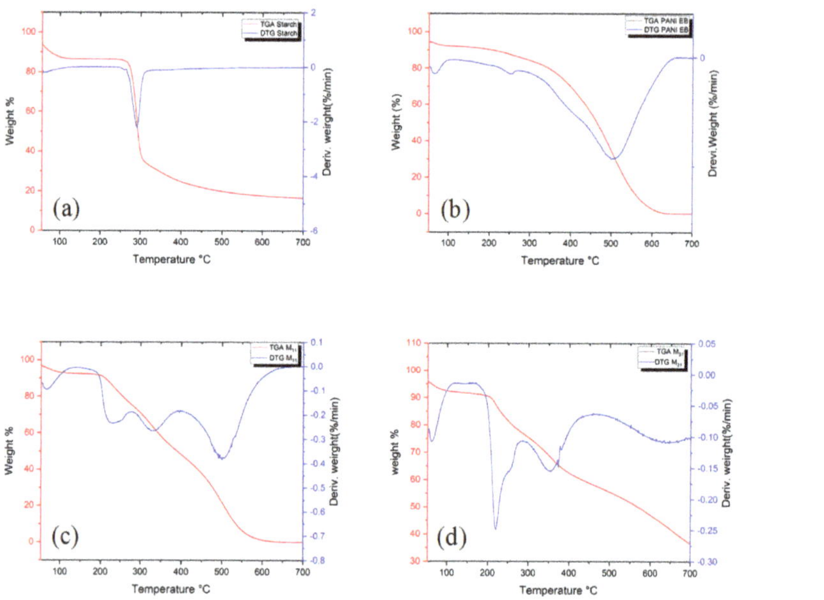

Figure 6. TGA and first derivatives curves of, (**a**): starch, (**b**): PANI EB and their biocomposites (**c**): M_{11}, (**d**): M_{31}.

Table 6. Thermal characteristics of PANI, starch, and PANI/starch biocomposites (% weight loss at specific temperatures).

Materials	T 5% (°C)	T 10% (°C)	T 50% (°C)	T 95% (°C)	Residue (%) at 700 °C
PANI EB	50	200	461.3	580.3	0
Starch	50.2	70.2	289.6	-	16.5
M_{11}	66.6	206.9	384.6	552.9	0
M_{21}	53.8	199.3	518.7	-	29.2
M_{31}	56.6	200.1	568.0	-	36.0

Detailed data can be derived from the TGA thermograms, which are related to the thermal stability such as T5, T50, and T95, defined as the temperature at which 5, 50 and 95% of the mass is volatilized, respectively. As observed in Table 6, PANI-EB and biocomposites are remarkably more stable than starch.

Figure 7 displays the DSC profiles of polyaniline, starch, and their biocomposites. Broad endothermic peaks are perceived in the temperature range of 60–120 °C, which may correspond to the evaporation of water as supported by the TGA and DTG data. Moreover, the PANI thermogram shows another endothermic step starting from 200 °C, which could be attributed to the glass transition temperature (Tg), where the movement of small chain segments starts [87,88]. On the other hand, the thermogram of starch reveals two narrow endothermic peaks centered at 266 °C and 278 °C that are attributed to the thermal decomposition of starch chains [89]. This result corroborates with the TGA obtained data. Regarding PANI/starch materials, we observe the presence of the endothermic peak related to water evaporation in the same temperature range, but the second endothermic peak, related to the variation in the Tg of polyaniline, shifts to a lower temperature, especially for M_{11}, indicating structural modification in the biocomposites owing to the interactions between components. The results of DSC measurement are reported in Table 7.

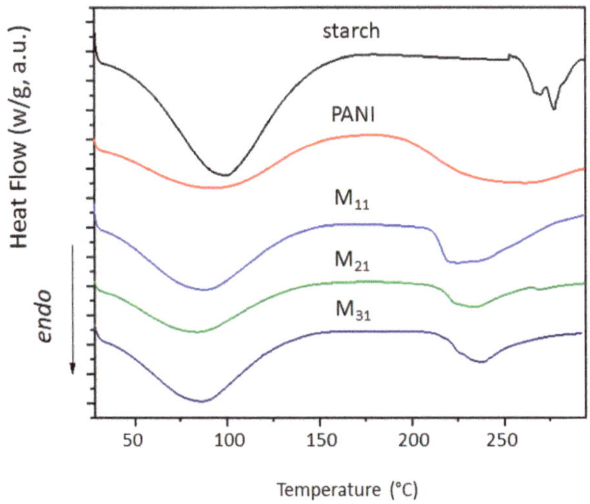

Figure 7. DSC thermograms of starch, PANI, and PANI–starch biocomposites.

Table 7. Data obtained from DSC thermograms of starch, PANI, and PANI–starch biocomposites.

Material	1st Peak			2nd Peak		
	Peak (°C)	Onset (°C)	ΔH (j/g)	Peak (°C)	Onset (°C)	ΔH (j/g)
M_{11}	88.19	36.89	254.4	255.65	211.94	121.0
M_{21}	84.49	34.28	231.5	233.19 270.08	215.19 266.57	52.58 1.854
M_{31}	68.83	39.02	249.2	237.73	218.07	60.89
PANI EB	94.69	35.33	203.5	234.72	195.81	113.6
Starch	100.01	50.45	439.7	266.36 278.10	261.71 273.84	9.058 15.82

4.5. Cyclic Voltammetry

The electroactivity of PANI and PANI-EB/starch composites was studied in the supporting aqueous electrolyte (KCl) by cyclic voltammetry, and the recorded voltammograms are shown in Figure 8. All the curves displayed a pair of oxidation and reduction peaks, arising from the redox behavior of PANI-EB. Indeed, the peaks appearing for each oxidation and reduction process for PANI and PANI-based composites can be assigned to leucoemeraldine/emeraldine transition states of polyaniline [44,90]. In most cases, voltammograms exhibit a well-defined oxidation peak at about 120 mV that appears to be associated with a less-evident reduction peak at about 110 mV, highlighting the occurrence of a redox electron transfer [91].

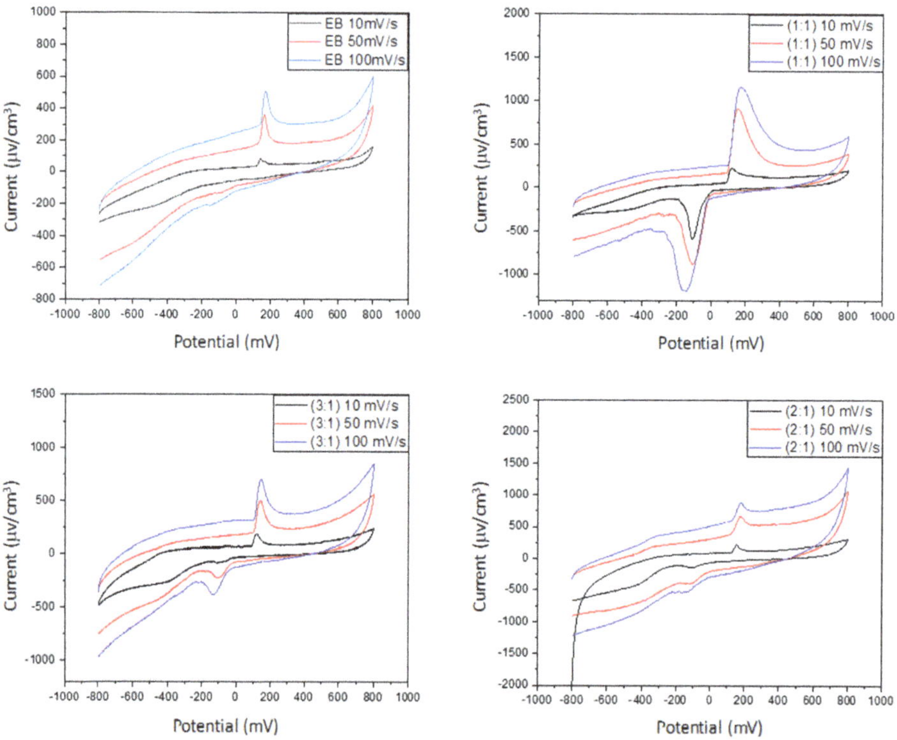

Figure 8. Cyclic voltammograms of PANI and PANI/starch biocomposites.

The species mobility within the composite was evaluated from the dependence of the redox peaks intensity on the scan rate, from 10 to 100 mV/s. As reported in Table 8, the current density increases with the increasing scan rate value, and the waves show an increase in their areas with increasing scan rates, revealing the good electroactivity as well as the electrochemical capacitive behavior of our materials [44,92].

Table 8. Redox parameters from cyclic voltammogram peaks of PANI and PANI/starch biocomposites.

Materials	Scan Rate (mV/s)	Ox	Red
PANI EB	10	84.71	−75.19
	50	352.12	−132.77
	100	506.45	−210.18
M_{11}	10	222.24	−587.66
	50	905.14	−879.00
	100	1158.75	−1179.44
M_{21}	10	118.56	−146.93
	50	669.08	−399.07
	100	866.21	−548.06
M_{31}	10	186.53	−81.03
	50	505.03	−226.46
	100	703.62	−380.19

In order to better understand the electronic behavior of PANI and its composites with starch, we have plotted the variations in the current density of the oxidation peak for each sample as a function of the square root of the scan rate $\sqrt{v}$.

We can observe in Figure 9 that the current density is linear with $\sqrt{v}$ for all the synthesized materials. The oxidation phenomenon is therefore limited by the ion diffusion, and the transport of charged species to the electrode surface is ensured by this transport mode.

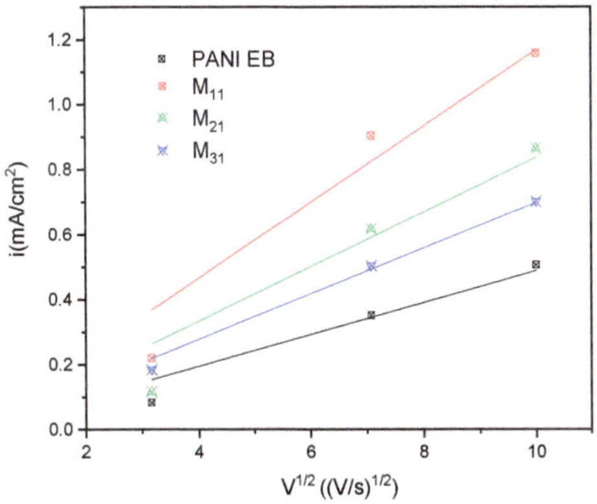

Figure 9. Peak current variation with scan rates.

In fact, the Randles Sevcik relation gives a straight line when pure diffusion controls the electrochemical reaction.

$$I_{pa} = 2.69 \times 10^5\, n^{3/2} A\, D^{1/2}\, C\, v^{1/2} \quad (7)$$

ip = current maximum in amps
n = number of electrons transferred in the redox event
A = electrode area in cm^2
D = diffusion coefficient in cm^2/s
C = concentration in mol/cm^3
v = scan rate in V/s

Therefore, it is clear that the PANI synthesized by chemical oxidation polymerization is a conductor material. Note that according to our results, the M11 sample is the best conductor among these materials.

4.6. Computational Assessment

The main electronic properties calculated for the aniline, PANI, and starch are summarized in Table 9. As amylopectin is the most abundant component (85%) in the starch under investigation, it was considered a suitable model. According to the energy values, comparable values of the aniline isomers ξ_{HOMO} and ξ_{LUMO} indicate a quasi-equivalent tendency to donate or accept electrons. That said, isomers I1 and I2 have the highest values, showing that they are the more probable molecules participating in the polymerization process to produce the PANI chain (this is confirmed by the highest electron escaping tendency from an equilibrium system, $\mu = -3.027$ eV). Both PANI and amylopectin are more reactive than the PANI parent species (isomers I1 or I2) (see Table 9). Regarding the gap values in Table 10, the amylopectin is the most stable in this set of polymers.

Table 9. ξ_{HOMO}, ξ_{LUMO} and related molecular properties of the studied compounds.

Parameter	An	An+	I1	I2	I3	PANI	Amylopectin
ξ_{HOMO} (eV)	−5.105	−5.268	−5.736	−5.736	−5.489	−4.770	−7.056
ξ_{LUMO} (eV)	−1.190	−0.1687	−0.318	−0.318	−0.261	−2.980	0.395
Gap ΔE (eV)	3.916	5.099	5.418	5.418	5.227	1.791	7.450
Electronegativity, $\chi = -\mu$	3.147	2.718	3.027	3.027	2.875	3.875	3.331
Hardness, η	1.958	2.550	2.709	2.709	2.614	0.895	3.725
Electrophilicity index, ω	2.530	1.450	1.691	1.692	1.581	8.386	1.489
Fractions of transferred electrons, ΔN	1.608	1.066	1.118	1.118	1.100	4.328	0.894

Table 10. The HOMO and LUMO electron density distributions of aniline, polyaniline, and amylopectin.

	An	PANI	Amylopectin
HOMO	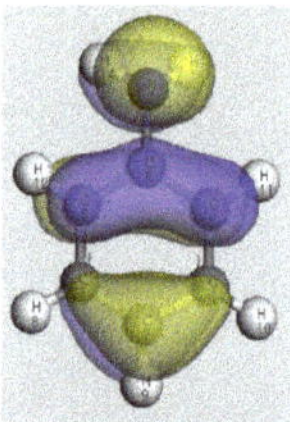	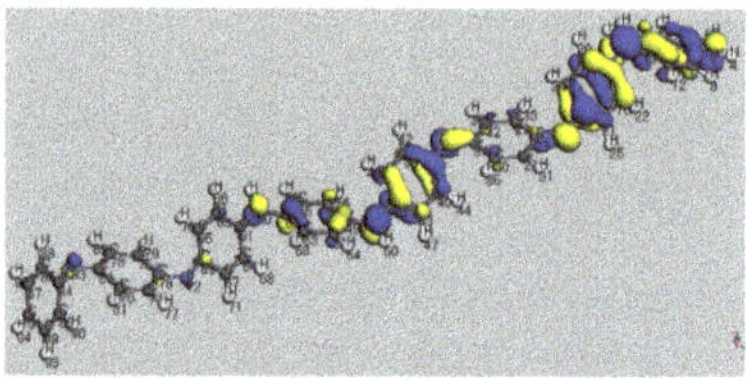	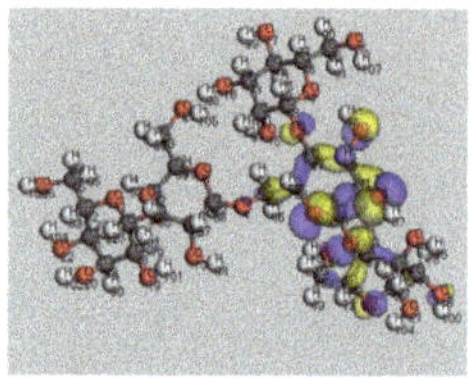
LUMO	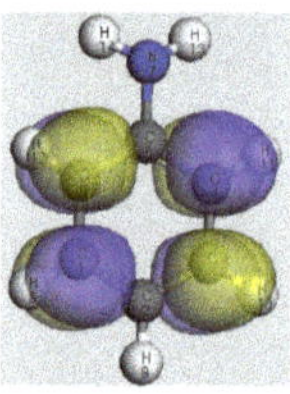	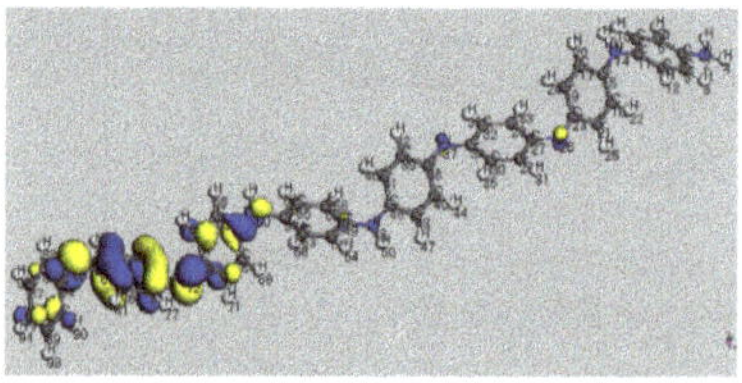	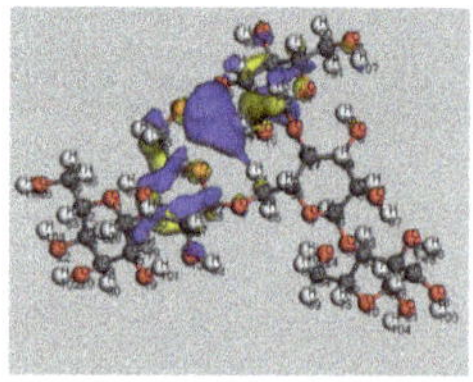

In conclusion, PANI has the highest ability to acquire an extra electron charge (highest ω value: 8.386 eV) and the lowest resistance to charge transfer (the lowest hardness η value, 0.895) (LUMO of PANI, see Table 10). It is thus concluded that amylopectin is the electron donor with the highest $\xi_{HOMO} = -7.056$ eV (see Table 2). The results for ΔN are 4.328 electrons (acquired) for PANI and 0.894 electrons (released) for amylopectin.

The Fukui results shown Table 11 led to the same conclusion as HOMO and LUMO. The PANI's f+ (LUMO) sites are localized on the N (72) (f+ = 0.076) and N (83) (f+ = 0.074) while the amylopectin's f- (HOMO) sites are localized on the O (1) (f− = 0.105) and O (9) (f− = 0.115).

Table 11. Fukui analysis of aniline, polyaniline, and amylopectin: The nucleophilic term f_k^+; the electrophilic term f_k^- values of the Fukui function.

	An	**PANI**	**Amylopectin**
f+	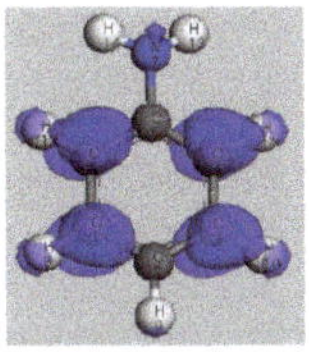 C(6) = 0.130; (4) = 0.130 C(3) = 0.127; C(1) = 0.127	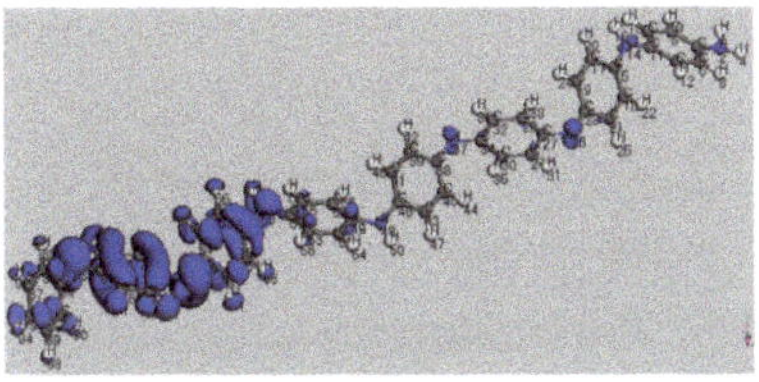 N(72) 0.076; N(83) 0.074	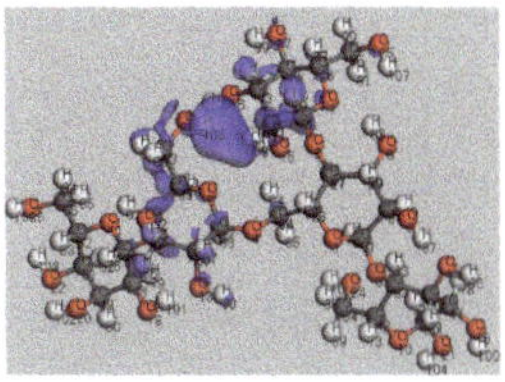 H(72) 0.107; H(105) 0.194
f−	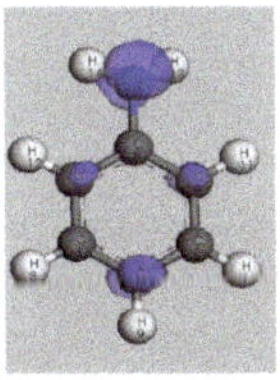 N(7) 0.193; C(2) 0.110	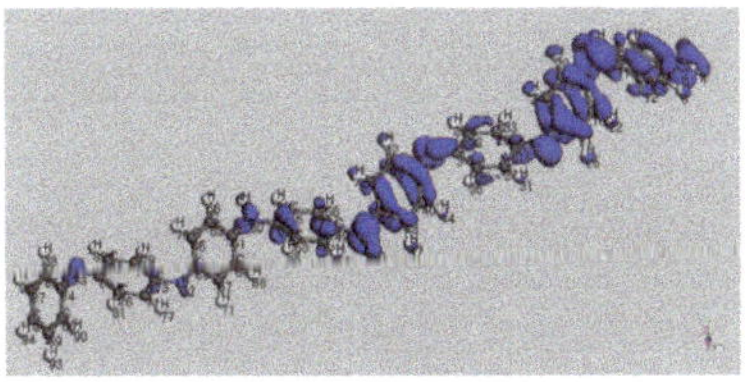 N(14) 0.037; N(2) 0.029; N(26) 0.027; N(37) 0.030; N(48) 0.031	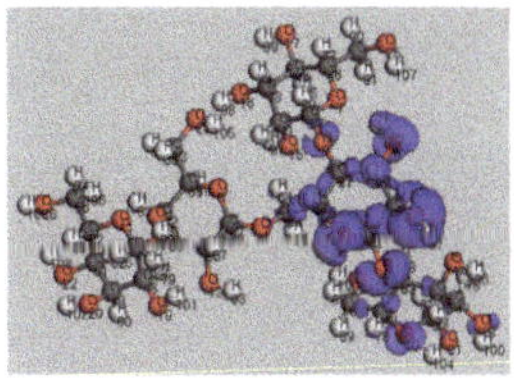 O(1) 0.105; O(9) 0.115

The sigma profile above is divided into three distinct regions:

- HBD Region—Hydrogen bond donor region: the sigma values are less than −0.01 eÅ^{-2}. The negative sigma values mean positive polarities.
- Non-Polar Region—σ values are given in the interval −0.01 eÅ^{-2} to +0.01 eÅ^{-2}.
- HBA Region—Hydrogen bond acceptor region: the σ values are greater than 0.01 eÅ^{-2}. Positive sigma values represent negative polarities.

Figure 10a shows that the highest peaks for all the selected molecules are in the non-polar region. They are showing the tremendous non-polar character of the molecule surfaces. Amylopectin has a net peak in the HBA region, while PANI is in the HBD region. In addition, PANI has the highest affinity to HBA, and amylopectin has the highest affinity to HBD (Figure 10b). This confirms the frontier molecular orbital study results.

The interaction energies reported in Table 12, reveal that the HB type contributes the most to the mixture's energy (−31.33496 and −40.29382 Ha). The PANI–amylopectin pair (ratio 1:1) achieves the highest value, while I2-amylopectin (ratio 3:1) has the lowest. The second contribution to the mixture's energy is Van der Waals (between −26.70916 and −27.83387 Ha), with the minimum value for the PANI–amylopectin couple (1:1) and the maximum value for I2-amylopectin (3:1). The mixture's energy ranges from −55.25390 for I2-amylopectin (3:1) to −63.51928 Ha for PANI–amylopectin (Emisfit).

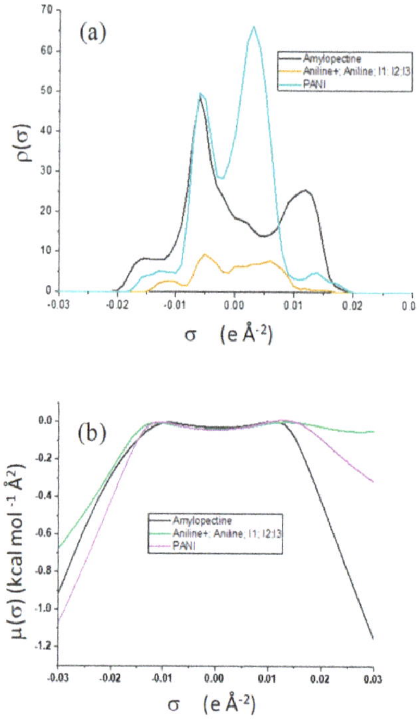

Figure 10. Interactions in PANI/starch biocomposites using COSMO-RS.

Table 12. Interaction energies of amylopectin with the various compounds, which are depicted in Figure 11.

Mixture	Mass Ratio	E_{VdW}	E_{Misfit}	E_{HB}	E_{mix}
Aniline + Amylopectin	1:1	−27.27875	8.19827	−37.56368	−61.26681
	2:1	−27.60525	8.45979	−34.01977	−57.78789
	3:1	−27.74166	8.61834	−31.70186	−55.44783
Aniline_i2 Amylopectin	1:1	−27.34835	8.15238	−37.28638	−61.10501
	2:1	−27.69050	8.39477	−33.66549	−57.58388
	3:1	−27.83387	8.53758	−31.33496	−55.25390
PANI Amylopectin	1:1	−26.70916	8.10636	−40.29382	−63.51928
	2:1	−26.94796	8.36426	−38.53551	−61.74186
	3:1	−27.02828	8.50672	−37.60637	−60.75059
	3:1	−27.02828	8.50672	−37.60637	−60.75059

Energies are given in Hartree.

Figure 12 shows a molecular graph of the optimized PANI–amylopectin complex, while Tables 13 and 14 give the topological characteristics at the BCP of interaction contacts in the PANI–amylopectin complex. The computed values of ρ(r) in BCPs are in the range [2.21×10^{-3}–7.83×10^{-3}] eV (Table 13), which are relatively low values, and $\nabla^2 2\rho(r)$ is positive, indicating the existence of hydrogen bond interactions in BCPs 231, 236, 239, 323, 324, 361, and 363 (BCPs are given in Table 13). The positive values of $\nabla^2 2\rho(r)$ and H demonstrate the H-bonding connection between the BCPs atoms (see Figure 12). Weak hydrogen bonding is characterized by $E_{HB} < 0$. The interaction between PANI and amylopectin is thus considered to be physical (with no covalent bonding).

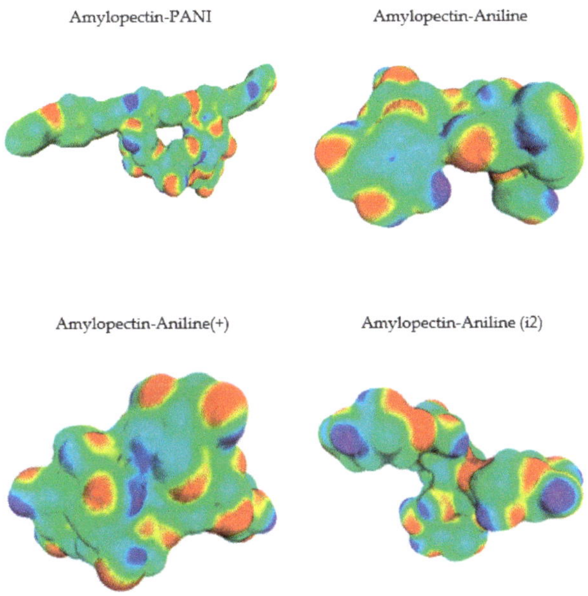

Figure 11. Molecular graph of the optimized PANI–amylopectin complex.

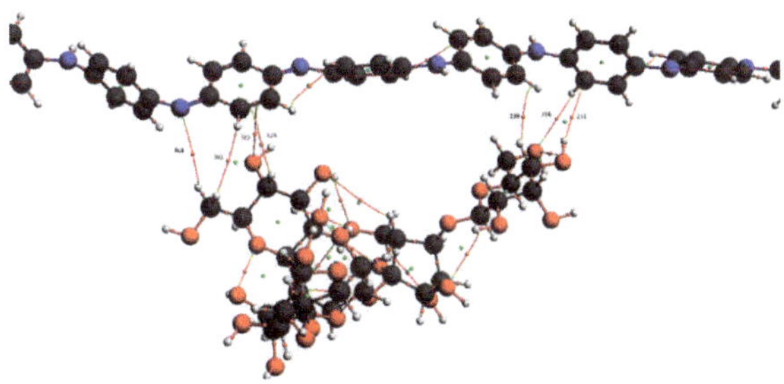

Figure 12. σ-profile (**a**) and σ-potential (**b**) distributions of aniline, polyaniline, and amylopectin.

Table 13. The topological characteristics at BCP of interaction contacts.

BCP	Amylopectin	PANI
231	103 (H)	171 (C)
236	26 (O)	174 (H)
239	82 (H)	166 (H)
323	17 (O)	138 (C)
324	75 (H)	136 (C)
361	91 (H)	139 (H)
363	90 (H)	134 (N)

Table 14. The topological characteristics at BCP of interaction contacts in the PANI–amylopectin complex.

BCP	$\rho(r)$	$\nabla^2\rho(r)$	$G(r)$	$V(r)$	E_{HB} (eV)	$H = G + V$	$\mid V \mid /G$
231	2.32×10^{-3}	7.35×10^{-3}	1.34×10^{-3}	-8.45×10^{-4}	-4.23×10^{-4}	4.96×10^{-4}	0.63
236	3.48×10^{-3}	1.28×10^{-2}	2.36×10^{-3}	-1.52×10^{-3}	-7.61×10^{-4}	8.33×10^{-4}	0.65
239	6.91×10^{-3}	2.28×10^{-2}	4.52×10^{-3}	-3.34×10^{-3}	-1.67×10^{-3}	1.18×10^{-3}	0.74
323	7.83×10^{-3}	2.72×10^{-2}	5.42×10^{-3}	-4.04×10^{-3}	-2.02×10^{-3}	1.38×10^{-3}	0.75
324	5.60×10^{-3}	2.10×10^{-2}	4.01×10^{-3}	-2.76×10^{-3}	-1.38×10^{-3}	1.25×10^{-3}	0.69
361	2.21×10^{-3}	7.58×10^{-3}	1.37×10^{-3}	-8.46×10^{-4}	-4.23×10^{-4}	5.24×10^{-4}	0.62
363	2.32×10^{-3}	7.35×10^{-3}	1.34×10^{-3}	-8.45×10^{-3}	-4.23×10^{-4}	4.96×10^{-4}	0.63

5. Conclusions

The results of this work demonstrate a facile method for the synthesis of polyaniline colloids. In particular, potato starch was employed as a steric stabilizer to obtain stable dispersions of polyaniline (emeraldine base). Polyaniline/starch biocomposites with various aniline/starch ratios were thus synthesized via an oxidative polymerization and characterized. To evaluate the effect of aniline/starch ratios variation, several tests (UV-vis, FTIR-ATR, Raman, XRD, TGA, DSC), as well as computational calculation (DFT, COSMO-RS, AIM), were performed. This study demonstrates the possibility of using starch as a steric stabilizer, and so to empower the processability of the polyaniline by improving the solubility in DMSO and DMF and the dispersion behavior in water, as well as in glycerol and chloroform, at higher temperatures.

The growth of polyaniline is observed on the starch particles' surfaces. The particle dimension and the degree of crystallinity appear to be strongly affected in the composite material, as demonstrated by the formation of smaller particles and the loss of the original starch crystalline lattice. Although less thermally stable, M11 appears to be the best in terms of morphology and retention of crystallinity degree. The cyclic voltammetry test shows better electro-activity in terms of the electron exchange and capacitive behavior of the composite materials, with the ratio of 1:1 being the best composition among these materials. Finally, the strong interactions between the PANI and amylopectin (the main starch component) were investigated by the DFT, COSMO-RS, and AIM methods, demonstrating their purely physical nature.

These results attest that the prepared composites as promising versatile materials, which are expected to broaden the electrochemical applications of PANI.

Author Contributions: Conceptualization, methodology, and investigation: S.B.; writing and supervision: S.D. and H.F.; writing-review and editing: M.C.; software and data correction: Y.B. and I.C. All authors have read and agreed to the published version of the manuscript.

Funding: This work was supported by the Department of Chemical Sciences of Padova University (grant number: P-DiSC#11NExuS_BIRD2019-UNIPD).

Acknowledgments: We acknowledge Nicola Dengo for his assistance with Raman analysis.

Conflicts of Interest: The authors declare no conflict of interest.

Abbreviations

(PANI-EB) polyaniline (emeraldine base), (FTIR) Fourier transform infrared, (XRD) X-ray diffraction analysis, (TGA) thermogravimetric analysis, (DTG) Derivative Thermogravimetry, (DSC) Differential scanning calorimetry, (SEM) Scanning Electron Microscopy, (DFT) Density-functional theory, (COSMO-RS) Conductor-like Screening Model for Real Solvents, (AIM methods) Atoms in Molecule, (DMSO) Dimethylsulfoxide, (DMF) Dimethylformamide, (ADF) The Amsterdam Density Functional.

References

1. Armes, S.P. Conducting polymer colloids. *Curr. Opin. Colloid Interface Sci.* **1996**, *1*, 214–220. [CrossRef]
2. Marcilla, R.; Ochoteco, E.; Pozo-Gonzalo, C.; Grande, H.; Pomposo, J.A.; Mecerreyes, D. New organic dispersions of conducting polymers using polymeric ionic liquids as stabilizers. *Macromol. Rapid Commun.* **2005**, *26*, 1122–1126. [CrossRef]
3. Bjorklund, R.B.; Liedberg, B. Electrically conducting composites of colloidal polypyrrole and methylcellulose. *J. Chem. Soc. Chem. Commun.* **1986**, *16*, 1293–1295. [CrossRef]
4. Mandal, T.K.; Mandal, B.M. Ethylhydroxyethylcellulose stabilized polypyrrole dispersions. *Polymer* **1995**, *36*, 1911–1913. [CrossRef]
5. Armes, S.P.; Miller, J.F.; Vincent, B. Aqueous dispersions of electrically conducting monodisperse polypyrrole particles. *J. Colloid Interface Sci.* **1987**, *118*, 410–416. [CrossRef]
6. Cawdery, N.; Obey, T.M.; Vincent, B. Colloidal dispersions of electrically conducting polypyrrole particles in various media. *J. Chem. Soc. Chem. Commun.* **1988**, *17*, 1189–1190. [CrossRef]
7. Digar, M.L.; Bhattacharyya, S.N.; Mandal, B.M. Conducting polypyrrole particles dispersible in both aqueous and non-aqueous media. *J. Chem. Soc. Chem. Commun.* **1992**, *1*, 18–20. [CrossRef]
8. Ohsawa, T.; Kabata, T.; Kimura, O.; Nakajima, S.; Nishihara, H.; Yoshi, K. Non-linear electric properties of polyaniline doped with organic acceptors. *Synth. Met.* **1993**, *57*, 4842–4847. [CrossRef]
9. Gospodinova, N.; Terlemezyan, L. Conducting polymers prepared by oxidative polymerization: Polyaniline. *Prog. Polym. Sci.* **1998**, *23*, 1443–1484. [CrossRef]
10. Aldissi, M. Is there a colloid in every solution-processable conducting polymer? *Adv. Mater.* **1993**, *5*, 60–62. [CrossRef]
11. Dong, H.; Nyame, V.; MacDiarmid, A.G.; Jones, W.E., Jr. Polyaniline/poly (methyl methacrylate) coaxial fibers: The fabrication and effects of the solution properties on the morphology of electrospun core fibers. *J. Polym. Sci. Part B Polym. Phys.* **2004**, *42*, 3934–3942. [CrossRef]
12. Swapna, P.; Subrahmanya, R.S.; Sathyanarayana, D. Water-soluble conductive blends of polyaniline and poly (vinyl alcohol) synthesized by two emulsion pathways. *J. Appl. Polym. Sci.* **2005**, *98*, 583–590. [CrossRef]
13. Milakin, K.A.; Morávková, Z.; Acharya, U.; Kašparová, M.; Breitenbach, S.; Taboubi, O.; Hodan, J.; Hromádková, J.; Unterweger, C.; Humpolíček, P.; et al. Enhancement of conductivity, mechanical and biological properties of polyaniline-poly (N-vinylpyrrolidone) cryogels by phytic acid. *Polymer* **2021**, *217*, 12345. [CrossRef]
14. Huang, Y.; Li, J.; Chen, X.; Wang, X. Applications of conjugated polymer based composites in wastewater purification. *RSC Adv.* **2014**, *4*, 62160–62178. [CrossRef]
15. Azmi, N.S.; Bahsa, R.; Othman, S.H.; Mohammed, M.A.P. Characterization of antioxidant tapioca starch/polyaniline composites film prepared using solution casting method. *Food Res.* **2019**, *3*, 317–324. [CrossRef]
16. Sarma, T.K.; Chattopadhyay, A. Reversible Encapsulation of Nanometer-Size Polyaniline and Polyaniline—Au-Nanoparticle Composite in Starch. *Langmuir* **2004**, *20*, 4733–4737. [CrossRef] [PubMed]
17. Liu, D.Y.; Sui, G.; Bhattacharyya, D. Synthesis and characterisation of nanocellulose-based polyaniline conducting films. *Compos. Sci. Technol.* **2014**, *99*, 31–36. [CrossRef]
18. Noreen, S.; Bhatti, H.N.; Iqbal, M.; Hussain, F.; Sarim, F.M. Chitosan, starch, polyaniline and polypyrrole biocomposite with sugarcane bagasse for the efficient removal of Acid Black dye. *Int. J. Biol. Macromol.* **2020**, *147*, 439–452. [CrossRef] [PubMed]
19. Cruz-Silva, R.; Arizmendi, L.; Del-Angel, M.; Romero-Garcia, J. pH-and thermosensitive polyaniline colloidal particles prepared by enzymatic polymerization. *Langmuir* **2007**, *23*, 8–12. [CrossRef] [PubMed]
20. Cruz-Silva, R.; Escamilla, A.; Nicho, M.E.; Padron, G.; Ledezma-Perez, A.; Arias-Marin, E.; Moggio, I.; Romero-Garcia, J. Enzymatic synthesis of pH-responsive polyaniline colloids by using chitosan as steric stabilizer. *Eur. Polym. J.* **2007**, *43*, 3471–3479. [CrossRef]
21. Chattopadhyay, D.; Mandal, B.M. Methyl cellulose stabilized polyaniline dispersions. *Langmuir* **1996**, *12*, 1585–1588. [CrossRef]
22. Luong, N.D.; Korhonen, J.T.; Soininen, A.J.; Ruokolainen, J.; Johansson, L.-S.; Seppälä, J. Processable polyaniline suspensions through in situ polymerization onto nanocellulose. *Eur. Polym. J.* **2013**, *49*, 335–344. [CrossRef]
23. Amarnath, C.A.; Venkatesan, N.; Doble, M.; Sawant, S.N. Water dispersible Ag@ polyaniline-pectin as supercapacitor electrode for physiological environment. *J. Mater. Chem. B* **2014**, *2*, 5012–5019. [CrossRef] [PubMed]
24. Anbalagan, A.C.; Sawant, S.N. Biopolymer stabilized water dispersible polyaniline for supercapacitor electrodes. *AIP Conf. Proc.* **2018**, *1942*, 140054.
25. Ibarra, L.E.; Tarres, L.; Bongiovanni, S.; Barbero, C.A.; Kogan, M.J.; Rivarola, V.A.; Bertuzzi, M.L.; Yslas, E.I. Assessment of polyaniline nanoparticles toxicity and teratogenicity in aquatic environment using Rhinella arenarum model. *Ecotoxicol. Environ. Saf.* **2015**, *114*, 84–92. [CrossRef] [PubMed]
26. Lai, J.; Yi, Y.; Zhu, P.; Shen, J.; Wu, K.; Zhang, L.; Liu, J. Polyaniline-based glucose biosensor: A review. *J. Electroanal. Chem.* **2016**, *782*, 138–153. [CrossRef]
27. Dhand, C.; Das, M.; Datta, M.; Malhotra, B. Recent advances in polyaniline based biosensors. *Biosens. Bioelectron.* **2011**, *26*, 2811–2821. [CrossRef]
28. Kuwahara, T.; Ogawa, K.; Sumita, D.; Kondo, M.; Shimomura, M. Amperometric glucose sensing with polyaniline/poly (acrylic acid) composite film bearing glucose oxidase and catalase based on competitive oxygen consumption reactions. *J. Electroanal. Chem.* **2018**, *811*, 62–67. [CrossRef]
29. Kamalesh, S.; Tan, P.; Wang, J.; Lee, T.; Kang, E.; Wang, C. Biocompatibility of electroactive polymers in tissues. *J. Biomed. Mater. Res.* **2000**, *52*, 467–478. [CrossRef]

30. Chen, Y.; Kang, E.; Neoh, K.; Tan, K. Oxidative graft polymerization of aniline on modified Si (100) surface. *Macromolecules* **2001**, *34*, 3133–3141. [CrossRef]
31. Humpolicek, P.; Kasparkova, V.; Saha, P.; Stejskal, J. Biocompatibility of polyaniline. *Synth. Met.* **2012**, *162*, 722–727. [CrossRef]
32. Xiao, C.; Tan, J.; Xue, G. Synthesis and properties of starch-g-poly (maleic anhydride-co-vinyl acetate). *Express Polym. Lett.* **2010**, *4*, 9–16. [CrossRef]
33. Ma, X.; Chang, P.R.; Yu, J.; Wang, N. Preparation and properties of biodegradable poly (propylene carbonate)/thermoplastic dried starch composites. *Carbohydr. Polym.* **2008**, *71*, 229–234. [CrossRef]
34. Predeep, P.; Sreeja, R.; Mazur, M.; Sharma, P. Role of starch as a steric stabilizer for doping natural rubber: Effect on optical, electrical, mechanical, and morphological properties. *J. Elastomers Plast.* **2006**, *38*, 333–348. [CrossRef]
35. De Bruyn, H.; Sprong, E.; Gaborieau, M.; David, G.; Roper, J.A., III; Gilbert, R.G. Starch-graft-copolymer latexes initiated and stabilized by ozonolyzed amylopectin. *J. Polym. Sci. Part A Polym. Chem.* **2006**, *44*, 5832–5845. [CrossRef]
36. Gugliemelli, L.; Swanson, C.; Baker, F.; Doane, W.; Russell, C. Cationic starch–polyacrylonitrile graft copolymer latexes. *J. Polym. Sci. Polym. Chem. Ed.* **1974**, *12*, 2683–2692. [CrossRef]
37. He, F.; Zhao, D. Preparation and characterization of a new class of starch-stabilized bimetallic nanoparticles for degradation of chlorinated hydrocarbons in water. *Environ. Sci. Technol.* **2005**, *39*, 3314–3320. [CrossRef]
38. Dong, H.; Cheng, Y.; Lu, Y.; Hou, K.; Zhang, L.; Li, L.; Wang, B.; Wang, Y.; Ning, Q.; Zeng, G. Comparison of toxicity of Fe/Ni and starch-stabilized Fe/Ni nanoparticles toward Escherichia coli. *Sep. Purif. Technol.* **2019**, *210*, 504–510. [CrossRef]
39. Wang, X.; Le, L.; Alvarez, P.J.; Li, F.; Liu, K. Synthesis and characterization of green agents coated Pd/Fe bimetallic nanoparticles. *J. Taiwan Inst. Chem. Eng.* **2015**, *50*, 297–305. [CrossRef]
40. Besheer, A.; Vogel, J.; Glanz, D.; Kressler, J.; Groth, T.; Mäder, K. Characterization of PLGA nanospheres stabilized with amphiphilic polymers: Hydrophobically modified hydroxyethyl starch vs. pluronics. *Mol. Pharm.* **2009**, *6*, 407–415. [CrossRef] [PubMed]
41. Gaborieau, M.; de Bruyn, H.; Mange, S.; Castignolles, P.; Brockmeyer, A.; Gilbert, R.G. Synthesis and characterization of synthetic polymer colloids colloidally stabilized by cationized starch oligomers. *J. Polym. Sci. Part A Polym. Chem.* **2009**, *47*, 1836–1852. [CrossRef]
42. Saikia, J.; Banerjee, S.; Konwar, B.; Kumar, A. Biocompatible novel starch/polyaniline composites: Characterization, anti-cytotoxicity and antioxidant activity. *Colloids Surf. B Biointerfaces* **2010**, *81*, 158–164. [CrossRef]
43. Lukasiewicz, M.; Ptaszek, P.; Ptaszek, A.; Bednarz, S. Polyaniline–starch blends: Synthesis, rheological, and electrical properties. *Starch-Stärke* **2014**, *66*, 583–594. [CrossRef]
44. Pandi, N.; Sonawane, S.H.; Gumfekar, S.P.; Kola, A.K.; Borse, P.; Ambade, S.B.; Guptha, S.; Ashokkumar, M. Electrochemical Performance of Starch-Polyaniline Nanocomposites Synthesized by Sonochemical Process Intensification. *J. Renew. Mater.* **2019**, *7*, 1279–1293. [CrossRef]
45. Janaki, V.; Vijayaraghavan, K.; Oh, B.-T.; Lee, K.-J.; Muthuchelian, K.; Ramasamy, A.; Kamala-Kannan, S. Starch/polyaniline nanocomposite for enhanced removal of reactive dyes from synthetic effluent. *Carbohydr. Polym.* **2012**, *90*, 1437–1444. [CrossRef]
46. Gautam, V.; Srivastava, A.; Singh, K.P.; Yadav, V.L. Preparation and characterization of polyaniline, multiwall carbon nanotubes, and starch bionanocomposite material for potential bioanalytical applications. *Polym. Compos.* **2017**, *38*, 496–506. [CrossRef]
47. Nazarzadeh, Z.E.; Najafi, M.P.; Azariyan, E.; Sharifian, I. Conductive and biodegradable polyaniline/starch blends and their composites with polystyrene. *Iran. Polym. J.* **2011**, *20*, 319–328.
48. Evers, F.; Weigend, F.; Koento, M. Coherent transport through a molecular wire: DFT calculation. *Phys. E Low-Dimens. Syst. Nanostruct.* **2003**, *18*, 255–257. [CrossRef]
49. Elsayed, A.; Elsayed, E.; Raya, A.-D.; Mahmoud, S.; Elshaer, A.; Kaialy, W. Thermal energy storage using metal–organic framework materials. *Appl. Energy* **2017**, *186*, 509–519. [CrossRef]
50. Lu, L.; Hu, H.; Hou, H.; Wang, B. An improved B3LYP method in the calculation of organic thermochemistry and reactivity. *Comput. Theor. Chem.* **2013**, *1015*, 64–71. [CrossRef]
51. Delley, B. An all-electron numerical method for solving the local density functional for polyatomic molecules. *J. Chem. Phys.* **1990**, *92*, 508–517. [CrossRef]
52. Kanouni, K.E.; Benguerba, Y.; Erto, A. Theoretical investigation of the solubility of some antiemetic drugs. *J. Mol. Liq.* **2019**, *282*, 626–632. [CrossRef]
53. Hellal, A.; Rachida, D.; Zaout, S.; Elkolli, M.; Chafaa, S.; Touafri, L.; Nadjib, C.; Mouna, M.; Khalissa, B. Structural, electronic, vibrational, optical and thermodynamic properties of 3-Oxo-3-p-tolylpropylphosphonic acid and 4-Oxo-4-p-tolyl-butyric acid: Density functional theory study. *J. Mol. Struct.* **2018**, *1171*, 527–540. [CrossRef]
54. Parr, R.; Szentpaly, L.; Liu, S. Electrophilicity Index. *J. Am. Chem. Soc.* **1999**, *121*, 1922–1924. [CrossRef]
55. Pearson, R.G. Absolute electronegativity and hardness: Application to inorganic chemistry. *Inorg. Chem.* **1988**, *27*, 734–740. [CrossRef]
56. Parr, R.G.; Pearson, R.G. Absolute hardness: Companion parameter to absolute electronegativity. *J. Am. Chem. Soc.* **1983**, *105*, 7512–7516. [CrossRef]
57. Domingo, L.R.; Arnó, M.; Contreras, R.; Pérez, P. Density functional theory study for the cycloaddition of 1, 3-butadienes with dimethyl acetylenedicarboxylate. Polar stepwise vs. concerted mechanisms. *J. Phys. Chem. A* **2002**, *106*, 952–961. [CrossRef]
58. Pérez, P.; Domingo, L.R.; Aizman, A.; Contreras, R. The electrophilicity index in organic chemistry. In *Theoretical and Computational Chemistry*; Elsevier: Amsterdam, The Netherlands, 2007; Volume 19, pp. 139–201.

59. Belakhdar, A.; Ferkous, H.; Djellali, S.; Sahraoui, R.; Lahbib, H.; Ben Amor, Y.; Erto, A.; Balsamo, M.; Benguerba, Y. Computational and experimental studies on the efficiency of Rosmarinus officinalis polyphenols as green corrosion inhibitors for XC48 steel in acidic medium. *Colloids Surf. A Physicochem. Eng. Asp.* **2020**, *606*, 125458. [CrossRef]
60. Klamt, A. Conductor-like screening model for real solvents: A new approach to the quantitative calculation of solvation phenomena. *J. Phys. Chem.* **1995**, *99*, 2224–2235. [CrossRef]
61. Miller, M.B.; Chen, D.-L.; Xie, H.-B.; Luebke, D.R.; Johnson, J.K.; Enick, R.M. Solubility of CO_2 in CO_2-philic oligomers; COSMOtherm predictions and experimental results. *Fluid Phase Equilibria* **2009**, *287*, 26–32. [CrossRef]
62. Macchi, P.; Sironi, A. Chemical bonding in transition metal carbonyl clusters: Complementary analysis of theoretical and experimental electron densities. *Coord. Chem. Rev.* **2003**, *238*, 383–412. [CrossRef]
63. Salehzadeh, S.; Gholiee, Y. A theoretical study on the encapsulation of halide anions by hexaprotonated form of aliphatic azacryptand 1, 4, 8, 11, 14, 18, 23, 27-octaazabicyclo [9.9.9] nonacosane in both the gas phase and solution. *Comput. Theor. Chem.* **2015**, *1060*, 43–51. [CrossRef]
64. Grabowski, S.J.; Ugalde, J.M. Bond paths show preferable interactions: Ab initio and QTAIM studies on the X−H$\cdots\pi$ hydrogen bond. *J. Phys. Chem. A* **2010**, *114*, 7223–7229. [CrossRef] [PubMed]
65. Rozas, I.; Alkorta, I.; Elguero, J. Behavior of ylides containing N, O, and C atoms as hydrogen bond acceptors. *J. Am. Chem. Soc.* **2000**, *122*, 11154–11161. [CrossRef]
66. Te Velde, G.T.; Bickelhaupt, F.M.; Baerends, E.J.; Fonseca Guerra, C.; van Gisbergen, S.J.; Snijders, J.G.; Ziegler, T. Chemistry with ADF. *J. Comput. Chem.* **2001**, *22*, 931–967. [CrossRef]
67. Bououden, W.; Benguerba, Y.; Darwish, A.S.; Attoui, A.; Lemaoui, T.; Balsamo, M.; Erto, A.; Alnashef, I.M. Surface adsorption of Crizotinib on carbon and boron nitride nanotubes as Anti-Cancer drug Carriers: COSMO-RS and DFT molecular insights. *J. Mol. Liq.* **2021**, *338*, 116666. [CrossRef]
68. Beck, A.D. Density-functional thermochemistry. III. The role of exact exchange. *J. Chem. Phys.* **1993**, *98*, 5648. [CrossRef]
69. Sulimenko, T.; Stejskal, J.; Křivka, I.; Prokeš, J. Conductivity of colloidal polyaniline dispersions. *Eur. Polym. J.* **2001**, *37*, 219–226. [CrossRef]
70. Do Nascimento, G.M.; Silva, C.H.; Temperini, M.L. Spectroscopic characterization of the structural changes of polyaniline nanofibers after heating. *Polym. Degrad. Stab.* **2008**, *93*, 291–297. [CrossRef]
71. Furukawa, Y.; Ueda, F.; Hyodo, Y.; Harada, I.; Nakajima, T.; Kawagoe, T.M. Vibrational spectra and structure of polyaniline. *Macromolecules* **1988**, *21*, 1297–1305. [CrossRef]
72. Skotheim, T.A.; Reynolds, J. *Conjugated Polymers: Theory, Synthesis, Properties, and Characterization*; CRC Press: Boca Raton, FL, USA, 2006.
73. Kang, E.; Neoh, K.; Tan, K. Polyaniline: A polymer with many interesting intrinsic redox states. *Prog. Polym. Sci.* **1998**, *23*, 277–324. [CrossRef]
74. Goswami, S.; Nandy, S.; Calmeiro, T.R.; Igreja, R.; Martins, R.; Fortunato, E. Stress induced mechano-electrical writing-reading of polymer film powered by contact electrification mechanism. *Sci. Rep.* **2016**, *6*, 19514. [CrossRef] [PubMed]
75. Li, D.; Jiang, Y.; Wu, Z.; Chen, X.; Li, Y. Fabrication of self-assembled polyaniline films by doping-induced deposition. *Thin Solid Films* **2000**, *360*, 24–27. [CrossRef]
76. Liu, H.; Chaudhary, D.; Yusa, S.; Tadé, M.O. Glycerol/starch/Na+-montmorillonite nanocomposites: A XRD, FTIR, DSC and 1H NMR study. *Carbohydr. Polym.* **2011**, *83*, 1591–1597. [CrossRef]
77. Gautam, V.; Srivastava, A.; Singh, K.P.; Yadav, V.L. Vibrational and gravimetric analysis of polyaniline/polysaccharide composite materials. *Polym. Sci. Ser. A* **2016**, *58*, 206–219. [CrossRef]
78. Lindfors, T.; Ivaska, A. Raman based pH measurements with polyaniline. *J. Electroanal. Chem.* **2005**, *580*, 320–329. [CrossRef]
79. Santha, N.; Sudha, K.; Vijayakumari, K.; Nayar, V.; Moorthy, A.S. Raman and infrared spectra of starch samples of sweet potato and cassava. *J. Chem. Sci.* **1990**, *102*, 705–712. [CrossRef]
80. Sovány, T.; Nikowitz, K.; Regdon, G., Jr.; Kása, P., Jr.; Pintye-Hódi, K. Raman spectroscopic investigation of film thickness. *Polym. Test.* **2009**, *28*, 770–772. [CrossRef]
81. Scharf, T.; Singer, I. Thickness of diamond-like carbon coatings quantified with Raman spectroscopy. *Thin Solid Films* **2003**, *440*, 138–144. [CrossRef]
82. Tegge, G. *Starch: Properties and Potential*; Galliard, T., Ed.; Volume 13 of Critical Reports on Applied Chemistry; John Wiley & Sons: Chichester, UK, 1987.
83. Dome, K.; Podgorbunskikh, E.; Bychkov, A.; Lomovsky, O. Changes in the crystallinity degree of starch having different types of crystal structure after mechanical pretreatment. *Polymers* **2020**, *12*, 641. [CrossRef]
84. Łużny, W.; Śniechowski, M.; Laska, J. Structural properties of emeraldine base and the role of water contents: X-ray diffraction and computer modelling study. *Synth. Met.* **2002**, *126*, 27–35. [CrossRef]
85. Doumeng, M.; Makhlouf, L.; Berthet, F.; Marsan, O.; Delbé, K.; Denape, J.; Chabert, F. A comparative study of the crystallinity of polyetheretherketone by using density, DSC, XRD, and Raman spectroscopy techniques. *Polym. Test.* **2021**, *93*, 106878. [CrossRef]
86. Lee, D.; Char, K. Thermal degradation behavior of polyaniline in polyaniline/Na^+-montmorillonite nanocomposites. *Polym. Degrad. Stab.* **2002**, *75*, 555–560. [CrossRef]
87. Wei, Y.; Jang, G.-W.; Hsueh, K.F.; Scherr, E.M.; MacDiarmid, A.G.; Epstein, A.J. Thermal transitions and mechanical properties of films of chemically prepared polyaniline. *Polymer* **1992**, *33*, 314–322. [CrossRef]

88. Ding, L.; Wang, X.; Gregory, R. Thermal properties of chemically synthesized polyaniline (EB) powder. *Synth. Met.* **1999**, *104*, 73–78. [CrossRef]
89. Hejna, A.; Lenża, J.; Formela, K.; Korol, J. Studies on the combined impact of starch source and multiple processing on selected properties of thermoplastic starch/ethylene-vinyl acetate blends. *J. Polym. Environ.* **2019**, *27*, 1112–1126. [CrossRef]
90. Htut, K.Z.; Kim, M.; Lee, E.; Lee, G.; Baeck, S.H.; Shim, S.E. Biodegradable polymer-modified graphene/polyaniline electrodes for supercapacitors. *Synth. Met.* **2017**, *227*, 61–70. [CrossRef]
91. Aoki, A.; Umehara, R.; Banba, K. Conductive Polyaniline/Surfactant Ion Complex Langmuir—Blodgett Films. *Langmuir* **2009**, *25*, 1169–1174. [CrossRef] [PubMed]
92. Aydinli, A.; Yuksel, R.; Unalan, H.E. Vertically aligned carbon nanotube–Polyaniline nanocomposite supercapacitor electrodes. *Int. J. Hydrogen Energy* **2018**, *43*, 18617–18625. [CrossRef]

polymers

MDPI

Article

Crystallization Behavior and Electrical Properties of Nanoparticle-Reinforced Poly(lactic Acid)-Based Films

Mei-Xian Li [1], Yu Ren [1,2], Dasom Lee [3] and Sung-Woong Choi [4,*]

1. School of Textile and Clothing, Nantong University, Nantong 226019, China; lmx321@ntu.edu.cn (M.-X.L.); ren.y@ntu.edu.cn (Y.R.)
2. Xinfengming Group Huzhou Zhongshi Technology Co., Ltd., Huzhou 313000, China
3. School of Mechanical and Aerospace Engineering, Seoul National University, Seoul 08826, Korea; dslee07@snu.ac.kr
4. Department of Mechanical System Engineering, Gyeongsang National University, Tongyeong-si 53064, Korea

* Correspondence: younhulje@gmail.com; Tel.: +82-55-772-9103; Fax: +82-50-4099-8373

Abstract: Graphene oxide (GO) and multiwalled carbon nanotubes with silver particles (MWNT-Ag) of different concentrations were used as nanofillers to prepare poly(lactic acid) (PLA) nanoparticle films through the solvent casting method. In this study, the effects of nanoparticles on the crystallization behavior, relationships between the dispersion and electrical properties, and hydrolytic degradation behaviors were investigated for the PLA/MWNT-Ag and PLA/rGO films. Differential scanning calorimetry was used to evaluate the crystallization behaviors of the PLA/MWNT-Ag and PLA/reduced GO (rGO) films. Electron probe microanalysis was performed to characterize the dispersion of MWNT-Ag, and X-ray diffraction and Raman spectroscopy were used to determine the degree of dispersion of rGO in the PLA matrix. The results showed that nanoparticles enhanced the crystallization kinetics of PLA as well as the hydrolytic degradation rate. From the measurement of electrical properties, the electrical conductivity of PLA/MWNT-Ag 1.0 wt% was much higher than that of the pure PLA and PLA/rGO films, showing that MANT and Ag nanoparticles contribute greatly to enhancing the electrical conductivity of the PLA/MWNT-Ag films.

Keywords: poly(lactic acid); MWNT-Ag; GO; dispersion; electrical conductivity

Citation: Li, M.-X.; Ren, Y.; Lee, D.; Choi, S.-W. Crystallization Behavior and Electrical Properties of Nanoparticle-Reinforced Poly(lactic Acid)-Based Films. *Polymers* **2022**, *14*, 177. https://doi.org/10.3390/polym14010177

Academic Editor: Evgenia G. Korzhikova-Vlakh

Received: 5 December 2021
Accepted: 29 December 2021
Published: 2 January 2022

1. Introduction

Poly(lactic acid) (PLA), as one of the environment-friendly biodegradable polymers, has been in the spotlight owing to its excellent mechanical properties [1–4], and because its hydrolysates are not harmful to the human body or to the environment [5–7]. PLA has been used in several industries, such as food packing [8,9], biomedicine [7,10–12], and electronics [13,14]. With properties being an issue for PLA, many research groups have made lots of effort to improve the mechanical and electrical conductivity of PLA-based composites by adding various carbon nanofillers, such as multiwalled carbon nanotubes (MWNT), carbon black, graphite, and graphene [1,15–19].

Among these carbon nanofillers, MWNT and graphene are the most popular owing to their superior electrical properties as well as their high aspect ratio. Qian et al. [20] designed PLA-based 3D composites with graphene oxide (GO) nanosheets which were aligned in the filament axial direction. The mechanical properties of these aligned GO nanosheets were outstanding when compared to that of nanosheets with random GO distributions. Dil et al. [21] studied the electrical conductivity of PLA/poly(butylene adipate-co-terephthalate)/MWNT, and their results showed that the MWNTs performed interface bridging, increasing the electrical conductivity of the samples. Guo et al. [22] proposed that a combination of reduced GO (rGO) and graphite could effectively improve the electrical and thermal properties.

Although the addition of nanofillers can enhance various properties of the polymer matrix because of their good mechanical, thermal, and electrical properties, it is difficult to disperse nanofillers in polymer matrixes because of their high aspect ratios. Therefore, dispersion of nanofillers is a crucial parameter in improving the mechanical, thermal, and electrical properties. Regarding the dispersion method for the surface treatment of nanofillers, lots of studies were examined to deal with the dispersion problem using chemical, thermal, and plasma methods. Rong et al. [23] investigated the chemical synthesis of silver nanoparticles by means of microemulsion techniques. Conductive networks can be established by freezing the nanoparticle aggregation. Ma et al. [24] studied the synthesis of MWCNTs using chemical vapor deposition (CVD). They found significant enhancements in dispersion quality and alignment stability for oxidized MWCNTs as compared to pristine MWCNTs. Stankovich et al. [25] examined the reduction of a colloidal suspension of exfoliated graphene oxide sheets in water with hydrazine hydrate. Tiwari et al. [26] has attempted to synthesize metal oxide nanoparticles in the presence of organic reagents, where the hydrothermally synthesized metal oxide nanoparticles can be covered with a hydroxyl group or alkyl chains, which made the surface of the nanoparticles hydrophilic or hydrophobic according to the acquired applications. Allen et al. [27] have undertaken a study on a comparison of the thermal and photochemical behaviors of a selection of nano-versus micron-sized (pigmentary) anatase and rutile titania pigments in polyethylene film and PVC-alkyd, acrylic, and isocyanate-paint films.

To improve the dispersibility of nanofillers in solvents and polymers, various methods have been used to modify their surface chemistry using plasma treatment [28] and plasma oxidation [29] to disentangle the nanofibers. Shi et al. [30] demonstrated the uniform deposition of ultrathin polymer films of 2 nm on the surfaces of Al_2O_3 nanoparticles by a plasma treatment. Yu et al. [31] employed plasma treatment method on diamond nanoparticles to improve dispersion in water. They found that the plasma treatment significantly reduced the water-contact angle of diamond nanoparticles and rendered the nanoparticles with a strong water affinity for dispersion enhancement in polar media such as water. Lee et al. [32] has observed plasma treatment of multiwalled carbon nanotubes (MWCNTs) using an atmospheric pressure hydroxyl radical (OH) source with micro-Raman spectroscopy. They found that the dominant effect of OH plasma on MWCNTs was reduced π-conjugated states due to the creation of structural defects and the attachment of oxygen-containing functional groups. Although several studies on the surface treatment of nanofillers using representative chemical, thermal, or plasma methods have been performed [33,34], the data remain insufficient. Moreover, most of the studies with polymer materials focused only on the dispersion phenomenon and results, not on the investigation of the relationship with crystallization behavior.

The objective of our study was to observe and analyze the overall crystallization behavior of PLA/MWNT-Ag and PLA/rGO films with respect to the dispersion behavior of MWNT-Ag and rGO. In this study, the effects of nanoparticles on the crystallization behavior, dispersion, electrical properties, and hydrolytic degradation behavior are investigated for the MWNT-Ag and GO. The crystallization behavior of PLA/MWNT-Ag and PLA/rGO films is evaluated by differential scanning calorimetry (DSC). Electron probe microanalysis (EPMA) is used to characterize the dispersion of MWNT-Ag. The degree of dispersion of rGO in the PLA matrix is determined via X-ray diffraction (XRD) and Raman spectroscopy. Lastly, the electrical conductivity and hydrolytic degradation behavior of PLA/MWNT-Ag and PLA/GO was measured with different concentrations of nanoparticles.

2. Experimental Section

2.1. Materials

PLA 4032D (NatureWorks, USA; $M_n \approx 155{,}715$, $M_w \approx 212{,}921$) was used as the matrix in this study. MWNT-Ag (Bioneer, Daejeon, Korea) and GO aqueous solution (Graphene Supermarket, Calverton, USA) were used as conducting nanofillers, and the details of the basic properties of these conducting nanofillers are summarized in Tables 1 and 2, respectively. Chloroform was used as a solvent to prepare the PLA/MWNT-Ag and PLA/GO films.

Table 1. Features of MWNT-Ag.

Property	MWNT-Ag Filler
Mass ratio	MWNT:Ag = 3:7
Average diameter of Ag (nm)	100
Diameter of MWNT (nm)	10–15
Length of MWNT (nm)	10–20

Table 2. Features of GO aqueous solution.

Property	MWNT-Ag Filler
Composition of carbon and oxygen (%)	Carbon (79), Oxygen (20)
Flake size (μm)	0.5–5
Thickness	1 atomic layer—at least 60%
Color	Brown
Concentration (g/L)	6.2 (Aqueous GO)
Single layer (%)	>80

2.2. Preparation of the PLA/MWNT-Ag Films

PLA was dissolved in chloroform using a magnetic stirrer with a stirring speed of 200 rpm at room temperature for 24 h, and MWNT-Ag was dispersed in chloroform through bath-type ultrasonication at room temperature for 3 h. During this process, the weight percentage of MWNT-Ag was calculated from that of MWNT (0.5 wt%, 1.0 wt%, and 1.5 wt%) according to the ratio of MWNT and Ag. After that, the MWNT-Ag/chloroform solution was added to the PLA/chloroform solution by mixing with a magnetic stirrer for 3 h to disperse the MWNT-Ag in the PLA solution. The PLA/MWNT-Ag films were created by spreading the solution on a glass Petri dish. Films were exposed to ambient conditions for 24 h, followed by evaporation under vacuum at 60 °C for 24 h (Figure 1a). Finally, the PLA/MWNT-Ag film samples with a size of 0.1 m × 0.1 m size were prepared (Figure 1b).

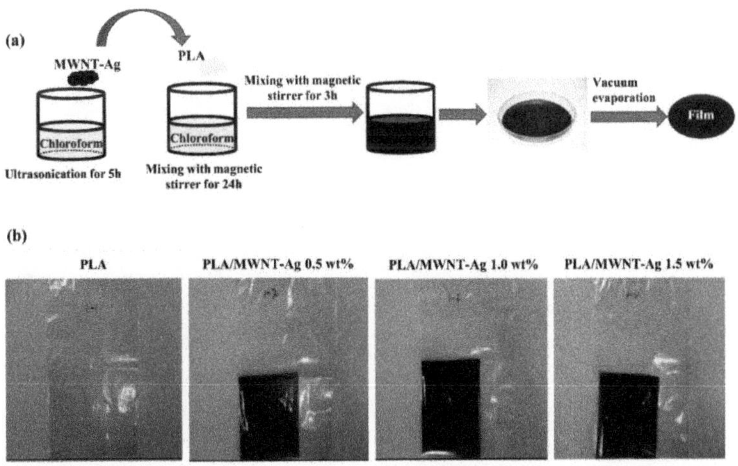

Figure 1. PLA/MWNT-Ag films prepared by solvent casting: (**a**) schematic illustration of the sample preparation; (**b**) PLA/MWNT-Ag film samples with different MWNT-Ag concentrations.

2.3. Preparation of the PLA/rGO Films

PLA was dissolved in chloroform using a magnetic stirrer with a stirring speed of 200 rpm at room temperature for 24 h, followed by adding GO into the PLA/chloroform solution by mixing with a magnetic stirrer at 60 °C for 24 h to disperse the GO in the PLA solution. In this step, the color of the GO solution was changed gradually from brown to dark brown according to the GO weight percentage and reduction of GO (Figure 2b,c). The PLA/GO solution was casted on a glass Petri dish, and the solvent was evaporated under vacuum at 60 °C for 24 h (Figure 2a). Finally, PLA/rGO film samples of 0.1 m × 0.1 m size were prepared (Figure 2d).

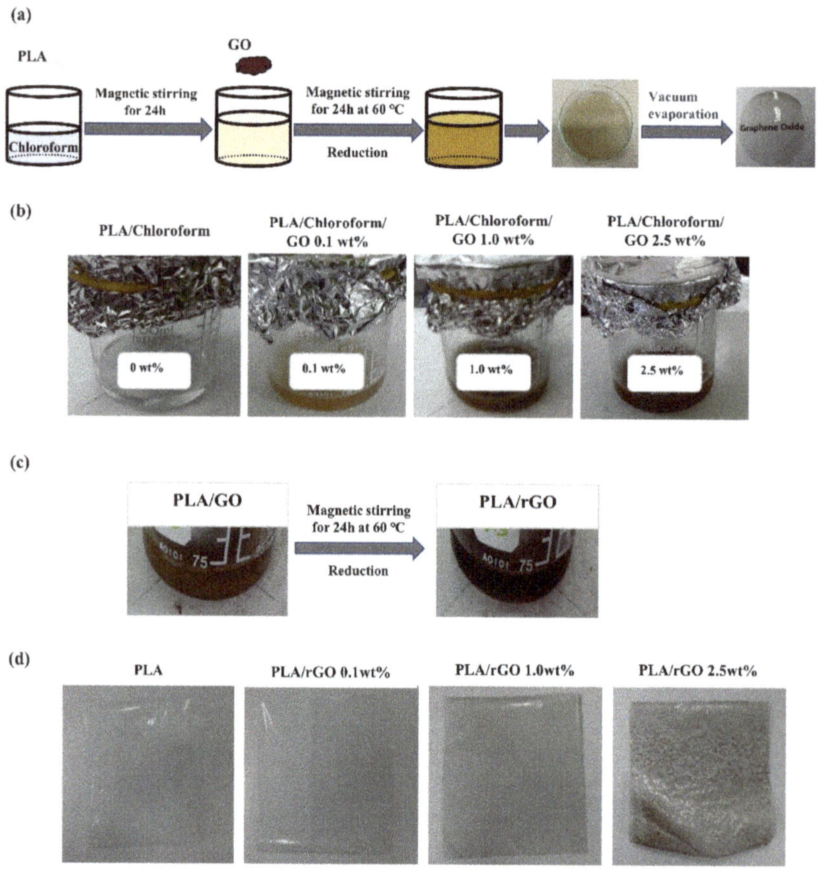

Figure 2. PLA/rGO films prepared by solvent casting: (**a**) schematic illustration of the sample preparation; (**b**) PLA/GO solutions with different GO concentrations; (**c**) color change of PLA/GO solution before and after reduction; and (**d**) PLA/rGO film samples with different rGO concentrations.

2.4. Characterizations

The visual comparison of the transparency for PLA, PLA/rGO, and PLA/MWNT-Ag films was performed. The thermal performance of the PLA-based films was evaluated using DSC (DSC-Q 1000, TA Instrument, New Castle, US) to evaluate the crystallization kinetics and crystallinity. Each film was cut into a small piece weighing approximately 10 mg, and the samples were heated from 30 to 200 °C at a rate of 20 °C/min. The samples were maintained at each temperature for 5 min to erase the previous thermal history. The samples were then cooled down to 110 °C and immediately subjected to isothermal scanning at 110 °C for 20 min. The samples were subsequently cooled to room temperature as fast as possible. Finally, they were heated again from room temperature to 200 °C at a heating rate of 10 °C/min.

EPMA (JXA-8900R, JEOL, Tokyo, Japan) was performed to examine the degree of dispersion in the PLA/MWNT-Ag films. Samples with a size of 5 mm × 5 mm × 0.03 mm were prepared according to our previous work [35], with a beam diameter of 300 μm × 300 μm. The weight ratio of MWNT and Ag, which are the components of the MWNT-Ag nanoparticles, is 3:7, and the degree of dispersion of MWNT can be estimated from that of the Ag element. In this study, both mapping and quantitative analyses were performed.

XRD analysis (D5005 X-ray diffractometry, Bruker Corporation, Germany) was performed under a voltage of 40 kV and current of 45 mA with Cu Ka radiation (1.5406 Å) with a scanning range of 5 to 30°. The Raman spectra of the samples were measured with a 532 nm wavelength laser (Confocal Micro-Raman Spectrometer, NRS-3100, Japan). The electrical conductivities of the PLA-based films were measured using high-resistance meters (Agilent, 4339B, Santa Clara, USA) with an applied voltage of 100 V (according to ASTM D257).

The degradation behaviors of the samples were evaluated according to our previously reported method [36]. The samples were immersed in a sodium hydroxide (NaOH) solution (PH = 13) at 37 °C for a predetermined period, and then rinsed with distilled water until the pH approached 7. Finally, the samples were dried in a vacuum oven at 70 °C for 48 h. The weight loss (W_{loss}) was estimated using the following equation:

$$W_{loss}(\%) = 100\% \times (W_0 - W_t)/W_0 \quad (1)$$

where W_0 is the initial weight of the specimen, W_t is the weight of the specimen dried in a vacuum oven at 70 °C for 48 h, and W_{loss} is the weight loss of the specimen.

3. Results and Discussion

3.1. Transparency

All PLA/MWNT-Ag films were opaque with a low-weight percentage of MWNT-Ag (Figure 1b), whereas, the PLA/rGO films were transparent at lower rGO concentrations (Figure 2d). However, the PLA/rGO films became slightly opaque when the concentration of rGO was higher than 2.5 wt%, owing to the uneven dispersion of the rGO particles. As shown in Figure 3, a highly transparent sample of the PLA/rGO 1.0 wt% film was observed, similar to the PLA film.

Figure 3. Transparency of PLA, PLA/rGO 1.0 wt%, and PLA/MWNT-Ag 1.0 wt% films.

3.2. DSC Characterization

The crystallization kinetics of the PLA-based films were evaluated. Figure 4a shows the overall heating–cooling–heating cycles for the crystallization behaviors of the PLA, PLA/MWNT-Ag, and PLA/rGO films. The isothermal scanning results of the PLA-based films are presented in Figure 4b, and it can be observed that it takes more than 15 min to fully crystallize pure PLA, while the crystallization time is reduced by almost one third for the PLA/MWNT-Ag and PLA/rGO films. The crystallization rate increases with the addition of nanoparticles, which can react as nucleation agents. The Avrami equation was used to evaluate the isothermal crystallization mechanisms of the PLA, PLA/MWNT-Ag, and PLA/rGO films. The relative crystallinity could be calculated from Equation (2):

$$X_t = \frac{\int_0^t (\mathrm{d}H/\mathrm{d}t)\mathrm{d}t}{\int_0^\infty (\mathrm{d}H/\mathrm{d}t)\mathrm{d}t} \quad (2)$$

where $\mathrm{d}H/\mathrm{d}t$ is the respective heat flow; the sum of $\mathrm{d}H/\mathrm{d}t$ from 0 to t is the enthalpy at time t; and the sum of $\mathrm{d}H/\mathrm{d}t$ from 0 to t_∞ is the enthalpy at time t_∞, which can be obtained as the total area under the crystallization DSC curve. Generally, the Avrami equation can be converted to the following linear form:

$$\log[-\ln(1 - X_t)] = \log K + n \log t \quad (3)$$

where n is the Avrami exponent that reflects the nucleation mechanism and dimension of crystal growth [37,38]. The temperature-dependent parameter K is the kinetic rate constant. By plotting the Avrami equation, log [ln(1−X_t)] versus log t, the values of n and K can be obtained from the intercept and slope, respectively (Figure 4c).

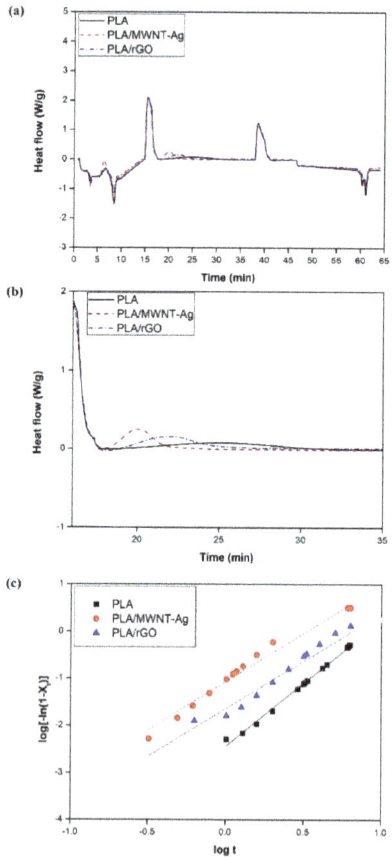

Figure 4. DSC thermograms for samples with different types of nanoparticles: (**a**) results of heating–cooling–heating cycles for the crystallization behavior of PLA-based films; (**b**) isothermal DSC thermograms at 110 °C; and (**c**) plots of log [ln(1−X_t)] versus log t for the PLA-based films at 110 °C.

The crystallization half-time ($t_{1/2}$), which is the time required to reach 50% crystallinity and indicates the crystallization rate, can be defined by Equation (4).

$$t_{1/2} = (\ln 2/K)^{1/n} \tag{4}$$

The thermal characteristic parameters such as the crystallization temperature (T_c), glass transition temperature (T_g), melting temperature (T_m), n, K, and $t_{1/2}$, are listed in Table 3.

Table 3. Thermal characteristic parameters of PLA-based films, obtained from DSC.

	T_c (°C)	T_g (°C)	T_m (°C)	n	K (min−n)	$t_{1/2}$ (min)
PLA	110	64	171	2.18	0.007	8.28
PLA/MWNT-Ag	110	64	171	2.19	0.105	2.37
PLA/rGO	110	64	171	2.07	0.023	5.13

For the degree of dispersion, XRD and Raman spectroscopy were used to evaluate the dispersion degree for the PLA/rGO, and EPMA was used for the PLA/MWNT-Ag with different concentration of 0.5–1.5 wt.%. Among them, a concentration of 1.0 wt.% was found to be well dispersed. Therefore, among the samples with different nanoparticle concentrations, PLA, PLA/MWNT-Ag 1.0 wt%, and PLA/rGO 1.0 wt% samples were prepared. The samples were dispersed to evaluate their crystallization kinetics from a DSC thermogram. The results show that the glass transition temperature and melting temperature of the PLA-based films remained almost same, without being affected by the nanoparticles, while the crystallization rate increased significantly with the addition of nanoparticles that acted as nucleating agents. Meanwhile, the crystallization rate of PLA/MWNT-Ag was faster than that of PLA/rGO. The reason could be that both MWNT and Ag nanoparticles acted as nucleating agents and affected the crystal formation and growth. Avrami exponent values of n for pure PLA, PLA/MWNT-Ag, and PLA/rGO were observed in the vicinity of 2, indicating that the crystallization mechanism was two-dimensional growth. Therefore, the addition of MWNT-Ag or rGO into the PLA matrix did not affect the geometric dimensions of PLA crystal growth during isothermal crystallization.

With the increasing of the concentration of the nanoparticles, the isothermal crystallization rate was found to be increased. However, after some point, the tendency to increase was reduced. This is because too many nanoparticles could hinder crystal growth. In sum, the crystallization rate was increased in early time with increasing concentrations of the nanoparticles. After some point, the crystallization rate was shown to decrease.

3.3. *Dispersion of Nanoparticles in PLA*

Both EPMA mapping and quantitative analysis were performed to evaluate the degree of dispersion of the MWNT-Ag nanoparticles in the PLA matrix. The degree of dispersion of MWNT could be estimated from that of the Ag elements, because of the grafting of Ag on MWNT.

Figure 5 shows the EPMA mapping images of PLA/MWNT-Ag films with different concentrations of MWNT-Ag nanoparticles. Ag is observed to be relatively well dispersed in PLA/MWNT-Ag films with 0.5 wt% and 1.0 wt% MWNT-Ag, especially. On the other hand, there are some distinct aggregations in the PLA/MWNT-Ag films with 1.5 wt% MWNT-Ag.

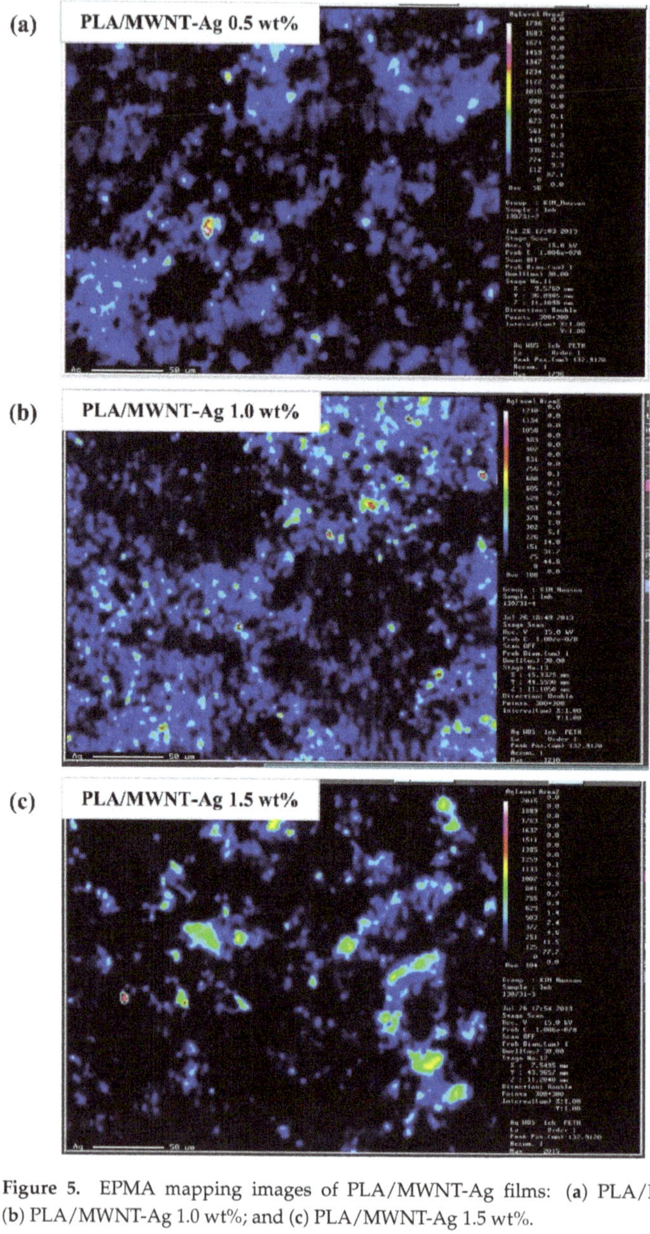

Figure 5. EPMA mapping images of PLA/MWNT-Ag films: (**a**) PLA/MWNT-Ag 0.5 wt%; (**b**) PLA/MWNT-Ag 1.0 wt%; and (**c**) PLA/MWNT-Ag 1.5 wt%.

Quantitative analysis was conducted with the same samples in order to measure the degree of dispersion more precisely. In Figure 6, the black dot indicates the weight percentage of Ag added to the PLA matrix, while the red dot represents that of Ag measured by the EPMA equipment. The measured weight percentage of Ag is slightly lower than the added Ag weight percentage. The reason may be that there still exist some aggregation parts of MWNT-Ag after 24 h sonication. Additionally, the relatively wide range of standard deviations is due to more nanoparticle aggregation.

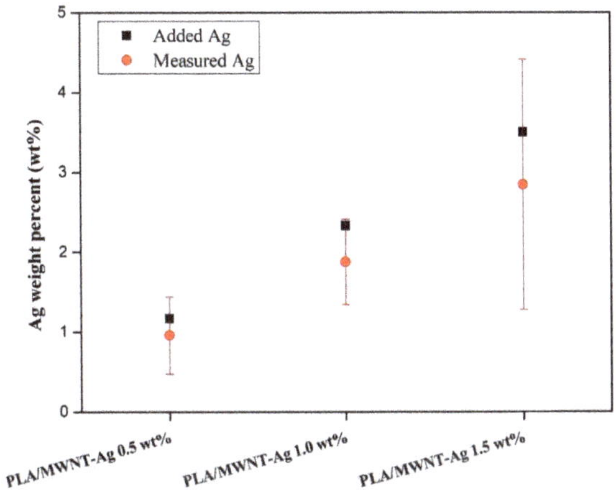

Figure 6. EPMA quantitative analysis of PLA/MWNT-Ag films.

For the PLA/rGO samples, it is difficult to evaluate the degree of dispersion of rGO using EPMA because both PLA and rGO are composed of only C, H, and O elements. Therefore, to roughly estimate the degree of dispersion of rGO in the PLA matrix, measurements using XRD and Raman spectroscopy were conducted. PLA with 1.0 wt% rGO film did not show the characteristic peak of rGO, indicating that rGO was relatively well dispersed in the PLA matrix.

The XRD patterns of pure rGO, pure PLA, PLA/rGO 0.5 wt%, and PLA/rGO 1.0 wt% are shown in Figure 7. Pure PLA exhibits two sharp peaks at $2\theta = 16.58°$ and 18.95°, which reflects crystal planes (200)/(110) and (203) indicating α- and α′- crystal formation. These results are in accordance with the results of a study by Kalani and Yunus [39]. As can be seen in Figure 7, there is a strong peak around 10.63°, which corresponds to the rGO structure. However, the peak associated with rGO disappeared for the PLA/rGO 0.5 wt% and PLA/rGO 1.0 wt% samples. It could be inferred that rGO was evenly dispersed in the PLA matrix. Meanwhile, the peaks around 16.58° and 18.95° for the PLA/rGO samples were more prominent than those of pure PLA, indicating more crystallinity owing to the rGO acting as a nucleating agent.

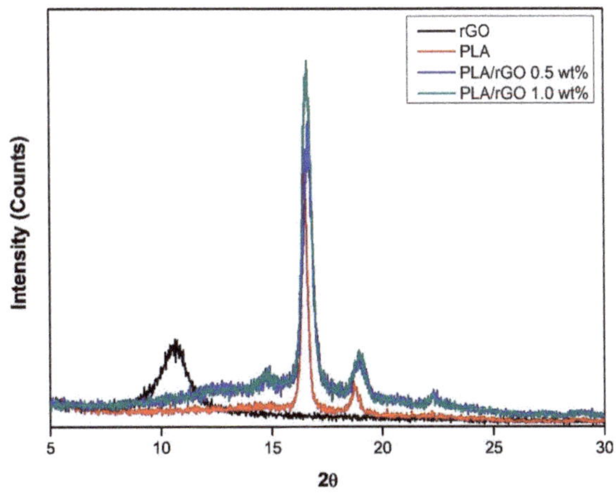

Figure 7. XRD patterns of rGO, pure PLA, PLA/rGO 0.5wt%, and PLA/rGO 1.0 wt%.

The Raman spectra of PLA/rGO 0.5 wt% and PLA/rGO 1.0 wt% are shown in Figure 8. The intensity of the D band (around 1400 cm^{-1}) and G band (around 1600 cm^{-1}) corresponded to the vibrations of the sp^3-bonded carbon atoms and sp^2-bonded carbon atoms, respectively. As already known, the D band reflects disordered structures (such as defects, crystal boundaries, and so on), while the G band reflects the crystalline structures [40]. The intensity ratios of the D band and G band (I_D/I_G) for PLA/rGO 0.5 wt% and PLA/rGO 1.0 wt% were 0.93 and 0.99, respectively, which were significantly higher than that of pure graphite [41], because GO was successfully exfoliated after reduction. In addition, there was no 2D band between 2600 cm^{-1} and 2800 cm^{-1}, indicating no stacked graphene layers.

3.4. Degradation Properties

The weight losses of pure PLA, PLA/MWNT-Ag, and PLA/rGO with degradation time are shown in Figure 10. The weight losses of all samples exceeded 50% within 10 days, and then, the degradation rate was decreased. The degradation rates of PLA/MWNT-Ag and PLA/rGO were higher than that of pure PLA because of the interface between the particles and the PLA matrix, which served as a leading path for hydrolytic degradation. In this study, the same weight percentage of MWNT and rGO were used in the PLA matrix. However, the Ag nanoparticles were grafted on MWNT; thus, the total weight percentage of the nanoparticles for the PLA/MWNT-Ag was higher than that of PLA/rGO. It means that the surface areas of the interfaces of PLA/MWNT-Ag were much larger than that of PLA/rGO. Therefore, the degradation rate of PLA/MWNT-Ag was higher than that of PLA/rGO. Another perspective of the above phenomenon was that increasing the crystallinity of the samples was attributed to making it more difficult to degrade parts than the amorphous parts for the PLA/rGO films prepared at 60 °C. Therefore, the degradation rate of PLA/rGO decreases compared to that of PLA/MWNT-Ag. These observed tendencies were identical to those of our previous study [42].

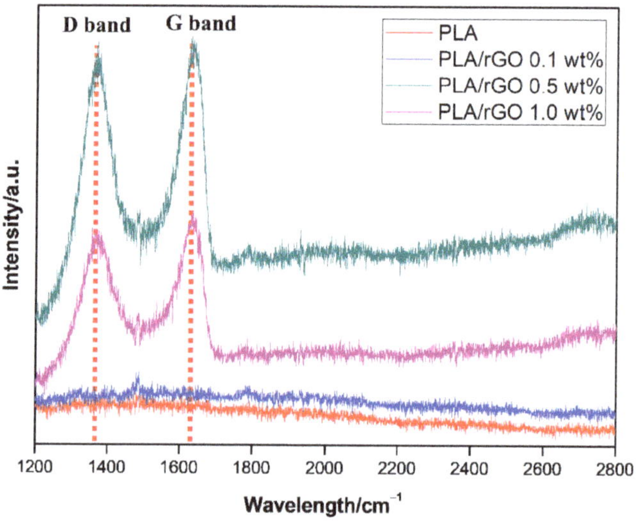

Figure 8. Raman spectra of PLA/rGO 0.5 wt% and PLA/rGO 1.0 wt%.

3.5. Electrical Properties

The electrical properties were analyzed for different concentrations of MWNT-Ag and/or rGO (Figure 9). The results showed that the electrical conductivity of PLA/MWNT-Ag and PLA/GO increased as the concentration of nanoparticles increased. PLA films with more than 1.0 wt% nanoparticles showed significantly higher electrical conductivities than pure PLA and PLA films with less than 0.5wt% nanoparticles. In addition, the electrical conductivity of PLA/MWNT-Ag 1.0wt%, which exhibited the best dispersion, was shown to be remarkably higher than that of pure PLA and other PLA films. Finally, PLA/MWNT-Ag films showed much higher electrical conductivities than PLA/rGO films for the same nanoparticle concentration. Ag nanoparticles also contributed greatly to enhancing the electrical conductivity of the samples; therefore, relatively higher electrical conductivities were observed with PLA/MWNT-Ag films.

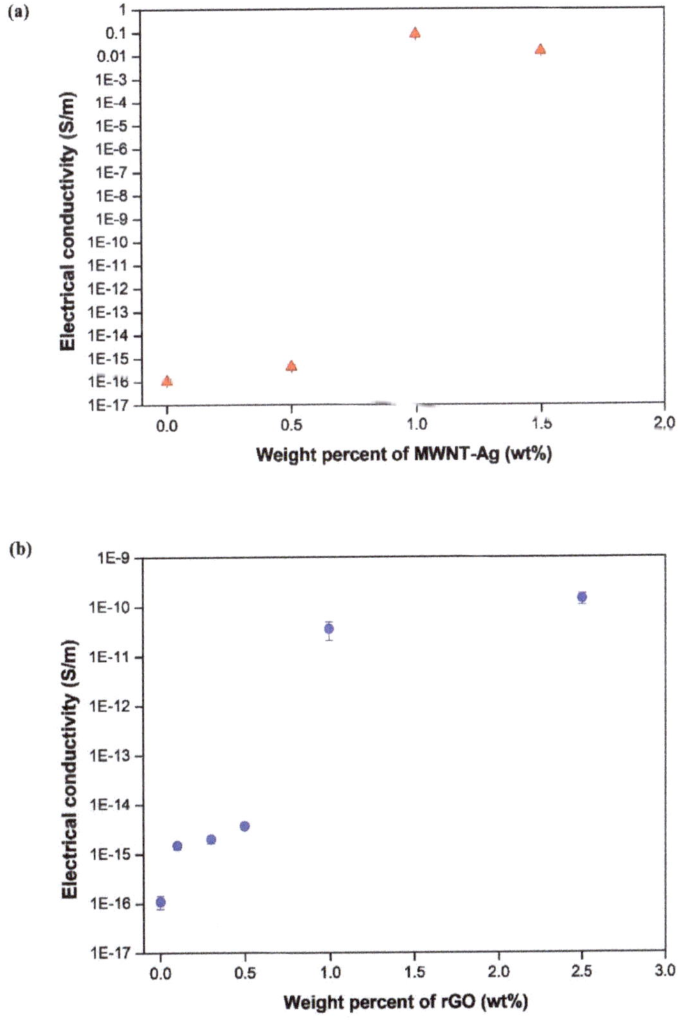

Figure 9. Electrical conductivity of: (**a**) PLA/MWNT-Ag; (**b**) PLA/rGO films.

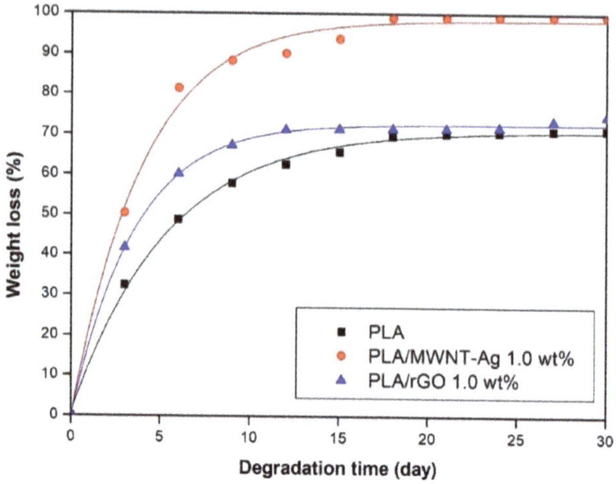

Figure 10. Degradation of PLA-based films.

Overall, transparent and flexible PLA/rGO thin films were prepared via a solvent casting method which overcame the brittle samples prepared in previous works. PLA/MWNT-Ag thin films prepared in this work were flexible and had high electrical conductivity. In this study quantitative and qualitative assessment could be conducted at the same time due to the Ag nanoparticles grafted on the MWNT which could easily evaluate the distribution of nanoparticles visually comparted to other research.

4. Conclusions

In this study, PLA/MWNT-Ag and PLA/rGO films were prepared to investigate the effects of nanoparticles on crystallization behavior, dispersion, electrical conductivity, and hydrolytic degradation behavior. For the PLA-based film of PLA/MWNT-Ag and PLA/rGO, there was no significant difference between the glass transition temperature and the melting temperature, while the crystallization rate increased significantly with the addition of nanoparticles that acted as nucleating agents. The crystallization rate of PLA/MWNT-Ag was faster than that of PLA/rGO. The results showed that nanoparticles such as MWNT-Ag and rGO acted as nucleating agents to enhance the crystallization rate and crystallinity of the samples. From the XRD patterns of pure rGO, pure PLA, PLA/rGO 0.5 wt%, and PLA/rGO 1.0 wt%, PLA/rGO was shown to have more crystallinity, owing to the rGO acting as a nucleating agent. The electrical conductivity of PLA/MWNT-Ag and PLA/GO increased as the concentration of nanoparticles increased. The electrical properties of the PLA films were affected by the nanoparticles and their dispersion degrees. At the same nanoparticle concentration, the electrical conductivity of PLA/MWNT-Ag was much higher than that of PLA/rGO, because Ag nanoparticles also contribute greatly to enhancing the electrical conductivity of the samples. The degradation rates of PLA/MWNT-Ag and PLA/rGO were higher than that of pure PLA because of the interface between the particles and the PLA matrix, which served as a leading path for hydrolytic degradation. The degradation rate of PLA/MWNT-Ag was also higher than that of PLA/rGO because of the larger surface areas of the interface of PLA/MWNT-Ag, compared to that of PLA/rGO. As GO has good transparency, it can be utilized to make biodegradable polymer-based thin films with high transparency and high electrical conductivity. The results and method presented in this paper can trigger further research endeavors in the direction of fabricating transparent biodegradable films with high electrical conductivity.

Author Contributions: Data curation, M.-X.L.; Investigation, M.-X.L.; Methodology, Y.R.; Resources, Y.R.; Visualization, D.L.; Writing–original draft, M.-X.L.; Writing–review & editing, S.-W.C. All authors have read and agreed to the published version of the manuscript.

Funding: This work was supported in part by "Regional Innovation Strategy (RIS)" through the National Research Foundation of Korea (NRF) funded by the Ministry of Education (MOE)(2021RIS-003) and in part by financial supports from Nantong University (No. 135421615008) and the Large Instruments Open Foundation of Nantong University.

Institutional Review Board Statement: Not applicable.

Informed Consent Statement: Not applicable.

Data Availability Statement: The data presented in this study are available on request from the corresponding author.

Conflicts of Interest: The authors declare no conflict of interest.

References

1. Kim, M.; Jeong, J.H.; Lee, J.Y.; Capasso, A.; Bonaccorso, F.; Kang, S.H.; Lee, Y.K.; Lee, G.H. Electrically conducting and mechanically strong graphene-polylactic acid composites for 3D printing. *ACS Appl. Mater. Interfaces* **2019**, *11*, 11841–11848. [CrossRef] [PubMed]
2. Qi, F.; Wu, J.; Li, H.; Ma, G. Recent research and development of PLGA/PLA microspheres/nanoparticles: A review in scientific and industrial aspects. *Front. Chem. Sci. Eng.* **2019**, *13*, 14–27. [CrossRef]
3. Wu, B.; Geng, B.; Chen, Y.; Liu, H.; Li, G.; Wu, Q. Preparation and characteristics of TEMPO-oxidized cellulose nanofibrils from bamboo pulp and their oxygen-barrier application in PLA films. *Front. Chem. Sci. Eng.* **2017**, *11*, 554–563. [CrossRef]
4. Xiao, H.; Xue, S.; Zhang, J.; Zhao, M.; Ma, J.; Chen, S.; Zheng, Z.; Jia, J.; Wu, H. Facile electrolytic synthesis of Pt and carbon quantum dots coloaded multiwall carbon nanotube as highly efficient electrocatalyst for hydrogen evolution and ethanol oxidation. *Chem. Eng. J.* **2021**, *408*, 127271. [CrossRef]
5. Schädlich, A.; Caysa, H.; Mueller, T.; Tenambergen, F.; Rose, C.; Göpferich, A.; Kuntsche, J.; Mäder, K. Tumor Accumulation of NIR Fluorescent PEG–PLA Nanoparticles: Impact of Particle Size and Human Xenograft Tumor Model. *ACS Nano* **2011**, *5*, 8710–8720. [CrossRef] [PubMed]
6. Wang, M.; Favi, P.; Cheng, X.; Golshan, N.H.; Ziemer, K.S.; Keidar, M.; Webster, T.J. Cold atmospheric plasma (CAP) surface nanomodified 3D printed polylactic acid (PLA) scaffolds for bone regeneration. *Acta Biomater.* **2016**, *46*, 256–265. [CrossRef]
7. Xu, W.; Pranovich, A.; Uppstu, P.; Wang, X.; Kronlund, D.; Hemming, J.; Öblom, H.; Moritz, N.; Preis, M.; Sandler, N.; et al. Novel biorenewable composite of wood polysaccharide and polylactic acid for three dimensional printing. *Carbohydr. Polym.* **2018**, *187*, 51–58. [CrossRef]
8. Bill, M.; Pillai, S.K.; Tinyane, P.; Ray, S.S.; Sivakumar, D. The Effect of Thyme Oil Low-Density Polyethylene Impregnated Pellets in Polylactic Acid Sachets on Storage Quality of Ready-to-Eat Avocado. *Food Bioprocess Technol.* **2017**, *11*, 141–151. [CrossRef]
9. Zhou, X.; Cheng, R.; Wang, B.; Zeng, J.; Xu, J.; Li, J.; Kang, L.; Cheng, Z.; Gao, W.; Chen, K. Biodegradable sandwich-architectured films derived from pea starch and polylactic acid with enhanced shelf-life for fruit preservation. *Carbohydr. Polym.* **2021**, *251*, 117117. [CrossRef] [PubMed]
10. Fairag, R.; Rosenzweig, D.H.; Garcialuna, J.L.R.; Weber, M.H.; Haglund, L. Three-Dimensional Printed Polylactic Acid Scaffolds Promote Bone-like Matrix Deposition in Vitro. *ACS Appl. Mater. Interfaces* **2019**, *11*, 15306–15315. [CrossRef]
11. Fujimoto, K.L.; Yamawaki-Ogata, A.; Uto, K.; Usui, A.; Narita, Y.; Ebara, M. Long term efficacy and fate of a right ventricular outflow tract replacement using an elastomeric cardiac patch consisting of caprolactone and D,L-lactide copolymers. *Acta Biomater.* **2021**, *123*, 222–229. [CrossRef] [PubMed]
12. Ng, H.-M.; Bee, S.-T.; Sin, L.T.; Ratnam, C.T.; Rahmat, A. Effect of electron beam irradiation sterilization on biomedical polylactic acid composite filled with Scomberomorus Guttatus-derived hydroxyapatite. *Compos. Part B: Eng.* **2019**, *176*, 107273. [CrossRef]
13. Paula, K.T.; Gaal, G.; Almeida, G.; Andrade, M.; Facure, M.H.M.; Correa, D.; Riul, A.; Rodrigues, V.; Mendonca, C. Femtosecond laser micromachining of polylactic acid/graphene composites for designing interdigitated microelectrodes for sensor applications. *Opt. Laser Technol.* **2018**, *101*, 74–79. [CrossRef]
14. Wang, Y.; Fan, Z.-W.; Zhang, H.; Guo, J.; Yan, D.-X.; Wang, S.; Dai, K.; Li, Z.-M. 3D-printing of segregated carbon nanotube/polylactic acid composite with enhanced electromagnetic interference shielding and mechanical performance. *Mater. Des.* **2021**, *197*, 109222. [CrossRef]
15. Laredo, E.; Bello, A.; Diaz, J.; Grimau, M.; Martinez-Tong, D.; Wu, D.; Wu, L. Effect of cold-crystallization on the AC and DC conductive properties of polylactide biocomposites with carboxylic or neat large aspect ratio MWCNT. *Polym. Compos.* **2013**, *34*, 67–76. [CrossRef]
16. Shen, Y.; Jing, T.; Ren, W.; Zhang, J.; Jiang, Z.-G.; Yu, Z.-Z.; Dasari, A. Chemical and thermal reduction of graphene oxide and its electrically conductive polylactic acid nanocomposites. *Compos. Sci. Technol.* **2012**, *72*, 1430–1435. [CrossRef]

17. Sullivan, E.M.; Gerhardt, R.A.; Wang, B.; Kalaitzidou, K. Effect of compounding method and processing conditions on the electrical response of exfoliated graphite nanoplatelet/polylactic acid nanocomposite films. *J. Mater. Sci.* **2016**, *51*, 2980–2990. [CrossRef]
18. Tambe, P.B.; Bhattacharyya, A.R.; Kulkarni, A.R. The influence of melt-mixing process conditions on electrical conductivity of polypropylene/multiwall carbon nanotubes composites. *J. Appl. Polym. Sci.* **2013**, *127*, 1017–1026. [CrossRef]
19. Tirado-Garcia, I.; Garcia-Gonzalez, D.; Garzon-Hernandez, S.; Rusinek, A.; Robles, G.; Martinez-Tarifa, J.; Arias, A. Conductive 3D printed PLA composites: On the interplay of mechanical, electrical and thermal behaviours. *Compos. Struct.* **2021**, *265*, 113744. [CrossRef]
20. Qian, Y.; Li, C.; Qi, Y.; Zhong, J. 3D printing of graphene oxide composites with well controlled alignment. *Carbon* **2021**, *171*, 777–784. [CrossRef]
21. Dil, E.J.; Arjmand, M.; Navas, I.O.; Sundararaj, U.; Favis, B.D. Interface Bridging of Multiwalled Carbon Nanotubes in Polylactic Acid/Poly(butylene adipate-co-terephthalate): Morphology, Rheology, and Electrical Conductivity. *Macromolecules* **2020**, *53*, 10267–10277. [CrossRef]
22. Guo, R.; Ren, Z.; Bi, H.; Xu, M.; Cai, L. Electrical and Thermal Conductivity of Polylactic Acid (PLA)-Based Biocomposites by Incorporation of Nano-Graphite Fabricated with Fused Deposition Modeling. *Polymers* **2019**, *11*, 549. [CrossRef]
23. Rong, M.; Zhang, M.; Zeng, H. Synthesis of silver nanoparticles and their self-organization behavior in epoxy resin. *Polymer* **1999**, *40*, 6169–6178. [CrossRef]
24. Ma, C.; Zhang, W.; Zhu, Y.; Ji, L.; Zhang, R.; Koratkar, N.; Liang, J. Alignment and dispersion of functionalized carbon nanotubes in polymer composites induced by an electric field. *Carbon* **2008**, *46*, 706–710. [CrossRef]
25. Stankovich, S.; Dikin, D.A.; Piner, R.D.; Kohlhaas, K.A.; Kleinhammes, A.; Jia, Y.; Wu, Y.; Nguyen, S.; Ruoff, R.S. Synthesis of graphene-based nanosheets via chemical reduction of exfoliated graphite oxide. *Carbon* **2007**, *45*, 1558–1565. [CrossRef]
26. Tiwari, S.; Srivastava, K.; Gehlot, C.L.; Srivastava, D. Epoxy/fly ash from thermal power plant/nanofiller nanocomposite: Studies on mechanical and thermal properties: A review. *Int. J. Waste Resour.* **2020**, *10*, 1–16. [CrossRef]
27. Allen, N.S.; Edge, M.; Ortega, A.; Sandoval, G.; Liauw, C.M.; Verran, J.; Stratton, J.; McIntyre, R.B. Degradation and stabilisation of polymers and coatings: Nano versus pigmentary titania particles. *Polym. Degrad. Stab.* **2004**, *85*, 927–946. [CrossRef]
28. Li, D.; Müller, M.B.; Gilje, S.; Kaner, R.B.; Wallace, G.G. Processable aqueous dispersions of graphene nanosheets. *Nat. Nanotechnol.* **2008**, *3*, 101–105. [CrossRef]
29. Brandl, W.; Marginean, G.; Chirila, V.; Warschewski, W. Production and characterisation of vapour grown carbon fiber/polypropylene composites. *Carbon* **2004**, *42*, 5–9. [CrossRef]
30. Shi, D.; Wang, S.X.; Van Ooij, W.J.; Wang, L.M.; Zhao, J.; Yu, Z. Uniform deposition of ultrathin polymer films on the surfaces of Al2O3 nanoparticles by a plasma treatment. *Appl. Phys. Lett.* **2001**, *78*, 1243–1245. [CrossRef]
31. Yu, Q.; Kim, Y.J.; Ma, H. Plasma treatment of diamond nanoparticles for dispersion improvement in water. *Appl. Phys. Lett.* **2006**, *88*, 231503. [CrossRef]
32. Lee, S.; Peng, J.-W.; Liu, C.-H. Probing plasma-induced defect formation and oxidation in carbon nanotubes by Raman dispersion spectroscopy. *Carbon* **2009**, *47*, 3488–3497. [CrossRef]
33. Dang, Z.M.; Zheng, M.S.; Zha, J.W. 1D/2D carbon nanomaterial-polymer dielectric composites with high permittivity for power energy storage applications. *Small* **2016**, *12*, 1688–1701. [CrossRef] [PubMed]
34. Hu, C.; Li, Z.; Wang, Y.; Gao, J.; Dai, K.; Zheng, G.; Liu, C.; Shen, C.; Song, H.; Guo, Z. Comparative assessment of the strain-sensing behaviors of polylactic acid nanocomposites: Reduced graphene oxide or carbon nanotubes. *J. Mater. Chem. C* **2017**, *5*, 2318–2328. [CrossRef]
35. Kahnt, A.; Flyunt, R.; Naumov, S.; Knolle, W.; Eigler, S.; Hermann, R.; Abel, B. Shedding light on the soft and efficient free radical induced reduction of graphene oxide: Hidden mechanisms and energetics. *RSC Adv.* **2016**, *6*, 68835–68845. [CrossRef]
36. Li, M.-X.; Lee, D.; Choi, S. Evaluation of the nanoparticle dispersion and the thermal behaviors of polyamide-6/silica composites. *J. Korean Phys. Soc.* **2020**, *77*, 530–536. [CrossRef]
37. Li, M.-X.; Kim, S.-H.; Choi, S.-W.; Goda, K.; Lee, W.-I. Effect of reinforcing particles on hydrolytic degradation behavior of poly (lactic acid) composites. *Compos. Part B—Eng.* **2016**, *96*, 248–254. [CrossRef]
38. Avrami, M. Kinetics of phase change. I General theory. *J. Chem. Phys.* **1939**, *7*, 1103–1112. [CrossRef]
39. Avrami, M. Kinetics of phase change. II Transformation-time relations for random distribution of nuclei. *J. Chem. Phys.* **1940**, *8*, 212–224. [CrossRef]
40. Kalani, M.; Yunus, R. Effect of supercritical fluid density on nanoencapsulated drug particle size using the supercritical antisolvent method. *Int. J. Nanomed.* **2012**, *7*, 2165–2172. [CrossRef]
41. Botta, L.; Scaffaro, R.; Sutera, F.; Mistretta, M.C. Reprocessing of PLA/graphene nanoplatelets nanocomposites. *Polymers* **2018**, *10*, 18. [CrossRef] [PubMed]
42. Zhang, C.; Wang, L.; Zhai, T.; Wang, X.; Dan, Y.; Turng, L.S. The surface grafting of graphene oxide with poly(ethylene glycol) as a reinforcement for poly(lactic acid) nanocomposite scaffolds for potential tissue engineering applications. *J. Mech. Behav. Biomed. Mater.* **2016**, *53*, 403–413. [CrossRef] [PubMed]

polymers

MDPI

Article

Characterization of Chitosan Extracted from Fish Scales of the Colombian Endemic Species *Prochilodus magdalenae* as a Novel Source for Antibacterial Starch-Based Films

Carlos Molina-Ramírez [1,2,*], Paulina Mazo [3], Robin Zuluaga [3], Piedad Gañán [2] and Juan Álvarez-Caballero [1]

[1] Grupo Química y Bioprospección de Productos Naturales, Universidad del Magdalena, Santa Marta 470004, Colombia; jalvarez@unimagdalena.edu.co
[2] Grupo de Investigación Sobre Nuevos Materiales (GINUMA), Universidad Pontificia Bolivariana, Medellín 050004, Colombia; piedad.ganan@upb.edu.co
[3] Grupo de Investigaciones Agroindustriales (GRAIN), Universidad Pontificia Bolivariana, Medellín 050004, Colombia; paulina.mazo@upb.edu.co (P.M.); robin.zuluaga@upb.edu.co (R.Z.)
* Correspondence: carlosan.molina@upb.edu.co

Citation: Molina-Ramírez, C.; Mazo, P.; Zuluaga, R.; Gañán, P.; Álvarez-Caballero, J. Characterization of Chitosan Extracted from Fish Scales of the Colombian Endemic Species *Prochilodus magdalenae* as a Novel Source for Antibacterial Starch-Based Films. *Polymers* **2021**, *13*, 2079. https://doi.org/10.3390/polym13132079

Academic Editor: Evgenia G. Korzhikova-Vlakh

Received: 3 June 2021
Accepted: 23 June 2021
Published: 24 June 2021

Abstract: Scales of *Prochilodus magdalenae*, a Colombian endemic fish species, were used to obtain chitosan for application as an antibacterial agent integrated into starch-based films. Analysis of its composition during the demineralization and deproteinization process indicated that minerals and protein were both removed successfully. At this point, mild conditions for the deacetylation process were employed, namely, 2, 4, and 6 wt.% NaOH at room temperature for 16 h. Chitosan processed under 2 wt.% NaOH had low molecular weight, with the lowest value of 107.18 ± 24.99 kDa, which was closely related to its antibacterial activity. Finally, this chitosan was integrated into a banana starch-based film, and its antibacterial activity was assayed in *Escherichia coli* and *Staphylococcus aureus* cultures, with positive results in the former culture, especially due to the low-molecular-weight characteristic of chitosan.

Keywords: chitosan; fish scales; *Prochilodus magdalenae*; antibacterial agent

1. Introduction

Chitosan is a derivative of chitin, which is one of the three most important and abundant polysaccharides on the planet, together with cellulose and starch [1,2]. Chitin is a structural polysaccharide similar to white, inelastic, and nitrogenous cellulose consisting of 2-acetamido-2 deoxy β-D-glucose linked together by β-1,4 linkages conventionally extracted from insects, fungi, annelids, mollusks, and crustaceans, although industrial production occurs particularly from shrimp and crab crustaceans [2–4].

As mentioned above, chitosan can be obtained from different natural sources, including fish scales, which are composed mainly of collagen and carbonates/hydroxyapatite [5], and minimally explored components such as chitin, which can be converted into chitosan [6]. Fish scales can be used in many applications such as filling material for paper [7], biomass for energy generation [5], or removing heavy metals [8]. However, fish scales are discarded because they are considered waste produced on a large scale, constituting 1 wt.% of whole weight in fish, which makes it one of the major sources of contamination in riverine regions in many parts of the world [9]. *P. magdalenae*, commonly called "bocachico" in Colombia, is an endemic species grown naturally mainly in the Magdalena River and Cauca River and can be used as a source to obtain scales to extract chitosan. "Bocachico" is highly commercialized informally by street vendors, as exhibited in Figure 1, because it constitutes an important food source for riverine populations [10,11]. Likewise, the importance of using local sources to obtain chitosan is due to the significant effect on its properties such as solubility, reactivity, affinity for solvents, and swelling [3].

Figure 1. "Bocachico" commercialized informally in the municipality of Plato, Magdalena, Colombia.

Chitosan is a nontoxic, biocompatible, biodegradable and polyelectrolyte material that inhibits the growth of a wide variety of fungi, bacteria, and yeasts, which has allowed it to have applications in different areas such as cosmetics, agriculture, food, biomedicals, textiles [12,13]. It also has been used for the removal of emerging pollutants such as antibiotics [14] and as a flocculant for treatment of wastewater [15]. Additionally, it is considered generally recognized as safe (GRAS) [16], which has allowed applications through combination with other polymers such as starch for the development of biodegradable packaging for food [13]. The antimicrobial activity of chitosan depends on various conditions such as the pH, type of microorganism, pKa, and degree of deacetylation, among others [13]; however, chitosan is an alternative as a natural preservative against food spoilation by microorganisms [17], replacing conventional preservatives such as potassium sorbate or sodium benzoate [18].

Therefore, this paper proposes a protocol to extract chitosan from highly polluting riverbanks, such as that from the scales of *Prochilodus magdalenae*, an endemic species in Colombia, as a novel source for this material and to characterize its use as a possible antimicrobial agent integrated into starch-based films, with a high potential to produce biodegradable food packaging.

2. Materials and Methods

2.1. Feedstock and Reagents

P. magdalenae scales were collected in the fish market square of Plato, Magdalena-Colombia, after the flaking process by artisanal fishermen of the city. Then, the scales were washed with plenty of tap water to clean organic debris and were stored at −20 °C until reaching the required quantity. Once the collection was completed, the scales were sent to Medellin, Colombia, to be processed, and chitosan was obtained.

Low-molecular-weight chitosan was purchased from Sigma-Aldrich (Sigma-Aldrich Co., Milwaukee, WI, USA) for comparison to chitosan obtained in this research, and sodium hydroxide and hydrochloric acid were purchased from Merck (Merck KGaA, Darmstadt, Germany). All the reactives used in this research were reagent grade.

2.2. Sample Preparation

The scales were dried as a previous step for chitosan extraction. The drying process was carried out in an oven at 60 °C until a constant weight was reached. Then, the scale size was reduced using a blender at 3000 rpm for 3 min.

2.3. Chitosan Extraction

Chitosan extraction involved three main stages: demineralization, deproteinization, and deacetylation. In Figure 2, the process to obtain chitosan from *P. magdalenae* scales is described. Deacetylation was performed with sodium hydroxide (NaOH) at a concentration of 50% wt./v and a temperature in the range of 60–90 °C.

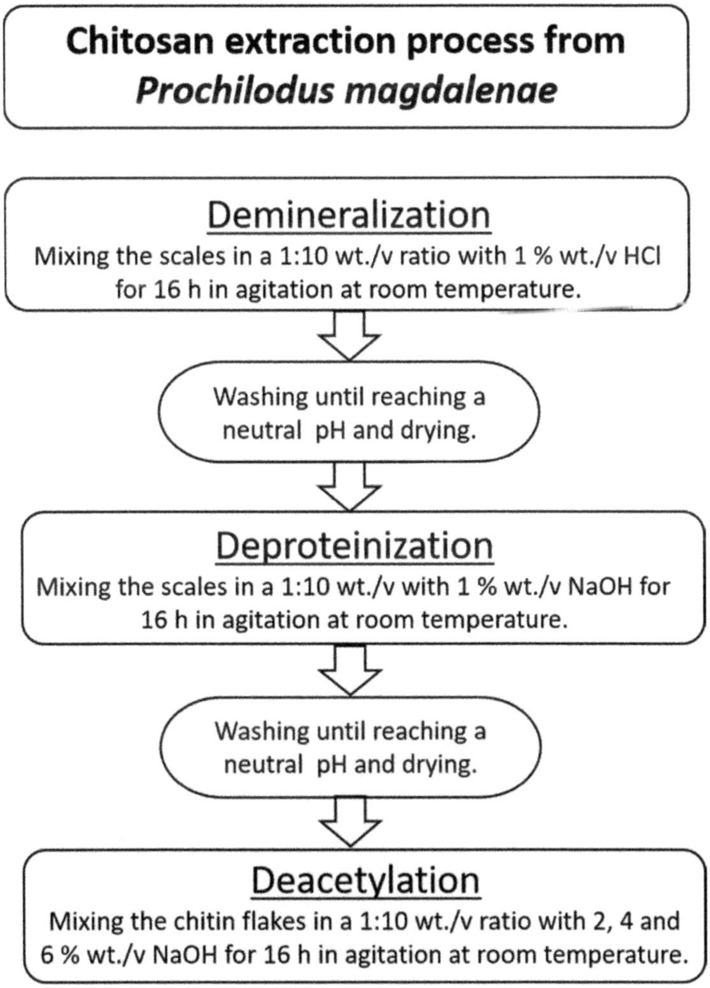

Figure 2. Chitosan extraction process from *P. magdalenae* scales.

2.4. Chitosan Characterization

The properties and characteristics of the chitosan samples obtained after deacetylation at different NaOH concentrations were analyzed using the following techniques. Before analysis, the samples were dried in an oven at 50 °C until a reaching constant weight, milled, and sieved through 100 mesh.

2.4.1. Protein Determination

To quantify protein removal after chemical treatment with NaOH, Biuret's method, modified and adapted by Gornall et al. (1949), was used [19]. All samples (0.1 g) were weighed and dispersed in 100 mL of distilled water and then vortexed and sonicated for 10 min to facilitate protein solubilization. Next, the samples were centrifuged at 6000 rpm, and 1 mL of supernatant was withdrawn from the samples to quantify the protein concentration. Finally, protein calculations were performed based on the dry weight of the samples. All analyses were performed in triplicate.

2.4.2. Mineral Quantification (Phosphorous and Calcium)

The mineral content in the *P. magdalenae* scales and chitin was determined using an ICE® 3500 atomic absorption spectrometer according to EPA method EPA-3050-B for digestion of the sample and Standard Method SM-3111-D to calculate the Ca content by atomic absorption spectroscopy (AAS). For phosphorous determination, the standard methods, SM-4500-P.B and SM-4500-P.E, were employed for acid digestion and P quantification by the spectrophotometric method, respectively. All values have a deviation of less than 5%.

2.4.3. Attenuated Total Reflection Fourier Transform Infrared (FTIR–ATR) Spectroscopy

The spectra were obtained on a Nicolet iS50 spectrophotometer (Thermo Scientific, Waltham, MA, USA) in the range of 40,000–400 cm^{-1} using a diamond ATR crystal. The spectra were recorded with a resolution of 4 cm^{-1} and an accumulation of 64 scans.

ATR correction was performed by Omnic software (Thermo Fisher Scientific, Waltham, MA, USA), and the baseline was defined manually.

The samples were dried at 50 °C for 2 d before the assay.

2.4.4. Scanning Electron Microscopy (SEM)

Chitosan samples were coated with gold/palladium using ion sputtering. The samples were observed with a JEOL® JSM 5910 LV microscope (JEOL Ltd., Tokyo, Japan) operated at 15 kV under high vacuum.

The samples were dried at 50 °C for 2 d and milled before the assay.

2.4.5. Thermogravimetric Analysis (TGA)

The thermal degradation of the samples was evaluated in a Mettler Toledo TGA/SDTA 851E thermogravimetric analyzer (Mettler Toledo, Columbus, OH, USA). Dried and ground samples were weighed and heated in a nitrogen atmosphere from room temperature to 800 °C at a heating rate of 10 °C min^{-1}.

2.4.6. Differential Scanning Calorimetry (DSC)

A DSC TA Q2000 Series system (TA Instruments, New Castle, DE, USA) was employed to carry out the differential scanning calorimetry analyses. Five milligrams of sample were placed into aluminum sealed pans, and scans were made from 0 °C to 250 °C at a 10 °C/min rate in a nitrogen atmosphere.

2.4.7. Viscosity Molecular Weight

The viscosity average molecular weight (MW) was determined by using a #75 Ostwald viscometer operated at a temperature of 25 °C. Chitosan was dissolved in 0.30 M acetic acid/0.2 M sodium acetate buffer, pH 4.5. Intrinsic viscosities were determined by extrapolation of the average values of flow time obtained for five solutions of different concentrations. The viscosity average molecular weight is calculated according to the Mark–Houwink–Sakurada equation (Equation (1)).

$$[\eta] = K(\overline{M}_v)^a \quad (1)$$

where $[\eta]$ is the intrinsic viscosity, $\overline{M}_v$ is the viscosity average molecular weight, and K and a are constant parameters equivalent to 74×10^{-3} mL/g and 0.76, respectively [20].

2.4.8. Deacetylation Degree

For the deacetylation degree, 0.1 g of sample was added to 10 mL of 0.2 M standard HCl solution. Then, the solutions were brought to 100 mL with distilled water and 0.746 g of KCl to adjust the ionic strength to 0.1. The titration processes were carried out with 0.1 M NaOH solution. A pH meter (SI Analytics LAB 850, Xylem Analytics, Weilheim in Oberbayern, Germany) was used for pH measurements under stirring conditions. The assays started at approximately pH 2 and went to pH 11. The method to determine the deacetylation degree (*DD* %) was the potentiometric titration method, and Equation (2) was used.

$$DD(\%) = 2.03\frac{V_2 - V_1}{m + 0.0042(V_2 - V_1)} \quad (2)$$

where m is the weight of the sample; V_1 and V_2 are the volume of NaOH solution at the inflection points; 2.03 is the coefficient correlated to the molecular weight of the monomer of chitin; and 0.0042 is the difference between monomers of chitin and chitosan [21].

2.4.9. Evaluation of Antibacterial Effect in the Starch-Based Film

To assess the antibacterial activity of chitosan in food packaging, a banana starch-based film was prepared according to a slightly modified method developed by Leites et al. (2018) [22]. Before film manufacturing, a 4.5 wt.% chitosan solution was prepared in 2 v. % acetic acid. The film was obtained by casting after adding chitosan (20 wt.%) into the mix of starch (4 wt.%), glycerol (1.5 wt.%), and water (74.5 wt.%). Once the film was dried, a sample of 1 cm diameter was taken and placed on separate *Escherichia coli* and *Staphylococcus aureus* cultures to evaluate the effect of chitosan as an antibacterial agent. The culture was grown for 3 d, and the effect of chitosan was measured by the formation of an inhibition halo around the film sample.

3. Results and Discussion

The extraction of chitosan from *P. magdalenae* was performed under the conditions shown in Figure 2. First, 45.2% chitin was recovered, which was slightly higher than the amount obtained by Lavall et al. (2007), who achieved a 40–42% recovery [23]. After the demineralization and deproteinization treatments, some chemical changes occurred on the scales, as shown in Table 1. These results show that a major quantity of non-chitosan components was removed from the scales; in the case of minerals, 99.99% of phosphorous and 98.51% of calcium were removed, and for protein, the removal yield was approximately 75.91%, which must be eliminated before to facilitate the deacetylation process. Then, the next step was deacetylation under three different concentrations of NaOH, namely, 2% (Chi-2), 4% (Chi-4), and 6% (Chi-6), to determine their effect on the chitosan properties. In this regard, in previous experiments, information was collected that pointed out the reason for using mild conditions during the chitin deacetylation process in this research: when 50 wt.% NaOH and 80–90 °C conditions were applied, the isolated chitin was degraded because it was β-chitin (data not shown). The above phenomenon was observed by Lavall et al. (2007), who obtained β-chitin samples that were more deacetylated due to the high susceptibility of this polymorph to the treatments employed during chitosan extraction from squid pens [23].

Table 1. Quantification of components other than chitin in *P. magdalenae* scales.

Component	*P. magdalenae* Scales (mg/kg)	Chitin (mg/kg)
Phosphorous	$109.34 \times 10^3 \pm 8638.30$	7.64 ± 0.61
Calcium	$131.79 \times 10^3 \pm 4.372$	$1.95 \times 10^3 \pm 4.37$
Protein	$314.10 \times 10^3 \pm 9081.99$	$75.65 \times 10^3 \pm 1454.28$

To determine which treatment is more convenient to obtain chitosan from *P. magdalenae* scales for use as an antimicrobial agent, the first properties evaluated were the deacetylation degree (*DD*) and molecular weight (MW), which are reported in Table 2. Regarding this topic, significant differences were not found in the MW of chitosan obtained by changing the NaOH concentration, which could be related to the high reactivity of β-chitin [24]. According to the classification of chitosan, it can exist in three categories: low-molecular-weight (less than 150 kDa), medium-molecular-weight (from 150 kDa to 700 kDa), and high-molecular-weight (higher than 700 kDa) chitosan [25]. In Table 2, it can be observed that the MW in the three samples obtained is consistent with low-molecular-weight chitosan. This type of chitosan is very useful as an antimicrobial agent due to its interaction with phospholipids in membranes by a chelation mechanism [26,27]. In this research, the sample Chi-2% was selected for use as the antimicrobial component in a starch-based film because it implies a lower quantity of NaOH used.

Table 2. Molecular weight and deacetylation degree of chitosan extracted from *P. magdalenae* scales.

Sample	Deacetylation Degree (%)	Molecular Weight (kDa)
Chi-2%	94.91 ± 1.35	107.18 ± 24.99
Chi-4%	100.06 ± 4.60	134.58 ± 24.39
Chi-6%	100.99 ± 0.00	240.3 ± 134.02
Chi-commercial (control)	72.00 ± 2.98	151.11 ± 4.47

Once the 2 wt.% NaOH treatment was selected, FTIR–ATR spectra were recorded for Chi-2% and Chi-commercial to study their chemical composition, as shown in Figure 3. The FTIR spectra of Chi-2% and Chi-commercial are shown in Figure 3. In the spectra, the peak at 564 cm^{-1} was assigned to NH out-of-plane bending in amide groups, and the 1160 cm^{-1} peak was assigned to C–O–C stretching [28]. Shown also is a 1026 cm^{-1} band corresponding to the free amino group of glucosamine at C_2 871 cm^{-1} related to C–N stretching, a 1545 cm^{-1} band assigned to the coupling of N–H bending and C–N stretching in amide II [6,23,29] and a band at 1640 cm^{-1} corresponding to amide I [30]. Likewise, the spectrum of Chi-2% shows the presence of certain bands such as that at 3073 cm^{-1}, which is assigned to amide B [31] related to proteins remaining in the sample, as shown in Table 1, and, unlike Kumari et al. (2017), they remain as β-chitosan because bands at 1604 or 1423 cm^{-1} were not formed [32]. In addition, a 3360 cm^{-1} band is observed in the Chi-commercial spectrum, which is not observed in Chi-2%; this finding is related to the lower intra- and inter-molecular hydrogen bonding characteristics of β-chitosan derived from β-chitin [33].

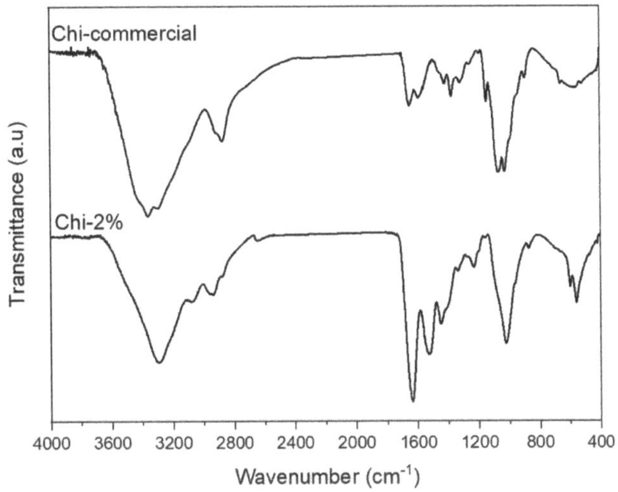

Figure 3. FTIR–ATR spectra recorded for Chi-commercial and chitosan (Chi-2%) extracted from *P. magdalenae* scales.

After evaluation using ATR–FTIR, SEM images of Chi-commercial and Chi-2% were captured to compare their surface characteristics and the effect of the 2% NaOH treatment, as shown in Figure 4. Figure 4a,c shows the morphology of the Chi-commercial and Chi-2% chitosan flakes, respectively, and both showed an irregular shape, as was reported by Kim et al. [34]. In Figure 4b,d, a magnified image is reported. These results demonstrate the fibrillar nature of chitosan, especially in Chi-2%, as was found by Kumari and Rath in 2014 [6]. Likewise, Chi-2% exhibits a cracked surface compared to the smoother surface of Chi-commercial, which is because β-chitin is more sensitive to the chemical treatments used, as mentioned above.

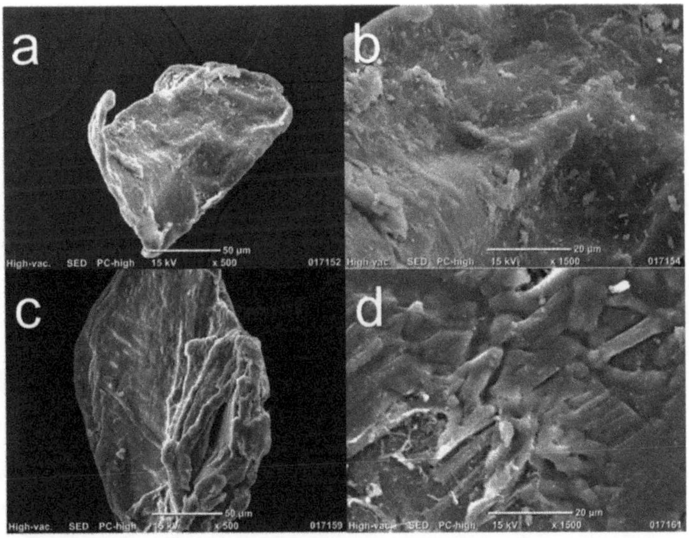

Figure 4. SEM images of chitosan flakes (**a**) and (**b**) are Chi-commercial (control) images, and (**c**) and (**d**) are Chi-2% images.

Once the surface and chemical characteristics of the chitosan obtained by the mild condition treatment of 2% NaOH were determined, thermal analyses with DTG and DSC were carried out, and the results are shown in Figure 5a,b. The results showed that the sample of chitosan extracted from *P. magdalenae* scales was very similar to commercial low-molecular-weight chitosan. Thermal analysis showed a sharp peak for Chi-commercial in Figure 5, which indicates that this sample is purer than Chi-2%. In addition, DTG of Chi-commercial presents major degradation at approximately 295 °C, unlike Chi-2%, which shows a broad peak of maximum degradation at approximately 330 °C, reported to be from the thermal rupture of glycosidic linkages in pyranose rings [35]. This result suggests a variable molecular weight in Chi-2%, which is correlated to the three different endothermic events presented in the DSC thermogram. Table 3 shows the value of the main thermal transformation extracted from the DSC thermogram. It is important to highlight that for polymers, the melting phenomenon (peak temperature) is an average of the melting temperature of crystallites, which is not a characteristic value for the sample but rather, depends on the MW [36,37]. Therefore, in Table 3, it can be appreciated that both samples evaluated present different values for onset temperature. Likewise, chitosan needs high thermal energy to initiate its thermal decomposition [38], and for the samples evaluated, a value of 195.66 J/g dry weight and 143.23 J/g dry weight was determined for Chi-commercial and Chi-2%, respectively, which is related to the MW previously calculated for each sample.

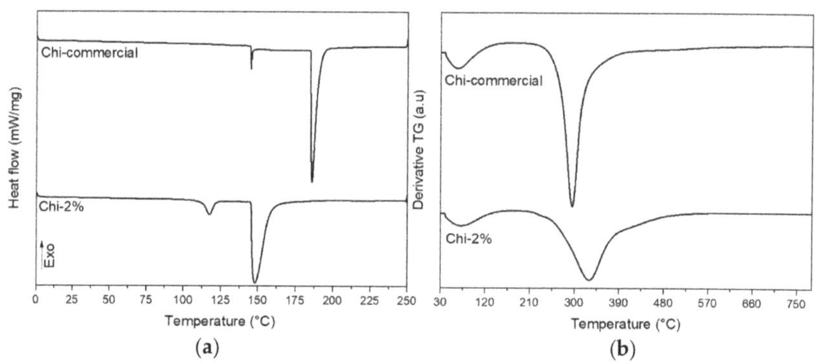

Figure 5. (**a**) DSC and (**b**) DTG thermograms of commercial chitosan and chitosan obtained from *P. magdalenae* scales.

Table 3. Parameters extracted from DSC analysis of commercial chitosan and chitosan obtained from *P. magdalenae* scales.

Sample	T_O (°C, Onset)	T_P (°C, Peak)	T_C (°C, Completion)	ΔH (J/g Dry Weight, Enthalpy)
Chi-commercial	182.85	186.25	199.00	195.66
Chi-2%	144.82	148.42	170.51	143.23

Finally, the Chi-2% sample was assayed in vitro to determine its antimicrobial effect against *Escherichia coli* (Gram-negative) and *Staphylococcus aureus* (Gram-positive), and the results are shown in Figure 6. Figure 6a,d shows the response of bacteria to the control film (without chitosan), which had no effect on the cultures. Figure 6b shows an inhibition halo of approximately 1 mm around the starch-based film with Chi-commercial added; however, this did not happen with *S. aureus* (Figure 6e). Finally, Figure 6c,f shows the effect of the starch-based film containing chitosan obtained in this research (Chi-2%), with an inhibition halo around the film in *E. coli* culture similar to that with Chi-commercial. As discussed above, low-molecular-weight (low MW) chitosan can enhance the antimicrobial effect;

however, according to Zheng and Zhu (2003), this does not apply to all microorganisms; i.e., in their research, they found that high MW improved the inhibitory effect on *S. aureus* because, in this way, it inhibited nutrient absorption, but for *E. coli*, low MW chitosan was more efficient than high MW chitosan in this application because lower MW chitosan can penetrate microbial cells more easily and disrupt cell metabolism [39].

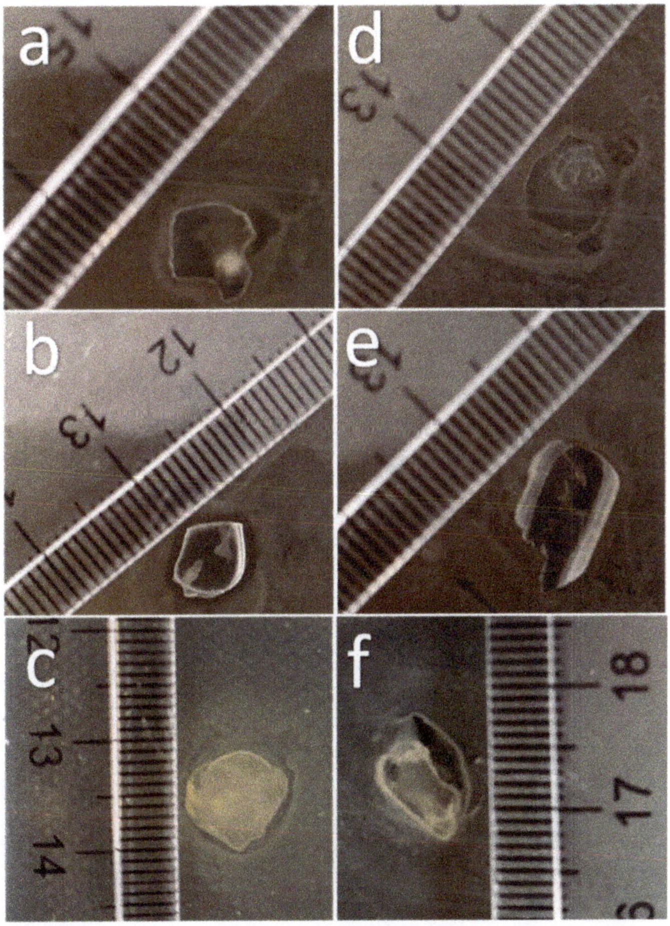

Figure 6. Images of the effect of starch-based films reinforced with chitosan extracted from *P. magdalenae* scales on *E. coli* culture (**a**) without chitosan, (**b**) with Chi-commercial and (**c**) with Chi-2%; and on *S. aureus* culture (**d**) without chitosan, (**e**) with Chi-commercial and (**f**) with Chi-2%.

4. Conclusions

Prochilodus magdalenae scales, a contaminant of riparian zones of rivers in Colombia, were proven to be a novel feedstock to obtain chitosan such as commercial chitosan of low molecular weight, which reduced Gram-negative microorganism growth. However, the extreme conditions of temperature (>80 °C) and NaOH concentration (50%) reported in the literature were not suitable for extraction of chitosan from *P. magdalenae* scales; therefore, it was necessary to develop a new method to obtain it because chitin extracted from these scales was β-type at room temperature and 2 wt.% NaOH conditions, as shown in this work. These mild conditions allowed chitosan to be obtained with desirable physical and chemical characteristics, such as lower molecular weight at 107.18 kDa than that of commercial chitosan at 151.11 kDa, which had effects on the thermal properties (lower T_P, 148.82 °C) but allowed its expected effect as an antibacterial agent to be achieved, owing to its low molecular weight.

Author Contributions: Conceptualization, C.M.-R. and R.Z.; methodology, C.M.-R.; validation, C.M.-R.; formal analysis, C.M.-R. and J.Á.-C.; investigation, C.M.-R.; data curation, P.G.; writing—original draft preparation, C.M.-R. and P.M.; writing—review and editing, J.Á.-C. and P.G.; supervision, R.Z.; project administration, J.Á.-C. All authors have read and agreed to the published version of the manuscript.

Funding: This research was funded by Ministerio de Ciencia, Tecnología e Innovación de Colombia (MINCIENCIAS) grant number 848 and Universidad Pontificia Bolivariana, and the APC was funded by Universidad del Magdalena.

Institutional Review Board Statement: Not applicable.

Informed Consent Statement: Not applicable.

Data Availability Statement: The data presented in this study are available on request from the corresponding author.

Conflicts of Interest: The authors declare no conflict of interest.

References

1. Rinaudo, M. Chitin and Chitosan: Properties and Applications. *Prog. Polym. Sci.* **2006**, *31*, 603–632. [CrossRef]
2. Ravi Kumar, M.N.V. A Review of Chitin and Chitosan Applications. *React. Funct. Polym.* **2000**, *46*, 1–27. [CrossRef]
3. Abdou, E.S.; Nagy, K.S.A.; Elsabee, M.Z. Extraction and Characterization of Chitin and Chitosan from Local Sources. *Bioresour. Technol.* **2008**, *99*, 1359–1367. [CrossRef] [PubMed]
4. Shepherd, R.; Reader, S.; Falshaw, A. Chitosan Functional Properties. *Glycoconj. J.* **1997**, *14*, 535–542. [CrossRef] [PubMed]
5. Silva, A.V.S.; Torquato, L.D.M.; Cruz, G. Potential Application of Fish Scales as Feedstock in Thermochemical Processes for the Clean Energy Generation. *Waste Manag.* **2019**, *100*, 91–100. [CrossRef] [PubMed]
6. Kumari, S.; Rath, P.K. Extraction and Characterization of Chitin and Chitosan from (Labeo Rohit) Fish Scales. *Procedia Mater. Sci.* **2014**, *6*, 482–489. [CrossRef]
7. Ural, E.; Kandirmaz, E.A. Potential of Fish Scales as a Filling Material in Surface Coating of Cellulosic Paper. *J. Appl. Biomater. Funct. Mater.* **2018**, *16*, 23–27. [CrossRef]
8. Mustafiz, S. The Application of Fish Scales in Removing Heavy, Metals from Energy-Produced Waste Streams: The Role of Microbes. *Energy Sources* **2003**, *25*, 905–916. [CrossRef]
9. Dutta, P.K.; Dutta, J.; Tripathi, V.S. Chitin and Chitosan: Chemistry, Properties and Applications. *J. Sci. Ind. Res.* **2004**, *63*, 20–31.
10. Jaramillo-Villa, U.; Jiménez-Segura, L.F. Algunos aspectos biológicos de la población de *Prochilodus magdalenae* en las ciénagas de Tumaradó (Río Atrato), Colombia. *Actual. Biológicas* **2008**, *30*, 55–66.
11. Berdugo, G.O.; Narváez Barandica, J.C. Genetic Diversity and Population Structure of Bocachico Prochilodus Magdalenae (Pisces, Prochilodontidae) in the Magdalena River Basin and Its Tributaries, Colombia. *Genet. Mol. Biol.* **2014**, *37*, 37–45. [CrossRef]
12. Sun, K.; Li, Z.H. Preparations, Properties and Applications of Chitosan Based Nanofibers Fabricated by Electrospinning. *Express Polym. Lett.* **2011**, *5*, 342–361. [CrossRef]
13. Kong, M.; Chen, X.G.; Xing, K.; Park, H.J. Antimicrobial Properties of Chitosan and Mode of Action: A State of the Art Review. *Int. J. Food Microbiol.* **2010**, *144*, 51–63. [CrossRef]
14. Truong, T.T.T.; Vu, T.N.; Dinh, T.D.; Pham, T.T.; Nguyen, T.A.H.; Nguyen, M.H.; Nguyen, T.D.; Yusa, S.; Pham, T.D. Adsorptive Removal of Cefixime Using a Novel Adsorbent Based on Synthesized Polycation Coated Nanosilica Rice Husk. *Prog. Org. Coat.* **2021**, *158*, 106361. [CrossRef]
15. Ariffin, M.; Hassan, M.; Li, T.; Zainon Noor, Z. Coagulation and Flocculation Treatment of Wastewater in Textile Industry Using Chitosan. *J. Chem. Nat. Resour. Eng.* **2009**, *4*, 43–53.
16. Leonida, M.; Ispas-Szabo, P.; Mateescu, M.A. Self-Stabilized Chitosan and Its Complexes with Carboxymethyl Starch as Excipients in Drug Delivery. *Bioact. Mater.* **2018**, *3*, 334–340. [CrossRef]
17. Shahidi, F.; Arachchi, J.K.V.; Jeon, Y.-J. Food Applications of Chitin and Chitosans. *Trends Food Sci. Technol.* **1999**, *10*, 37–51. [CrossRef]
18. Kaczmarek, M.; Avery, S.V.; Singleton, I. Microbes Associated with Fresh Produce: Sources, Types and Methods to Reduce Spoilage and Contamination. *Adv. Appl. Microbiol.* **2019**, *107*, 29–82. [CrossRef]
19. Gornall, A.G.; Bardawill, C.J.; David, M.M. Determination of Serum Proteins by Means of the Biuret Reaction. *J. Biol. Chem.* **1949**, *177*, 751–766. [CrossRef]
20. Rinaudo, M.; Milas, M.; Dung, P.L. Characterization of Chitosan. Influence of Ionic Strength and Degree of Acetylation on Chain Expansion. *Int. J. Biol. Macromol.* **1993**, *15*, 281–285. [CrossRef]

21. Czechowska-Biskup, R.; Jarosińska, D.; Rokita, B.; Ulański, P.; Rosiak, J.M. Determination of Degree of Deacetylation of Chitosan—Comparison of methods. *Prog. Chem. Appl. Chitin Its Deriv.* **2012**, *17*, 5–20.
22. Luchese, C.L.; Pavoni, J.M.F.; dos Santos, N.Z.; Quines, L.K.; Pollo, L.D.; Spada, J.C.; Tessaro, I.C. Effect of Chitosan Addition on the Properties of Films Prepared with Corn and Cassava Starches. *J. Food Sci. Technol.* **2018**, *55*, 2963–2973. [CrossRef]
23. Lavall, R.L.; Assis, O.B.G.; Campana-Filho, S.P. β-Chitin from the Pens of Loligo Sp.: Extraction and Characterization. *Bioresour. Technol.* **2007**, *98*, 2465–2472. [CrossRef]
24. Kurita, K.; Ishii, S.; Tomita, K.; Nishimura, S.-I.; Shimoda, K. Reactivity Characteristics of Squid β-Chitin as Compared with Those of Shrimp Chitin: High Potentials of Squid Chitin as a Starting Material for Facile Chemical Modifications. *J. Polym. Sci. Part A Polym. Chem.* **1994**, *32*, 1027–1032. [CrossRef]
25. Minh, N.C.; Van Hoa, N.; Trung, T.S. Preparation, Properties, and Application of Low-Molecular-Weight Chitosan. In *Handbook of Chitin and Chitosan*; Elsevier: Amsterdam, The Netherlands, 2020; pp. 453–471.
26. Liu, H.; Du, Y.; Wang, X.; Sun, L. Chitosan Kills Bacteria through Cell Membrane Damage. *Int. J. Food Microbiol.* **2004**, *95*, 147–155. [CrossRef]
27. Je, J.-Y.; Kim, S.-K.; Byun, H.-G.; Moon, S.-H. Antimicrobial Activity of Hetero-Chitosans and Their Oligosaccharides with Different Molecular Weights. *J. Microbiol. Biotechnol.* **2004**, *14*, 317–323.
28. Varma, R.; Vasudevan, S. Extraction, Characterization, and Antimicrobial Activity of Chitosan from Horse Mussel Modiolus Modiolus. *ACS Omega* **2020**, *5*, 20224–20230. [CrossRef] [PubMed]
29. Stuart, B.H. *Infrared Spectroscopy: Fundamentals and Applications*; Analytical Techniques in the Sciences; John Wiley & Sons, Ltd.: Chichester, UK, 2004; ISBN 978-0-470-01114-0.
30. Varun, T.K.; Senani, S.; Jayapal, N.; Chikkerur, J.; Roy, S.; Tekulapally, V.B.; Gautam, M.; Kumar, N. Extraction of Chitosan and Its Oligomers from Shrimp Shell Waste, Their Characterization and Antimicrobial Effect. *Vet. World* **2017**, *10*, 170–175. [CrossRef]
31. Riaz, T.; Zeeshan, R.; Zarif, F.; Ilyas, K.; Muhammad, N.; Safi, S.Z.; Rahim, A.; Rizvi, S.A.A.; Rehman, I.U. FTIR Analysis of Natural and Synthetic Collagen. *Appl. Spectrosc. Rev.* **2018**, *53*, 703–746. [CrossRef]
32. Kumari, S.; Kumar Annamareddy, S.H.; Abanti, S.; Kumar Rath, P. Physicochemical Properties and Characterization of Chitosan Synthesized from Fish Scales, Crab and Shrimp Shells. *Int. J. Biol. Macromol.* **2017**, *104*, 1697–1705. [CrossRef] [PubMed]
33. Kumari, S.; Kishor, R. Chapter 1—Chitin and chitosan: Origin, properties, and applications. In *Handbook of Chitin and Chitosan*; Gopi, S., Thomas, S., Pius, A., Eds.; Elsevier: Amsterdam, The Netherlands, 2020; pp. 1–33. ISBN 978-0-12-817970-3.
34. Kim, H.-S.; Lee, S.-H.; Eun, C.-J.; Yoo, J.; Seo, Y.-S. Dispersion of Chitosan Nanoparticles Stable over a Wide PH Range by Adsorption of Polyglycerol Monostearate. *Nanomater. Nanotechnol.* **2020**, *10*, 184798042091726. [CrossRef]
35. Pereira, F.S.; Da Silva Agostini, D.L.; Job, A.E.; González, E.R.P. Thermal Studies of Chitin-Chitosan Derivatives. *J. Therm. Anal. Calorim.* **2013**, *114*, 321–327. [CrossRef]
36. Bair, H.E.; Salovey, R. The Effect of Molecular Weight on the Structure and Thermal Properties of Polyethylene. *J. Macromol. Sci. Part B* **1969**, *3*, 3–18. [CrossRef]
37. Kawai, T. Melting Points of Polyethylene Crystals and the Relation between Molecular Length and Chain Folding. *Kolloid-Zeitschrift Zeitschrift für Polymere* **1965**, *201*, 15–20. [CrossRef]
38. Subhapradha, N.; Ramasamy, P.; Shanmugam, V.; Madeswaran, P.; Srinivasan, A.; Shanmugam, A. Physicochemical Characterisation of β-Chitosan from Sepioteuthis Lessoniana Gladius. *Food Chem.* **2013**, *141*, 907–913. [CrossRef]
39. Zheng, L.Y.; Zhu, J.F. Study on Antimicrobial Activity of Chitosan with Different Molecular Weights. *Carbohydr. Polym.* **2003**, *54*, 527–530. [CrossRef]

polymers

MDPI

Article

Development of Edible Coating from Gelatin Composites with the Addition of Black Tea Extract (*Camellia sinensis*) on Minimally Processed Watermelon (*Citrullus lanatus*)

Salwa Salsabiela [1], Ambar Sukma Sekarina [1], Hanifa Bagus [1], Aulia Audiensi [1], Farah Azizah [1], Windy Heristika [1], Manikharda [1], Eko Susanto [2], Heli Siti Halimatul Munawaroh [3], Pau Loke Show [4] and Andriati Ningrum [1,*]

1 Department of Food and Agricultural Product Technology, Faculty of Agricultural Technology, Universitas Gadjah Mada, Flora Street No. 1, Bulaksumur, Yogyakarta 55281, Indonesia; salwasalsabiela@ugm.ac.id (S.S.); ambarnana@gmail.com (A.S.S.); hanifabagus@gmail.com (H.B.); auliaaudiensi@gmail.com (A.A.); farahazizah@gmail.com (F.A.); windyheristika@gmail.com (W.H.); manikharda@gmail.com (M.)

2 Department of Fish Products Technology, Faculty of Fisheries and Marine Science, Universitas Diponegoro, Jl. Prof. Soedarto SH Kampus Tembalang, Semarang 50275, Indonesia; ekothp@live.undip.ac.id

3 Study Program of Chemistry, Department of Chemistry Education, Faculty of Mathematics and Science Education, Universitas Pendidikan Indonesia, Bandung 40154, Indonesia; heli@upi.edu

4 Department of Chemical and Environmental Engineering, University of Nottingham Malaysia, Jalan Broga 43500, Selangor, Malaysia; pauloke.show@nottingham.edu.my

* Correspondence: andriati_ningrum@ugm.ac.id

Citation: Salsabiela, S.; Sukma Sekarina, A.; Bagus, H.; Audiensi, A.; Azizah, F.; Heristika, W.; Manikharda; Susanto, E.; Munawaroh, H.S.H.; Show, P.L.; et al. Development of Edible Coating from Gelatin Composites with the Addition of Black Tea Extract (*Camellia sinensis*) on Minimally Processed Watermelon (*Citrullus lanatus*). *Polymers* **2022**, *14*, 2628. https://doi.org/10.3390/polym14132628

Academic Editor: Evgenia G. Korzhikova-Vlakh

Received: 9 June 2022
Accepted: 22 June 2022
Published: 28 June 2022

Abstract: The purpose of this research was to determine the effect of composite fish gelatin–chitosan edible coatings enriched with black tea extract on the physical, chemical, and fungal decay properties of minimally processed watermelons stored at ±4 °C for 13 days. In this study, tuna skin gelatin was extracted and used to prepare edible coating solutions which comprised 4% gelatin, 2% chitosan, 2% calcium lactate, 2% glycerol, and black tea extract (0%; 0.25%; 0.50%; 0.75%; 1%). The samples were coated using the layer-by-layer dipping technique. This study showed that composite fish gelatin–chitosan edible coating enriched with black tea extract maintained and improved weight loss, texture (hardness), color, pH, and total soluble solid antioxidant activity and prevented fungal decay on minimally processed watermelons stored at ±4 °C for 13 days. The development in this study of edible film and a coating prepared from fish gelatin–chitosan and the incorporation of black tea extract as an antioxidant or antimicrobial agent can be a new approach to preventing postharvest loss and increasing the shelf life of minimally processed watermelon.

Keywords: tuna skin by-product; fresh-cut watermelon; extract tea; edible coating

1. Introduction

It is hoped that the consumption of sufficient fruits and vegetables during the current pandemic can have many functional effects on human health [1,2]. In Indonesia, fruit consumption is less than vegetable consumption. One of the reasons for the low consumption of fruit is the necessity to peel it before consumption. Therefore, there is the potential to develop minimally processed fruit because it only involves washing, peeling, cutting, packaging, and storing the fruit at low temperatures to maintain the freshness and nutritional content [2]. An example of minimally processed fruit that is often found in the market is watermelon (*Citrullus lanatus*).

Watermelon is very popular with the public because of its sweet taste and high water content, which give it a freshness when consumed. Watermelon is rich in vitamins A, B6, C, and K and antioxidants, which are very good for maintaining a healthy body [1,3]. Currently, many watermelons are sold in a minimally processed form. Minimally processed watermelon, in the form of slices without the skin, is more in demand by the public because of the convenience and practicality of consuming it. However, minimally processed watermelon easily loses weight through the evaporation of its water content, the growth of spoilage microbes, and the several enzymatic reactions that cause changes in the texture, color, taste, and nutritional content [3]. One way to maintain quality and freshness and extend the shelf life of minimally processed watermelon is to coat the pieces of fruit using a coating solution that is safe for consumption, commonly called edible coating [4,5].The edible coating can maintain quality and extend the shelf life of minimally processed fruit. The requirements for the components that can be used as edible coatings include those that do not affect the smell and taste of the food used and are easy to obtain, easy to digest, and non-toxic [5,6]. The edible coating could also contain several polymers from volatile or non-volatile parts [7]. Edible coating is used for the coating of fresh-cut fruits and vegetables such as watermelon. The sample is immersed in the film-forming solution, and it creates a protective coating directly on the surface of food such as minimally processed fruits [8]. The polymers that are usually used for edible coating could be from polymers of protein, carbohydrates, and lipid derivatives [8,9]. Gelatin is one of the ingredients that is often used as a component of edible coatings. One type of fish that is effective in producing gelatin is tuna. Gelatin from tuna fish skin as a food by-product has a good gel strength and viscosity [8,9]. Several studies have succeeded in combining gelatin with chitosan as a composite edible coating. Chitosan is one of the polysaccharides that is widely used as a constituent of edible coatings because it has antimicrobial properties. The combination of gelatin and chitosan can increase the antimicrobial properties of chitosan [10,11].

The addition of natural extracts can also improve the physical and functional properties of edible coatings, especially natural extracts that are rich in bioactive compounds [12]. Tea (*Camellia sinensis*) is a plant that has both antimicrobial and antioxidant activity [13]. Black tea contains tannins such as thearubigins, which have antimicrobial activities. On the other hand, several polyphenolic compounds, such as catechins, theaflavins, and methyl jasmonate, also have antimicrobial and antioxidant activity [14,15]. The fermentation process in black tea converts polyphenolic compounds (catechins and their derivatives) into theaflavins and thearubigins so that the content of catechin compounds decreases. However, many studies have shown that the antioxidant activity of black tea is comparable to that of green tea [12,14,15]. The theaflavins in black tea have the same antioxidant potential as the catechins in green tea [14,16]. The theaflavins in black tea extract are more effective in capturing free radicals than the catechins found in green tea. Therefore, based on the description above, in our study the aim was to determine the effect of using tuna skin gelatin composite and chitosan enriched with the addition of black tea extract as an edible coating on minimally processed fruits, e.g., fresh-cut watermelon. The samples were stored at ±4 °C for 13 days at low-temperature storage (±4 °C), and the physicochemical characterization and the fungal decay of the minimally processed watermelon was evaluated.

2. Material and Methods

2.1. Materials

The ingredients used in this study include red seeded watermelon 7–8 days after harvest (Berkat fruit shop, Yogyakarta, Indonesia), black tea (Tong Tji Super), yellowfin tuna (*Thunnus albacares*) skin (Omah Tuna, Yogyakarta, Indonesia), chitosan with a 20–100 mPas viscosity and a min 94% degree of acetylation (Phy Edumedia, Malang, Indonesia), glycerol (Progo Mulyo, Yogyakarta, Indonesia), aquades (Progo Mulyo, Yogyakarta, Indonesia), sodium hydroxide (Merck, Darmstadt, Germany), sulfuric acid (Merck, Darmstadt, Germany), citric acid (Merck, Darmstadt, Germany), acetic acid (Merck, Darmstadt, Germany), methanol (Merck, Darmstadt, Germany), and DPPH (2,2-diphenyl-1-picrylhydrazyl) (Sigma Aldrich, Singapore).

2.2. Extraction of Gelatin from Yellowfin Tuna Fish Skin

The extraction of the gelatin began with the separation of the tuna skin from the scales, bones, and nonskin components, followed by the cutting of the tuna skin to a size of $\pm 5 \times 5$ cm. The tuna skin was soaked in NaOH solution with a 0.2% ratio of 1:6 (w/v) for 2 h and neutralized to pH 6–7 with water. The fish skin was then immersed in a 0.2% H_2SO_4 solution at a ratio of 1:6 (w/v) for 2 h, washed to neutralize, soaked in citric acid ($C_6H_8O_7$) 0.1%, with a ratio of 1:6 (w/v) for 2 h, and washed to neutral pH (6–7). Continued extraction was performed with distilled water in a 60 °C water bath shaker for 6 h at a ratio of 1:3 (w/v) and filtered twice before drying at a temperature of 50 °C for 24 h [17]. The gelatin sheet was then milled to obtain the gelatin powder [17]. The specification of gelatin viscosity is approximately 25 cPs.

2.3. Preparation of Edible Coating Solution

The process of applying the edible coating solution onto the watermelon (*Citrullus lanatus*) was achieved using a layer-by-layer technique, with three different types of edible coating solutions. The first solution was a gelatin–chitosan composite solution and glycerol; the second solution was a black tea extract at five concentration variations (0%; 0.25%; 0.50%; 0.75%; and 1%); and the third solution was a 2% calcium lactate solution.

2.4. Coating of Edible Coating Solution on Minimally Processed Watermelon (Citrullus lanatus)

The process of applying the edible coating solution to the minimally processed watermelon (*Citrullus lanatus*) was carried out using the layer-by-layer immersion technique. First, a sample of watermelon that had been minimally processed with a size of $\pm(3 \times 3$ cm) was immersed in a 2% calcium lactate solution for two minutes, then drained and dried for two minutes. Next, the sample was immersed in a gelatin–chitosan composite solution and glycerol for two minutes, then drained and dried for two minutes. After that, the sample was again immersed in a 2% calcium lactate solution for two minutes, then drained and dried for two minutes. Furthermore, the packaging was carried out in plastic cups for samples treated with 0% black tea extract, while the other samples were immersed in a black tea extract solution with different concentrations, namely 0.25%, 0.50%, 0.75%, and 1% each for two minutes, then drained and dried for two minutes. After that, the sample was again immersed in a 2% calcium lactate solution for two minutes, then drained and dried for two minutes. Then, the sample was put into a plastic cup and stored in the refrigerator at a temperature of $\pm$ 4 °C. The sample was stored for 13 days of storage, and the physicochemical characterization and fungal decay evaluation were investigated at 4, 7, 10, and 13 days of storage. The following is a schematic of the edible coating solution coating process on watermelons that are minimally processed with the layer-by-layer technique. Further processes of the layer-by-layer technique can be seen in Figure 1.

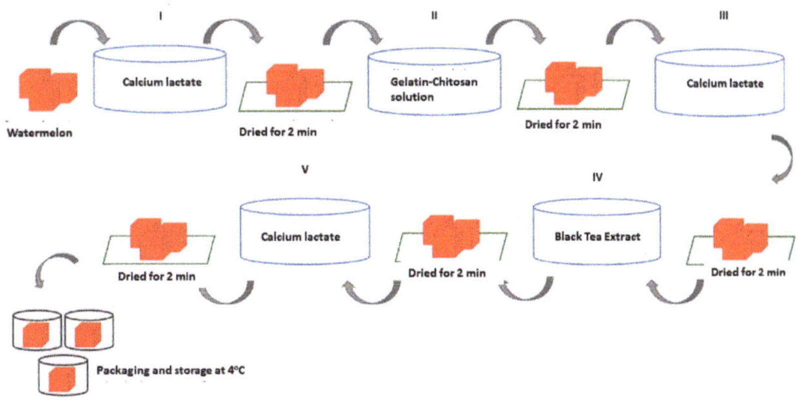

Figure 1. Layer-by-layer technique of edible coating.

2.5. Weight Loss Analysis

Weight loss analysis was carried out gravimetrically, namely by calculating the difference in weight before and after storage [18,19]. The measurement of the sample weight was carried out using an analytical balance every three days for 13 days of storage.

$$\text{Weight loss } (\%) = \frac{(\text{initial weight (grams)} - \text{final weight (grams)})}{(\text{initial weight (grams)})} \times 100\% \quad (1)$$

2.6. Texture Analysis

The texture parameter observed was hardness. Texture analysis was performed using a Universal Testing Machine (UTM) with a pre-load of 0.02 N; a pre-load speed of 50 mm/min; and a test speed of 50 mm/min. The analysis of sample hardness was interpreted by the maximum force (Fmax) required to pierce 30% of the sample height. Higher Fmax (Newton) values indicate that the texture of the sample is becoming harder [17].

2.7. Color Analysis

Color intensity analysis was carried out using a Minolta CR-400 Chroma Meter (Konica Minolta, Inc., Tokyo, Japan). The sample was placed on top of the chroma meter sensor, then a light was fired at the part to be measured so that the values of L (lightness), a (green-red chromaticity), and b (yellow-blue chromaticity) would appear on the chroma meter display [17].

2.8. pH Analysis

The pH analysis was performed using a pH meter previously calibrated with standard buffer solutions of pH 4 and 7. A total of 10 g of the sample was mashed using a blender and then centrifuged for one hour at a speed of 1000× *g* at 4 °C until separation between the natant and the supernatant occurred. The supernatant was used for pH analysis. To measure the pH value, the pH meter probe was immersed in the sample supernatant to obtain the pH value directly [20].

2.9. Total Dissolved Solids Analysis

Total dissolved solids analysis was performed using an Atago Master-53M refractometer (Atago Co., Ltd., Fukuoka, Japan). Sample preparation for the total dissolved solids analysis was carried out as in step 2.8. One to two drops of the sample supernatant at room temperature were placed on the prism of the refractometer, then the Brix percentage was read through the eyepiece of the refractometer [21].

2.10. Antioxidant Activity Analysis

A total of 15 g of watermelon samples were mashed using a blender for 2 min, then centrifuged for one hour at a speed of 1000× *g* at 4 °C until the separation between the natant and the supernatant occurred. Then, the natant or solid particles resulting from the centrifugation were filtered using a vacuum filter.

For the analysis of the antioxidant activity, a sample solution with a concentration of 100,000 ppm was made by dissolving 2.5 g of the sample, which had been filtered with a vacuum filter in 25 mL of methanol, then stirred with a magnetic stirrer for 90 min. After that, the samples were filtered with filter paper; then, 1 mL of each was taken and put into a test tube. Each sample had 7 mL of 0.1 mM DPPH solution added and homogenized using a vortex. The samples were then incubated for 30 min in the dark. After that, the absorbance value was measured using a UV-Vis spectrophotometer at a wavelength of 517 nm.

The antioxidant activity of the sample is interpreted in terms of the percentage of radical scavenging activity, namely the ability of the sample to capture free radical compounds. The more free radical compounds that can be captured, the more the antioxidant activity content in the sample [17].

$$\text{DPPH radical scavenging activity } (\%) = \frac{(\text{A blank} - \text{A sample})}{(\text{A blank})} \times 100\% \quad (2)$$

where A = Absorbance.

2.11. Fungal Decay

Determination of fungal decay was performed by observing the presence or absence of the fungi that grow on the surface of the sample, then calculating the percentage of the sample that was overgrown with fungus (Equation (3)). The sample is considered damaged if there is fungal mycelium on its surface. The results of these observations are expressed as the percentage of samples contaminated with fungi [22].

$$\text{Fungal contamination } (\%) = \frac{(\text{Number of samples contaminated with fungus})}{(\text{Total number of samples})} \times 100\% \quad (3)$$

2.12. Statistical Analysis

The statistical analysis was performed using Minitab v. 19 statistical software (State College, PA, USA). Analysis of variance (ANOVA) and Tukey's pairwise comparisons were carried out using a level of 95% confidence. The experiments were performed at least in triplicate, and the data were reported as mean ± standard deviation.

3. Results and Discussion

3.1. Weight Loss

Figure 2 shows that the application of an edible coating of gelatin composite enriched with black tea extract can effectively reduce the percentage of weight loss in minimally processed watermelon stored at ±4 °C for 13 days. The samples not treated with an edible coating (control) had a significantly higher percentage of weight loss ($p < 0.05$) than the samples treated with an edible coating (Figure 1). Weight loss is generally solely attributed to water loss, although the loss of some other components may also contribute to this problem. Nevertheless, except for water loss, the contribution from other components is considered negligible. This loss of water decreases the turgor and firmness of fruits. It can cause acceleration of surface depression and deformation of the produce. Water loss is associated with several other changes occurring in fruits and can act as a trigger to initiate these changes [23]. The edible coatings cover the surface layer of the fruit and inhibit the processes of respiration, transpiration, and syneresis [21,24,25]. This finding is evidence for the benefits of applying the edible coating to fresh-cut watermelon pieces, mainly due to the formation of a polymeric barrier that can reduce the water loss from fresh-cut samples, as is found with other fruits [3,22].

Based on Figure 2, the higher the concentration of black tea extract added, the lower the percentage of weight loss. The addition of black tea extract as a natural bioactive compound can act as a cross-linking agent in edible coating components to increase the barrier against water vapor because gelatin is sensitive to water vapor when used as the sole constituent of edible coatings [24]. Although there was a decrease in the percentage of weight loss, there was no significant difference ($p > 0.05$) in the addition of 0.25–0.75% black tea extract. However, there was an increase in weight loss with the addition of 1% black tea extract. This could be caused by the addition of phenolic compounds to a certain level that can reduce the gel properties of gelatin. Phenolic compounds are more likely to interact with gelatin as an aggregate, causing an irregular gel structure [20,24].

Based on Figure 2, the concentration of black tea extract, storage time, and the interaction between the two treatments had a significant effect ($p < 0.05$) on the percentage of minimally processed watermelon weight loss during storage. Our results showed that there was no significant difference in the percentage of weight loss ($p > 0.05$) in the treatment with the addition of 0.25% black tea extract or with 0.50% and 0.75% on the 13th day. Therefore the treatment of adding black tea extract with a concentration of 0.25% to the edible coating solution of the gelatin composite was the best in reducing the percentage of weight loss of watermelons processed for at least 13 days of storage at a temperature of ±4 °C.

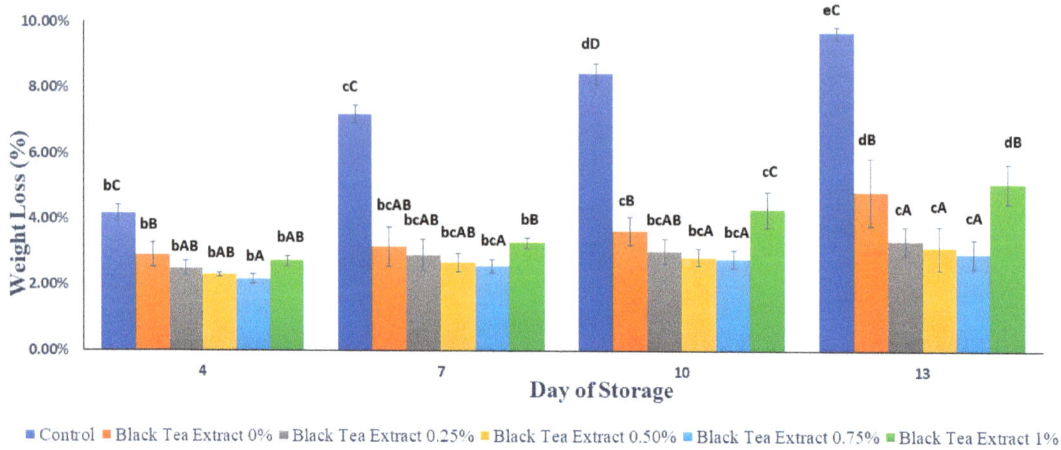

Figure 2. Effect of an edible coating of gelatin composite enriched with black tea extract on weight loss (%) of minimally processed watermelon during storage. The value shown is the average of the three experimental replications. [A–D] Values followed by the same capital letters showed no significant difference ($p > 0.05$) between treatments on the same day. [b–e] Values followed by the same non-capital letters showed no significant difference ($p > 0.05$) between storage times in the same treatment.

3.2. Texture

Figure 3 shows the results of texture analysis (firmness) on minimally processed watermelon coated with an edible coating of a gelatin composite enriched with black tea extract. The higher the Fmax (Newton) value, the harder the sample texture. Based on Figure 3, the longer the storage time, the lower the Fmax value, but the edible coating treatment can reduce the loss of firmness. Firmness is one of the most important quality attributes for all fruits, although it may be of more critical importance to some fruits (such as berries) than others. Tissue softening is thus a severe concern that affects visual quality and reduces the shelf life of fruits, especially in fresh-cut products [23]. The control sample had a significantly lower firmness ($p < 0.05$) than the sample treated with an edible coating. The hardness of the watermelon samples was minimally reduced during storage, but the samples treated with an edible coating of sodium alginate had a significantly higher hardness ($p < 0.05$) than the control. This is because during storage there are still metabolic processes in fruit, such as respiration and transpiration, as well as the activity of pectinase enzymes such as polygalacturonase and pectin methylesterase, which can convert insoluble pectin into water-soluble pectin so that the fruit texture becomes soft [3,22].

Based on Figure 3, the higher the concentration of black tea extract added, the greater the Fmax value, but there was a decrease with the addition of 1% black tea extract. The addition of black tea extract can act as a cross-linking agent that can increase hardness at certain concentrations. The hardness of the fruit can be related to the percentage of weight loss because the higher the percentage of weight loss (loss of water), the softer the texture of the fruit. The incorporation of tea extract into coating formulations for fruits and vegetables may have a remarkable effect on the firmness of the tissue. The presence of some compounds in tea extract may have some activity that promotes alterations in firmness. In addition, the use of calcium lactate also plays a role in maintaining fruit texture, because calcium ions can strengthen cell walls. It can be seen in Figure 3 that on days 10 and 13 there was no significant difference in hardness ($p > 0.05$) with the addition of 0.50% and 0.75% black tea extract.

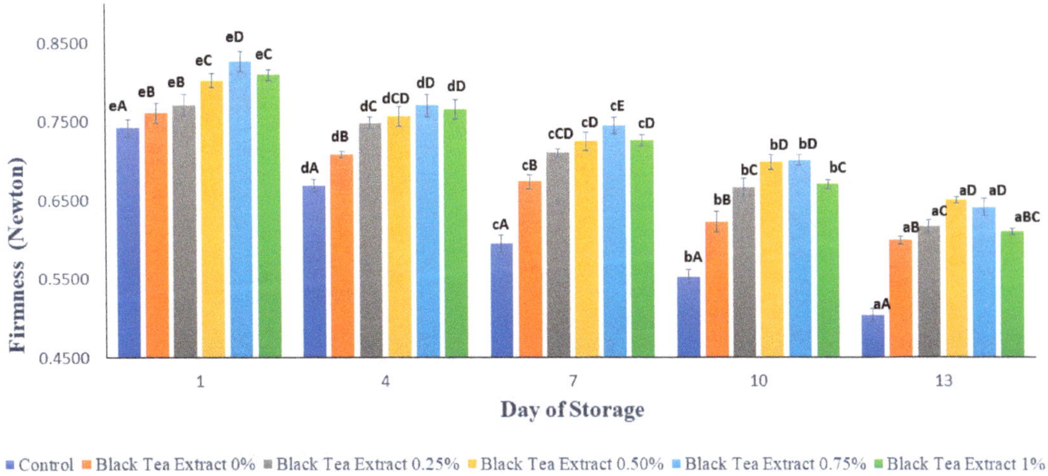

Figure 3. Effect of edible coating of gelatin composite enriched with black tea extract on the hardness (Newton) of minimally processed watermelon during storage. The value shown is the average of the three experimental replications. $^{A–E}$ Values followed by the same capital letters showed no significant difference ($p > 0.05$) between treatments on the same day. $^{a–e}$ Values followed by the same non-capital letters showed no significant difference ($p > 0.05$) between storage times in the same treatment.

3.3. *Color*

3.3.1. Lightness

Visual appearance and color are important quality criteria that directly influence the customer's perception of quality and are two of the most crucial quality attributes in fruits. The color of fruits particularly changes in their storage; when transformed into fresh-cut products they become highly prone to developing a browning discoloration. This discoloration in fruits is thus a critical issue that may render the product unacceptable to the consumers [23].

Figure 4 shows the results of the brightness analysis (L*) of minimally processed watermelon coated with an edible coating of a gelatin composite enriched with black tea extract. Based on Figure 3, the surface brightness value of the processed watermelon samples minimally decreased during storage. This can be caused by the activity of the PPO (polyphenol oxidase) enzyme which causes a decrease in brightness on the surface of the fruit flesh and tends to brown [25]. Browning reactions can occur because oxygen reacts directly with the polyphenol compounds catalyzed by the polyphenol oxidase enzyme to form a brown melanin compound. Oxygen can react directly with polyphenol compounds if some cells or tissues are open due to wounds [26]. Enzymatic browning diminishes visual appearance, makes undesirable changes in flavor, and causes a loss of nutrients in fresh/fresh-cut fruits and vegetables. Such changes make the product less desirable to the consumers. The development of browning degrades the original color of the product [2]. In the control, however, the decrease in brightness value tends to be greater than the sample treated with the edible coating because the edible coating can reduce the contact between oxygen and the polyphenol compounds. Gelatin and chitosan, as components of edible coatings, have good barrier properties against oxygen [4,27].

Based on Figure 4, on the first day, the higher the concentration of black tea extract added, the darker the color of the black tea extract solution used as an edible coating and therefore the lower the brightness value of the sample. However, during storage, the edible coating treatment enriched with black tea extract was better able to maintain brightness than the control. It can be seen in Figure 4 that on day 13 there was no significant difference in brightness value ($p > 0.05$) in the control treatment or in the edible coatings with the addition of 0% and 1% black tea extract. The application of an edible coating incorporated with green tea has been applied to fresh lettuce. The treatments prevented ascorbic acid and carotenoid loss. Green tea has also been incorporated into the coating solution applied to peach slices to enhance their shelf life [2].

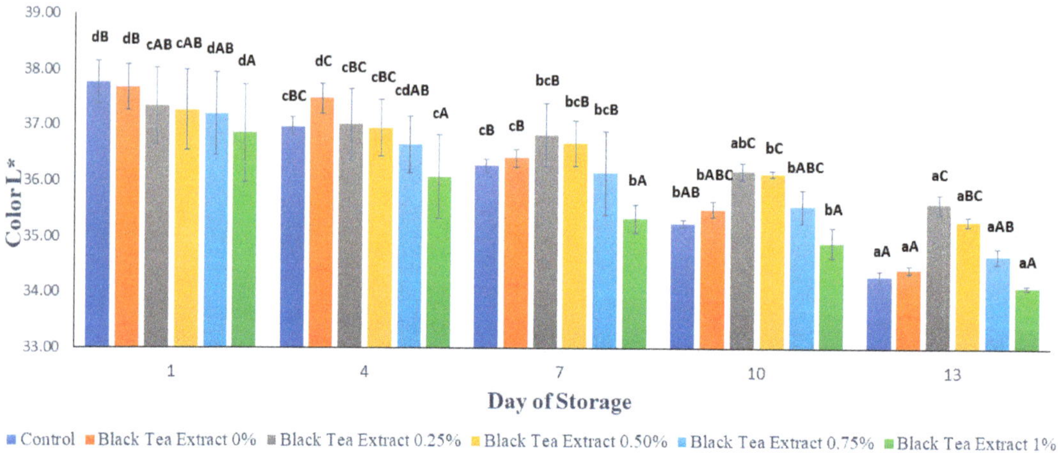

Figure 4. Effect of edible coating of gelatin composite enriched with black tea extract on the color (L*) of minimally processed watermelon during storage. The value shown is the average of the three experimental replications. [A–C] Values followed by the same capital letters showed no significant difference ($p > 0.05$) between treatments on the same day. [a–d] Values followed by the same non-capital letters showed no significant difference ($p > 0.05$) between storage times in the same treatment.

The results of the statistical analysis showed that the concentration of black tea extract, storage time, and the interaction between the two treatments had a significant effect ($p < 0.05$) on the brightness of the watermelon flesh that was minimally processed during storage. Further Duncan's test results showed that there was no significant difference in brightness ($p > 0.05$) with the addition of 0.25% and 0.50% black tea extract treatment on day 13. Therefore, the addition of black tea extract with a concentration of 0.25% in the edible coating solution of gelatin composite is the best treatment for maintaining the brightness of the color of watermelons processed for at least 13 days of storage at a temperature of ± 4 °C.

3.3.2. a (Degree of Redness)

Figure 5 shows the results of the analysis of the reddish color (a*) of minimally processed watermelon coated with an edible coating of a gelatin composite enriched with black tea extract. Based on Figure 5, the red color of all the samples of minimally processed watermelon experienced a significant decrease ($p < 0.05$) during the storage period. Lycopene is a red pigment in watermelon that has the ability to be an antioxidant; so, it is sensitive and easily damaged when exposed to light and oxygen during storage [1]. Samples without treatment (control) had a higher intensity of red color than samples treated with an edible coating. This can be caused by the red pigment in watermelon dissolving during the drying process in an edible coating solution. There was a significant difference ($p < 0.05$) for each addition of black tea extract concentration to the red color of minimally processed watermelon samples. Black tea contains theaflavin and thearubigin compounds, which give the tea a brown color so that the higher the concentration of black tea extract added, the more theaflavin and thearubigin compounds are present, which will affect the appearance of color in minimally processed watermelon samples [14].

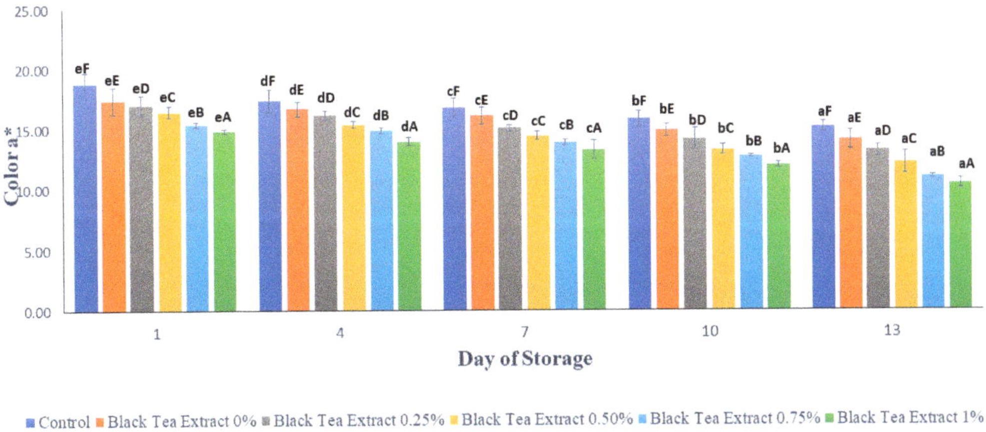

Figure 5. Effect of edible coating of gelatin composite enriched with black tea extract on the color (a*) of cut watermelon during storage. The value shown is the average of the three experimental replications. A–F Values followed by the same capital letters showed no significant difference ($p > 0.05$) between treatments on the same day. a–e Values followed by the same non-capital letters showed no significant difference ($p > 0.05$) between storage times in the same treatment.

3.3.3. Degree of Yellowness

Figure 6 shows the results of the analysis of the yellowish color (b*) on minimally processed watermelon coated with an edible coating of a gelatin composite enriched with black tea extract. Based on Figure 6, the yellow color in all the samples of minimally processed watermelon tends to decrease during the storage period. This can be seen in Figure 6, where, on the first day, the higher the concentration of the addition of black tea extract, the lower the intensity of the yellow color. On the first day, the sample without treatment (control) had a higher yellow color intensity than the sample treated with an edible coating, but there was a significant decrease ($p < 0.05$) during storage so that the control had the lowest yellow color intensity on the 13th day. However, on day 13 there was no significant difference ($p > 0.05$) between the control and the treatment with the addition of 1% black tea extract.

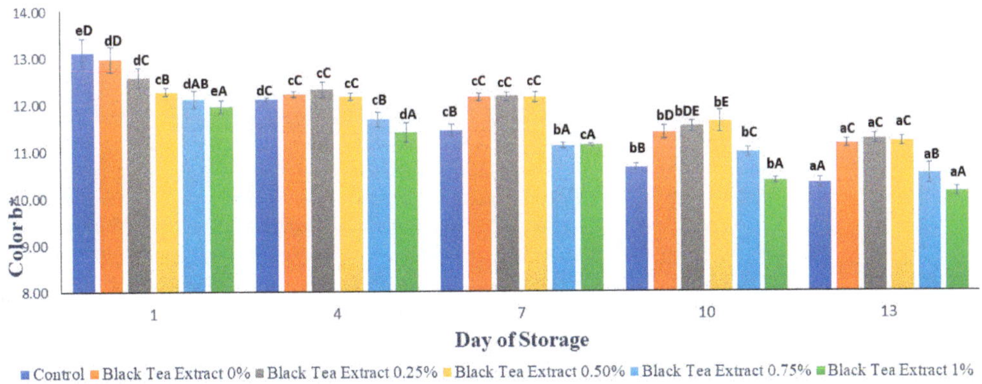

Figure 6. Effect of edible coating of gelatin composite enriched with black tea extract on the color (b*) of minimally processed watermelon during storage. The value shown is the average of the three experimental replications. A–E Values followed by the same capital letters showed no significant difference ($p > 0.05$) between treatments on the same day. a–e Values followed by the same non-capital letters showed no significant difference ($p > 0.05$) between storage times in the same treatment.

3.4. pH

Figure 7 shows the results of the pH analysis on minimally processed watermelon coated with an edible coating of a gelatin composite enriched with black tea extract. Based on Figure 7, on the first day the sample without treatment (control) had the highest pH value compared to the sample with the edible coating treatment, and the higher the black tea extract added to the edible coating solution, the lower the pH of the sample. This could be caused by the use of 1% acetic acid to dissolve chitosan as a component of the edible coating composites and by the presence of organic acids in tea, such as malic acid, citric acid, oxalic acid, and succinic acid, which affect the pH of the sample [15].

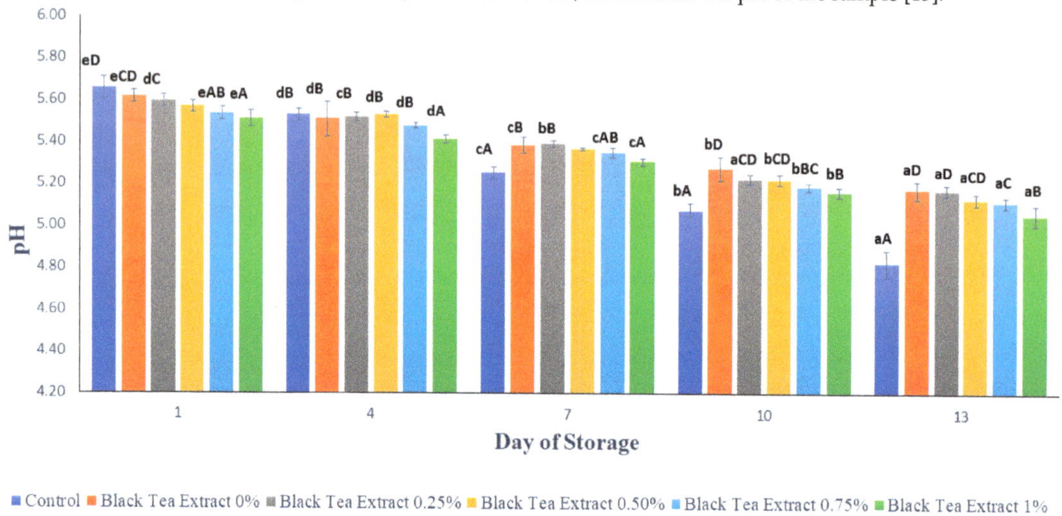

Figure 7. The effect of edible coating from gelatin composite enriched with black tea extract on the pH of watermelon was minimally processed during the storage period. The value shown is the average of the three experimental replications. [A–D] Values followed by the same capital letters showed no significant difference ($p > 0.05$) between treatments on the same day. [a–e] Values followed by the same non-capital letters showed no significant difference ($p > 0.05$) between storage times in the same treatment.

There was a decrease in pH in all the treatments during the storage period. This was presumably due to the activity of the microorganisms that can convert sugar into organic acids. It can be seen in Figure 6 that the decrease in pH in samples treated with an edible coating enriched with the addition of black tea extract was not as sharp as in the control, because chitosan and black tea, as components of edible coatings, have antimicrobial activity, and the combination of gelatin and chitosan can increase the antimicrobial properties of chitosan [22]. Fruits are a naturally existing source of several organic acids such as Vitamin C. Vitamin C (ascorbic acid) degradation occurs during storage because it is highly sensitive and is lost upon exposure to heat, light, oxygen metals, and enzymes. Coating treatments have been shown to reduce the loss of organic acids such as ascorbic acid in fresh/fresh-cut fruits. Usually, ascorbic acid content decreases with the storage of fruits, irrespective of whether the fruit is covered with a coating or not, but the extent of the loss may be less in fruits that are coated [23].

3.5. Total Soluble Solid

Figure 8 shows the results of the analysis of total soluble solids in minimally processed watermelon coated with an edible coating of a gelatin composite enriched with black tea extract. Total dissolved solids can be used to interpret the amount of sugar in the material and can be used as an indicator of the level of fruit maturity. The total dissolved solids values in all the treatments tended to increase significantly during the storage period ($p < 0.05$), and on the 13th day of storage, there was a slight decrease, as shown in Figure 8. However, the increase in the total dissolved solids values in the treated sample's edible coating was not as high as in the sample without treatment (control). This could be because the coating can inhibit the rate of metabolism of the polysaccharides, allowing the sugar content in the fruit to be maintained [18].

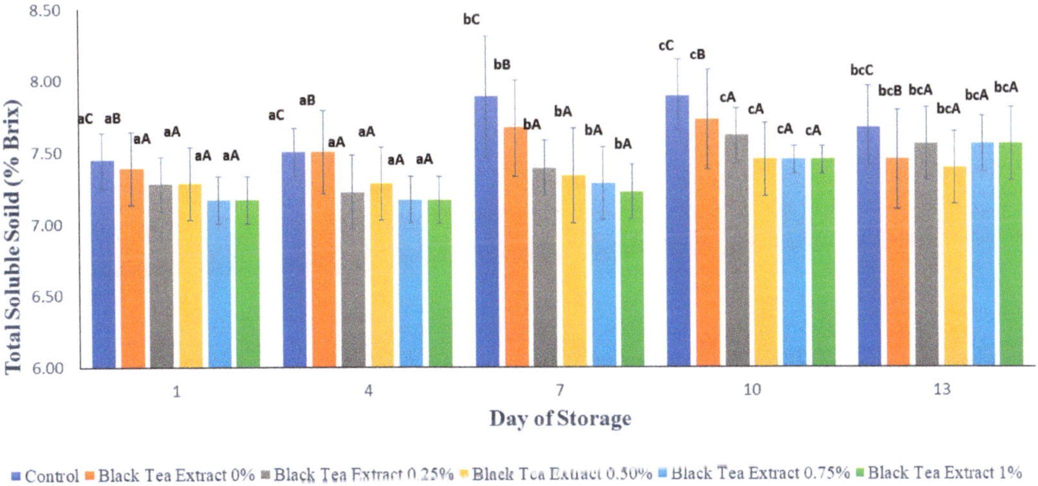

Figure 8. Effect of edible coating of gelatin composite enriched with black tea extract on total soluble solids (% Brix) of minimally processed watermelon during storage. The value shown is the average of the three experimental replications. [A–C] Values followed by the same capital letters showed no significant difference ($p > 0.05$) between treatments on the same day. [a–c] Values followed by the same non-capital letters showed no significant difference ($p > 0.05$) between storage times in the same treatment.

During storage, the sugar content in fruit tends to increase due to the fruit ripening process; however, there is also a decrease in sugar content due to the use of sugar for fruit respiration and the activity of microorganisms. Chitosan has antimicrobial properties, and the combination of gelatin with chitosan can increase the antimicrobial properties of chitosan [27–29]. In addition, the addition of black tea extract can also increase the antimicrobial properties of the edible coating because tea has antimicrobial activity. Figure 7 shows that the addition of black tea extract to the edible coating solution affected the total soluble solids value in the minimally processed watermelon samples stored at a temperature of ±4 °C for 13 days, but there was no significant difference ($p > 0.05$) with the addition of 0.25–1% black tea extract. Often the studies involving the coating of fruit indicate that the TSS increases over time. One obvious reason for this increase is the progress in the ripening process leading to increased TSS. However, another justification for this increase is associated with the fact that water loss happens during the storage, which raises the TSS; this is a reason which has been cited in several studies [23].

3.6. Antioxidant Activity

Figure 9 shows the results of the analysis of the antioxidant activity in minimally processed watermelon coated with an edible coating of a gelatin composite enriched with black tea extract. Based on Figure 9, there was a significant decrease in antioxidant activity ($p < 0.05$) in all the treatments during the storage period. The storage duration of fruits and vegetables affects their antioxidant activity: the longer the storage period, the lower the antioxidant activity due to the respiration processes [12]. The preparation and storage process can affect antioxidant activity in minimally processed watermelon samples because antioxidant compounds are sensitive and easily damaged when exposed to oxygen and light. In this study, the samples were stored for 13 days in transparent plastic cups that were translucent so that they could damage the antioxidant compounds contained in them.

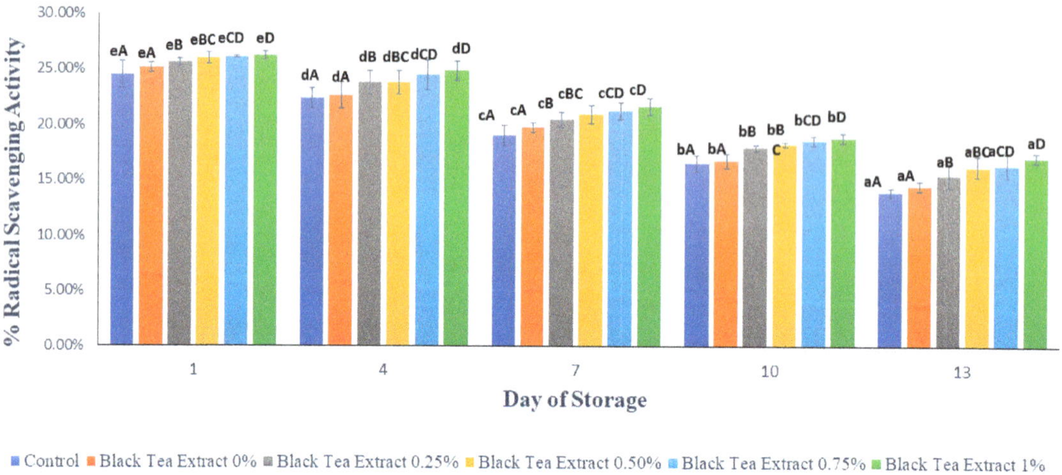

Figure 9. Effect of edible coating of gelatin composite enriched with black tea extract on antioxidant activity (%) of sliced watermelon during storage. The value shown is the average of the three experimental replications. $^{A–D}$ Values followed by the same capital letters showed no significant difference ($p > 0.05$) between treatments on the same day. $^{a–e}$ Values followed by the same non-capital letters showed no significant difference ($p > 0.05$) between storage times in the same treatment.

In the samples without treatment (the control), the decrease in antioxidant activity was higher than that of the samples treated with the edible coating because the diffusion of O_2 into the tissue could not be inhibited. Antioxidant compounds are sensitive to oxygen, light, and high temperatures. Gelatin as a component of edible coatings has good barrier properties against oxygen and light; so, it can reduce the decrease in antioxidant activity during storage.

The addition of black tea extract to the edible coating solution can increase the antioxidant activity of the sample, as shown in Figure 9, because tea is rich in antioxidants. One of the hurdles in the food industry is the limited shelf life of foods due to the oxidation or degradation phenomenon. The addition of tea extract provides natural antioxidant compounds, and their addition to coatings could improve their functional characteristics, making them more effective in protecting fruits [23]. Many researchers have studied how edible coatings incorporated with antioxidants affect the quality and preservation of fresh/fresh-cut fruits. The addition of tea extracts is one of the several types of antioxidants used in coatings for fresh-cut fruits. The catechin, theaflavin, and thearubigin compounds in black tea have strong antioxidant abilities so that the higher the concentration of black tea extract added, the greater the antioxidant activity [14]. Antioxidant compounds can fight free radicals and increase tissue resistance so that they can reduce damage and decay due to fungal growth. Thus, the content of antioxidant compounds can affect the shelf life of the fruit. High antioxidant activity can increase the shelf life of fruit [5].

3.7. Fungal Decay

Our investigation showed that in all the treatments no samples were found that were damaged by fungal contamination during 13 days of storage at a temperature of ±4 °C. In this study, the selection of low temperature (±4 °C) and storage time for 13 days was motivated by previous experiments, which can be seen in Figure 10. Figure 10 shows that there was damage to the sample due to fungal growth on the 2nd day of storage at room temperature (±25 °C) and on the 15th day of storage at low temperature (±4 °C). Storage at low temperatures (±4 °C) can inhibit fungal growth. Previous research showed that ciplukan fruit (Physalis peruviana), coated with an edible coating of gelatin and stored at ±5 °C, had a shelf life of up to 21 days and a lower percentage of damage due to fungal growth when compared to fruit storage at ±20 °C [30]. Storage at a temperature of ± 4 °C in minimally processed watermelon coated with an edible coating of sodium alginate can effectively reduce fungal growth until the 15th day [3].

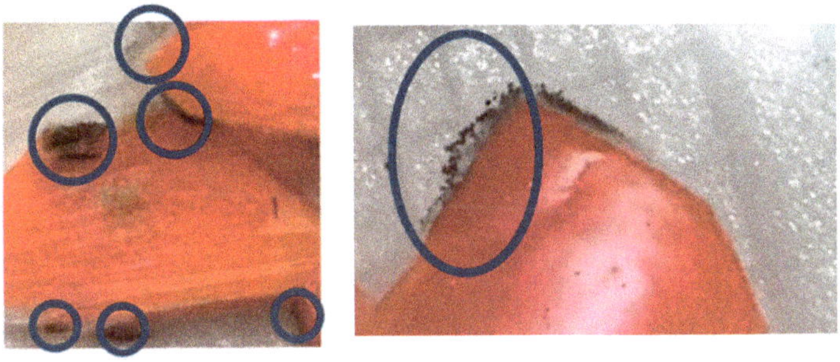

(a) (b)

Figure 10. Fungal decay on the 2nd day of storage at room temperature. (±25 °C) (**a**) and the 15th day of storage at low temperature (±4 °C) (**b**).

The antimicrobial nature of the edible coating may be due either to its inherent features or to an antimicrobial agent that has been added to the coating formulation, such as chitosan [29,31]. With reference to the literature, the antimicrobial potential of essential oils is the major reason for the incorporation of essential oils into edible coatings applied to fruits. The increased consumption of fresh-cut produce in recent times has raised the interest in antimicrobial coatings. The addition of plant extracts, such as tea extract, to edible coating formulations can improve the functionality of coatings in safeguarding fruits and vegetables from microbiological spoilage and hence extend their shelf life [23].

Fresh-cut fruits such as watermelon are convenient and ready-to-eat products that deliver advantages to consumers. However, the development of novel technologies to maintain the quality and extend the shelf life of fresh-cut fruits is a major challenge for the food industry and an issue of concern for future research. The use of edible coatings helps to maintain the quality of fresh-cut watermelon and extends its shelf life. In addition, coatings can be used to incorporate active/functional ingredients such as black tea into the fresh-cut produce. They can also perform effectively as carriers of bioactive compounds.

4. Conclusions

Treatment with an edible coating of tuna fish skin gelatin composite and chitosan enriched with the addition of black tea extract can inhibit weight loss and texture softening, as well as provide a protective effect against the discoloration of minimally processed watermelon which is stored for 13 days at ±4 °C. An edible coating treatment of tuna skin gelatin composite and chitosan enriched with the addition of black tea extract can inhibit acidity reduction and provide a protective effect against the changes in sugar content and the antioxidant activity of minimally processed watermelon which is stored for 13 days at a temperature of ±4 °C. Treatment with an edible coating of tuna fish skin gelatin composite and chitosan enriched with the addition of black tea extract can provide a protective effect against damage due to fungal contamination in minimally processed watermelon which is stored for 13 days at a temperature of ±4 °C.

Author Contributions: Investigation, S.S., A.S.S., H.B., A.A., F.A. and W.H.; Resources, E.S. and H.S.H.M.; Supervision, M., H.S.H.M., P.L.S. and A.N. All authors have read and agreed to the published version of the manuscript.

Funding: The authors would like Riset Kolaborasi Indonesia Program (RKI) for fiscal year of 2022 (NOMOR: 1575/UN1/DITLIT/Dit-Lit/PT.01.03/2022).

Institutional Review Board Statement: Not applicable.

Informed Consent Statement: Not applicable.

Data Availability Statement: Not applicable.

Acknowledgments: The authors would like to thank Riset Kolaborasi Indonesia Program (RKI) for fiscal year of 2022 (NOMOR: 1575/UN1/DITLIT/Dit-Lit/PT.01.03/2022).

Conflicts of Interest: The authors declare no conflict of interest.

References

1. Isabelle, M.; Lee, B.L.; Lim, M.T.; Koh, W.P.; Huang, D.; Ong, C.N. Antioxidant activity and profiles of common fruits in Singapore. *Food Chem.* **2010**, *123*, 77–84. [CrossRef]
2. Yousuf, B.; Qadri, O.S.; Srivastava, A.K. Recent developments in shelf-life extension of fresh-cut fruits and vegetables by application of different edible coatings: A review. *LWT-Food Sci. Technol.* **2018**, *89*, 198–209. [CrossRef]
3. Sipahi, R.E.; Castell-Perez, M.E.; Moreira, R.G.; Gomes, C.; Castillo, A. Improved multilayered antimicrobial alginate-based edible coating extends the shelf life of fresh-cut watermelon (Citrullus lanatus). *LWT-Food Sci. Technol.* **2013**, *51*, 9–15. [CrossRef]
4. Rodrigues, M.Á.V.; Bertolo, M.R.V.; Marangon, C.A.; da Conceição Amaro Martins, V.; de Guzzi Plepis, A.M. Chitosan and gelatin materials incorporated with phenolic extracts of grape seed and jabuticaba peel: Rheological, physicochemical, antioxidant, antimicrobial and barrier properties. *Int. J. Biol. Macromol.* **2020**, *160*, 769–779. [CrossRef] [PubMed]
5. Yang, H.J.; Lee, J.H.; Lee, K.Y.; Song, K.B. Application of gelatin film and coating prepared from dried alaska pollock by-product in quality maintanance of grape berries. *J. Food Process. Preserv.* **2017**, *41*, e13228. [CrossRef]
6. da Silva Pereira, G.V.; de Oliveira, L.C.; Cardoso, D.N.P.; Calado, V.; de Fatima Lourenço Henriques, L. Rheological characterization and influence of different biodegradable and edible coatings on postharvest quality of guava. *J. Food Process. Preserv.* **2021**, *45*, e15335. [CrossRef]
7. Teacă, C.A.; Roşu, D.; Mustaţă, F.; Rusu, T.; Roşu, L.; Roşca, I.; Varganici, C.D. Natural bio-based products for wood coating and protection against degradation: A review. *BioResources* **2019**, *14*, 4873–4901. [CrossRef]
8. Jancikova, S.; Dordevic, D.; Tesikova, K.; Antonic, B.; Tremlova, B. Active edible films fortified with natural extracts: Case study with fresh-cut apple pieces. *Membranes* **2021**, *11*, 684. [CrossRef]
9. Hassan, B.; Chatha, S.A.S.; Hussain, A.I.; Zia, K.M.; Akhtar, N. Recent advances on polysaccharides, lipids and protein based edible films and coatings: A review. *Int. J. Biol. Macromol.* **2018**, *109*, 1095–1107. [CrossRef]
10. Ehsani, A.; Hashemi, M.; Afshari, A.; Aminzare, M.; Raeisi, M.; Zeinali, T. Effect of different types of active biodegradable films containing lactoperoxidase system or sage essential oil on the shelf life of fish burger during refrigerated storage. *LWT* **2020**, *117*, 108633. [CrossRef]
11. Benítez, S.; Achaerandio, I.; Pujolà, M.; Sepulcre, F. Aloe vera as an alternative to traditional edible coatings used in fresh-cut fruits: A case of study with kiwifruit slices. *LWT-Food Sci. Technol.* **2015**, *61*, 184–193. [CrossRef]
12. López-Palestina, C.U.; Aguirre-Mancilla, C.L.; Raya-Pérez, J.C.; Ramírez-Pimentel, J.G.; Gutiérrez-Tlahque, J.; Hernández-Fuentes, A.D. The effect of an edible coating with tomato oily extract on the physicochemical and antioxidant properties of garambullo (myrtillocactus geometrizans) fruits. *Agronomy* **2018**, *8*, 248. [CrossRef]
13. Jamróz, E.; Kulawik, P.; Krzyściak, P.; Talaga-Ćwiertnia, K.; Juszczak, L. Intelligent and active furcellaran-gelatin films containing green or pu-erh tea extracts: Characterization, antioxidant and antimicrobial potential. *Int. J. Biol. Macromol.* **2019**, *122*, 745–757. [CrossRef]
14. Ngure, F.M.; Wanyoko, J.K.; Mahungu, S.M.; Shitandi, A.A. Catechins depletion patterns in relation to theaflavin and thearubigins formation. *Food Chem.* **2009**, *115*, 8–14. [CrossRef]
15. Anggraini, T.; Febrianti, F.; Aisman; Ismanto, S.D. Black Tea with *Averrhoa bilimbi* L. Extract: A Healthy Beverage. *Agric. Agric. Sci. Procedia* **2016**, *9*, 241–252. [CrossRef]
16. Kim, A.N.; Kim, H.J.; Chun, J.; Heo, H.J.; Kerr, W.L.; Choi, S.G. Degradation kinetics of phenolic content and antioxidant activity of hardy kiwifruit (Actinidia arguta) puree at different storage temperatures. *LWT-Food Sci. Technol.* **2018**, *89*, 535–541. [CrossRef]
17. Ningrum, A.; Perdani, A.W.; Supriyadi; Munawaroh, H.S.H.; Aisyah, S.; Susanto, E. Characterization of Tuna Skin Gelatin Edible Films with Various Plasticizers-Essential Oils and Their Effect on Beef Appearance. *J. Food Process. Preserv.* **2021**, *45*, e15701. [CrossRef]
18. Ganduri, V.S.R. Evaluation of pullulan-based edible active coating methods on Rastali and Chakkarakeli bananas and their shelf-life extension parameters studies. *J. Food Process. Preserv.* **2020**, *44*, e14378. [CrossRef]

19. Medina-Jaramillo, C.; Quintero-Pimiento, C.; Gómez-Hoyos, C.; Zuluaga-Gallego, R.; López-Córdoba, A. Alginate-edible coatings for application on wild andean blueberries (Vaccinium meridionale swartz): Effect of the addition of nanofibrils isolated from cocoa by-products. *Polymers* **2020**, *12*, 824. [CrossRef]
20. Ningrum, A.; Hapsari, M.W.; Nisa, A.A.; Munawaroh, H.S.H. Edible film characteristic from yellowfin skin tuna (Thunnus albacares) gelatin enriched with cinnamon (cinnamomum zeylanicum) and roselle (hibiscus sabdariffa). *Food Res.* **2020**, *4*, 1646–1652. [CrossRef]
21. Yıldırım-Yalçın, M.; Sadıkoğlu, H.; Şeker, M. Characterization of edible film based on grape juice and cross-linked maize starch and its effects on the storage quality of chicken breast fillets. *LWT* **2021**, *142*, 111012. [CrossRef]
22. Badawy, M.E.I.; Rabea, E.I.; El-Nouby, M.A.M.; Ismail, R.I.A.; Taktak, N.E.M. Strawberry Shelf Life, Composition, and Enzymes Activity in Response to Edible Chitosan Coatings. *Int. J. Fruit Sci.* **2017**, *17*, 117–136. [CrossRef]
23. Yousuf, B.; Wu, S.; Siddiqui, M.W. Incorporating essential oils or compounds derived thereof into edible coatings: Effect on quality and shelf life of fresh/fresh-cut produce. *Trends Food Sci. Technol.* **2021**, *108*, 245–257. [CrossRef]
24. Huang, T.; Tu, Z.c.; Shangguan, X.; Sha, X.; Wang, H.; Zhang, L.; Bansal, N. Fish gelatin modifications: A comprehensive review. *Trends Food Sci. Technol.* **2019**, *86*, 260–269. [CrossRef]
25. Nagarajan, M.; Rajasekaran, B.; Benjakul, S.; Venkatachalam, K. Influence of chitosan-gelatin edible coating incorporated with longkong pericarp extract on refrigerated black tiger Shrimp (Penaeus monodon). *Curr. Res. Food Sci.* **2021**, *4*, 345–353. [CrossRef]
26. Leong, S.Y.; Oey, I. Effect of endogenous ascorbic acid oxidase activity and stability on vitamin C in carrots (Daucus carota subsp. sativus) during thermal treatment. *Food Chem.* **2012**, *134*, 2075–2085. [CrossRef]
27. Sahraee, S.; Milani, J.M.; Regenstein, J.M.; Kafil, H.S. Protection of foods against oxidative deterioration using edible films and coatings: A review. *Food Biosci.* **2019**, *32*, 100451. [CrossRef]
28. Hanani, Z.A.N.; Husna, A.B.A.; Syahida, S.N.; Khaizura, M.A.B.N.; Jamilah, B. Effect of different fruit peels on the functional properties of gelatin/polyethylene bilayer films for active packaging. *Food Packag. Shelf Life* **2018**, *18*, 201–211. [CrossRef]
29. Poverenov, E.; Zaitsev, Y.; Arnon, H.; Granit, R.; Alkalai-Tuvia, S.; Perzelan, Y.; Weinberg, T.; Fallik, E. Effects of a composite chitosan-gelatin edible coating on postharvest quality and storability of red bell peppers. *Postharvest Biol. Technol.* **2014**, *96*, 106–109. [CrossRef]
30. Licodiedoff, S.; Koslowski, L.A.D.; Scartazzini, L.; Monteiro, A.R.; Ninow, J.L.; Borges, C.D. Conservation of physalis by edible coating of gelatin and calcium chloride. *Int. Food Res. J.* **2016**, *23*, 1629–1634.
31. Mohammad, S.; Hashemi, B.; Mousavi, A.; Ghaderi, M.; Ismail, E. Postharvest Biology and Technology Basil-seed gum containing Origanum vulgare subsp. viride essential oil as edible coating for fresh cut apricots. *Postharvest Biol. Technol.* **2017**, *125*, 26–34. [CrossRef]

polymers

Communication

A Potential of New Untreated Bio-Reinforcement from *Caesalpinia sappan* L. Wood Fiber for Polybutylene Succinate Composite Film

Ekkachai Martwong [1], Yvette Tran [2], Nattawadee Natsrita [3], Chaithip Kaewpang [3], Kittisak Kongsuk [3], Yeampon Nakaramontri [4] and Nathapong Sukhawipat [3,*]

[1] Division of Science, Faculty of Science and Technology, Rajamangala University of Technology Suvarnabhumi, Phra Nakhon Si Ayutthaya 13000, Thailand; ekkachai.m@rmutsb.ac.th
[2] Soft Matter Sciences and Engineering, ESPCI Paris, PSL University, CNRS, F-75005 Paris, France; yvette.tran@espci.fr
[3] Division of Polymer Engineering Technology, Department of Mechanical Engineering Technology, College of Industrial Technology, King Mongkut's University of Technology North Bangkok, Bangkok 10800, Thailand; natthwd41@gmail.com (N.N.); chaithip4@gmail.com (C.K.); kittisak1487@gmail.com (K.K.)
[4] Sustainable Polymer & Innovative Composite Materials Research Group, Department of Chemistry, Faculty of Science, King Mongkut's University of Technology Thonburi, Bangkok 10140, Thailand; y.nakaramontri@gmail.com
* Correspondence: Nathapong.s@cit.kmutnb.ac.th; Tel.: +66-824846930

Citation: Martwong, E.; Tran, Y.; Natsrita, N.; Kaewpang, C.; Kongsuk, K.; Nakaramontri, Y.; Sukhawipat, N. A Potential of New Untreated Bio-Reinforcement from *Caesalpinia sappan* L. Wood Fiber for Polybutylene Succinate Composite Film. *Polymers* **2022**, *14*, 499. https://doi.org/10.3390/polym14030499

Academic Editor: Evgenia G. Korzhikova-Vlakh

Received: 6 January 2022
Accepted: 24 January 2022
Published: 26 January 2022

Abstract: Natural cellulose-based *Caesalpinia sappan* L. wood fiber (CSWF) has been demonstrated to have significant promise as a new untreated bio-reinforcement of the polybutylene succinate (PBS) composite film. The morphology, mechanical characteristics, and biodegradation were investigated. The morphology, the fiber distribution, and the fiber aggregation has been discussed. The properties of the composite have been improved by the addition of CSWF from 5 phr to 10 phr, while with the addition of 15 phr, the properties were dropped. The result showed that CSWF could be used as a new reinforcement without any treatment, and 10 phr of CSWF was the best formulation of a new biocomposite film. The PBS/CSWF10 composite film had the highest mechanical strength, with a tensile strength of 12.21 N/mm^2 and an elongation at break of 21.01%, respectively. It was completely degraded by soil bury in three months. Therefore, the PBS/CSWF10 composite film has the potential to be a green with a promising short-term degradation.

Keywords: biocomposites; polybutylene succinate; polymer films; fiber; cellulose; bio-reinforcement

1. Introduction

Petroleum-based packaging films, such as polyethylene and polypropylene, are a major source of pollution and have a negative impact on the environment [1,2]. Almost all plastic films are designed to be used once before being destroyed, resulting in a large volume of plastic film waste that poses a direct threat to the environment [3]. Thus, several researchers have attempted to develop a new solution to replace conventional plastic film with a fully biodegradable one.

Biodegradable plastic, also known as bioplastic, is defined by its ability to degrade naturally, such as polybutylene succinate (PBS) and poly(lactic acid) (PLA). PBS is mentioned as one of the possibilities appropriate for utilization as a bioplastic because of its outstanding mechanical characteristics, heat resistance, and processibility [4,5]. However, its properties are insufficient for daily use.

PBS has recently been enhanced by the use of inorganic materials [6]. However, it has a direct impact on humans, such as health problem explanation. Many research groups utilize cellulose in order to prevent these issues. It is a well-known natural-based additive used to improve the properties and biodegradability of several substances [7].

However, in order to improve the polymer matrix properties, cellulose-based fiber must first be treated before use, either chemically [8] or physically [9]. Almost every piece of research [10,11] has reported that treating cellulose with a chemical reagent or adding a compatibilizer enhanced the interfacial interaction between the polymer and natural-based cellulose. Finding a new bio-reinforcement without treating it or adding compatibilizers is a promising and desired alternative to save time and cost.

Caesalpinia sappan L. wood fiber (CSWF) is cellulose derived naturally from a Caesalpiniaceae plant. The mechanical characteristics are comparable to those of cellulosic materials. Brazilin and its derivatives, a heterotetracyclic structure with a hydroxyl functional group, are found in CSWF [12–14]. Thus, we hypothesized that CSWF may be a good reinforcement, even without any surface treatment or compatibilizers. Untreated CSWF is focus on being used as a bio-reinforcement for the PBS composite film to examine this hypothesis.

Therefore, the potential of untreated CSWF as a new reinforcement in a novel bio-composite film was investigated. The influences of CSWF content on the morphology, mechanical properties and biodegradability of PBS/CSWF composite films were studied.

2. Materials and Methods

2.1. Materials

PBS grade FZ71PM with a density of 1.24 g/cm^3 and MFR (190 °C, 2.16 kg.) of 22 g/10 min was purchased from PTTMCC Biochem in Rayong, Thailand.

The CSWF originated from the OTOP center in Nan, Thailand. To obtain mesh size 100, CSWF was ground and sieved by mesh sieve size number 100. CSWF was less than 150 m long and 30 m in diameter. All of the components were dried for 12 h at 80 °C before use.

2.2. Method

2.2.1. Preparation of PBS/CSWF Composite Films

To control the mixing conditions, a mixing temperature of 160 °C and a rotor speed of 50 rpm were utilized. The formula for PBS/CSWF composites is shown in Table 1. An additional minute was spent in an internal mixer preheating a PBS pellet. The CSWF was then added to the melted PBS at 0, 5, 10, and 15 phr and continuously mixed for 3 min. After that, the PBS/CSWF composite was dumped and rapidly cooled to ambient temperature. Finally, the composite was crushed by a granulator machine before being blown up.

Table 1. Compositions of the composite films.

Sample Name	PBS (phr)	CSWF (phr)
Neat film	100	0
PBS/CSWF5	100	5
PBS/CSWF10	100	10
PBS/CSWF15	100	15

The PBS/CSWF composite film was formed by blow film extrusion (type LF-400, Labtech, Sorisole (BG), Italy). Extruder temperature profiles from zone 1/zone 2/ zone 3/zone 4 were 130/135/135/140 °C, respectively. The die temperature and screw rotational speed were 145 °C and 40 rpm, respectively. The nip-roll and windup rates were 4.2 and 4.4 mm/min, respectively. The lay flat width of composite films was fixed at 13 cm. To prevent humidity absorption, all of the composite films were kept in a desiccator.

An overview of PBS/CSWF composite preparation process is illustrated in Figure 1.

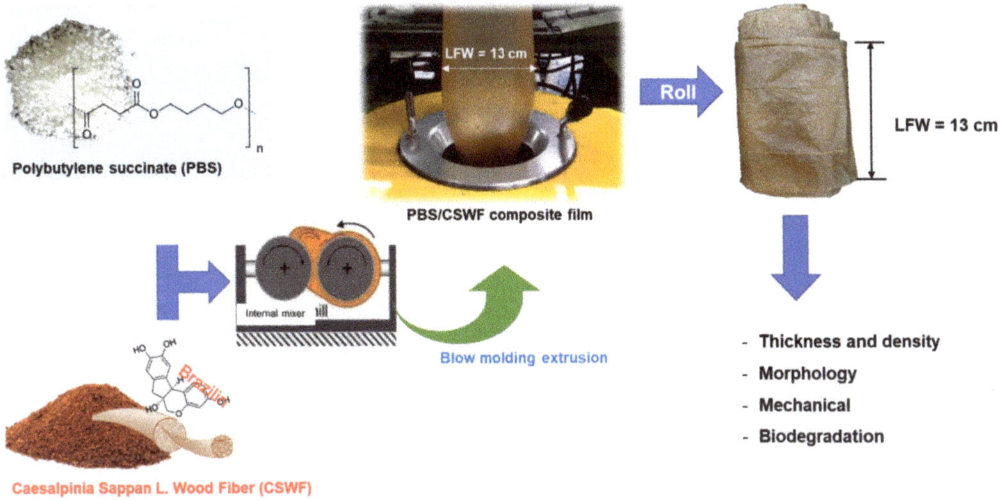

Figure 1. A short overview of PBS/CSWF composite preparation process.

2.2.2. Characterization

The density of the composite was determined using ASTM D4635-16. A Vernier calliper was used to accurately measure the length and thickness of composite films. The specimens were precisely weighed in order to calculate the density using the equation below.

$$\text{Density } (g/cm^3) = \text{Mass } (g)/\text{Volume } (cm^3)$$

The morphology of PBS/CSWF composite films was studied with an optical microscope at a magnification of 10× and 40×, using Xenon (DN-117M, Nanjing Jiangnan Novel Optics, Nanjing, China).

Following imaging, measurements of each sample were completed using ImageJ. The fiber size, aggregation area of fiber in PBS film composite and total area were analyzed. The statistical analysis of aggregation area, total area, and % area of fiber has been accomplished and reported.

The mechanical properties of the composite films were measured using Testometric (M500-25AT) (Testometric, Rochdale, UK) on a universal testing machine in accordance with ASTM D638. The size of the specimen was 130 × 10 mm^2. The instrument included a 1 kN load cell and a 25 mm gauge length extensometer with a crosshead speed of 10 mm/min. Tensile strength and elongation at were reported.

Biodegradation was studied via soil burying in an ambient environment, and the weight change before and after soil burying was measured. The weight change was measured once a month for three months under ambient conditions. The Equation (1) was used to determine the film degradation.

$$W_{loss} (\%) = (W_f - W_i)/W_i \times 100 \quad (1)$$

where: W_{loss} (%) is weight loss percentage, W_i is the weight of the composite film before the burial test, and W_f is the weight of the composite film after the soil burial test.

3. Results and Discussion

3.1. Properties of PBS/CSWF Composite Films

The thicknesses of PBS/CSWF composite films were determined by setting the flat lay width to 13 cm as shown in Figure 2. It was discovered that the thicknesses of composite films for neat film, PBS/CSWF5, PBS/CSWF10, and PBS/CSWF15 were 0.031, 0.043, 0.052, and 0.069 mm, respectively. By increasing the CSWF component, the thickness of the PBS/CSWF composite film was marginally increased. It was attributed to agglomeration effects and limited dispersion in CSWF with a higher content.

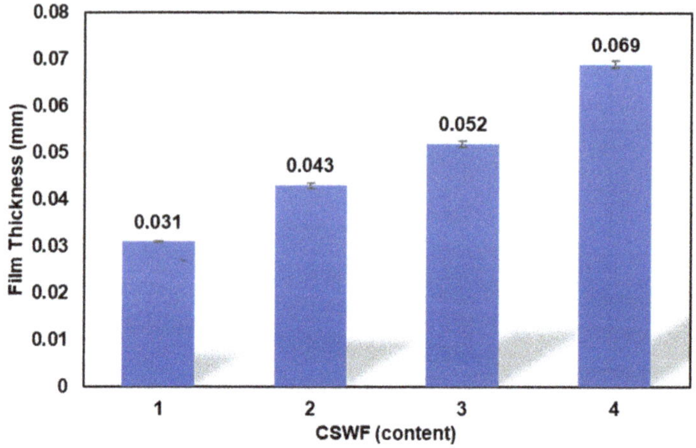

Figure 2. Thickness of PBS/CSWF composite films.

Figure 3 illustrates the density of PBS/CSWF composite films. It was marginally reduced from 1.24, 1.20, 1.19, and 1.16 g/cm^3 by increasing the CSWF content from 0, 5, 10, and 15 phr, respectively. Because CSWF has a lower density than neat PBS, the density tendency of PBS/CSWF composite film was therefore reduced in the formulation with a higher content of CSWF. As a result, by increasing the CSWF content, the density of the PBS/CSWF composite in this study was reduced significantly.

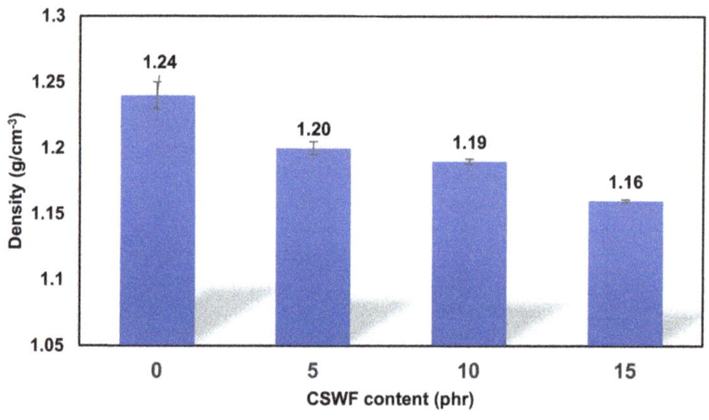

Figure 3. Density of PBS/CSWF composite films.

3.2. Morphology of PBS/CSWF Composite Films

Optical microscopy was used to examine the morphologies of PBS/CSWF composite films. Figure 4a shows OM images of PBS/CSWF composite films at 10× and 40× magnification. It was determined that CSWF was well-dispersed in the PBS matrix at 5 and 10 phr, with some CSWF aggregated at 15 phr. Furthermore, the results indicated that the composite film had no flaws between the PBS and CSWF interfacial surfaces. It was due to the polar–polar interaction and hydrogen bonding between PBS and Brazilin derivatives, an organic heterotetracyclic molecule found in CSWF. It was considered to be one of the reasons that enhanced the interfacial contact between PBS and CSWF. FTIR and TGA were two further techniques used to validate the interaction of PBS and fiber, as shown in Figures S1 and S2, respectively. Moreover, the ImageJ analysis findings were provided to access the behavior of the fiber aggregation.

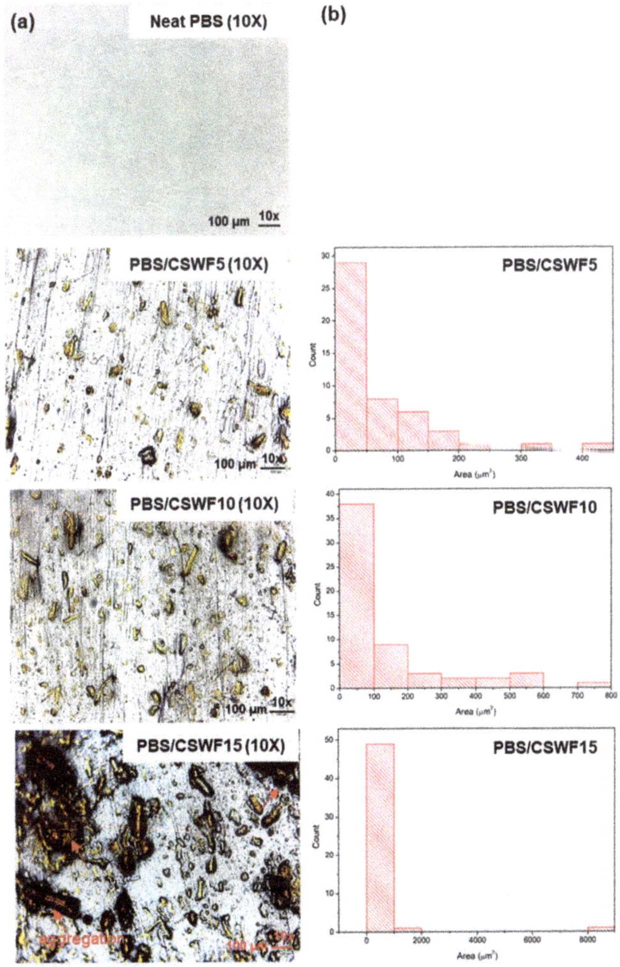

Figure 4. (**a**) Optical microscope images of PBS composite films at magnification of 10× and 40×, and (**b**) fiber diameter and its aggregation area analyzed by ImageJ technique.

Figure 4b presents the outcomes of fiber diameter and aggregation area, which are summarized in Table 2. The counts of fiber and aggregation point were 49, 58 and 51 for PBS/CSWF5, PBS/CSWF10 and PBS/CSWF15, respectively. By increasing the CSWF content from 5 to 15 phr, the overall aggregation area and average size of CSWF in PBS matrix rose from 3644 to 14,273 μm^2 and 74.37 to 279.86 μm^2, respectively. It was discovered that the area distribution was too high, and the SD value of all samples was greater than the average size of area. At this point, it was established that the distribution of fiber area was dramatically different, with PBS/CSWF15 showing the highest distribution of area. Furthermore, by increasing the fiber content from 5 to 15 phr, the cover area of the fiber in the film increases from 10.17% to 40.03%. The CSWF was entirely dispersed in the PBS matrix for the PBS/CSWF5 composite film, which included the least amount of CSWF and low area distribution. In contrast, the dispersion in the PBS matrix was low in the PBS/CSWF15 film, and the fiber was tightly aggregated in the PBS matrix confirmed by their statistical analysis of fiber area.

Table 2. Summary results from ImageJ analysis.

Sample Name	Average Size (μm^2)	Total Area (μm^2)	Area (%)
PBS/CSWF5	74.37 ± 79.91	3644	10.17
PBS/CSWF10	126.26 ± 163.40	7323	20.22
PBS/CSWF15	279.86 ± 1145.97	14,273	40.03

3.3. *Mechanical Properties of PBS/CSWF Composite Films*

The mechanical characteristics of the PBS/CSWF composite films were evaluated using tensile strength and elongation at break, as shown in Figure 5. They were studied in two directions: machine direction (MD) and transverse direction (TD). Figure 5a presents the tensile strength of composite films. The tensile strengths of composite films with 0, 5, 10 and 15 phr are 4.77, 5.69, 12.21 and 2.16 N/mm^2 in MD and 2.21, 3.91, 6.51 and 0.43 N/mm^2 in TD, respectively. Tensile strength tends to improve with increasing CSWF content. PBS/CSWF10 had the highest tensile strength in both directions, reaching 12.21 N/mm^2 in MD and 6.51 N/mm^2 in TD. In contrast, the mechanical properties of the PBS/CSWF15 composite film were significantly reduced in both directions.

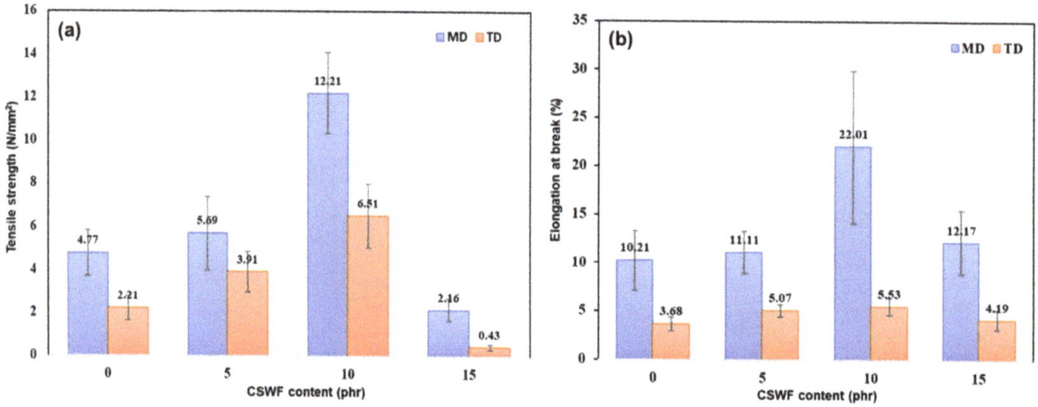

Figure 5. Mechanical properties of PBS/CSWF composite films—(**a**) tensile strength and (**b**) elongation at break.

Figure 5b presents the elongation at break results of the PBS/CSWF composite films. The elongations at break of the composite films with 0, 5, 10 and 15 phr are 10.21%, 11.11% 22.01% and 12.17% in MD and 3.68%, 5.07%, 5.53% and 4.19% in TD respectively. It resembled the tensile strength trend. PBS/CSWF10 had the highest elongation at break, with 22.01% of MD and 5.53% of TD, respectively. It could be described as a decrease in the PBS-to-fiber ratio and an increase in the aggregation area due to an increase in CSWF content from the OM image and ImageJ results. The optimal ratio and aggregation area of CSWF is 10 phr. As a result of CSWF overload and aggregation, the mechanical characteristics of PBS/CSWF15 were reduced. It is reasonable to conclude that an untreated CSWF has a high potential to be a bio-reinforcement.

3.4. Biodegradation Properties of PBS/CSWF Composite Films

The biodegradation process of the PBS/CSWF composite was monitored by soil burying for three months. The weight reduction of the composite films is shown in Figure 6a. The weights reductions of the composite films with 0, 5, 10 and 15 phr are 1.28%, 6.63%, 37.12% and 2.53% for the first month, 6.25%, 16.51%, 51.97% and 12.14% for 2 months and 19.33%, 42.85%, 100% and 13.1% for 3 months respectively. It was found that PBS/CSWF10 is the highest biodegradation rate. However, it was a surprise that the degradation rate of PBS/CSWF15 is lowest. This finding is supported and consistent with the mechanical properties. It might be because of the antibacterial abilities of Brazilin derivatives found in CSWF [12–14]. Composite films with CSWF contents ranging from 0 to 10 phr that released Brazilin derivatives onto the surface of the PBS/CSWF composite film were insufficient to inhibit the activity of bacteria attached to the composite film. Meanwhile, the PBS/CSWF15 film delivers sufficient Brazilin derivatives to the surface to inhibit bacterial activity. Figure 6b displays the appearance of composite films two months after degradation. According to the findings, PBS/CSWF10 degraded the fastest, whereas PBS/CSWF15 film degraded the slowest.

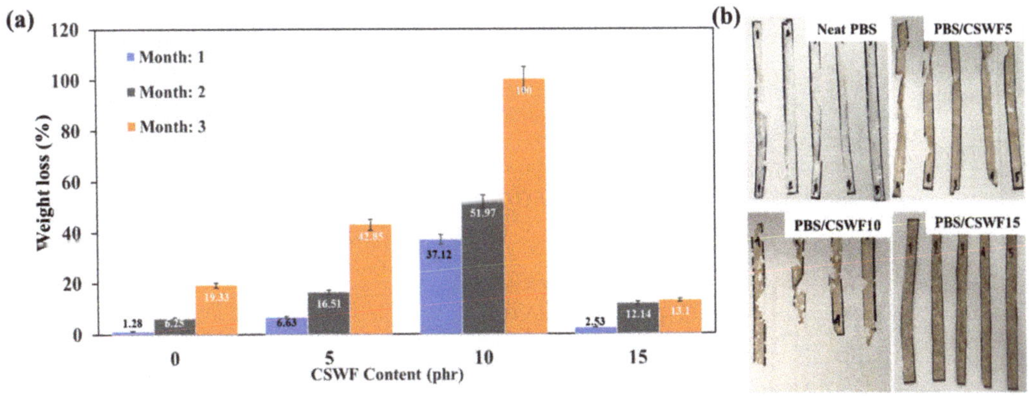

Figure 6. (**a**) The biodegradation progress of PBS/CSWF composite films with different burial times, and (**b**) The appearance of PBS/CSWF composite film after a 2-month biodegradation test.

4. Conclusions

In the absence of treatment, CSWF was well-dispersed in the PBS matrix and could be employed as a new bio-reinforcement. The morphology, the dispersion, and the aggregation of fibers have all been validated and studied. The addition of CSWF increased the properties of the composite film from 5 to 10 phr, whereas the addition of 15 phr decreased the properties. After three months, the PBS/CSWF10 had completely degraded. PBS/CSWF10 is the best composite film, according to the results. Therefore, CSWF has the potential to be a promising alternative addition for the industrial sector of green plastic film.

Supplementary Materials: The following are available online at https://www.mdpi.com/article/10.3390/polym14030499/s1, Figure S1: FTIR spectra of CSWF (orange), PBS (bleu), and PBS/CSWF composite films with 5 (green), 10 (black) and 15 (red) phr, Figure S2: Thermal degradation characteristics of PBS/CSWF composite film by TGA and DTG.

Author Contributions: E.M.: Writing—Review and Editing, Y.T.: Writing—Review and Editing, N.N.: Investigation, C.K.: Investigation, K.K.: Investigation, Y.N.: Conceptualization, Writing—Review and Editing, N.S.: Conceptualization, Investigation, Writing—Original Draft, Writing—Review and Editing, Supervision, Visualization. All authors have read and agreed to the published version of the manuscript.

Funding: This research received no external funding.

Institutional Review Board Statement: Not applicable.

Informed Consent Statement: Not applicable.

Data Availability Statement: The study did not report any data.

Acknowledgments: The authors thanks to College of Industrial Technology, King Mongkut's University of Technology North Bangkok for providing the facilities.

Conflicts of Interest: The authors declare no conflict of interest.

References

1. Patil, S.S.; Jena, H.M. Performance assessment of polyvinyl chloride films plasticized with Citrullus lanatus seed oil based novel plasticizer. *Polym. Test.* **2021**, *101*, 107271. [CrossRef]
2. Sundar, N.; Kumar, S.A.; Keerthana, P.; Stanley, S.J.; Kumar, G.A. Schiff's base (SB) modified zirconium dioxide reinforced PLA bio-composite film for industrial packaging applications. *Compos. Commun.* **2021**, *25*, 100750. [CrossRef]
3. Kalogerakis, N.; Karkanorachaki, K.; Kalogerakis, G.C.; Triantafyllidi, E.I.; Gotsis, A.D.; Partsinevelos, P.; Fava, F. Microplastics Generation: Onset of Fragmentation of Polyethylene Films in Marine Environment Mesocosms. *Front. Mar. Sci.* **2017**, *4*, 84. [CrossRef]
4. Rafiqah, S.A.; Khalina, A.; Harmaen, A.S.; Tawakkal, I.A.; Zaman, K.; Asim, M.; Nurrazi, M.N.; Lee, C.H. A Review on Properties and Application of Bio-Based Poly(Butylene Succinate). *Polymers* **2021**, *13*, 1436. [CrossRef] [PubMed]
5. Jost, V. Packaging related properties of commercially available biopolymers—An overview of the status quo. *Express Polym. Lett.* **2018**, *12*, 429–435. [CrossRef]
6. Grigoriadou, I.; Nianias, N.; Hoppe, A.; Terzopoulou, Z.; Bikiaris, D.; Will, J.; Hum, J.; Roether, J.A.; Detsch, R.; Boccaccini, A.R. Evaluation of silica-nanotubes and strontium hydroxyapatite nanorods as appropriate nanoadditives for poly(butylene succinate) biodegradable polyester for biomedical applications. *Compos. Part B Eng.* **2014**, *60*, 49–59. [CrossRef]
7. Goutianos, S.; Arévalo, R.; Sørensen, B.F.; Peijs, T. Effect of Processing Conditions on Fracture Resistance and Cohesive Laws of Binderfree All-Cellulose Composites. *Appl. Compos. Mater.* **2014**, *21*, 805–825. [CrossRef]
8. Tarchoun, A.F.; Trache, D.; Klapötke, T.M.; Abdelaziz, A.; Derradji, M.; Bekhouche, S. Chemical design and characterization of cellulosic derivatives containing high-nitrogen functional groups: Towards the next generation of energetic biopolymers. *Def. Technol.* **2021**. [CrossRef]
9. Mohammed, A.A.; Bachtiar, D.; Rejab, M.R.M.; Siregar, J.P. Effect of microwave treatment on tensile properties of sugar palm fibre reinforced thermoplastic polyurethane composites. *Def. Technol.* **2018**, *14*, 287–290. [CrossRef]
10. Baley, C.; Kervoëlen, A.; Le Duigou, A.; Goudenhooft, C.; Bourmaud, A. Is the low shear modulus of flax fibres an advantage for polymer reinforcement? *Mater. Lett.* **2016**, *185*, 534–536. [CrossRef]
11. Sun, M.; Wang, H.; Li, X. Modification of cellulose microfibers by polyglutamic acid and mesoporous silica nanoparticles for Enterovirus 71 adsorption. *Mater. Lett.* **2020**, *277*, 128320. [CrossRef] [PubMed]
12. Xu, H.-X.; Lee, S.F. The antibacterial principle of Caesalpina sappan. *Phyther. Res.* **2004**, *18*, 647–651. [CrossRef] [PubMed]
13. Tewtrakul, S.; Tungcharoen, P.; Sudsai, T.; Karalai, C.; Ponglimanont, C.; Yodsaoue, O. Antiinflammatory and Wound Healing Effects of Caesalpinia sappan L. *Phyther. Res.* **2015**, *29*, 850–856. [CrossRef] [PubMed]
14. Ye, M.; Xie, W.D.; Lei, F.; Meng, Z.; Zhao, Y.N.; Su, H.; Du, L.J. Important immunosuppressive component from *Caesalpinia sappan* L. *Int. Immunopharmacol.* **2006**, *6*, 426–432. [CrossRef] [PubMed]

MDPI
St. Alban-Anlage 66
4052 Basel
Switzerland
Tel. +41 61 683 77 34
Fax +41 61 302 89 18
www.mdpi.com

Polymers Editorial Office
E-mail: polymers@mdpi.com
www.mdpi.com/journal/polymers

www.ingramcontent.com/pod-product-compliance
Lightning Source LLC
Chambersburg PA
CBHW061457220926
51850CB00004B/111
* 9 7 8 3 0 3 6 5 6 4 9 2 0 *